Berichte der Internationalen Symposien
der Internationalen Vereinigung für Vegetationskunde
Herausgegeben von Reinhold Tüxen

Gesellschaftsentwicklung (Syndynamik)

(Rinteln, 20.-23. März 1967)

Redaktion

Reinhold Tüxen
und
Wolf-Hildebrand Sommer

Springer-Science+Business Media, B.V.
1979

Additional material to this book can be downloaded from http://extras.springer.com

ISBN 978-94-011-7594-4 ISBN 978-94-011-7593-7 (eBook)
DOI 10.1007/978-94-011-7593-7

Originally published by A.R. Gantner Verlag K.G., FL-9490 Vaduz in 1979
Softcover reprint of the hardcover 1st edition 1979

Dem wegweisenden russischen Geobotaniker
V.N.SUKATSCHEV
zum Gedenken

Die Teilnehmer am Internationalen Symposion
über Gesellschaftsentwicklung
(Syndynamik)

VORBEMERKUNG

Eine Reihe widriger Umstände, vor allem aber die Überlastung des Herausgebers, machten das frühere Erscheinen dieses Berichtes leider unmöglich. Wenn wir ihn dennoch jetzt der Öffentlichkeit übergeben, so geschieht das einmal, um die Berichte der Symposien der Internationalen Vereinigung vollständig zu veröffentlichen, zweitens, weil die in Vorträgen und Diskussionen vorgetragenen Äußerungen u.E. keineswegs überholt sind, in jedem Falle aber für den Zeitpunkt des Symposion festgehalten zu werden verdienen, und nicht zuletzt, weil wir jetzt die bedeutenden Beiträge russischer Gäste endlich veröffentlichen können.

R.Tüxen

INHALT

TEILNEHMERVERZEICHNIS

Argentinien

ESKUCHE,Prof.Dr.U., Corrientes,C.C.326

Canada

Le BLANC,Prof.Dr.F., Ottawa, Dept.of Biology, University of Ottawa, Ont.Kin 6N5

Chile

KOHLER,Prof.Dr.A., Santiago de Chile (jetzige Anschrift): 7 Stuttgart-Hohenheim,Universität,Abt.Landeskultur

ČSSR

BALÁTOVÁ-TULÁČKOVÁ,Frau Dr.Emilie, Brno 16, Stará 18, Geobotanicka laborat.ČSAV

HEJNÝ,Dr.S., Průhonice u Prahy,Botanický ustav

KÜHN,Dr.F., Brno, Zemedelska Katedra Botanický

MORAVEC,Dr.J., Průhonice u Prahy, Geobot.Institut

NEUHÄUSL,Dr.R., Průhonice u Prahy, Geobot.Institut

DEUTSCHLAND

ALTEHAGE,C.†, Realschullehrer i.R., 45 Osnabrück.

BAUCH,Frau Dr.Erika, 2 Hamburg 71, Benzstr.3

BIERHALS,Dipl.rer.hort.E., 3 Hannover

BLOMBERG,Freiherr v.H.G., 3251 Pohle, Gut Nienfeld

BRACKER,D.H.-H., 2381 Bollingstedt, am Mühlenteich

BURRICHTER,Prof.Dr.E., 44 Münster i.W.,Schloßgarten 3,Bot.Inst. der Universität

DAMMANN,Dr.Hildegard,Studienrätin, 314 Lüneburg, Wilschenbrucher Weg 87

DIETZ,H.-D.,cand.rer.hort., 3 Hannover-Herrenhausen, Dorotheenstr.7

DIERSCHKE,Prof.Dr.H., 34 Göttingen, Untere Karspüle 2

DUTHWEILER,Dr.H., 3011 Hannover-Laatzen,Kronsbergerstr.80A

ELLENBERG,Prof.Drs.h.c.H., 34 Göttingen, Untere Karspüle 2 System.Geobot.Institut d.Universität

ELLENBERG,Frau Charlotte, 34 Göttingen-Hetjershausen-Amsterdam

FEISE ,Dr.J.,Oberlandw.Rat i.R., 29 Oldenburg,Sodenstich 115

FEISE,Frau 29 Oldenburg,Sodenstich 115

GALONSKE,D.,Dipl.hort., 3151 Equort, Schulhaus

GRAČANIN,Prof.Dr.Z., 463 Bochum, Ruhr-Universität,Postf.2148

GRIES,Dr.Brunhild, 44 Münster i.W.,Himmelreichallee 50 Landesmuseum

GROSSE-BRAUCKMANN,Prof.Dr.G., 61 Darmstadt, Bot. Inst.d.TU.

HABER,Prof.Dr.W., 805 Freising-Weihenstephan, Inst.f.Landespflege der Universität München

HACKER,E.,Dipl.Landw., 3 Hannover, Nieders.Landesamt f.Bodenforschung, Stilleweg

HARTMANN,Prof.Dr.F.K. , 351 Hann.-Münden.

HARTMANN,W.,Dipl.hort., 3 Hannover, Institut f.Landespflege der TU Hannover

HERMS,R.,Landschaftsarchitekt, 2 Hamburg 13, Klosterallee 53

HOFMEISTER,Dr.H., 33 Braunschweig, Hainburgstr.42

HOLTZ,Dr.Fr., 34 Göttingen, Untere Karspüle 2

HOLTZ,Frau

HORST,Dr.K., 314 Lüneburg, Sültenweg 23, Päd.Hochschule

HORSTMEYER,D.,Lehrer, 4831 Verl, Wiesenweg 31

HORSTMEYER,Frau

JANIESCH,Dr.P., 44 Münster i.W., Botanisches Museum

KNAUER,Prof.Dr.N., 23 Kiel 16, Buschberg 8

KRUKENBERG,Dr.med.†, 326 Rinteln

LEIPPERT,Dr.H., 3 Hannover,Herrenhäuserstr.2, Institut für Vegetationskunde

LUCHTERHANDT,Dipl.Ing.Bdbs.Abt.Präs.i.R. , 56 W.-Elberfeld.

MÜLLER,G.,Ober-Studienrat, 325 Rinteln, Eichendorffstr.14

PFADENHAUER,Dr.Dozent,J., 7 Stuttgart, Univ.Hohenheim, Abt. Landeskultur

PREISING,Prof.Dr.E., 3 Hannover, Altenbekener Damm 30

REHAGEN,Dr.H., 415 Krefeld, Steinstr.203-205, Geol.Landesamt

RÖDEL,Prof.Dr.H., 75 Karlsruhe, Kolbergerstr.11

RUNGE,Dr.F., 44 Münster i.W. Kinderhaus, Diesterwegstr.63

SCHMITHÜSEN,Prof.Dr.J., 66 Saarbrücken, Trillerweg 40, Geogr. Institut der Universität

SCHMITHÜSEN,Frau

SCHWERDTFEGER,Prof.Dr., 3113 Suderburg Krs.Uelzen,Staatl.Ing.-Schule

SPEIDEL,Prof.Dr.B., 643 Bad Hersfeld, Kolbergerstr.16

STIEF,J. 4960 Stadthagen, Gartenstr.24

STOEVESANDT,W. 326 Rinteln, Westcontrescarpe

SUKOPP,Prof.Dr.H., Berlin 33, Thaerweg 4, Inst.f.Ökologie

TÜXEN,Dr.J.,Oberreg.Rat, 3 Hannover, Nieders.Landesamt f.Bodenforschung, Stilleweg

TÜXEN,Prof.Dr.Drs.h.c., 326 Rinteln I Todenmann, Arbeitsstelle für Theoretische und Angewandte Pflanzensoziologie

ULLRICH,Prof.Dr.C., 58 Hagen, Stirnband, 10, Päd.Hochschule

WEBER-OLDECOP,Dr.D., 3011 Gehrden, Lindenweg 5

WEBER,Frau

WILDENHAIN,Dr.J., 5 Köln, Brüderstr.3,IV

WILMANNS,Prof.Dr.Otti, 78 Freiburg, Schänzlestr.9-11, Biol.Inst. II der Universität

WINTERHOFF,Prof.Dr.W., 6902 Sandhausen, Bunsenstr.24

WOLFF-STRAUB,Dr.Rotraud, 4 Düsseldorf, Landesanst.f.Ökologie

ZEIDLER,Prof.Dr.H., 3 Hannover, Herrenhäuserstr.2, Institut für Vegetationskunde der TU Hannover

England

BELLAMY,Prof.Dr.D., Durham, Department of Botany, University

Frankreich

CARBIENER,Prof.Dr.R., Strasbourg, 2 rue Saint Georges, Labor de Botanique de l'Université

CARBIENER,Mme

GARONNE,B., 34 Montpellier, 5 Auguste Broussonet, Institut de Botanique de l'Université

LINDER,Prof.Dr.R., Lille, 14 rue Malus, Institut de Botanique

POISSONET,Mme Dr.Paule, 34 Montpellier, B.P.1018

TRABAUD,Dr.J., 34 Montpellier, B.P.1018

WURCH,Dr.Th., 67 Selestat, 12 rue de la paix

WURCH,Mme, 67 Selestat, 12 rue de la paix

Griechenland

LAWRENTIADES,Prof.Dr.G., Thessaloniki,Institut f.System.Botanik J.Universität

LAWRENTIADES,Frau Carola

Irland

MOORE,Prof.Dr.J.J., Dublin 4, University College, Dept.of Botany

Israel

LITAV,Dr.M., Jerusalem, Hebrew University, Dept.of Botany

Italien

LAUSI,Dr.D., Trieste,Istituto ed Orto Botanico, Via Cumano 2

PIGNATTI,Prof.Dr.S., Trieste,Istituto ed Orto Botanico, Via Cumano 2

POLDINI,Dr.L., Trieste,Istituto ed Orto Botanico, Via Cumano 2

Japan

SUZUKI-TOKIO,Prof.Dr.† Ooita.

USUI,Prof.Dr.H., Utsunomiya Universität, Tochigi-Pref.

Jugoslavien

KOŠIR,Z., Dipl.-Ing.,Direktor, Ljubljana, Casta na loke 4, Biro za gozdarsko naortovanje
SERCELJ,Dr.A., Lubljana, Slowen.Akad., Novi trg 3
WRABER,Prof.Dr.M.† Ljubljana.

Niederlande

BARKMAN,Prof.Dr.J.J., Wijster, Dr.Kampsweg 29, Biologige Station
BEEFTINK,Dr.W.G., Yerseke, Vierstraat 28, Hydrobiol.Institut
BLIJSWIJK,M., Den Haag, van Stolkweg, Verlag Dr.Junk
BUIL,Mejuffrouw, Zeist, R.I.V.O.N. Laan van Beek en Royen 40
DE BOER, Th.Ir., Wageningen, Duivendaal 10, Proefstation voor akker-en weidebouw
DIEMONT,Dr.H.,Landforstm. , Maastricht.
ERNST,Prof.Dr.W., Amsterdam-Buitenveldert, Vrieje Universiteit, De Boelelaan 1087
FREIJSEN,Dr.A., Oostvoorne, Biologisch Station Weevers Duin
FRESCO,W.F.M., Haren (Gn) Rijkstraatweg 78 D
JOENJE,W., Haren (Gn) Rijkstraatweg 78 D
PLASTERK,K., Direktor, Den Haag, Uitgeeststraat 32
SEGAL,Dr.S., Amsterdam 4, Keizersgracht 776
SISSINGH,Dr.G., Olfm., Wageningen, Stichting Bosbouwproefstation "De Dorschkamp"
SMIDT, de Th., Utrecht, Heidelberglaan
VAN LEEUWEN,Dr.Chr., Bilthoven, Grote Beer 99, R.I.V.O.N.
VAN DER MAAREL, Dr.E., Nijmegen, Abt. Geobotanik, Universiteit, 6525 FD
WESTHOFF,Prof.Dr.V., Nijmegen, Bot.Inst.d. Kath.Universiteit
WESTHOFF,Frau, Nijmegen
WOLTERSON,K., Direktor, Bennekom, Hullenbergweg 16
ZONNEVELD,Prof.Dr.I., Enschede, Vaarwerkshorst 63

Österreich

HÜBL,Prof.Dr.E., Wien XVIII, Bot.Inst.der Hochschule f.Bodenkultur, Greg.Mendelstr.33
THIELE,Kl., Wien XIX, Peter Jordanstr.82, Waldbauinstitut

Polen

WOLAK,Dr.J., Warszawa, Institut Badawczj, Leśnictwa

Schweden

MERKER,Dr.H., Hälsingborg, Sköldenborgsgatan 5 c, Bot.Trädgarden

Schweiz

FURRER,Dr.E.†, Zürich.
HEGG,Dr.O., Bern, Humboldtstr.53, Bot.Institut d.Universität

UdSSR

GORCHAKOVSKY,Prof.D.P., Sverdlovsk 8, Biol.Inst.Ural Branch of Academy of Science
KORCHAGIN, Prof.Dr.A., Leningrad, Bot.Inst.der Universität, Prof.Popovastr.2
MATVJEJEWA, Frau Prof. Dr.Elsabeth, Leningrad, Bot.Inst.der Universität, Prof.Popovastr.2
RABOTNOW,Prof.Dr.T., Moskau B 234, Lehrstuhl f.Geobotanik der Universität
ZAKIROV,Prof.Dr.K., Taschkent, Schelkovitschnaja 24

USA

KÜCHLER,Prof.Dr.A.W., Lawrence, Kansas, Dept. of Geography,University
MUELLER-DOMBOIS,Prof.Dr.D., Honolulu, Hawaii, St.John Plant Science Lab.101

ERÖFFNUNG DES SYMPOSION

R e i n h o l d   T ü x e n

Meine sehr verehrten Damen und Herren!

Ich habe die Ehre, im Auftrage unseres hochverehrten Herrn Präsidenten, Prof.LEBRUN, unser 11.Symposion, das syndynamische Probleme behandeln soll, zu eröffnen. Sie alle, die Sie aus 20 verschiedenen Ländern heute zu uns in unser schönes, altes Rinteln gekommen sind, möchte ich herzlich willkommen heißen. Ich muß leider davon absehen,Sie einzeln zu begrüßen. Aber ich kann mir nicht versagen, einer ganz großen Freude Ausdruck zu geben, die mir eine Erfüllung eines sehr alten Wunsches bedeutet. Seit mehr als drei Jahrzehnten habe ich meine Aufgabe nicht nur als Sekretär der Internationalen Vereinigung darin gesehen, Gegensätze, Unterschiede und Schwierigkeiten, die auftauchten in unserer Wissenschaft (Phytozönologie, Pflanzensoziologie, Vegetationskunde, Geobotanik, oder wie Sie sie nennen wollen), auszugleichen. Ich habe versucht, die Unterschiede zwischen Nord und Süd, zwischen der skandinavischen und der mitteleuropäischen Schule zu verringern. Das ist gelungen, wenn Sie sich an eines unserer letzten Symposien in Stolzenau erinnern, wo wir über die nicht mehr bestehenden gegensätzlichen Auffassungen lebhaft diskutiert haben. Damals dachte ich allerdings auch an die Querrichtung Ost und West.

Heute ist es nun endlich wahr geworden, daß von unseren verehrten russischen Kolleginnen und Kollegen hier in diesem Saale nicht weniger als vier hervorragende Vertreter anwesend sind. Es ist mir leider nicht möglich, Sie auf Russisch zu begrüßen. Aber ich habe in Hannover einen Freund, der gut russisch spricht, und von dem habe ich gelernt: Dobro pojaluwat! Soyez les bienvenues, darf ich meinen französischen Freunden sagen, and welcome to Rinteln, my english friends! Und die deutsch-sprechenden verstehen mich ja ohnehin!

Meine Damen und Herren, das wissenschaftliche Problem, das uns hier beschäftigen soll, ist die Dynamik in der Vegetation. Aber es ist nicht nur dieses, sondern es gibt noch ein anderes, ein menschliches. Alle, die früher hier gewesen sind - ich glaube, es sind einige von Ihnen heute zum 11. Male hier - verstehen, was ich damit meine. Wir freuen uns, daß einige "baliveaux",das

sind 'Überhälter' im forstlichen Sinne, bei uns sind. Vor allem einer der Begründer der dynamischen Pflanzensoziologie. Es sind gerade 40 Jahre her, als ich in Zürich zu Füßen von Ihnen, Herr Dr.FURRER, zum ersten Male etwas von dynamischer Pflanzensoziologie gehört habe. Es bedeutet mir eine außerordentliche Freude, Sie heute unter uns wiederzusehen, und wir hoffen, auch heute wieder von Ihnen lernen zu dürfen. Wir bedauern aber, daß viele andere alte Freunde nicht kommen konnten. Wir hoffen, daß sie das nächste Jahr wieder bei uns sein werden.
Für drei unserer alten verehrten Freunde können wir diese Hoffnung nicht mehr hegen. Ich habe Sie daran zu erinnern, das Madame Gabriele BRAUN-BLANQUET, die Gattin unseres Lehrers und Freundes Josias BRAUN-BLANQUET vor einigen Monaten von uns gegangen ist. Es gibt wenige Frauen, die so der Wissenschaft und dem Lebenswerk ihres Lebensgefährten ergeben durch viele Jahrzehnte gefolgt sind, wie diese bescheidene, grazile Frau, die immer um ihn und mit ihm war. Viele von uns haben sich ihrer Fröhlichkeit, ihrer Kameradschaft,Fürsorge und Freundschaft erfreut, die sie gegen uns alle, sei es in der "Sigma" oder auf Exkursionen und Reisen immer gezeigt hat. Wir beklagen mit ihrem Gatten und seiner Tochter ihr Hinscheiden. Ihnen gilt unsere ganze Teilnahme.

Wir haben weiter einen schweren Verlust zu beklagen. Den unerwarteten Tod von Professor SUKATSCHEV, eines der bedeutendsten weltweit anerkannten russischen Forschers auf dem Gebiete der Phytocoenologie. Sein Lebenswerk hier zu würdigen, kann heute nicht meine Aufgabe sein. Wir wissen, wie viel er über methodische Fragen, über systematische, über waldtypologische und über andere Probleme publiziert hat, und wie ungeheuer anregend seine in mehreren Sprachen erschienenen Arbeiten gewesen sind. Ich brauche nur daran zu erinnern, daß der Begriff der Geobiozönose von SUKATSCHEV stammt, ein Wort, das wir täglich, wenn vielleicht auch nicht aussprechen, so doch denken.
Es gehört zu den bemerkenswerten Tatsachen, daß in einem anderen Lande, wo dieses Wort geprägt wurde, die Vegetation in derselben umfassenden Weise betrachtet, wie sie bei uns studiert wird.
Auch wir bemühen uns, nicht Spezialisten zu sein, sondern das Ganze im Sinne der Geobiozönose im Sinne von SUKATSCHEV zu sehen, auch wenn wir von Biozönose, von Lebensgemeinschaft, von Biochore oder auch nur von Pflanzengesellschaften sprechen. Wir meinen das Gleiche. Und mir ist das wieder eine Bestätigung des-

sen, daß die Unterschiede in der Auffassung, die scheinbar in den verschiedenen Teilen der Welt bestehen, gar nicht so bedeutend sind, daß wir uns bei gutem Willen nicht verständigen und verstehen könnten. Auch aus diesem Grunde freue ich mich, daß unsere russischen Kollegen zu uns gekommen sind, und daß sie wissen, wie hoch unsere Achtung vor der russischen Forschung und vor dem verstorbenen Professor SUKATSCHEV ist.
Noch einen schmerzlichen Verlust habe ich Ihnen mitzuteilen: vor wenigen Wochen starb in Uppsala unser alter Freund Prof. Gunnar Einar DU RIETZ. Er kam aus seinem Laboratorium, in dem er wie immer gearbeitet hatte, und ist auf dem Heimwege an einer Hirnblutung gestorben. Mitten aus der Arbeit herausgerissen: einen glücklicheren Tod könnten wir uns wohl nicht wünschen! Dieser Mann hat uns vor wenigen Jahren in Stolzenau in unserem Kreise seine ganze dynamische Ausstrahlung und seine ganze Vitalität vorgeführt. Sie alle wissen, daß auch er zu den Begründern unserer Wissenschaft gehört, die er mit Rolf NORDHAGEN und anderen im Norden ins Leben rief, während in den Alpen und im westlichen Mittelmeer-Raum Josias BRAUN-BLANQUET seine Lehre entwickelte, und in Rußland unter dem führenden Einfluß von SUKATSCHEV eine eigene große Schule entstand. Wir erinnern uns auch, wie sehr die nordische und mittel-europäische Richtung rang, um den fruchtbarsten Weg zu finden für die Erforschung der Vegetation, und wie dann Einar DU RIETZ, Rolf NORDHAGEN und Josias BRAUN-BLANQUET 1935 während des Internationalen Botaniker-Kongresses in Amsterdam zu einer Synthese ihrer Begriffe gelangten.
Bevor wir uns unseren wissenschaftlichen Problmen voll widmen, will ich die andere Seite, die menschliche, noch einmal kurz beleuchten. Erlauben Sie mir, daß ich Ihnen aus der "Universitas" eine halbe Seite vorlese, die ganz und gar dem alten Stolzenauer, jetzt Rintelner Symposion-Geist entspricht. Ich nenne nicht den Autor, um nur die Worte wirken zu lassen:
"Der moderne Mensch hat neue und hervorragende Gipfel wissenschaftlichen Erfolges erreicht. Er hat Maschinen erbaut, die denken und Instrumente ersonnen, die unermeßlichen Weiten des interplanetaren Raumes erforschen. Er hat gigantische Brücken geschlagen, die die Meere überspannen und riesige Gebäude errichtet, die in den Himmel ragen. Seine Flugzeuge und Raumschiffe haben die Entfernungen zusammenschrumpfen lassen, die Zeit in Ketten gelegt und Wege durch die Stratosphäre gebahnt. Uns

bietet sich ein blendendes Bild vom wissenschaftlichen und technischen Fortschritt des modernen Menschen. Und trotz dieser offensichtlichen und noch kommenden Fortschritte in Wissenschaft und Technik, deren Grenzen nicht abzusehen sind, fehlt etwas Grundlegendes. Je reicher wir in materieller Hinsicht geworden sind, desto bedrohter wurden wir innerlich. Wir haben gelernt, die Luft zu durchfliegen wie die Vögel und das Meer zu durchschwimmen wie die Fische, aber nicht die einfache Kunst, als Brüder zusammen zu leben. Jeder Mensch lebt in zwei Reichen, dem inneren und dem äußeren. Das innere ist auf das geistige Ziel gerichtet, das äußere ist jene Gesamtheit von Apparaten, technischen Anlagen, Mechanismen und Werkzeugen, durch die sich unser Dasein gestaltet. Unser gegenwärtiges Problem ist, daß wir das innere Reich vor dem äußeren verloren gehen ließen. Hier liegt die ernste Gefahr, das tiefe und quälende Problem für den modernen Menschen. Wenn wir heute überleben wollen, müssen wir diesen Rückstand aufholen. Eine Erweiterung der materiellen Kräfte zieht erhöhte Gefahren nach sich, wenn damit nicht eine entsprechende Stärkung der Seele einhergeht. Wenn das Äußere der menschlichen Natur das Innere unterjocht, beginnen sich in der Welt dunkle Gewitterwolken zusammenzuziehen. Dies ist das gewaltige neue Problem der Menschheit. Wir haben ein stattliches Haus geerbt, ein großes Welthaus, in dem wir zusammen leben müssen, weil wir nie mehr ohne einander leben können, und darum irgendwie lernen müssen, in dieser großen Welt miteinander zu leben."

Meine Damen und Herren, wir haben in unseren Symposien seit jeher versucht, in dieser Weise in gegenseitiger Achtung, Hilfsbereitschaft und in immer wachsender Freundschaft miteinander zu leben und miteinander zu sprechen über die Aufgaben, die unser Leben ausmachen, nämlich die Erforschung der Vegetation. Ich bin ganz sicher, daß auch dieses heutige Symposion im selben schönen freundschaftlichen Geiste gegenseitiger Achtung und Liebe sich abspielen wird, und ich möchte Sie noch einmal herzlich in diesem Geiste willkommen heißen.

Meinen Mitarbeitern, die mir geholfen haben, dieses Symposion vorzubereiten, danke ich herzlich. Ich bitte alle Kritik, die nicht ausbleiben wird, an mich und nicht an sie zu richten.

Ich bitte nun Herrn Prof. KÜCHLER das Präsidium zu übernehmen.

BEGRIFFE DER SUKZESSIONSFORSCHUNG

Diskussionsgrundlage

H. E l l e n b e r g

1. Hauptbegriffe

Abfolge - Reihe - Serie von Vegetationseinheiten:

Räumliche oder (und) zeitliche Aneinanderreihung verschiedener Pflanzenbestände, die meist auch im Hinblick auf ihre Standorte eine Reihe bilden.

Zonierung = Ökologische Reihe = Sequenz:

Räumliches Nebeneinander verschiedener Pflanzenbestände bei kontinuierlicher oder abgestufter Änderung eines oder mehrerer Standortfaktoren.

Sukzession = Entwicklungsreihe:

Zeitliches Nacheinander von verschieden zusammengesetzten Pflanzenbeständen an demselben Wuchsort.

2. Nach ihrer Richtung ist die Sukzession:

primär progressiv(oder kurz primär):

Erstbesiedlung eines vorher von Pflanzen unbesiedelten Rohbodens und darauf folgende natürliche Sukzession, verbunden mit entsprechender Bodenreifung (z.B. auf neuer Insel, frischem Lavastrom, Erdrutschmaterial, Gletschervorfeld).

regressiv:

In entgegengesetzter Richtung verlaufend, also meist von einer kompliziert gebauten und hoch entwickelten Formation zu einer einfacher gebauten und niedrigeren hinführend. Häufig vom Menschen ausgelöst (z.B. Degradation eines Waldes zu Magerrasen oder Heide durch Beweidung).

sekundär progressiv:

Wiederbesiedlung eines mehr oder minder reifen Bodens nach teilweise oder gänzlicher Zerstörung des natürlichen Pflanzenbestandes (z.B. nach Kahlschlag, nach Wirbelsturm,nach Brand, nach Aufhören der Beweidung oder der Beackerung).

discessiv = richtungsneutral:

Innerhalb der gleichen Formation verlaufend, also im Sinne der soziologischen Progression weder progressiv noch regressiv (z.B. bei Düngung oder bei Entwässerung einer Wiese, bei Fruchtwechsel auf dem Acker).

3. Nach der Geschwindigkeit der Sukzessionen unterscheidet man:

mehr oder minder <u>rasche</u>:

durch direkte Beobachtungen in wenigen Jahren oder Jahrzehnten zu verfolgende.

<u>säkulare</u>:

in größeren Zeiträumen verlaufende und bisher noch nicht direkt registrierte.

Viele hypothetisch angenommene säkulare Sukzessionen sind inzwischen als äußerst unwahrscheinlich oder unmöglich erkannt worden!

4. Die auslösende Wirkung kann erfolgen:

<u>stetig</u>:

Durch stetig wirkende (wie Allgemeinklima, Ausgangsgestein) oder sich sehr rasch wiederholende Einflüsse (z.B. Beweidung, regelmäßiges Mähen, öfteres Betreten, alljährliches Brennen) entsteht schließlich ein Gleichgewichtszustand, der als Endstadium der Sukzession gelten darf.

<u>rhythmisch = zyklisch</u>:

nach mehr oder minder regelmäßigen, aber in Abständen von Jahren oder Jahrzehnten erfolgenden Veränderungen wiederholt sich die gleiche Vegetationsabfolge (z.B. Kahlschlag-Krautflur, Gebüsch, Weichholzwald, Hartholzwald).

<u>katastrophal</u>:

Nach plötzlichem Eingreifen äußerer, nicht rhythmischer Faktoren (z.B. Vulkanausbruch, Erdrutsch, Brand, Windbruch, Überstauung mit Wasser, Umpflügen einer Wiese) beginnende Sukzession.

5. Nach der Rolle der Pflanzen bei der Sukzession sollte man auseinanderhalten:

<u>phytogene (autogene) Sukzession = Entwicklung</u>:

Nur durch die Lebenstätigkeit, das Vorhandensein und die Abfälle von Pflanzen bewirkte Sukzession, ohne daß der Standort von außen her verändert wird (z.B. Verlandung eines Stillwassers, das keine sedimentierenden Zuflüsse erhält, Wiederbesiedlung eines Brachlandes).

<u>ökogene (allogene) Sukzession = Verdrängung = Standorts- und Vegetationsüberlagerung</u>:

Durch Änderung der Standortsbedingungen bewirkte Sukzession, ohne daß diese durch die am Ort wachsenden Pflanzen ausgelöst wurde (z.B. Umwandlung eines Bruchwaldes in einen Hartholzwald infolge Grundwassersenkung, Ablösung eines Weiden- oder Grauerlen-Auenwaldes durch einen Kiefern-Trockenwald infolge fortschreitender Flußerosion).

gemischte, phyto-ökogene (auto-allogene) Sukzession:

Sowohl von der Vegetation selbst als auch durch Änderung des Standortes von außen her bewirkte Sukzession (z.B. Dünenbildung, Marschenbildung, Verlandung eines Stillwassers mit sedimentierendem Zufluß, die meisten Sukzessionen in Flußtälern).

6. Nach dem hauptsächlichen kontrollierenden Faktor kann die Sukzession bezeichnet werden als:

allgemeinklimatisch:

Auf tiefgründigen, grundwasserfreien Böden von mittlerer Beschaffenheit in mehr oder minder ebener Lage ablaufende primäre Sukzession. Führt zur "zonalen" Vegetation.

lokalklimatisch:

Primäre Sukzession unter Klimaverhältnissen, die erheblich von denen des betreffenden Gebietes abweichen (z.B. an einem Südhang oder Nordhang). Führt zu "extrazonaler" Vegetation.

edaphisch:

Primäre Sukzession an Sonderstandorten, an denen die zonale Vegetation unter natürlichen Verhältnissen niemals erreicht wird (z.B. Verlandung eines Sees, Vegetationsabfolge in Flußauen, Dünenbildung). Führt zu "azonaler" Vegetation.

brandbedingt:

weidebedingt:

Regressive Sukzessionen, die durch natürlichen oder von Menschen veranlaßten Brand oder Verbiß ausgelöst werden.

mahdbedingt:

düngungsbedingt:

bewässerungsbedingt:

entwässerungsbedingt:

Regressive oder richtungsneutrale Sukzessionen, die durch die entsprechenden Faktoren ausgelöst werden.

aufforstungsbedingt:

Durch künstliche Waldbegründung ausgelöste, mehr oder minder richtungsneutrale oder progressive Sukzession.

<u>ackerbaubedingt</u>:

Nach Umbruch von Grünland oder Rodung von Wald beginnende Entwicklung von Ackerunkraut-Gesellschaften.

<u>usw</u>.

7. Sukzessionseinheiten

<u>Stadium</u>:

Physiognomisch und floristisch deutlich abgrenzbarer Sukzessionsschritt, der im Verhältnis zum Gesamtverlauf der Sukzession eine gewisse Dauer hat (z.B. Besiedlung nackter Kalkfelsen: Krustenflechten-Stadium, Blattflechten-Stadium, Moos-Stadium, Polsterphanerogamen-Stadium usw.).

<u>Phase</u>:

Kleinerer Sukzessionsschritt von verhältnismäßig kurzer Dauer; meist nur mit Bezug auf eine bestimmte, floristisch definierte Vegetationseinheit (z.B. Initial-, Optimal-, Degenerations- und Schlußphase einer Assoziation).

<u>Serie</u>:

Reihe von aufeinander folgenden Stadien, die unter dem Einfluß eines und desselben kontrollierenden Faktors stehen (z.B. Verlandungsserie eines eutrophen Stillwassers, Brachlandserie vom verlassenen Acker bis zum zonalen Wald).

Innerhalb der Serie kann man unterscheiden:

<u>Initial</u>stadium = Anfangsstadium,

<u>Zwischen</u>stadien = Übergangsstadien,

<u>Terminal</u>stadium = Endstadium - Schlußstadium.

Vorschläge zur einheitlichen Darstellung von Vegetationsabfolgen

(1) —— Zonierung: Strich <u>ohne</u> Pfeilspitze
⟶ Sukzession: Pfeil

(2) ↑ primäre progressive S.: dicker Pfeil nach oben
↓ regressive S.: dünner Pfeil nach unten
↑ sekundär progressive S.: dünner Pfeil nach oben
↘ discessive S.: dünner Pfeil zur Seite

(3) ⟶ rasche S.: einfacher Pfeil
–S→ säkulare S.: Pfeilstrich mit S

(4) ⟶ stetig: normaler Pfeil
o⟶ rhythmisch: Kreis am Beginn des Pfeils
▌⟶ katastrophal: dicker Querstrich am Beginn des Pfeils

(5) ⟶ phytogene S.:V-förmige Pfeilspitze
⟶ gemischte (phyto-ökogene) S.: Zwischenform
—▷ ökogene S.: Dreieck als Pfeilspitze

(6) ⟶» allgemeinklimatische S.: doppelte Pfeilspitze
⟶ lokalklimatische S.: teiweise doppelte Spitze
⟶ edaphische S.: einfache Pfeilspitze
∿> brandbedingte S.: geschlängelter Strich
·····> weidebedingte S.: punktierter Strich
-·-·-> mahdbedingte S.: kurzgestrichelter Strich
D düngungsbedingte S.
+W bewässerungsbedingte S.
-W entwässerungsbedingte S. } Buchstaben neben dem Pfeilstrich
F aufforstungsbedingte S.
++++++> ackerbaubedingte S.: durchstrichelter Strich
⋰↓ Kombinationen sinngemäß (z.B.durch Brand und Weide bedingte Regression)

(7) <u>Initialstadium</u>: einfach unterstrichen
Zwischenstadien: nicht unterstrichen
<u>Terminalstadium</u>: doppelt unterstrichen

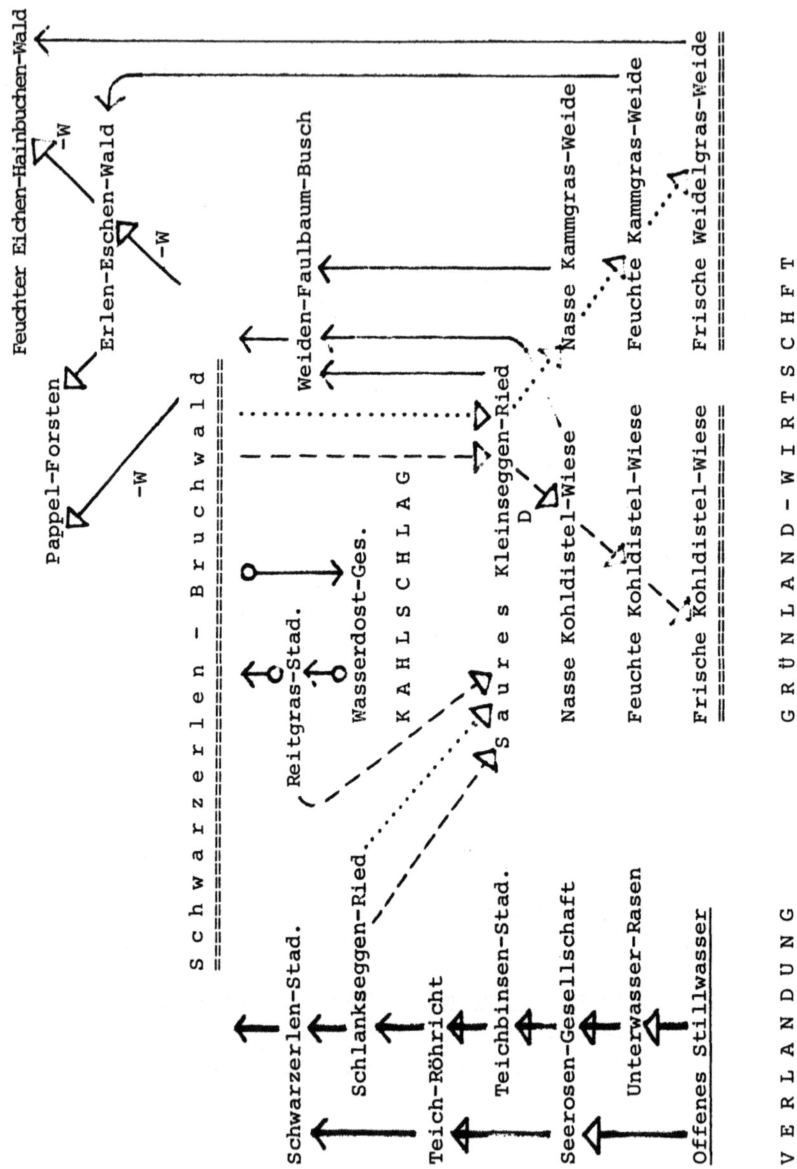

Beispiel eines Sukzessions-Schemas

ON GENERAL TRENDS DURING SUCCESSION IN BIOTIC COMMUNITIES, WITH SPECIAL REFERENCE TO PLANT ECOLOGY

S. S e g a l

An exhaustive discussion of succession in vegetation was given by CLEMENTS (1928), who also reviewed previous opinions on the subject. CLEMENTS (1904) was probably the first biologist who tried to ascertain if some general principles regarding the succession of plant communities can be recognised. Another useful attempt was made by WARMING & GRAEBNER (1918). In recent times this subject received more attention from, e.g. DANSEREAU (1957), ODUM (1963) and KERSHAW (1964), who also tried to find some causal relations and 'laws' in succession. After the appearance of a special publication on succession by CLEMENTS (1928), the knowledge of the subject matter has been considerably augmented and the gradually acquired insights into the structure of biotic communities have furthered the understanding of succession concerning many of the relevant problems.
The purpose of this publication is to give a survey of the most promising methods of approach to the various problems and to summarise the hypotheses concerning the general trends, principles and 'laws' which apply in successions of biotic communities, with special reference to plant ecology. This survey, which is partly based on a study of the pertaining literature, may serve as a starting-point for future discussions. The time is not ripe yet for a general synthesis, but a critical review of the different problems may perhaps stimulate future research on the subject. I shall attempt to incorporate several aspects of novel ideas. Presumably it will be necessary to accumulate more factual data before a more generally applicable theory can be conceived and the various hypotheses can be judged on their merits.

Before one can discuss general trends in succession, it is necessary to consider three major points:
(a) To what level of organisation of the biotic communities must these trends be related?
(b) What is succession, do all trends and associated phenomena occur in all types of succession and are the trends invariable irrespective of the type of succession? and
(c) What must be our conception of "general" tendencies" ?

Each of these questions is of sufficient importance to deserve a lengthy discussion, but this is not the purpose of the present paper. I shall restrict myself to those general remarks necessary for an understanding of the subsequent analysis of the problems to be stated.

Ad (a): If succession is regarded as the sum total of a number of processes with a certain degree of direction, we must rid ourselves of a purely static analysis of the concept of a biotic community (which is, incidentally, already inevitable if all kinds of short-term, periodic an non-periodic changes are to be accounted for). It is of course rather inane if one speaks of 'the succession of a species' (because a species appears and goes, which is one of the essences of succession), but strictly speaking, it is not quite correct to speak of a 'succession of plant communities', as we are wont to do. What changes is a functional system of relations in space and time, in other words: an ecosystem. An ecosystem is an 'open' system and can, on the one hand, be considered to represent a subordinate element of a system of a higher level of organisation (e.g., the whole biosphere) and, on the other hand, as a complex of integrated systems of a lower order right down to the submolecular level of organisation. Accordingly, the circumscription and delimitation of an ecosystem is by no means simple. In a number of instances different workers will easily reach an agreement as to what may or may not be referred to as a 'plant community', a 'biotic community' or an 'ecosystem'. However, the less conspicuous the functional relations appear (as is quite often the case in pioneer communities), the greater the chance of alternative interpretations and the greater the tendency towards a subdivision of the whole into a number of small systems.

Ad (b): Several classifications of successions have been proposed, such as the distinction of autogenic and allogenic successions (TANSLEY, 1926) primary and secondary successsions, succeesions of rather short durations and seculares ones, progressive and regressive or ' discessive ' and cyclic succession (ELLENBERG, 1967, symposium paper). Within the limited scope of the present paper, especially the progressive primary successions will be discussed in more detail. As far as the distinction between autogenic and allogenic successions is concerned, one should bear in mind that these terms refer to the two extremes of a series of cases, between which all possible transitions may occur. In reality every type of successions is, to a certain extent, always of a 'mixed' character, because the autogenic and allogenic factors interact frequently (compare WESTHOFF, 1965).

Ad (c): An objective yard-stick for a certain trend is, particularly, a quantification (at least some quantitative expression of the phenomena). Flotted against the time-axis, this results in a diagram, for which a mathematical expression can be sought which describes the trend at least during a part of the process. The rising or falling graph can, theoretically, continue in the same direction, change into a horizontal line after a certain level has been attained (in the case of a saturation diagram), reach a certain optimum or minimum value subsequently to change from positive to negative (or vice versa), or show a regularly periodic character (an undulation or regular fluctuation). In a succession series the second and the third possibilities are the most important.

At this stage of our knowledge only working hypothesis can be proposed, which should be put to the test. All trends to be discussed presently should in point of fact be provides with a question mark. Furthermore, we must consider in how far certain trends are more or less consistently valid throughout or to what extent 'exceptions to the rule' may occur, and of what kind these special cases probably are. The subject matter discussed in the present paper emanated from studies of plant communities, more particularly of communities of bryophytes and of vascular plants. There are sufficient indications suggesting a more general validity of the preliminary conclusions to render their applicability to thallophytes, micro-organism and at least certain groups of animals rather likely.

The following survey is by no means intended to be exhaustive. In a forthcoming publication the various points will be discussed in greater detail, so as to lay the foundations of a more comprehensive theory.

1. Number of species

During succession, generally speaking, the number of species gradually increases, at least, during several of its stages, but often starts decreasing again after some time before a terminal stage is reached (cf. WARMING & GRAEBNER, 1918). Pioneer communities frequently exhibit dominance of only a single species or co-dominance of a few. The number of individual taxa of macro-organism may eventually rise to several scores or even several hundreds of species (as in a tropical rain forest) on a representative surface area. The increase in the number of species apparently does not invariably continue till the final phase of succession series has been attained, but may decrease to a certain extent in respect of preceding stages, as can be observed in, (e.g., the succession series of beach forests in Central Europe, in the development of a tropical rain forest and (manifestly so) during the formation of ombrotrophic peat bogs. In a few cases the number of species occuring during the pioneer stages can be relatively high. Presumably in such cases the environment is inhomogeneous as, for instance, or the rocky slopes and ledges of river valleys and glens. Such habitats are rich in gradients and ecotones. During the process of succession the environment as a whole may converge towards the same general end phase,which in the abovementioned example may be a kind of seasonally flooded riverine forest. (see addition 1,p.19).

2. Taxonomic groups

One may consider whether during succession certain changes occur in the relations between different idiosystematic groups of species. Changes definitely take place in every individual succession series and are associated with the structural composition as associated with the various growth habits and life forms (see under 4). In haloseries in Europe and elsewhere, for instance, the relative frequency of *Chenopodiaceae* is hight, in other pioneer vegetation there may be a strong representation of *Asteraceae*, *Brassicaceae*, *Poaceae* or *Alismatales* sensu *lato*. The possible incidence of a more general trend involving taxa of a higher rank, especially thallophytic groups, should be studied.

Another interesting question is the relation between Monocots and Dicots. In Europe the number of dicotyledonous taxa shows a relative increase in a number of cases, at least in xero- and hygroseries. The epiphytic orchids of the tropical rain forest, on the other hand, are probably forms only appearing during the terminal phases of the succession series.

3. Structure in a quantitative sense.

In many cases during succession, the number of physiognomic and functional structural elements increases during several stages, but often decreases again before a terminal stage is reached. This development is, generally speaking, concomitant with the increase in the number of species, and is partly to be ascribed to the fact, that the vegetation by itself causes a progressive differentiation of the environment. The increasing structural complexity is especially niticeable in the rising number of growth- and life-forms and in the higher degree of stratification. In pioneer communities usually only a single physiognomic type prevails, or only a few types are co-dominant, whereas other types gradually become relatively more important as succession proceeds, a least in some of the vegetation layers. The number of layers also tends to rise, but the boundaries between the various superimposed structural components may gradually become more obscure. Also in the horizontal direction the structural patterns often become less conspicuous by the gradual fading of originally sharper boundaries. Although the overall structure may gain in complexity, at least during several stages, the patterns become less surveyable. MARGALEF (1958) distinguishes coarsely and finely patterned biotic communities: "The structure of a community can take various forms. There are "fine grain" ones, that is, with small heterogeneous nuclei, which give rise to many similar structures, which are repeated indefinitely. There are also relatively large surfaces or volumes occupied by individuals distributed in a relatively uniform manner, but differently than in other contiguous places. Finally, the heterogeneity may pass from one plane to another in a gradual way, so that "grains" or structures are not distinguishable". In this terminology (which was adopted by VAN LEEUWEN,1966) one night say that coarse-grained structure or structures with a coarse lattice pattern are gradually replaced by structures with a finer grain or finer lattice. According to ODUM (1963) the climax community exhibits the highest degree of differentiation. There are arguments, however, which render it rather doubtful whether this helds in general, and suggest rather that the highest degree of differentiation is reached before the stages of stabilization in the ecosystem have developed.

4. Structure in a qualitative sense

When special ecological conditions prevail, the physiognomic structure apparently alters itself according to certain causal relations; in other words, it is, up to a point at least, pre-

dictable and, accordingly, there is a fixed sequence of the first appearance of the various growth- and life forms involved and also of their respective biomass relations. In addition, the dimensions can be predicted within certain limits. The pioneer phases at least may be of a different nature in different series. Some basic patterns are presumably recognisable amomg the variants. It is, for instance, very striking that the majority of the pioneer species belongs either to thallophytic groups or to graminoid or therophytic spermatophates. The questions which of these growth forms prevails is mainly decided by the habitat, but it is probably, among other things, also dependent on the time-scale during which the environmental factors are operative. One gets the impression that in many instances lichens are rather typical of solid substrates such as rocks and the barks of trees. The favoured habitats of algae are water an (in the second place) relatively soft mineral subtrates. Pioneers among the Bryophytes often grow on soft humose or completely organic substrates, frequently in secundary successions, e.g. in humid places after fires, on sides of ditches, on dug-op peat or on decaying wood. (Dr.J.J.BARKMAN, personal communication).Cormophytes are often pioneers on more or less mobile substrates. Therophytes occur more often than not in situations which have a more ephemeral character, or are consistently of short duration (e.g.,during the rather brief rainy periods in desert areas). Graminoid pioneer forms presumably prefer a habitat with fairly strong, recurring or more lasting changes in the environmental factors, usually concerning the substratum (such as changes in the water level or a shifting of soil particles).
In all instances, usually a phase occurs in which geophytes and hemicryptophytes start plying an important part, the succession often proceeding via brushwood or scrub vegetation in the direction of a forest and showing an increasing number of life- and growth forms. In a well-established forest the aspect it not only determined by the trees, but also by lianas and epiphytes. There is evidently a convergence of physiognomic features and this served as one of the foundations of the climax theory of CLEMENTS (1928) which, in its original form, is no longer accepted as generally valid.

5. Dimensions and life cycles

In the course of a succession series gradually species appear with increasingly greater dimensions and life cycles of longer duration. Pioneer species are often small and short-lived,which holds especially for therophytes. Cormophytic pioneers are usually preceded by micro-organism, such as bacteria, with a generation cycle lasting from a few hours to a few days. The largest forms we know, the trees, are not only of a considerable size, but also have a life cycle of long duration. In the case of peatmoor development, the ultimate stage is built up chiefly by relatively small organism (*Sphagnum* species and other Bryophytes).

6. Morphological structures

Frequently, in pioneer stages, the number of species with a simple morphological structure is relatively high and the number of species with a relatively more complicated structure may increase as the succession series proceeds. The question of the primary or the secondary nature of the simplicity of structure is not considered here. Pioneer taxa show, generally speaking, little ramification. Of general occurrence in initial vegetation are also species with simple leaves, which are often long and narrow (grasslike, linear, strap-shaped,etc.) or very much reduced (to spines or scales) to absent (in stem succulents).

7. Symmetry relations

During succession, the symmetry relations between species often appear to be changing. This is another aspect which up to now has received hardly any attention, but is may well prove to have an essential bearing upon our understanding of vegetational ecology. There are reasons to suppose that in several cases, as an example, the number of species with a more complicated phyllotaxis (e.g.,3/8 or 5/13 instead 1/2 or 1/3 increases). It may be of interest to verify if there will be any relation with efficiency in the utilisation of energy. Similarly, it may be of interest to study the number of planes of symmetry and especially to include the subterranean organs in the comparison. In pioneer habitats bilateral symmetry (or even asymmetry) is of fairly common occurrence, e.g., in species of ferns in rock crevices, among graminoid forms and in species with rhizomes or runners. Many ecological factors acting vectorially, a polarity is not at all surprising. Preliminary studies clearly point to the frequent relation between the direction of growth of rhizomes and the gradient along which environmental factors change.

Such "polarities" will probably decrease during succession.

8. Changes in the habitat

As has been discussed under (1), the vegetation cover by itself contributes towards a change of the ecological habitat. The following interactions between plant cover and habitat, partly already mentioned by CLEMENTS (1928), are the most important: the effect of vegetation on the weathering of the substratum; the inhibition of soil erosion; catching of soil and dust particles transported by winds, deposition of silt suspensions in water, etc.; silting-up, formation of litter, accumulation of humus, enrichment or exhaustion of mineral nutrients in the subsoil; changes in microclimate by an influence on the temperature (shading etc.), on the relative humidity (by evaporation, but also by protection of the soil against an excessive loss of water), on the quantity of light (diminishing towards the ground level) and on its spectral composition; on the effect of the wind or the dash of waves; on the oxygen metabolism in the soil (aeration, etc.); and biotic reactions and relations.

A vegetation cover is even the cause of an increased differentiation within the environment: as soon as a single plant has established itself, the habitat has become somewhat altered in the immediate vicinity of that lone specimen in respect of the neighbouring places, in which no vegetation has developed as yet.

Whenever the habitat is temporarily submitted to extreme conditions, such as a severe and long-lasting period of frost, a very dry summer, or an exceptionally high (or low) water level, the degree of organisation may show a temporary fall-off to an earlier phase, especially in less stable systems (see under 19).

9. Complex water relations

In both xero- and hygroseries the micro-environment becomes less extreme as far as the factor 'water' is concerned. This is also apparent from the specific composition: gradually more species with a mesophytic character appear.

10. Ecological amplitudes

Certain changes also occur in the manner, in which the various species react on their changing environment, but such changes can not always be detected, simply because we but seldom have a thorough knowledge of the environmental factors and of the alterations to which they are submitted. In the present context it seems best to restrict the discussion to those factors or complexes of factors which determine more or less clearly and directly the progressive trend in a succession series. I have the impression, that many pioneer species exhibit a relatively broad ecological amplitude in respect of these vectorial environmental factors (but not necessarily in respect of other environmental conditions). We know as yet comparatively little about these directional factors, but we may presumably accept that such factors as the progressive development of the substratum (in all sorts of succession series), the displacement of soil particles in psammo-series, the amount of water in lithoseries, and the dash of waves in hydroseries belong here. In this connexion, a confusion may be caused by the use of such terms as euryecious and stenoecious, if the latter are not related to the prevalence of varying conditions ("*Wechselbedingungen*"). A plant series which is tolerant of, say, considerable fluctuations in salinity can be called 'stenoecious', but perhaps one should rather apply the term 'poikiloecious' to it, or preferably *poikilohaline*, if one knows only very little of the possible alterations of other environmental conditions. This does not imply that pioneer species are euryecious (compare VAN STEENIS, 1941), nor that they are stenoecious. One should avoid the use of such terms as much as possible. The so-called 'Law of Ecesis' of DANSEREAU (1957) can, therefore, not be accepted without certain restrictions. DANSEREAU expressed his 'Law' as follows: "The resources of an unoccupied environment will first be exploited by organism with high tolerance and generally with low requirements".

The degree of fluctuation of environmental condition during the vegetation period is usually of essential importance for the ecological interpretation. The extreme values (the maxima and the minima) are often much more important than the 'average' prevailing conditions in determining the composition and the structure of the community (SEGAL, 1965). In the initial phase the minima and maxima of essential environmental factors are often relatively far apart (KERSHAW, 1965). Pioneers, the early settlers, are quite often specialized in that they are adapted to very special environmental factors, the 'master factors', which in many instances implies that they are tolerant of the extreme values of those factors.

11. Habitat of the ultimate stage

The ultimate stage of the succession series is determined by the degree of extremity of the habitat. The 'extremity' does not only include very high or exceptionally low values of master factors, but also fluctuations with a considerable amplitude in a certain lapse of time. The vegetation as such can level out the fluctuations in the environmental conditions, among other things, by its heat-insulating capacities (also by the confinement of air masses between the plants or parts of the plants), by the locally high relatively humidity (partly caused by stomatal evaporation), by shading, by providing shelter against the wind, by producing litter and humus etc. In very extreme habitats, e.g., in arid zones, in south-facing exposed rocks in warmer regions, on beaches, but also in extremely fertile or very poor environments, in very hot or very cold climates, in sites strongly influenced by winds of considerable force or by wave-dash, and, generally speaking, also in extreme unstable environments (such as sites with a considerable degree of fluctuation of the water level), the vegetation cover does not develop beyond a certain stage in the succession series (see under 4). In the most extreme cases no vegetation establishes itself at all, or at best can only maintain itself at the initial level.

12. Secundary succession

Generally speaking, secundary succession does not proceed in exactly the same way as in a primary series. Several forms of disturbance cause a reversion to a coarser lattice pattern,for instance to a situation of (co-)dominance of one species or of a few, but without a complete restauration of the composition of the comparable structural organisation prevailing in one of the phases of the original succession series. After the disturbance has set in, the environment is basically different from what it was at the initiation of the primary series, because the vegetation cover had previously developed in that habitat and changed the environment, for instance by the production of litter and humus, which also retains water. In several respects the environment is not quite so 'extreme' as it was in the beginning. Only in some extreme forms of disturbance verging upon calamities can a situation develop which corresponds with the

first stage of a primary succession. After the disturbance, the changed environment is also different from all subsequent phases of the primary succession, because every disturbance of any importance destroys at least a part of the vegetation cover,whilst creating new favourable sites for the germination of diaspores, and results in a new combination of environmental factors hardly ever corresponding with any stage of the primary series. The tangible result is that quite often plants can settle whose adaptive capacities differ appreciably from those of the species normally occurring in the primary succession series. Vegetation established in places where a patch of forest has been felled or burnt down, for instance, if often composed of tall, mostly perennial herbs ('Hochstauden'), a structural type characteristic of several forms of natural and anthropogenic disturbance.
In other cases vegetation consisting of tussock-forming species appear, e.g., *Carex hudsonii* vegetation after felling of alder carr in fen regions in the Low Countries, and *Molinia coerulea* after the sudden drainage of peat bogs which are rather poor in nutrients. (Incidentally 'Hochstauden' are often also typical of natural or semi-natural ecotones, for instance of vegetation of stream banks, and of margins of lakes and ditches. They usually appear in sites where sharp boundaries, both in time and in space, prevail. The same applies to plants with a tussock habit).
Jungle vegetation developing after forest fires likewise specific of secondary succession series. In the tropics and subtropics, certain shrubs and trees such as species of *Melaleuca* and *Eucalyptus* appear to be typical of secondary vegetation in areas where forest fires are of common occurrence, the bark of their trunks being sufficiently resistent to survive an ordeal by fire. The bark of such forms does not burn, but only becomes scorched (VAN STEENIS,1941).

13. Range of processes

In the coarse of succession series a number of stages take place, which have been accurately described by CLEMENTS (1928). According to this author, the stages passed through by each individual species can be described as migration, ecesis (colonisation), aggregation, competition, and reaction, to which stabilisation may be added. The range for the community can be described, partly in accordance with CLEMENTS, as follows: initiation, settlement (selection), interference, consolidation (development, regulation, controll: DANSEREAU,1957) and stabilisation.

14. Interference and interdependence

Little is known about interference phenomena and mutual despendence relations between organism. After the appearance of CLEMENTS' publication it has generally been assumed that the effect of competition is not inconsiderable in or after a certain successional stage. Apart from competition, there are related phenomena, such as allelopathy, a number of forms of symbiosis, heterothrophy, etc. According to ODUM (1963), the number of autotrophic and, at least during certain successional phases, also the number of heterotrophic organism increases. Many higher plants are not obligate autotrophs and it is feasible that particularly in pioneer situations or in other extreme environments the number of true autotrophs is relatively high and heterotrophy becomes more important in later stages. Forms of symbiosis between tress or shrubs and soil fungi are most probably very common. In a number of cases an increased rate of litter decomposition can be observed, leading to remineralisation of the accumulated organic matter, which thus becomes available again for resorption by higher plants. This decomposition process retards the exhaustion of the soil (WESTHOFF,1965). The relatively little studied subject of antibiosis is much better known from some lower fungi than among higher plants (see,e.g.,FUNKE, 1943; GRÜMMER,1955). The question in how far species exhibiting allelopathy under cultural conditions show this phenomenon in natural circumstances has hardly been studied.

15. Dominance

Dominance of single species may prevail both in pioneer habitats and in terminal stages. Dominance of the pioneer species is usually associated with the rapid development and the rate of growth of these forms, notably of the more ephemerous ones.More instantaneous reactions upon the environmental conditions gain the upper hand over the slower ones. The broader range of tolerance in fluctuations of the master factors of many pioneers is of course important in this connection. At a later stage other reactions become accelerated as the external conditions change in the habitat. In spite of their broad range of tolerance (and, hence, adaptibility) and rapid rate of development, the competitive capacities of pioneer species are, generally speaking, not very strong (cf.IVERSEN,1960). The seeds of most of the pioneer species requiring high light intensities for their germination, a rejuvenation is hampered in later stages by the much denser vegetation cover. The dominance of trees in the terminal stages need not be impeded by prerequisite germination conditions for maintaining a population equilibrium.

16. Dissemination

Upon the whole, pioneer species have a high dissemination capacity (cf.IVERSEN,1960), either by producing a large number of diaspores, or by producing runners. Among the early settlers, anemochorous species (e.g.,*Asteraceae* and 'tumble weeds') and hydrochorous ones (e.g., in haloseries and hydroseries) are quite common. Autosamy is probably also relatively common in pioneer stages. Species with large and heavy seeds (or fruits), and more particularly the arborescent forms, are more common in the terminal succession stages.

17. Neophytes

Neophytes can usually only establish themselves in pioneer habitats, or in unstable (frequently much disturbed) environments, generally speaking, in 'open' situations (cf.SUKOPP,1962). In later successional stages the ecological niches have all been occupied already and each species plays its own part in the more stabilised ecosystem. In newly colonised regions such neophytes initially are not necessarily controlled by their natural enemies (if the latter were not introduced simultaneously and their competitive power may thus be considerably higher than in their country of origin.

18. Rate of Succession

The velocity of succession may be initially rather high or rather low. In the most extreme habitats the rate of succession is practically nil. This rate also falls to zero as soon as the structure reaches its upper limit. In some cases the velocity gradually accelerates during several stages, as may be observed in the hydroseries in a lake. This is partly attributable to the accelerated accumulation of dead vegetable matter, from which most of the silt is derived and which, by raising the soil level, creates favourable conditions for the establishment of new forms and of more and larger individuals, which in turn more and more level out the differences in environmental conditions and form new niches until the highest attainable level of differentiation has developed. The acceleration of succession in its later stages may perhaps also be the result of the increased competitive capacities of species appearing during the later stages (see also SEGAL 1966).

A succession probably proceeds more rapidly in secundary series than in comparable primary series. On the whole, only few exact data on the velocity of succession are known, so that every conclusion at this stage of our knowledge is decidedly premature. The role of geomorphological conditions is discussed by SEGAL (1970).

19. Stability

The progress of a succession series is generally concomitant with a greater stability of the ecosystem (cf.MARGELEFF,1958; 1963). This means that the extremes of many of the fluctuating environmental factors approach one another more closely and that sudden changes in the habitat are more readily levelled out.[1] The greater stability is partly the result of the higher degree of differentiation and of the increasing number of infra- and syn-specific interrelations. In a pioneer situation an ecosystem is almost non-existent. However, the plants themselves contribute towards an increase in stability of the habitat, so that they function to a certain extent as 'regulators'. In proportion to the increase in the number of species and the complexity of the structure, more specialised regulators will appear.

The accumulation of litter and humus also contributes towards an increase in stability. The development of micro-organism and of mycorrhizal symbiosis inhibits exhaustion of the soil.

20. Biomass and production

According to ODUM (1963), during succession normally an increase of the biomass and of the dead vegetable matter can be observed, which is concomitant with an increase of the total production per unit of surface area (or of volume). The rate of metabolism per unit of biomass is suppoded gradually to fall off. However, the general validity of these principles is open to doubt. The experiences gained in agricultural practice are certainly not confirmative: quite often there is an increase in rate of the metabolic processes per unit of biomass resulting in a decrease of litter and humus.

[1] that species composition, number of organisms, and number relation between organisms of different species, only fluctuate little in time.

The various ecological relations and sequences under discussion can also occur in a smaller compass. During the establishment of bryophyte communities, for instance, it is a rule of the thumb that acrocarpic mosses are gradually replaced by pleurocarpic forms when the habitat becomes less extreme. Sometimes even 'dendroid' mosses may follow. In this case we again find an increase in, e.g.,morphological complexity, in mode of branching and in dimensions. Something similar occurs during the succession in lichen vegetation, in which the appearance of the dendroid forms frequently follows after the establishment of the relatively little differentiated crustaceous species and of the frondose ones, respectively.
There are several points of contact between information theory and some of the above-mentioned principles, provided the latter are valid. In this respect, the publications of PATTEN (1959) and of MARGALEFF (1958) are of interest. WHITTAKER (1953) already expressed the view that succession can be regarded as a complex of phenomena proceeding towards a maximum of complexity, of differentiation (diversity), of production, and of stability, i.e., of all characteristics, that enhance regulation and various interrelations and thus lead to a higher degree of 'information'. The disorder decreases and consequently the 'entropy' of the system. A rising degree of order or of information can also be expressed as an increase of the negative entropy. PATTEN says: "Entropy is continually created from the information stored. To offset this tendency to lose order, the system continually accumulated negentropy from the environment. The balance between information gained and entropy generated in a unit of time determines the dynamic state of the ecosystem. If more information is accumulated than issued, the excess is converted to biomass. Such a condition of positive balance represents a growing or successional stage....". A conclusion emanating from VAN LEEUWEN's (1966) relation theory is that during succession the variation in time is replaced by a variation in spational distribution.An investigation into succession in aquatic communities yielded the same conclusion (SEGAL,1970).

ZUSAMMENFASSUNG

Es wird eine Inventarisation von Tendenzen gegeben, die möglicherweise bei autogenen und allogenen primären (und manchmal anderen) Sukzessionsreihen eine Rolle spielen könnten. Diese Inventarisation kann vielleicht zu einer Diskussionsbasis führen. Sie muß aber nicht unbedingt als eine Reihe von Gesetzmäßigkeiten gesehen werden. Die Hypothesen sind teilweise der Literatur entnommen.

1. Die Zahl der Arten nimmt zu; wenigstens während einiger Stadien, aber in Endstadien kann diese Zahl verhältnismäßig niedrig sein.
2. Es kann einen Wechsel von bestimmten taxonomischen Gruppen geben.
3. Wenigstens während einiger Stadien findet eine zahlenmäßige Zunahme von Strukturelementen statt. Die Grenzen zwischen den Strukturelementen verwischen sich. Grobe Muster werden durch feinere ersetzt.
4. Die physiognomische Struktur verändert sich gesetzmäßig. Allgemeine ökologische Bedingungen werden für Thallophyten, Therophyten und Graminoiden an verschiedenen Pionierstandorten erwähnt.
5. Während der Sukzession treten anfangs allmählich immer mehr Arten auf mit größeren Ausmaßen und längeren Generationsperioden.
6. Öfter nimmt die Zahl der Arten mit einer relativ komplizierten morphologischen Struktur zu.
7. In den Symmetrieverhältnissen der aufeinander folgenden Arten können sich Veränderungen zeigen.
8. Die Vegetation ändert das Milieu, die Differenzierung nimmt in horizontaler und vertikaler Richtung zu.
9. Das System bekommt einen immer mehr mesophytischen Charakter.
10. Die ökologischen Amplituden ändern sich mit den Wechselbedingungen, insbesondere in Bezug auf Maximum- und Minimumfaktoren.
11. Der Endpunkt der Sukzession wird bestimmt von dem Grad der Eigenschaften, in welchem sich der Standort extrem zeigt (nicht nur in der Bedeutung von hohen und niederen Intensitätswerten, sondern auch von Wechselbedingungen).
12. Die sekundäre Sukzession verläuft nicht analog der primären. Bestimmte Wuchsformen sind spezifisch für sekundäre Reihen (z.B. Hochstauden, rasenbildende Graminoiden).

13. Die Reihenfolge der Prozesse umfaßt folgende Stadien: Initiation, Ansiedlung (Selektion), Interferenz, Konsolidation (Entwicklung, Regulation) und Stabilisation. Für die Arten: Migration, Kolonisation, Aggregation, Interferenz, Reaktion und Stabilisation.
14. Über die biologischen Interferenzeinflüsse ist wenig bekannt.
15. Dominanz kann sowohl in Pionier- wie in Endstadien eine wichtige Rolle spielen. Die Konkurrenzkraft der Pionierarten ist öfter niedrig.
16. Pionierarten haben öfter eine verhältnismäßig große Disseminationskapazität.
17. Neophyten siedeln sich öfter auf Pionierstandorten, auf gestörten oder auf anderen Standorten mit starken Wechselbedingungen an.
18. Die Schnelligkeit der Sukzession ist abhängig von dem Grad der Eigenschaften, in welchem sich der Standort extrem zeigt.
19. Sukzession geht mit größerer Stabilität des Ökosystems zusammen.
20. Die Sukzession braucht nicht notwendig eine Zunahme der Biomasse oder der totalen Produktion pro Oberflächeneinheit mit sich zu bringen.

Wenn einige allgemeine Prinzipien vorhanden sind, können sie wahrscheinlich auf der Basis der Informationstheorie formuliert werden.

SUMMARY

The purpose of this paper is to give a survey of the various possible changes in the ecosystem during succession. General trends, apart from other phenomena, include a number of groups of patterns and processes: number of species, taxonomic groups, structure in a quantitative and qualitative sense, dimensions and duration of life cycles, morphological structures, symmetry relations, changes in the habitat (e.g., complex water relations), ecological amplitudes, range of processes, interference and interdependance, dominance, dissemination, rate of succession, stability, biomass and production.

Special attention is focussed on progressive primary succession, and to the determination of the ultimate stage. A number of general trends are tentatively enumerated. The role of neophytes in succession and some principles in secundary succession have also been taken into account. It is suggested that, if some general trends can be indicated and if needed, can be quantified, there are several points of contact with information theory.

LITERATUR

CLEMENTS,F.E. -1904- The development and structure of vegetation.- Bot. Survey Nebraska 13. Lincoln.

-- -- -1928- Plant succession and indicators.- New York. 453 pp.

DANSEREAU,P. -1957- Biogeography, and ecological perspective.- New York.394 pp.

FUNKE,G.L. -1943- Experimenteele plantensociologie.- Gorinchem. 224 pp.

GRÜMMER,G. -1955- Die gegenseitige Beeinflussung höherer Pflanzen.- Jena. 162 pp.

IVERSEN,J. -1960- Problems of the early post-glacial forst development in Denmark.- Denm.Geol.Unders.IV, Rackke 4 (3): 6-32.

KERSHAW,K.A. -1964- Quantitative and dynamic ecology.- London. 183 pp.

LEEUWEN,C.G.van -1966- A relation theoretical approach to pattern and process in vegetation.- Wentia 15:25-46. Amsterdam.

MARGALEF,R. -1958- Information theory in ecology. General Systems, Yearb. Soc.Gen.Systems 3:36-71.

-- -- -1963- On certain unifying principles in ecology.- Am.Natural. 97:357-374. Lancaster,Pa.

ODUM,E.P. -1963- Ecology.- New York. 152 pp.

PATTEN,B.C. -1959- An introduction to the cybernetics of the ecosystem: The tropic-dynamic aspect.- Ecol.40. New York.

SEGAL,S. -1966- Ecological studies of peat-bog vegetation in the northwestern part of the province of Overijsel (The Netherlands).- Wentia 15:109-141. Amsterdam.

-- -- -1968- On structure, zonation and succession in vegetation of higher aquatics.- Meeting of Experts on Ecological and Control of Aquatic vegetation. Unesco,Paris,16-18.Dez.1968. (Mskr.vervielf.).

-- -- -1970- Strukturen und Wasserpflanzen.- In: Tüxen,R.(Ed.): Gesellschaftsmorphologie. Ber.Intern.Sympos.Rinteln 1966: 157-171. Den Haag.

STEENIS,C.G.G.J.van -1941- Oekologische eigenschappen van pionierplanten.- Verslag 28e Verg.Veren.Proefstation Personeel Buitenzorg: 195-205. Buitenzorg.

SUKOPP,H. -1962- Neophyten in natürlichen Pflanzengesellschaften Mitteleuropas.- Ber.Dtsch.Bot.Ges.75:193-205. Stuttgart.

WARMING,E. & GRAEBNER,P. -1918- Lehrbuch der ökologischen Pflanzengeographie.- Berlin. 988 pp., 3.Aufl.

WESTHOFF,V. -1965- Plantengemeenschappen.- In:"Uit de plantenwereld", Arnhem. 288-349 pp.

WHITTAKER,R.H. -1953- A consideration of climax theory: the climax as a population and pattern.- Ecol.Monogr.23: 41-78. Durham,N.C.

Addition to p.14.

The fall of the species number is partly due by concurrence during the interference- and consodilations phases (see under 13). These phenomena particularly held for the primary producers. Generally, the species number of consumers and decomposers also increase, and the total sum of all species, including fungi, bacteria and animals, may possibly be highest in the end phase of progressive series.

Addition to p.18.

It is impossible to summarize the literature on the subject published after my paper was finished in 1967. One of the most related papers was E.P.ODUM,1969: The strategy of ecosystem development, Science 164: 262-270. Our knowledge has increased considerably in the meantime, but for the moment only two notes are added.

1. During succession we often find an increase in homogeneity, and a decrease of overdispersion in the middle stages.
2. Discontinuities in space and time may arise if the change in a specific habitat factor steps across a threshold, by which a complex of factors is changed all at once. An example is the first phase of hygroserie continuing from a hydrosere, about or behind the contact zone of the surface water and the soil surface.Here we may find an apparent impoverishment in species composition and structure of the shore belt, which is, however, only manifest for the primary producers.

DISKUSSION

A.W.KÜCHLER

Ich möchte diesen Vortrag sehr begrüßen, weil er großen Wert auf die Struktur gelegt hat. Und das ist besonders wichtig, wenn es sich um Vegetationen handelt, mit denen man nicht sehr vertraut ist. Das ist natürlich beinahe immer der Fall, wenn man die Literatur aus anderen Ländern liest, die man nicht kennt. Es kommt sehr häufig vor, daß man die Struktur für selbstverständlich hält, weil man die Arten kennt. Wenn man die Arten kennt, dann glaubt man, man kenne auch die Struktur. Das braucht aber nicht der Fall zu sein, und kann sogar ganz irreführend sein. Außerdem, wenn man mit der Vegetation nicht vertraut ist, kann man sich überhaupt nicht vorstellen, um was es sich handelt, selbst wenn man die Artennamen vor sich hat. Bei einer Tagung der UNESCO im vorigen Jahr kam dieser Standpunkt auch wieder zutage: "Wir kennen ja die Arten, da brauchen Sie ja keine Struktur zu bezeichnen". Ich habe aber dann eine der größten und ausgedehntesten Formen der Vegetation in den USA rein floristisch beschrieben und habe die Gruppe gebeten, mir zu erklären, wie die Struktur aussieht. Das konnte natürlich niemand, weil man nicht wußte, um was es sich handelt, und die Arten haben nicht genug gesagt, daß man sich ein Bild hätte machen können. Eins ohne das andere ist m.E. immer etwas sehr unvollkommenes, und es ist wichtig, daß man beides macht. Ich darf in diesem Zusammenhang auch auf das Buch von ELLENBERG hinweisen, das vor 10 Jahren veröffentlicht wurde. In diesem hat er die verschiedenen Arten der Vegetations-Analyse und -Gliederung beschrieben, und die Struktur immer zusammen mit der floristischen Analyse gesehen.

H.ELLENBERG

Das, was Herr SEGAL uns hier zu geben versucht hat, ist sozusagen eine allgemeine Sukzessionslehre, oder eine Philosophie der Sukzession, oder letzten Endes eine mathematische Fassung der Sukzession. So wertvoll und notwendig auch in der Sukzessionsforschung eine synthetische Betrachtung ist, so problematisch ist doch eine zu frühe Verallgemeinerung. Manche der von SEGAL aufgestellten "Gesetzmäßigkeiten" sind nicht allgemein gültig, wie an vielen Beispielen zu zeigen ist. Man sollte nur einige wenige der von SEGAL aufgezeigten Probleme herausgreifen, diese aber konkret an zahlreichen Einzelfällen gründlich prüfen. Nur dann lassen sich die allgemeinen Schlußfolgerungen absichern.

A.W.KÜCHLER

Die Bemerkungen von Herrn ELLENBERG waren sehr interessant,weil ja die Verallgemeinerung der Sukzession schon lange vor Herrn SEGAL klar und zusammenfassend dargelegt wurde, und infolgedessen stehen wir in keiner Weise am Anfang, wir haben die Verallgemeinerung schon vor Jahrzehnten vor uns gehabt und so ist es vielleicht geradezu begrüßenswert, wenn wir weitere Verallgemeinerungen haben, die die Schwächen aufzeigen und die Lücken, die man dann ausfüllen kann.

Aber ich glaube doch, gerade auf grund der amerikanischen Arbeiten ist das Studium der Sukzession so weit fortgeschritten, daß Verallgemeinerungen heutzutage nicht nur unannehmbar, sondern geradezu veraltet sind. Es kommt mir vor wie Herr GAUSSEN das seinerzeit von einer Vegetationskarte beschrieb, als er seine erste Karte von Perpignan machte: die Karte mußte vollkommen werden. Es war eine wunderschöne fehlerlose Karte. Schließlich kam er aber zu dem Schluß, wenn er die Karte vollkommen machen müßte, dann würde sie nie veröffentlicht, weil immer wieder ein Fehler, etwas Neues entdeckt werden würde, außerdem, wenn er sie endlich vollkommen findet, dann kann es ja immer noch sein, daß jemand anders sie nicht vollkommen findet. Es ist also besser, man veröffentlicht seine Idee, und dann kann man anhand der Kritik der Kollegen feststellen, in welcher Weise die Arbeit verbessert werden könnte.

S.SEGAL

Die Frage ist wohl eher, ob es Zweck hat, schon frühzeitig einen Überblick über Prinzipien zu geben, die es geben kann. Ich habe auch anfangs schon gesagt, daß es nur um die Diskussion geht, und ich bin nicht sicher, daß alle Prinzipien gelten. Aber dessen bin ich sicher, daß von verschiedenen Prinzipien noch vieles anderes zu sagen wäre. Aus Zeitgründen habe ich bei meinem Vortrag gerade die Ausnahmen nicht erwähnen können.

BEZIEHUNGEN ZWISCHEN STRUKTUR UND DYNAMIK IN ÖKOSYSTEMEN

E. van der M a a r e l[1], C.G. van L e e u w e n
Mitteilung des Reichsinstitutes für Naturverwaltung, Leersum (NL)

1. Einführung

In diesem Beitrag versuchen wir, Struktur und Dynamik in der Vegetation in einem Zusammenhang zu betrachten. Diese Betrachtung stützt sich auf die System- und die Informationstheorie, wie sie z.B. von ROSS ASHBY (1964) und von MARGALEF (1958,1961) schon in die Biologie eingeführt worden sind. Der Zweite von uns hat eine Relationstheorie aufgestellt, die sich als Grundlage für Untersuchungen der Beziehungen zwischen Struktur und Dynamik, wohl auch "pattern and process" genannt, eignet. Er hat darüber im letzten Symposion schon in seinem Vortrag "Raum-Zeit-Beziehungen in der Vegetation" (siehe VAN LEEUWEN 1966) berichtet.
Wir wissen, daß diese neuen Disziplinen schwierig sind, jedenfalls bislang in der Pflanzensoziologie und der allgemeinen Ökologie noch schwer anwendbar sind. Doch sehen wir in dieser Richtung Perspektiven für die Entwicklung einer allgemeinen Theorie der Vegetationsdynamik, was uns gerade im Rahmen dieses Symposion über Syndynamik interessiert. Dabei suchen wir Anschluß bei zwei schon weiter entwickelten Betrachtungsweisen, der phytosoziologischen Sukzessionstheorie einerseits, und der Lehre von "biomass and energy flow" in Ökosystem - Syndynamik, bzw. Symphysiologie andererseits. Wir fassen ein Ökosystem im üblichen Sinne als eine funktionelle Einheit von Vegetation, Milieu und damit verknüpften Fauna-Elementen auf, die etwa dem mehr geographisch gefaßten Begriff Biogeozönose der russischen Ökologen parallel ist. Die Hauptthese der Relationstheorie ist für uns die gegenseitige Bedingtheit von räumlicher und zeitlicher Variation innerhalb eines Ökosystems, wie sich aus dem folgenden Schema erkennen läßt:

$$a \ (s) \rightleftarrows \bar{a} \ (t)$$
$$\bar{a} \ (s) \rightleftarrows a \ (t)$$

a = Variation; $\bar{a}$ = "Nicht-Variation"; (s) = Raum; (t) = Zeit.

[1] Der Autor war zur Zeit des Symposiums am Laboratorium für Pflanzenökologie der Universität Groningen und hat seinen Anteil an diesem Referat in diesem Institut vorbereitet.

Diese Beziehungen können wir auch folgendermaßen andeuten: Großräumliche Variation hängt mit großer Stabilität, großzeitliche Variation (Instabilität) mit kleinräumlicherer Vegetation, also Uniformität zusammen. Die beiderseits gerichteten Pfeile bedeuten, daß eine Änderung in der räumlichen Situation sich in einer Änderung der temporellen Beziehungen äußert. Aber auch umgekehrt wird eine Stabilisierung, bzw. Instabilisierung auf die räumliche Struktur einwirken.
Selbstverständlich dürfen wir die gegebene Ausgangssituation nur als einen Rahmen sehen, in dem jedes Ökosystem eine bestimmte Raum-Zeit bedingte Stellung einnehmen kann.
Besonders in verhältnismäßig jungen Systemen können wir uns Einwirkungen zwischen Vegetation und Milieu in erster Linie von Boden und Klima denken.
Einerseits ist die Vegetation in ihrer Variation und Stabilität bedingt durch die raum-zeitlichen Eigenschaften des Milieus, andererseits wirkt die Vegetation auf das Milieu ein und zwar in einer Art Rückkoppelung, wobei die räumliche Variation im Milieu vergrößert wird, und die im Milieu gegebenen Instabilitäten abgeschwächt werden. Wir können dabei an die Entwicklung von Boden-Horizonten, die Stratifikation in der Vegetation, die Bildung von Bulten und Schlenken in bestimmten Hochmoortypen usw., denken.
Trotz dieser Verknüpfungen zwischen Vegetation und Milieu ist doch die primär gegebene räumliche und zeitliche Variation in Boden und Klima für die Gestalt des ganzen Ökosystems maßgebend.

Diese Tatsache weist darauf hin, daß wir unsere Untersuchungen gerade auf das Milieu konzentrieren sollten. Aber bisher ist es noch kaum gelungen, diese Eigenschaften des Milieus effektiv zu beschreiben. Wir wissen oft nicht, welche Milieukomponenten von direkter Bedeutung für eine Differenzierung der Vegetation sind; bisher hat man noch sehr wenig quantitative Beschreibungen der Variation von bestimmten Komponenten durchgeführt. Die statistische Varianzanalyse ist in dieser Hinsicht natürlich von Bedeutung, doch genügt sie unseres Erachtens nicht. VAN DER MAAREL (1966b) hat zwar versucht, die Varianten mittels einer der Informationstheorie entlehnten Formel zu bestimmen, aber bisher ist diese Bestimmung nicht auf Milieuvariation angewandt worden.

Das schwierigste Problem ist wohl, die Instabilität quantitativ zu betrachten. In welchem Maße sollte man z.B. die jährliche Schwankung einer Grundwasserganglinie ausdrücken? Noch schwieriger ist es, den Einfluß einmaliger großer Änderungen zu beschreiben, z.B. die Abschnürung einer Salzwiese vom Meer. Diese Schwierigkeiten können wir vorläufig besser vermeiden; wir wollen unsere Untersuchungen vor allem der Vegetation widmen, wie es gerade in der europäischen Pflanzensoziologie besonders seit BRAUN-BLANQUET Tradition geworden ist. Dabei gehen wir von der allgemeinen Erfahrung aus, daß Struktur und Dynamik in den Vegetationsverhältnissen im Milieu widergespiegelt werden.

2. Bestimmung der räumlichen und zeitlichen Variation in der Vegetation

Die Variation in einer Vegetation kann zuerst auf die Arten bezogen werden. Hier sind die wertvollen Arbeiten von C.B.WILLIAMS (1964) und MARGALEF (1958,1961) zu erwähnen. Beide Ökologen haben einen Index vorgeschlagen, der auf der Verteilung von Pflanzenindividuen (Arten) beruht (Index of Diversity or Index of Biotic Diversity). PIELOU (1966a,b) hat einige von diesen Indices in einer Sukzessionsuntersuchung angewandt und dabei auch einen neuen Index of pattern diversity entwickelt. Die Indices von MARGALEF und PIELOU basieren völlig auf der Informationstheorie. Bei MARGALEF geht es um Individuen-Verteilung von Arten, bei PIELOU auch über Individuen-Verteilung über Oberflächen.

Neben Art-Diversität und Dispersions-Diversität könnte man auch die Struktur-Diversität zu quantifizieren versuchen. Bisher ist uns aber ein diesbezüglicher Index nicht bekannt geworden.
Die zeitlichen Änderungen in einer Vegetation könnte man parallel zur räumlichen Variation betrachten und zwar in der Weise, daß man die Raum-Indices in der Zeit verfolgt. Dabei kann man die Veränderlichkeit auf drei Niveaus beschreiben: Art-, Individuen- und Struktur-Veränderlichkeit. Ob man gewisse Änderungen in einem Stabilitätsrahmen zuläßt, hängt vom Zeitmaß, in welchem man diese Änderungen betrachtet, oder von der Definition der Stabilität ab. Das irregulär-ephemere und zyklische Vorkommen von Arten ist in einem stabilen System immerhin möglich - in Definitionen der Stabilität wie z.B. derjenigen von ASHBY (1964) sind allerdings zyklische Prozeßketten einbegriffen; man vergleiche auch zyklische Sukzessionen, wie sie z.B. in vorbildlicher Weise von WATT (1947) in seinem "pattern and process in plant communities" beschrieben worden sind.- In unserem Beitrag lassen

wir die Begriffe Stabilität und Instabilität außer acht und beschreiben jede Änderung als "zeitliche Variation".
Ein erstes Niveau der Stabilität oder Unveränderlichkeit ist erreicht, wenn die Arten-Garnitur von Jahr zu Jahr keine Änderungen zeigt. Das zweite Niveau der Unveränderlichkeit ist erreicht, wenn auch in der Individuen-Zusammensetzung keine großen Änderungen stattfinden. Es gibt schöne Beispiele von einer großen Diskrepanz zwischen diesen beiden Niveaus: So ist in einem reinen Salicornietum die Art-Unveränderlichkeit maximal, aber die Individuen-Unveränderlichkeit minimal; denn es handelt sich hier um eine annuelle Art, deren Population sich von Jahr zu Jahr völlig erneuert. Im allgemeinen könnte man vielleicht sagen, daß eine große Diskrepanz zwischen den genannten Stabilitätsniveaus darauf hindeutet, daß im Milieu eine sehr starke Veränderlichkeit herrscht, aber mit einem zyklischen Charakter, so daß im großen und ganzen doch eine gewisse Stabilität vorhanden ist.
Auf dem dritten Niveau der Unveränderlichkeit, also in den Individuen einer solchen Gemeinschaft, gibt es keine Änderungen in der Struktur. Unveränderlichkeit im absoluten Sinne kommt in einer Lebensgemeinschaft überhaupt nicht vor. Auch Struktur-Stabilität in einem systemtheoretischen Sinne, also auf der Basis von zyklischen oder rhythmischen Änderungen, ist wohl kaum zu erwarten. Hier denkt man wohl eher an das Bild des "dynamischen Gleichgewichtes".
In unseren hier zu erwähnenden kurzjährigen Beobachtungen haben wir uns also nicht um Stabilitätsbetrachtungen bemüht, sondern uns auf die unmittelbaren Beziehungen zwischen räumlicher und zeitlicher Variation beschränkt.
Die räumliche Variation haben wir dabei gemessen als Artenzahl bezogen auf die Flächengröße: $\alpha_v = G / \ln A$.
In Untersuchungen eines Dünen-Trockenrasen-Komplexes (VAN DER MAAREL 1966b) hat sich erwiesen, daß α_v deutlich mit dem Index α von WILLIAMS und auch mit dem Index von MARGALEF korreliert ist.
Die zeitliche Variation ist in einer einfachen Weise gemessen: $I_T = G_2 - G_c + G_1 - G_c$; G_2 und G_1 sind die Artenzahlen in zwei aufeinanderfolgenden Jahren, 2 und 1, G_c ist die Zahl der für die beiden Jahre gemeinsamen Arten: die zeitliche Variation ist also die Summe der nicht gemeinsamen Arten in zwei Artenlisten einer Vegetation in aufeinanderfolgenden Jahren. Dieser Index ist iden-

tisch mit einem Index I, der auch für vegetationssynthetische Studien schon angewandt worden ist, und der dem Index I von WILLIAMS, LAMBERT & LANCE (1966) sehr ähnlich ist.
Außerdem ist ein relativer Index I_T/I_T max erprobt worden. Dieser Index ist zurückzuführen auf die bekannte JACCARD-Formel für Similarität zwischen Artenlisten:

$$I \,/\, I_{max} = 1 - C_J, \text{ wo } C_J = \frac{G_c}{G_1 + G_2 - G_c}$$

Es ist allerdings noch fraglich, inwieweit man derartige Similaritäts- und Dissimilaritätskoeffizienten sowohl für räumliche als für zeitliche Beziehungen verwenden darf. Jedenfalls kommt in dem relativen Instabilitätsmaß neben der Änderungskomponente auch die Konstanzkomponente zum Ausdruck.
Im Großen verlaufen absolute und relative zeitliche Variation in ähnlicher Weise. Nur in Fällen, wo eine erhebliche Variation in der Artenzahl auftritt, ergibt sich ein regelmäßiger und besser zu interpretierender Verlauf der relativen Änderung.
In unseren Beispielen geben wir hauptsächlich Angaben über $^{I}\text{rel.}$

3. Diskussion über eine Landschaft mit großer zeitlicher Variation. (Untersuchungen von VAN LEEUWEN)

Unser erstes Beispiel ist die junge und dynamische Landschaft des ehemaligen Naturschutzgebietes De Beer. Dieses Gebiet lag an der Mündung des Rotterdamse Waterweg, südlich von Hoek van Holland und ist der Ausbreitung des Hafen- und Industriekomplexes Europoort zum Opfer gefallen, also eliminiert von einer "Dynamik höherer Ordnung".
1944 wurde in diesem hervorragendem Küstengebiet mit gelben und grünen Stränden, Salzwiesen, jungen primären Dünentälern und Sanddorn-Gebüschen ein Deich gebaut. Dieser Deich hat einen Teil dieses Gebietes von der Nordsee abgeschlossen.
Ab 1946 wurden 26 Dauerquadrate festgelegt, teilweise hinter diesem neuen Deich, teilweise im Vorland, wo die Gezeitenbewegung des Meerwassers ihre Wirkung immer noch ausüben konnte. Ab 1952 sind einige Quadrate teilweise eingezäunt worden, um den offenbar großen Einfluß der Kaninchen auf die Vegetationsentwicklung detailliert untersuchen zu können.
Neben diesen beiden erheblichen Einflüssen, Entsalzung und Begrasung, konnte ab 1955 noch ein dritter Einfluß, das Ausbrechen der Myxomatosis, studiert werden.
Zuerst vergleichen wir die Quadrate 2,7 und 18.

Tab. 1. Das Auftreten dominanter Arten in der Vegetationsentwicklung im Quadrat 2.- The occurrence of dominant species in the vegetation development of quadrat 2.

Aüfnahme-Nr.:	1	2	3	4	5	6	7	8	9	10	11	12	13
Chenopodium rubrum	3	.	1	.	.	.	.	.	.	.	.	.	.
Plantago coronopus	+	+	2	1	2	3	4	3	2	2	1	+	.
Spergularia salina	.	2	3	.	+	.	.	.	.	.	.	.	.
Agrostis stolonifera	.	+	+	+	1	1	1	1	2	3	2	2	2
Calamagrostis epigeios	.	.	.	.	+	.	.	+	1	2	3	4	4
Phragmites communis	.	.	.	.	.	.	.	.	.	.	2	2	5

Quadrat 2 liegt innerhalb des Deiches in einer feuchten, sandigen Ebene, die vor der Eindeichung völlig unter dem Einfluß des Meerwassers stand und wahrscheinlich kaum bewachsen war. In diesem Quadrat sehen wir rasche Änderungen. In Abb. 1 ist der Verlauf der Vegetationsbedeckung und der Höhe seit 1946 wiedergegeben. Abb. 2 zeigt den Verlauf der Artenzahl, der relativen zeitlichen Änderung in der Arten-Garnitur, bezogen auf die floristische Zusammensetzung im Jahr 1946, und die zeitliche Änderung von Jahr zu Jahr. Die Vegetation hat sich von einer offenen Chenopodium rubrum - Atriplex hastata - Gesellschaft im Jahre 1946, die wahrscheinlich schon die zweite Entwicklungsphase nach der Bedeichung war, über niedrige, artenreiche, mehr und mehr geschlossene Phasen mit *Spergularia*-Arten ond vor allem *Plantago coronopus*, in den Jahren 1947-1953 bis zu geschlossenen hohen Rasen mit rasch wechselnden Massenverhältnissen der Gras-Arten *Agrostis stolonifera*, *Calamagrostis epigeios* und *Phragmites communis* geändert.

Tab. 2 Das Verhalten der Armerion - und Agropyro - Rumicion - Arten in Quadrat 7.- The behaviour of Armerion and Agropyro - Rumicion species in Quadrat 7.

Aufnahme-Nr.:	1	2	3	4	5	6	7	8	9	10	11
Triglochin maritimum	1	.	+	.	.	.	.	.	.	.	.
Aster tripolium	+	+	.	.	.	.	.	.	.	.	.
Juncus gerardii	3	5	3	3	3	3	3	3	3	4	3
Glaux maritima	2	3	2	2	2	2	2	2	2	1	+
Plantago maritima	2	+	1	+	+	+	.	.	.	.	.
Festuca rubra f.litoralis	+	1	1	2	+	1	1	1	1	+	r
Agrostis stolonifera f. salina	3	3	3	3	2	2	2	2	3	4	5
Potentilla anserina	4	4	4	4	4	3	3	3	2	1	+
Carex distans	.	+	+	+	1	+	+	+	+	+	.
Plantago major	.	.	+	+	+	+	+	+	+	r	.

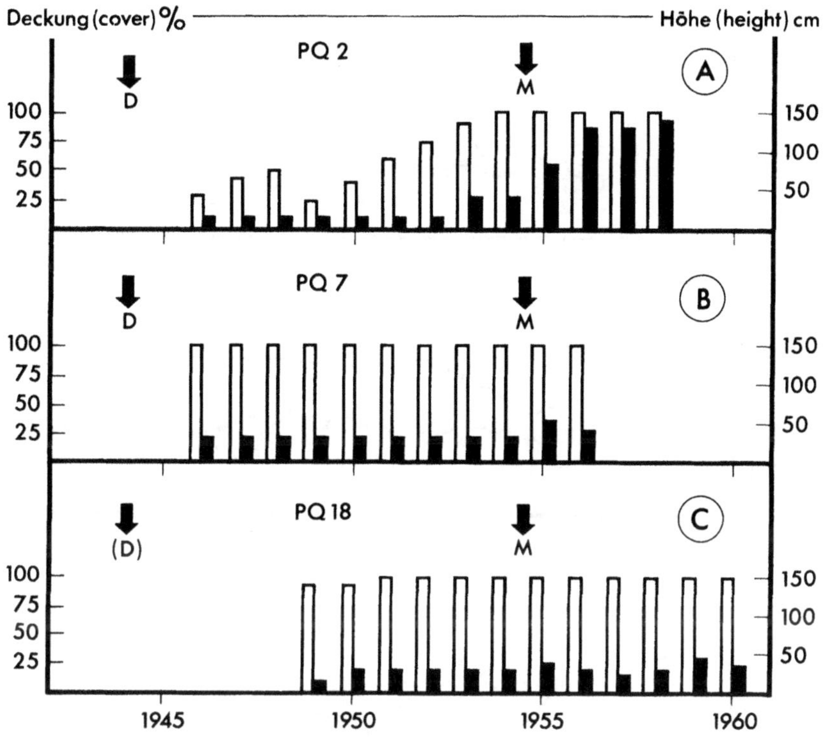

Abb.1 Deckung und Höhe der Vegetation in den Quadraten 2, 7 und 18 auf De Beer. D=Bedeichung, M=Myxomatosis
Cover and height of vegetation in quadrats 2, 7 and 18 on De Beer D=embankment, M=Myxomatosis

Quadrat 7 liegt ebenfalls innerhalb des Deiches in einer etwas mehr tonigen Ebene in größerer Entfernung vom Meer. Hier war schon eine Vegetation entwickelt, die von Fluktuationen sowohl in der Feuchtigkeit als im Salzgehalt des Bodenwassers beeinflußt war und zwar ein Juncetum gerardii mit einem deutlichen Agropyro-Rumicion-Anteil.
Tabelle 2 zeigt, wie sich allmählich einige weitere salztolerante Agropyro-Rumicion-Arten einfügen, weil die Arten-Garnitur im großen Ganzen konstant ist. Erst nach dem Eintritt der Myxomatosis ändert sich die Situation in einer leicht zu interpretierenden Weise: *Agrostis stolonifera* kommt zur Dominanz und die Artenzahl nimmt ab. Abb.1 und 2 zeigen die relativ große Konstanz in diesem Quadrat.

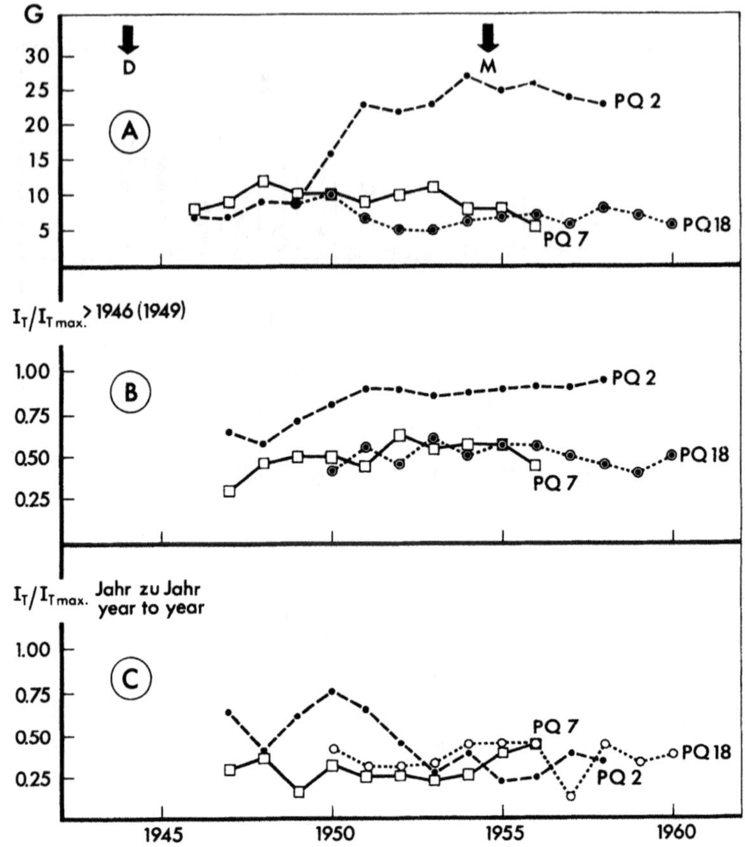

Abb.2. Artenzahl G, zeitliche Variation der Artengarnitur I_T/I_{max}, bezogen auf die Anfangssituation und die jahreszeitliche Variation in den Quadraten 2, 7 und 18.
Species number G, temporal variation I_T/I_{max}, in species composition compared with the original situation and yearly temporal variation in quadrats 2, 7 and 18.

Tab.3. Das Verhalten einiger konstanter A r m e r i o n - Arten im Quadrat 18.
The behaviour of some constant A r m e r i o n species in quadrat 18.

Aufnahme-Nr.:	1	2	3	4	5	6	7	8	9	10	11	12
Juncus gerardii	3	5	3	3	3	3	5	5	5	5	5	5
Glaux maritima	3	3	3	3	3	3	2	2	2	2	2	2
Plantago maritima	1	2	2	2	.	+	r	r	+	+	+	r
Agrostis stolonifera-salina	1	1	1	2	2	3	1	r	1	1	+	+

Quadrat 18 liegt im grünen Strand vor dem Deich und zwar in einem Priel, der unregelmäßig vom Meerwasser überflutet wird. Die Vegetation gehört zum J u n c e t u m g e r a r d i i und verhält sich ähnlich wie im Quadrat 7. Die Arten-Garnitur variiert auf dem gleichen niedrigen Niveau, die Massenverhältnisse sind hier im Anfang noch etwas konstanter. Die charakteristischen Arten wie *Juncus gerardii* und *Glaux maritima* sind im Jahre 1960 immer noch da, aber auch hier ist der Einfluß der Myxomatosis ganz deutlich: *Juncus gerardii* - nicht *Agrostis stolonifera* kommt zu völliger Dominanz. (Siehe Tabelle 3.)

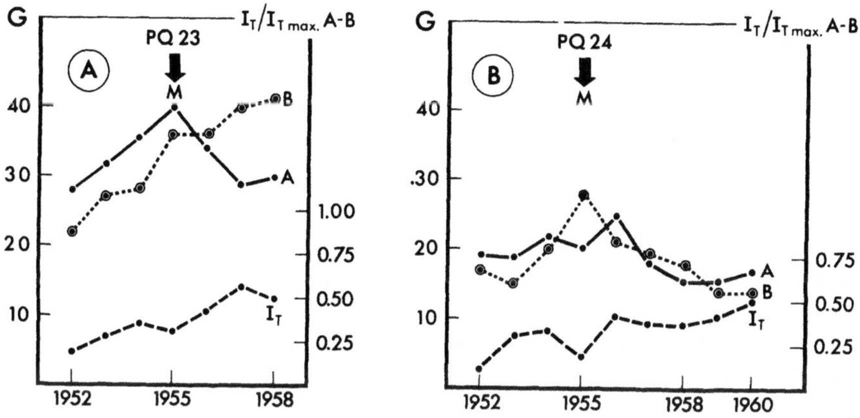

Abb.3. Artenzahl und floristische Unterschiede zwischen von Kaninchen beweideter und unbeweideter Vegetation in den Quadraten 23 und 24 auf De Beer. Species number and floristic difference between rabbit-grazed and ungrazed vegetation in quadrats 23 and 24 on De Beer.

Wir sehen also, wie unter bestimmten Verhältnissen eine Aussüßung einer Salzwiese überraschend geringe Änderungen in der Arten-Garnitur hervorruft. Über ähnliche Ergebnisse hat WESTHOFF (1969) in einem früheren Symposion-Beitrag berichtet. Es handelt sich dort um den Einfluß der Rinder auf die Entwicklung der Vegetation am Ufer der ehemaligen Zuidersee nach 1933. Auch hier war ein J u n c e t u m g e r a r d i i , das sich im großen und ganzen im beweideten Gebiet erhalten hat, anzutreffen, während im unbeweideten Abschnitt sich ein mit *Phragmites* durchstellter, an Hochstauden reicher Gürtel entwickelte. Auf De Beer verläuft die Entwicklung in ähnlicher Weise.

Zunächst werden zwei Quadrate besprochen, in denen der Einfluß einer Umzäunung gegen Kaninchen genau verfolgt werden konnte. Dies sind Quadrat 23, hinter dem Deich in einem feuchten Dünental neben einem Sanddorn-Gebüsch gelegen und Quadrat 24 vor dem Deich, ebenfalls neben *Hippophae* in einem hochliegenden Priel, der nur selten vom Meerwasser erreicht wird. Die Vegetation im Quadrat 23 war bereits schon vor Beginn dieser Untersuchung von WESTHOFF et al. (1961) (acht Jahre nach der Eindeichung) als Loto-Trifolion, vor allem durch Agropyro-Rumicion-Arten charakterisiert, untersucht worden; so liegen gleichermaßen vor den grünen Stränden der Insel Goeree Caricion davallianae und Centaurio-Saginetum-Gesellschaften in grasreichen Ausbildungen vor.

Die Tabelle 4 beweist die durch die Umzäunung bedingte Vegetationsänderung; Deckungsgrad und Höhe der Gräser nehmen zu, wogegen die salztoleranten Arten und die aus den letzten Syntaxa genannten deutlich zurücktreten.

Abb.3 zeigt, wie sich die Artenzahl widerspiegelt: In dem nicht umzäunten Teil ist die Artenzahl im Anfang niedriger, steigt aber dann nach sechs Jahren weit über die Artenzahl im umzäunten Teil an. Die Myxomatosis scheint auf diese Entwicklung nur geringen Einfluß auszuüben.

Im Quadrat 24 war die Vegetation im Anfang durch eine Mischung von Loto-Trifolion und Juncetum gerardii-Arten, ebenfalls unter Vorherrschaft eines Grases, *Festuca rubra* var. *litoralis* gekennzeichnet.

Wie die I_T-Linie in Abb.3b zeigt, nimmt der Unterschied zwischen den beiden Teilen rasch zu. Hier wird aber diese Differenzierung deutlich von der Myxomatosis getrennt: die Artenzahl im freien Teil, die über die Zahl im umzäunten Teil angestiegen war, fällt nach 1955 wieder auf das Anfangsniveau zurück.

Fassen wir nun diese Ergebnisse zusammen: Die meisten Vegetationstypen ändern sich ziemlich schnell, nachdem sie dem Einfluß des Gezeiten-Rhythmus entzogen sind, besonders die Typen auf relativ stabilen trockenen und nassen Stellen. Die Agropyro-Rumicion-Gesellschaften aber, die schon auf eine Instabilität eingestellt waren, fangen die plötzliche Änderung im Milieu gut auf und ändern sich wenig.

Der Einfluß der Beweidung durch Kaninchen ist im allgemeinen sehr deutlich, vor allem in den halbgeschlossenen "natürlichen" Vegetationen, die einem regelmäßigen Aussüßungsprozeß unterworfen sind. Auch in den höheren Gebieten vor dem Deich ist dieser Einfluß deutlich.

Im Bereich des Juncetum gerardii, des Saginion maritimae und Loto-Trifolion hat derartige Beweidung eine deutlich stabilisierende Funktion. So spielt beispielsweise die Beweidung in Salicornieten, die regelmäßig vom Meer-Wasser überspült werden, gar keine Rolle.

4. Diskussion über eine Landschaft mit großer räumlicher Variation.
- Untersuchungen von VAN DER MAAREL -

Das zweite Beispiel ist die verhältnismäßig alte und stabile Landschaft der Altdünen auf Voorne, südlich vom schon beschriebenen Gebiet De Beer. Diese Dünen sind über 2000 Jahre alt und zeigen das zugehörige abgerundete Relief. Eine allgemeine Beschreibung dieser Dünenlandschaft gab VAN DER MAAREL (1966a). Es gibt in diesem Gebiet ein Mosaik von Trockenrasen (Tortulo-Phleetum, Festuco-Galietum), feuchte Rasen (Agropyro-Rumicion, Poo-Lolietum) und offene niedrige Rasen auf karbonatreichen Trittstellen (Nanocyperion, Lolio-Plantagion). Dieses Mosaik ist eingehend beschrieben worden in VAN DER MAAREL (1966b). Während des Symposion über Gesellschaftsmorphologie ist über die Struktur dieser Vegetation schon etwas mitgeteilt worden (VAN DER MAAREL 1970b); 1963 wurde eine Serie von Karten angefertigt von einem äußerst variierenden Teil dieses Gebietes von etwa 2000 qm. Die Vegetationsdecke wurde in über 1000 homogene Flecken eingeteilt, die insgesamt 15 Arten von Gefäßpflanzen enthielten. Eine Vegetationstypologie dieser Flecken ergab 14 lokale Vegetationstypen. Von jedem Fleck wurde außer der floristischen Komposition die Struktur und die Art-Diversität bestimmt. Einige Teile dieses Gebietes wurden in den nachfolgenden Jahren in sogenannten permanenten Transsekten weiter studiert und zwar bis 1966 auf der Grundlage des 1963 aufgestellten Musters der homogenen Flekken. In diesem Jahre wurde dieses Muster von einem Netzwerk von Quadraten ersetzt.

Abb.4 zeigt einige Ergebnisse dieser Untersuchungen und zwar von Transsekt A, etwa 40 x 4. Die Höhenlinien sind gezeichnet in Abb.4a, der Unterschied zwischen Dünenkopf und Tal beträgt etwa 100 cm. Das Grundwasser steigt im Tal im Winter bis zu einigen dm unter die Oberfläche.

Abb. 4a zeigt die 7 Vegetationstypen, die in diesem Teil zu unterscheiden waren (siehe VAN DER MAAREL 1966b).

Abb. 4b zeigt die Strukturverhältnisse im Transsekt. Die Strukturtypen sind auf Grund der Höhe, der Bedeckung und der vorherrschenden Wuchsformen unterschieden worden.

Abb.4c zeigt das Muster der Diversität. Man sieht, wie die Flekke mit hohen Diversitätswerten in Zonen liegen, die etwa mit den Zonen der intermediären Strukturen zusammenfallen.

Tab.4, Divergierende Entwicklung einer Dünental-Vegetation im Quadrat 23 unter Einfluß der Beweidung durch Kaninchen.

Divergent development of a dune slack vegetation in quadrat 23 under influence of rabbit grazing.

A: Unbeweidet/ungrazed
B: Beweidet/grazed

	A							B						
Aufnahmen:	1	2	3	4	5	6	7	8	9	10	11	12	13	14
Deckung %:	100	98	98	100	100	100	100	98	98	98	100	100	98	98
Höhe der Krautschicht cm:	25	45	70	90	80	80	100	10	15	15	40	40	40	30
a. Wesentliche konstante Arten														
Holcus lanatus	3	3	3	3	4	3	3	3	3	3	3	3	3	3
Agrostis stolonifera	2	2	2	2	2	2	2	2	2	2	2	2	2	2
Carex flacca	2	2	2	2	2	1	1	2	2	2	2	2	2	2
Parnassia palustris	2	1	1	1	1	1	1	2	2	1	2	2	2	2
Rubus caesius	1	1	1	2	3	2	1	1	1	+	2	2	2	2
Prunella vulagaris	1	1	1	1	1	+	1	2	2	2	2	2	2	2
Galium uliginosum	1	1	2	1	1	2	1	1	2	2	2	1	2	2
Linum catharticum	1	2	1	1	+	.	1	1	1	2	2	+	+	1
b. Verschwindende Art														
Glaux maritima	+	.	.	.	.	.	.	+	.	.	.	.	.	.
c. Durch Beweidung begünstigt														
Sagina nodosa	1	+	.	.	.	.	.	.	1	+	1	+	+	+
Ranunculus sardous	+	+	1	1	+	.	.	+	+	+	+	+	+	+
Gentiana amarella-uligonosa	.	.	+	r	r	.	.	.	+	+	1	1	2	2
d. Durch Beweidung unterdrückt														
Calamagrostis epigeios	.	r	+	2	1	1	1	.	.	.	.	.	+	r
Pulicaria dysenterica	.	r	+	+	+	+	1	.	.	.	.	.	.	.
Festuca arundinacea	.	.	+	+	+	+	+	.	.	.	.	.	.	.
e. Eintretende Arten														
Juncus articulatus	.	.	.	+	r	.	.	.	.	.	2	1	1	2
Taraxacum palustre	.	.	.	r	+	.	.	.	.	.	r	r	1	+
Ranunculus repens	.	.	.	.	+	+	+	.	.	.	.	.	.	+
Trifolium repens	.	.	.	.	.	.	+	.	.	.	r	.	+	2

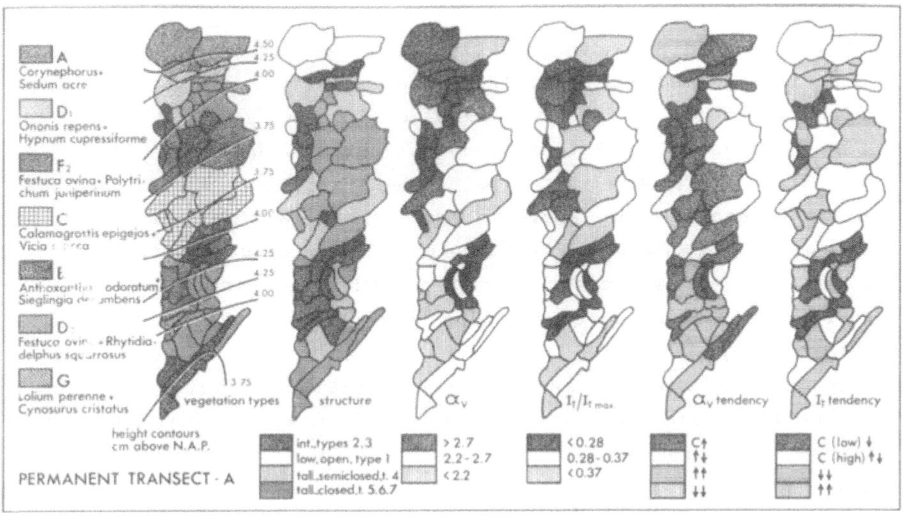

Abb.4. Vegetationstypen, Struktur, Diversität und Stabilität in einem Dünenrasen-Transsekt auf Voorne.
Vegetation types, structure, diversity and stability in a dune grassland transect on Voorne. A. Vegetation types (after VAN DER MAAREL 1966 b): A= Corynephorus-Sedum acre type; F.2= Festuca ovina-Polytrichum juniperinum; D.1= Ononis repens-Hypnum cupressiforme; D.2= Festuca ovina-Rhytidiadelphus squarrosus; C= Calamgrostis epigejos-Vicia cracca; B= Anthoxanthum odoratum-Sieglingia decumbens; G= Lolium perenne-Cynosurus cristatus.
B. Structural types after VAN DER MAAREL 1966 b: a: intermediate; b: low,open; c: tall,semiclosed; d: tall,closed.
C: Diversity (VAN DER MAAREL 1966 b). D: Instability. E: Tendency of diversity development. F: Tendency of instabilty changes (See text).

Abb.4d zeigt das Bild der zeitlichen Variation, gemessen als Durchschnittswert über die Untersuchungsjahre. Deutlich sehen wir wieder eine Korrelation zwischen Zonen mit hoher Diversität und hoher Stabilität. Diese Korrelationen sind statistisch überprüft mittels eines χ^2-Tests und erwiesen sich als signifikant. (P 0,05). Soweit wir wissen, ist dieses Ergebnis eines der ersten exakten Belege für eine direkte Beziehung zwischen Diversität und Stabilität in Pflanzengesellschaften. MAC ARTHUR (1955) hat allerdings in einem system-theoretischen Modell schon eine derartige Beziehung für eine ganze Lebensgemeinschaft aufgestellt, und zwar auf Grundlage der Nahrungsketten im Ökosystem. MARGALEF (1958) gibt Andeutungen in dieser Richtung für Plankton-Gemeinschaften.

Abb.4e zeigt die Tendenz der Diversität. Auch hier gibt es wieder ein regelmäßiges Bild. In einigen relativ armen Zonen steigt die Diversität, im feuchten Rasen des A g r o p y r o - R u m i c i o n wechseln die Artenzahlen von Jahr zu Jahr stark.

Das Bild der Tendenz in der zeitlichen Variation (Abb.4f) ist etwas unregelmäßiger, aber man erkennt ein bandförmiges Muster. Auf die Erklärung dieser Änderungen und die pflanzensoziologischen Verhältnisse in der Zeit können wir hier nicht eingehen. Inzwischen ist schon etwas darüber publiziert (siehe VAN DER MAAREL 1969).

Nach nur 4 Jahren kann man in der Interpretation der hier erwähnten Beziehungen selbstverständlich nicht zu weit gehen. Immerhin können wir zu dem Schluß kommen, daß die Untersuchung von permanenten Transsekten gewisse Vorteile haben, vor allem aber die Möglichkeit bietet, zeitliche auf räumliche Variation unmittelbar zu beziehen.

5. Allgemeine Schlußbetrachtung

Zuerst möchten wir betonen, daß ungestörte langjährige Untersuchungen an Dauer-Quadraten und besonders an Dauer-Transsekten von großer Bedeutung sind. In dieser Weise können wir nicht nur verfolgen, wie sich bestimmte Pflanzengesellschaften entwickeln, sondern auch die dynamischen Verhältnisse informationstheoretisch beschreiben.

Die folgende Behauptung kann dabei wie eine allgemeine Theorie der Vegetationsdynamik dienen. Jede Landschaft hat eine bestimmte Verteilung der Vegetationsstabilität. Diese Verteilung fällt zusammen mit einer Verteilung über die Landschaft der Variation in Struktur und Diversität und auch der Instabilität in den Milieuverhältnissen.

In jungen, dynamischen Landschaften - mit einem limes convergens oder ecotonen Charakter - sind alle Vegetationstypen mehr oder weniger instabil, aber innerhalb großer Gradienten gibt es ein Maximum der Instabilität in der Mitte der Gradienten.

In diesen Landschaften können große und plötzliche Änderungen im Milieu sehr starke Änderungen in der Vegetation hervorrufen und zwar in der Weise, daß die relativ stabilen Zonen sich nun am stärksten ändern. Das ist also eine Umkehrung des Stabilitätsmusters.

In alten, stabilen Landschaften - limes divergens oder ecocline Landschaften - sehen wir, daß die Zonen mit hoher Stabilität in der Mitte der Gradienten auftreten und dort mit intermediären Strukturen und hoher Diversität zusammengehen.

Das führt zu einer Hypothese über die Stabilitätsverhältnisse in Klimagebieten mit einer völligen Entfaltung der Strukturmöglichkeiten, also von Gras-Fluren über Saum- und Mantel-Gesellschaften zu Wäldern.

In derartigen Landschaften sind es vielleicht nicht die Wälder, sondern die Komplexe von Säumen, Mänteln, Gebüschen und Wäldern, die die höchste Stabilität erreichen.

SUMMARY

Relations between structure and dynamics in ecosystems.

1. Introduction. Structure and dynamics are considered in a general way (Relation Theory of VAN LEEUWEN). An immediate relation exists between the spatial and temporal variation in ecosystems: spatial variety tends to coincide with stability, instability tends to lead to spatial uniformity. The actual situation of an ecosystem will be in between these two extremes. This relation between pattern and process should be approached in a general quantitative way by measuring vegetational and environmental variety in space and time. Such a general approach may be a helpful addition to

a. the general approach on the biomass and energy level;

b. the phytosociological approach to the succession of particular plant communities.

2. Measurement of vegetational variety in space and time. Spatial variety can be expressed as species diversity (e.g. index of WILLIAMS), species pattern diversity (e.g. index D of PIELOU) and structure diversity. These diversities can be studied both in permanent quadrats and permanent transects, the latter comprising vegetation complexes. Changes in vegetation can be expressed as changes in the above mentioned diversities. They may be considered on three levels: species level, individual level and structure level.

In this study two simple parameters were used. Spatial variety is measured as G / ln A, G = number of species, A = area (unit of area depends on density of individuals). The index is correlated with WILLIAMS' index of diversity (VAN DER MAAREL). With seris of equal areas simply G is taken. Temporal variety is measured as $I_T = G_2 - G_c + G_1 - G_c$. G_1 and G_2 are numbers of species in successive years, G_c = number of constant species. This index is related to the "I" of WILLIAMS, LAMBERT & LANCE. Also a relative index Irel = I_I/Imax was used.

3. Discussion of a landscape with a great temporal variety.
Some results are discussed of a 10-15 years study in permanent plots on the dune and salt marshes area of De Beer (South of Hook of Holland, The Netherlands) by VAN LEEUWEN. The vegetation was described since 1947, three years after a dike was built, which cut off sea part of the area. Most plots behind the dike show very rapid changes in composition and diversity, e.g. *Salicornia* and *Chenopodium rubrum* vegetations, which developed into dense *Calamagrostis epigeios* grasslands within 4-6 years. Relatively few changes occurred in A g r o p y r o - R u m i c i o n - communities.
In general the effect of the stopping of the salt water inundation rhythm was great on places with originally a "cyclic instability" which were under direct influence of the sea and it was small on places which were already rather instable. The influence of rabbit grazing was great on places with a regular desalinisation development and small on places outside the dike, which remained under direct influence of the sea.

4. Discussion of a landscape with a great spatial variety.
The results are discussed of a 4 years study in permanent transects in a dune grassland complex on Voorne (South of De Beer, The Netherlands). This complex shows a great variety in topography, structure and species diversity. Zones with high diversity (up to 28 species/sq.m.) were found in the middle parts of topographical dry-wet gradients and gradients of treading intensity. High diversity tends to coincide with intermediate vegetational structures, in between low and open vegetations respectively tall and dense grasslands. Clear patterns of vegetational instability could be detected, in most cases in close relation with structure and diversity, high diversity and intermediate structure being found together with high stability.

5. General remarks.

The study of permanent transects covering representative parts of the local variation, over long periods and with both phytosociological and informations theoretical methods, is emphasised as one of the future plant ecology approaches promising. The following statement can be considered as a general theory of vegetation dynamics, among other things to be tested in future work. Each landscape has a definite pattern of vegetational stability, which coincides with the pattern of structure and diversity, as well as with the pattern of environmental instability. In young landscapes with a coarse-grained pattern (limes convergens or ecotone landscapes) zones with relatively high species stability are found at the ends of gradients. Such landscapes are sensitive to major environmental changes, which may invert the original pattern of stability. In old landscapes with a fine-grained pattern (limes divergens or ecocline landscapes) the zonation of stability is much more pronounced, the zones with high species stability being found in the middle parts of environmental gradients in coicindence with intermediate structures and high species diversity.

In climate zones with an intermediate position, such as found in Europe, structural gradients are well developed. Borderline and mantle communities, e.g. Prunetalia spinosae, may represent the regional optimum of stability rather than forest communities, e.g. Querco-Fagetea. This view may be extended to other climatic regions and will have repercussions for a general theory of succession.

LITERATUR

ASHBY,W.Ross -1964- An Introduction to Cybernetics.- London. 295 pp.

LEEUWEN,C.G.van -1966- A relation-theoretical approach to pattern and process in vegetation.- Wentia 15: 25-46. Amsterdam.

-- -- -1970- Raum-zeitliche Beziehungen in der Vegetation.- In: Tüxen,R. (Ed.):Gesellschaftsmorphologie. Den Haag.

MAAREL,E.van der -1966a- Dutch studies in coastal sand dune vegetation, especially in the Delta region.- Wentia 15: 47-82. Amsterdam.

-- -- -1966b- On vegetational structures, relations and systems, with special reference to the dune grasslands of Voorne the Netherlands.- Thesis. Utrecht. 170 pp.

-- -- -1969- On the use of ordination models in phytosociology.- Vegetatio 19: 21-46. Den Haag.

-- -- -1970- Vegetationsstruktur und Minimumareal in einem Dünentrockenrasen.- In: Tüxen,R. (Ed.): Gesellschaftsmorphologie. Den Haag. p.218-230.

MAC ARTHUR,R.H. -1955- Fluctuations of animal populations, and a measure of community stability.- Ecology 36: 533-536. Durham,N.C.

MARGALEF,R. -1958- Information theory in ecology.- Gen.Syst.3: 36-71.
-- -- -1961- Communication of structure in planktonic populations.- Limnology and Oceanography 6: 124-128.
PIELOU,E.C. -1966a- Species diversity and patterns diversity in the study of ecological succession.- J.theor.Biol.10: 370-383.
-- -- -1966b- The measurement of diversity in different types of biological collections.- J.theor.Biol.13: 131-144.
WATT,A.S. -1947- Pattern and process in the plant community.- J.Ecol.35: 1-22. Cambridge.
WESTHOFF,V. -1969- Langjährige Beobachtungen an Aussüßungs-Dauer-Probeflächen beweideter und unbeweideter Vegetation an der ehemaligen Zuidersee.- In: Tüxen,R. (Ed.): Experimentelle Pflanzensoziologie. Den Haag. p.246-253.
-- -- & LEEUWEN,C.G.van -1966- Ökologische und systematische Beziehungen zwischen natürlicher und anthropogener Vegetation.- In Tüxen,R. (Ed.): Anthropogene Vegetation. Den Haag. p.156-172.
-- -- -- -- & ADRIANI,M.J. -1961- Enkele aspecten van vegetatie en bodem der duinen van Goeree, in het bijzonder de contactgordels tussen zoet en salt milieu.- Jb.Wet.Gen.Goeree-Overflakkee, p.46-92.
WILLIAMS,C.B. -1964- Patterns in the balance of nature.- London. 324 pp.
-- -- LAMBERT,J.M. & LANCE,G.N. -1966- Multivariate methods in plant ecology.V. Similarity analyses and informations analysis.- J.Ecol.54: 437-445. Oxford.

SPEZIALMETHODEN UND SONDERFRAGEN BEI DER SUKZESSIONSFORSCHUNG VON EPIPHYTEN.

J.J. B a r k m a n

Manuskript nicht eingegangen.

SYNDYNAMISCHE BEZIEHUNGEN ZWISCHEN EINIGEN JAPANISCHEN DAUERGESELLSCHAFTEN ERKLÄRT AUS DER GESELLSCHAFTSSCHICHTUNG

S u z u k i - T o k i o †

In der Hochgebirgsvegetation verläuft die Sukzession langsam, so daß dort die Dauergesellschaften fast unverändert bleiben.

I. Alpine Stufe Mitteljapans

In den japanischen Nordalpen, wo aus makroklimatischen Ursachen das V a c c i n i o - P i n e t u m p u m i l a e die Klimaxgesellschaft ist, wirkt der Wintermonsun auf die Gesellschaftsverbreitung bestimmend ein. Auf den Windseiten, wo die mechanische Wirkung am stärksten ist, wachsen nur Flechtengesellschaften. Leider sind sie noch nicht genau pflanzensoziologisch charakterisiert.
Die ersten Besiedler, aufbauende Arten, sind *Rhizocarpon geographicum* und *Gyrophora leiocarpa*, die beide allmählich günstige Standorte für einige Laub- und Strauchflechten, wie *Parmelia stygia, Parmelia diffugiens, Alectoria ochroleuca, Stereocaulon paschale, Stereocaulon dendroides* u.a. bilden. Ein Laubmoos, *Rhacomitrium lanuginosum*, begleitet sie. Hier kommen oft Keimlinge von *Pinus pumila* auf.
Wo die Windwirkung auf den losen Böden stark ist, kommen alpine Windheiden, wie das A r c t e r i o - L o i s e l e u r i e t u m vor. Diese Assoziation ist eine Zwergstrauch-Strauchflechten-Gesellschaft. Sehr dichte Matten von Zwergsträuchern werden unmittelbar der Winderosion ausgesetzt, denn die Schneedecke fehlt dort fast ganz. Sie ertragen gut die Windwirkung, werden aber oft teilweise erodiert. Strauchflechten, wie *Cetraria crispa* var. *japonica* und *Cetraria islandica* fo. *angustifolia* wachsen üppig zwischen den Zwergsträuchern. Auch sie zeigen Windresistenz, denn sie können, wenn sie auch teilweise geschädigt werden, immer wieder regenerieren. Im Sommer tritt die dritte Lebensform, die der Schaftpflanzen auf. Dazu gehören *Gentiana algida, Geum calthaefolium, Potentilla matsumurae* u.a. Im Winter hat die Gesellschaft nur eine Schicht, in der alle drei Lebensformen überwintern, während im Sommer die Schaftpflanzen über 15 cm Höhe mit ihren Blättern erreichen, also zwei Schichten ausgebildet sind. Windstärke und Schneedecke sind antagonistische Erscheinungen im Winter. Wenn die Windstärke etwas geschwächt ist, dringt *Pinus pumila* mit einigen Begleitern sofort ein. Dann kommt die Subassoziation c e t r a r i e t o s u m des V a c c i n i o - P i n e t u m p u m i l a e

als das dritte Stadium zur Ausbildung. Die Zwergkiefer wächst hier zerstreut und niedrig, aber die Zwergsträucher und die Strauchflechten vom A r c t e r i o - L o i s e l e u r i e t u m bleiben noch. Die Assoziation erreicht ihre Initialphase.
Die Optimalphase vertreten drei Varianten des V a c c i n i o - P i n e t u m r u b e t o s u m p e d a t i. Die R h o d o d e n d r o n a u r e u m - und die R h o d o d e n d r o n f a u r i e i- Variante sind die Optimalphase der Trockenserie. Die Varianten trennen sich gegeneinander durch die Höhenlage. Es gibt auch eine Torfserie (S a s a k u r i l e n s i s - Variante), denn das Schmelzwasser verursacht Vermoorung auf den Leeseiten, wo mehr Schnee abgelagert wird. Die vierte Variante ist die von *Sorbus matsumurana*, die sich entwickelt, wo eine mächtige Schneedecke ohne Vermoorung liegt.

Die Gesellschaften der Optimalphase haben folgende vier Schichten:
In der ersten Strauchschicht (F 1) dominiert *Pinus pumila*, sie ist kontinuierlich 60-100 cm hoch.
Die zweite Strauchschicht (F 2) wird dominiert von immergrünem *Rhododendron* und begleitet von sommergrünen *Vaccinien*; sie tritt zerstreut auf.
Die Krautschicht (H), beherrscht von *Vaccinium vitis-idaea* oder *Gaultheria migueliana*, ist immergrün, kontinuierlich oder gruppenweise wachsend.
Die Moosschicht(M) wird von *Pleurozium schreberi* u.a. gebildet, sie ist kontinuierlich 3-5 m hoch.
In der S a s a k u r i l e n s i s -Variante dominiert der Zwergbambus *Sasa kurilensis*, anstelle von *Rhododendron*, während *Sorbus matsumurana* in der danach genannten Variante vorherrscht.

Die Terminalphase ist durch das Auftreten einer niedrigen Baumschicht (As) ausgezeichnet. Sie kommt aus dem A b i e t u m m a r i e s i i und wird von vereinzelt wachsender, deformierter Tanne gebildet. Wenn die Tannen höher wachsen,könnte vielleicht das A b i e t u m m a r i e s i i entstehen, aber die vorherrschende Windwirkung hindert die Sukzession weiter fortzuschreiten. Das V a c c i n i o - P i n e t u m bleibt daher das Endstadium dieser Serie. Man kann also zusammenfassen:
In die alpine Stufe Japans teilen sich vier Vegetationsschichten; die fünfte Schicht von *Abies mariesii* spielt die Rolle eines Zerstörers. Weil die Assoziation des V a c c i n i o - P i n e t u m eine zonale Gesellschaft in großen Zügen ist, muß der Temperatur-

Faktor entscheidend sein. In einem Gebiet, wo der Wintermonsun unmittelbar auf die Gebirgshänge stößt, ist die Assoziation streifenartig an den Graten entlang verteilt. Die entscheidende Wirkung des Relieffaktors der Exposition erscheint dabei sehr klar, doch beeinflussen in Wirklichkeit die mechanische Windwirkung auf den Windseiten und die Schneeauflagerung auf den Leeseiten die Entstehung der Assoziation.

II. Gipfelstufe der Mittelkyûsû Gebirgsregion

Die in der südwestlichen Insel Kyûsyû erreichte Höhenlage ist niedriger als in Mittel-Japan. Dort ist der höchste Gipfel Kuzyû nur 1788 m N.N. Wenn man die Temperaturen aus der Meereshöhe berechnet, wären sie noch hoch genug für Buchenwälder, aber die Landschaft erscheint ähnlich der alpinen. An einigen windoffenen Stellen tritt eine den alpinen Heiden ähnliche Gesellschaft auf, und der Wuchs des angrenzenden Gesträuchs wird durch Kälte und Windwirkung stark gehemmt.

Die hier wachsende Assoziation ist das Maianthemo-Rhododendretum kiusiani. Das schön blühende *Rhododendron kiusianum* ist ein Sohayki-Element, d.h. eine Art des japanischen äußeren Bogens und in der hochmontanen Stufe von Sikoku und Kyûshû heimisch.

Maianthemum dilatatum gehört dagegen in die Klasse Vaccinio-Piceetea japonica, des japanischen Teiles der panborealen *Ericaceen*- und *Pinaceen*-Nadelwälder und Heiden. Diese Assoziation ist eine der südlichsten Vertreter der Klasse, soweit sie auf den Japanischen Inselbogen beschränkt ist. Sie fehlt schon den *Pinaceen*-Gesellschaften.

Sie gliedert sich in zwei Subassoziationen: alnetosum und vaccinietosum. Die Trennarten der ersten sind *Alnus firma*, *Acer micranthum*, *Weigela japonica* und *Athyrium tashiroi*, während *Vaccinium vitis-idaea*, *Carex chrysolepis* und *Shortia soldanelloides* als diejenigen der zweiten gelten können. Die Subassoziation alnetosum ist in zwei besondere und eine Typische Variante zu trennen. Sonst dominieren an einigen Standorten drei Zwergbambus-Arten, wo die entsprechenden Fazies von *Arundinaria pygmaea* im alnetosum, von *Sasa hikosanensis* in der Typischen Subass. und von *Sasa quelpaertensis* im vaccinietosum beherrscht werden.

Alle diese Untereinheiten sind Dauergesellschaften und bleiben unverändert, so lange der heutige Standort in der Gipfelstufe existiert. Man kann jedoch die syndynamischen Zusammenhänge in der Schichtenstruktur erkennen.

Folgende Schichten sind ausgebildet:

Unterbaumschicht (As): Hier dominiert *Alnus firma*, 2 m hoch, diskontinuierlich oder truppweise entwickelt, nur in der Subassoziation a l n e t o s u m vorhanden.

Erste Strauchschicht (F 1): Dominant sind *Hydrangea paniculata*, *Viburnum furcatum* oder *Enkianthus cernuus*, 1 m hoch, kontinuierlich, nur in der Subassoziation a l n e t o s u m vorhanden.

Zweite Strauchschicht (F 2): Dominiert von *Rhododendron kiusianum*, 30-50 cm hoch, kontinuierlich, mit F 1-Schicht vereinigt in der typischen Fazies.

Erste Krautschicht (H 1): Dominiert von *Miscanthus sinensis* und *Miscanthus oligostachyus*, 50 cm hoch, truppweise entwickelt.

Zweite Krautschicht (H 2): Dominiert von *Maianthemum dilatatum*. *Shortia solanelloides*, *Vaccinium vitis-idaea* über 10 cm hoch, diskontinuierlich.

Moosschicht (M): Flechten und Moose umfassend, nur stellenweise entwickelt, F_2-Schicht kommt mit *Sasa* und *Arundinaria* (in der danach genannten Fazies) vor.

Außerdem kommt oft eine F_2-H_2-M-Verbindung in einer Gesellschaft von halbwintergrünen Strauch-Horstpflanzen und Deckenmoosen oder Strauchflechten vor, die zur Subassoziation v a c c i n i e t o s u m gehört. Sie ist sehr widerstandsfähig gegen Wind und plötzliche Temperaturänderungen. Hohe Temperatur und die starken Regengüsse im Sommer fördern aber die Wald-Entstehung, so daß die Unterbaumschicht von *Alnus firma* vordringen kann. Sie wird oft begleitet von *Sorbus commixta*, *Acer micranthum* und *Enkianthus cernuus* und in der Strauchschicht tritt *Viburnum furcatum* auf. Diese Art wird als eine der Kennarten japanischer Buchenwälder bewertet.

Beide Subassoziationen bleiben gewöhnlich als Dauergesellschaften bestehen, aber die syndynamischen Zusammenhänge sind durch die Schichtenstruktur der Untereinheiten erklärbar.

ZUSAMMENFASSUNG

Jedes Stadium der alpinen Serie hat in der japanischen alpinen Region seinen eigenen Verein. So kommen im ersten Stadium die Krustenflechten vor, gefolgt von Laub- und Strauchflechten und Deckenmoosen. Das nächste, das Windheiden-Stadium, ist aus drei Gruppen,geordnet nach der Lebensform: Zwergsträucher, Schaftpflanzen und Strauchflechten, zusammengesetzt, und kann als Dauergesellschaft angesehen werden. Ein Weiterschreiten der Sukzession wird durch den vorherrschenden Wintermonsun verhindert. Die Zwergstrauch-Heide und die Strauchflechten-Gesellschaften sind windhart; die Schaftpflanzen entwickeln sich nur im Sommer.

Die Klimax-Gesellschaft in der sub-alpinen Stufe, das Vaccinio-Pinetum pumilae, ist aus der Schichtenverbindung zweier Strauchschichten, einer Kraut- und einer Moos-Schicht aufgebaut. Jede Phase hat ihre eigenen Dominanten in den untergeordneten Schichten, aber die aufbauende Art bleibt in der Strauchschicht *Pinus pumila*, solange sich diese Klimaxgesellschaft erhält. Bei nachlassender Windgeschwindigkeit wandert *Abies mariesii* als "störende" Pflanze ein und bildet über der Zwergkiefer eine entwicklungshemmende Baumschicht.
Auf der Kyûsyû-Insel in West-Japan, wo die alpine Region gar nicht erreicht wird, wächst als Strauch-Gesellschaft das Maianthemo-Rhododendretum kiusiani in der Gipfel-Stufe mit einem Lokalklima und einer Landschaft ähnlich der alpinen Stufe. Diese Assoziation erweist sich hier mit ihren beiden Subassoziationen als Dauer-Gesellschaft. Ihr Aufbau zeigt eine Baumschicht, zwei Strauchschichten, zwei Krautschichten und eine Moosschicht. Die Temperaturen würden in diesem Bereich eine Wald-Entwicklung zulassen, aber das Lokalklima im Winter erlaubt eine Fortsetzung der Sukzession zur Schluß-Gesellschaft nicht. Sonst würde hier unter anderen *Alnus firma* zerstörend eindringen. Durch den Vulkanismus, der sich bis heute erhalten hat und die Frühjahrs-Brände, die früher wegen der Jagd, neuerdings zur Weide-Pflege angelegt werden, sind die Natur-Wälder im Kuzyû-Vulkangebiet vernichtet worden. Dennoch läßt die anthropogen beeinflußte Waldvegetation pflanzensoziologisch drei höhere Einheiten und mehrere Assoziationen unterscheiden. Obwohl für deren Kennzeichnung Trennarten zu finden sind, ist es leichter, die Regressionsstadien durch das Erscheinen und Vergehen der Schichten-Gesellschaften zu verstehen.

DISKUSSION

D.MUELLER-DOMBOIS

Eine Frage zum Monsun-Einfluß: Die *Pinus*-Art ist in den Alpen mosaikartig verteilt. Wie kann man das in Zusammenhang bringen? Auf der anderen Seite würde es mich auch interessieren, wie Herr SUZUKI-TOKIO zu diesen Ergebnissen gekommen ist. Es ist mir nicht ganz klar, wie er anzunehmen scheint, daß auf Standorten, wo heute *Rhizocarpon geographicum* erscheint, nachher *Sasa* aufkommt. Oder welche Methode wurde angewandt?

R.TÜXEN

Ich kann nur wenig antworten, weil ich diese Vegetation nur ganz kurz gesehen habe. Professor SUZUKI arbeitet nach der Methode von BRAUN-BLANQUET. Er hat Aufnahmen und Tabellen gemacht, diese verglichen und dabei Kenn- und Trennarten gefunden; er hat die Sukzession ebenso studiert, wie wir es tun.
Pinus pumila sieht fast genau so aus wie unsere Latschen, *Pinus montana*. Beide leben unter sehr ähnlichen Bedingungen.
Die Vegetation, die hier behandelt worden ist, ist physiognomisch und auch in der Artenkombination der unseren in gleicher Lage so ähnlich, daß man als Europäer 60-80% der Arten sofort aufschreiben kann und nur die eine oder andere Art erfragen muß. Die sehr markanten, wie z.B. *Shortia*, erkennt man aber sofort wieder, wenn man sie einmal gesehen hat. Flechten und Moose sind meist dieselben wie bei uns. Und viele Gattungen, *Arctostaphylus* z.B., sind die gleichen.

J.SCHMITHÜSEN

Zur Frage der Bedeutung des Wintermonsuns kann ich noch sagen, daß der japanische Wintermonsun ein niederschlagsreicher Wind ist, der zu ganz besonders hohen Schneemengen führen kann. So ist also hier dieser Gegensatz von Luv und Lee einerseits ein Gegensatz der Windstärke auf der Luvseite und der windschwachen Seite auf der Leeseite, und zugleich ein Gegensatz der hohen Schneelagen dieser selben Luvseite und der geringeren Schneelagen in diesem Fall auf der Leeseite, bezogen auf den Gesamtkamm des Gebirges. Die beiden Faktoren spielen also zusammen, einmal die Windwirkung direkt, die verhindern kann, daß das *Pinus pumila*-Gebüsch in den windexponiertesten Lagen entsteht und zum anderen die Höhe der Schneedecke, die verhindert, daß sich in den *Pinus pumila*-Gebüschen der *Abies mariesii*-Wald bilden kann.

R.CARBIENER

Ich möchte nur kurz darauf hinweisen, wie außerordentlich die Analogien zwischen gewissen japanischen und westeuropäischen Verhältnissen sind. Und diese Analogien zeigen sich auch in diesen starken ozeanischen Einflüssen ausgesetzten Gebirgen. Besonders drängen sich Vergleiche mit den Vogesen auf, wo ganz ähnliche Verhältnisse vorliegen. Natürlich, der "raisonnement par analogie" hat bestimmt auch Nachteile, ist aber immerhin in der Naturwissenschaft etwas, das sehr nützlich ist.

Ich möchte noch zu einem besonderen Problem Stellung nehmen.Man hat öfter behauptet, daß ein vorhandener geschlossener Waldbestand für Neuankömmlinge ein Ausbreitungs-Hindernis darstellt. Man hat behauptet, daß das Fehlen der Fichte in den Vogesen auf solche Ursachen zurückzuführen sein könnte. Das wird sogar heute noch in der Literatur angegeben: Die Buche hat das Terrain schon beherrscht, als die Fichte kam, und deswegen konnte sich die Fichte nicht mehr ausbreiten. Das ist eine ziemlich anthropomorphe Erklärung, und in Japan sehen wir, obwohl mit anderen Arten, auch sehr schöne ähnliche Verhältnisse. Es ist doch bemerkenswert, daß über einem Gürtel von *Abies mariesii* ein *Abies marisii*-Gürtel mit *Betula ermani*, wie in West-Skandinavien mit *Betula pubescens* folgt. Darüber findet sich ein *Betula ermani*-Busch, der physiognomisch viel mit dem subalpinen Buchenkrüppelwald, z.B. der Vogesen gemein hat. In diesem *Betula ermani*-Gebüsch findet man keine *Abies mariesii* mehr, die aus klimatischen Ursachen weiter unter sehr gut gedeiht. Und so ist es auch in den Vogesen höchstwahrscheinlich eine klimatische Ursache, daß die Fichte oben nicht mehr wächst. Wenn die Buche nicht da wäre, könnte sie wohl durch die Fichte teilweise ersetzt werden, denn die Konkurrenz spielt bei den Grenzen zwischen den Pflanzengesellschaften immer eine große Rolle. Aber sie allein als Faktor anzugeben, wäre doch wohl übertrieben.

DYNAMIK DER SUKZESSION

D.J. Bellamy and D.W. Shimwell

Ein Ökosystem kann als Zwischenakt der zufälligen, der abgestorbenen und der lebenden Energie in einem gewissen Flächenraum, über eine gewisse Zeitspanne betrachet werden.
Ein Ökologe ist mit der Erforschung dieses Zwischenaktes beschäftigt. Wir dürfen den abgestorbenen Bestandteil des Ökosystems als Vorbild für den Aufbau betrachten, von dem die Zusammensetzung des lebenden Bestandteiles abhängt, wenn wir LUNDEGÅRD's Theorie über die Faktoren, welche die Pflanzenwelt beeinflussen, ausdehnen.
Der Verlauf der Sukzession darf deshalb als eine langsame Änderung dieses Vorbildes betrachtet werden, das durch den Lebensablauf der organischen Bestandteile verursacht wurde, und die Aufspeicherung von Energie in Gestalt von wachsenden Pflanzen, Humus und Torf mit umfaßt.
Der Lebenslauf der Biota beeinflußt den geochemischen Zyklus, besonders den von Wasser, Stickstoff und Kohlenstoff. Aufspeicherung von Energie in Gestalt von wachsenden Pflanzen hat bedeutenden Einfluß auf das Klima in der Nähe des Bodens. Die Schwankungen von Temperatur und Feuchtigkeit werden ausgeglichen, wodurch Nischen für Schattenpflanzen entstehen, die das Ökosystem vertikal ausweiten, und dadurch mehr Raum und größere Vorräte von Rohmaterialien ausnützen.
Aufspeicherung von Energie in Gestalt von Humus bildet die anfänglichen xerachen Sukzessionen, wobei die Wasserspeicherung in dem sich neu bildenden Boden zunimmt, und dadurch den austauschbaren zählbaren Bruch modifiziert.
Energiespeicherung in Gestalt von Torf kann Bodenbildung unter Wasser verursachen, wodurch eine hydrache Sukzession entsteht. Die während dieser Zeit vorliegende überschüssige Energie, die durch das Ökosystem bestimmt ist, wird in den wachsenden Pflanzen als Humus oder als Torf gespeichert. Das Ökosystem hat ein positives Energie-Gleichgewicht, das Vorbild wird verändert,neue Arten wandern ein, und eine Sukzession findet statt. Sukzession ist die zusammenhängende Verwandlung der lebenden und der abgestorbenen Bestandteile eines Ökosystems, die durch die Aufspeicherung von Energie innerhalb des Ökosystems hervorgebracht wird.

Diese Entwicklung wird weitergehen, bis irgend ein Umstand den jährlichen Energie-Vorrat stabilisiert, so daß alle Energie, die durch die Photosynthese bestimmt ist, durch die Erhaltung des Ökosystems verbraucht wird. Zu diesem Zeitpunkt kann es keine weitere Zunahme in der Menge an gespeicherter Energie geben; das Vorbild und die lebende Gemeinschaft sind stabilisiert, und die Sukzession ist zu Ende. Bild 1 zeigt dies im Diagramm.

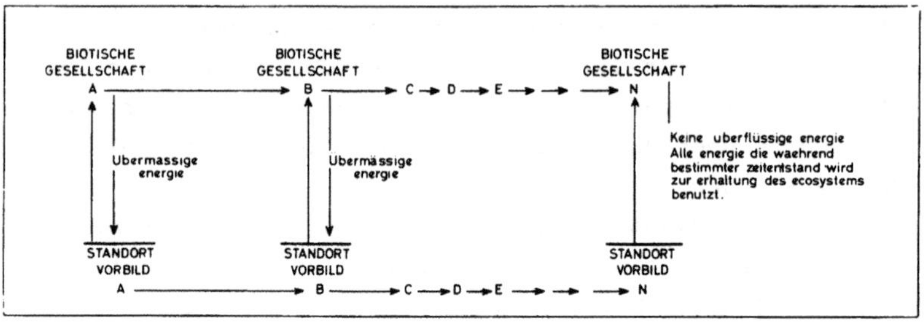

Abb.1

Der genaue Zeitpunkt, an dem die Stabilisierung stattfindet, hängt von kontrollierenden, edaphischen, biotischen oder anthropogenen Faktoren ab.

Normales phytosoziologisches Verfahren macht eine genaue Beschreibung des lebenden Wesens möglich, wobei es an seiner floristischen Gleichartigkeit und/oder seiner Unveränderlichkeit der Bestandteile erkannt wird.

Die Phytosoziologie sucht nach Abänderungen und benutzt sie als Maßstab in der Einteilung. Eine statistische Analyse von Lebewesen, die aufs Geratewohl gesammelt wurden, kann für den gleichen Prozeß benutzt werden, indem man alle, die gewisse ähnliche floristische Merkmale haben, zusammennimmt.

Mit Hilfe von zahlreichen Beispielen und der richtigen Analyse werden die Abänderungen unklarer, und sie erscheinen als ununterbrochen. Auf diese Weise ergibt sich mehr eine Ordination als eine Klassifikation. Wenn eine solche Analyse auf Grund der floristischen Gleichartigkeit ausgeführt wird, muß eine der Hauptsachen der sich ergebenden Ordination mit Sukzession verbunden sein. Es ist deshalb wohl anzunehmen, daß eine Zusammensetzung der phytosoziologischen Daten ein klares Bild der biotischen Veränderungen der Sukzession gibt. Durch Erwägung normaler, phytosoziologischer Eigenschaften sollte es möglich sein, jede Aufnahme in ihre theoretische und ihre richtige sukzessionelle Reihenfolge einzuordnen.

Bei Anwendung trophisch-dynamischer Ökologie ist es unter anderem möglich, die folgenden Attribute eines Ökosystems zu ermessen:

1. Verhältnis von Energie zu Nutzbarkeit; die gesamte anfängliche Produktion.
2. Verhältnis von Energie zu Verbrauch; Pflanzenatmung und Weitergabe an Pflanzenfresser und Zersetzung.
3. Verhältnis von Energie zu Aufbewahrung.

Wenn man die von Phytosoziologen anerkannten floristisch verschiedenen Einheiten in ihrer sinnvollen Anordnung unterbringt, sie als ein bestimmtes Ökosystem betrachtet, und die angeführten Eigenschaften erwägt, bekommt man ein klares Bild von der Dynamik der Sukzession. Auf ähnliche Weise könnten Messungen der Umgebung innerhalb jedes bestimmten Ökosystems das Bild vervollständigen, und Daten über das Ausmaß der Unterschiede zum Vorbild zur Verfügung stellen, das mit jedem erkennbaren Unterschied des lebenden Bestandteiles verbunden ist.

Fortdauer von Information - das Pflanzenleben der Welt -, dessen Beschreibung und Klassifizierung das Ziel der Phytosoziologen ist, wird dann nicht nur die Arena sein, in der der Ökologe seine Probleme findet, sondern auch das Werkzeug, mit dem er sie lösen kann.

Ich möchte dies durch Hinweis auf zwei kurzfristige und unvollständige Untersuchungen dieser Art illustrieren, deren Ergebnis in jedem Fall nur als Wegweiser gelten kann.

Das erste Beispiel ist die Sukzession vom E u s c o r p i d i o n-zum E u f u s c i o n - Verband bei DU RIETZ.

Ein Bult von *Sphagnum fuscum* aus einem E u s c o r p i d i o n-Niedermoor wurde für dieses Experiment benutzt. Ein Stück Torf von gewisser Größe wurde aus diesem Bult entnommen und in Stükke von vier Zentimeter Länge geteilt. Jeder Teil wurde auf seinen floristischen, Wasser-, Kalorien- und Mineralien-Gehalt untersucht.Die Produktion von *Sphagnum fuscum* wurde durch Ernte und Wiegen des jährlichen Zuwachses ermittelt.

Das Ergebnis ist in Abb.2 zu sehen.

Ersichtlich wird, daß die ganze Verwandlung durch die Bildung von nur acht Zentimeter Torf verursacht wurde. Man kann hieraus entnehmen, daß das Wachstum der Pflanzen, die in der tiefsten Strata zu finden sind, nämlich *Cratoneuron falcatum* und *Campylium stellatum*, die beide in großen, schwellenden Büschen

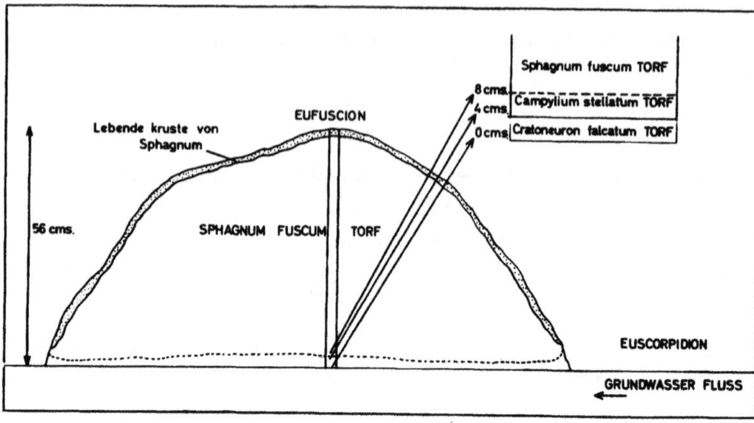

Abb.2.

wachsen können, das Grundwasser um sich herum ableiten und auf diese Weise ein kleines Gebiet schaffen, wo *Sphagna* sich ansiedeln können. Sobald diese Pflanzen sich angesiedelt hatten, fand eine sehr schnelle Veränderung statt, und die Ablagerung von nur acht Zentimeter Torf verhinderte jeden Einfluß des Grundwassers.

Das Ergebnis ist in Abb.3 zusammengefaßt, worin MAUCHA Ionenfelder - Diagramme für das Grundwasser und Kerngehäusewasser dargestellt sind. Die Diagramme zeigen einen Wechsel von Wasser und dessen Eisengehalt, das von Calcium und Bicarbonaten, zu solchem, das von gelösten Hydrogen und Sulfaten beherrscht ist. Ähnliche Diagramme für Grundwasser, das von klassischen Beispielen von Euscorpidion, Subsecundo-Apiculation, Apiculation und Eufuscion von DU RIETZ gesammelt wurden, werden zum Vergleich gezeigt. Die Gleichartigkeit ist auffallend. Es darf deshalb angenommen werden, daß in dem untersuchten Schlamm eine Bildung von acht cm Torf, das sind 8326 Kcal. der vom Ökosystem aufgespeicherten Energie, die vollständige Transformation eines Quadratmeters von Euscorpidion zu Eufuscion mit allen Merkmalen verursachen kann. Eine Verwandlung der hydrologischen, edaphischen und floristischen Eigenschaften wird durch die Aufspeicherung von Energie in Gestalt von Torf verursacht. Bei gegenwärtiger jährlicher Produktion durch *Sphagnum fuscum* würde dieses sieben Jahre dauern.

Das zweite Beispiel ist das der Sukzession des Genistetum

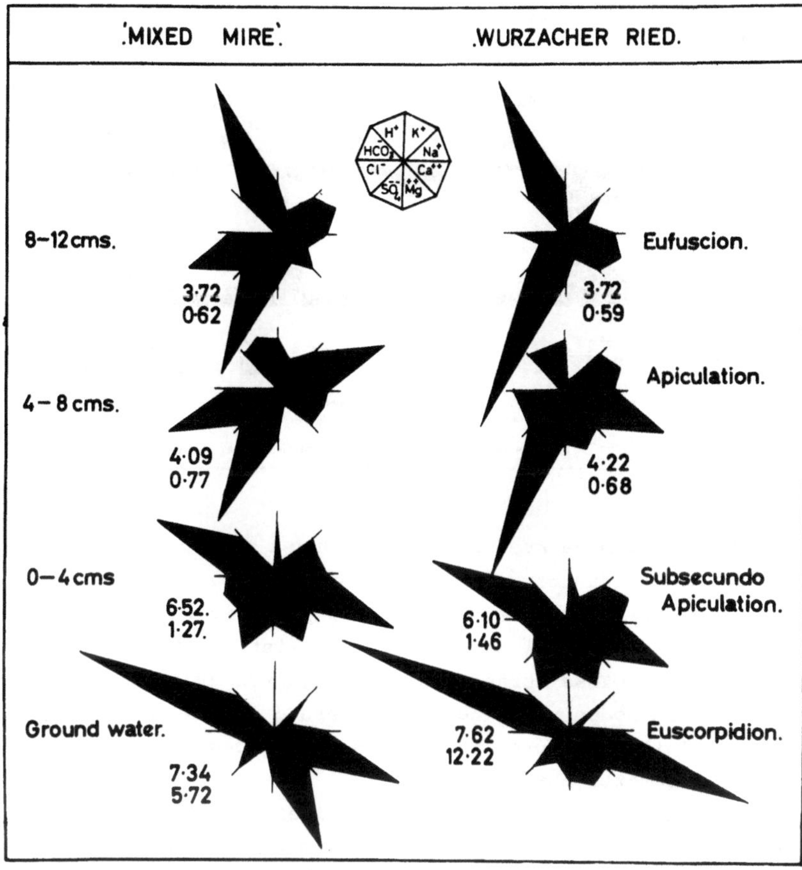

Abb.3.

l e p t o c l a d a e (anthropogene Variante) zum Q u e r c e t u m p y r e n a i c a e. Die Sukzession wurde in diesem Fall durch landwirtschaftliche Maßnahmen, und zwar hauptsächlich durch Abbrennen und übermäßige Beweidung beeinflußt. Jeder Fortschritt in der Sukzessions-Serie wird von nur einer Art einer perennierenden holzigen Pflanze bestimmt. Wachsende Pflanzen und ihre Produktion wurden gemessen, wobei normale Erntemethoden benutzt wurden. Einfache klimatische Daten, Temperatur und Feuchtigkeit, wurden innerhalb jeder Zone während der Untersuchung festgestellt. Der Anteil organischen Materials jeden Bodentyps und sein Wassergehalt bei vollständiger Sättigung wurden gemessen.
Das Ergebnis ist aus Abb.4 zu ersehen.

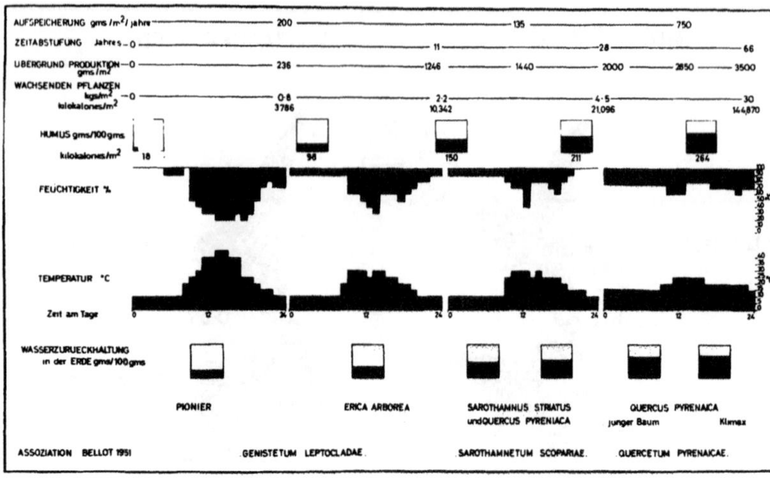

Abb.4

Die Sukzession vom nackten Boden zum Quercetum pyrenaicae verläuft über vier sich floristisch unterscheidende Phasen. Die Affinität jeder Phase zur phytosoziologischen Einheit, wie sie von BELLOT dargestellt ist, ist unten auf dem Diagramm zu sehen. Diese Veränderungen werden von größerer Wasser-Rückhaltung im Boden und einer Verbesserung des Klimas in Bodennähe begleitet, wodurch tägliche Schwankungen von Feuchtigkeit und Temperatur herrühren.

Diese Änderungen am Standort sind mit einer allmählichen Zunahme der vom Ökosystem aufgespeicherten Energie verbunden, sowohl in Form von Humus als auch von wachsenden Pflanzen.

Die Zunahme von jährlichem Wachstum über der Erde während der Sukzession ist bemerkenswert. Die Zeitabstufung wird von den Altersgruppen der perennierenden vorherrschenden Pflanzen im Endstadium jedes bestimmten Fortschrittes kalkuliert. Hieraus kann man eine Versuchszahl für jährliche Speicherung in Gestalt von wachsenden Pflanzen berechnen; dies ist oben auf dem Diagramm zu sehen. Es ist anzunehmen, daß während eines langfristigen Projektes das Verhältnis der Energie-Speicherung als Teil der Untersuchung gemessen werden wird. Diese Arbeit wurde während einer vierwöchigen Exkursion nach Nordwest-Spanien unternommen. Das Ergebnis ist auf verschiedene Annahmen aufgebaut, und kann bestenfalls als unvollständige Schätzung betrachtet werden. Es weist jedoch auf die Möglichkeit langfristiger, ein-

gehender Untersuchungen hin.
Unser Forschungsprogramm in Durham ist auf langfristige, eingehende Untersuchungen dieser Art konzentriert, die alle auf feste phytosoziologische Daten begründet sind. Diese Untersuchungen haben gerade erst angefangen, und wir möchten den phytosoziologischen Hintergrund eines unserer Forschungsprogramme, die Untersuchung der xerach Sukzession von Kohle enthaltendem Kalkstein in Zentralengland darstellen.
Abb.5 zeigt die theoretische Sukzessions-Folge.

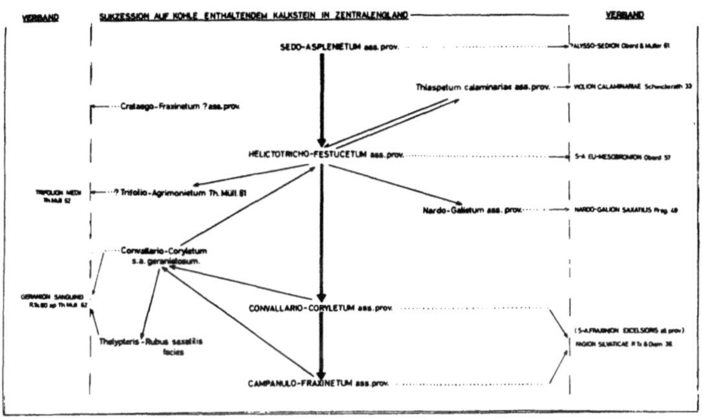

Abb.5.

Diese Bezeichnungen sind zuerst nur provisorisch, und ihre Affinität mit kontinentalen Verbänden ist auf jeder Seite des Diagramms zu sehen. Abb.6 zeigt die wichtigen Kenn- und Trennarten, die anscheinend die Hauptstufen der ursprünglichen Sukzession darstellen: S e d o - A s p l e n i e t u m , H e l i c t o t r i c h o - F e s t u c e t u m , C o n v a l l a r i o - C o r y l e t u m und C a m p a n u l o - F r a x i n e t u m . Diese müssen als die Hauptarten betrachtet werden, die sich den Änderungen am Standort anpassen.
Abb.7 macht auf die Entwicklung vom S e d o - A s p l e n i e t u m zum H e l i c t o t r i c h o - F e s t u c e t u m aufmerksam. Alle Aufnahmen sind in theoretischer Anordnung der Sukzession dargestellt, die auf den Deckungswert der Kenn- und Trennarten und den Pflanzenarten, die oft vorkommen und dominieren, aufgebaut ist. Der Deckungswert ist durch verschiedene Schattierungen dargestellt, je dunkler die Schattierung, um so höher der Deckungswert.

	(a)	(b)	(c)	(d)	
Zahl d. Aufnahmen.	10	20	12	12	
Sedum acre	IV				(a) KENNARTEN und TRENNARTEN SEDO-ASPLENIETUM
Asplenium ruta-muraria	V				
A. trichomanes	IV				
Saxifraga tridactylites	IV				
Erophila verna	IV				
Cardamine hirsuta	V				
Encalypta streptocarpa	IV				
Carlina vulgaris	+	IV			(b) KENNARTEN und TRENNARTEN HELICTOTRICHO-FESTUCETUM
Leontodon taraxacoides		V			
Carex caryophyllea		V			
Galium sterneri		IV			
Gentianella amarella		IV			
Helictotrichon pratense	+	V			
Poterium sanguisorba		V			
Helianthemum chamaecistus		V			
Thymus drucei	+	V			
Convallaria majalis			IV	+	(c) KENNARTEN und TRENNARTEN CONVALLARIO-CORYLETUM
Epipactis helleborine			V		
Geranium robertianum			III		
Pimpinella major			III		
Trollius europaeus			III		
Thalictrum minus			III		
Origanum vulgare			IV		
Clinopodium vulgare			IV		
Cruciata chersonensis			IV		
Teucrium scorodonia			V	1	
Mycelis muralis				V	(d) KENNARTEN und TRENNARTEN CAMPANULO-FRAXINETUM
Asperula odorata				III	
Polystichum aculeatum				IV	
Campanula latifolium				V	
Arum maculatum				IV	
Moehringia trinevia				V	
Lapsana communis				V	

Abb.6.

Die floristischen Änderungen der sukzessionellen Prozesse sind sehr offenkundig, und es ist einfach, Bestände für weitere Untersuchungen auszulesen.

Für alle Erdteile mit einer annehmbaren Phytosoziologie ist das Gerüst der grundlegenden, erläuternden Daten bereits fertig,und eine Untersuchung der Dynamik der Sukzession könnte sofort beginnen. Außerdem besteht in diesen Gebieten die Möglichkeit, zu irgendeinem Bestand, der nicht zerstört wurde, zurückzukehren, und ihn aufs neue zu beschreiben. Die durch Sukzession veranlaßten floristischen Änderungen werden sofort offenbar, und ein genaues Zeitmaß für diesen Prozeß kann ermittelt werden. Auf die-

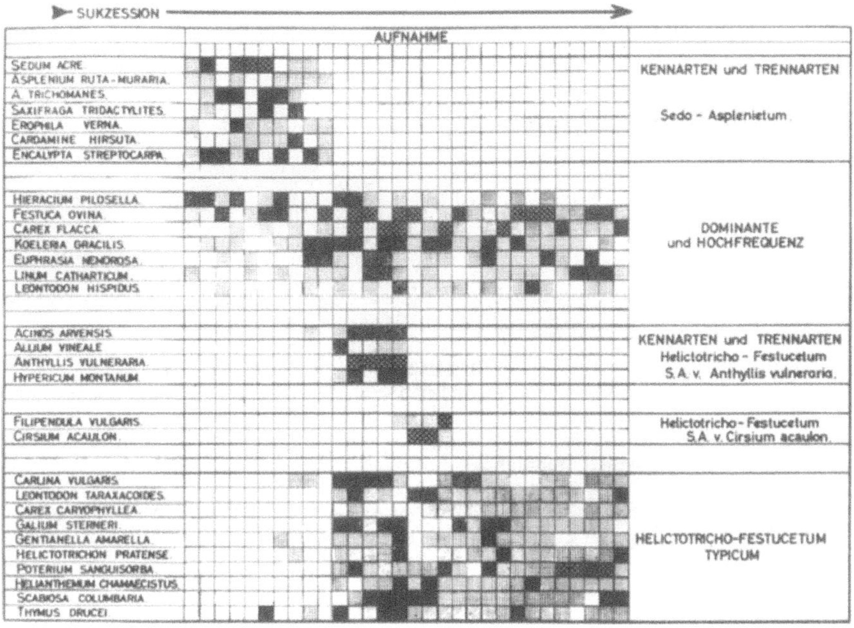

Abb.7.

se Weise wird den Untersuchungen dieser Art wichtige Information neu zugefügt.

Auf Abb.8 sind Daten dieser Art dargestellt.

Drei Flächen von Querco-Ilicetum aquifoliae, eine provisorische Assoziation in Südwest-Irland, die 1936 von TURNER & WATT zuerst als Quercetum sessiliflorae beschrieben worden ist, wurden 1966 wieder besichtigt. Das Diagramm zeigt die Veränderungen in Umfang und Art der vorherrschenden Bäume. Es ist anzunehmen, daß die drei Arten von Waldland, die zuerst beschrieben wurden, das Endstadium der Sukzession darstellen.

Man kann hieraus ersehen, daß allein auf Grund der Zunahme des Stamm-Durchmessers innerhalb von dreißig Jahren die Typ III-Fläche sich in Typ II, und die Typ II-Fläche sich in Typ I verwandelte. Die ursprüngliche Typ I-Fläche zeigte nur eine Ausbreitung der Alters- und Umfangsstruktur. Diese Umfangsmessungen können sehr einfach auf Menge umgerechnet werden, von hier auf Biomasse, und schließlich auf Daten der Energie-Speicherung. Das einzige, was uns fehlt, ist eine genaue phytosoziologische Beschreibung

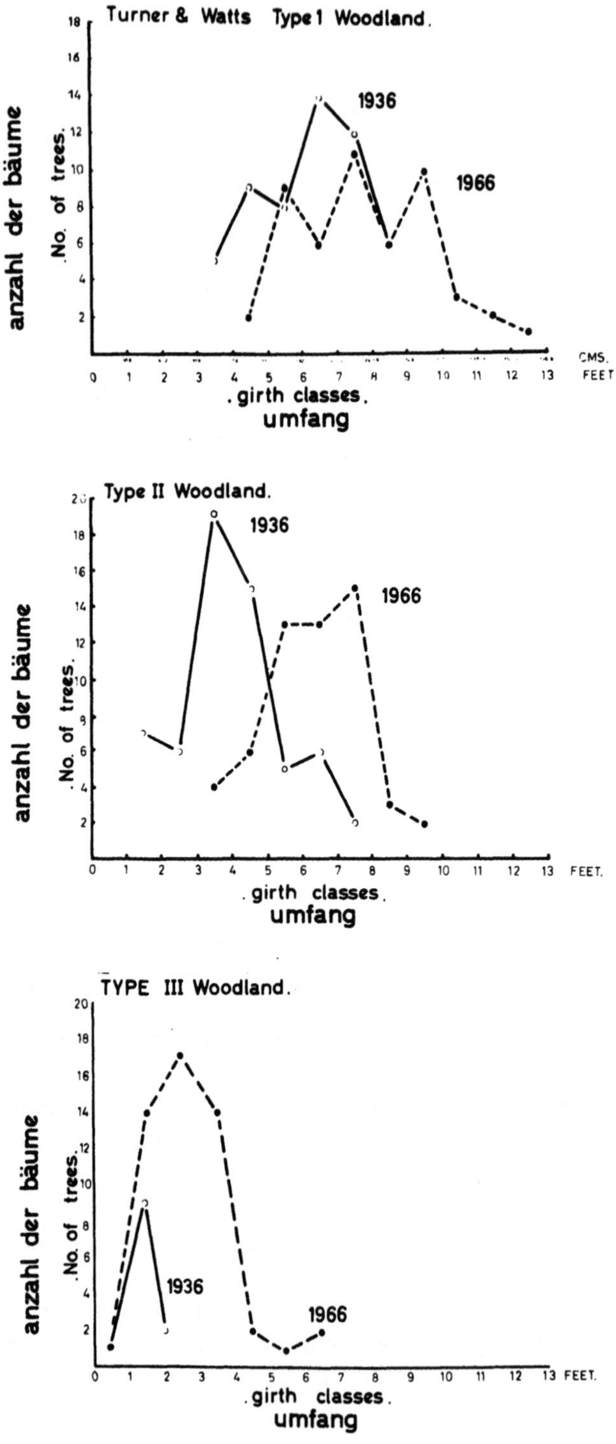

Turner & Watts Type 1 Woodland.
1936
1966
anzahl der bäume
No. of trees.
CMS.
FEET
girth classes.
umfang
Type II Woodland.
1936
1966
anzahl der bäume
No. of trees.
FEET.
girth classes.
umfang
TYPE III Woodland.
1936
1966
anzahl der bäume
No. of trees.
FEET.
girth classes.
umfang

Abb.8

der Bestände von 1936.
Zum Schluß möchte ich um Entschuldigung bitten, weil diese Vorlesung nichts Neues enthält, und alle Ideen schon vor langem von BRAUN-BLANQUET und anderen erläutert wurden und in phytosoziologischen Kreisen allgemein bekannt sind. Wir glauben jedoch, daß eine nochmalige Zusammenfassung dieser Ideen angebracht ist, da wir davon überzeugt sind, daß die Zukunft unserer Wissenschaft in der Verbindung von phytosoziologischer und dynamischer Betrachtungsweise liegt, um grundlegende ökologische Fragen beantworten zu können. Welche Zeit könnte sich besser dazu eignen, als die Anfangsphase des Internationalen Biologischen Programmes.

SUMMARY
Succession is a linked change of the biotic and abiotic components of an ecosystem brought about by the storage of energy (in the form of standing crop, humus, peat or extracellular product) within the ecosystem. It is suggested that by linking the phytosociological and dynamic approaches, the continuum of information - the world vegetation - will not only be the arena in which the ecologist finds his problems, but the tool with which he solves them. Four examples of the study of the application of these concepts are described.

DISKUSSION

A.W.KÜCHLER
You seem to say that succession implies a surplus of energy and that succession continues until there is a balance. And when there is no more surplus of energy the community becomes stabilized. Can you visualize that there is the opposite happening too, so that instead of having a surplus of energy you have a lack of energy and consequently we get what was mentioned this morning, a retrogressive succession which some people felt was no succession?

D.J.BELLAMY
I must agree that both processes are succession. But I believe that in vegetation which is undisturbed, the natural process of the evolution of an ecosystem is with an increasing energy storage. I will agree that man can affect the ecosystem, that we can have changes in climate, changes in the biotic component,

new species, new herbivores coming and help to break this ecosystem down. The basic drive behind the evolution of this thing which we call an ecosystem is this gradual build-up. I think that ecology is lacking a basic principle. We have been groping in the dark, trying to find a basic principle to which we can work. I believe that this principle of succession, as first perhaps set up by CURTIS & LUNDEGÅRDH, is the light one, but there are retrogressive successions and on my diagram for Central England some of those of course were retrogressive steps. This will not go hand in hand with increased storage.

D.MUELLER-DOMBOIS
To the question of continous energy increase upto the state of equilibrium I would say that this does not apply in Hawaii. For example on ash soils it is like this: We have raw ash first. If it is in the humid tropical climate, a forest eventually develops and we have a state whereby the organic matter content in the soils is very high. But afterwards, due to the laterization process, organic matter becomes completely depauperated, and the old soils, e.g. physiographically older soils, show very little organic matter. So the energy gets lost. And we come to an older type of vegetation, an older type of forest which is also in equilibrium.

A.W.KÜCHLER
How can it build up first, and then get lost?

D.MUELLER-DOMBOIS
I haven't studied this, it is just phenomena observation, and I can't tell the reason as yet, but in the ash we may assume a high component of nutrients originally and these get gradually washed out in the tropical humid climate. Therefore the nutrient storage in the soil is eventually so much decreased that it is entirely tied up in the living vegetation only.

D.J.BELLAMY
I do not know the vegetation in Hawaii, I wish I did! When one looks at energy storage, we must not only think of that under the ground, but also on that above ground. The standing crop must be included in any measurement of this point. Now, what happened? Your soil begins to get much humus and you have a standing crop of vegetation. Does this vegetation also disappear? In your process of succession you say the soil slowly builds up

until you get much humus, and then the humus disappears, but what happened to the standing crop?

D.MUELLER-DOMBOIS

You mean how the vegetation changes. During the time when you have a maximum of humus in the soil and the soil is not there until you have a brown latozolic forest soil, and at that time it is very rich and the vegetation population is also of high value. And later on, as the organic matter is removed gradually by the laterization process we still have a forest, but it is not a productive forest. We get more species as a rule, the stand becomes more diversified, the volume production, the standing crop, e.g.the energy in the vegetation will decrease, so it is not only in the soil but it is also in the standing crop.

D.J.BELLAMY

Than I must stand corrected. I have measurements for tropical vegetation, in Assam and in West Africa, and I have found nothing which goes against this gradual development. So I must come to Hawaii.

H.SUKOPP

I only want to tell you that the last publication of Professor BERGER-LANDEFELDT, who died last month, on "Energieumsatz am Standort" (Bot.Jb.1967) gives some detailed examples on the use of energy coming in at the habitat, by different types of vegetation, although not in combination with succession.

A.W.KÜCHLER

If I understood Dr.MUELLER-DOMBOIS correctly, we have here precisely a case of retrogressive succession, because, if normal succession builds the matter up and then breaks down again, that would be retrogressive. And apparently we have a building up here, based on the rich nutrients in the soil and a break down as they are leached out.

J.J.MOORE

Was your idea that, once the storage of energy reached equilibrium - no build up, no loss - you have a stable community? No further succession, whereas when energy storage was building up, the succession of vegetation was in progress? Now in your own study on bogs, on mires, on moors, does the accumulation of humus not go on with a constant vegetation cover? With no real change in the type of vegetation?

D.J.BELLAMY

I don't think so. In the peat bogs in Ireland and also in England we do not have the "Stillstand-Komplex". It is when we come to the "Stillstand-Komplex" that the succession is slowing down, is coming to climax. I still do not understand the peat bogs of Ireland, they are something very peculiar, both phytosociologically and ecologically, and there is much work to be done on them, but I still think, peat bogs must come to this stage of equilibrium. Professor Stanislav KULCZYNSKI says in his monographion the "Moore des Polesie-Gebietes" that we pass west across Europe, we find more and more erosion and the end product of the "Hochmoor" is a cupola with a certain amount of growth and a certain amount of erosion brought about by rainfall and I think at this point the energy is stable and we can go and we describe the same stand which LLOYD PRAEGER described many years before and they are the same type of vegetation. I think in this case succession is not taking place and we are in equilibrium.

J.J.MOORE

I think i will disagree as regards Ireland. I believe that there is a climax, a steady state of vegetation with a continous accumulation of peat.

H.ELLENBERG

I would like to support Father MOORE, because when you see the large bogs, e.g. here in northwestern Germany, these bogs have many meters of mass and they are growing slowly, but for a long time. In the studies of OVERBECK, e.g., they had the same association, even the same mosaic, for hundreds of years, and the accumulation of peat still went on. So is it not probable, at least for these types of mires, that you are right. There still is an accumulation of humus and the type of vegetation is still remaining the same. I think also, e.g., that a climax forest still gains some material, it is growing, slowly, but it still grows. I think it is too extreme to imagine that there should be no accumulation at all after having reaches some climax stage. I think you are a bit theoretical in this point, and it is merely theory but not proved.

D.J.BELLAMY

I said, our work is just beginning. We are studying climax woodland. Now, we do not have climax woodland in the British Isles, it was all gone, but we are studying something which we think may be climax woodland. But it takes much time to cut down 100 trees and dry them and weigh them. We are still doing this, and when we have finished, I will come back and answer your question.

H.ELLENBERG

When I understood your slide well, you demonstrated that within 750 years, there has grown an oak forest out of bare stony ground in Spain, by succession. I don't believe this. I think this is an example of zonation which you interpreted as succession. In reality the stony ground is developing very slowly, and it is very improbable that a stony ground becomes a brown earth or something like that within some hundred years. This has been believed by pedologists in earlier times but more and more they come to the conviction that this is not correct. The development of soil is a very slow one, and on a stony ground, e.g. a calcareaous stony ground, there never develops a brown earth. When you find a brown earth there, there is a cover of loamy soil which is formed by glaciation, or by loessic material, or colluvial material or other things and not formed by the same spot: it is not autochthonous but allochthonous. What is your argument to prove that only 750 years are necessary to complete this succession?

D.J.BELLAMY

I'm afraid it is even worse, because I said 75 (which I can't say it in German).

H.ELLENBERG

Oh, even less? I think 7000, or 70000!

D.J.BELLAMY

There was a valley high in the mountains, 1400 m above sea level, and this valley almost within living memory had been forest with *Quercus pyrenaica*. And then the local peasants had cut all the forest. There was an old man in the village in this valley who said he can remember when there were no trees. We found trees growing here, a forest, which was 67 years old. And this must have grown in the last 75 years. When man cut the forest down here, he turned it over to a desert, and *Erica arborea* started to grow, when the *Erica* is 11 years old, and we have a standing crop of 2 kg/m^2, then the next stage of succession can begin.

The climate must be changed to a certain extent, before seedlings of *Quercus* and *Sarothamnus* can begin to grow. When *Sarothamnus* is 17 years old, *Quercus* can grow into adult trees.

H.ELLENBERG

Now I understand! This is a secondary succession, not a primary one, of course, then it is possible so quickly.

In any case it is dangerous to interpret this in terms of energy, because the various stands you compared perhaps are not linked in a succession really all, and the soils you found form a mosaic in this case.

DYNAMIK DER PFLANZENGESELLSCHAFTEN UND PFLANZENSOZIOLOGISCHE SYSTEMATIK

R. C a r b i e n e r

ZUSAMMENFASSUNG:
An Hand einiger Beispiele von strittigen Fragen in der heutigen pflanzensoziologischen Systematik versucht der Referent Anregungen zu bringen zu stärkerer Berücksichtigung zonal-klimatischer und allgemein-dynamischer Gesichtspunkte (natürliche und potentielle Sukzession) der Eingliederung verschiedener pflanzensoziologischer Einheiten in das System, sowie bei der Fassung der höheren Einheiten selbst.
Es wird vorausgesetzt, daß es dabei grundsätzlich nicht um eine Änderung des floristischen Gliederungsprinzips nach BRAUN-BLANQUET gehen kann. Die von BRAUN-BLANQUET gegründete Methodik zur Erforschung der Vegetation hat trotz der bekannten und immer wieder auftretenden theoretischen Einwände, einen nicht bestreitbaren, wesentlichen, ausschlaggebenden Fortschritt in der exakten großmaßstäblichen Kenntnis der Vegetation all derer Gebiete, in denen sie gewissenhaft gehandhabt wurde, zur Folge gehabt. Die großen synthetischen Übersichts- und Sammelwerke der letzten Jahre, in welchen eine noch nie erreichte Feinheit der Analyse sich mit klarer überregionaler Übersichtlichkeit verbindet, stellen dafür den eklatantesten Beweis dar. Die Bewährung in der Praxis liefert den Prüfstein einer jeden Methodik.
Es geht also nur darum, gewisse Mängel und Schwierigkeiten zu verringern. Zu diesen Mängeln gehört die Überwindung der Gegensätze zwischen Pflanzensoziologie einerseits und der Pflanzengeographie und Ökologie andererseits, die von vielen Autoren hervorgehoben werden. Diese Gegensätze sind z.T. normal, z.T. maßstabsbedingt. Andere Mängel sollten jedoch im Interesse einer besseren interdisziplinären Verständlichkeit behoben werden. Zu diesen Mängeln gehört die große Uneinheitlichkeit der praktischen, räumlichen und und ökologischen Bedeutung der Vegetationseinheiten höheren systematischen Ranges. Es ist unbestreitbar, daß bei der Abgrenzung derselben ein subjektives Moment nicht vermieden werden kann, und Zweck und Ziel des Bearbeiters mitspielen. Es wäre wünschenswert, wenn bei der Erarbeitung der höheren Einheiten großklimatische, d.h. zonal-dynamische Gesichtspunkte (gegen die Arealgrenzen in lokalklimatische über-

gehend) den Vorrang gegenüber edaphischen oder biotischen erhalten würden. Damit dürfte der nicht immer unberechtigte Vorwurf vieler Pflanzengeographen und Ökologen einer zu großen Künstlichkeit des Systems entkräftigt werden, und z.B. der Feststellung von C.TROLL, daß in gewissen Fällen "die pflanzensoziologische Systematik einheitliche Landschaftszellen, die ökologische und physiognomische Ganzheiten sind, völlig zerreißen kann" abgeholfen werden. Bei unmittelbar sukzessionsbedingten Einheiten, wie z.B. den Schlagfluren, Gebüsch- und Saumgesellschaften usw., wäre es günstig, die wirkliche oder potentielle dynamische und ökologische Verbindung mit den meist unmittelbar benachbarten klimatischen oder klimatisch-edaphischen Waldgesellschaften im System zu berücksichtigen. Dies, z.B. indem sie vielleicht nicht als selbständige Klassen gefaßt werden, sondern nur zu eigenen Ordnungen zusammengestellt werden, Ordnungen die z.T. parallel in verschiedenen Klassen sich wiederfinden können. Dies würde eine Annäherung an ein mehrdimensionales, dynamisch-ökologisches Gliederungsprinzip darstellen. Durch Maßnahmen dieser Art wäre der zur Zeit bestehende Streit zwischen den Anhängern eines auf ökologische Zeigergruppen aufgebautem Systems,oder den Befürwortern eines dynamischen auf Sukzession und Klimax geeichten Systems und den Getreuen des Kenn- und Trennarten-Systems nach BRAUN-BLANQUET, wenigstens entschärft.

Der Referent erläutert dann durch Beispiele Möglichkeiten einer Verbesserung der pflanzensoziologischen Systematik durch eine dynamische und zonale Betrachtungsweise. So dürften z.B. die oligotrophen bodensauren Buchenwälder nicht an die Q u e r c e - t a l i a angegliedert werden, weil solche Wälder weit über das Eichenklimaxgebiet heraus verbreitet sind und oft in engstem räumlichen Kontakt zu "echten" F a g e t a l i a -Gesellschaften stehen, was von ELLENBERG schon betont wurde. Desgleichen dürfte das Problem der subalpinen Krüppelbuchen-Wälder (A c e r i - F a - g e t u m) nur in zonal-klimatischer und dynamischer (verminderter Vitalität der Holzgewächse) Sicht zu lösen sein (eventuelle Angliederung an die B e t u l o - A d e n o s t y l e t e a, mit welchen sie in äußerst engem dynamischen Kontakt stehen). In der subalpinen und alpinen Stufe stellt sich die Frage der dynamischen Beziehungen zwischen der Zwergstrauchheiden, den N a r d e t e n und pro parte den Koniferen-Wäldern. Hier stört der Gegensatz zwischen mehr edaphisch (N a r d e t a l i a) und mehr klimatisch ge-

faßten Einheiten (V a c c i n i o - P i c e e t a l i a). Es dürfte sich eine neue klimatisch-zonale, d.h. höhenstufenbedingte Klassifikation als zweckmäßiger erweisen, wie sie z.B. in Skandinavien gebraucht wird (D e s c h a m p s i o - M y r t i l l e t a l i a). Was die Schlag-, Mantel- und Saumgesellschaften betrifft, dürfte den dynamischen Zeigerarten eine größere systematische Bedeutung angemessen werden, was eine schnellere und bessere Eingliederung in landschaftsökologische Systeme erlauben würde.
Ob sich solche Vorschläge verwirklichen lassen, wird nur neue Tabellenarbeit erweisen.

H.ELLENBERG
Dieser Vortrag erinnert uns daran, daß man die Sukzession nicht um ihrer selbst willen betrachten kann, sondern als Hilfsmittel benutzen muß, um in gewissen Fällen der Systematik, gerade in der Einordnung schwieriger Klassen, zu einer guten Lösung zu kommen.

R.TÜXEN
Es war von *Geranium sanguineum* die Rede, das man etwa an die Q u e r c e t a l i a p u b e s c e n t i s annähern könnte. Ich denke daran, daß WESTHOFF und ich eine solche Gesellschaft in der Nähe von Helsinki aufgenommen haben, daß um Uppsala und nördlich davon in den Säumen keine Pflanze häufiger ist als *Geranium sanguineum*, daß es sie in England, in Irland, an den Küsten der Bretagne, und daß es sie weit und breit in Jütland gibt, daß dort aber weit und breit keine Spur von Q u e r c e t a l i a p u b e s c e n t i s ist.

R.CARBIENER
Es steht doch fest, daß auch in Norwegen, oder England, oder Finnland diese Gesellschaften doch in engem syndynamischen Verhältnis zu den thermophilen Wäldern stehen. Natürlich sind wir da außerhalb des Areals der *Quercus pubescens*, und das ist eben auch etwa die Klasse der Q u e r c e t a l i a p u b e s c e n t i s. Wenn man die Klasse enger fassen würde, also sie mehr an die mediterrane Vegetation anschließen würde, und wenn man eine besondere, zonale oder interzonale Klasse der thermophilen Eichenwälder bilden würde, dann könnte man da teilweise die acidophilen Eichenwälder hineinbringen, und dann wäre das Problem der Q u e r c e t e a r o b o r i - p e t r a e a e gelöst, indem diese Klasse sich eben in verschiedene Klassen auflösen würde.

EXEMPLE D'APPLICATION DU TRANSECT LINEAIRE (BISECT) DANS L'ETUDE DE LA DYNAMIQUE DE LA VEGETATION.

J. W o l a k

Au Symposion précédent, qui était consacré à la synmorphologie nous avons présenté un rapport traitant des complexes de groupements végétaux. Ce rapport était basé sur les matériaux de travail, dont le but essentiel était l'étude de la succession des groupements.

C'est la succession secondaire des anciennes prairies du bord, du Parc National de Białowieza qui a été examiné. Le terrain s'étand le long des rives des rivières Narewka et Hwoźna qui constituent les limites ouest et nord de la réserve.

Dès la creation du Parc National ce terrain s'est trouvé dans la réserve proprement dite. Ce qui a entrainé le processus de la succession régénérative. Nous connaissons approxymativement le stade intial et la stade final de cette succession. Le premier ressemble aus prairies de fauche et patûrages des bords opposés. Le stade final représent les groupements forestiers contigus.

Pourtant l'image de la végétation sur les anciennes prairies est plus compliqué qu'on puisse penser. Il y a deux causes qui provoquent ces complications. La première c'est que l'inclusion des terrains riverains à la réserve ne se fit pas d'un seul coup. Elle a duré de 1929 à 1939. Et ensuite en temps de guerre, et même plus tard, certaines parties ont été fauchées de nouveau. Alors la succession n'a pas duré le même temps sur tout le territoire considéré, et comme résultat on trouve différents chainons de la même série régénérative l'un à côté de l'autre.

La seconde cause qui entraine les perturbations dans le processus de la succession est due aux anciennes régularisations des rivières. Ces régularisations ont eu lieu sous le règne du dernier roi polonais pendant la deuxième moitié de XVIII siècle. Puis on a renouvelé ces régularisations encore quelques fois. Jusqu'en 1930 les rivières Narewka et Hwoźna étaient navigables.

Sur le profil transversal de la valée de la Narewka on voit que sur 500 mètres de tourbière basse la différence des niveaux est de 5 mètres. L'inclinaison sur 150 mètres de la rivière morte est de l'ordre de 5°.

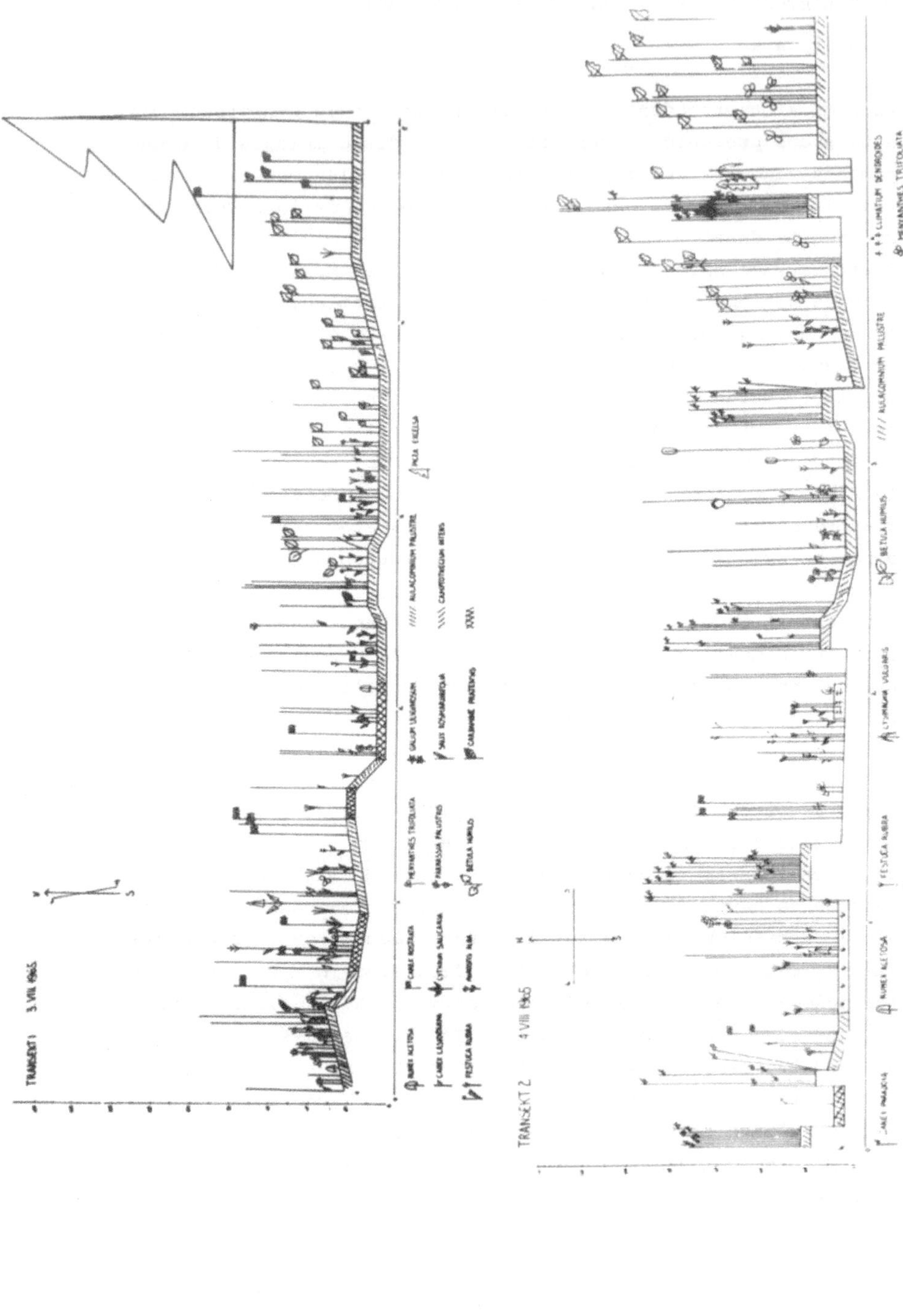
TRANSEKT 1 3. VIII 1965
RUMEX ACETOSA
CAREX LASIOCARPA
FESTUCA RUBRA
CAREX ROSTRATA
LYTHRUM SALICARIA
AGROSTIS ALBA
MENYANTHES TRIFOLIATA
PARNASSIA PALUSTRIS
BETULA HUMILIS
GALIUM ULIGINOSUM
SALIX ROSMARINIFOLIA
CARDAMINE PRATENSIS
AULACOMNIUM PALUSTRE
CAMPTOTHECIUM NITENS
PICEA EXCELSA
TRANSEKT 2 4 VIII 1965
CAREX PANICEA
CAREX ROSTRATA
RUMEX ACETOSA
POA PRATENSIS
FESTUCA RUBRA
PARNASSIA PALUSTRIS
POLYGONUM BISTORTA
LYSIMACHIA VULGARIS
CIRSIUM RIVULARE
ERIOPHORUM ANGUSTIFOLIUM
BETULA HUMILIS
GALIUM ULIGINOSUM
AULACOMNIUM PALUSTRE
CAMPTOTHECIUM NITENS
ACROCLADIUM CUSPIDATUM
CLIMATIUM DENDROIDES
MENYANTHES TRIFOLIATA

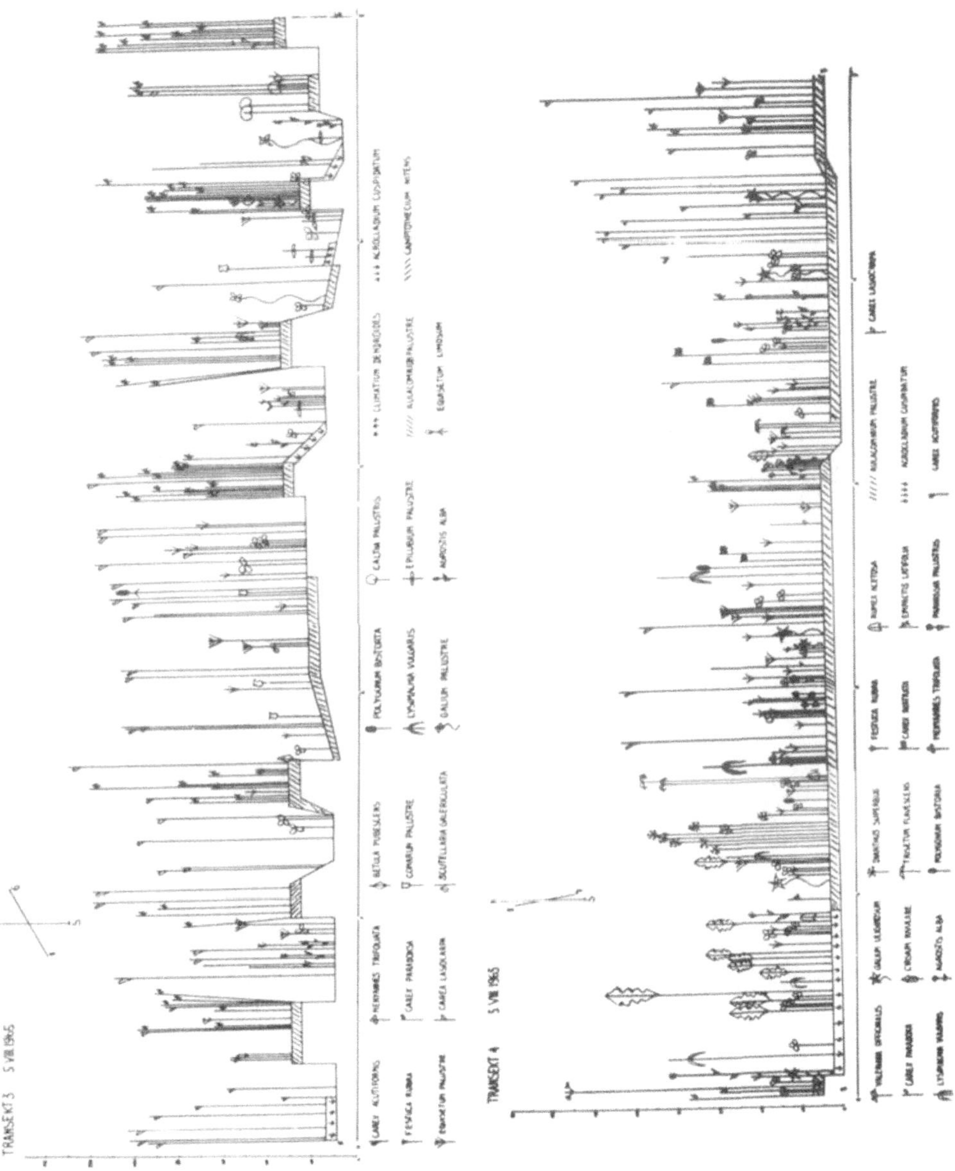
TRANSEKT 3 5 VIII 1965
CAREX ACUTIFORMIS
FESTUCA RUBRA
EQUISETUM PALUSTRE
MENYANTHES TRIFOLIATA
CAREX PARADOXA
CAREX LASIOCARPA
BETULA PUBESCENS
COMARUM PALUSTRE
SCUTELLARIA GALERICULATA
POLYGONUM BISTORTA
LYSIMACHIA VULGARIS
GALIUM PALUSTRE
CALTHA PALUSTRIS
EPILOBIUM PALUSTRE
AGROSTIS ALBA
CLIMATIUM DENDROIDES
AULACOMNIUM PALUSTRE
EQUISETUM LIMOSUM
ACROCLADIUM CUSPIDATUM
CAMPTOTHECIUM NITENS
TRANSEKT 4 5 VIII 1965
VALERIANA OFFICINALIS
CAREX PARADOXA
LYSIMACHIA VULGARIS
GALIUM ULIGINOSUM
CIRSIUM RIVULARE
AGROSTIS ALBA
DIANTHUS SUPERBUS
TRISETUM FLAVESCENS
POLYGONUM BISTORTA
FESTUCA RUBRA
CAREX ROSTRATA
MENYANTHES TRIFOLIATA
RUMEX ACETOSA
EPIPACTIS LATIFOLIA
PARNASSIA PALUSTRIS
AULACOMNIUM PALUSTRE
ACROCLADIUM CUSPIDATUM
CAREX ACUTIFORMIS
CAREX LASIOCARPA

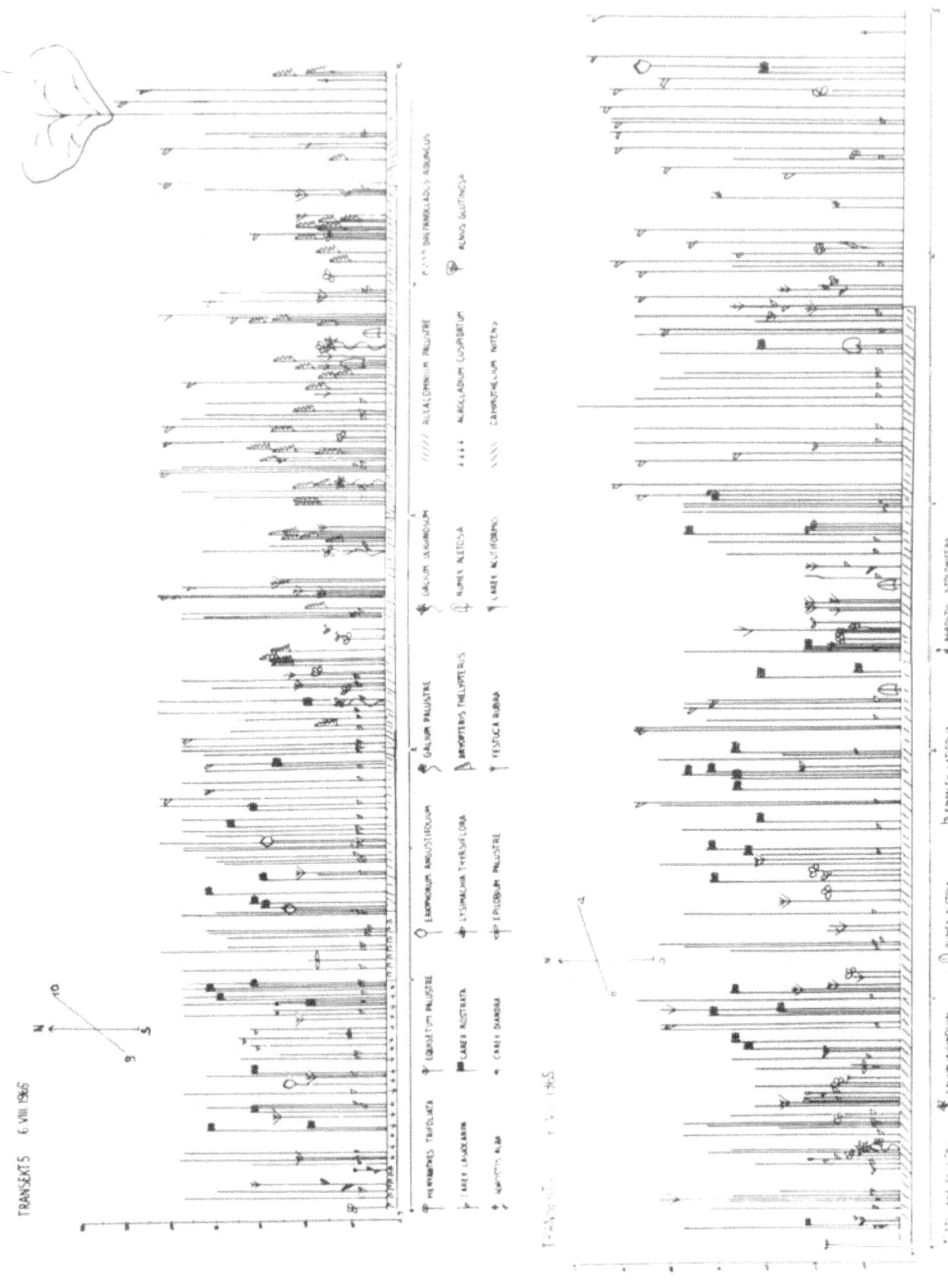
TRANSEKT 5 6 VIII 1966
N
MENYANTHES TRIFOLIATA
EQUISETUM PALUSTRE
ERIOPHORUM ANGUSTIFOLIUM
GALIUM PALUSTRE
GALIUM ULIGINOSUM
AULACOMNIUM PALUSTRE
DREPANOCLADUS ADUNCUS
CAREX LASIOCARPA
CAREX ROSTRATA
LYSIMACHIA THYRSIFLORA
DRYOPTERIS THELYPTERIS
RUMEX ACETOSA
ACROCLADIUM CUSPIDATUM
ALNUS GLUTINOSA
AGROSTIS ALBA
CAREX DIANDRA
EPILOBIUM PALUSTRE
FESTUCA RUBRA
CAREX ACUTIFORMIS
CAMPTOTHECIUM NITENS
GALIUM ULIGINOSUM
MENYANTHES TRIFOLIATA
RUMEX ACETOSA
CALTHA PALUSTRIS
AGROSTIS ALBA
AGROSTIS STOLONIFERA
AULACOMNIUM PALUSTRE

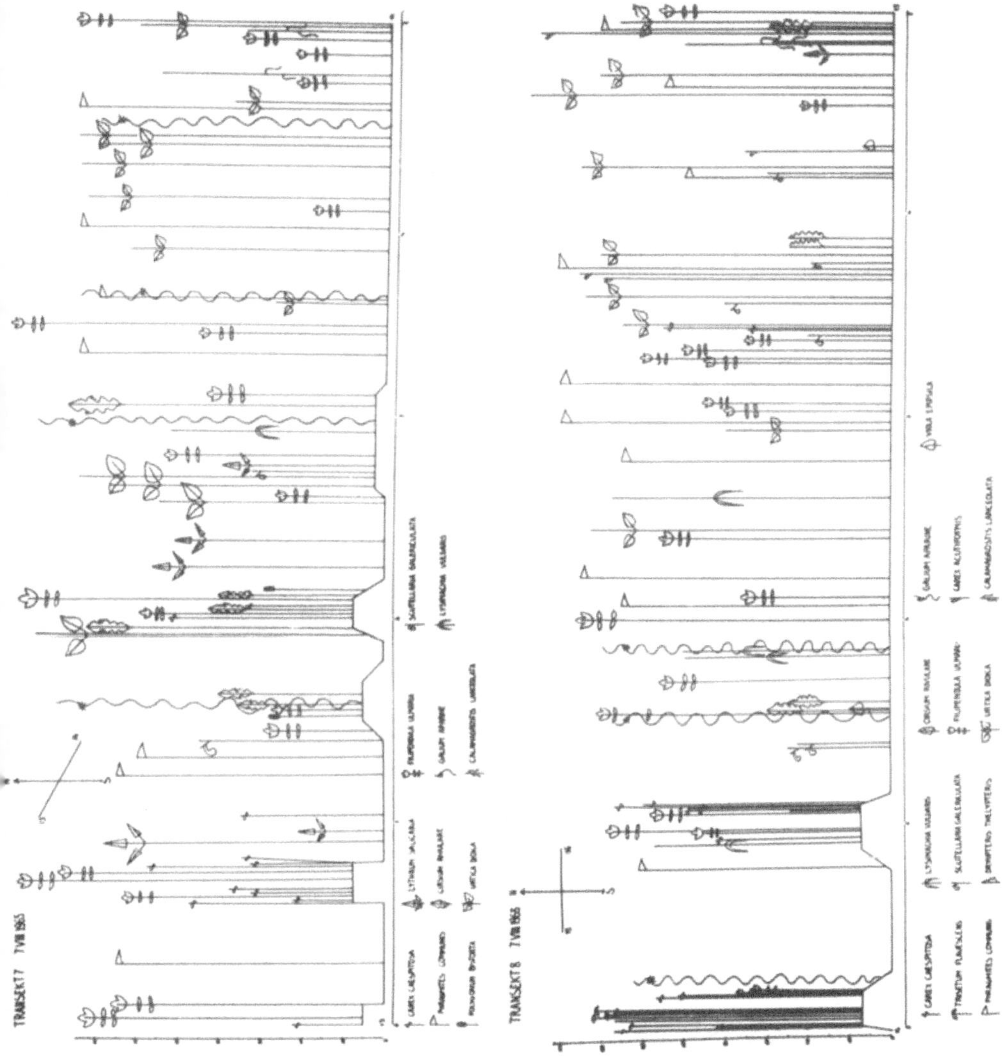
TRANSEKT 7
TRANSEKT 8

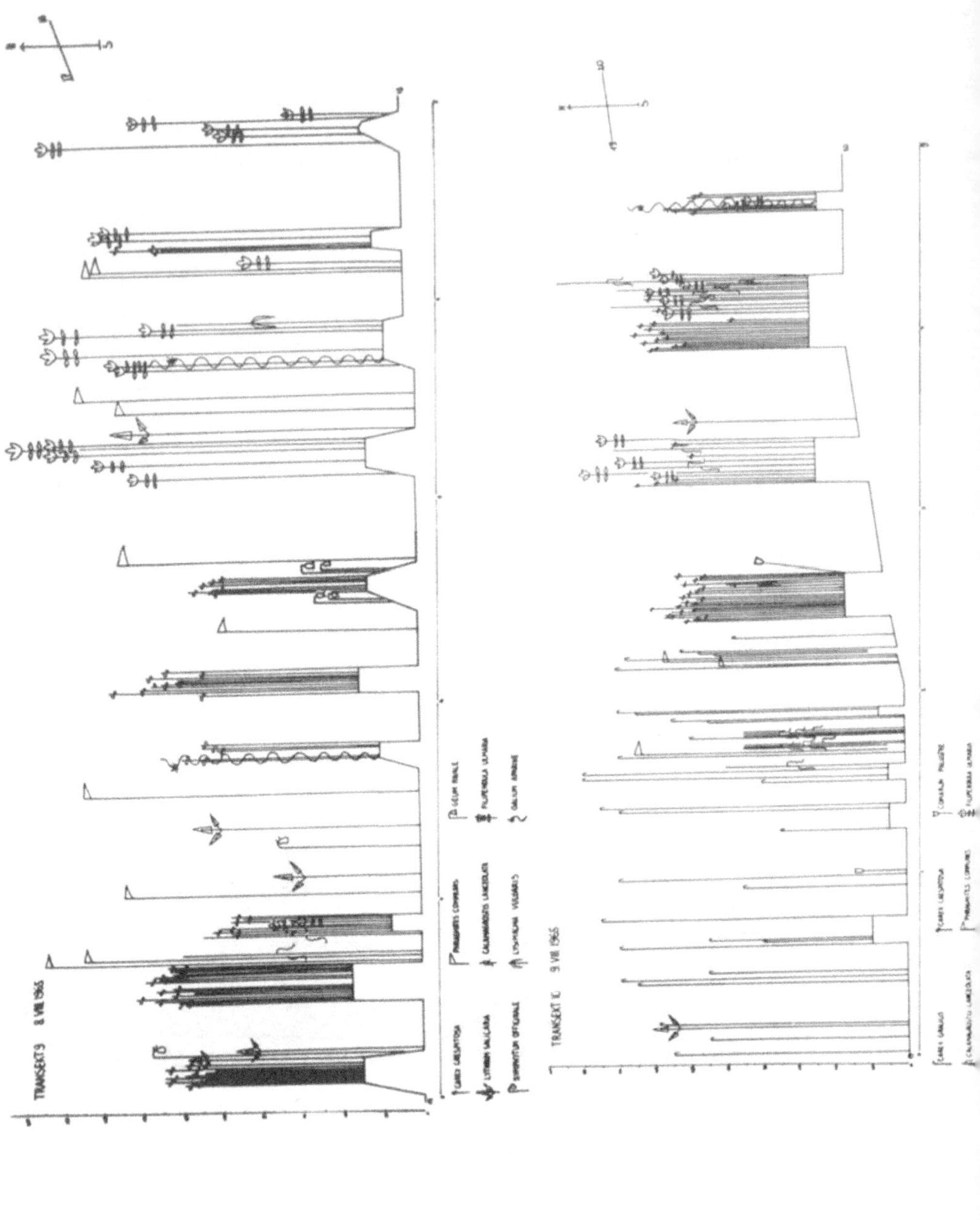
TRANSEKT 9 8 VIII 1965
TRANSEKT 10 9 VIII 1965

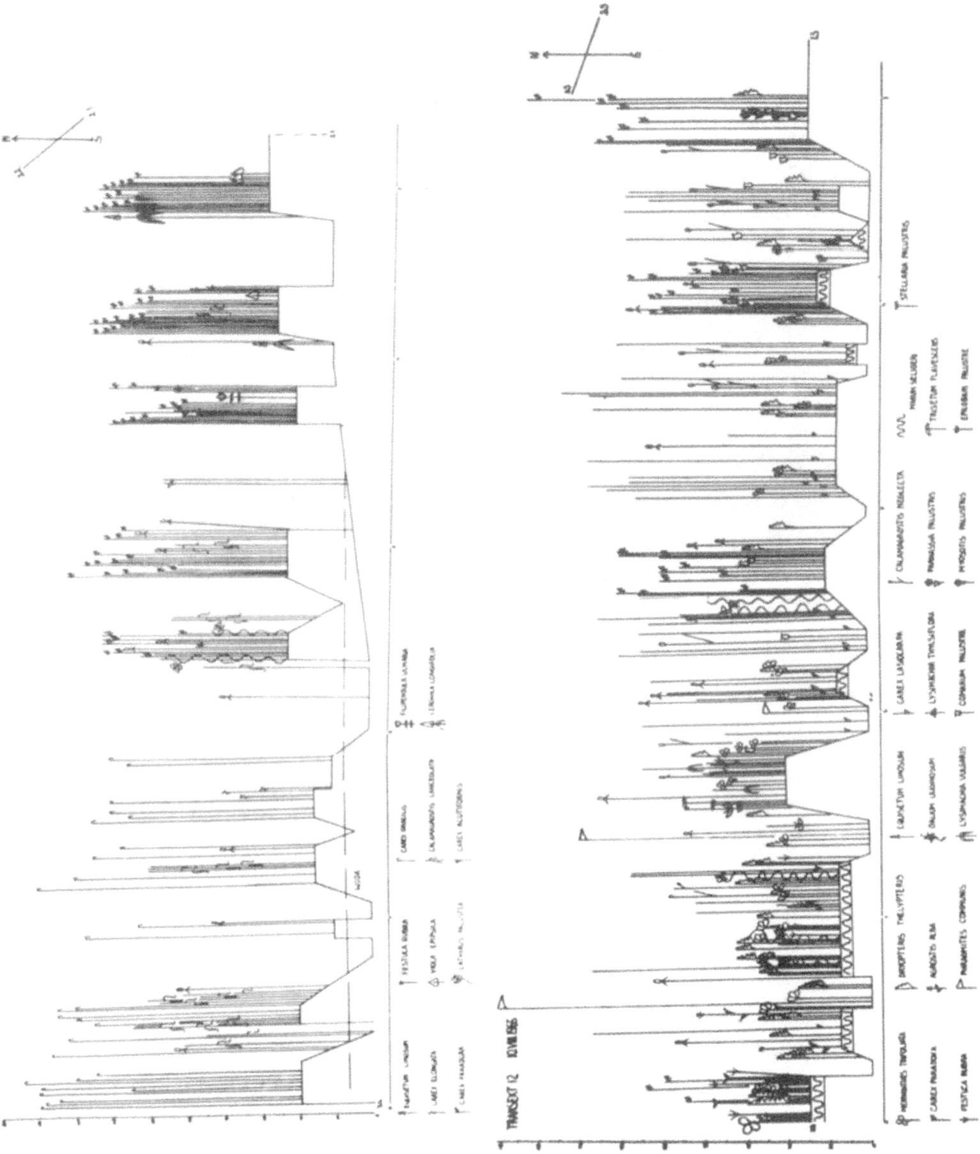

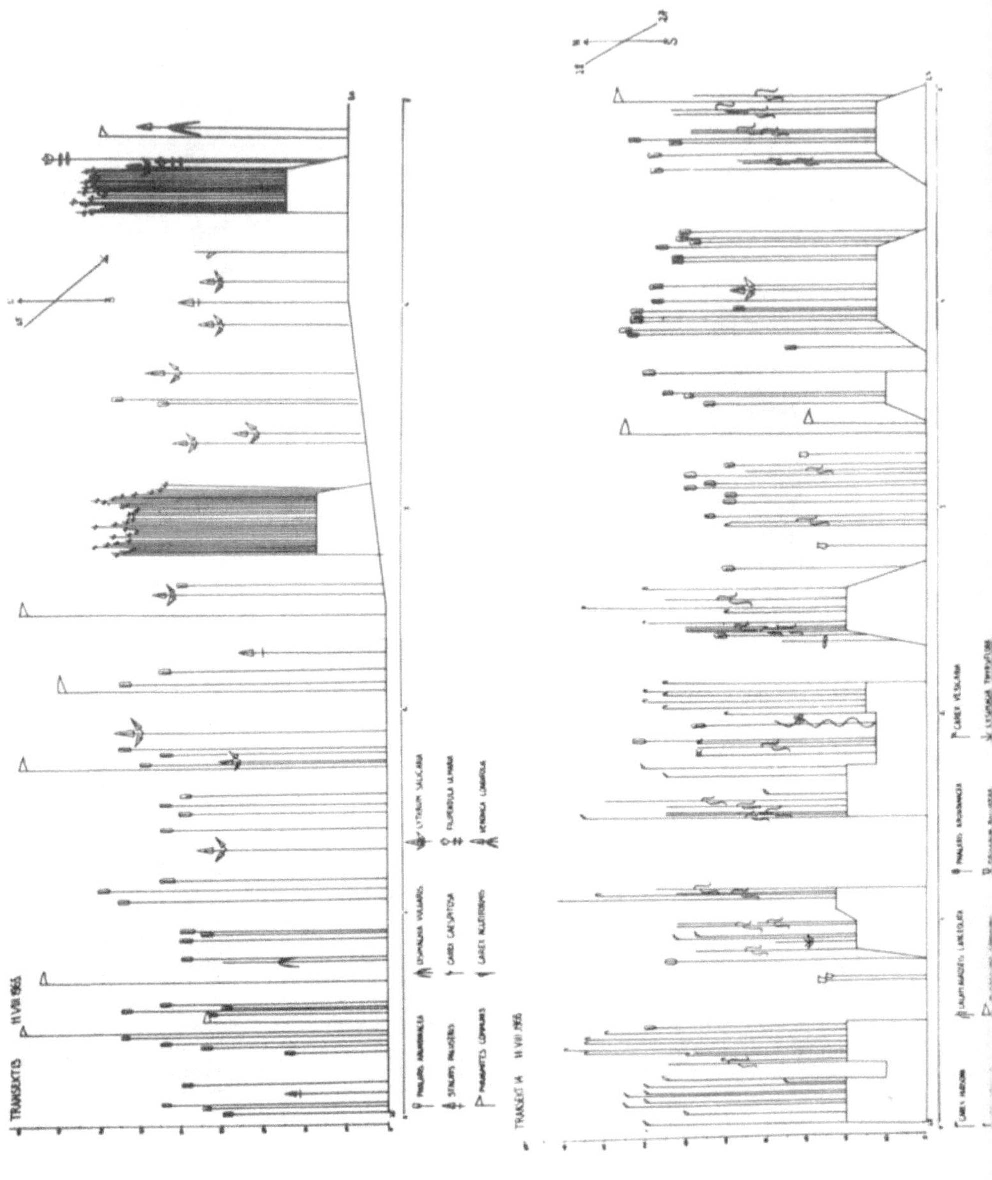
TRANSEKT 13 11 VIII 1965
TRANSEKT 14 11 VIII 1965

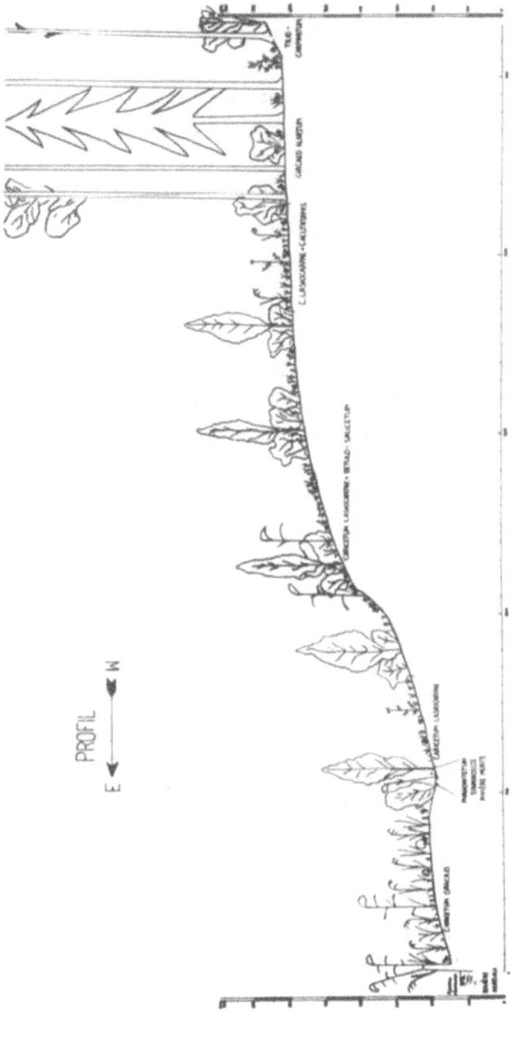

D'aussi grandes inclinaisons sur la tourbière basse ne peuvent que tirer leur origine de la minéralisation de la tourbe à la suite de ces anciennes régularisations. Ce facteur semble jouer encore une certain rôle dans les localités plus élévées et éloignées de la rivière. Mais il y a plusieurs signes qui indiquent qu'à présent les deux rivières se trouvent de nouveau en phase d'accumulation. Ce qui entraine l'élévation de la nappe d'eau dans le voisinage. Les deux facteurs mentionnés étant les facteurs écologiques dominants, poussent la succession en sens contraires.

Le but essentiel dans les recherches sur la succession végétale est la détermination de l'ordre de succession des groupements. En principe seul la méthode des surfaces permanentes est estimée comme précise et exacte. Les résultats, bases sur des observations qui n'ont eu lieu qu'une fois, seront considéres comme plus ou moins hypothétiques et provisoires.

Nous voudrions présenter ici la méthode, qui bien qu'elle soit basée sur une observation unique est néanmoins apte à fournir des informations valables sur l'ordre de succession des groupements. La méthode se fonde sur l' application des transects linéaires (bisects) sur des espaces où les processus dynamiques sont spécialement visibles. En cas de changement linéaire c'est la zone transitoire entre deux groupements.

Les transects ont été réalisés sur une longeur de 5 m. On a dressé le diagramme des espèces à l'échelle 1:10 en utilisant le papier quadrillé d'un milimètre et en admettant pour chaque espèce les symboles schématisés.

En voici quelques exemples:

Les transects 7,8,9 montrent la succession du Filipendulo-Geranietum à partir du Caricetum caespitosae. On voit qu'ici le Filipendulo-Geranietum est le groupement succédant au Caricetum caespitosae. Des espèces de Molinietalia s'installent sur les touffes de *Carex*, puis elles dominent complètement tandis que les espèces de Phragmitetalia disparaîssent. Ce changement n'a lieu que dans les localités constamment émergées. D'ailleurs les deux groupements peuvent former un complexe vertical sinaptique assez persistant.

Dans les transects 5 et 6 le Caricetum acutiformis prend le terrain du Caricetum lasiocarpae. Ce changement est provoqué probablement par la mobilisation de l'eau dans la tourbière comme conséquence de l'élévation du niveau de l'eau au voisinage de la rivière. Il nous semble qu'une telle succession demeure aussi dans la dynamique endogène du Caricetum lasiocarpae. Comme résultat, il se forme complexe vertical cryptodisaptique, assez répandu comme stade de succession.

Les transects 3 et 4 représentent aussi l'extension de *Carex acutiformis:* 3-e sur le Caricetum appropinquatae, 4-e sur le Cirsietum rivularis. La situation du transect 3-e possède un caractère semblable aus transects 5 et 6. Dans le transect 4-e seule l'élévation de la nappe d'eau semble être décisive.

Sur le transect 14-e nous avons une situation où comme résultat de l'élévation de la nappe d'eau et en conséquence de sa mobilisation et de son enrichissement à proximité de la rivière, le Phalaridetum arundinaceae envahit le Caricetum elatae. En effet on a saisi ici deux phases de ce processus, toutes deux étant assez répandues. En première phase C. gracilis supplante C. elatae, en deuxième C. gracilis est supplanté par le Phalaridetum arundinaceae. Même dans cette dernière phase les touffes formées par *Carex elata* sont bien conservées, bien que cette espèce y n'existe plus.

Le transect 12-e présente une situation toute différente causée par l'élévation du niveau d'eau. On voit ici la succession du Caricetum appropinquatae vers le Caricetum lasiocarpae. C'est un exemple de changement sur une surface entière d'individu d'association. Une observation superficielle donne seulement l'impression d'un complexe vertical disaptique de ces deux groupements. D'ailleurs un complexe pareil existe réellement comme stade durable. Le transect révèle de nombreuses touffes mortes de *C.paradoxa* qui indique la régression de cette espèce. Ce phénomène a lieu dans une localité assez éloignée de la rivière où l'élévation de la nappe d'eau provoque seulement de plus grandes oscillations, sans mobilisation. Dans le complexe disaptique ces mouvements verticaux sont favorables pour le groupement émersif et peuvent être néfastes pour le groupement immersif.

Les transects 1 et 2 illustrent la succession vers le Betulo-

S a l i c e t u m r e p e n t i s; 1-e à partir du C a r i c e t u m l a s i o c a r p a e, 2-e à partir d'un complexe vertical disaptique de C. l a s i o c a r p a e et C. a p p r o p i n q u a t a e. C'est l'exemple de la succession endogène sur la tourbière basse constamment émergée.

Enfin les transects 10,11,13 illustrent les transitions entre deux groupements sur les bords des eaux, où existent des limites écologiques nettes. Les zones de transition sont ici extrêmement étroites. Le sens de la succession est dirigé par l'élévation de la nappe d'eau.

CONCLUSIONS

Dans ce rapport on a présenté une partie du travail sur la succession secondaire des groupements végétaux de tourbière basse. La succession végétale se réalise par des changements progressifs qualitatifs, quantitatifs et structuraux créant une phase intermédiaire entre les groupements en cause. Un transect détaillé à grande échelle réalisé dans une telle phase révèle les différents stades du processus de succession, l'un à côté de l'autre. Ces stades sont fixés comme sur un film. Cela permet de déterminer la direction de la succession avec certitude, même si les causes de ce processus sont discutables.

ZUSAMMENFASSUNG

Der Verfasser stellt die Methode des Linientransekts (Bisekts) und deren Anwendung bei den Untersuchungen der Pflanzensukzession dar. Versuchsobjekt waren Niedermoore an Flüssen, welche die West- und Nordgrenze des Białowieża-Nationalparkes bilden. Die am Ufer gelegenen Wiesen wurden in das strenge Reservat in den Jahren 1929-1937 eingegliedert, jedoch wurden noch manche Partien nach dem zweiten Weltkrieg gemäht. Aus diesem Grunde gibt es im Untersuchungsgebiet nebeneinander verschiedene Stadien derselben sekundären Regenerations-Sukzession.

Der Verlauf der Sukzession wird gleichzeitig durch Veränderungen der Wasserverhältnisse, die im Untersuchungsgebiet stattfinden, kompliziert. Die Richtungen dieser Veränderungen sind in Abhängigkeit vom Flußabschnitt und von der Entfernung vom Wasserlauf entgegengesetzt.

Einerseits wirkt noch die alte Flußregulierung ein, welche die Senkung des Wasserspiegels und die Mineralisierung des Torfes in immer größerer Entfernung vom Wasserlauf verursacht. Andererseits

beobachtet man in direkter Nachbarschaft als gegenteilige Tendenz die Erhöhung des Wasserspiegels als Folge des Überganges der Flüsse in die Akkumulationsphase.
Die in der Übergangszone zwischen den Gemeinschaften durchgeführten Transekte ließen die Richtung der Sukzession der Gesellschaften von Molinietalia nach Phragmitetalia, von Phragmitetalia nach Scheuchzerietalia palustris und umgekehrt, die gegenseitige Ersetzung der Schilfgesellschaften, endlich das Vorschreiten der Gebüschgesellschaften in die Schilfgesellschaften nachweisen.
Die Methode des Linientransektes erwies sich als eine genaue Methode der Bioindikation der Veränderungen der Wasserverhältnisse im Moor und der Entwicklungsphase der anliegenden Wasserläufe.

SUMMARY

The author presents the technique of linear transect (bisection) and its application in studies on plant succession. Low peatlands along rivers constituting the western and northern boundary of the Białowieża National Park provided the study area. Riverside meadows were included into the strict reserve during the years 1929-1937. Some parts of them were mown still following to the World War II. Hence various stages of the same secondary regenerative succession are to be found side by side in the study area.

The course of succession is at the same time disturbed by alterations in water relations occurring in the study area. Directions of these alterations are contrary, depending upon the river section and the distance from watercourse.
On the one hand the previous river regulation, causing lowering of water level and mineralization of peat at ever increasing distance from the watercourse, is still in operation. On the other hand a contradictory trend - the rise in water level as a result of entering accumulation phase by rivers, is to be noted in an immediate neighbourhood of river.
Transects established in the transitional zone between associations made it possible to indicate the direction of the sequence of associations from Molinietalia to Phragmitetalia, Phragmitetalia to Scheuchzerietalia palustris and vice versa, substition of individual rush associations, finally the encroachment of shrubby assosiations into rush ones.

The technique of linear transect appeared to be an accurate method of bioindication for alterations of water relations in a peatland and of the developmental phase of adjacent watercourses.

METHODES D'ETUDE DES FORMATIONS HERBACEES DENSES. ESSAIS D'APPLICATION A L'ETUDE DE LA DYNAMIQUE DE LA VEGETATION

M. Godron, J. Poissonet, P. Poissonet.

Les formations herbacées denses occupent une grande partie du territoire de certaines régions françaises. Or, ces formations présentent deux caractères qui compliquent singulièrement leur étude:

- les individus de la plupart des espèces ne peuvent pas y être dénombrés exactement,
- la biomasse présente une fluctuation saisonnière très grande, et elle est constituée d'un grand nombre d'espèces.

De très nombreuses méthodes qualitatives et quantitatives d'analyse de la végétation sont préconisées par les spécialistes. Pour être en mesure de choisir judicieusement celles qui sont le mieux adaptées au cas particulier des formations herbacées denses, nous nous sommes proposés de comparer les méthodes les plus couramment employées. A cet effet, nous avons mis en place *un dispositif comparatif d'échantillonnage des formations herbacées denses*.

Ce dispositif permet:

- d'étudier quelques caractères de plusieurs méthodes (rapidité d'exécution, précision des mesures, intégration des données dans l'étude floristico-écologique de la station, etc....)
- d'envisager l'application des résultats à l'étude de la dynamique de la végétation des formations herbacées denses.

Ces différents points sont examinés dans les paragraphes qui suivent.

1.- Dispositif comparatif d'échantillonnage des formations herbacées denses

11. Principes généraux

Les principes généraux se réfèrent aux notions suivantes:

-le mode de répartition des échantillons,

-le nombre d'échantillons,

-les méthodes d'observation.

111. Le mode de répartition des échantillons

Du mode de répartition des échantillons dépendent les calculs qui permettront d'interpréter les observations faites sur le terrain.

La parcelle étudiée a été choisie dans une formation aussi homogène que possible, pour réduire les variations dues aux facteurs du milieu. En conséquence, il était logique d'y faire un échantillonnage au hasard et un échantillonnage systématique; mais aucun caractère ne permettait d'y faire un échantillonnage "stratifié".

112. Le nombre d'échantillons

Il influe sur la précision des données. Il fallait donc rechercher quel était le nombre optimal d'échantillons qui, avec le moindre temps de travail, peut apporter la précision désirée.

113. Les méthodes d'observation

Pour le phytosociologue, une bonne méthode d'observation de la végétation doit être *rapide*, *objective* et *reproductible*, tout en donnant le *maximum d'informations* sur la végétation.
En outre, pour le phyto-écologue, une bonne méthode doit s'adapter aux exigences d'une étude des relations *végétation-milieu-productivité*.
Dans notre essai, les observations floristiques ont été effectuées:

- ponctuellement (points-quadrats),
- sous des segments linéaires,
- dans des surfaces de formes et de dimensions variables.

12. Dispositif comparatif d'échantillonnage des formations herbacées denses

Le dispositif que nous avons utilisé (voir figure 1) est un carré de 64 m x 64 m qui présente une grille avec des mailles de différents modules (4 m x 4 m; 1 m x 1 m; 0,50 m x 0,50 m). Cette grille est matérialisée sur le terrain par des fils colorés qui servent au repérage.
Ce dispositif a été installé dans un herbage du Plateau de Salers (Cantal). Après avoir été essayées rapidement sur d'autres formations herbacées, toutes les observations ont été faites en juin 1965, dans un laps de temps très court (5 jours).

112. Les observations

Elles peuvent être divisées en neuf opérations, que nous résumons sommairement ci-après:

Opération I - Relevés des espèces par poignées

Etablissement de la liste floristique, et estimation visuelle des abondances relatives des espèces dans chaque poignée.

Opération 2 - Etude des carrés de 1 m x 1 m

a. Etablissement de la liste floristique,et estimation visuelle des abondances relatives des espèces, dans chaque carré.

b. Lecture des contacts avec une bayonette, en un point.

Opération 3 - Etude des carrés de 0.50 m x 0.50 m

Les observations sont les mêmes que celles de l'opération 2.

Opération 4 - Comparaison de surfaces égales (carré, rectangle et cercle)

Etablissement de la liste floristique et estimation visuelle des abondances relatives des espèces pour chaque type de surface.

Opération 5 - Relevés des espèces par points, sur des lignes, et mesure des hauteurs

a. Recensement des présences des espèces et lecture des contacts tous les 2,5 cm le long d'une ligne de 4 mètres (points alignés sous un bâti) [1)]

b. Recensement des espèces et lecture des contacts tous les 4 cm le long d'une ligne de 4 mètres (points alignés sur la tranche graduée d'un double-mètre).

c. Mesure de la hauteur de la végétation de certaines espèces dominantes (10 espèces).

Opération 6 - Recensement des espèces et lecture des contacts aux points échantillonnés systématiquement

Opération 7 - Recensement des espèces et lecture des contacts aux points échantillonnés au hasard

Opération 8 - Récolte des échantillons destinés à la détermination des poids secs après séparation des espèces en sept catégories (3 catégories de Graminées, les Légumineuses fourragères, les diverses espèces non fourragères, les espèces "refusées").

Opération 9 - Lecture des présences des espèces sous des segments successifs de 0,25 m

(La localisation de chaque opération est indiquée sur la figure 1).

122. La répartition des échantillons

La répartition des échantillons, et leur nombre sont indiqués succinctement ci-après:

. échantillonnage au hasard

- opération 7: 289 échantillons

[1)] Les définitions relatives aux "présences" et aux "contacts" sont données au § 226.2

. échantillonnage systématique

- opération 1: 145 échantillons
- opération 2: 4 groupes de 64 échantillons chacun
- opération 3: 4 groupes de 64 échantillons chacun
- opération 4: 16 échantillons
- opération 5: 12 échantiloons
- opération 6: 289 échantillons
- opération 8: 48 échantillons
- opération 9: 4 lignes de 256 échantillons chacune

. échantillonnage à différentes mailles, mais avec une même forme de relevé

+ échantillons individuels

- opérations 1 et 4
- opérations 2a, 3a et 4
- opérations 5a et 5b
- opérations 2b, 3b, 6 et 7

+ échantillons obtenus par regroupement

- opérations 2a, 3a, 8 et 9

+ échantillons obtenus par sous-échantillonnage

- opérations 1, 2b, 3b, 5a, 5b, 6,7,8,9

. échantillonnage à la même maille, mais avec des formes différentes

- opérations 2a, 3a et 4

Dans l'échantillonnage systématique, pour connaître le degré d'objectivité des différentes méthodes d'observation, des répétitions ont été faites par division du dispositif en carré latin.

123. Le matérial utilisé

Les mesures ponctuelles des opérations 5a et 5b ont été effectuées avec un appareil voisin de celui décrit par N.S.MOUNTIER & J.E.RADCLIFFE (1964); celles des opérations 2b, 3b, 6 et 7 avec un appareil qui ressemble à une "bayonette". Ce dernier se compose d'une lame de section triangulaire dont un des angles est très aigu; les lectures sont effectuées le long du fil de la lame, la bayonette étant plantée verticalement dans le sol.

Les autres mesures n'ont pas exigé de matérial spécial.

2. Comparaison des résultats obtenus

21. Comparaison des temps d'exécution des observations en fonction de la répartition des échantillons

L'échantillonnage au hasard est plus long à effectuer que l'échantillonnage systématique.

Ex: il a fallu 16 heures de travail pour examiner 289 points répartis au hasard sur toute la parcelle (opération 7) contre 12 heures pour examiner le *même nombre de points* répartis systéma-

Additional material from Gesellschaftsentwicklung (Syndynamik),
ISBN 978-94-011-7594-4 (978-94-011-7594-4_OSFO1),
is available at http://extras.springer.com

tiquement sur la même surface (opération 6).
L'échantillonnage systématique est d'autant plus rapide que les échantillons sont rapprochés; il est particulièrement rapide quand les échantillons sont contigus (segments linéaires consécutifs).
Ex: 5 heures environ ont été nécessaires pour examiner 300 segments consécutifs de 0,25 m (opération 9), alors qu'il a fallu 12 heures pour observer 289 points répartis systématiquement (opération 7). Le temps passé à recenser un point ou un segment est sensiblement le même.

22. Comparaison des résultats phytosociologiques

En raison du grand nombre de données recueillies, les calculs ont été automatisés.
La comparaison est faite à partir d'un schéma général qui permet d'étudier:

1 - la présence des espèces (courbes "aires-espèces"),
2 - la fréquence des espèces,
3 - la répartition qualitative et quantitative des espèces,
4 - les variations de l'abondance des espèces dues à l'hétérogénéité de la surface et aux observateurs,
5 - les corrélations entre les espèces,
6 - les corrélations entre la présence ou l'abondance des espèces et les mesures pondérales effectuées sur les catégories floristiques.

Il n'est pas possible d'entrer ici dans le détail des résultats obtenus. Nous les résumerons:

221. L'exploitation des courbes "aires-espèces", "points-espèces", "poignées-espèces" ou "segment-espèces" permet:

- de mettre en évidence l'existence de plusieurs zones légèrement différentes à l'intérieur de la parcelle.
- de classer les différentes zones les unes par rapport aux autres, en fonction des espèces présentes,
- de déceler l'hétérogénéité floristique, qui correspond à des variations des conditions écologiques (sol et influence de l'homme),
- de constater que l'ajustement des courbes à une équation générale peut être obtenu avec précision.

222. L'étude de la fréquence des espèces montre que les fréquences obtenues peuvent varier considérablement selon les méthodes d'observation. surtout pour certaines espèces.
Ex: *Galium verum* est recontré dans 75% des poignées de l'opération 1, et seulement dans 17% des points des opérations 6 et 7. Les courbes de RAUNKIAER obtenues sont cependant toujours en forme de L.

223. La répartition des espèces

Les "diagrams" (de VRIES 1937) ne semblent avoir d'intérêt que pour visualiser schématiquement cette répartition, car ils ne permettent pas de contrôler l'homogénéité avec précision.
La répartition des espèces a aussi été examinée à l'aide du test statistique du nombre de suites, qui permet d'ébaucher quelques hypothèses sur la biologie de plusieurs espèces.

224. L'étude de l'abondance numérique et de l'abondance estimée des espèces a permis de connaître les variations d'abondance dans certaines directions et, surtout, entre les observateurs. Ce sont malheureusement les espèces qui constituent la majeure partie de la masse végétale qui sont, le plus souvent, quantifiées différemment par les divers observateurs.

225. L'étude des corrélations entre espèces a permis de mettre en évidence des groupes d'espèces statistiquement liées entre elles. Mais ces groupes varient suivant les méthodes d'observation; ils varient aussi, pour une même méthode, suivant les emplacement examinés. Ces résultats pourraient être la base d'hypothèses sur les relations inter-spécifiques. Ainsi, par exemple, *Festuca rubra* est presque toujours en corrélation négative ou nulle avec toutes les autres espèces de la communauté; cependant, dans un cas, *Festuca rubra, Agrostis vulgaris* et *Trifolium repens* sont liés positivement entre eux, mais sont en corrélation négative ou non significative avec les autres espèces.

226. L'utilisation d'observations ponctuelles a permis de mettre en évidence plusieurs relations qui peuvent être un point de départ pour l'étude de la dynamique de la végétation des formations herbacées:

226.1. Dans la formation herbacée étudiée, il existe une relation entre la somme des présences des espèces d'une catégorie "fourragère" et la somme des contacts de ces mêmes espèces.
Cette relation est présentée en coordonnées semi-logarithmiques (figure 2) et en coordonnées logarithmiques (figure 3). Dans ce dernier cas, la relation devient presque linéaire.

226.2. Pour apprécier la fraction de la masse végétale qui est occupée par une espèce, les méthodes ponctuelles permettent de définir:

226.21. La fréquence spécifique

C'est le nombre (A) de points où une espèce (a) a été rencontrée; dans le cas de l'opération 5b où il y a 100 points échantillonnées, le nombre (A) devient une fréquence "relative", au sens précis de ce mot (LAMOTTE, 1948); elle peut s'exprimer directement en pourcentage, et peut aussi être assimilée à une *probabilité de présence*.

En supposant que les points au niveau desquels les lectures sont faites sont tellement petits qu'ils n'ont pas de dimension, et si leur nombre est suffisamment grand, la probabilité de présence est une expression du *recouvrement* (GREIG-SMITH, 1964).

226.22. La somme des fréquences spécifiques

C'est la somme des fréquences spécifiques de toutes les espèces.

226.23. La contribution spécifique d'une espèce

La contribution spécifique, due à la présence d'une espèce, est le rapport de la fréquence spécifique d'une espèce à la somme des fréquences spécifiques de toutes les espèces.

Le symbole de la "contribution spécifique due à la fréquence d'une espèce (a)" est $C.S.P._a$:

$$C.S.P_a = \frac{A}{A + B + C \ldots M}$$

Si B,C, ... M sont les fréquences spécifiques des espèces b,c, ... m. Le symbole de la "contribution spécifique, due au nombre de contacts, d'une espèce (a)" est $C.S.C._a$:

$$C.S.C._a \quad \frac{\sum_{i=1}^{i=n} \alpha\ ia}{\sum_{i=1}^{i=n} (\alpha\ ia + \alpha\ ib + \ldots \alpha\ im)}$$

avec α ia: le nombre de contacts de l'espèce a au point observé i

α ib: " " " " " " b " " " i

α im: " " " " " " m " " " i

n: le nombre de points observés

226.24. La contribution spécifique d'un groupe d'espèces

Pour le groupe d'espèces formé par les espèces a,b et c, la contribution spécifique due à la présence de ces espèces est:

$$C.S.P_{a,b,c} = \frac{A + B + C}{A + B + C \ldots M}$$

celle due au nombre de contacts de ces espèces est:

$$C.S.C._{a,b,c} = \frac{\sum_{i=1}^{i=n} (\alpha_{ia} + \alpha_{ib} + \alpha_{ic})}{\sum_{i=1}^{i=n} (\alpha_{ia} + \alpha_{ib} + \alpha_{ic} + \ldots \alpha_{im})}$$

La relation entre C.S.C et C.S.P. pour les sept catégories d'espèces indiquées à la p.101 est présentée sur la figure 4, en coordonnées logarithmiques. La relation est alors linéaire dans la portion considérée, mais il est certain qu'elle ne reste pas linéaire dans ses parties extrêmes.

226.3. La comparaison de l'opération 5 et de l'opération 8 permet de relier les poids de matière sèche de chaque catégorie "fourragère" - exprimés en pourcentage du poids total - à la C.S.P. et à la C.S.C. de ces groupes. Seule, la relation avec la C.S.P. peut se mettre sous une forme assez simple en coordonnées semi-logarithmiques (voir figure 5).

226.4. L'absence de relation nette entre le nombre de présences (ou de contacts) des deux méthodes ponctuelles (opérations 5a et 5b), mais l'existence de relations entre les C.S.P. et C.S.C. obtenues par ces deux méthodes (voir figure 6) met en évidence l'intérêt et la fidélité de la notion de contribution spécifique.

226.5. Il semble donc qu'il existe des relations très générales:

- entre le nombre de présences et le nombre de contacts des espèces (ou des groupes d'espèces),
- entre les contributions spécifiques dues à la présence, ou au nombre de contacts des espèces (ou des groupes d'espèces),

- entre leur contribution relative et leur poids relatif.

Il reste à déterminer comment varient ces relations dans d'autres milieux.

3.- Essais d'application à l'étude de la dynamique de la végétation

31. La méthode utilisée

311. Description

Les résultats exposés précédemment conduisent à utiliser une méthode d'observation ponctuelle, pour "suivre" la dynamique de la végétation, dans les formations herbacées denses, en adoptant deux processus légèrement différents, selon que la formation herbacée est basse ou haute.

311.1. Dans les formations herbacées basses, 100 lectures ponctuelles sont effectuées le long de deux double-mètres (cf opération 5b, p.[illegible]) - DAGET,Ph. & POISSONET,J. 1966-.

La mise en place du dispositif de lecture se fait de la manière suivante:

- le premier double-mètre est placé parallèlement au petit côté de la parcelle: le double-mètre déployé est glissé à la surface du sol, à travers la végétation de manière que la ligne des graduations soit aussi proche que possible du sol.
- les lectures sont faites le long du bord gradué, tous les 4 cm. L'opérateur vise la graduation retenue et note toutes les espèces qui sont présentes à la verticale de cette graduation.
  50 observations sont ainsi effectuées.
- Le deuxième double-mètre est placé perpendiculairement au premier. La graduation "1 mètre" correspond, dans les deux mesures, au centre géométrique de la parcelle élémentaire.
- les mesures sont toujours effectuées dans le même sens.

311.2. Dans les formations herbacées hautes (herbages de fauche) les lectures à l'aide du double-mètre seraient très difficiles à faire; aussi utilise-t-on la "bayonette" (cf p. 98), en notant toutes les espèces qui touchent le fil de la lame.

312. Protocole d'expérimentation

Cette étude a été entreprise en 1966 dans le CANTAL, sur des parcelles d'expérimentation; elle sera développée en 1967 et 1968 dans quatre autres régions de France.

312.1. But

Le protocole a deux buts:

- suivre qualitativement et quantitativement, pendant 5 ans les variations de la liste floristique et des contributions spécifiques dans une formation herbacée sous l'action de quelques traitements.
- exprimer le potentiel de production en matière sèche et suivre des modifications de ce potentiel sous l'action de deux facteurs écologiques.

Cette expérimentation est faite en collaboration avec d'autres organismes (Institut National de la Recherche Agronomique et Institut Technique des Céréales et des Fourrages).

312.2. Dispositif

Le dispositif adopté est un dispositif factoriel à 4 répétitions, avec 3 dates de premiere coupe, et 4 doses d'éléments fertilisants, soit 12 "traitements" et 48 parcelles élémentaires, avec distribution au hasard des traitements dans les quatre blocs.

Des témoins supplémentaires, sans fertilisation minérale, seront installés chaque année, dans la partie précédemment soumise au pâturage, pour éviter des erreurs d'interprétation dues à l'influence de la fauche sur la flore.

312.3. Traitements

Les facteurs écologiques dont l'action est étudiée concernent les dates d'exploitation et les fertilisation azotée (N) et phosphatée (P).

312.31. Dates d'exploitation

Les premières coupes ont lieu à trois dates différentes décalées chacune d'une semaine (c'est-à-dire aux jours J, J + 7 et J + 14). La date J + 7 correspond à la date normale de mise à l'herbe des animaux dans la région, ou à une hauteur d'herbe d'environ 15 cm.

Dans la placette qui a été coupée la première - au jour J - les repousses après la première coupe sont récoltées chaque fois que la hauteur de l'herbe atteint 20 cm, ou, au plus tard, au bout de six semaines. Les autres placettes, où les premières coupes ont été effectuées respectivement au jour J + 7 et au jour J + 14, sont récoltées au même rythme que la placette J: en conséquence, l'intervalle entre deux coupes successives dans les traitements J + 7 et J + 14 est égal à l'intervalle de temps qui sépare deux coupes successives dans le traitement J.

312.32 Fertilisation minérale

Les 4 doses de fertilisants, exprimées en kg/hectare, sont:

Type de fertilisation	N	$P^2 O^5$
F_1	0	0
F_2	80	0
F_3	0	100
F_4	80	100

32. Etude des formations herbacées denses à l'échelle de la parcelle

321. Observations recueillies

Deux séries d'observations sont effectuées chaque année dans deux répétitions de chaque traitement (soit 24 parcelles);l'une d'elles a lieu au départ de la végétation, avant la première coupe; l'autre, en pleine période de végétation (floraison des Graminées précoces).

Après chaque analyse floristique, la végétation est coupée sur 400 cm^2 (80 cm x 5 cm), au centre géométrique de la parcelle,la plus grande longueur du prélèvement étant parallèle au plus petit côté de la parcelle élémentaire. Ce prélèvement est trié en trois catégories floristiques: Graminées, Légumineuses, espèces diverses, afin de préciser les liaisons qui existent entre les mesures pondérales et les mesures de fréquences.

322. Prémices des résultats

Une première année d'observation (deux séries de mesures en mai et juillet 1966) sur la première parcelle d'expérimentation mise en place (SALERS, dans le CANTAL) a permis de dénombrer 70 espèces dont seulement 5 interviennent d'une façon importante dans la biomasse (contribution spécifique supérieure ou égale à 5%); trois de ces espèces *Festuca rubra, Agrostis vulgaris, Trifolium repens* ont une contribution spécifique supérieure à 50% (tableau I).

Cette première année d'étude a permis de mettre en évidence que l'influence de l'observateur était pratiquement nulle pour les espèces qui interviennent d'une façon significative dans la biomasse (fréquence spécifique supérieure à 10%) mais que, par contre, cette influence n'est pas négligeable et souvent plus si-

gnificative que "l'effet du traitement" (dans le cadre du protocole adopté) pour les espèces ayant une faible fréquence spécifique (inférieure à 10%). Quelques résultats sommaires sont exprimés dans le tableau II.
Après une seule année d'étude, l'évolution est peu visible (tableau I), car la fertilisation initiale n'a été apportée que le 20 mars 1966.

Espèces	mai 1966		juillet 1966	
	avec N	sans N	avec N	sans N
Festuca rubra	24,5	25	29,5	27,5
Agrostis vulgaris	21	18	23,5	21
Trifolium repens	10,5	13,5	12,5	16,5
Luzula vulgaris	7,5	8,5	5,5	6
Poa pratensis	7,5	5,5	6	4,5
Galium verum	2,5	3,5	3,5	4,5
Mousses diverses	2,5	3	2	1,5
Trisetum flavescens	4	3	1,5	1,5
Ranunculus bulbosus	1,5	2	1,5	1,5
Anthoxanthum odoratum	2	1	1,5	1,5
Cerastium caespitosum	1,5	2	1,5	1,5
Cynosurus cristatus	1,5	1,5	1,5	1,5
Phleum pratense	2	2	1	1

Tableau I. Contribution des espèces (en%) intervenant dans la biomasse, en fonction de la fertilisation azotée

(Résultats d'une première année d'étude sur la parcelle expérimentale de Salers, Cantal)

33. Etude des formations herbacées denses à l'echelle régionale

L'emploi d'observations ponctuelles telles qu'elles sont effectuées dans l'opération 5b, pour l'étude des formations herbacées denses de deux régions naturelles (Cantal et Margeride), a permis de mettre en évidence que la somme des fréquences spécifiques de quelques espèces semble être relativement constante.
Ces espèces (que l'on peut appeler "le fond prairial") constitu-

Espèces	Observateur A	Observateur B
Festuca rubra	82	84
Agrostis vulgaris	70	68
Trifolium repens	37	36
Luzula vulgaris	23	26
Poa pratensis	27	27
Trisetum flavescens	16	13
Phleum pratense	1	7
Cynosurus cristatus	9	3
Anthoxanthum odoratum	8	4

Tableau II Influence de l'observateur sur les fréquences spécifiques

ent plus de 50% de la biomasse.

Ex: dans les herbages pâturés, sur sol sain des plateaux basaltiques de l'ouest du Cantal, la somme des fréquences de *Festuca rubra, Agrostis vulgaris, Trifolium repens*, qui était comprise entre 161 et 188 (moyenne 172) sur 100 points échantillonnées, a été trouvée indépendante de la somme des fréquences de toutes les espèces, qui était comprise entre 203 et 593.

4- Conclusions

41: Le dispositif de comparaison des méthodes d'étude de la végétation qui a été placé dans une formation herbacée dense du Plateau de Salers (Cantal) montre ou confirme que:

- la fréquence des espèces varie en fonction de la méthode d'observation,
- l'estimation de l'abondance des espèces dépend beaucoup de l'observateur,
- des phénomènes de concurrence interspécifique peuvent être mis en évidence à partir des méthodes d'observation utilisées,
- les observations ponctuelles paraissent réunir le plus d'avantages: rapidité, sécurité, précision, signification écologique et agronomique,
- l'observation de la présence des espèces sous des segments est rapide et bien adaptée à l'étude de la structure de la végétation.

42. Les résultats obtenus grâce à ce dispositif de comparaison de méthodes servent de base pour un protocole d'étude expérimentale de la dynamique de la végétation et pour l'étude écologique générale de l'ensemble des formations herbacées denses de deux régions naturelles. Ces deux études sont le point de départ de recherches approfondies sur les formations herbacées denses.

SUMMARY

The aim of this study is:

1. to compare methods of vegetation analysis in order to determine those the most suitable for dense herbaceous formations
2. to apply the results obtained to the study of vegetation dynamics.

For that purpose a comparative sampling device was conceived. It was placed in position on the Plateau de Salers (Cantal, France). It allows the study of:

1. the mode of repartition of the samples
2. the influence of the number of samples
3. various observation methods, taking in account their speed, their objectivity, their reproductibility, the quantity of collected information, the relations vegetation - environment - productivity.

The results obtained show or confirm that:

1. the systematic sampling is speedier that the sampling "at random"
2. the circumstantial study of the "area-species" curves allows to show up the heterogeneity of the plant community
3. the frequency of a species may vary considerably according to the observation method utilized
4. the visual estimation of the abundance of certain species hangs on the observer
5. the study of interspecific correlations can allow to outline hypothesis on the interspecific concurrence
6. in the parcel under study, the relations between the presence of a species (or a group of species) and the number of contacts of this species (or group of species) - these relations being obtained by punctual methods - are significant. It remains to determine how these relations vary according to the environment.

Working from these results, the experimental study of the vegetation dynamics as a function of the mineral fertilization and of the harvesting dates was undertaken in various regions. The observation method utilized is a practical adaption of the punctual methods.

Moreover the general ecological study of the dense herbaceous formations of two natural regions is in progress.

BIBLIOGRAPHIE

ANDERSON,K.L. -1942- A comparison of line transects and permanent quadrats in evaluating composition and density of pasture vegetation on the tall-prairie grass type.- J.Amer.Soc.Agron.34: 805-822. Washington.

ARCHIBALD,E.A. -1952- A possible method for estimating the area covered by the basal parts of plants.- S.Afr.J.Sci.45: 286-292. Johannesburg.

ARNY,A.C. & SCHMID,A.R. -1942- A study of the inclined point-quadrat method of botanical analysis of pastures mixtures.-J.Amer.Soc.Agron.34: 238-247. Washington.

BLACKMAN,C.V. -1933- A statistical study of the distribution of species in grassland associations.- Proc.Linn.Soc.Lond.146: 31-32. London.

BLOUARD,R. & DESCHUYTENER,G. -1962- L'èchantillonnage dans les prairies par la méthode du poids sec.- Bull.Inst.Agron.Gembloux 30 (1-2): 8-29. Gembloux.

BOUDET,G. & Bayens,F. -1963- Une méthode d'étude et de cartographie des pâturages tropicaux.- Rev.d'élevage et de médecine vétér.des Pays tropicaux 16 (2).

BRAUN-BLANQUET,J. -1932- Plant sociology.- Traduit par Fuller,G.D. & Conard,H.S. New-York-London. 439 pp.

CANFIELD,R.H. -1941- Application of the line intercept method in sampling range vegetation.- J.For.39: 388-394. Washington,D.C.

CLARKE,S.E., CAMPBELL,J.A. & CAMPBELL,J.B. -1942- An ecological and grazing capacity study of the nature grass pastures in southern Alberta, Saskatchewan.- Publ.738, Techn.Bull.46, Canad.Dep.Agrig. 55 pp.

C.N.R.S., C.E.P.E. -1965- Séminaire sur les formations herbacées denses, document de travail ronéo. 111 pp.

COTTAM,G.& Curtis,J.T. -1956- The use of distance measures in phytosociological sampling.- Ecol.37 (3): 451-460. Lancaster,Pa.

DAGET,Ph. & POISSONET,J. -1965- Contribution à l'étude des herbages des plateaux basaltiques de l'Ouest du Cantal. Série des Etudes locales n° 16, C.N.R.S., C.E.P.E. Montpellier. 120 pp. Ronéo.

-- -- -- -- -1966- Etude phyto-écologique de la commune de THORAS (Haute Loire). Etude locale n° 28, 620 pp. (en 5 fascicules).

DANSEREAU,P. -1951- Description et relevé de la végétation sur une base structurale.- Ecol.32 (2): 172-229. Lancaster,Pa.

DE VRIES,D.M. -1937- Bestandsuntersuchungsmethoden von Wiesen und Weiden.- Rap.4° Congr.Internat.Herbages. Aberystwyth, Grande Bretagne: 474-480.

-- -- -1938- The plant sociologica combined specific frequency and order method.- Chronica Botanica 4 (2): 115-117. Waltham,Mass.

-- -- & DE BOER,Th. -1959- Methods used in botanical grassland research in the Netherland and their application.- Herbage Abstract 24 (1): 1-7. Aberystwith.

DIX,R. -1961- An application of the point-centered-quarter method to the sampling of grassland vegetation.- J.Range Mangt.14 (2): 63-69. Portland,Ore.

GOUNOT,M. -1960- Méthodes d'étude et d'inventaire de la végétation pastorale et prairiale.- Rapport présenté au Colloque sur l'étude des prairies, organisé par la Société Botanique de France, Paris, 8 mai 1960. Paris.

-- -- -1962- Etude statistique d'une pelouse à Brachypodium ramosum.Etude de la distribution des espèces au moyen d'un test non paramètrique.- Bull.serv.Carte Phytogéographique, série B 7 (1): 65-81. Paris.

GREIG-SMITH,P. -1952- The use of random and contiguous quadrats in the study of the structure of plant communities.- Ann.Bot.16: 293-316. London-Oxford.

-- -- -1957- Quantitative plant ecology.- Butterworths scientific publications. 198 pp.

-- -- -1961- Data on pattern within plant communities.- J.Ecol.49 (3): 695-708. Cambridge.

JOHNSTON,A. -1957- A comparison of the line interception, vertical point-quadrat and loop methods.- Canad.J.Plant Sci.37 (1): 34-42. Ottawa.
KERGUELEN,M. -1954- Etude de quelques herbages des Vosges.- Ann.de l'I.N.R.A., série B 1: 33-50. Paris.
KERSHAW,K.A. -1957- The use of cover and frequency in the detection of pattern in plant communities.- Ecology 38 (2): 291-299. Lancaster,Pa.
-- -- -1958-1959- An investigation of the structure of a grassland community.- J.Ecol.46-47: 571-592; 31-55. Cambridge.
-- -- -1960- The detection of pattern and association.- J.Ecol.48 (1): 233-242. Cambridge.
LAMOTTE,M. -1962- Initiation aux méthodes statistiques en biologie.- Paris. 144 pp.
LEVY,E.A. & MADDEN,E.A. -1933- The point method of pasture analysis.- N.Zeal.J.Agric.46: 267-279. Wellington.
MOUNTIER,N.S. & RADCLIFFE,J.E. -1964- Problems in measuring pasture composition in the field.- N.Zeal.J.Bot.2 (1-2): 90-105, 131-142. Wellington.
NAVEH,Z. -1963- A comparison of several methods of determining the botanical composition of mediterranean Grassland.- Israel J.Agric.Res. 13 (2): 79-91.
NICHOLSON,I.A. & ROY HUGUES -1963- A method for caracterisation of range type vegetation.- J.Range Mangt. 16 (6): 293-299. Portland,Ore.
PARKER,K.N. & SAVAGE,D.A. -1944- Reliability of the line interception method in measuring vegetation on the southern great Plains.- J.Amer. Soc.Agron.36: 97-110. Washington.
PECHANEC,J.F. & STEWART,G. -1940- Sagebrush-grass range sampling studies: size and structure of sampling unit.- J.Amer.Soc.Agron.32: 669-682. Washington.
PIELOU,E.C. -1962- The use of plant to-neighbour distance for the detection of competition.- J.Ecol.50 (2): 357-368. Cambridge.
PORYATOWSKA,V.M. -1958- On the method of studying the structure of plant communities.- Bot.Zhur.43: 585-605. Leningrad.
RAY,R.J. -1959- A phytosociological analysis of the tallgrass prairie in Northeastern Oklahoma.- Ecology 40 (2): 255-261- Lancaster,Pa.
SHOOP,M.C. & Mc ILVAIN,E.H. -1963- The micro-unit forage inventory method.- J.Range Mangt.16 (4): 172-179. Portland,Ore.
SPEDDING,C.R.W. & LARGE,R.V. -1957- A point-quadrat method for the description of pasture in terms of height and density.- J.Brit.Grassl.Soc. 12: 229-234. Aberystwyth.
STAPLEDON,R.G. -1913- Pasture problems. Drought resistance.- J.Agr.Sci.5: 129-151.
STEPHENS,J.L. -1942- Pasture of the coastal plain of Georgia.- Bull.27, Georgia Coastal Plain Exp.Sta. 57 pp.
STRICKLER,G.S. -1961- A grid method for obtaining loop readings on small plots.- J.Range Mangt.14 (5): 261-263. Portland,Ore.
SUMMERHAYES,V.S. -1941- The effect of voles (Microtus agrestis) on vegetation.- J.Ecol.29: 14-48. Cambridge.
THOMPSON,H.B. -1958- The statistical study of plant distribution patterns.- Austral.J.Bot.6: 322-342. Melbourne.
TIDMARSH,C.E.M. & HAVENGA,C.M. -1955- The wheel-point method of survey and measurement of semi-open grasslands and Karoo vegetation in South Africa.- Bot.Survey of South Africa, Mêmoire no 29: 49.
TINNEY,F.W., AAMODT,O.S. & AHLGREN,H.L. -1937- Preliminary report of pasture swards.- J.Amer.Soc.Agron.29: 835-840. Washington.
VAN DYNE,G.M. -1960- A procedure for rapid collection, processing and analysis of line intercept date.- J.Range Mangt.13 (5): 247-251. Portland, Ore.
-- -- -1963- Influence of small plot size and shape on range herbage production estimates.- Ecology 44 (4): 746-759. Lancaster,Pa.

WAGNER,R.O. -1960- Distribution patterns of some prairie plants and their relationships to methods of reproduction.- Ph.D.Thesis Univ., Wisconsin Madison. 157 pp.

WARREN-WILSON,J. -1959- Analysis of the spatial distribution of foliage by two dimensional point quadrats.- New Phytol.58 (1): 92-101. London.

WAYNE COOK,C. & BOX,Th.W. -1961- A comparison of the loop and point methods of analysing vegetation.- J.Range Mangt.14 (1): 22-27. Portland,Ore.

-- -- & REX HURST -1962- A quantitative measure of plant association on ranges in good and poor conditions.- J.Range Mangt.15 (5): 266-273. Portland,Ore.

WEAVER,J.E. & DARLAND,R.W. -1947- A method of measuring vigour of range grasses.- Ecology 28: 146-162- Lancaster,Pa.

YAROSHENTO,P.D. -1958- On the methods of studying the pattern of vegetation cover.- Bot.Zhur.43: 381-387. Leningrad.

DISCUSSION

H.ELLENBERG

Nous avons effectué des relevés comparés avec plusieurs méthodes en Argentine. J'ai étudié une pelouse dense avec plusieurs méthodes, p.e. la méthode de plusieurs points et d'autres méthodes que vous avez mentionnées. Nous avons aussi usé la méthode classique de BRAUN-BLANQUET. Un relevé selon BRAUN-BLANQUET donne dans 1/2 heure toutes les espèces présentes; nous avons travaillé 1 1/2 jours avec des points selon GOODALL. Nous avons eu besoin de plus 10 000 points pour attrapper 80% des espèces présentes, c.à.d. toutes les espèces rares ne s'attrappent pas par les méthodes plus ou moins statistiques. Voilà l'avantage que donne un relevé selon BRAUN-BLANQUET, parce que la plupart des espèces rares sont les espèces les plus valables indicatrices. Selon nos expériences, on voit que ces espèces rares donnent une réaction claire, p.e. lorsqu'on fertilise avec de l'azote, tandis que les espèces dominantes, comme *Festuca rubra* et *Agrostis* et *Trifolium*, donnent une réaction quantitative mais pas claire.

P.POISSONET

Nous avons fait la méthode des relevés complets; c'était sur les tout petits carrés de 1 m x 1 m. Je suis entièrement d'accord avec Monsieur ELLENBERG que c'est essentiellement la méthode de BRAUN-BLANQUET qui donne, au point de floristique,les indications les plus rapides. Seulement, nous avons aussi un problème quantitatif et c'est pourquoi, - je ne l'ai pas ainsi specifié - nous faisons d'abord, pour l'étude de la région, le relevé floristique, et à côté nous étudions une ligne pour avoir une appréciation quantitative, meilleure que l'estimation habituelle.

H.ELLENBERG

Je proposerais d'étudier la succession en utilisant le relevé selon BRAUN-BLANQUET sur la surface minimum, pour l'étude comparée de divers traitements sur cette pelouse, et vous verrez que l'on peut assez exactement calculer quantitativement les données sur les diverses espèces; on a l'estimation très exacte p.e. selon KLAPP, qui donne des données assez exactes sur les quantités.

R.CARBIENER

Permettez-moi que je commente un certain nombre de données sur les corrélations qu'il y a entre la productivité d'un groupement et le système de BRAUN-BLANQUET. On commence à avoir du matérial, qui montre qu'il y a une corrélation très belle entre la productivité des associations, sous-associations et variantes, et que cette corrélation est relativement facile à utiliser aussi. L'inconvénient de la méthode concrète c'est justement celui-ci qu'elle met en excès un certain nombre d'espèces d'amplitude écologique assez basse, qui ont encore une vitalité assez grande et qui de ce fait ont une grande chance d'être rencontrées par les dispositifs, alors que les espèces d'amplitude et d'écologie étroite, qui sont les plus intéressantes de point de vue de la valeur d'indicateur au point de vue écologie ont très peu de chance d'être rencontrées.

R.LINDER

Est-ce que c'est une pelouse naturelle? Pourriez-vous nous donner quelques précisions sur la nature de la pelouse étudiée, est-elle fumée par engrais chimique?

P.POISSONET

C'est une étude sur l'évolution de cette prairie sous l'influence d'une part de la fertilisation par engrais chimique et d'autre part l'influence de l'homme.

R.TÜXEN

Je suis botaniste et pas mathématicien. Vous avez étudié les espèces *Festuca rubra* et *Agrostis vulgaris*, *Trifolium repens*, mais j'aimerais à savoir le nom de votre association.

P.POISSONET

Cette association est celle à F e s t u c a r u b r a et C y n o s u r u s c r i s t a t u s . Seulement, *Cynosurus cristatus* n'est pas très développée dans l'association qui est paturée. Au point de vue de la biomasse, elle n'a pas beaucoup d'importance.

G.LAVRENTIADES

Etudiez-vous aussi la vitalité des espèces par cette méthode?

P.POISSONET

La vitalité de chaque espèce sera étudiée plus tard.

H.ELLENBERG

KLAPP a additionné à la méthode de BRAUN-BLANQUET cette méthode d'estimation quantitative de la biomasse de chaque espèce. Il fait, presque la même chose que vous, en séparant, par croix, les différents parties, les graminées, et les légumineuses et tout ça, mais il fait une autre chose additive: il estime la quantité de chaque espèce dans cette biomasse. Alors, c'est une combinaison de la méthode de BRAUN-BLANQUET de relevés complets et de la méthode de peser chaque groupe d'espèces. Il estime cela en 100%, c.à.d. il reçoit des dates assez exactes et assez rapidement. Nous avons trouvé un relevé selon KLAPP en seulement 1/2 heure en plus sur la relevé selon BRAUN-BLANQUET.

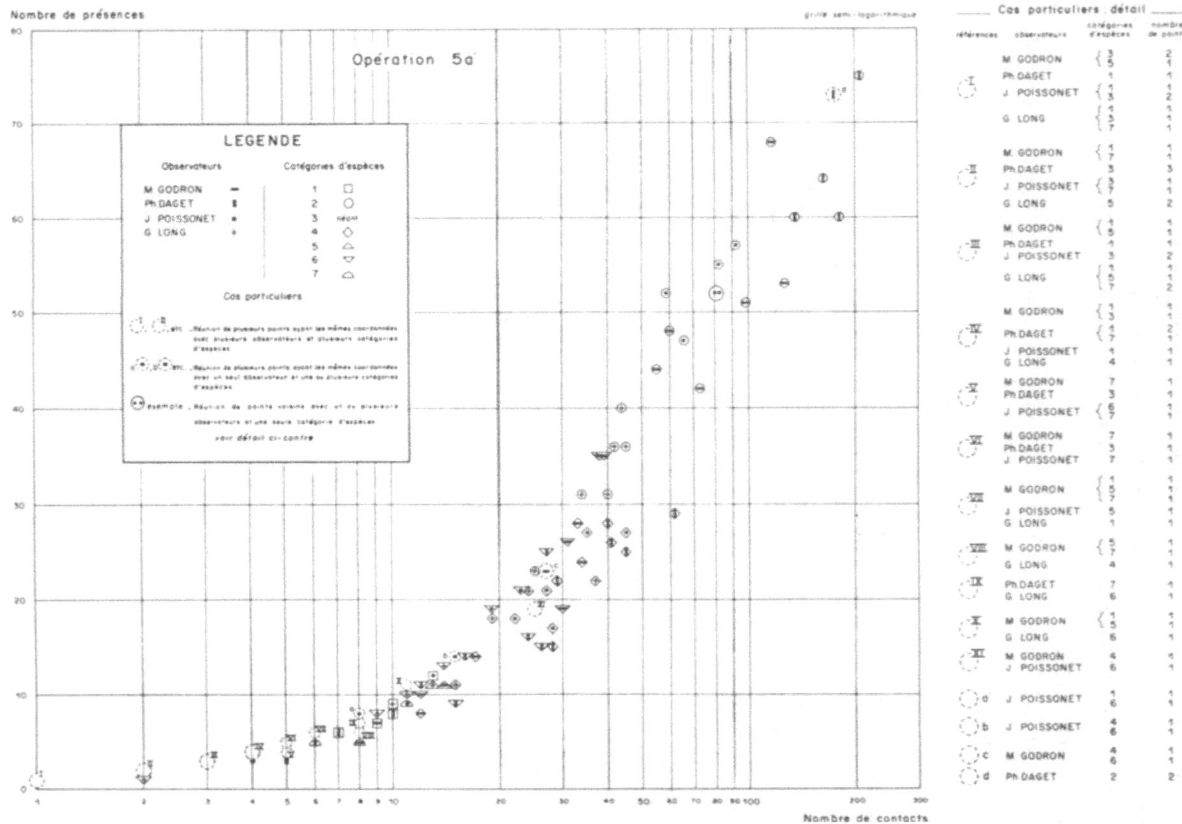

Figure 2 _ Relation entre le nombre de présences et le nombre de contacts à partir de mesures ponctuelles, par catégories d'espèces

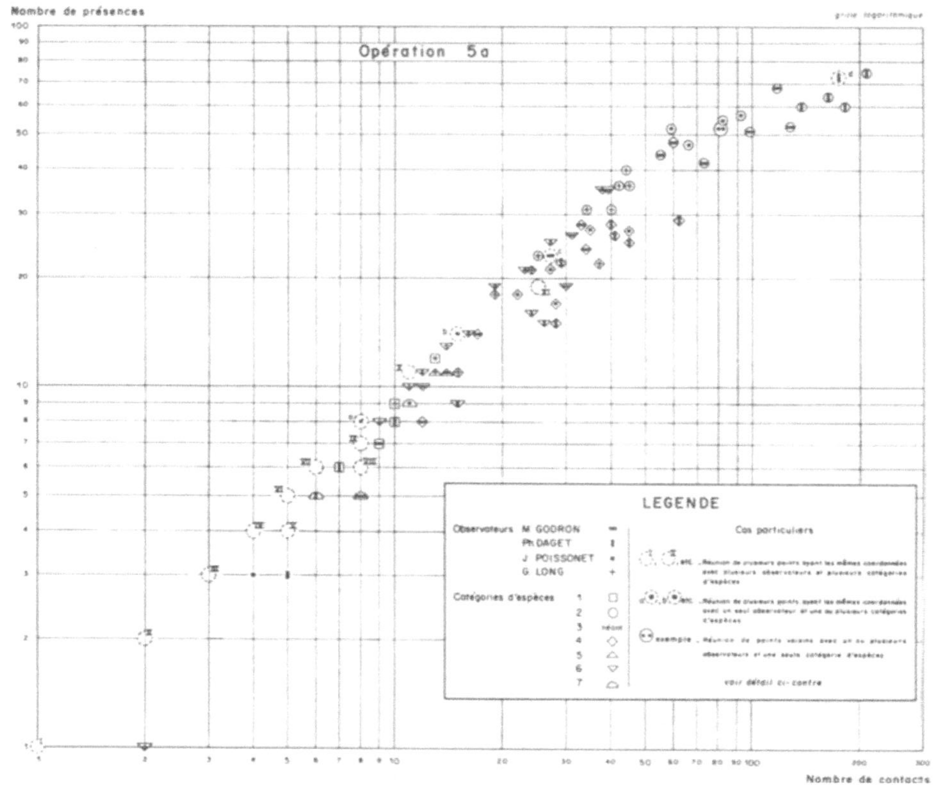

Cas particuliers : détail

références	observateurs	catégories d'espèces	nombre de points
I	M. GODRON	3, 5	2, 1
	Ph. DAGET	1	1
	J. POISSONET	1, 3	1, 2
	G. LONG	1, 3, 7	1, 1, 1
II	M. GODRON	1, 7	1, 1
	Ph. DAGET	3	3
	J. POISSONET	3, 7	1, 1
	G. LONG	5	2
III	M. GODRON	1, 5	1, 1
	Ph. DAGET	1	1
	J. POISSONET	3	2
	G. LONG	1, 5, 7	1, 1, 2
IV	M. GODRON	1, 3	1, 1
	Ph. DAGET	1, 7	2, 1
	J. POISSONET	1	1
	G. LONG	4	1
V	M. GODRON	7	1
	Ph. DAGET	3	1
	J. POISSONET	6, 7	1, 1
VI	M. GODRON	7	1
	Ph. DAGET	3	1
	J. POISSONET	7	1
VII	M. GODRON	1, 5, 7	1, 1, 1
	J. POISSONET	5	1
	G. LONG	1	1
VIII	M. GODRON	5, 7	1, 1
	G. LONG	4	1
IX	Ph. DAGET	7	1
	G. LONG	6	1
X	M. GODRON	1, 5	1, 1
	G. LONG	6	1
XI	M. GODRON	4	1
	J. POISSONET	6	1
a	J. POISSONET	1, 6	1, 1
b	J. POISSONET	4, 6	1, 1
c	M. GODRON	4, 6	1, 1
d	Ph. DAGET	2	2

Figure 3 _ Relation entre le nombre de présences et le nombre de contacts à partir de mesures ponctuelles, par catégories d'espèces

Inv. 1748

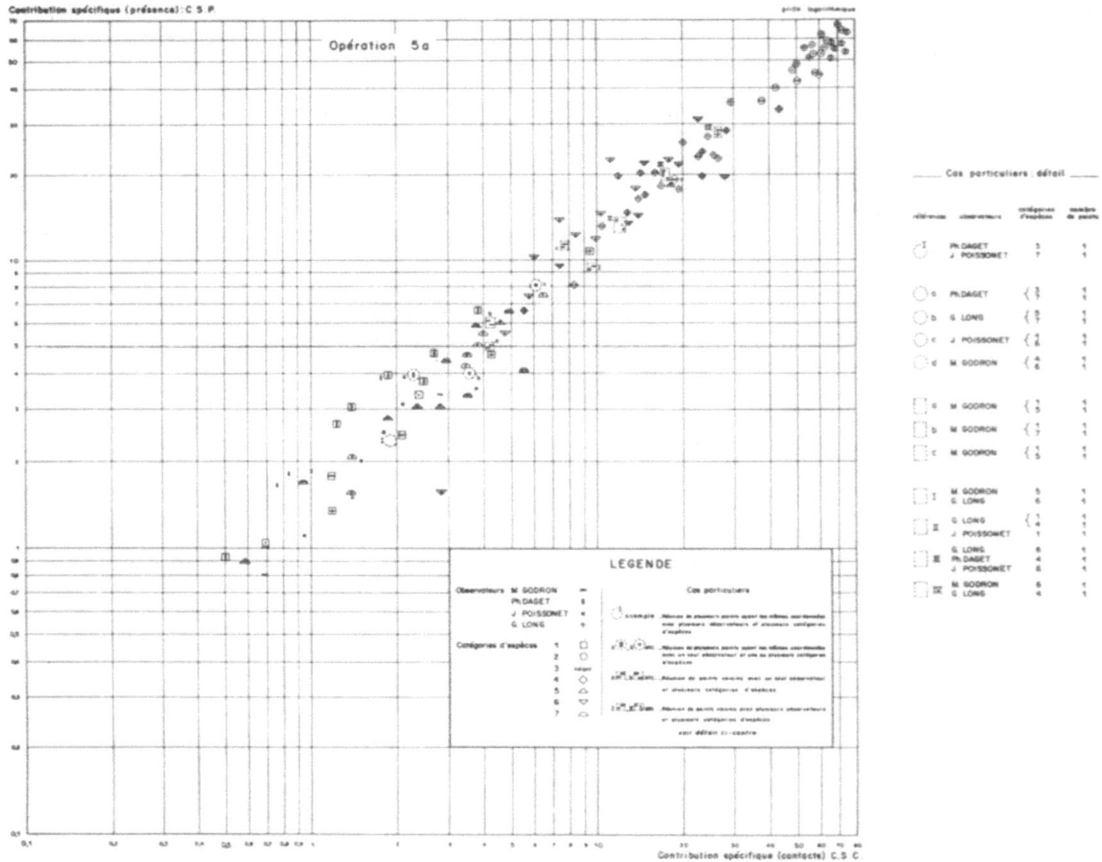

Figure 4 _ Relation entre la contribution spécifique (présence): C.S.P. et la contribution spécifique (contacts): C.S.C. de catégories d'espèces, à partir de mesures ponctuelles

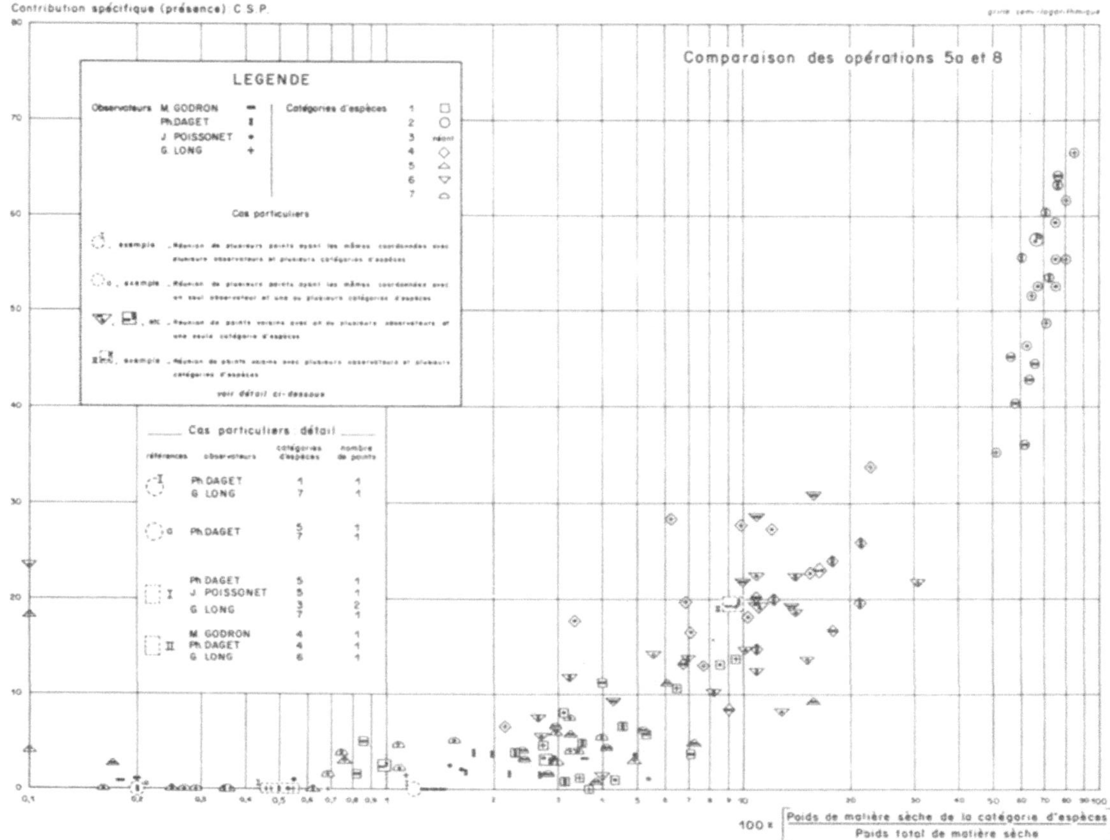

Figure 5_Relation entre la contribution spécifique (présence): C.S.P. de catégories d'espèces et leur poids

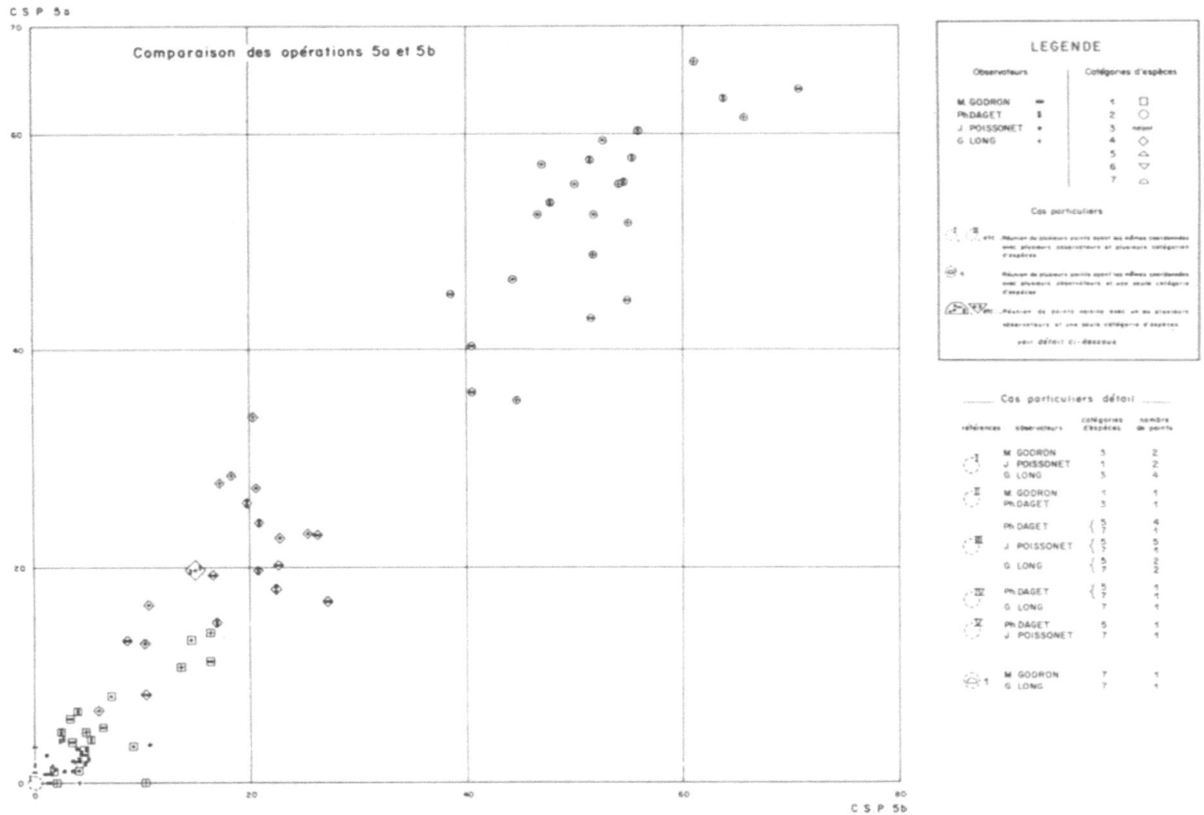

Figure 6 _ Relation entre la contribution spécifique (présence) C.S.P. observée par deux méthodes ponctuelles, par catégories d'espèces

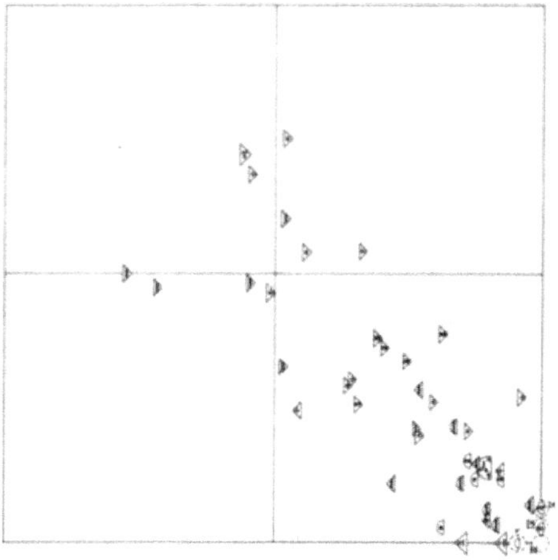

EVOLUTION DE LA MASSE VEGETALE EN MILIEU DE GARRIGUE SOUMISE A DIVERSES INTERVENTIONS HUMAINES

F. Fay, G. Long, M. Thiault, L. Trabaud[1]

I- INTRODUCTION

L'étendue considérable des garrigues de la région méditerranéenne française menace, à brève échéance, de correspondre à un vide économique total, face à une bordure littorale en pleine expansion.

Plus de 80% de la surface de cette région sont des terres incultes; la garrigue elle-même couvrirait de 360 000 à 500 000 hectares.

La dentité moyenne de la population est inférieur à 10 habitants au kilomètre carré. L'utilisation traditionelle des terres par le parcours et l'exploitation du bois de chauffe y est en pleine régression. Une telle situation peut être estimée comme étant normale dans un milieu ingrat, sous un climat difficile. Elle peut, au contraire, être la conséquence d'une mauvaise utilisation du milieu, et de la rupture d'un équilibre naturel. Il importait de vérifier ces hypothèses.

C'est pour cette raison, que depuis 1957, certaines études théoriques ont été entreprises par le CENTRE D'ETUDES PHYTOSOCIOLOGIQUES ET ECOLOGIQUES du C.N.R.S. à Montpellier, pour connaître les conditions de l'évolution et de l'amélioration de la végétation naturelle de la zone des garrigues. Ces travaux ont permis de dégager assez rapidement des conceptions d'application intéressantes sur le plan pastoral.

Les garrigues sont depuis très longtemps utilisées comme terrains de parcours pour ovins. Les traitements couramment appliqués par les utilisateurs se limitent à une exploitation extensive du terrain, et, parfois encore, à la pratique de l'incendie. Les coupes de nettoyage (coupes à blanc) deviennent exceptionnelles à cause de la mévente du bois de chauffage.

L'évolution actuelle de la végétation tend, de ce fait, vers un envahissement de plus en plus important par le Chêne kermès, incompatible avec l'exploitation rationnelle d'un troupeau, sans pour cela apporter d'améliorations tangibles sur le plan fore-

(1) Phyto-écologistes au Centre d'Etudes Phytosociologiques et Ecologiques du C.N.R.S., Montpellier, France

stier,les risques d'incendies subsistant intégralement.

II- SITUATION ECOLOGIQUE DE LA ZONE EXPERIMENTALE

La zone choisie pour l'expérimentation est une garrigue typique de Chêne kermès. Ce type de milieu correspond à l'un des stades les plus pauvres de dégradation en garrigue; des formations semblables couvrent près de 100 000 hectares en France méditerranéenne.

L'ensemble du périmètre disponible dépasse 50 hectares et se trouve à une dizaine de kilomètres au Nord de Montpellier, commune de Saint Gély du Fesc (Hérault), au lieu dit Puech du Mas du Juge.

La zone expérimentale a été implantée sur le versant Sud-Ouest de cette colline. La pente générale est de 5 à 10%, assez régulière; l'altitude est comprise entre 100 et 175 mètres.

La pluviométrie moyenne annuelle, sur cinquante ans, est de 770 mm à Montpellier, mais la moyenne des dix dernières années est de 920 mm à Saint Gély du Fesc, l'automne et le printemps étant les saisons les plus pluvieuses.

Hiver	22%
Printemps	25,5%
Eté	15,8%
Automne	36,7%

La moyenne des températures minima (m) du mois le plus froid (janvier) est, à Montpellier, de 09°C, et la moyenne des températures maxima (M) du mois le plus chaud (juillet) est de 30° C; à Saint Gély de Fesc m doit être voisin de 0° C. C'est entre ces températures moyennes extrêmes que se déroule la vie végétale.

Le sol est un sol brun très squelettique formé sur un calcaire marneux dur, d'âge miocène.

Autrefois, le Puech du Mas du Juge devait être couvert par la forêt de Chênes verts et de Chênes pubescents, comme toute la région environnante. Actuellement, la couverture végétale est une garrigue typique à Chêne Kermès; les espèces les plus fréquemment rencontrées sont:

Quercus coccifera	Rubia peregrina
Brachypodium ramosum	Teucrium chamaedrys
Brachypodium phoenicoides	Carex halleriana
Dorycnium suffruticosum	Thymus vulgaris
Carex humilis	Sanguisorba minor
Stachys silvatica	Euphorbia characias

La communauté végétale appartient à l'association C o c c i f e re t u m (b r a c h y p o d i e t o s u m) de BRAUN-BLANQUET. L'extension du Chêne kermès est due aux incendies, car seules les parties aériennes sont détruites par le feu et il émet, ensuite, de nombreux rejets. Le dernier incendie connu au Puech du Mas du Juge date de septembre 1951.

III- ETUDES PRELIMINAIRES

A - Problème de la destruction du Chêne kermès

Trois procédés de destruction ont été envisagés:

1) Eradication mécanique: c'est, en effet, le seul procédé permettant d'éliminer directement la masse végétale incriminée, c'est-à-dire, d'atteindre le plus rapidement le but recherché. De plus, il permet l'obtention de renseignements préliminaires indispensables sur la masse végétale à traiter qualitivement et quantitativement.

Plusieurs types de "débroussailleuses" ont été utilisés pour détruire le Chêne kermès: le simple sécateur à main, la scie circulaire portée à dos d'homme et actionnée par un moteur, les broyeurs tractés et actionnés par un tracteur agricole, utilisant des principes de travail différents, parties travaillantes fonctionnant dans un plan horizontal ou dans un plan vertical.

Quant à l'époque d'utilisation de ce procédé, elle a été choisie pendant la phase d'activité photosynthétique considérée comme maximum, c'est-à-dire au printemps.

2) Destruction chimique: c'est la destruction des rejets de Chêne kermès par la pulvérisation de certains produits phytocides. Après étude bibliographique, seules les phytohormones ont été retenues comme produits susceptibles d'être utilisés. En effet, les autres produits sont soit inefficaces, soit trop peu sélectifs et risqueraient d'entraîner la destruction totale de la végétation, ce qui n'est pas le but recherché.

Du point de vue du mode d'application, ces hormones phytocides peuvent être utilisées soit sur la végétation dans son état initial, soit sur les souches ou sur les repousses, après débroussaillement mécanique. Les souches du Chêne kermès étant d'un diamètre trop faible pour permettre une opération sur une superficie assez grande, seul le traitement sur les repousses a été retenu.

3) Lutte biologique: elle n'a été envisagée que sous l'effet de la concurrence entre espèces végétales, pour remplacer le Chêne kermès défaillant, et non sur le plan de l'introduction dans le

biotope d'un ou plusieurs parasites spécifiques du Chêne kermès, procédé qui mériterait d'être étudié spécialement.
La concurrence sur le plan nutritif doit nécessairement intervenir après l'application de l'un ou des deux procédés de lutte précédents. En effet, la rupture de l'équilibre biologique, au détriment du Chêne kermès, provoquée par la destruction mécanique ou la destruction chimique appliquées séparément ou simultanément, offre la possibilité de favoriser le développement de l'ensemble des espèces herbacées ou semi-ligneuses présentes dans la garrigue à Chêne kermès: *Arrhenatherum elatius*, *Brachypodium* spp., *Dorycnium suffruticosum*, etc.

B - Essai d'orientation

Les premiers travaux entrepris en 1957 et en 1958 ont eu pour but principal, d'une part, de déterminer les conditions dans lesquelles devaient s'opérer la lutte mécanique et, d'autre part, de permettre un premier choix parmi les produits phytotoxiques proposés, spécifiques du Chêne kermès.
Le premier essai d'orientation réalisé au printemps 1957 a porté sur une surface de 200 mètres carrés. Le protocole mis en place comportait:

1) Une coupe à blanc de l'ensemble de la végétation ligneuse arbustive. La totalité du matériel végétal coupé sur la parcelle a été pesé, le poids total était de 172 kg pour 200 m^2 de surface, dont 107 kg de matière sèche (Tableau I).

Tableau I
Poids et pourcentages respectifs des matières sèches et de l'eau.

	sur 200 m^2	par ha	%
Matière sèche feuilles	31 kg	1550 kg	29%
Matière sèche tiges,rameaux	76 kg	3800 kg	71%
Matière sèche totale	107 kg	5350 kg	62%
Eau totale	65 kg	3250 kg	38%

Ce qui donne un poids de masse végétale verte de 8600 kg par hectare, pour une garrigue de 5 ans 1/2.

2) Division de la parcelle en trois parties: une zone centrale, le témoin, deux zones latérales qui reçurent, chacune, un trai-

tement chimique différent.
L'année suivante, en 1958, à la même époque, le Chêne kermès, de la garrigue environnante, non coupée, non traitée, atteignait 1 m de haut et recouvrait 100% de la surface, il avait une teinte vert clair, avec de jeunes pousses sur les rameaux.
Sur les deux zones traitées par les phytocides, le *Quercus coccifera* ne recouvrait que 10 à 70% de la surface, il présentait une teinte vert roux, avec les repousses issues des souches souterraines.

IV- MISE EN PLACE DE L'EXPERIMENTATION PROPREMENT DITE

L'expérimentation comportait trois essais:
A) La première expérimentation (parcelle I) portait sur une surface de 1000 m^2.
En avril 1959, coupe à blanc de la végétation ligneuse sur l'ensemble de la surface au moyen d'une petite débroussailleuse à scie circulaire portée à dos d'homme et actionnée par un moteur à essence. La masse végétale ainsi coupée fut pesée: 1385 kg de matière verte, soit 13850 kg/ha pour une garrigue de 7 ans 1/2, ont été ramassés er pesés. Cette masse végétale était en grande partie ligneuse. La parcelle fut divisée en 3 parties, une bande centrale et deux bandes latérales.
Au mois de juillet de la même année, les deux bandes latérales reçurent un traitement chimique; les rejets de Chêne kermès atteignaient en moyenne 20 à 25 cm de hauteur. L'herbicide utilisé, à base de 2,4,5-T, est le même sur les deux bandes, mais à deux concentrations différentes, et avec deux supports mouillants différents: eau et fuel. Les modes d'application furent différents: pulvérisation pour l'eau, atomisation pour le fuel.
Au début septembre 1959, la bande n° 1 (traitée avec l'eau) présentait moins de 5% de la masse végétale atteinte, la bande n° 3 (traitée au fuel) présentait moins de 10% de la masse végétale atteinte.
En octobre 1959, essai d'introduction d'espèces herbacées par semis à la volée sur la bande n° 1, et en poquets sur la bande n° 3. Une fumure y fut apportée. Huit jours après, la levée s'était produite.
En juillet 1960, un nouveau traitement chimique, analogue au précédent, a été effectué. Les rejets de *Quercus coccifera* atteignaient en moyenne 25 cm de hauteur; les Graminées étaient de belle venue.

En août 1960, les effets des phytocides sont sensibles sur les repousses de l'année, l'aspect général est brunâtre, les extrémités sont brûlées.
En décembre 1960, la bande n° 1 présente peu de rejets, le traitement chimique a donc donné de bons résultats. Très bon développement des espèces herbacées, malgré la période froide et la présence de lapins. La bande n° 3 présente une très bonne levée, malgré les semis tardifs et le froid. L'effet du traitement chimique est assez superficiel, en général beaucoup de rejets et de feuilles ne sont pas touchés.
En avril 1962, après un hiver froid et long, pas de départ de *Quercus coccifera*. Par contre, le développement des Graminées locales ou introduites est bon: hauteur 15 à 20 cm.
Au mois de septembre 1962, un débroussaillement mécanique d'entretien a été effectué sur la totalité de la parcelle avec broyage des débris, laissés sur place.

b) La deuxième expérimentation, qui a débuté en 1960 concernait deux parcelles (parcelles II et III).
La parcelle II, de 1300 m^2, fut débroussaillée avec un gyrobroyeur Gard type CFT. Le Chêne kermès atteignait 80 à 150 cm de hauteur. La végétation broyée fut laissée sur place, en litière.

Au mois de juillet 1960, la parcelle était divisée en 7 bandes dont une bande témoin. Le produit utilisé était toujours à base de 2,4,5-T, à des concentrations différentes et avec des supports différents. Deux bandes de la parcelle reçurent, en outre, un mélange équivalent de 50% du produit précédent et 50% de débroussaillant concentré, dans deux supports différents.
Au mois d'octobre 1960, la parcelle II présentait une destruction du feuillage de l'ordre de 75 à 95%. L'effet phytocide se caractérise par le dessèchement des feuilles, plus ou moins complet sur toute la longueur des tiges. Le sommet des tiges est desséché, mais leur base reste verte.
Le même mois, des espèces herbacées étaient semées; suivait une fertilisation.
La parcelle III, de 700 m^2, fut débroussaillée avec un appareil "Forage Harvester H 1 Gloster Bamfords". Le Chêne kermès atteignait 80 à 100 cm de hauteur, la végétation broyée fut laissée sur place.
Au mois de juillet 1960, la parcelle était divisée en deux ban-

des et un témoin. Le produit utilisé était le même que pour la parcelle II à des concentrations différentes et avec des supports différents.

Au mois d'octobre 1960, le Chêne kermès de la parcelle présentait une destruction du feuillage de 50 à 60%. Le même mois, des espèces herbacées étaient semées, suivait une fertilisation différente de l'autre parcelle.

En septembre 1962, les deux parcelles (II et III) étaient débroussaillées mécaniquement, avec broyage des débris, laissés sur place.

C) Enfin, à titre de comparaison, un *essai de débroussaillement chimique*, sans débroussaillement mécanique préalable a été réalisé au mois de juin 1962.

Trois parcelles de 1000 m^2 chacune, avec deux traitements (A et B) par parcelle, sur la moitié de la surface. De plus, sur la moitié de la surface de chacun des six traitements (parties A1 et B1 de chaque traitement A et B), soit 250 m^2, il a été effectué un traitement chimique supplémentaire, en vue de détruire les Graminées.

A la fin octobre de la même année, les phytocides avaient atteint leur but. La végétation présentait un aspect souffreteux.

V- LES RESULTATS OBTENUS

A) Sur la parcelle des essais préliminaires (1957) aucun prélèvement, aucun traitement n'ont été effectués; on note un retour progressif mais rapide à la situation primitive au bout de 6 ans (1963). Il n'y a, à cette date, pratiquement aucune différence avec le stade initial ou avec la végétation environnante.

B) En 1963, sur les trois parcelles I,II,III, des prélèvements, en vue de connaître la variation de la masse végétale, étaient effectués, selon un rythme saisonnier, correspondant à des périodes de pâturage possible, c'est-à-dire, printemps, été, automne. La masse végétale était coupée et recueillie, sur trois échantillons de 2 m^2 chacun, soit 6 m^2 et ceci sur chaque parcelle. Après un tri manuel, la partie herbacée de la masse végétale, était pesée à l'était frais, puis séchée à l'étuve à 110°C et pesée à nouveau (Tableau II).

La parcelle I a donné un rendement théorique annuel de matière verte totale de 10 890 kg/ha; la parcelle II, un rendement théorique de 12 950 kg/ha; la parcelle III, un rendement théorique

TABLEAU II

RESULTATS PAR PARCELLE DES 3 PERIODES DE COUPES

ANNEE 1963

Date des prélèvements	PARCELLE I			PARCELLE II			PARCELLE III		
	Poids total de matière verte	Poids de la partie herbacée		Poids total de matière verte	Poids de la partie herbacée		Poids total de matière verte	Poids de la partie herbacée	
		Etat frais	Etat sec		Etat frais	Etat sec		Etat frais	Etat sec
6/4/63 Poids total/6m2	2830 g	2680 g	835 g	3130 g	2685 g	924 g	2610 g	2340 g	816 g
Poids/m2	471 g	446 g	139 g	521 g	447 g	154 g	435 g	390 g	136 g
Rdt théorique/ha	4710kg	4460kg	1390 kg	5210kg	4470 kg	1540 kg	4350 kg	3900 kg	1360 kg
10/7/63 Poids total/6m2	1580 g	629 g	255 g	2486 g	1959 g	932 g	2640 g	1686 g	754 g
Poids/m2	263 g	105 g	42,5 g	414 g	326 g	155 g	440 g	281 g	125 g
Rdt théorique/ha	2630 kg	1050 kg	425 kg	4140 kg	3260 kg	1550 kg	4400 kg	2810 kg	1250 kg
1/10/63 Poids total/6m2	2132 g	770 g	300 g	2162 g	1810 g	687 g	2074 g	1628 g	568 g
Poids/m2	355 g	128 g	50 g	360 g	300 g	114,5g	345 g	271 g	94 g
Rdt théorique/ha	3550 kg	1280 kg	500 kg	3600 kg	3000 kg	1145 kg	3450 kg	2710 kg	940 kg
Rdt théorique total des 3 coupes/ha	10890 kg	6790 kg	2315 kg	12950 kg	10730 kg	4235 kg	12700 kg	9420 kg	3550 kg

TABLEAU III

RESULTATS PAR PARCELLE DES 3 PERIODES DE COUPES

ANNEE 1964

Date des prélèvements	PARCELLE I			PARCELLE II			PARCELLE III		
	Poids total de matière verte	Poids de la partie herbacée		Poids total de matière verte	Poids de la partie herbacée		Poids total de matière verte	Poids de la partie herbacée	
		Etat frais	Etat sec		Etat frais	Etat sec		Etat frais	Etat sec
24/4/64									
Poids total/6m2	2596 g	1957 g	630 g	3733 g	2805 g	960 g	3822 g	3364 g	980 g
Poids/m2	432,6g	326,1g	105 g	622 g	467,5 g	160 g	637 g	560,6g	165 g
Rdt théorique/ha	4326 kg	3261 kg	1050 kg	6270 kg	4675 kg	1600 kg	6370 kg	5606 kg	1650 kg
10/6/64									
Poids total/6m2	1575 g	736 g	250 g	2815 g	2178 g	877 g	1910 g	1470 g	510 g
Poids/m2	226 g	122 g	41 g	469 g	363 g	146 g	318 g	245 g	85 g
Rdt théorique/ha	2260 kg	1220 g	410 kg	4690 kg	3630 kg	1460 kg	3180 kg	2450 kg	850 kg
27/10/64									
Poids total/6m2	3368 g	741 g	388 g	2118 g	723 g	375 g	1444 g	500 g	202 g
Poids/m2	561 g	123 g	64,5g	353 g	120,5g	62,5g	240,6g	83 g	33,6g
Rdt théorique/ha	5610 kg	1230 kg	645 kg	3530 kg	1205 kg	625 kg	2406 kg	830 kg	336 kg
Rdt théorique total des 3 coupes/ha	12196 kg	5711 kg	2105 kg	14440 kg	9510 kg	3685 kg	11956 kg	8886 kg	2836 kg

TABLEAU IV

RESULTATS PAR PARCELLE

(kg/ha)

1963	Poids total de matière verte	Poids de la partie herbacée	
		Poids en matière verte	Poids en matière sèche
Parcelle I	10 890	6 790	2 315
Parcelle II	12 950	10 730	4 235
Parcelle III	12 200	9 420	3 550
Total	36 040	26 940	10 100
Rendement moyen	12 013	8 980	3 366

1964	Poids total de matière verte	Poids de la partie herbacée	
		Poids en matière verte	Poids en matière sèche
Parcelle I	12 196	5 711	2 106
Parcelle II	14 440	9 510	3 685
Parcelle III	11 956	8 886	2 836
Total	38 492	24 107	8 627
Rendement moyen	12 830	8 036	2 876

de 12 200 kg/ha, soit une moyenne de 12 013 kg/ha de matière verte pour l'ensemble des parcelles. Cette production est fournie par des espèces graminéennes presque exclusivement, car il y a appauvrissement en Chêne kermès dû à la destruction et à la concurrence.

En 1964 des échantillons ont été prélevés sur les trois parcelles I,II et III pesés et triés (Tableau III). Le rendement théorique de matière verte de la parcelle I est de 12 196 kg/ha,celui de la parcelle II, de 14 440 kg/ha et celui de la parcelle III de 11 956 kg/ha, au cours de l'année. Les rendements sont, en général, plus élevés que ceux de 1963. Le rendement théorique moyen est de 12 830 kg/ha. Il y a donc un accroissement de la masse végétale par rapport à 1963, ceci est dû à l'abondance des Graminées qui sont toujours dominantes, mais aussi à un embuissonnement bas, mais effectif, du Chêne kermès qui repousse. En effet,le poids de la partie herbacée, constitué aux 3/4 par les Graminées ou des Légumineuses herbacées, est en nette régression.

La première coupe apporte, toujours, le poids de masse végétale, tant matière verte totale, que matière verte assimilable, le plus élévé.

Le poids de matière verte totale a augmenté entre 1963 et 1964 (Tableau IV) passant de 12 013 kg/ha en moyenne, à 12 830 kg/ha, mais la partie herbacée de matière verte a diminué de 8 980 kg/ha en moyenne, à 8 036 kg/ha, ce qui prouve, malgré l'augmentation de la masse végétale verte, un accroissement de l'embuissonnement; le Chêne kermès regagne, peu à peu, le terrain perdu. Or, il ne faut pas oublier que la végétation des trois parcelles a été coupée en septembre 1962. Si nous calculons, le rendement moyen annuel de production de matière verte totale est de 12 421 kg/ha et an. Or, une garrigue typique, en couvert continu de Chêne kermès donnerait, en moyenne par an, 1580 kg/ha de matière verte (cf chiffres de 1957 à 1959) et 2900 kg/ha de litière (comm.verbale de M.RAPP,1966) soit 4500 kg de masse végétale totale par hectare et par an. Ainsi, l'ensemencement par espèces graminéennes et la fertilisation déterminent une production végétale annuelle plus élevée. Sous l'action de l'homme, la garrigue à Chêne kermès acquiert un potentiel de production plus important.

C) Le traitement chimique seul, par phytocides aux hormones, sur les trois dernières parcelles, n'avait produit, en 1963, aucun effet sur la végétation des trois parcelles. Seul *Quercus coccifera* présentait un aspect jaunâtre, les extrémités des rameaux étaient dénudés. En 1964, il n'y avait plus aucun effet visible.

VI- CONCLUSIONS

Par des coupes, des traitements chimiques et des semis, l'évolution de la végétation naturelle et la masse végétale d'une garrigue typique à Chêne kermès ont été modifiées. Lors des premières coupes (1957 ,1959) la masse végétale était constituée, en majeure partie, par une masse ligneuse; après les interventions, en 1963, cette masse végétale est constituée aux trois quarts par une masse herbacée, et, en 1964, celle-ci ne représente plus que les deux tiers de la masse totale. Il y a donc embuissonnement progressif.

Avec ou sans traitement chimique, en l'absence de traitement mécanique d'entretien type fauche ou débroussaillement par machine, le retour au stade initial est très rapide (reconstitution de la garrigue à 75% au bout de 3 à 4 ans). Si l'on veut améliorer la végétation naturelle, telle qu'elle se présente dans la garrigue à Chêne kermès, il faut faire appel, obligatoirement, à un traitement mécanique ou à des animaux pâturants.

La masse végétale annuelle produite est importante, elle serait de l'ordre de 4500 kg/ha dans le cas d'une garrigue typique utilisée de façon traditionnelle,elle est de 12 470 kg/ha dans une garrigue soumise à des traitements.

Compte tenu de ce potentiel de production, les traitements appliqués à une formation végétale de ce type, en considérant le contexte écologique pris dans son sens le plus large, conduiraient actuellement à envisager une évolution de la végétation orientée:

- soit vers une amélioration des possibilités pastorales du milieu, en facilitant les espèces herbacées, ou en introduisant de nouvelles espèces dites "fourragères"
- soit vers une amélioration des possibilités forestières, en favorisant le développement des espèces existantes présentant quelque intérêt: Pin d'Alep, Chêne vert, ou en introduisant de nouvelles espèces: Cèdre, Cyprès.

Le terme de comparaison est constitué par une mise en défens où la végétation évolue naturellement en l'absence de toute intervention humaine.

RESUME

La garrigue est un milieu naturel couramment utilisé d'une façon abusive par surpâturage, coupe à blanc des arbres qui peuvent s'y développer (Chênes verts, Pins d'Alep) et surtout par la pratique des incendies répétés. L'évolution actuelle de la végétation tend, de ce fait, vers un embroussaillement de plus en plus important.

Les expérimentations entreprises depuis 1957, par le Centre d' Etudes Phytosociologiques et Ecologiques, tendent à comparer l' action de cette utilisation sur l'évolution de la végétation, avec l'application de techniques moins défavorables à un bon équilibre du milieu. Toutes les expériences ont été effectuées sur une garrigue typique à Chêne kermès.

Le traitement favorable à une évolution de la garrigue vers un stade herbacé s'est révélé efficace, le Chêne kermès a été rapidement concurrencé par des graminées, telles que: *Arrhenatherum elatius, Brachypodium phoenicoides, Avena bromoides*, etc.

Plusieurs essais de coupe, traitement chimique, fertilisation et semis d'espèces ont été réalisés pendant les années 1957 à 1960.

Resultats obtenus

Le poids de masse végétale annuelle est important, de l'ordre de 4 500 kg/ha théoriques dans une garrigue typique, de 12 420 kg/ha sous l'intervention humaine.

Par divers traitements on note une amélioration de la garrigue, le poids de masse végétale herbacée devient prépondérant; cependant, malgré cela, un nouvel embroussaillement se manifeste très vite.

SUMMARY

The "garrigue" constitutes a natural environment generally misused by over-grazing, clear-cutting of the trees which can live in it *(Quercus ilex, Pinus halepensis)* and more particularly by the fire practice. The present evolution of the vegetation shows a trend towards an increasing colonization by bushes.

The experimentation undertaken since 1957 by the Centre d'Etudes Phytosociologiques et Ecologiques aims to compare the action of this utilization on the evolution of the garrigue, with the application of techniques favorable to a good equilibrium of the environment. All the experimentation was carried out on a typical *Quercus coccifera* garrigue.

ZUSAMMENFASSUNG

Die Garrigue ist eine natürliche Vegetation, die durch übermäßige Beweidung, Schlag der dort wachsenden Bäume *(Quercus ilex, Pinus halepensis)* und Anwendung von Brand mißbräuchlich verwendet ist. Die aktuelle Entwicklung der Vegetation strebt zu einem mehr und mehr zunehmenden Heranwachsen der Gesträuche.

Die seit 1957 begonnenen Freilandversuche bei dem "Centre d'Etudes Phytosociologiques et Ecologiques" bezwecken einen Vergleich zwischen der Wirkung der Nutzung mit der Dynamik der Vegetation und der Anwendung weniger störender Eingriffe für ein gutes Gleichgewicht. Alle Versuche sind in einer typischen Garrigue von *Quercus coccifera* gemacht worden.

Die vorgeschlagene Behandlung für eine Etnwicklung der Garrigue zu einem krautigen Zustand war wirksam; grasartige Pflanzen wie *Arrhenatherum elatius, Brachypodium phoenicioides, Avena bromoides* etc. haben der *Quercus coccifera* schnell Konkurrenz gemacht.

Mehrere Kahlschlag-Versuche, chemische Behandlung, Düngung und Saat von krautigen Pflanzenarten wurden während der Jahre 1957 bis 1960 gemacht.

Ergebnisse

Das Gewicht der jährlich erlangten Grünmasse betrug ungefähr 4500 kg/ha in einer typischen Garrigue, bis 1240 kg/ha für eine Garrigue mit menschlichen Einflüssen. Bei verschiedenen Behandlungen bemerkt man eine Verbesserung der Garrigue; das Gewicht der krautigen Pflanzen-Masse steigt. Indessen erfolgt ein neues Anwachsen des Buschwerks schnell.

SUMMARY

The treatment aiming to the evolution of the garrigue towards an herbaceous stage was efficient. *Gramineae* such as *Arrhenatherum elatius, Brachypodium phoenicoides, Avena bromoides*, etc. were soon able to compete with *Quercus coccifera*.

Several trials of clear-cutting, chemical treatment, fertilization and sowing of different herbaceous species have been made during the 1957 to 1960 years.

Results

The weight of annual plant mass is considerable, amounting to about 4 500 kg per hectare, theoretical, in a typical garrigue, and 12 420 kg per hectare in a garrigue submitted to the human intervention.

After the use of different treatments some improvements can be seen, the weight of the herbaceous plant mass becomes preponderant, nevertheless and in spite of all these treatments a new bush-cover appears quickly.

BIBLIOGRAPHIE

ARBONNIER, FONDEUX, VIGIER & CHEVALIER -1966- Résultats et enseignements de chantiers expérimentaux de dégagements chimiques suivis pendant cinq ans par la Recherche Forestière.- Bull.Vulgarisation Forestière 66 (4) 9 pp.

BRAUN-BLANQUET,J. -1935- Un problème économique et forestier de la garrigue languedocienne.- SIGMA 35: 11-22. Montpellier.

HARRANGIER,J. & VARLET,G. -1961- l'amélioration des herbages d'altitude moyenne.- B.T.I.163, octobre 1961: 859-868.

HEDIN,H, COMMEAU,J., HARRANGER,J., VARLET,G., POIGNANT,P., CHAFFARD,M. & THELLOT,B. -1956- Le rôle et les possibilités d'emploi des desherbants et des débroussaillants dans le cadre de l'amélioration des prairies et des pâturages. Journées françaises d'information de la Fédération Nationale de protection des cultures sur les désherbants sélectife et les régulateurs de croissance, Paris 28-29 novembre 1956, Paris. 26 pp.ronéo.

KUHNHOLTZ-LORDAT,G. -1958- L'Ecran vert.- Mém.Mus.Nat.Hist.Nat., n.s.,s.B, botanique 9: 276 pp.

LEE LING -1951- Hormones et Herbicides.- F.A.O. Etude agricole 13,avril 1951.

LONG,G. -1958- Points de vue de l'écologie et la mise en valeur de la garrigue.- Ann.Soc.Hort.Hist.Nat.de l'Hérault: 11-34.

-- -- FAY,F. & THIAULT,M. -1964- Possibilité d'utilisation de la garrigue par le mouton.- Journées CETA, décembre 1964. Mouton. Etude n° 982.

-- -- RAMI,J. & VISONA,L. -1958- La végétation du domaine de Coulondres (Herault). Relations avec les problèmes de mise en valeur. Série des Etudes locales, n° 6, C.N.R.S., C.G.V., Montpellier. 48 pp.

RAUCOURT,M. & VENTURA,E. -1953- Une opération à grande échelle de destruction chimique des broussailles.- B.T.I.85: 875-881.

TRABAUD,L. -1962- Monographie phytosociologique et écologique de la région de Grabels-St Gély du Fesc.- Thèse de 3ème Cycle (Ecologie). Faculté des Sciences, Montpellier. 131 pp.

Les questions qui sont traitées dans cette communication ont fait l'object de nombreux développements depuis 1967. Il convient, par conséquent, d'être très prudent sur l'utilisation des informations contenues dans cette note (L.TRABAUD 1978).

DISCUSSION

J.WOLAK
Vous avez ici donné une estimation de la production de matière verte. Est-ce que dans vos échantillons vous avez distingué la matière morte?

L.TRABAUD
Nous avons récolté le matériel vivant et le matériel mort sur pied.

J.BARKMAN
Vous avez bien démontré qu'on peut augmenter la production de la biomasse du Quercetum cocciferae par sa transformation en pelouse. Cependant, avant d'en tirer des conclusions, il faudrait savoir la production annuelle de la formation climacique, le Quercetum ilicis. Quelles sont vos connaissances actuelles concernant ce problème?

L.TRABAUD
Actuellement, nous ne connaissons pas la production de biomasse du Quercetum ilicis, mais des recherches sont en cours à Montpellier (depuis 1967, des études ont été entreprises concernant ce sujet: voir les travaux récent de M.RAPP 1969,1971).

R.CARBIENER
Ce sont les problèmes les plus intéressants de ces traveaux, à savoir le type d'évolution qui serait préférable de soutenir. Au fond, les expérimentations, comme elles ont été faites, consistent à précipiter une évolution régressive,telle que la nature la réalise, sous l'effet des incendies. Je me demande si vous n'auriez pas obtenu exactement les mêmes résultats en brûlant le chêne kermès et en appliquant des engrais après, vous auriez également obtenu une pelouse à *Brachypodium phoenicoides* d'une productivité tout à fait analogue à celle que vous avez constatée.

L.TRABAUD
Nous venons d'entreprendre une expérimentation afin de connaître l'influence des feux sur l'évolution de la végétation (voir les travaux de L.TRABAUD 1973,1977).

R. CARBIENER
Une autre chose extrêmement importante à considérer c'est la question de substrat. Selon les cas que nous avons étudies il est très net que le substrat est extrêmement favorable à la végétation graminéenne, parce que la pelouse à *Brachypodium phoenicoides* demande un sol rélativement lourd. Dans ce cas là il est évidemment préferable sur le plan économique de transformer une chênaie à *Quercus coccifera* en un pelouse à *Brachypodium phoenicoides*.

R. LINDER
Est-ce cette expérimentation est poursuivie dans le cadre d'un plan d'aménagement du territoire? Si oui, la garrigue est-elle condamnée? Qu'en disent les organismes de protection de la nature?

L. TRABAUD
L'étude entreprise n'est pas inscrite dans un plan d'aménagement, mais il est possible que dans l'avenir, les résultats puissent être utilisés pour un aménagement pastoral. Ces garrigues ne seront pas toutes transformées en terrains de parcours améliorés. Il y a plus de 400 000 hectares de garrigues dans la région méditerranéenne française.

GEGENWÄRTIGE DYNAMIK DER WALD-VEGETATION IM EUROPÄISCHEN NORDEN DER SOWJET-UNION

A.A. K o r c h a g i n

Die Wälder des europäischen Nordens der Sowjet-Union sind gegenwärtig in verschiedenen Formationen sowohl aus Nadel- wie auch aus Laubgehölzen vertreten. Aber Klimax oder plakor sind nur die dunklen aus *Picea abies, P.obovata, Abies sibirica* und *Pinus sibirica* zusammengesetzten Nadel-Wälder.

Botaniker der Sowjet-Union, die der Lehre von WYSOTZKY (1909) folgen, verstehen unter dem Terminus "Plakorvegetation" eine natürliche Klimax-Pflanzengesellschaft, die mehr oder weniger ebene, genügend dränierte, aus lehmigen oder tonhaltigen Böden zusammengesetzte Wasserscheiden besiedelt. Klimabedingte zonale Besonderheiten der Pflanzendecke treten unter solchen Verhältnissen am schärfsten hervor.

Die Wälder des europäischen Nordens gliedern sich in zwei botanisch-geographische Provinzen: eine westliche Provinz der europäischen Fichte *Picea abies*, und eine östliche Provinz der sibirischen Nadelgehölze - *Picea obovata, Abies sibirica, Pinus sibirica, Larix sukaczewii*. Die Kiefer - *Pinus sylvestris* - wächst überall. Die Grenze dieser Provinzen zieht sich als ein breiter Übergangsstreifen vom Nordosten nach Südwesten dem Flusse Nördliche Dwina entlang.

In diesen Wäldern entwickeln sich gegenwärtig sehr mannigfaltige Vegetationsfolgen, die von verschiedenen Faktoren hervorgerufen werden, und mit verschiedener Geschwindigkeit sich vollziehen. Man kann hier nur einige Sukzessionen nennen, die von größtem Interesse und von größter Bedeutung für die Erklärung des jetzigen Bestandes und Zustandes dieser Wälder sind.

Die erste Gruppe - Säkulare (GAMS 1918) oder hologenetische (SUKASZEW 1954) Sukzessionen, die sich im laufe mehrerer Jahrhunderte und Jahrtausende als Folge physikalisch-geographischer Abänderungen der Umweltbedingungen vollziehen. Im europäischen Norden gehören hierzu wohl die nacheiszeitlichen durch mehrfache Klima-Änderungen bedingten Waldvegetations-Sukzessionen.

Die dunkle Taiga, die ihren Ursprung wahrscheinlich noch in der ersten Hälfte der Tertiärzeit in den Bergwäldern Europas fand (TOLMATSHEV 1943,1954; KORCHAGIN 1945) stieg am Ende des Terti-

ärs von den Bergen herab und nahm schnell große Landflächen ein, einschließlich fast das ganze Territorium des europäischen Nordens der Sowjet-Union. Das mächtige Inlandeis,welches im Pleistozän entstand und den ganzen Norden Europas und den europäischen Teil der Sowjet-Union überzog, hatte auf der gesamten Fläche die Vegetation vollständig vernichtet. Das wurde zweimal, nach Meinung mancher Geologen auch dreimal, von Interglazialperioden abgelöst, wobei die freigewordenen Regionen sich wieder mit Vegetation bedeckten. Demgemäß wurde der laufende Prozeß der Vegetations-Phylogenese durch jede neue Vereisung zum Stehen gebracht(SUKATSHEV 1944,1954),um in jeder Interglazialperiode und im darauf folgenden Holozän mit Hilfe der am nächsten liegenden und aus den unvereisten Refugien in die vom Eis befreiten Territorien migrierenden Florenkomplexe von Neuem zu beginnen.
Im Holozän hat sich der Vegetationscharakter des Europäischen Nordens mehrfach infolge Klimawechsels geändert. Die günstigsten Verhältnisse für die Waldvegetation haben sich im mittleren Holozän (während der atlantischen und subborealen Periode gebildet (NEISTADT 1957) für das ein Klimaoptimum nachgewiesen ist.
Dunkelnadelwälder drangen in dieser Zeit hoch nach Norden, wo sie die ganze Tundrazone vom Kontinent beinahe verdrängend, das Barenzowo Meer erreichten. Dieser ziemlich schmale Streifen der Dunkelnadelwälder wurde südlich der Breite von Archangelsk von Dunkelnadelwäldern mit Beimischung breitblättriger Gehölze (Eiche, Linde, Ulmus usw.) abgelöst, deren Menge in südlicher Richtung immer mehr zunahm. Als Folge hatten sich die Wälder in gemischte Dunkelnadel-Breitblatt-, und noch südlicher (ab 57-58° nördl.Breite) in reine Breitblatt-Wälder verwandelt.
Während der letzten Etappe des Holozän (der Subatlantischen Periode oder im Späten Holozän nach NEISTADT), die im europäischen Norden etwa vor 2500-3000 Jahren begann, fand eine allmähliche Abkühlung und Feuchtigkeitserhöhung statt, was eine damit zusammenhängende allmähliche Verdrängung aller Vegetationszonen nach Süden hervorrief. Im hohen Norden wurden die Dunkelnadelwälder von der Tundra abgelöst. In dem gemischten Dunkelnadel-Breitblatt-Waldbestand begann der Ausfall der breitblättrigen Elemente und deren Ablösung durch reine Nadelwälder.
Diese Sukzessionen dauerten fast bis zur Gegenwart und sind wahrscheinlich erst in den letzten vier oder fünf Jahrzehnten als

Folge der begonnenen Klimaerwärmung der nördlichen Teile Eurasiens gehemmt oder ganz abgebrochen worden.

Auf solchem Wege durch jahrhundertlange Abfolgen haben sich die gegenwärtigen dunkelnadeligen primären Wälder des europäischen Nordens gebildet. Die in ihrem Bestande vorhandenen breitblättrigen Gehölze und nemoralen Kräuter kann man darum als Relikte des mittleren Holozän (Subboreale Periode) auffassen. Solche dynamischen Folgen wiederholen sich auch gegenwärtig.

Zweite Gruppe - Florengenetische Sukzessionen

In den Wäldern des europäischen Nordens vollzieht sich gleichzeitig ein anderer schon mehrere Jahrhunderte dauernder Prozeß in der Veränderung der Dunkelnadelwälder, die durch eine gegenseitige Einwirkung verschiedener Nadelholzarten hervorgerufen worden ist.

Die Besiedlung des Gebietes des europäischen Nordens vollzog sich in der Nacheiszeit aus zwei Zentren (KORSHINSKY 1899). Aus südwestlicher Richtung (aus dem südeuropäischen Zentrum, wahrscheinlich aus den Karpaten) verbreitete sich die europäische Art - *Picea abies*, und aus südöstlicher Richtung (aus dem Altai und vielleicht auch aus dem Süd- und Mittel-Ural) drangen die sibirischen Arten - *Picea obovata*, *Abies sibirica*, *Pinus sibirica* und andere Gehölze vor.

In ihrem Marsch durch den europäischen Norden trafen diese beiden Pflanzen-Legionen im Bassin der Nördlichen Dwina zusammen und bildeten somit zwei botanisch-geographische Provinzen.

Die Sibirische Fichte verbreitete sich schneller als die Tanne, und besiedelte früher das Territorium des europäischen Nordens. Das war durch die Tatsache bedingt, daß die Fichtensamen, die nur im Winter und Frühling aus den sich dann öffnenden noch auf den Bäumen hängenden Zapfen herausfallen, und vom Winde auf der Schneedecke weiter von den Mutterpflanzen weggetrieben werden als die Tannensamen, die schon im Herbst mit den Schuppen zerfallender Zapfen zusammen auf die Kräuterdecke niederfallen (KORCHAGIN 1940a).

Europäische und Sibirische Fichten sind junge Arten, die während des Quartär aus einer der Sibirischen Fichte näher stehenden präglazialen Art durch die Zergliederung deren Areals in zwei isolierte Waldinseln (Europäische und Uralisch-Altaiische Refugialgebiete entstanden sind.

Sehr aufschlußreich ist die Aufklärung der gegenseitigen Beziehungen beider Arten und der Richtung, in welcher das Ablösen der einen Art durch die andere im Gebiete ihres Zusammentreffens im Bassin von Nord-Dwina sich entwickelte. Leider ist diese Frage beinahe noch unerforscht. Manche Beobachtungen zeigen, daß in der Berührungszone beider Arten oft mehrere hybride Übergangsformen vorkommen, und daß es einige Gesetzmäßigkeiten der Verteilung dieser Arten in verschiedenen Lebensräumen gibt, da diese Arten in ökologischer Beziehung verschieden sind. Die Sibirische Fichte kann sich auch in minder günstigen Verhältnissen entwikkeln, als die zartere europäische Fichte, die wärmere und nährstoffreichere Böden beansprucht. Die Sibirische Fichte kann man darum hauptsächlich unter weniger günstigen Umweltbedingungen (Wasserscheiden, Nordhänge, feuchte Böden) als die europäische Fichte (Flußtäler, Südhänge usw.) antreffen (KORCHAGIN 1940b, 1954). Die Geschichte der Tannenverbreitung in Gebieten, die schon von der Fichte besiedelt sind, kann man auf Grund der Ökologie beider Arten verstehen.

Während die Sibirische Tanne für ihre normale Entwicklung genügend nahrhafte und feuchte, gut durchlüftete Böden beansprucht, kann die Sibirische Fichte sowohl auf reichen, wie auch auf sehr armen, dürren und versumpften, verhältnismäßig schlecht durchlüfteten Böden wachsen. Gleichzeitig ist die Sibirische Tanne mehr schattenertragend als die Sibirische Fichte; darum kann der Tannenwuchs sich auch gut im Fichtenwald entwickeln, aber der Fichtennachwuchs wird im Tannenwald äußerst unterdrückt und fällt aus. Darum entwickelt sich die Tanne, die in Fichtenwälder eindringt, auf nährstoffreichen Böden ganz normal und kann nach 2-4 Generationen die Fichte vollkommen verdrängen und einen Tannenbestand bilden. Auf armen, trockenen oder versumpften Böden wird die Tanne, wenn sie unter die Kronendecke des Fichtenwaldes gelangt, sehr unterdrückt und ist nicht imstande, unter diesen Verhältnissen die Fichte zu verdrängen. In Übergangslagen ist ein Zusammenleben der Fichte und der Tanne möglich (KORCHAGIN 1940a).

Die Tanne, die nur auf reichen Böden vorwärts dringt und sich sehr langsam verbreitet, vermochte darum bis jetzt noch nicht, alle Fichten aus den für sie günstigen Lebensräumen zu verdrängen. Tannen- und Fichten-Tannen-Wälder sind weit im Vor-Uralgebiet verbreitet, und westliche Tannenvorposten haben schon die Eisenbahnlinie Wolodga-Archangelsk erreicht (40° östl.Breite).

Die Geschichte der Lärchenwälder des europäischen Nordens ist wenig bekannt. Die letzte Vereisung haben sie wahrscheinlich in den Refugien des Süd- und Mittel-Urals (ILJISKY 1929, NEISTADT 1957) oder auch am Südtiman (DYLIS 1940) überlebt, um dann ziemlich schnell am Anfang des Holozän sich nach Westen zu verbreiten. Lärchenreste wurden im Torf im Pskovdistrikt gefunden (NEISTADT 1957). Dann aber wurde die Lärche fast allerorts in der Ebene durch dunkle Nadelwälder verdrängt. Primäre Lärchenwälder sind im europäischen Norden nur noch am Mittel- und Nord-Ural anzutreffen. Die Lärchenwälder des Südtimans sind wahrscheinlich auch Relikte (DYLIS 1940).

Dritte Gruppe- Anthropogene,exogene (nach SUKATSHEV 1928,1954) oder katastrophale (nach WYSOTSKY 1915, GAMS 1918) Sukzessionen. Diese oben genannten jahrhundertelangen Vegetationsfolgen, die sich jetzt im europäischen Norden abspielen, werden durch höchst mannigfaltige Waldvegetationsfolgen kompliziert, die durch Schlag und Brand hervorgerufen sind. Die unter dem Einfluß dieser katastrophalen Faktoren enstandenen sekundären oder Übergangs-Wälder unterscheiden sich von den primären Stammwäldern in ihrem Bestand und in der Struktur. Wenn sie sich selbst überlassen werden, beginnen sie, in ihren primären Zustand zurückzukehren. Unter verschiedenen Umweltbedingungen nimmt die Waldvegetationsfolge recht unterschiedlichen Charakter an.

I) Primäre Fichten, Fichten-Tannen- und Tannenwälder auf tonhaltigen und lehmigen Böden werden nach Schlag und Brand von sekundären Birken- oder Aspenwäldern abgelöst. Das hängt damit zusammen, daß auf Schlagflächen der Wasserscheiden die Keimpflanzen der Fichte späte Nachtfröste und ein Ausfrieren nicht ertragen können und dadurch absterben. Die Fichte verbreitet sich hier nur unter einem Kronendach, gewöhnlich von kleinblättrigen Gehölzen, wo sie eine ganze Schicht des jungen Nachwuchses bildet und die Laubgehölze allmählich verdrängt. In 80-100 Jahren vollzieht sich eine Wiederherstellung des Primärwaldes (kurzzeitliche Sukzession).

Nach dem Brand und Schlag der Tannen- und Fichten-Tannen-Wälder werden sehr oft auf großen entwaldeten Flächen die ehemaligen Waldbestände gar nicht wieder hergestellt, sondern es entwickeln sich zuerst kleinblättrige Gehölze, dann ein sekundärer Fichtenwald, da die Fichte sich schneller als die Tanne ausbreitet. Erst später beginnt die Tanne langsam in diese sekundären Fichtenwälder einzudringen und allmählich stellt sich wieder der primäre Tannen- oder Fichten-Tannenbestand ein (langzeitliche Sukzession). In ein und demselben Lebensraum und geographischem Bezirk kann man darum Kleinblatt- und Fichten-, Fichten-Tannen- und Tannenbestände als verschiedene Wiederherstellung des primären Waldes beobachten.

2) Fichtenwälder und Böden, die sich auf Kalkstein, Mergel oder Gips bildeten, sind im europäischen Norden verbreitet. Die Brandflächen werden gewöhnlich nicht von der Fichte, sondern von der Lärche (*Larix sukaczewii*) allein, oder in Verbindung mit kleinblättrigen Gehölzen besiedelt. Unter ihrem Kronendach erscheint später Nachwuchs der Fichte und allmählich bilden sich gemischte Fichten-Lärchen- oder Fichten-Birken-Lärchen-Wälder, in welchen die obere Schicht von der Lärche gebildet wird, da diese schneller als die Fichte und die Birke wächst. Die Lärche kann sich als Lichtpflanze nicht in diesen schattigen Wäldern erneuern. Darum kehren diese Wälder nach dem Absterben der Lärchen-Überhälter (nach 400-500 Jahren) zum primären Zustand - dem Fichtenwald - zurück.

Im europäischen Norden wiederholen sich Brände periodisch alle 100-300 Jahre. Die Fichte, welche eine dünne Borke und ein oberflächennahes Wurzelsystem besitzt, kann die Brände nicht ertragen und stirbt ab. Die Lärche dagegen,mit ihrem tiefliegenden Wurzelsystem und dicker Borke hält diese Brände aus. Infolgedessen verwandeln sich die gemischten reifen Fichten-Lärchen-Wälder wieder in reine Lärchenbestände, in welchen sich später Fichten- und Lärchennachwuchs wieder ansiedeln. Die Brände verhelfen also im europäischen Norden, sekundäre Lärchen- und Fichten-Lärchenwälder zu erhalten (langzeitliche Sukzession).

3) Auf armen Sandböden kommen im europäischen Norden Fichten- (hauptsächlich *Picea obovata*), Fichten-Kiefer- und Kiefernwälder vor. Ob diese Kiefernwälder unter solchen Umweltbedingungen als primär anzusehen sind, oder ob sie auch erst anstatt der Fichtenwälder nach dem Brande erschienen sind, so wie auch die Probleme der gegenseitigen Beziehungen der Kiefer zu der Fichte, und die Frage nach der Richtung der Dynamik in gemischten Fichten-Kiefer-Wäldern - alle diese Fragen werden in der Weltliteratur beinahe schon hundert Jahre diskutiert (BLOMQUIST 1881; KIHLMAN 1890 u.a.). In dieser Hinsicht besteht eine große Meinungsverschiedenheit. Ohne auf diese Diskussion näher einzuge-

hen (GORDJAGIN 1901,1902; MOROZOV 1924; SUKATSHEV 1928,1938; SOKOLOV 1928; SAIZEV 1933; KORCHAGIN 1940,1954; MELECHEV 1924 u.a.), möchte ich hier nur meinen eigenen Standpunkt klarlegen.

Beinahe alle Botaniker stimmen darin überein, daß auf genügend feuchten Sandböden die Kiefernwälder allmählich von Fichtenwäldern abgelöst werden, da die Fichte unter solchen Bedingungen ziemlich dichten Nachwuchs unter dem Kronendach der lichtbedürftigen Kiefer bilden kann. In 50-80 Jahren wächst der Fichtennachwuchs empor, dringt in die Baumschicht der Kiefer ein, einen gemischten Kiefer-Fichtenbestand bildend, der sich nach dem Absterben der Kiefer in einen reinen Fichtenbestand umwandelt, da unter der dichten, von der Fichte gebildeten Schicht die Kiefer sich nicht anzusiedeln vermag.

Auf versumpften Böden aber kann die Fichte die Kiefer nicht ablösen, da sie der steigend wachsenden Torfmächtigkeit und der damit verbundenen schlechten Durchlüftung nicht gewachsen ist.

Auch auf sehr trockenen und armen sandigen Böden fehlt die Fichte in Kieferwäldern, und nur manchmal kommt sie als niedriger verkrümmter Baum vor. Diese Kiefernwälder (P i n e t u m c l a d i n o s u m , P i n e t u m v a c c i n i o s u m) werden von vielen Gelehrten (TKACHENKO 1911, MOROSOV 1924, EUKATSHEV 1928,1938 u.a.) als primäre angesehen, die durch die Fichte nicht ersetzt werden können. Aber auch die Ansicht wird vertreten, daß diese Kiefernwälder sekundär und aus ehemaligen Fichtenwäldern entstanden sind (GORDJAGIN 1901,1922, SKOLOV 1928, KORCHAGIN 1940,1954 u.a.). Nach meiner Meinung kann man diese Frage für *Picea abies* und *Picea obovata* nicht gleich beantworten.

Auf armen, trockenen Böden, besonders in Gegenden, die südlicher liegen, kann die europäische Fichte, da sie größere Ansprüche an Edaphon und Bodenfeuchtigkeit stellt, nicht leben und die Kiefer dort vertreiben. In den Grenzen des Areales der europäischen Fichte sind diese Kiefernwälder also primär. Die sibirische Fichte kann auf trockenen, armen Böden wachsen und darum besiedelt sie besonders in nördlicher liegenden Gegenden, wo das Klima feuchter ist, diese trockenen Kiefernwälder. Wenn sie sich allmählich vermehrt, kann sie nach der 3.,4.,5.Generation die Kiefer verdrängen und einen Fichtenbestand bilden, der von minderwertiger Bonität vielleicht noch eine geringe Kiefernbeimischung enthält. Eine solche theoretisch denkbare Folge wird jedoch kaum jemals verwirklicht, da in den trockenen Kiefernwäldern sich die

Brände periodisch alle 50-100-150 Jahre wiederholen und alle erschienenen Fichten vernichten, während die Kiefer mit ihrer dikken Borke und ihren tiefgreifenden Wurzeln erhalten bleibt. Darum verhindern die Brände unter diesen Bedingungen ein Aufkommen der Fichte (KORCHAGIN 1940,1954). Nur sehr selten kann man jetzt im europäischen Norden auf Sand trockene Flechten-Kiefernwälder mit ziemlich großer Beimischung von unterdrückter Fichte antreffen (KORCHAGIN 1954, MELECHOV 1944).

Das gegenwärtige, sehr komplizierte und bunte Bild der Waldstruktur und der Waldbestände im europäischen Norden ist also das Resultat des Zusammenwirkens von langsam, aber unaufhaltsam sich entwickelnden jahrhundertelangen Wald-Vegetationsabfolgen und von den katastrophalen Ereignissen, die von Zeit zu Zeit in das Waldleben grob eingreifen, Folgen, die von Brand und Schlag hervorgerufen werden, und die sekundären Baumbestände äußerst mannigfaltige Stufen der Wiederherstellung des primären Waldzustandes erzeugen.

LITERATUR

BLOMQUIST,A.G. -1881- Finlands tradslag i forstligt hänseende beskrifna.- Finska forstfören meddel.3 (1).

DYLISS,N.V. -1940- Lärchenwaldtypen des Südlichen Timan.- Trudy Bot.Inst. AN SSSR, Ser.III, Geobotanika 4.

GAMS,H. -1918- Prinzipienfragen der Vegetationsforschung.- Vierteljschr. Naturf.Ges.Zürich 63.

GORDJAGIN,A.J. -1901- Materialien zur Kenntnis der Böden und der Vegetation Westsibiriens.- Trudy Obschtsch.Jesstjesstvoisspt.Kasansk.Univ. 35 (2).

-- -- -1922- Vegetation der Tatarischen Republik. Geographische Beschreibung der Tatarischen Republik,Tl.1 (Natur des Gebietes).-

ILJINSSKIJ,A.P. -1929- Methoden und Aufgaben der Untersuchung der geographischen Verbreitung der Gehölze in der UdSSR.- Trudy Prikl.Bot.Gen. Selekz.21 (3).

JAROSHENKO,P.D. -1961- Geobotanik. Grundzüge des Begriffs, der Richtung und der Methode.- AN SSSR.

KIHLMAN,A.O. -1890- Pflanzenbiologische Studien aus Russisch Lappland.- Acta Soc.fauna et flora Fennica 6 (3).

KORSHINSSKIJ,SS.I. -1899- Die Pflanzenwelt Rußlands.- Brockhausen und Ephron-Lexikon 27 (1).

KORCHAGIN,A.A. -1940a- Zur Frage der Beziehungen zwischen Fichte und Tanne.- Trudy Leningradssk.Obschtsch.Jesstjesstvoissp. 68 (3).

-- -- 1940b- Pflanzendecke der Nordhälfte des Pjetschore-Ilytsch-Naturschutzgebietes.- Trudy Pjetsch.- II.Naturschutzgeb.2.

-- -- -1945 - Die Herkunft der Dunkelnadelwälder Eurasiens.- Tes.Dokl. Ssekz.Geogr.Nauk Nautschn. Ssess.Leningradssk.Univ.1945.

-- -- -1954 - Der Einfluß von Bränden auf die Waldvegetation und ihre Regeneration nach Brand im europäischen Norden.- Trudy Bot.Inst.AN SSSR, Ser.III, Geobotanika 9.

NJEJSCHTADT,M.I. -1957- Waldgeschichte und Paläogeographie der UdSSR im Holozän.- Moskau (AN SSSR).

MOROSOV,G.F. -1924- Die Lehre vom Walde.- Leningrad.

SAJZEV,B.D. -1933- Über die Verbreitung von Föhren- und Fichtenbeständen des Nordbereichs in Abhängigkeit von den Bodenbedingungen.- Mater. 2.Konfer.Stud.Produktionskräfte d.Nordbereichs 2.
SSOKOLOV,SS.Ja. -1928- Waldpflanzenassoziationen und Moore der Datscha Ossinoroschtschin im Lehr- und Versuchsrevier Pargolov des Leningrader Waldinstitutes.- In: "Natur und Wirtschaft des Lehr- und Versuchsreviers des Leningrader Waldinstitutes".-
SUKATSHEV,V.N. -1938a- Pflanzengesellschaften. Einführung in die Phytosoziologie. 4.Aufl.- Leningrad.
-- -- -1938b- Dendrologie mit den Grundzügen der Waldgeobotanik.- Leningrad.
-- -- -1944 - Über die Grundsätze der genetischen Klassifikation in der Biologie.- Shurn.Obschtsch.Biol.5 (4).
-- -- -1954 - Einige allgemeine theoretische Fragen der Phytozönologie.- In: "Voprossy Botaniki" 1.
TKATSHENKO,M.E. -1911- Die Wälder des Nordens.- Trudy Ljessn.Optyn. Djelu v Rossii 25.
TOLMATSHEV,A.I. -1943- Zur Frage der Entstehung der Taiga als zonale Landschaft.- Sovjetssk.Bot.4.
-- -- -1954- Zur Geschichte der Herkunft und Entwicklung der Dunkelnadelwaldtajga.- AN SSSR 1954.
VYSSOZKIJ,G.N. -1909- Phyto-topologische Karten, Verfahren zu ihrer Anfertigung und ihre praktische Bedeutung.- Potschvovjedjenje 2.
-- -- -1915- Jergenij, kulturell-phytologische Skizze.- Trudy Bjuro Prikl.Bot.8 (10-11).

DISKUSSION

A.W.KÜCHLER

Was sind nemorale Gehölze?

Was Sie über das Feuer sagten, interessiert mich sehr wegen der Parallelen, die ich in den nordöstlichen Vereinigten Staaten (Maine) erlebt habe. Hier läßt sich ein großer Unterschied in der Wirkungsweise des Feuers feststellen, je nachdem, welche Gehölze brennen. Bei einem Waldbrand überleben Tannen und Kiefern, wogegen Fichten vollständig vernichtet werden. Wenn der Wald abgebrannt ist, folgt ein Birken-, manchmal auch ein Aspen-Bestand, und im Schutze der Birken erst kommen die Koniferen zurück, aber nie direkt. Manchmal kommen sie gleichzeitig mit den *Betula*-Arten. Diese wachsen aber derart rasch, daß der Aspekt eines reinen Birken-Waldes entsteht. Die kleinen Tannen stehen jedoch schon darin.

Kann man derartige Beobachtungen auch bei Jhnen machen?

A.A.KORCHAGIN

Bei uns ist es die Eiche. Es handelt sich um boreale Eichen-Wälder mit *Quercus robur*. Die genannte Sukzession,mit der Rückkehr der Fichte, bzw. der Nadelhölzer überhaupt zusammen mit den kleinblättrigen Laubbäumen wie Birke und Aspe erfolgt die Zuwanderung aller dieser Arten nur sehr selten gleichzeitig. Im Gebiet der UdSSR erscheinen zuerst die Laub- und dann erst sekundär die Nadel-Gehölze. Dazu kommt noch, daß die russischen Wälder auf völlig ebenem Boden wachsen, und daß infolgedessen die Versorgung mit Nährstoffen, Wasser etc.gut ist.

J.SCHMITHÜSEN
Es würde mich interessieren, ob man darüber etwas aussagen kann, auch zahlenmäßig, in welchem Ausmaß die Brände anthropogen oder durch Blitz ausgelöst werden.

A.A.KORCHAGIN
Die Brände können sowohl durch Gewitter, als auch durch den Menschen verursacht sein. Zahlenmäßig ist nichts bekannt.

D.MUELLER-DOMBOIS
Ich möchte die Beobachtungen von Herrn KÜCHLER noch dahingehend ergänzen, daß im südöstlichen Manitoba beide Formen der Sukzession vorkommen: auf oligotrophen Sanden mit hohem Wasserspiegel geht die Sukzession von *Pinus banksiana* direkt zu *Picea mariana*, auf mesotrophen Böden (mit organischen oder mineralischen Kolloiden) oftmals von *Pinus banksiana* über *Populus tremuloides*, *Betula papyrifera* zu *Picea mariana* und *P.glauca* (und auch *Abies balsamea*).
Weiter scheinen sich die zwei Fichtenarten *Picea mariana* und *P.glauca* ökologisch zu ergänzen, indem *P.mariana* nur ein kurzes Wurzelsystem auszubilden vermag, sogar bei häufiger Irrigation, wie ich experimentell feststellen konnte, während *P.glauca* bei derselben Irrigation ein tiefgehendes Wurzelsystem entwikkeln kann. Deshalb kann man die letztgenannte Fichte auch auf Sandböden finden.

DIE WICHTIGSTEN TYPEN ANTHROPOGENER SUKZESSIONEN VON WIESENGESELLSCHAFTEN DER SOWJETISCHEN OSTSEELÄNDER[1)]

E.P. Matvjejeva

1) Sowjetische Ostseeländer heißt das Territorium an der Ostsee und ihren Meerbusen innerhalb der Verwaltungsgrenzen der Litauischen, Lettischen, Estnischen Sozialistischen Sowjetrepublik, der Gebiete Leningrad und Kaliningrad RSFSR (UdSSR).
Die Untersuchung der Dynamik in den Jahreszeiten und verschiedenen Jahren, sowie der Sukzession von Wiesengesellschaften wurde von uns vorwiegend in Wasserscheidenbereichen der sowjetischen Ostseeländer von 1949 bis 1963 durchgeführt (LARIN, MATVJEJEVA und SSROKOMSSKAJA-TJAN-SCHANSSKAJA 1960).
Die sowjetischen Ostseeländer liegen in den Unterzonen der südlichen Tajga und der Laub-Nadel-Mischwälder. Die Waldbedeckung des Gebietes nimmt von Nordosten nach Südwesten im Durchschnitt von 50 auf 16-13% ab. In gleicher Richtung wächst, von 9 auf 33% die Fläche, die von Wiesen eingenommen wird, als Ergebnis der Abholzung von Wäldern oder der Kultivierung von Mooren (MATVJEJEVA 1953,1960,1965,1967; ERINGISS 1964).
Die verschiedenen Wiesentypen dieses Territoriums sind in 7 Formationsklassen vereinigt (MATVJEJEVA 1953,1960,1965,1967): Prata genuina eumesophytica, Prata oligo- et psychromesophytica et psammomesophytica, Prata oligomesophytica, Prata steppo- sa euxeromesophytica, Prata alvaro-carbonatomesophytica, Prata hydromesophytica-minerale, Prata oxilomesophytica-turfosa.

Die herrschenden Wiesentypen sind die 3 erstgenannten; im folgenden wird im wesentlichen von ihnen die Rede sein.
Unsere Untersuchungen der Dynamik und der Sukzessionen von Wiesengesellschaften wurden sowohl auf Exkursionen als auch stationär ausgeführt. Angewandt wurden unmittelbare Methoden und Vergleiche in verschiedenen Jahren, ferner mittelbare unter Heranziehen von Pionier-, Differential- und Rest-Indikatoren verschiedener Phasen und Stadien der Sukzession.
Ohne allgemeine Fragen der Dynamik von Phytozönosen und Pflan-

[1)] Deutsche Übersetzung von Prof.Dr.H.ZEIDLER, Hannover

zendecke insgesamt im Sinne von SUKATSCHEVs (1942) u.a. zu berühren, die genügend ausreichend von ALJEKSSANDROVA (1964) behandelt wurden, verweilen wir bei einigen Gesetzmäßigkeiten und Beispielen für Dynamik und Sukzessionen von Wiesengesellschaften des Untersuchungsgebietes.

Auf das heutige Bild der Wiesenvegetation drückte die lange landwirtschaftliche Kultur, besonders die Grünlandwirtschaft, einen tiefen Stempel. Unter den sekundären natürlichen Wiesengesellschaften herrschen solche aus verschieden alten Brachen vor, die sich in einem labilen Zustand befinden und ziemlich leicht zu beobachtende, schnell wechselnde Sukzessionsstadien durchlaufen. Ebenso sind weit verbreitet mehrjährige Grasgemische und einartige Wiesengras-Saaten, die sowohl auf verschiedenen Stufen der Agrotechnik angebaut sind, wie auch ohne entsprechende Pflege, Düngung und rationelle Nutzung degradieren. Dynamik und Sukzessionen des Bestandes, der Struktur und der Produktivität der letzteren sind besonders klar ausgeprägt und lassen sich im Lauf von verhältnismäßig wenigen Jahren verfolgen.

Sukzessionen durchlaufen sich ablösende Anfangs- (Initial-), Übergangs- (Optimal-) und Schluß- (Terminal-) Phasen im Sinne von LÜDI (1932) und BRAUN-BLANQUET (1951). Ebensolche Phasen unterschied auch H.ELLENBERG auf dem Symposium über Vegetationsdynamik in Rinteln/Weser (Bundesrepublik Deutschland) 1967. Außer den Stadien unterscheidet H.ELLENBERG nur kurze Zeit dauernde Sukzessionsphasen mit noch ungefestigten (noch nicht fertig ausgebildeten) Gesellschaften. An diese Einteilung halten auch wir uns, indem wir beispielsweise der Phase des Initialstadiums die Gesellschaften aus Unkräutern und Ruderalpflanzen auf den Brachen zuteilen.

Manchmal können auch sich rasch ablösende Gesellschaften von *Agrostis tenuis* Sibth. zur selben Phase führen, wenn sie auch in anderen Fällen ein verhältnismäßig langlebiges Anfangsstadium einer Wiesengesellschaft darstellen.

In der Waldzone entstehen die allerersten Anfangsphasen der Bildung natürlicher Vorgänger der Wiesengesellschaften und dieser selbst noch unter einem aufgelichteten Walddach, an seinen Rändern und auf den Lichtungen (POLJANSSKAJA, KORNILOV, TRUTNJEV 1937; NIZENKO 1961). Je nach dem Ausgangstyp des Waldes und der Art des menschlichen Einflusses auf ihn bilden sich entsprechende Wiesen-Wald-, dann Wald-Wiesen- und Wiesen-Phytozönosen.

Nach vor längerer Zeit geschlagenen Beständen ursprünglicher Fichtenwaldtypen und an Stelle von Misch- und kleinblättrigen Wäldern entwickeln sich großblättrige (*Aegopodium podagraria* L., *Anthriscu sylvestris* (L.) Hoffm., *Geranium sylvaticum* L. u.a.) und Hochgras- (*Calamagrostis arundinacea* Roth., *Dactylis glomerata* L., *Festuca pratensis* Huds. u.a.) Wiesen-Wald- und Wald-Wiesengesellschaften aus mesophytischen Kräutern, die in ihrer Artenzusammensetzng und ihrem Aufbau durch die Mahd erhalten werden.

Dies sind Vorwiesen- und initiale Wiesen-Phasen oder -stadien der Sukzessionsreihe, die kürzere oder längere Zeit ohne einschneidende Veränderungen des Artenbestandes und der Struktur bestehen können. Bei längerer systematischer Mahd oder Weide und gemischter Nutzung fallen die erwähnten und andere hochwüchsige Pflanzen, die hinsichtlich der Mineralernährung anspruchsvoll sind oder keinen systematischen Schnitt, Verbiß und Tritt durch die Tiere aushalten, allmählich aus der Krautschicht aus. Ihren Platz nehmen niederwüchsige Pflanzen ein, die sich mit einer begrenzten Menge von Nährstoffen begnügen und an das Leben auf Mähwiesen und Weiden angepaßt sind. Unter ihnen erlangen eine dominierende Rolle *Agrostis tenuis* Sibth. *Anthoxanthum odoratum* L., *Briza media* L., *Achillea millefolium* L., *Alchemilla vulgaris* L., *Prunella vulgaris* L.

Häufiger als anderswo nimmt auf sauren Rasenpodsolböden im zweiten Optimalstadium der Entwicklung auf lange Dauer *Anthoxanthum odoratum* eine führende Stellung ein. Auf weniger sauren Böden der westlichen Bezirke der betrachteten Landschaft bekommt nicht selten *Briza media*, in besser dränierten Lagen *Helictotrichon pubescens* (L.) Bess. Bedeutung als Dominante.

Auf Weiden liegt häufiger als sonst in diesem Stadium die Dominanz bei *Alchemilla vulgaris*.

Dieses zweite Optimalstadium kann sich über 10-20 Jahre und länger in Abhängigkeit von einer Reihe von Bedingungen, in erster Linie der Nutzung, hinziehen.

Viele Jahre später kommt, je nach dem Ausmaß des Rasenbildungsprozesses, in Verbindung mit Verdichtung und Verschlechterung der Durchlüftung des Bodens und Anhäufung sich nur wenig zersetzender Pflanzenreste, das Endstadium der Wiesenentwicklung. Auf Mähwiesen erringt meist *Nardus stricta* L., auf Weiden *Deschampsia cespitosa* (L.) Beauv. die Vorrherschaft. Diese Stadi-

Schema I

Phases and stages of development of meadow communities, formed on the soddy-podsol soils on the place of coniferous and broadleaved-coniferous forests under the influence of different catastrophic and regular anthropogenous factors

Meadow communities successions under the conditions of mowing and pasturage in the parkland forst (Anthriscus silvestris, Aegopodium podagraria, Solidago virgaurea, Alchemilla vulgaris, Anthoxanthum odoratum, Nardus stricta, Deschampsia cespitosa, Musci		Cereal cultures -fallow- -regular- -mowing- (without fertilization)	Field of Phleum pratense + Trifolium pratense - many years regular mowing (without fertilization)	Field of Phleum pratense + Trifolium pratense irregular pasturage (without fertilization) ↓
Periodical mowing (without fertilization)	Irregular pasturage (without fertilization)	Agrostis tenuis + Herbae ruderale ↓	Phleum pratense + Trifolium pratense ↓	Phleum pratense + Trifolium pratense + Trifolium repens + Deschampsia cespitosa ↓
Anthoxanthum odoratum + Herbae diversae + Musci ↓	Alchemilla vulgaris + Anthoxanthum odoratum + Musci ↓	Agrostis tenuis + Herbae diversae + Anthoxanthum odoratum ↓	Phleum pratense + Agrostis tenuis ↓	Phleum pratense + Deschampsia cespitosa + Trifclium repens + Herbae diversae ↓
Anthoxanthum odoratum + Nardus stricta + Herbae diversae + Musci ↓	Alchemilla vulgaris + Deschampsia cespitosa ↓ Deschampsia cespitosa + Alchemilla vulgaris ↓	Anthoxanthum odoratum + Herbae diversae + Nardus stricta ↓	Agrostis tenuis + Phleum pratense + Herbae diversae ↓	

↓
Nardus stricta + Anthoxanthum odoratum + Herbae diversae + Musci
↓
Nardus stricta + Musci + Herbae diversae

↓
Deschampsia cespitosa

↓
Anthoxanthum odoratum + Nardus stricta + Herbae diversae + Musci
↓
Nardus stricta + Anthoxanthum odoratum + Herbae diversae + Musci
↓
Nardus stricta + Musci + Herbae diversae

↓
Agrostis tenuis + Anthoxanthum odoratum + Herbae diversae
↓
Anthoxanthum odoratum + Herbae diversae + Musci
↓
Anthoxanthum odoratum + Nardus stricta + Herbae diversae + Musci
↓
Nardus stricta + Anthoxanthum odoratum + Musci + Herbae diversae
↓
Nardus stricta + Musci + Herbae diversae

↓
Deschampsia cespitosa + Trifolium repens + Herbae diversae
↓
Deschampsia cespitosa + Trifolium repens

en können auch sehr ausgedehnt oder verhältnismäßig kurz sein in Abhängigkeit von den Lebensbedingungen und der Nutzung.
In den Endstadien der Sukzession der Wiesenentwicklung hört die systematische Wegnahme der oberirdischen Phytomasse der krautigen Pflanzen (*Nardus*, *Deschampsia*) wegen ihres niedrigen Futterwertes, und weil sich auf dem Boden Horstwuchs einstellt (dichte Bulten), oft auf.
Dann ergeben sich günstige Bedingungen für Samenkeimung und Aufkommen anderer Jungpflanzen, für Ausschlagen und Entwickeln vegetativer Vermehrungskörper und Ausläufer der ursprünglich heimischen Sträucher und Bäume. In einiger Zeit bedeckt sich die Wiese wieder mit Jungpflanzen von Sträuchern, mit einem Buschwald, dann mit einem gegenüber dem ursprünglichen etwas veränderten Waldtyp.
Auf armen Sandböden kann das Zuwachsen auf nicht gemähter Brache mit Jungpflanzen von *Pinus sylvestris* L. und *Juniperus communis* L. schon in der Initialphase des Entstehens einer Wiesengesellschaft mit Vorherrschen von *Agrostis tenuis* beginnen.
Faßt man das oben Dargelegte zusammen, kann man sagen, daß folgende Sukzessionsstadien betrachtet wurden: hochwüchsige Krautpflanzen(*Aegopodium*, *Anthriscus*, *Geranium*, *Calamagrostis*), *Agrostis tenuis*, *Anthoxanthum odoratum*, *Nardus stricta* mit den Weide-Varianten von *Alchemilla vulgaris* und *Deschampsia cespitosa* stellen sich als die Grundstadien der Sukzession der Entwicklung der Wiesenvegetation in den Wasserscheidenbereichen (Plakor) der Sowjetischen Ostseeländer nach dem Schlagen der Wälder, dem Umbrechen der Böden unter ihnen und dem Aufgeben der Felder ein. Außer den oben erwähnten Grundstadien der Sukzession der Wiesenvegetation gibt es in den Sowjetischen Ostseeländern jedoch eine Menge Varianten und Abweichungen von dem betrachteten Schema I (die 3 ersten Reihen links). So entwickeln sich z.B. bei hoher Bodenfruchtbarkeit auf Wiesenbrachen in den ersten Jahren Gesellschaften von *Agropyron repens* (L.) P.B., darauf von *Agrostis alba* L. und schon dann, je nach der Verarmung der Böden, von *Agrostis tenuis*, danach von *Anthoxanthum odoratum* usw.
Noch verschiedenartiger ist die Sukzessszionsdynamik bei der Degradation von mehr- oder einartigen Wiesensaaten, wenn man sie ohne Pflege und Düngung läßt.
Am weitesten verbreitet sind Sukzessionen aus degenerierten An-

saaten von *Phleum pratense* und *Trifolium pratense*, die zusammen oder einzeln für sich ausgesät wurden, wie im Schema I gezeigt ist (in den 2 letzten Sukzessionsreihen von rechts). Sie unterscheiden sich merklich von den früher betrachteten nur in den ersten Phasen (oder Stadien) der Degradation der Saatwiese. Schon die Initial-, ebenso die Optimal- und Terminalstadien des Wiederaufbaues natürlicher Wiesen sind dieselben wie in den früher betrachteten Fällen, die in Schema I in den ersten 3 Reihen von links dargestellt sind.

Bei der Degeneration von mehrartigen Grasmischungen ausdauernder Mähwiesensaaten treten viele Varianten der Anfangsphasen der Sukzession auf, deren Lebensdauer je nach der Geschwindigkeit der Degradationsvorgänge der Kulturböden, die ohne Pflege und Düngung gelassen sind, verschieden ist. Aber von dem Augenblick an, wo die Fruchtbarkeit der Kulturböden unter die Stufe der natürlichen der Zone sinkt, beginnen die Grundstadien der Sukzession des Wiederaufbaues und der Entwicklung der natürlichen Wiesengesellschaften. In diese Vorgänge kann das Stadium von *Agrostis tenuis* fallen, und die Sukzessionen des Wiederaufbaues werden beispielsweise mit dem Stadium von *Anthoxanthum odoratum* beginnen.

Im Falle der Sukzessionen der Degradation von Dauermischsaat kann sich unter dem Einfluß ungeregelter Beweidung, auch ohne schnelles Nachlassen der Bodenfruchtbarkeit, sofort die Weide-Variante des Sukzessions-Endstadiums mit Vorherrschen von *Deschampsia cespitosa* in der Krautschicht einstellen.

Als häufiger Fall der Degradation ausdauernder Grünlandgesellschaften auf leichten Böden, die mit K_2O überdüngt wurden, erweist sich das Endstadium mit herrschend *Taraxacum officinale* Web. Als eine Besonderheit ist dieses Stadium im Schema I nicht eingezeichnet.

LITERATUR

ALJEKSSANDROVA,V.D. -1964- Dynamik der Pflanzendecke. Feldgeobotanik III.1964.

ERINGISS,K.I. -1964- Dauer-Kulturweiden Litauens, ihre Düngung und Nutzung.- Vilnjus (Wilna).

LARIN,I.V., MATVJEJEVA,E.P. & SSYROKOMSSKAJA,I.V. -1955-1956- Dynamik der Entwicklung der Wiesenvegetation des Gebietes Kaliningrad.- Trudy BIN AN SSSR, Ser.III, Geobotanika 1.

MATVJEJEVA,E.P. -1953- Mähwiesen und Weiden und Maßnahmen zu ihrer Verbesserung.- Moskau-Leningrad.

-- -- -1960- Phytozönotische Klassifikation und Produktivitäts-Typologie der krautigen Pflanzendecke von Mäh- und Weidegrünland der Sowjetischen Ostseeländer.- Trudy BIN AN SSSR,Der.III, Geobotanik 12. Moskau-Leningrad.

MATVJEJEVA,E.P. -1965- Vergleichende Analyse der krautigen Natur- und Kulturpflanzendecke von Mäh- und Weidegrünland der Sowjetischen Ostseeländer. Autorreferat der Doktor-Dissertation.- Leningrad.

-- -- - - Wiesen der Sowjetischen Ostseeländer.- Leningrad (Nauka).

-- -- & SSEMJENOVA-TJAN-SCHANSSKAJA,A.M. -1960- Charakteristik der Futterpflanzen der natürlichen Landschaften des Gebietes Leningrad.- Trudy BIN AN SSSR, Ser.III, Geobotanik 12. Moskau-Leningrad.

NIZENKO,A.A. -1961- Veränderung der natürlichen Pflanzendecke des Gebietes Leningrad unter dem Einfluß des Menschen.- LGU 1961.

POLJANSSKAJA,O.SS., KORNILOV,M.F., TRUTNJEV,A.G. -1937- Bodenkundlich-geobotanische Untersuchungen im Bereich der Kolchosen des Rayon Krjestezk im Gebiet Leningrad.- Trudy BIN AN SSSR, Ser.III, Geobotanik 2.

SSUKATSCHEV,V.N. -1942- Der Entwicklungsgedanke in der Phytozönologie.- Sovjetsk.Bot.1-3.

SUMMARY

The Soviet Pribaltic[1] is situated in the southern taiga and mixed with broadleaved-coniferous forest subzones of the forest zone. At present this territory is covered with forest rather disproportionally (16% in the south-west and up to 50% in the north-east region). The territory under grasslands increases from 9% to 33% in the opposite direction.

To the diversity of natural conditions and history of cultivation corresponds the high degree of diversity of the meadow types of Pribaltic. One distinguishes here 7 classes of formations (MATVJEJEVA,1965): true meadows, depoverished ones, poor, steppizated, alvar and overmoistened meadows and meadows on the peat soils. The secondary anthropogenous meadow types predominate. The first among them are perennial standing depoverished oligo- and psychromesophytic herb-short-grass meadows of fallow origin covering vast areas; the true cultivated mesophytic grass, clover-grass and leguminous-herb-grass meadows do not lay far behind them.

The above mentioned categories of meadows are characterized by their strongly expressed dynamics of structure, composition and seasonal and annual productivity. The chronological reestablishing successions on fallow and the degenerative and secondary reestablishing successions on the cultivated hay-meadows and pastures may be clearly distinguished.

Everyone can easily trace the main stages of the reestablishing

1) Under the Soviet Pribaltic we like to understand the territory adjoining to the Baltic sea and its bays in the administrative limits of Lithuanian, Latvian, Estonian SSR, Leningrad and Kalinin regions of RSFSR. The total area of it reaches 275 000 km^2.

successions according to the conceptions of LÜDI (1932) and BRAUN-BLANQUET (1951).

Not touching on the general problems of the dynamics and the successions of plant communities and the vegetative cover as a whole, which have been examined in details by V.D.ALEXANDROVA (1964), I shall mark only the most characteristic features of the anthropogenous successions of the area under investigation.

The most typical and widely spread **stages of the fallow reestablishing** successions of the hay-meadow type are the following ones: the initial one - *Agrostis tenuis*, the optimal one - *Anthoxanthum odoratum*, the terminal one - *Nardus stricta*.

The pasture type is represented by the stages of *Taraxacum officinale*, *Alchemilla pastoralis* and *Deschampsia caespitosa*.

In its generalized shape the above mentioned stages of successions with their transitional phases are submitted on the scheme 1.

It does not show the place of *Taraxacum officinale* stage, which has the special meaning on the cultivated pastures with the pure potassium fertilizer.

If there has been other initial composition of seed-mixture of the cultivated *Phleum* or *Trifolium pratense* meadows, the degenerative successions (before the beginning of the reestablishing stages) may have others, more compose ways (MATVEJEVA,1967).

The profound study of the meadow vegetation and the stages of its succession allows us to prevent the appearance of the least desirable of them and to control the composition, structure and productivity of meadow communities with help of the modern methods and means of agronomy.

DISKUSSION

H.ELLENBERG

I would ask which method you use to study the succession to be sure that there is a succession and not a zonation, that there is a real development and not a zonation of ecologically different sites?

E.MATVJEJEVA(translated by P.L.Gortchakovsky)

The spot, where the investigation took place, is not very large; it is the Soviet part of the pre-baltic region. During many years I have been able to make some experiments on permanent plots and I am sure that this fact shows the real stages of development.

I.ZONNEVELD
As far as I understood, you use also a method of estimation comparable with the BRAUN-BLANQUET method, but you also use quantitative methods. What is the main method in Russia, a very statistically mathematical method or something in the middle of both?

E.MATVJEJEVA
The investigations were carried out on the basis on the method of the late Professor SUKATCHEV, and also the works of Professor SHENEKEL were taken into consideration. The method BRAUN-BLANQUET was not used, but some effects of his method, because the Soviet method is based both on the study of dominance, the structure and the floristic composition of the society. So some methods of the school of BRAUN-BLANQUET were taken into consideration.

I.ZONNEVELD
I am not very acquainted with the Russian methods, but how far are pure measurements included in this method, weighing, counting or just estimation of dominants, and relating this to mass.

P.L.GORCHAKOVSKY
There are two principal schools of geobotany in our country.The school of Prof.SUKATCHEV,and the school of Prof.RAMENSKY. The method of Prof.SUKATCHEV,is based on the study of dominants and the floristic composition of the plant societies and the structure of societies and it is necessary to study the biomass, the ration between components, and so on. Another school of geobotany is the school of Prof.RAMENSKY. He studied meadows and he proposed an individualistic conception. His works were not much estimed at the time when he lived, but now WHITTAKER and some other botanists have found that their view has much in common with the conception of Dr.RAMENSKY.

H.SUKOPP
In ABDERHALDEN's Handbuch der Biologischen Arbeitsmethoden two articles, one of the method of SUKATCHEV, and the other on the method of RAMENSKY are to be found.

FACTS AND IDEAS ON SUCCESSION IN THE MEDITERRANEAN TERRITORY OF ISRAEL

M. L i t a v

This review constitutes an attempt a reevaluation of the accepted ideas on plant succession in limestone areas of the Mediterranean territory of Israel. Limestone covers the greatest portion of this territory in Israel as well as extensive areas in other Near Eastern countries. The following ideas and suggestions are based on data acquired mostly during the last two decades in Israel but they should be applicable to neighbouring countries having similar climate and vegetation.

The prevalent opinion regarding the course of primary succession on hard limestone (lithosere) is as follows (ZOHARY 1962): Lichens ⟶ mosses ⟶ shallow-rooted rock plants ⟶ deep rooted rock plants ⟶ annuals in shallow soil ⟶ annuals in deep soil ⟶ perennial grasses and other hemicryptophytes ⟶ dward shrubs ⟶ garigue ⟶ maquis ⟶ forest.

On abandoned land the process would commence at the annual plants stage and proceed as before to the climatic (or edaphic) climax. Such a scheme presupposes that each stage prepares the way in some manner for the next, either by providing essential shading for seedling establishment, by modifying the soil environment or by doing both things. While there is an extensive amount of descriptive literature on succession (BRAUN-BLANQUET 1964, WHITTAKER 1953) some of it supporting (openly or by inference) and some of it critisizing the foregoing assumption, markedly little effort has gone into its critical evaluation. This is especially true for the East Mediterranean region where conditions for such work are particularly difficult.

This is because:

a) The vegetation of this region has not been described in quantitative terms earlier than three or four decades ago.

b) There are practically no areas where the vegetation has been left to develop undisturbed for any appreciable length of time.

c) The rate of change in the successional sere is much slower than in wetter climates.

Thus, present ideas on the course of succession in the Eastern Mediterranean region rest alsmost exclusively on indirect evidence according to which various associations which occur side by side under similar habitat conditions are regarded as different stages of one successional sere (see however: ELLENBERG 1957). The similarity between the South of France and the East Mediterranean in climatic conditions and in history of land use (and abuse) led authoroties like EIG & ZOHARY (1962) to assume that succession follows simular patterns on both sides of the Mediterranean. The Western side of which was studied mainly by workers of the Monspellier school (e.g. BRAUN-BLANQUET 1936, BHARUCHA 1933, BRAUN-BLANQUET & DE BOLOS 1957). The above assumption was further supported by the strong physiognomic similarity between the two sub-regions, forming a sort of vegetational vicariism.

Difficulties arose, however, when it became increasingly apparent that in many cases presumably advanced successional stages may actually initiate the course of secondary succession, while in other cases succession seems to be at a standstill at an early annual plants stage.

The following parts will deal with: 1) a critical evaluation of the basic concepts of plant succession as applied to the limestone areas of the Mediterranean territory of Israel, 2) observations and data on the ecology of establishment of various 'Association builders'.

It is obvious that a predetermined sequence of associations could be the rule only if succession and soil development go 'hand in hand' (it is unlikely that light and shade relations alone would control an orderly course of succession), meaning that soil forming processes are constantly affected and modified by vegetation

The process of conversion of limestone rocks into soil entails the dissolution of carbonates and the release of clay minerals, iron oxides, etc. According to CORBEL's figures (1959) a 10 cm layer of limestone is removed each year under the Mediterranean climate of Marseille and about 2,5 times as much under the wet climate of Cambridge, England. YAALON (1959) who arrived at about the same figures, calculated that if the rock contains 1% of insoluble residues then it would take about 1 million years to form a 10 cm layer of soil! Owing to the presence of numerous marly intercalations in the limestone strata of Israel a more likely figure would be 10% of risidue. This will provide 1 m of soil in one million year or 1 mm of soil per 1000 years. This means that formation of a soil layer which will be of any significance takes incomparably longer than the time which passed since certain abandoned areas reverted to a climax vegetation e.g. CRISADER's strongholds deserted since the thirteenst century and now covered by dense maquis.

If we adopt the view that the existing soil mantle consists of a mixture of aeolian and aurochtonous material (VROMAN 1944, YAALON & GINSBOURG 1966, DAN & YAALON 1964) then the accumulation of soil depends in part on the frequency of dust-laden winds from adjacent deserts (GINZBOURG & YAALON 1963, TOKATLY 1960). A dense vegetation cover may conceivably act as a 'trap' for the dust particles. In addition, vegetation may be assumed to promote weathering of clay minerals, leaching of deposited carbonates and the creation of stable aggregates. These processes in so far as they migh affect the course of succession merit some examination.

It is true that carbonation is an important weathering agent in primary minerals. In the present case, however, the insoluble rock residue and the silicates within the wind-borne dust consist both of the same secondary clay minerals (mainly montmorilonite, kaolinite and illite) which undergo little additional change in the soil proper (YAALON et al.1966).[1)]

Leached carbonates, originally present in the deposited dust, form horizontal bands of lime-enriched soil which are visible in many soil profiles of the more arid parts of Israel. This leaching is due of course to the same process of bicarbonate formation which causes the wearing away of hard limestone strata. It can be expected however to be much faster than the latter owing to the highly dispersed state of the reactant(BOYNTON 1966). The depth at which the dissolved bicarbonates are re-precipitated will depend on rainfall and on the depth reached by plant roots (MARBUT 1928). This is because decomposing roots form channels along which lime-saturated solutions flow more frely and because of higher CO_2 concentration along them (JOFFE 1949,p.58).

[1)] Exceptions occur in the case of the unstable palygorskite and sepiolite which are absent in all but very young soils, and in the case of montmorilonite which is converted into kaolinite in strongly leaches terra rossa on calcareous limestone (YAALON 1955, KOYOUMDJINSKY, YAALON & DAN 1966).

It has long been established that the formation of stable aggregates is considerably promoted by the presence of plant roots, the degree of aggregation depending, among other things, on the density of the root system (BAVER 1956, for a detailed discussion). In fact, the best soil structure is found under a prairie type vegetation with its highly ramified fibrous root system. The latter is represented in our case by the annual-perennial grass stage which, according to prevalent ideas is but one step removed from the abandoned field! Indeed, preliminary observations indicate that there is less intermixing of roots and soil under 'batha' and garigue than under tall annual or perennial grasses.

Although these presumed effects of vegetation on soil development do not seem very far reaching compared to those occuring in more humid climates or on other parent materials, they might prove nevertheless indispensable for the establishment of certain successional stages. In what follows, available data on the ecology of establishment of leading plants in some of the main stages which lead to the *Quercus-Pistacia* climax is examined and discussed.

Climax communities - often much degraded - of the Quercus calliprinos-Pistacia palaestina association cover at present considerable areas in the Mediterranean territory of Israel (ZOHARY 1960,1962). Most of these areas are rocky and thinly covered with soil. Adult trees are often found growing straight into exposed rock crevices, in only slightly weathered marl and, occasionally, in strongly calcareous rendzina overlying soft chalk. Obviously, such sites cannot be regarded as edaphically mature. As all the tree and shrub species are able to regenerate from sprouts, it is impossible to ascertain the age of the stands nor to guess the habitat conditions which pravailed at the time of establishment.

Assuming for the sake of argument that establishment could have taken place in mature soil only, it may be suggested that erosion under conditions of excessive browsing and cutting removed the top soil "from under the feet" of the trees and exposed the parent rock or the subsoil. Such a process should habe resulted in at least part-exposure of root collars of adult trees. There is very little evidence to this affect; exposed tree roots can be seen in stream beds only where flowing water is cutting away the banks.

Recently established maquis is very rare, even in the vicinity of a copious seed supply. However, the few records on oak and *Pistacia* seedlings growing away from adult trees indicate preference for shade of some kind, but there is no evidence that their suvival is related to any particular type of existing plant cover or state of soil maturity. They have been found under *Poterium* plants, in the shade of olive trees, on relatively exposed soil, beside jutting rocks, etc. (HARIF,oral.comm.). Soil conditions range from well leached terra rossa to moderately calcareous rendzina overlying marly limestone.

Mediterranean garigue in Israel consists predominantly of *Calycotome villosa, Rhamnus palaestina,Phyllirea media, Pistacia lentiscus,* etc. usually mixed with oak and other maquis species in shrubby form. The former may also form a lower stratum inside well developed maquis. The ability of *Rhamnus, Phyllirea* and *Pistacia* to regenerate from seed under natural conditions has never been studied under natural conditions, but the extreme scarcity of young individuals or seedlings in open habitats indicates a very low capacity for territorial expansion. *Calycotome villosa* on the other hand, is spreading rapidly into new habitats which include: abandoned terraces on terra rossa, recently abandoned fields on red sandy loam in the coastal plain, erosion gullies, highly calcareous rendzina on marly chalk or on sandy limestone of the coast. *Calycotome* does not seem to require a previous plant cover for succesful establishment and poor edaphic conditions no more than retard its subsequent growth (LITAV & ORSHAN 1963a).

The next step down the presumed successional sere is the dwarf shrub ('batha') stage which covers extensive areas in the Near East. Its woody components include: *Poterium spinosum, Thymus capitatus, Majorana syriaca* and many others. *Poterium* is dominant in most batha areas while *Thymus* has a similar status in certain areas of calcareous rendzina overlying marly chalks (LITAV & ORSHAN 1963a). Observations as well as experimental data regarding the two species (LITAV & ORSHAN 1963b) indicate that both are successful invaders of habitats where the quick growing annual vegetation is lacking or poorly developed (*Thymus* is more susceptible than *Poterium* to competition). Such habitats include: shallow soils, rock outcrops, highly calcareous rendzina, erosion gullies and - occasionally - dumps of crushed rock and gravel. *Poterium* and sometimes *Thymus* too will thrive in deep and fertile soils, provided the annual vegetation has been kept back or else, they have somehow managed to survive the critical seedling stage in spite of the competing annuals. *Cistus* ssp. and other species which form what may be called low garigue are known to establish themselves in the shade of *Poterium* individuals, overgrowing and suppressing them subsequently (ORSHAN, oral comm.). They are however equally capable of germinating and growing on bare calcareous soil, where they often share dominance with *Thymus capitatus* (LITAV & ORSHAN 1963a).

It is not proposed to deal here with the ecological reqirements of the multitude of perennial grasses, hemicryptophytes and annuals - as yet unclasified into vegetational units - which constitute presumably the lowermost successional stages. Many of them for pioneer associations on abandoned lands, on eroded slopes and other disturbed sites. Others seem to be flexible in their demands, surviving dense shade, shallow soils and poor nutrients supply. There are, on the other hand, several widely occuring species which exhibit unequivocal preference for favourable habitats. A typical example is *Avena sterilis* which forms dense stands in alluvial mountain valleys or in pouches of deep residual soil between hard limestone outcrops (LITAV 1965). In fact, *Avena* with some of its associates (e.g. *Trifolium purpureum, Phalaris tuberosa*) belong to the rather uncommon type of plants whose area of distribution more or less overlaps their potential area (cf. KNAPP 1954).

I am aware that the picture presented here is far from complete and our information concerning the course of succession toward the Mediterranean climax and its causes has many serious gaps in it. It is remarkable, however, that this admittedly limited amount of information does not contain one single clear-cut case where a leading component of association depends in its establishment on the pre-existence of a specific plant community (shade requirements of maquis tree seedlings can be supplied by the canopy of adult trees as well as by north facing rock surfaces). It would seem that the only kind of relationship existing between one association and another is that of competition:the established community acting as an obstacle to any kind of invasion.

It may be argued that all the foregoing conclusions refer to secondary succession (a highly disturbed one that) and the course of primary succession may have proceeded in a much more orderly manner with one stage preparing the site for the next one. While such an assumption cannot be proved or disproved (even the remotest areas in the Near East have been cultivated or grazed at one time or another) it can be taken for granted that a plant community consisting of large-sized plants cannot maintain itself without an adequate soil volume. To the extent that the formation of this soil is related to biological processes, it can be said that the establishment of one plant community depends on the accumulated effects of previous generations of plants.It seems however that a prodigious number of generations will have passed before the amount of soil necessary for the successful establishment of deep rooted plants is produced. Once this stage is reached there is good reason to believe that the course of succession would hence-forward be determined predominantly by seed production, by efficiency of dispersal and by competitive relations.

SUMMARY

It has long been accepted that plant succession in the Mediterranean territory of Israel is closely connected with soil development, so that with each succeeding stage the soil becomes more mature and better suited for a more demanding community. Thus, the large sized components of the garigue and maquis associations require for their establishment better edaphic conditions than dwarf shrubs and these in turn are more exacting than components of the annual and perennial grass stages. The assumption is that a successional stage contributes to soil formation and maturation processes thereby preparing the ground for the next one.

This scheme was examined in the ligh of recent data concerning the rate of pedogenic processes on limestone. It was concluded that at the calculated rate of 1 mm layer of soil per 1000 years the addition of a quantity of soil material which will significantly affect the edaphic conditions should take incomparably longer than it took certain abandoned areas to revert to a climax vegetation.

Vegetation may however affect the soil by promoting maturation processes, notably those occuring in the aeolian material which is believed to be constantly deposited in considerable quanti-

ties on the existing soil mantle. Processes like the downward transport of lime particles and the formation of stable aggregates may create the necessary conditions for the establishment of certain successional stages.
It has been found, however, that the leading species of presumably advanced stages germinate and develop on shallow and immature soils, in rock crevices, in the shade as well as in full sunlight. Exceptions occur in the case of most maquis tree seedlings which require shade for successful establishment and in the case of some large sized annual grasses which form dense stands in deep residual soils, but are poorly developed or absent in strongly calcareous soils.
It is concluded that, once a certain minimum quantity of soil has been formed, succession will depend hence-forward on seed production, on efficiency of dispersal and on competitive relations.

BIBLIOGRAPHIE

BAVER,L.D. -1956- Soil physics (3rd.Ed.).- John Wiley & Sons,New York.

BHARUCHA,F.R. -1933- Etude écologique et phytosociologique de l'association à Brachypodium ramosum et Phlomis lychnitis des garigues languédociennes.- Beih.Bot.Zbl.50 B 247-379. Dresden.

BOYNTON,R.S. -1966- Chemistry and technology of lime and limestones.- John Wiley & Sons, New York.

BRAUN-BLANQUET,J. -1964- Pflanzensoziologie 3.Aufl.- Wien.

-- -- -1936- La chênaie d'Yeuse méditerranéenne (Quercion ilicis).- Mém. Soc.Etud.Sci.Nat.Nîmes 5. (Comm.S.I.G.M.A. 45) Nîmes.

-- -- & BOLOS,O.de -1957- Les groupements végétaux du bassin moyen de l'Ebre et leur dynamisme.- Ann.Estac.exp.Aula Dei 5. Zaragossa.

CORBEL,J. -1959- Vitesse de l'érosion.- Geomorphologie 3: 1-28.

DAN,J. & YAALON,D.H. -1964- The application of catena concept in studies of pedogenesis in Mediterranean and desert fringe regions.- Trans. Intern.Congr.Soil Sci., 8th. Bucarest.

ELLENBERG,H. -1953- Führt die alpine Vegetations- und Bodenentwicklung auch auf reinen Karbonatgesteinen zum Krummseggenrasen (Caricetum curvulae).- Ber.Deutsch.Bot.Ges.66: 241-246. Stuttgart.

GINZBOURG,D. & YAALON,D.H. -1963- Petrography and origin of the loess in the Beér Sheva Basin.- Israel J.Earth Sci. 12: 68-70.

KNAPP,R. 1954- Experimentelle Soziologie der höheren Pflanzen.I. Stuttgart.

KOYUMDJINSKY,H., YAALON,D.H. & DAN,J. -1966- Red and reddish brown terra rossa in Israel.- Conf.Medit.Soils Intern.Soc.Soil Sci. (Comm.V). Madrid (Mim.).

LITAV,M. -1965- Effects of soil type and competition on the occurence of Avena sterilis L in the Judean hills (Israel).- Israel J.Bot.14: 74-89.

-- -- KUPERNIK,G. & ORSHAN,G. 1-963- The role of competition as a factor in the distribution of dwarf shrub communities in the Mediterranean territory of Israel.- J.Ecol.51: 467-480. Cambridge.

-- -- & ORSHAN,G. -1963- Ecological studies on some sub-lithophytic communities in Israel.- Israel J.Bot.12: 41-54.

MARBUT,C.F. -1928- A scheme for soil classification.- Proc.1st.Intern. Congr.Soil Sci. 4: 1-31.
TOKATLY,Y.L. -1960- Easterly storms - November 1958.- Israel Expl.J.10: 112-117. Jerusalem.
VROMAN,J. -1944- The petrology of sandy sediments of Palestine.- Geol.Dept. Hebrew Univ.Bull.5: 1-11.
WHITTAKER,R.H. -1953- A consideration of climax theory: the climax as a population and pattern.- Ecol.Mon.23: 41-78. Durham,N.C.
YAALON,D.H. -1959- Dynamics of limestone solution and its application in the interpretation of geological processes.- Bull.Res.Council Israel Sect.G: Geo-sciences 8: 217-218.
-- -- & GINZBOURG,D. -1966- Sedimentary characteristics and climatic analysis of easterly dust storms in the Negev (Israel).- Sedimentology 6: 315-332.
-- -- NATHAN,Y, KOYUMDJINSKY,H. & DAN,J. -1966- Weathering and catenary differentiation of clay minerals in soils on various parent materials in Israel.- Proc.Intern.Clay Conf.1: 187-198. Jerusalem.
ZOHARY,M. -1960- The maquis of Quercus calliprinos in Israel and Jordan.- Bull.Res.Council of Israel 9 D: 51-72. Jerusalem.
-- -- -1962- Plant Life of Palestine.- New York.

DISKUSSION

G.LAVRENTIADES

It is interesting to notice here that there are many similarities between the garrigue of Israel and the Phrygana formation in Greece, well known by the ancient greek botanist THEOPHRASTOS. The Phrygana formation is also characterized by the same dominant species: *Quercus coccifera, Poterium spinosum, Thymus capitatus, Phyllirea media* and *Pistacia lentiscus*. However, in Greece the last two species also belong to Maquis formation. The difference is that, while the shrubs of *Phyllirea media* and *Pistacia lentiscus* are very low in the Phrygana formation, they are very tall in the Maquis and occur with higher vitality.

H.ELLENBERG

Bislang glaubte ich mit meiner Ansicht über Boden- und Vegetations-Entwicklung ein Extremist zu sein. Was hier jedoch demonstriert wurde, ist extremer als mein Skeptizismus.
Allein die Berechnung, wieviel Feinerde im Laufe der Zeit aus einem Kalkstein entstehen kann, nicht im Laufe von 1000 Jahren 1 mm, bei einem verhältnismäßig hohen Tongehalt, und auch die Feststellung, daß, wenn nur genügend Spalten im Gestein sind, bereits ein Wald sich ansiedeln kann, ohne daß vorher eine Feinerdebildung vorausgegangen ist, auch das muß man m.E. voll unterstützen. Wir sind gar nicht skeptisch genug gegenüber dem, was wir bisher als Entwicklung angesehen haben. Es tut gut, einmal extrem zu sein!

D.MUELLER-DOMBOIS

I would say that the last categorical statement shoud be extended by giving recognition to the fact that the competitive relations depend: (a) on the flora of the area which provides for the plants that may be able to compete with one another; (b) on the local environment (habitat) where the competition takes place; because the kind of competition is dependent on the physical factors of environment and competition is only (for the plants involved) a modification of the primary form of the environmental factors.

SUKZESSION UND BODENENTWICKLUNG AUF TROCKENEN SILIKATHÄNGEN

J. M o r a v e c

Die Sukzession der Biozönosen im engeren Sinne und die Bodenentwicklung stellen einen einheitlichen, autonomen, natürlichen Prozeß dar, der durch die Lebenstätigkeit der Biozönosen verursacht wird, und dessen Ergebnis das Entstehen des Bodens aus totem Gestein ist. Ohne diese Tätigkeit von Biozönosen könnte kein Boden entstehen. Die Biozönosen und der Boden gelangen im Verlaufe des Sukzessions- und Bodenentwicklungsprozesses in eine Beziehung einer rückwirkenden Bindung, welche die Treibkraft dieses Prozesses darstellt. Diese rückwirkende Bindung zwischen der Biozönose und dem Boden äußert sich in der durch die Lebenstätigkeit der Biozönose hervorgerufene Änderung des Bodens, die für diese Biozönose ungeeignet und für diejenige des nachfolgenden Stadiums geeigneter wird. Dieser Prozeß verläuft so lange, bis ein Endstadium erreicht ist, das dem Klima des gegebenen Gebietes entspricht. Dann wird der Boden durch die Lebenstätigkeit der Schlußbiozönose nicht mehr derart geändert, daß derselbe für die gegebene Biozönose ungeeignet wird. Die Phytozönose ist der wesentlichste Teil und aktivste Bestandteil einer Biozönose während des Sukzessions- und Bodenentwicklungsprozesses als Energieakkumulator und Produzent der Biomasse.

Im Verlaufe des Sukzessions- und Bodenentwicklungsprozesses mildern sich die Extreme des Standortes, der Einfluß des Muttergesteins wird unterdrückt, und es entstehen einige wenige konvergierende Klimaxbodentypen. Während die Biozönosen einen aktiven Bestandteil des Sukzessions- und Bodenentwicklungsprozesses bilden, stellt der Boden ein "Gedächtnis" dieses Systems dar. Im Boden verbleiben die durch die Biozönose fixierten und vor Erosion beschützten Verwitterungsprodukte, es sammeln sich hier der größte Teil des Tonhumuskomplexes und im Kreislauf zwischen der Biozönose und dem Boden verbleibt auch ein großer Teil der bereits durch die ersten Sukzessionsstadien freigewordenen mineralischen Nährstoffe. Dagegen bleiben in mehr vorgeschrittenen Sukzessionsstadien, bzw. sogar in Klimaxbiozönosen, fast keine Organismen aus den ersten Sukzessionsstadien erhalten.

Der Sukzessions- und Bodenentwicklungsprozeß wirkt "standortsbildend", d.h. im Komplex der Standortsfaktoren nehmen solche

Faktoren relativ zu, die durch die Lebenstätigkeit der vorhergehenden Biozönosen beeinflußt wurden. Solche Faktoren wirken auf die gegebene Biozönose nicht als ganz von der lebenden Welt unabhängige, sondern sie übertragen auch die Wirkung der früheren Biozönosen. Es ist dies vor allem die Gruppe edaphischer Faktoren, besonders die an den Tonhumuskomplex und an den Kreislauf der mineralischen Nährstoffe gebundenen Faktoren. Mann kann sagen, daß ein bestimmter Teil der Umwelt während des Sukzessions- und Bodenentwicklungsprozesses immer mehr und mehr in die Biozönose einbezogen wird, wo er als "Kanal" bei der Übertragung von Wirkungen zwischen bestimmten Organismen, bzw. ihren Gruppen, wirkt und zum Bestandteil der Biozönose wird.

Aus dem Angeführten folgt, daß man im Rahmen des Standortes, als des auf die Biozönose einwirkenden Faktorenkomplexes, folgende drei Gruppen von Faktoren unterscheiden kann, die an drei analoge Bestandteile der Umwelt gebunden sind:

1. Faktoren, die durch die Lebenstätigkeit der Biozönose unbeeinflußbar sind, z.B. die auf die Biozönose einfallende Strahlung, die Niederschlagsmenge u.a.
2. Faktoren, die von der Lebenstätigkeit der früheren Biozönosen gebildet oder beeinflußt wurden, z.B. die Sorptionskapazität des Bodens, die Wasserkapazität des Bodenprofils.
3. Faktoren, die von der Lebenstätigkeit der bestehenden Biozönose abhängen, z.B. Stickstoffkreislauf.

Geschichte der Erkenntnis der Sukzession und der Bodenentwicklung als einheitlicher Prozess.

KERNER (1863) beschrieb als erster diesen Prozeß in groben Umrissen, bereits mit genügend genau zutreffenden grundlegenden Prinzipien der Sukzession und der Bodenentwicklung mit der aktiven Rolle der Vegetation. Die kann man z.B. der Beschreibung der Sukzession (KERNER verwendet noch nicht diesen Ausdruck) von einer Kalkschuttgesellschaft bis zur Schlußformation mit *Azalea* (l.c.242-244) entnehmen.

CLEMENTS bearbeitete die Theorie der Sukzession und ihres Schlußstadiums (Klimax) vor allem vom Gesichtspunkt der Vegetation; die Veränderungen des Bodens behandelt er nur allgemein. Die Sukzession von CLEMENTS repräsentiert eine durch die Wirkung der Vegetation auf den Standort hervorgerufene Sukzession im engeren Sinne (vgl. CLEMENTS 1907, p.274). COWLES (1911) erweiterte den Begriff der Sukzession auch auf die durch Wirkung von

Außenfaktoren hervorgerufenen Änderungen der Vegetation (z.B. durch Änderung des Klimas - "regional successions" oder durch topographische Änderungen, im Grunde genommen Änderungen des Muttergesteins - "topographic successions"). Regionale Sukzessionen entsprechen der Migration von Pflanzen-Gesellschaften infolge Änderung des Klimas und topographischer Sukzessionen einer neuen Besiedlung der freien Erdoberfläche, die weiter zur Sukzession im engeren Sinne führen kann. Der Sukzession im engeren Sinne entsprechen nur die "biotic successions" von COWLES.

Zur Zeit der Arbeiten von CLEMENTS und COWLES konnten noch keine detaillierten Kenntnisse des Sukzessions- und Bodenentwicklungsprozesses vorliegen, weil hierzu noch zwei Voraussetzungen fehlten:

1. ein System der Pflanzengesellschaften als Schlüssel zur Erkenntnis der sich in der Sukzession geltend machenden Biozönosen und
2. eine detaillierte Boden-Klassifikation. Zu dieser Zeit beachteten die Pedologen vor allem klimazonale Bodentypen,und junge Entwicklungsstadien einzelner Bodenentwicklungsreihen bezeichneten sie allgemein als unentwickelte Böden.

Es ist ein Verdienst der Pflanzensoziologen, daß sie die Sukzession und die Bodenentwicklung im Zusammenhang zu untersuchen begannen. Die Alpen boten hierzu die vorzüglichsten Bedingungen. Diese Anfänge, bei denen BRAUN-BLANQUET mit JENNY, weiter PALLMANN und HAFTER standen, inspirierten die Pedologen zur Bildung einer Theorie der Bodenentwicklung, deren Grundlagen von PALLMANN stammen und die von KUBIENA (1948) in seiner "Entwicklungslehre des Bodens" zusammenfassend formuliert wurde.

Bearbeitung der Systeme der Pflanzengesellschaften und der Böden in vorigen Dezennien bieten uns definierte Stufen, auf denen man den Sukzessions- und Bodenentwicklungsprozeß verfolgen kann, um in die Gesetzmäßigkeiten einzelner Ökosysteme vorzudringen.

Die Sukzession und die Bodenentwicklung auf Silikathängen in Südwestböhmen

Mein weiterer Beitrag beschreibt eine vom Menschen beeinflußte Serie. Es ist dies eine Serie, deren fortgeschrittenere Stadien vom Menschen modifiziert wurden, und es ist sogar keine vollständige Serie. Diese Sukzessions- und Bodenentwicklungsserie wurde im Gebiet des mittleren Otavaflusses in Meereshöhen von

373-530 m untersucht. Die durchschnittliche Jahrestemperatur schwankt von 7,0 bis 7,6 °C, die verhältnismäßig niedrigen Niederschläge von 538 bis 595 mm und der daraus folgende LANG'sche Faktor von ca. 71 bis 85. Das Gebiet wird von sauren Silikatgesteinen gebildet, vor allem aus Gneisen, Migmatiten und Graniten. Es gehört zum Klimaxgebiet des Quercion robori-petraeae. Die Hänge der ca. 10-20 m tief eingeschnittenen breiten Täler der Otava und ihrer Nebenflüsse, an denen die Sukzession und die Bodenentwicklung verfolgt wurden, besitzen allerdings ein wärmeres und trockeneres Mikroklima.

Das Asplenietum septentrionalis - Protoranker-stadium

Abgesehen von der direkt auf der Felsoberfläche haftenden Krustenflechten-Gesellschaft, beginnt die Sukzession der Gesellschaften höherer Pflanzen mit dem Asplenietum septentrionalis Schwickerath 1944 (Asplenietia rupestria). Es ist ziemlich artenarm mit nur 4-14 Arten in der Krautschicht und 6-10 in der Moosschicht (Fig.1).

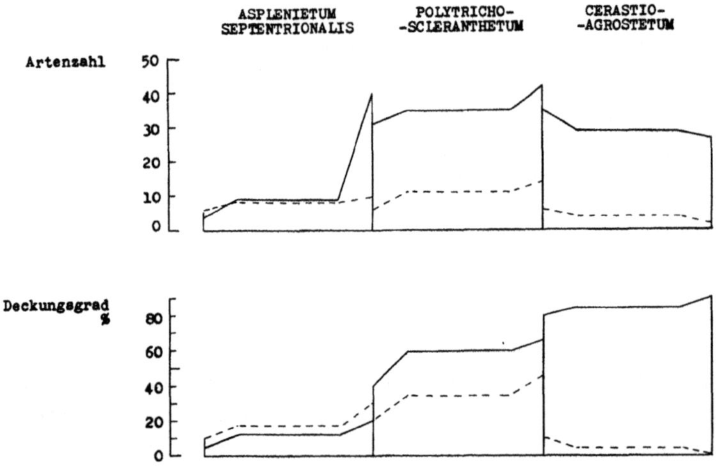

Fig.1. Artenzahl und Deckungsgrad einzelner Sukzessionsstadien; ——— Krautschicht, ------ Moosschicht.

Der Deckungsgrad der Pflanzengesellschaft ist sehr niedrig (5-30%), wobei die Moosschicht überwiegt. Die Pflanzen sind nur in Felsspalten, auf kleinen Felsvorsprüngen und Felsbändern konzentriert, die mit einer lückigen, maximal 5 cm mächtigen Feinerdeschicht bedeckt sind. Die Krautschicht wird praktisch nur von *Asplenium septentrionale* und *Festuca ovina* gebildet (Fig.1). In der Moosschicht dominiert *Polytrichum piliferum* (Fig.2). Der Standort wird von der Pflanzengesellschaft nur wenig beeinflußt. Er ist besonders physikalisch sehr extrem, was sich in der großen Schwankung der Temperatur der Oberfläche und der Feuchtigkeit im Wurzelraume äußert. Die Gesamtneigung der Felsen ist zwar ziemlich groß (38-60°), die Oberfläche ist jedoch in Felsbänder, -vorsprünge und -spalten mit viel kleinerer Neigung aufgegliedert. Ein Antritt eines weiteren Sukzessions- und Bodenentwicklungsstadiums erfordert erst einen Abbau des Gesteins und eine Verminderung der Hangneigung. Für das Studium der Sukzession und Bodenentwicklung kann uns jedoch die auf Felsbänder und -spalten konzentrierte Pflanzengesellschaft und der Boden genügende Auskunft geben.

Der Boden des A s p l e n i e t u m ist der Protoranker (KUBIËNA 1948, p.122, 1953, p.194) - ein primitiver A_0C- bzw. (A)C-Boden (Fig.3). Der flachgründige A -Horizont besteht aus einem losen, sandigen, ungleichmäßig mit Humus vermischten Boden und liegt auf einem festen Felsen. Das Mikroskelett ist meistens nicht verwittert und die Verwitterungsprodukte kleben noch an der Oberfläche fester Bodenteilchen. Der mineralische Bestandteil ist mit dem organischen nur mechanisch vermischt. Der Sorptionskomplex besitzt eine geringe Austauschkapazität und enthält einen verhältnismäßig hohen Anteil an Austauschwasserstoff (Fig.4), der wahrscheinlich durch den organischen Anteil gebunden wird. Ein so hoher Gehalt an Austauschwasserstoff kommt in vorgeschritteneren Bodenentwicklungsstadien nicht mehr vor. Verhältnismäßig hoch ist auch der Gehalt an Austausch-Aluminium, dagegen ist nur wenig Austausch-Kalzium, -Kalium und -Magnesium vorhanden. Daraus folgt eine hohe Bodenazidität. Da der Humus keine kittende Wirkung ausübt, wird der Boden nur durch das Wurzelsystem stabilisiert. Stirbt infolge einer übermäßigen Trockenheit ein Horst, so wird die Feinerde durch Wasser- und Winderosion auf tiefer gelegene Felsbänder verlagert und der organische Bestandteil vermischt sich auf mechanische Art mit dem mineralischen.

Das P o l y t r i c h o - S c l e r a n t h e t u m - Braunes Ranker-Stadium

Die Pflanzengesellschaft dieses Stadiums gehört zur Assoziation P o l y t r i c h o p i l i f e r i - S c l e r a n t h e t u m p e r e n n i s Moravec 1967 (S e d o - S c l e r a n t h e t e a). Diese Assoziation ist wesentlich artenreicher als das A s p l e n i e -

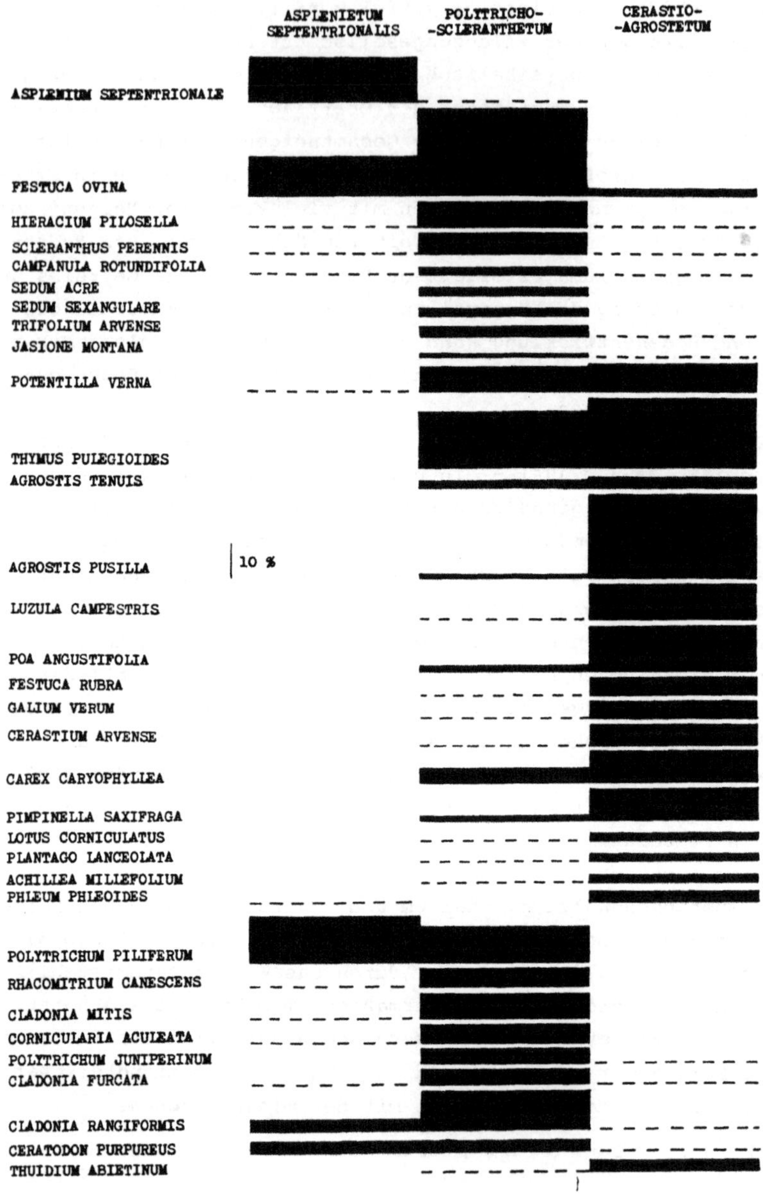

Fig.2. Durchschnittlicher Deckungsgrad der Arten mit dem Deckungsgrad 1 oder mehr% beim A s p l e n i e t u m aus 6, beim P o l y t r i c h o - S c l e r a n t h e t u m aus 10, beim C e r a s t i o - C e r a s t i o - A g r o s t e t u m aus 8 Aufnahmen errechnet.

t u m s e p t e n t r i o n a l i s : es enthält 30-42 Arten in der Krautschicht und 6-14 in der Moosschicht (s.Fig.1). Die Krautschicht setzt sich vor allem aus horst- und polsterbildenden,zum Teil sukkulenten Arten zusammen (Abb.2). Die Lücken sind teils von einigen ausläufertreibenden oder einjährigen Arten, teils von Moosen und Flechten ausgefüllt. An Stellen, wo die Felsunterlage bis zur Bodenoberfläche aufsteigt, wachsen noch vereinzelte Horste von *Asplenium septentrionale* als Zeugen des vorhergehenden Sukzessionsstadiums. Die wichtigste Dominante der Krautschicht bildet *Festuca ovina* (s.str.); in der Moosschicht dominiert *Polytrichum piliferum* mit *Cladonia rangiformis* (s.Fig.2).

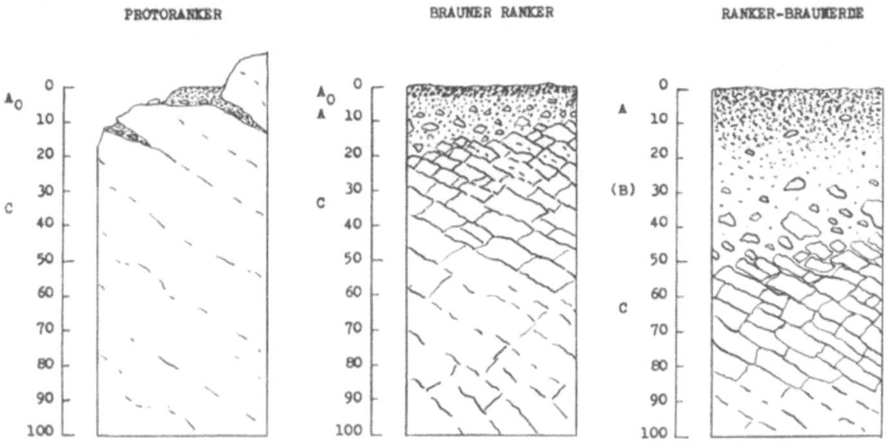

Fig.3. Profile einzelner Bodenentwicklungsstadien (schematisiert).

Der Deckungsgrad der Krautschicht beträgt 40-75%, der Moosschicht 20-45% (s.Abb.2). Beide Schichten wechseln mosaikartig ab.

Der Antritt des P o l y t r i c h o - S c l e r a n t h e t u m ist durch Anhäufung einer zusammenhängenden, mindestens 10 cm mächtigen Bodenschicht bedingt. Dieser Boden entspricht dem braunen Ranker (KUBIËNA 1953, MÜCKENHAUSEN 1962), der ein morphologisch etwas komplizierteres Profil besitzt (s. Fig.3 u.Abb.3).

In einigen Fällen entsteht teils eine Differenzierung eines 3 cm mächtigen humosen A_0Horizontes, teils ein schwacher Horizont eines mechanisch verwitterten Muttergesteins, der dem festen Fels aufsitzt. Andere Profile besitzen einen morphologisch nicht differenzierten A-Horizont, der auf einer angewitterten zersprungenen, sonst jedoch festen Felsunterlage (C-Horizont) liegt. Normalerweise zeigt das Profil des braunen Rankers folgenden Aufbau:

Horizont	Tiefe (cm)	Charakteristik
A_0	0 - 3	dunkelgrau bis graubraun, humos, schluffig,
A_1	3 - 10 (20)	grauockerbraun, schluffig-sandig, ziemlich skelettreich, oft filzartig durchwurzelt,
C	unter 10 (20)	zersprungener, angewitterter Gneis.

Theoretisch sollte dem Stadium des Protorankers ein Stadium des grauen Rankers folgen. Das Fehlen dieses Stadiums ist in unserem Falle dadurch bedingt, daß das feste Muttergestein schon im Laufe früherer geologischer Perioden ziemlich tief angewittert wurde, und deshalb die Eisen-Mineralien an der Oberfläche eine dünne Schicht von Eisen^{III}-Verwitterungsprodukten bilden, die beim Lockerwerden der Mineralkörner und ihrer Vermischung mit Humus, diesem einen braunen Farbton geben. Dieser Prozeß ist wahrscheinlich auch durch die hohe Azidität des Bodens bedingt.

Die chemische Verwitterung des Mikroskelettes des braunen Rankers ist nicht allzusehr vorangeschritten, und es bilden sich keine organmineralischen Komplexe, die eine kittenden Wirkung ausüben würden, was die staubartige Textur und Einzelkornstruktur des Bodens anzeigt. Der Antritt des braunen Rankers wird von einer Reihe chemischer Umwandlungen des Bodens begleitet (Fig.4).
Vor allem sinkt der Humusgehalt, auch im A_0-Horizont, der dem nichtdifferenzierten A_0-Horizont des Protorankers enstspricht. Dies ist auch vom Sinken der Sorptionskapazität begleitet, besonders im A_1-Horizont. Der Gehalt an Austausch-Kalzium erhöht sich im A_0-Horizont gegenüber dem Protoranker sogar auf das Zweifache. Diese Erhöhung erfolgt zum Nachteil von Austausch-Aluminium und -Wasserstoff. Ein höherer Gehalt an Austausch-Wasserstoff findet sich nur im A_0-Horizont.
Infolge der Lebenstätigkeit des Polytricho-Scleranthetum vertieft sich allmählich das Bodenprofil, weiter wird der Standort milder, was eine Invasion von ausläufertreibenden und lockerhorstigen Gräsern höheren Wuchses zur Folge hat. Diese Gräser beherrschen sodann den Bestand und bilden eine geschlossene Schicht; sie zerstören das mosaikartige Gefüge der Pflanzengesellschaft und unterdrücken die Moosschicht. Das Polytricho-Scleranthetum- Brauner-Ranker-Stadium wird auf diese Weise durch das nächste Stadium ersetzt.

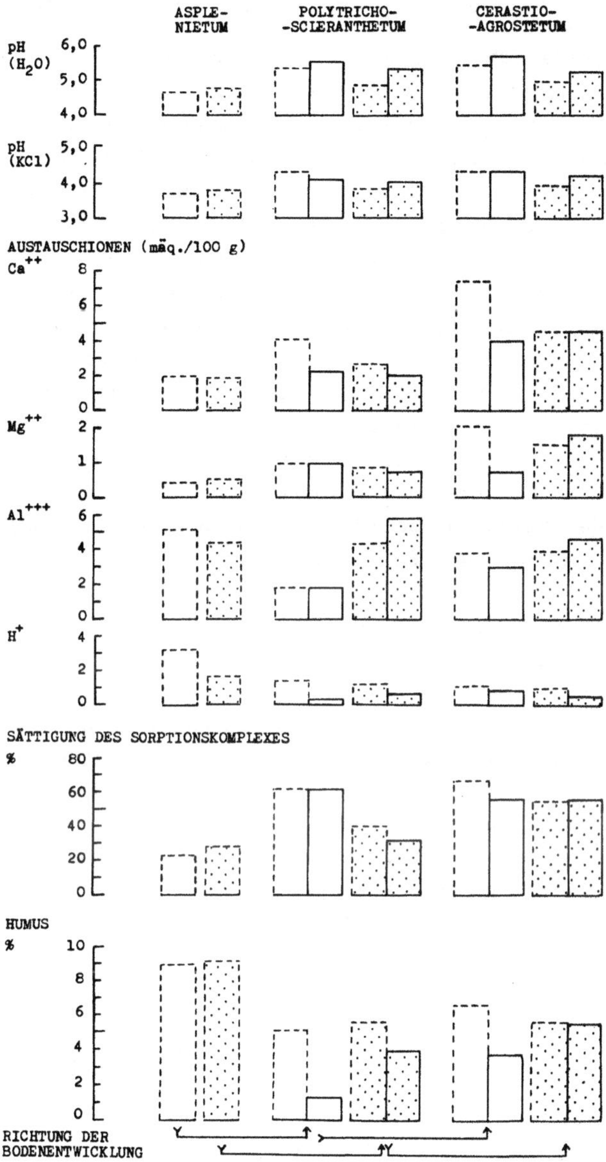

Fig.4. Einige chemische Eigenschaften der Rhizosphäre einzelner aneinander anknüpfender Sukzessionsstadien von zwei Lokalitäten: (1) S-Hänge über dem linken Otava-Ufer, w vom Dorf Horní Poríčí, ca. 420 m ü.d.M., Gneis (nicht punktiert); (2) SW-Hänge über dem Bach Rojický potok zwischen den Dörfern Velká Turná und Malá Turná, ca. 450 m ü.d.M., Gneis (punktiert).

Das Cerastio-Agrostietum pusillae-Ranker-Braunerde-Stadium

Die Pflanzengesellschaft dieses Stadiums bildet das Cerastio-arvensis-Agrostietum pusillae Moravec 1967 (Sedo-Scleranthetea). Diese Assoziation ist etwas artenärmer als das Polytricho-Scleranthetum, was sich besonders in der Unterdrückung der Moosschicht zeigt und nicht nur die Verringerung der Artenzahl (2-6), sondern vor allem des Deckungsgrades (1-10%) zur Folge hat (s.Abb.1). Aber auch in der Krautschicht sinkt trotz ihres hohen Deckungsgrades (75-90%) ein wenig die Artenzahl (27-35) (s.Abb.1). Zum Vorherrschen gelangt *Agrostis pusilla (= A.coarctata)*, die zusammen mit *Poa angustifolia* und *Festuca rubra* die obere Etage bilden (Fig.2). In der unteren Etage dominiert *Thymus pulegioides*. In der Moosschicht erreicht nur *Thuidium abietinum* eine höhere Dominanz; Flechten *(Cladonien)* treten nur sporadisch auf.

Der Wechsel der Pflanzengesellschaften wird durch die im Boden entstandenen Umwandlungen unterstützt. Das Bodenprofil vertieft sich von 10-20 auf 30-50 cm und differenziert sich weiter; es wird fähig, einen höheren Wassergehalt zu halten, es verringert die Schwankung der Bodenfeuchtigkeit und damit zugleich auch die der Bodentemperatur. Dies äußert sich in dem fast vollen Schluß der Krautschicht und durch Zurücktreten sukkulenter Arten.

Der Boden des Cerastio-Agrostietum ist nicht einheitlich und man kann zwei schwer unterscheidbare Entwicklungsstufen feststellen. Die Profile weisen meistens sehr undeutlich diffenzierte, dunkler oder heller ockerbraune Horizonte auf,was der Farbe angewitterter Silikatgesteine entspricht. Das Bodenprofil zeigt gewöhnlich folgenden Aufbau (Fig.3 u.Abb.4):

Horizont	Mächtigkeit (cm)	Charakteristik
A	0-14 (25)	Graubraun bis dunkel-ocker, meist feinkrümelig, sandig-schluffig,stark durchwurzelt, mit geringer Skelett-Beimischung,
A/C bzw.(B)	14 (25) -30 (37)	heller ocker-braun, mit vielen kleinen Skelettanteilen vermischt, sandig; mit der Tiefe nimmt das Skelett zu,
C	<30 (37)	hell-ockerbraun, sandig; Feinerde füllt die Lücken des angewitterten Gneises oder Granits.

Die Bestimmung des hellockerfarbenen Horizontes in der Tiefe von 14-30 (37) cm ist ziemlich schwierig und erfordert noch weitere spezielle Untersuchungen. In einigen Profilen ist dieser Horizont vom humusreichen A-Horizont teils durch seine hellere Farbe,teils durch einen andersartigen orange-ockerfarbenen Farbton deutlicher abgegrenzt. In einem solchen Fall kann man diesen Horizont als (B)-Horizont betrachten. In anderen Profilen ist der erwähnte Horizont jedoch vom A-Horizont nicht deutlich abgegrenzt, von dem er sich durch eine etwas hellere Farbe und einen größeren Skelettgehalt unterscheidet. In einem solchen Fall kann man diesen Horizont nur als A/C-Übergangshorizont ansehen. Die ockerbraune Farbe des Muttergesteins erschwert die korrekte Unterscheidung des (B)-Horizontes. Der AC-Boden gehört noch zum Ranker und zwar zu dem der Braunerde am nächsten stehenden Subtyp Braunerde-Ranker (MÜCKENHAUSEN 1962, p.54), der das letzte Stadium des Rankers in der Entwicklung der Braunerde darstellt. Der A(B)C-Boden gehört bereits zur Braunerde und zwar zum Subtyp Ranker-Braunerde (MÜCKENHAUSEN 1962, p.54 u.78), der in der weiteren Entwicklung zur mitteleuropäischen Braunerde dem Ranker folgt.

Die vorgeschrittene Bodenentwicklung beider angeführter Entwicklungsstufen äußert sich durch Bildung einer feinkrümeligen Struktur im A-Horizont. Der Gehalt an Austausch-Kalzium erhöht sich weiter (Fig.4) und man kann sagen, daß im Verlauf der Bodenentwicklung unter azidophilen Trockenrasen der Boden mit Austausch-Kalzium angereichert wird. Die Erhöhung des Austausch-Kalziumgehaltes wird jedoch durch eine Erhöhung des Austausch-Aluminiumgehaltes kompensiert, so daß die Böden des Polytricho-Scleranthetum und des Cerastio-Agrostietum in der Basensättigung und der Bodenreaktion fast übereinstimmen.

ZUSAMMENFASSUNG

Die Sukzession und Bodenentwicklung der auf trockenen gegen Süden exponierten Silikathängen des südwestböhmischen Hügellandes verläuft über folgende Stadien: Asplenietum septentrionalis- Protoranker-St., Polytricho-Scleranthetum- Brauner Ranker St., Cerastio-Agrostietum pusillae- Ranker-Braunerde-St. In ungestörten Verhältnissen wäre dieser Prozeß durch einen Quercion robori-petracae- Wald auf der reifen oligotrophen mitteleuropäischen Braunerde abgeschlossen. Die beschriebene Sukzession ist nicht in allen Stadien natürlich. Sie wird besonders in vorgeschritteneren Stadien vom Menschen in der Weise modifiziert, daß eine Invasion von Bäumen und Sträuchern durch extensive Beweidung verhindert wird. Das Asplenietum septentrionalis ist eine natürliche, primäre Pflanzengesellschaft. Für das Polytricho-Scleranthetum kann dies nicht mehr so eindeutig bestätigt werden. Sein Boden ist zwar nicht genügend tiefgründig, um einen Wald wachsen zu lassen, man kann jedoch nicht ausschließen, daß er das Wachsen einiger Sträucher oder auch vereinzelter kümmernder Bäume (*Quercus robur* oder *Pinus silvestris*) ermöglichen könnte. Die Böden des Cerastio-Agrostietum pusillae erlauben jedoch bereits das Bestehen eines Waldes, wenn auch mit niedrigerem Wuchs. Deshalb muß diese Pflanzengesellschaft als eine sekundäre Ersatzgesellschaft betrachtet werden. Die Bodenentwicklung würde anthropogen unbeeinflußt die gleichen Entwicklungsstadien durchlaufen und ist mit der oligotrophen mitteleuropäischen Braunerde (als Klimax-Bodentyp) abgeschlossen.

ANHANG

Viele Autoren werten verschiedene an einer bestimmten Stelle verlaufende Änderungen der Vegetation als Sukzession. Dadurch entsteht eine Vermischung und Verschleierung verschiedener Prozesse, die ganz verschiedene Ursachen und einen unterschiedlichen Charakter haben. Nach den Ursachen der hervorgerufenen Änderungen der Phytozönosen (bzw. Biozönosen) kann man folgende Gliederung vorschlagen:

A. Änderungen durch Faktoren die auf die Biozönose von außen wirken und die von der Lebenstätigkeit der Biozönose unabhängig sind - dies sind adaptive Änderungen (die Biozönose paßt sich den veränderten Umweltbedingungen passiv an).

1. Änderungen der Klimafaktoren - deren Folge eine Migration der Biozönosen (Symmigration) ist, entspricht den "regional successions" COWLES 1911).

2. Änderungen der edaphischen Faktoren durch Erosion und Sedimentation - diese Änderungen stören die Sukzession im engeren Sinne (entsprechen den "topographic successions" COWLES 1911).

3. Änderungen der Biozönosen selbst durch Eingriffe von außen - z.B. durch menschliche Eingriffe, spontaner Brand u.a. (Auch diese Änderungen stören die Sukzession im engeren Sinne).

B. Änderungen, hervorgerufen durch die Lebenstätigkeit der Biozönosen - S u k z e s s i o n im engeren Sinne (im Sinne CLEMENTS 1907).
Zu dieser Kategorie gehört der Sukzessions- und Bodenentwicklungsprozeß, und nur für diesen empfehle ich, den Ausdruck Sukzession zu verwenden.
1. progressive (primäre) Sukzession (beginnt auf totem Gestein oder im Wasser)
   a) natürliche progressive Sukzession - wo alle Sukzessionsstadien von natürlichen Biozönosen gebildet werden,
   b) anthropogen modifizierte progressive Sukzession - wo vorgeschrittenere Sukzessionsstadien durch anthropogen bedingte Ersatzbiozönosen gebildet werden, die jedoch die Funktion natürlicher Biozönosen übernehmen und eine Fortsetzung der Bodenentwicklung hervorrufen.
2. R e g e n e r a t i o n (regenerative Sukzession - sekundäre Sukzession im Sinne CLEMENTS) beginnt auf Böden bzw. auf Bodentorsen vorgeschrittenerer Sukzessionsstadien und regeneriert die durch einen äußeren Eingriff gestörte Sukzessionsserie.
3. D e g r a d a t i o n (degradative Sukzession) - ein Prozeß, bei dem die Sukzession durch anthropogen ungünstig beeinflußte Ersatzbiozönosen verläuft und eine weitere standortsverschlechternde Bodenentwicklung verursacht (z.B. die Verheidung und dadurch verursachte Ortstein-Bildung).

Die gesamte Richtung der Sukzession und der Bodenentwicklung und ihr Endstadium (Klimax) werden vom Makroklima bestimmt. Die Schnelligkeit dieses Prozesses und der spezifische Verlauf der einzelnen Serien werden von dem Relief der Landschaft und den Eigenschaften des Muttergesteins bestimmt. Die Sukzession und die Bodenentwicklung unter ungestörten natürlichen Bedingungen sind irreversibel. An Orten, die weder vom Grund- noch vom Überschwemmungswasser oder durch lokale klimatische Einflüsse beeinflußt werden, enden alle Sukzessionsserien eines klimatisch einheitlichen Gebietes in einem gleichen Biozönose-Typ - einer Klimax, die sich durch die Lebenstätigkeit der Biozönose nicht weiter ändern läßt.

SUMMARY

On southern slopes built of silicate rocks in SW Bohemia succession and soil development passes through the following stages: the Asplenietum septentrionalis - protoranker-st., the Polytricho-Scleranthetum-brown ranker-st., the Cerastio-Agrostietum pusillae-ranker-braunerde-st. Under undisturbed conditions this process would terminate in a Quercion robori-petraeae forest on mature oligotrophic Central European braunerde. Not all stages of this successional series are natural. The more advanced stages are modified by human interference so that the invasion of shrubs and trees is hindered by extensive grazing. The Asplenietum septentrionalis is a natural primary plant community. This cannot be said so definitely of the Polytricho-Scleranthetum; its soil is not deep enough for growing wood, however, it must be admitted that spread shrubs or even individual crippled trees *(Quercus robur* or *Pinus silvestris)* could grow here. The soil of the Cerastio-Agrostietum pusillae makes the growth of a low wood possible; this plant community must be therefore considered a secondary substitution plant community. Soil development without human interference can be supposed to pass through the same developmental stages terminating in the oligotrophic Central-European braunerde as soil climax.

APPENDIX

Many authors consider all gradual changes of vegetation on the same spot as succession. This means that processes of quite different causes and different nature are placed together and thus obscured. The following classification can be proposed, based on the causes evoking directional changes of phytocenoses (biocenoses) (see also MORAVEC 1969):

A. Changes induced by external factors independent on the life activity of the biocenosis (biocenoses adapt passively to changing environmental factors) - these are adaptive changes.
   1. Changes of climate - induce the migration of biocenoses - synmigration - (they correspond to the "regional successions" of COWLES 1911).
   2. Changes of edaphic factors (by erosion or sedimentation, change of water regime etc.) (they correspond to the "to-

pographic succession" of COWLES 1911).

3. Changes of the biocenoses themselves by direct external intervention (e.g. by human intervention).

B. Changes evoked by the life activity of the biocenoses - s u c c e s s i o n sensu stricto (in the sense of CLEMENTS 1907).

1. Primary succession (initiating on dead rock or in water)
   a) natural primary succession - with stages formed by natural biocenoses,
   b) anthropogenically modified primary succession - with stages partly formed by substitution biocenoses.
2. Regeneration (regenerative succession, secondary succession in the sense of CLEMENTS) - is a succession starting on stripped soils of more advanced successional stages and regenerating the biocenosis destroyed by external influence.
3. Degradation (degradative succession) - is a process where succession passes through anthropogenically unfavourably influenced substitution biocenoses and evokes further soil development deteriorating the properties of the climax soil (e.g. the formation of heath vegetation and the consequent formation of iron pan ("Ortstein") in the soil).

The general direction of succession and soil development and their terminal stage (the climax) are determined by macroclimate (except for parent material of extreme chemical and physical properties). The rate of this process and the specific course of individual series are determined by relief and by the properties of the parent material. In undisturbed natural conditions succession and soil development are irreversible. In territory of the same macroclimate all successional series on sites influenced neither by ground water nor by floods nor by local climate terminate in the same type of biocenosis - the climax that cannot be further changed by the life activity of the biocenosis.

LITERATUR

CLEMENTS,F.E. -1907- Plant physiology and ecology.- London.
COWLES,H.C. -1911- The causes of vegetation cycles.- Bot.Gaz.51: 161-183. Chicago.
KERNER,A. -1863- Das Pflanzenleben der Donauländer.- Innsbruck.
KUBIËNA,W.L. -1948- Entwicklungslehre des Bodens.- Wien.
-- -- -1953- Bestimmungsbuch und Systematik der Böden Europas.- Stuttgart.
MORAVEC,J. -1967- Zu den azidophilen Trockenrasengesellschaften Südwestböhmens mit Bemerkungen zur Syntaxonomie der Klasse Sedo-Scleranthetea.- Fol.geobot.et phytotax.2: 137-178. Praha.
MÜCKENHAUSEN,E. -1962- Entstehung, Eigenschaften und Systematik der Böden der Bundesrepublik Deutschland.- Frankfurt/M.
SCHWICKERATH,M. -1944- Das Hohe Venn und seine Randgebiete.- Pflanzensoz.6. Jena.

DISKUSSION

J.TÜXEN

Man sollte zweifellos nicht unnötig alte, eingebürgerte Begriffe ändern, es sei denn, sie könnten durch etwas Besseres ersetzt werden. Der Wert eines Begriffes steigt bekanntlich mit seinem Inhalt, seiner Aussagefähigkeit. Wir praktizieren so etwas in der Systematik seit langem. Die im Grund verschiedenen Vorgänge der Sukzession und Substitution sollten auseinandergehalten werden, auch wenn sie im konkreten Fall beide nebeneinander ablaufen, damit der Vorgang besser verständlich wird oder, wie Herr MORAVEC gesagt hat, eine Verminderung und Verkleinerung verschiedener Prozesse vermieden wird. Über die Bezeichnung der Vorgänge, die von mir als Substitution bezeichnet werden, müssen wir uns einigen.

H.ELLENBERG

Ich möchte zu dieser letzten, von Herrn JES TÜXEN angeschnittenen Frage nur sagen, daß die begriffliche Trennung bei dieser allogenen und autogenen Sukzession in jedem Falle vorliegt. Ob man einen anderen Ausdruck dafür unbedingt braucht oder nicht, müßte eigentlich die Praxis der Verwendung entscheiden. Beide Ausdrücke, die vorgeschlagen wurden, scheinen mir nicht glücklicher als die bisherigen. Auch der Ausdruck "adaptive Änderung" könnte mißverstanden werden, wenn er für sich allein verwendet wird.

Ich möchte aber auf das Referat von Herrn MORAVEC zurückkommen, weil mir da grundsätzliche Bedenken kommen. Die anschauliche und konkrete Darstellung der Bodenentwicklung oder des Vergleichs verschiedener Bodenzustände und der dazugehörigen Vegetation zeigte, daß es sich in dem Falle der Asplenium-Gesellschaft um eine Felsspaltengesellschaft handelt, die man mit BRAUN-BLANQUET seit langem als Dauergesellschaft betrachtet, die also keine weitere Entwicklung hat. Und auch als er von dem Polytricho-Scleranthetum sagte, daß infolge der Lebenstätigkeit dieser Gesellschaft das Bodenprofil sich vertiefe, konnte ich nicht mitmachen. Denn ich kann mir nicht vorstellen, wie eine Gesellschaft aus *Polytrichum* und *Scleranthus*, auch wenn man die Tiere, die darin leben, einbezieht, ein mehrere Dezimeter tiefes Bodenprofil aus einem steinigen Boden machen kann. Die Zweifel, ob es sich hier überhaupt um eine Bodenentwicklung handelt, wurden dann vollends verstärkt, und eigentlich bestätigt durch die letzten chemischen Bemerkungen. Denn es kam daraus klar zum Ausdruck, daß der Gehalt an austauschbarem Ca steigt, nicht fällt, und daß die Basensättigung steigt, nicht fällt, der Rohhumusgehalt sinkt, nicht steigt; und das widerspricht den Vorstellungen, die wir über Bodenbildung im humiden Klima haben. Ich möchte deshalb annehmen, daß es sich hier um eine typische Zonierung auf mehr oder wenigen felsigen Partien handelt. Andere tiefgründigere und auch sehr viel tiefgründige Partien bilden an diesen Hängen ein Mosaik, das wie die tiefe Granitverwitterung aus der langen Geschichte dieser Gegend zu erklären ist, insbesondere aus der diluvialen Geschichte, wo Bodenfließen und andere Vorgänge stattfanden. Die heutige Vegetation hat fertige, oder fast fertige, verschiedenartige Bodenzustände vorgefunden, die sie nur besiedelt hat. Wahrscheinlich ist der höhere Ca-Gehalt und die höhere Basensättigung gerade durch solche Fließerden bedingt, vielleicht sogar durch Anwehung von Material, also durch lößähnliche Bildungen.

J.MORAVEC

Es gibt keine Merkmale für Fließerden in diesen Profilen, es sind wirklich gewachsene Bodenprofile. Es gibt auch keine Erscheinungen von einem durch Wind gebrachten Bodenmaterial. Die Beispiele der Ca-Anreicherung liegen ganz dicht nebeneinander, einige Meter voneinander entfernt, und der ganze Hang ist von dem gleichen Gneis gebildet. Die Idee, daß der Humusgehalt während der Sukzession steigen muß, ist, glaube ich, nicht ganz richtig. Das Endstadium, die oligotrophe Braunerde, hat einen ziemlich flachgründigen A_0-Horizont, wo also die organische Substanz konzentriert ist, und dann ein ziemlich tiefes Profil fast aus Mineralsubstanz zusammengesetzt. Aber die vorherigen Stadien haben einen ziemlich tief reichenden Humushorizont, der weiter in der Klimaxgesellschaft zu einem 5 cm dicken Horizont verengt ist.

Das Diagramm des Humus ist nicht ganz richtig: es ist der relative Gehalt in Prozenz angegeben und nicht der Gesamtgehalt im Bodenprofil. Im Asplenietum ist der Gehalt an Humus auf eine 5 cm mächtige Feinerdeschicht beschränkt, im Polytricho-Scleranthetum schon auf 10-15 cm; man kann also sagen, daß der Gesamthumus steigt, prozentual aber sinkt.
Die Steigerung des Ca-Gehaltes habe ich schon öfter unter solchen xerophilen Vegetationen gefunden. Das sind eigentlich steppenartige Gesellschaften, keine Gesellschaften, die den Nardetalia angehören. Der Ca-Gehalt steigt immer allmählich, natürlich nicht sehr stark. Etwas ähnliches kann auch in Waldgesellschaften beobachtet werden, wo der Gehalt an Austauschbasen mit der Artenzahl steigt und dann wieder sinkt. In dem Stadium, das unter natürlichen Bedingungen dem Cerastio-Agrostietum entsprechen würde, würde der Gehalt wieder sinken, und in der Klimaxgesellschaft des Quercion robori -petraeae könnte er wieder niedriger sein.

R.CARBIENER
Aus den Bildern ist zu erkennen, daß das geomorphologische Ausgangsmaterial durchaus nicht einheitlich war. Man sieht, daß die Felsen dort, wo *Asplenium septentrion.* wächst, anders geartet sind als die sehr tief verwitterten Stellen, Das Ausgangsmaterial ist petrographisch vielleicht einheitlich, geomorphologisch aber nicht.

R.TÜXEN
Ich möchte auf eine merkwürdige Parallele hinweisen. Denken Sie an unser holländisch-nordwestdeutsches- nordostdeutsches Flachland, bis Polen und noch weiter, wo wir von Flugsand gebildete Rohböden finden. Dort ist der Erstbesiedler *Corynopherus*, und dann kommt *Polytrichum piliferum*, und dann *Agrostis arida*, alles genau wie bei Ihnen, und dann *Festuca ovina* und Begleiter. Vielleicht ist das letzte auch eine sekundäre, bzw. anthropogen geleitete, durch Schafweide beeinflußte Sukzession zum Rasen. Das Corynephoretum und auch der *Agrostis canina-arida*-Rasen können eine Flechten-Subassoziation haben mit *Cladonia mitis*, mit *Cornicularia aculeata*, mit *Cladonia furcata* und noch einigen anderen. Was mich höchst überrascht hat, war, daß diese flechten-

reiche Gesellschaft von Herrn MORAVEC nachher in einen flechtenarmen Rasen übergeht. Ich habe immer gemeint, bei uns wäre das ein Endstadium, und in diesem Punkte stimme ich mit ELLENBERG überein, daß von diesen flechtenreichen, sehr sauren und sehr armen eine Anreicherung ohne anthropogene Hilfe zu den geschlossenen Rasen bei uns schwer vorstellbar ist.
Noch eine kurze Bemerkung: Das erste Stadium mit *Asplenium septentrionale* und *Festuca* scheinen mir 2 Gesellschaften zu sein, eine Spaltengesellschaft und eine beginnende Initialphase der Rasen, die auf den kleinen Absätzen wächst.

D.MÜLLER-DOMBOIS
Selbst wenn man annimmt, daß das geologische Ausgangsmaterial ursprünglich gleicher Natur war, könnte man diese verschiedenen Stadien eigentlich nur in eine Sukzession, in eine Zeitreihe oder chronologische Folge bringen, wenn auch die einzelnen Standorte geologisch zeitlich unterschieden sind, und zwar eine solche mit dem tiefen Boden als älteste, und die mit dem flachen oder steinigen Boden als jüngste. Könnte man dafür irgendwelche Gründe annehmen? Ich nehme an, daß das nicht der Fall ist, sondern daß sich die Standorte durch topographische Unterschiede herausgebildet haben, also durch Unterschiede im Mikroklima.Dann ist es natürlich nur eine ökologische Reihe und keine zeitliche. Das würde m.E. nicht dazu berechtigen, eine Sukzessionsreihe aufzustellen, da sich der zeitliche Unterschied der einzelnen Entwicklungsgänge durchaus floristisch durch Verschiedenheiten auszeichnen kann.

H.MERKER
Was das Ausgangsgestein betrifft, ist zu bedenken, daß der Gneis eine ganz verschiedene Verklüftung aufweisen kann. Wir haben Gneise, z.B. in Schweden, die horizontal zerklüftet sind, die horizontal abbanken und demnach verwittern, während die Profile, die Herr MORAVEC gezeigt hat, immer schräg gerichtete Verklüftungen aufweisen, und da kann man sich sehr wohl denken, daß dann die Verwitterung und die Bodenbildung eine ganz andere ist als in den horizontal abplattenden Gneisen.
Was den Basen-Säure-Status betrifft, ist es ohne weiteres möglich, daß auch ein saures Gestein wie der Gneis je nach der Jahreszeit eine sehr große Schwankung in der Basen- und Säureführung aufweist. Wir haben es schon in einem früheren Symposium einmal

berührt, daß das pH durchaus nicht konstant ist, sondern während der Jahreszeit stark schwankt. Hätte Herr MORAVEC die Messung statt im Herbst im Juli oder Anfang August gemacht, hätte er wahrscheinlich noch viel höhere Schwankungen erreicht.
Was nun das Profil des Gesteins selbst betrifft, diese tiefe Ausfällung der Sesquioxyde, so handelt es sich ja in diesem Falle um ein Gestein, das niemals der Glaziation ausgesetzt gewesen ist, während wir doch, die Skandinavier, die Norddeutschen und teilweise die Holländer, aus dem Moränengebiet kommen, und für uns wirkt wohl so ein Bild sehr frappierend. Es ist aber in Wirklichkeit doch sehr leicht zu erklären, da ja diese Gesteine nun seit dem Tertiär mehr oder weniger ungestört verwittern konnten.

H.ELLENBERG

Zwar ist das Eis über diese Böden nicht hinweggegangen, wohl aber das Klima der Eiszeit, und das bedeutet, daß in dieser Zeit Solifluktion stattgefunden hat und damit eine örtliche Anreicherung. Ich möchte das Argument von MÜLLER-DOMBOIS unbedingt unterstützen. Es ist ja keinerlei Beweis gegeben, daß die als verschieden alt angegebenen Stadien tatsächlich verschieden alt sind. Sie sind nebeneinander und sind gleich alt, am gleichen Platz, sie sind topographisch herausgearbeitet.
Ich möchte noch zu der grundsätzlichen Frage Stellung nehmen,die JES TÜXEN angeschnitten hat. Es ist die Frage, ob ein neuer Begriff für die allogene Sukzession möglich ist oder nicht. Die Vorschläge von Herrn MORAVEC und die von Herrn TÜXEN sprechen beide für diese Richtung. Im Grunde auch mein Vorschlag. Meines Erachtens wird zu wenig unterschieden zwischen diesen exogenen und endogenen, oder allogenen und autogenen Vorgängen, d.h.zwischen der echten Entwicklung der Vegetation, die nur durch die Vegetation und ihre Tätigkeit ausgelöst wird, und Vorgängen,die durch Außenvorgänge entweder ausschließlich oder doch teilweise veranlaßt sind.
Was die Benennung betrifft, so würde ich es für unzweckmäßig halten, einen neuen Begriff international einzuführen. Man sollte bei der Unterscheidung autogen und allogen bleiben. Vielleicht könnte man aber im deutschen Sprachgebrauch den Unterschied deutlicher machen; denn in Deutsch gibt es keinen Begriff, der mit Sukzession wirklich synonym wäre. Und da könnte man die Sukzession in engem Sinne, die autogene Sukzession, als "Entwicklung" übersetzen, und die Substitution im Sinne von JES TÜXEN

als "Standorts- und Vegetationsüberlagerung", wie das vor einigen Jahren MOOR vorgeschlagen hat.

J.SCHMITHÜSEN

Das beste und solideste Argument, m.E. das ausschlaggebende,ist folgendes: Wir können aus der Beobachtung erkennen, daß sich irgendwo eine Sukzession abspielt, d.h. wir können das, was wir beobachten, als eine Veränderung der Vegetation, im alten Sinne eine Sukzession nennen. Wir wissen aber am Anfang niemals, ob es eine autochthone oder allochthone, oder eine autogene oder allogene, oder eine endogene oder exogene ist. Das müssen wir erst erforschen. Und wir könnten von den Dingen gar nicht reden,wenn wir den Begriff Sukzession so aufteilen; dann brauchten wir ja wieder ein neues Wort für das Allgemeine, einfach dafür, daß eine Entwicklung da ist, die sich in gewissen Stadien vollzieht.

J.BARKMAN

Herr Professor ELLENBERG schlägt da eine Art Kompromiß vor,indem man in der deutschen Sprache von Entwicklung und von Standortsüberlagerung spricht. Wir müssen uns doch ganz klar darüber sein, daß 90% der wissenschaftlichen Literatur, die wir durcharbeiten müssen, in anderen Sprachen geschrieben ist. In wissenschaftlichen Publikationen soll man sich doch so viel wie nur möglich wissenschaftlicher internationaler Begriffe bedienen. Wenn die Deutschen unter sich diese Termina verabreden würden, würde das zu großer Verwirrung führen. Die Worte "Entwicklung" und "Überlagerung" würden den Ausländern gar nicht einleuchten, und wir würden uns etwas ganz anderes darunter vorstellen. CLEMENTS hat damals gesagt: "Ein Klimax ist ein Organismus, der sich entwickelt". Wir müssen von dieser dogmatischen Theorie abkommen; das sind wir ja eigentlich alle schon. Wenn wir jetzt den Begriff "Entwicklung" wieder einführen, dann hat man wieder so eine Analogie zu der Entwicklung eines Organismus. Ich würde "Entwicklung", wenn man es unbedingt will, nur für die Entwicklung einer Pioniergesellschaft verwenden. Dann wird sie von einer anderen Gesellschaft abgelöst.Das ist keine Entwicklung mehr, auch nicht im deutschen Sprachgebrauch. Sogar die autogene, primäre natürliche Sukzession zeigt in sich eine Entwicklung von Stadien und Verdrängungen von einem Stadium zum anderen.Nur im dogmatischen Sinne CLEMENTS' ist das Ganze eine Entwicklung vom nackten Fels bis zum Klimaxwald.

E.BURRICHTER
Ist es möglich, daß eine Beeinflussung durch Schnecken vorliegt? Vielfach sind xerophile Rasengesellschaften sehr von Schnecken heimgesucht, was oftmals mit dem Ca-Gehalt des Bodens im Einklang steht.

J.MORAVEC
Zu der Ca-Anreicherung möchte ich sagen: Gegen Anreicherung aus einem anderen Ort, durch Wind oder Ähnliches spricht, daß sich der Gehalt an Austausch-Magnesium nicht ändert, der bleibt bei diesen beiden Böden derselbe in der Tiefe von 10 cm. In der Tiefe von 1-5 cm steigt hier natürlich der Mg-Gehalt, aber das ist auch durch Steigung der organischen Substanz verursacht; denn die organische Substanz gibt mehr Mg, weil dies im Chlorophyll vorhanden ist. Auch der Sättigungsgrad des Bodens ist fast gleich. Was die Schnecken betrifft: Ich habe dort keine Schnecken gesehen. Im Boden können die Schalen nicht erhalten bleiben, der ist zu sauer. Auch im Bestand habe ich keine beschalten Mollusken gesehen.

NACHTRAG (September 1978)
Die Sukzession und Bodenentwicklung vom festen Felsen zum reifen Boden kann wegen zwei großer Schwierigkeiten kaum direkt verfolgt werden: (1) diese Prozesse verlaufen so langsam, daß sie einer direkten Untersuchung unzugänglich sind, (2) es ist bisher unmöglich, das Alter der Bodenprofile bzw. einzelner Bodenhorizonte (ausgenommen der Humushorizonte) zu bestimmen. Deshalb ist man auf eine Rekonstruktion der Sukzession und Bodenentwicklung auf Grund nebeneinander vorkommender Phytozönosen und Böden angewiesen; diese Rekonstruktion bleibt jedoch nur eine Hypothese. Wie gefährlich diese Rekonstruktion sein kann, zeigt dieser Beitrag und besonders die Diskussion darüber.
Man kann annehmen, daß auf den untersuchten Silikathängen Verwitterung und Bodenbildung entsprechend dem Klima in der Richtung zur Braunerde verläuft. Das angeführte Schema der Sukzession und Bodenentwicklung bezieht sich auf die heutige landwirtschaftlich beeinflußte Landschaft und zeigt ihre Richtung und anthropogen modifizierte Stadien. Es bedeutet nicht, daß eine jede Stelle des Cerastio-Agrostietum-Braunerdestadiums alle vorherigen Stadien durchgemacht hat, und daß es dazu sogar in der kurzen Zeitspanne nach der Besiedlung der Landschaft durch den Menschen kam. Dasselbe gilt für das Po-

l y t r i c h o - S c l e r a n t h e t u m - Braun-Ranker-Stadium. In der natürlichen progressiven Sukzession, die sich hier vor der Kolonisation (also fast während des ganzen Holozäns) abspielte, war statt des C e r a s t i o - A g r o s t i e t u m eine primäre Gehölz-reichere Pflanzengesellschaft anwesend. Auch das P o l y - t r i c h o - S c l e r a n t h e t u m war wahrscheinlich durch das Vorkommen von Straucharten beeinflußt, und wohl nur das A s p l e - n i e t u m s e p t e n t r i o n a l i s konnte dem heutigen gleich sein. Das heutige Mosaik der Pflanzengesellschaften entstand teils durch Sukzession, teils durch die Symmigration an Hängen mit verschieden tief angewitterten Felsen und wahrscheinlich auch mit verschieden tiefen Decken pleistozäner Verwitterungsprodukte. Es ist nicht ausgeschlossen, daß die Bodenentwicklung polyzyklisch sein könnte, d.h., daß sie schon während des (der) letzen Interglazials (le) in ähnlicher Richtung, mit einer Unterbrechung während des Glazials, verlaufen ist.

Der Analyse der mit den Veränderungen der Vegetation verknüpften Begriffe wurde eine selbständige Arbeit gewidmet: MORAVEC,J. -1969- Succession of plant communities and soil development.- Fol.geobot.et phytotax.4: 133-164. Praha.

Abb.1. Aplenietum septentrionalis.

Abb.2. Polytricho (piliferi) - Scleranthetum perennis.

Abb.3. Profil des braunen Rankers (1 m tief) unter dem Polytricho-Scleranthetum.

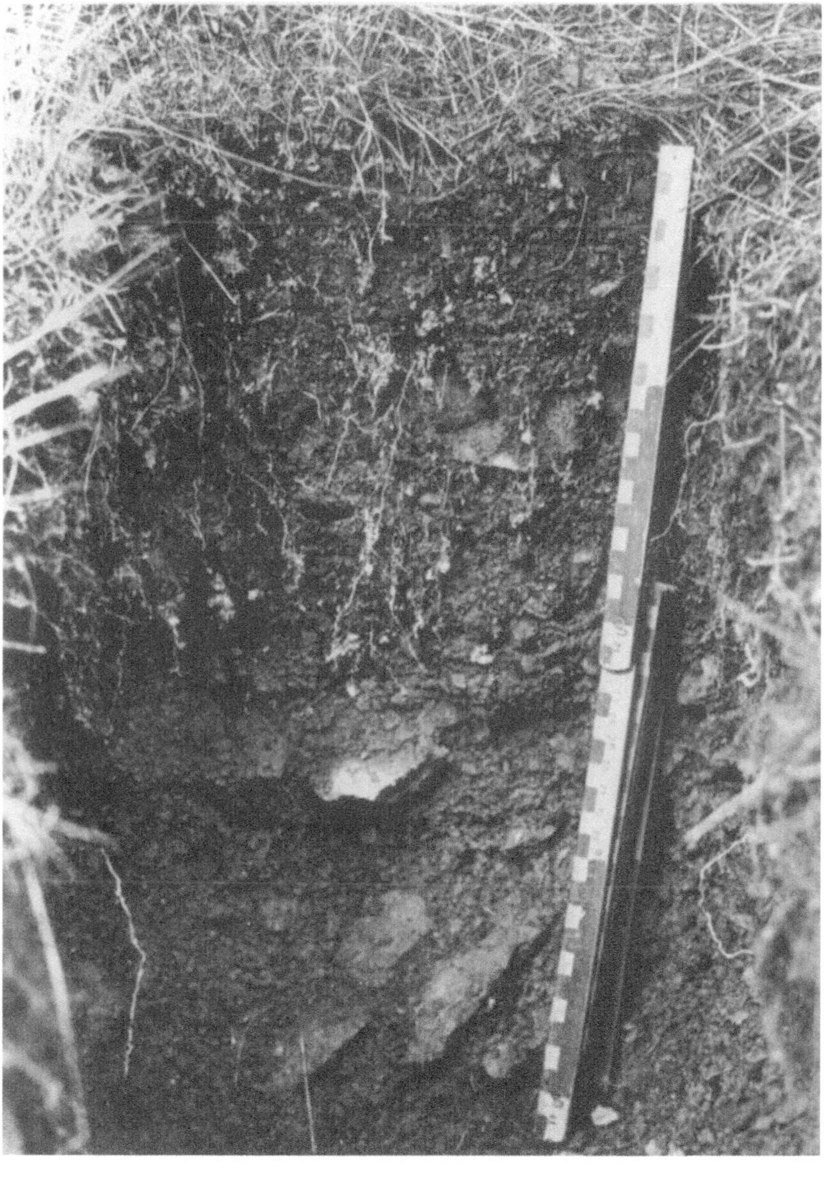

Abb.4.Profil der Ranker-Braunerde (40 cm tief).

BODEN- UND VEGETATIONSENTWICKLUNG AUF DEM HAUPTDOLOMIT IN DER ALPINEN RASENSTUFE DER ALLGÄUER UND LECHTALER ALPEN

Z. G r a č a n i n

Einleitung

Nachdem J.BRAUN-BLANQUET & H.JENNY (1926) die Boden- und Vegetationsentwicklung in der alpinen Stufe der Zentralalpen untersucht haben, und H.JENNY (1930) diese Ergebnisse auch in dem "Handbuch der Bodenlehre" von H.BLANCK dargestellt hat, galt in der Bodenkunde die Bodenentwicklung in der alpinen Stufe der Alpen lange Zeit als erforscht und wurde von bodenkundlicher Seite nur wenig behandelt. Die Allgäuer und die Lechtaler Alpen gehören aber nicht zu den Zentralalpen, sondern zu den nordalpinen Alpenketten, auf die sich die erwähnte Arbeit von J.BRAUN-BLANQUET & H.JENNY gar nicht bezog. J.BRAUN-BLANQUET & H.JENNY (1926 p.V) schrieben sogar ausdrücklich: "Eine vorsichtige Verallgemeinerung unserer Stufeneinteilung und Vegetationsbeschreibung ist also höchstens für die Gebirge des inneralpinen Föhrenbezirkes, niemals aber für die Außenketten der Nord- und Südalpen angängig. Dasselbe dürfte, wenn auch in abgeschwächtem Maß, für die Bodenbildung und Bodenverteilung gelten; doch ist man hierüber noch recht wenig unterrichtet".

In den Allgäuer Alpen wurden zwar die ersten Bodenuntersuchungen bereits 1854 von dem bayerischen Pflanzengeographen O.SENDTNER durchgeführt, seither ist aber in den mittleren und höheren Lagen der Allgäuer Alpen - außer nicht veröffentlichten Arbeiten der Reichsbodenschätzung - bodenkundlich nicht gearbeitet worden.

Die Lechtaler Alpen hat O.SENDTNER nicht besucht, und m.W. hat auch kein Bodenkundler aus ihnen etwas veröffentlicht. Dagegen hat H.ELLENBERG (1953) im Gebiet der Stuttgarter Hütte einige Bodenprofile untersucht und daraus bedeutende Schlußfolgerungen über die Boden- und Vegetationsentwicklung in der alpinen Stufe gezogen; eine Beschreibung der Böden selbst hat er aber nicht veröffentlicht.

Die Vegetation ist dagegen besser erforscht, wenn auch nicht nach einheitlichen Prinzipien. Für das Allgäu besteht eine ausführliche, obgleich nicht erschöpfende pflanzensoziologische Arbeit von E.OBERDORFER (1950), für den Vorarlberger Teil der Lechtaler Alpen dagegen nur eine kurze Beschreibung von H.GAMS (1961), die sich aber einer anderen Nomenklatur und Systematik bedient.

In diesen Gebieten untersuche ich seit 1964 die Bodenentwicklung auf verschiedenen Kalken und Dolomiten in verschiedenen Höhenstufen. Da die Vegetation immer ein wichtiger Indikator der Standortsverhältnisse ist, und erst recht im Gebirge am besten die stark wechselnden Standortsverhältnisse widerspiegelt, habe ich ihr natürlich große Aufmerksamkeit gewidmet. Dadurch erscheint es mir berechtigt,in diesem Kreise über meine Untersuchungen zu berichten. Im Rahmem der mir zur Verfügung stehenden Zeit sollen hier lediglich die Ergebnisse aus der alpinen Rasenstufe (= alpinen Stufe schlechthin), und zwar nur von dem Hauptdolomit kurz dargestellt werden.

Dies ist ein Teil der Untersuchungsergebnisse, die im Rahmen des Forschungsvorhabens "Bodenentwicklung auf Kalk und Dolomit im mittleren Abschnitt des Alpenmassivs" gewonnen wurden. Das Forschungsvorhaben kam auf Anregung von Herrn Prof. Dr.GANSSEN zustande und wurde von der *Deutschen Forschungsgemeinschaft* unterstützt, wofür ich ihnen auch an dieser Stelle meinen verbindlichsten Dank sagen möchte. Herrn Prof. Dr.GANSSEN verdanke ich auch die freundliche Durchsicht des Manuskriptes.Herzlich danke ich auch den Herren Prof. Dr. V.JACOBSHAGEN, Prof. Dr. R.HUCKRIEDE, Dipl.Geol. J.WESTRUP, Dipl.Geol. D.HELMCKE, Geologisch-Paläontologisches Institut der Universität Marburg, für die Auskünfte und Gewährung der Einsichtnahme in die nicht veröffentlichten Dissertationen und Diplomarbeiten; Herrn Prof. Dr.E.OBERDORFER und Herrn Dr.G.PHILIPPI, Landessammlungen für Naturkunde, Karlsruhe, für die Bestimmung der kritischen Pflanzen und Moose, sowie Frau R.GOEZE, Fräulein E.RÜHMEKORF und Frau B.RIESER, den technischen Assistentinnen des Instituts für die Durchführung der Untersuchungen im Labor.

Die alpine Rasenstufe

Im untersuchten Gebiet ist die alpine Rasenstufe an ihrer unteren Grenze durch Latschenbestände, an der oberen durch die subnivale Stufe begrenzt. Bei der Beurteilung der unteren Grenze der alpinen Stufe ist zu berücksichtigen, daß diese durch die menschliche Wirtschaft nach unten verlagert wurde. Wenn auch heute die Weidewirtschaft in den Alpen stark zurückgeht und die Vegetation sich vielfach wieder besser entwickeln kann als in den vergangenen Jahrhunderten, so konnten doch die Latschen ihren ganzen natürlichen Verbreitungsraum nicht zurückerobern. Vielmehr werden sie auch heute noch mancherorts vernichtet.
Die untere Grenze der alpinen Rasenstufe dürfte in den Allgäuer Alpen allgemein zwischen 1980 m (Nebelhorn) und 2050 m N.N. (Kanzelwand, Walser Gaißhorn) liegen. In den Lechtaler Alpen befindet sie sich in der Regel etwas höher und zwar zwischen

2100 und 2200 m.
Im Gebiet des Nebelhorn (2224 m, Bild 1) und auf den meisten Allgäuer Bergen wird die obere Grenze der alpinen Rasenstufe gar nicht oder gerade noch erreicht. Die kahlen Felsen täuschen zwar stellenweise schon die subnivale Stufe vor, doch bei näherer Betrachtung zeigt sich, daß der geringe Bewuchs auf die Steilheit des Geländes und nicht auf eine klimatische Höhengrenze zurückzuführen ist. So erscheint z.B. der Widderstein (2536 m NN)(Bild 2) von weitem bereits ab etwa 2130 m kahl, die nähere Betrachtung zeigt aber einen guten Bewuchs an geeigneten Stellen bis etwas über 2400 m (Bild 3) und darüber, in der subnivalen Stufe, reichen einzelne Vegetationspolster fast bis zum Gipfel. Das Walser Gaißhorn trägt bis zu seinem Gipfel 2366,4 m (Bild 4) eine nahezu geschlossene Rasendecke.
In den Lechtaler Alpen liegt die obere Grenze der alpinen Rasenstufe auf Hauptdolomit zwischen etwa 2400 m (auf dem Nordhang der Rüfi-Spitze 2632 m) und 2450-2500 m auf dem Südhang der Weißschrofen-Spitze. Auf den Substraten, die für die Bodenbildung günstiger sind als der Hauptdolomit, scheint allerdings diese Grenze etwas höher zu liegen, so z.B. im Gebiet der Rogg-Spitze (2747 m) bei etwa 2600 m.
Nach den Messungen der meteorologischen Station Nebelhorn (1929 m) ist in der alpinen Rasenstufe der Allgäuer Alpen mit einer mittleren Jahrestemperatur von weniger als 1,4°C und mit über 2500 mm Jahresniederschlag zu rechnen; in den Lechtaler Alpen dürften die entsprechenden Werte unter 0,5°C und über 1500 mm liegen.

Der Hauptdolomit

In den Nördlichen Kalkalpen, zu denen auch die Allgäuer und Lechtaler Alpen zählen, ist der Hauptdolomit stark verbreitet und baut zahlreiche Gipfel auf. Er ist bekannt durch seine bizarren Verwitterungsformen und ausgedehnten Schutthalden (Bild 5),doch weist er gelegentlich auch mildere Formen auf. Die Bezeichnung "Hauptdolomit" vereinigt eigentlich zwei verschiedene Begriffe: einerseits ist sie der Name einer geologischen Schicht-Serie in der Norischen Stufe der Trias, die auf großen Flächen eine ziemlich gleichmäßige Zusammensetzung aufweist und größtenteils aus Dolomit besteht, andererseits aber auch die Bezeichnung für dieses Gestein in seiner chemischen Zusammensetzung.
Von Zwischenschichten aus andersartigem Material (Mergel, Quarz,

Ton, Seefelder Schiefer u.a.) wird zwar berichtet (vgl.O.AMPFERER 1932, p.41, P.J.HAMANN 1956, p.14, V.JACOBSHAGEN 1957, p.9-10, B.SCHÜLER 1959, p.49, J.WESTRUP 1966 u.a.), sie sind aber verhältnismäßig selten. Böden, die vorwiegend aus den nicht-carbonatischen Zwischengebieten entwickelt wären, habe ich in der alpinen Stufe des untersuchten Gebietes nicht angetroffen.
Dennoch sind die Gesamtserie und das Gestein: Hauptdolomit als Muttergestein für die Bodenbildung nicht gleich zu bewerten. Die Gleichsetzung des namengebenden Hauptbestandteils mit der geologischen Gesamtserie hat bei der Beurteilung der Boden- und Vegetationsentwicklung aus Carbonatgesteinen schon genug Verwirrung gestiftet, zumal die nicht-carbonatischen oder mergeligen Teile solcher Serien oft viel stärker in Erscheinung treten, als dies beim Hauptdolomit der Fall ist.
Gerade bei sehr reinen Carbonatgesteinen, wie es der Hauptdolomit mit seinem Gehalt an nicht-carbonatischen Anteilen von vielfach nur 1% und darunter und selten bis über 5% ist, kann örtlich schon eine geringe Zwischenschicht andersartigen Materials die Boden- und Vegetationsentwicklung wesentlich beeinflussen.
Für den Pflanzenwuchs sind die nicht-carbonatischen Bestandteile unentbehrlich, doch für die Bodenbildung und den Wasserhaushalt ist auch der physikalische Detritus des Dolomits sehr wichtig, da er, besonders in den jüngeren Stadien der Bodenentwicklung, als schluffiger bis schluffig-lehmiger Sand den größten Teil der Feinerde aufbaut.
Die Schichtserie des Hauptdolomits schließt oft mit dem Plattenkalk, einem sehr reinen Kalkstein, ab. Daher muß stellenweise mit einer gewissen Beimischung des Calcits bzw. des Kalksteins im physikalischen Detritus des Hauptdolomits gerechnet werden.

Stadien der Bodenentwicklung und ihre Vegetation

Wenn wir von den Schneetälchen absehen, haben wir auf dem Hauptdolomit in der alpinen Stufe der Allgäuer und Lechtaler Alpen einige typische Stadien der Verwitterung und Bodenentwicklung, die wir immer wieder antreffen. Sie sollen im weiteren kurz erörtert werden.

1) Hauptdolomit-Felsen

Die Hauptdolomit-Felsen, die in den Allgäuer und Lechtaler Alpen in großen Mengen vorkommen, stehen dem Ausgangsstadium der Verwitterung und Bodenbildung am nächsten. Nähere Betrachtung zeigt

Tabelle 1. Chemische Zusammensetzung des Hauptdolomits von untersuchten Bodenprofilen.

Profil	Gesamt-carbonat %	$CaCO_3.MgCO_3$ (Dolomit) %	$CaCO_3$ (Calcit) %	R_2O_3 %	Unlösl. Rück-stand %	Summe %
60.Weißschrofen-spitze	98,64	96,27	2,37	0,83	0,42	99,89
62.Weißschrofen-spitze	99,42	98,85	0,57	0,54	0,39	100,35
61.Weißschrofen-spitze	98,34	95,16	3,18	0,93	0,53	99,80
63.Weißschrofen-spitze	99,69	98,11	1,58	0,26	0,18	100,13
38.Walser Gaißhorn	99,65	98,67	0,98	0,43	0,28	100,36
41.Widderstein	95,05	91,47	3,58	1,15	3,28	99,48
15.Koblat	98,74	97,74	1,00	0,80	0,43	99,97
35.Walser Gaißhorn	97,80	94,43	3,37	0,80	1,23	99,83

jedoch, daß auch sie selbst bereits in größerem oder kleinerem Maße angewittert sind. Ihre Verwitterbarkeit ist stark von der tektonischen Beanspruchung abhängig, der sie während der Gebirgsbildung ausgesetzt wurden. Stark beanspruchte Partien des Hauptdolomits vergrusen verhältnismäßig leicht, während wenig beanspruchte, als steile Felsen emporragend, bleiben. So zeigt sich, daß der Hauptdolomit - trotz seiner verhältnismäßig sehr gleichmäßigen Zusammensetzung - nicht überall die gleichen Voraussetzungen für die Boden- und Vegetationsentwicklung bietet.

Wenn auch von weitem kahl und grau aussehend (Bild 6), so sind die Hauptdolomit-Felsen der alpinen Stufe doch nicht ganz leblos.

Da der Hauptdolomit meist gut gebankt ist, bieten die Schichtfugen ebenso wie die durch die tektonische Beanspruchung und Verwitterung entstandenen Spalten gute Ansatzplätze für die Besiedlung. So finden sich auf den Hauptdolomit-Felsen oft nicht nur Pflanzen der typischen Felsbesiedler-Gesellschaft A n d r o - s a c e t u m h e l v e t i c a e Br.-Bl.1918, sondern auch einzelne Pflanzen und Fragmente der Alpenmatten-Gesellschaften; sie bewirken eine gewisse Ansammlung der organischen Substanz, die auch in die darunterliegenden Schutthalden gelangen kann.

2) Beweglicher Hauptdolomit-Schuttrohboden mit T h l a s p i e - t u m r o t u n d i f o l i i Br.-Bl.1926

Von den Felsen spaltet der Frost größere oder kleinere Steine und Blöcke ab, die beim Aufprallen am Fuße der Felsen weiter zerkleinert und in Form von Schutthalden angesammelt werden. Die von älteren Autoren (A.PIWOWAR 1903,p.359, u.a.) entwickelte und bis in die neuere Zeit vielfach übernommene Vorstellung, daß hierbei das feinere Material mehr im oberen Teil und der gröbere Schutt mehr im unteren Teil der Schutthalde angesammelt wird, ist nur zum Teil richtig. Auch bei ganz steilen Felsen gelangt der Schutt aus dem Fels in die Schutthalde meist nicht unmittelbar, sondern durch kleine Runsen. Unterhalb ihres Austrittes aus dem Fels bilden sich Schuttfächer, so daß größere Schutthalden meist in Schuttfächer gegliedert sind (die sich freilich überschneiden). Durch die Wirkung von Runsen entsteht auf den Schutthalden eine Zonierung derart, daß in gleicher Höhe die Partien mit gröberem und feinerem Material miteinander wechseln.
Ein Beispiel eines Rohbodens auf der Hauptdolomit-Schutthalde gibt das folgende Profil.

Profil Nr.60. 29.6.1965

Weißschrofenspitze, bewegliches Geröll am Südhang, auf dem oberen Weg zur Leutkircher Hütte, 2380 m über N.N., Exposition S, Neigung 62%. Muttergestein: Hangschutt aus Hauptdolomit. An der Oberfläche befinden sich sowohl große wie auch kleine Steine und Grus, selten etwas Sand. Im Hang sind einige Felsen, die zu der Schutt-Stauung etwas beitragen, doch die Oberfläche ist noch beweglich.
Die Pflanzen der alpinen Täschelkraut-Halde, T h l a s p e e t u m r o t u n d i f o l i i Br.-Bl.1926, decken nur 2% der Fläche. Auf 100 m^2 um das Profil sind es nur folgende vier Arten:

Assoziations-Kennart	Thlaspi rotundifolium	1.2
Verbands-Kennart	Hutschinsia alpina	+
Klassen-Kennart	Silene cucubalus ssp.prostrata	1.2
Begleiter	Galium anisophyllum	+.2

Der bewegliche Hauptdolomit-Schuttrohboden unter dieser Vegetation hat den folgenden Profilaufbau (Bild 7):

A_{sk} (0-2 cm) Grus und kleine Steine;

A_i (2-10 cm, z.T.bis 12 cm) skelettreicher, zu 80-90% aus Steinen und Grus bestehender Horizont mit hellgrauem (10YR 6/1, feucht 10YR 5/1) schluffigem Sand und einzelnen Wurzeln;

C_{v1} (10-25 cm) grober Schutt mit zahlreichen Hohlräumen, die praktisch ohne Feinboden sind;

C_{v2} (25-40 cm und tiefer) skelettreicher Horizont vorwiegend aus großen und kleinen Steinen aufgebaut, enthält außerdem hellgrauen (10YR 6,5/1, feucht 10YR 5/1), schluffigen Sand und einzelne Wurzeln

Steine brausen mit HCl 1:3 auf frischem Bruch nur schwach, die Feinerde dagegen intensiv auf.

3) Initiale Alpine Polsterrendzina über anstehendem Hauptdolomit mit D r y a s - Stadium des C a r i c e t u m f i r m a e (Kerner) Br.-Bl.1926

Durch rein abiotische, wohl vorwiegende physikalische Verwitterung entsteht aus den Hauptdolomit-Felsen nicht nur Schutt, sondern auch feines Verwitterungsmaterial, das aus den Felsspalten ausgeschwemmt wird. Wenn die Neigung des Geländes es zuläßt und die Zufuhr des groben Schuttes gering ist, sammelt sich dieses feine Material unter Mitwirkung der Pioniervegetation über anstehendem Hauptdolomit-Fels. Ein Beispiel dafür ist folgendes Profil.

Profil Nr.62. 30.6.1965

Weißschrofenspitze, Südhang, 2440 m über N.N., Exposition SSE, Neigung 72%. Muttergestein: Hauptdolomit, etwas Hangschutt über anstehendem Fels. Vegetation in Polstern, unterbrochen durch angeschwemmten und angerollten Schutt.

Ein Polster (1 m²) des D r y a s - Stadiums des C a r i c e t u m f i r m a e (Kerner) Br.-Bl.1926 hat folgende Zusammensetzung:

Assoziationskennarten	
Carex firma	1.2
Saxifraga caesia	1.2
Verbands-, Ordnungs-und Klassen-Kennarten	
Dryas octopetala	4.4
Festuca pumila	1.2
Sedum cf.atratum	+
Begleiter:	
Salix retusa.ssp. serpyllifolia	2.2
Primula auricula	+
Tortella tortulosa	1.2
Hypnum spec.	+

Moose (2% deckend).

Unter dem Schutz dieser Vegetation ist eine *Initiale Alpine Polsterrendzina* entwickelt. In der subnivalen Stufe stellen vereinzelte Vegetationspolster und alpine Polsterrendzinen das höchste Stadium der Boden- und Vegetationsentwicklung dar,hier aber in der alpinen Stufe ist es nur ein Initialstadium; die weitere Entwicklung führt zu einer nahezu zusammenhängenden Bodendecke. Zwischen den Vegetationspolstern befindet sich an der Bodenoberfläche der A_{sk}-Horizont, zum Teil aber auch kahle ungeschützte Feinerde. Im unteren Teil des untersuchten Vegetationspolsters, und zwar in der Mitte des Polsterquerschnittes, zeigt das Bodenprofil den folgenden Aufbau (Bild 8):

O_L (0-1 cm) trockene *Dryas*-Blätter zwischen lebenden Stengeln; dazwischen A_{sk} (ebenso etwa 0-1 cm): eingeschwemmter Grus und kleine Steine, zusammen etwa 10% deckend;

A_h (1-15 cm) grauer (10YR 4,5/1,5, feucht dunkelgrau: 10YR 3/1), humoser, lehmiger Sand, stark skeletthaltig, grusig, gut durchwurzelt. Struktur staubig bis krümelig. Schwaches Aufbrausen mit HCl 1:3(mehr hörbar als sichtbar). Ab 15 cm, an den stärker erodierten Rändern bereits ab 6 cm, beginnt

A_hC_v (15-30 cm) Felsen des Hauptdolomits deren Spalten mit hellbraunem (10YR 6,5/3, feucht 10YR 5/4), nur schwach humosem, sandig-schluffigem Lehm gefüllt sind. Er enthält noch viele Wurzeln. Mit HCl 1:3 reagiert die Feinerde intensiv, das Skelett nur schwach positiv;

C_v (30-80 cm und tiefer) durchaus skelettiger-felsiger Horizont, dessen Spalten ebenfalls noch etwas gelblich braunen feinsandig-schluffigen Feinboden und einzelne Wurzeln enthalten. In einer Spalte befindet sich auch etwas rostgelber bis gelblichbrauner geschieferter Lehm (Ursprung vermutlich synsedimentär). Mit HCl 1:3 reagiert auch hier der Hauptdolomit nur schwach, die Feinerde dagegen intensiv.

4) Initiale Alpine Polsterrendzina auf Hauptdolomit-Schutthalde mit Initialstadien des C a r i c e t u m f i r m a e (Kerner) Br.-Bl.1926

An Stellen, wo die Schuttzufuhr nachgelassen hat, oder wo die Stauung durch größere Felsblöcke die Ansammlung von Feinmaterial bewirkt, oder einfach an geschützteren Stellen der Schutthalde,siedelt sich *Dryas octopetala* L. an. Unter ihren Polstern entwickelt sich die *Initiale Alpine Polsterrendzina* . So lange die *Dryas*-Polster von feinem Schutt und Dolomit-Sand immer wieder überspült werden,kann sich kein reiner Humushorizont bilden. Neben der Silberwurz findet man hier weitere Pflanzen, die zwar meist mengenmäßig hinter der *Dryas* weit zurückstehen, trotzdem aber die Entwicklung zum C a r i c e t u m f i r m a e klar anzeigen.

Ein Beispiel für eine solche Entwicklung ist das nachstehende Profil.

Profil Nr.61. 29.6.1965

Weißschrofenspitze, wenig östlich vom Profil Nr.60, 2380 m N.N., Exposition S, Neigung 64%. Muttergestein: Hangschutt aus Hauptdolomit, immer noch etwas beweglich, doch durch große Felsblöcke (fraglich ob anstehend !) stärker gestaut als beim Profil Nr.60. Große Felsblöcke bedecken 25%, Steine und Grus 60% der Fläche.
Die Vegetation (vgl.Bild 9), ein Initialstadium des C a r i c e t u m f i r m a e (Kerner) Br.-Bl.1926 zeigt folgende Zusammensetzung:

Aufnahmefläche (m^2)	25
Deckungsgrad (%)	20
Assoziations-Kennarten	
Carex firma	2.2
Saxifraga caesia	1.2
Crepis jaquini	+
Seslerietalia variae (coeruleae)-Kennarten	
Festuca pumila	+.2
Minuartia verna	+.2
Elyno-Seslerietea-Kennarten	
Dryas octopetala	2.3
Begleiter	
Tortelle tortuosa	1.2
Primula auricula	1.1
Androsace helvetica (Auf Felsblock!)	+.2
Ditrichum flexicaule	+

Das Profil wurde unter einem Vegetationspolster von *Dryas octopetala, Carex firma, Saxifraga caesia* und *Tortella tortuosa* geöffnet. Dieses Vegetations-Polster, wie auch die übrigen auf der untersuchten Fläche, wird immer wieder von feinem Dolomit-Schutt überrieselt, so daß sich kein stark humoser Horizont bilden kann. Der Boden ist eine *Initiale Alpine Polsterrendzina*.

Profilaufbau:

O_L (0-1 cm) trockene, meist braunschwarze *Dryas*-Blätter, dazwischen etwas eingeschwemmter Dolomit-Grus. Ein dunkler O_HA_h-Horizont ist nur ganz schwach angedeutet;

A_h (1-7 cm) graubrauner (10YR 5,5/1,5, feucht 10YR 4/1,5), humoser, schluffiger Sand von staubig- bis krümeliger Struktur, skeletthaltig (Grus und kleine Steine). Durchwurzelung mittel;

A_hC_v (7-15/20 cm) skelettreicher Übergangshorizont mit hellbraungrauem (10YR 6,5/2, feucht 10YR 5/2) schluffigem Sand;

C_{v1} (20-50 cm) lockerer Hauptdolomit-Hangschutt mit etwas hellbraunem (10YR 7/2, feucht 10YR 6/2) schluffigem Sand. Ab etwa 50 cm

C_{v2} (50-80 cm +) größere Blöcke, aber nicht anstehend. Einzelne Wurzeln.

Mit HCl 1:3 braust das Skelett auf frischem Bruch schwach, die Feinerde dagegen stärker auf.

5) Alpine Rendzina unter Caricetum firmae (Kerner) Br.-Bl.1926

Alpine Rendzina unter gut entwickelten Beständen des Caricetum firmae weist verschiedene Oberflächen- und Ausbildungsformen auf (vgl.auch Z.GRACANIN 1968). Einige davon mögen durch die Profile 63,38,42 und 132 charakterisiert werden.

Profil Nr.63. 30.6.1965

Weißschrofenspitze oberhalb der Kreuzung des oberen Weges zur Leutkircher Hütte mit den Wegen zur Ulmer Hütte und zum Kapall. 2390 m N.N., Exposition S, Neigung 68%. Muttergestein: Hangschutt aus Hauptdolomit. Skelett bedeckt 30%, ungeschützte Feinerde 10% und die Vegetation 60% der untersuchten Fläche. Das Gelände ist leicht treppenartig geformt, die Böschungen der Stufen sind durch *Carex firma* und andere Pflanzen befestigt; auf den fast waagerechten bis schwach geneigten Treppenabsätzen befinden sich meist kleine Steine und Grus, zum Teil aber auch kahle ungeschützte Feinerde. Die untersuchte Fläche liegt zwischen zwei beweglichen Schutthalden; sie selbst ist durch die Lage der Felsen am Hang oberhalb des Profils vor einer stärkeren Schuttzufuhr weitgehend geschützt.
Das Caricetum firmae (Kerner) Br.-Bl.1926, der Polsterseggenrasen, kommt hier in einer Ausbildung mit reichlich *Erica carnea* vor, die ganze Vegetation ist aber jetzt, bei Beginn der Vegetationsperiode nur 1-3 cm hoch. Sie hat folgende Zusammensetzung:

Caricetum firmae

Aufnahmefläche (m^2)	25
Assoziations-Kennart	
Carex firma	3.3
Seslerion variae (=corueleae)-Kennarten:	
Gentiana clusii	1.1
Helianthemum alpestre	1.1
Pedicularis rostrato-capitata	+
Seslerietalia- und Elyno-Seslerietea-Kennarten:	
Dryas octopetala	1.2
Anthyllis vulneraria ssp.alpestris	1.2
Sesleria varia	+.2
Festuca pumila	+
Phyteuma orbiculare	+
Arabis cf.corymbifolia	+
Begleiter:	
Erica carnea	2.2
Primula auricula	1.1
Pinguicula alpina	1.1
Artostaphylos uva-ursi	+
Tortella tortuosa	+.2
Orthothecium rufescens	+
Campylium spec.	+

Ein unter dieser Vegetation geöffnetes Profil der *Alpinen Rendzina* zeigt folgende Horizonte:

O_L nur schwach angedeutet: trockene *Carex firma*-Blätter, meist noch an der lebenden Pflanze hängend;

A_h (0-20 cm) je nach Ab- und Auftrag durch Bodenfließen und Bodenerosion durch oberflächlich abfließendes Wasser schwankt die Mächtigkeit dieses Horizontes zwischen 14 und 24 cm, dunkelgrauer (10YR 4/1, feucht 3/1), stark humoser, lehmiger Sand, skeletthaltig: steinig und grusig, stark durchwachsen mit feinen Wurzeln. Gefüge im Grunde staubig, doch werden auch größere Aggregate durch Wurzeln zusammengehalten. Die Reaktion mit kalter HCl 1:3 ist in der Feinerde und im Skelett nur schwach positiv (hörbar!).
In ungleicher Tiefe von 14-24 cm scharfer Übergang in

C_{v1} (30-50 cm) lockerer, hohlraumreicher Horizont des Hauptdolomit-Schuttes mit nur noch ganz schwach humosem, hellbraungrauem (10YR 5,5/2, feucht 10YR 6,5/1,5) schluffigem Sand. Keine ausgeprägten Aggregate. Nur einzelne Wurzeln. Starkes Aufbrausen der Feinerde mit HCl 1:3, im Skelett ist die Reaktion nur hörbar. Ab etwa 50 cm beginnt

C_{v2} (50-80 cm) etwas dichter als der vorherige Horizont. Im gleichen Hauptdolomit-Hangschutt erscheinen größere Steinblöcke, dazwischen aber auch immer noch hellgrauer bis hellbraun-grauer (10YR 7/1,5, feucht 10YR 6/2) schluffiger Sand. Alle Steine in diesem Horizont sind stark angewittert. Nur vereinzelte Wurzeln reichen bis 80 cm. Die Reaktion mit HCl 1:3 ist in der Feinerde stark, im Skelett nur schwach.

Profil Nr.38. 25.8.1964

Walser Gaißhorn, westlich vom Gipfel, 2292 m N.N., Exposition W, Neigung 56%. Muttergestein: Hangschutt aus dem Hauptdolomit. Skelett, vorwiegend kleine Steine und Grus (=A_{sk}-Horizont), bedeckt 30-40% der Fläche.
Das C a r i c e t u m f i r m a e (Kerner) Br.-Bl.1926 ist in Polster aufgelöst, teilweise treppenartig, aber nicht ganz regelmäßig. Der Boden ist deutlich erodiert; kahle ungeschützte Feinerde nimmt 1-2% der Fläche ein.

Die Vegetation zeigt folgende Zusammensetzung:

Aufnahmefläche (m^2)	25
Assoziations-Kennarten:	
Carex firma	4.2
Saxifraga caesia	+.2
Seslerion variae-Kennarten:	
Sedum atratum	+
Biscutella laevigata	+
Seslerietalia- und Elyno-Seslerietea-Kennarten:	
Dryas octopetala	1.2
Festuca pumila	1.2
Androsace chamaejasma	+.2
Galium anisophyllum	+.2
Arabis corymbiflora	+
Anthyllis vulneraria ssp.alpestris	+
Begleiter:	
Salix retusa	2.2
Silene acaulis	1.2
Primula auricula	1.1
Polygonum viviparum	1.1
Ranunculus alpester	+

Bartsia alpina	+
Campanula cf scheuchzeri	+
Achillea spec.	+
Veronica aphylla	+
Rhododendron hirsutum	$+^{o}$

Die unter dieser Vegetation entwickelte alpine Rendzina zeigt folgenden Aufbau (Bild 10):

O_L (0-1 cm) unzersetzte organische Reste, vornehmlich *Carex firma*-Blätter. Auf 30-40% der Fläche statt dessen von 0-3 cm der A_{sk}-Horizont.

A_h (1-15 cm) sehr dunkler, grauer bis schwarzer (10YR 3/1), feucht 10YR 2/1), stark humoser, gut durchwurzelter toniger Lehm von staubiger bis krümeliger Struktur, skeletthaltig (Grus und kleine Steine des Hauptdolomits). Um 15 cm deutlicher Übergang in

A_hC_v (20-60 cm) absolut skeletthaltiger (99%), sehr hohlraumreicher, wenig zusammenhaltender, beim Graben ständig nachrutschender Horizont des hauptdolomitischen Grus mit kleinen Steinen, sehr wenig schluffigem Sand und einzelnen Wurzeln;

C_v (60-90 cm) wesentlich besser zusammenhaltender, aber auch noch sehr durchlässiger, skelettreicher und fein-grusiger Horizont mit hell (grau)braunem (10YR 7/3, feucht 10YR 5/2,5), dolomitischem, schluffigem Sand.

Die Reaktion mit HCl 1:3 ist im Skelett und in der Feinerde verzögert und schwach, aber anhaltend. Das schwächste Aufbrausen weist das Skelett aus frischem Bruch und die Feinerde des A_h-Horizontes auf.

Profil Nr.42. 28.8.1964

Widderstein, kleiner, weniger geneigter Absatz auf dem Hang zwischen steilen Felsen, 2380 m N.N., Exposition ESE, Neigung am Profil 61%. Muttergestein: Hauptdolomit-Hangschutt über anstehendem, leicht bergwärts geneigtem, vertikal zerklüftetem Hauptdolomit. Steine mit etwas Grus und Feinmaterial gelangen von den Felsen im Oberhang auf den untersuchten Felsabsatz hinunter, zum Teil als kleine Schuttzüge und Schuttrinnen. Um das Profil deckt das Skelett etwa 20%. Die Bodenoberfläche ist leicht getreppt, wohl auch durch Begehen.
Die Vegetationsdecke bildet hier ein C a r i c e t u m f i r m a e (Kerner) Br.-Bl.1926 mit Übergangscharakter zum S e s l e r i o - S e m p e r v i - r e t u m Br.-Bl.1926.

Aufnahmefläche (m^2):	6
Höhe der Vegetation (cm) Carex firma	3
Sonstige Rasen (cm):	10
Assoziations-Kennart:	
Carex firma	2.2
Seslerietalia-Kennarten:	
Sesleria varia	2.2
Androsace chamaejasme	1.2
Anthyllis vulneraria (verbissen)	+.2
Phyteuma orbiculare ssp.orbiculare	+
Galium anisophyllum	+
Carduus defloratus ssp.rhaeticus	+
Elyno-Seslerietea-Kennart:	
Festuca pumila	1.2
Begleiter:	
Carex sempervirens	2.2

Ligusticum mutellina	1.1
Salix retusa ssp.serpyllifolia	+.2
Campanula scheuchzeri	+
Bartsia alpina	+
Gentiana campestris	+
Thymus polytrichus	+
Homogyne alpina	+
Polygonum viviparum	+
Ranunculus alpester	+
Aster bellidiastrum	+
Leontodon hispidus	+
Euphrasia spec.	+
Soldanella spec.	+

Dank der allmählichen Materialzufuhr vom Oberhang wuchs das Profil nach oben und nicht wie üblich durch die Verwitterung des anstehenden Muttergesteins nach unten. So ist eine besonders mächtige, bis zu 50 cm Tiefe humusreiche und dennoch bis zur Bodenoberfläche skelettreiche und stark carbonathaltige Alpine Rendzina entstanden. Ihr Aufbau:

O_L meist nur 1 cm, in *Carex firma*-Polstern jedoch bis 5 cm mächtiger Horizont der unzersetzten organischen Reste. Wo Skelett die Bodenoberfläche bedeckt, ist der A_{sk}-Horizont 2-5 cm mächtig. Auf dem größten Teil der Fläche beginnt aber ab 1 cm

A_{h1} (1-20 cm) skelettreicher (>60%), steiniger und grusiger Horizont mit sehr dunkelgrauem bis dunkelgraubraunem (10YR 3,5/1, feucht 10YR 2/2) carbonatreichem, stark humosem tonigem Lehm krümeliger Struktur, stark durchwurzelt. Skelett besteht hier ebenso wie im ganzen Profil, ausschließlich aus Hauptdolomit. Die Reaktion mit kalter HCl 1:3 ist in der feuchten Feinerde kaum wahrnehmbar, in der trockenen Feinerde dagegen deutlich hörbar (wenn auch kaum sichtbar) und lange andauernd. Der Hauptdolomit reagiert auf frischem Bruch schwach positiv, auf den angewitterten Außenflächen etwas deutlicher.

A_{h2} (20-35 cm) skelettreicher (ca 50-60%), steiniger und grusiger Horizont mit sehr dunklem graubraunem (10YR 3,5/2, feucht 10YR 2,5/2), humosem, lehmigem Ton von krümeliger bis bröckeliger Struktur. Mittlere Durchwurzelung. Am Skelett befinden sich helle Verwitterungsrinden, die nach der Entnahme des Skeletts als helle Abdrücke an der sonst dunklen Feinerde haften bleiben. Die Reaktion mit HCl 1:3 ist in der Feinerde schwach positiv (zunächst nur hörbar!) aber lange andauernd. Übergang kaum erkenntlich.

A_{h3} (38-50 cm) skeletthaltiger Horizont (ca 50%) mit dunkelgraubraunem (10YR 3,5/2, feucht 10YR 2,5/2) lehmigem Ton, krümeliger Struktur, noch reich an Wurzeln. Helle Verwitterungsabdrücke an der Feinerde neben Skelett. Der Gehalt an Carbonaten erreicht in diesem Horizont mit 35,7% (SCHEIBLER) sein Minimum und deutet damit eine Periode langsamerer Carbonatzufuhr und intensiverer Entcarbonatisierung an. In ungleicher Tiefe von 45-50 cm Übergang in

A_hC_v (50-80 cm) stark skelettreicher (ca 90%) steiniger und grusiger Horizont mit braunem (10YR 5,5/3, feucht 10YR 4/3), dolomitischem schluffig-lehmigem Fein - Sand ., in welchem noch einzelne Wurzeln vorhanden sind. Die Feinerde reagiert mit HCL 1:3 stärker als in den oberen Horizonten und im Skelett. Ab 80 cm Übergang in

C_n anstehende Hauptdolomit-Felsen.

6) Alpine Rendzina unter Seslerio-Semperviretum Br.-Bl.1926

Die mächtigsten Profile der Alpinen Rendzina findet man unter der alpinen Blaugras-Horstseggen-Halde (Seslerio-Semperviretum Br.-Bl.1926). Öfter als bei der Alpinen Rendzina unter Caricetum firmae kann man hier eine stärkere Entcarbonatisierung oder gar Carbonatfreiheit des A_h-Horizontes feststellen; bei den meisten Profilen weist aber der A_h-Horizont auch unter Seslerio-Semperviretum einen erheblichen Carbonatgehalt auf.
Die nur in den obersten Horizonten entcarbonatisierten und schwach sauren Bodenprofile der alpinen Rendzina stellen die am weitest entwickelten Stadien der Bodenentwicklung auf Hauptdolomit dar, die in der alpinen Rasenstufe der Allgäuer und Lechtaler Alpen anzutreffen sind. Stark saure Böden und ein Caricetum curvulae sind hier auf Hauptdolomit nicht entwickelt. Im unteren Teil der alpinen Rasenstufe konnte sich auf Hauptdolomit auch kein Nardetum entwickeln, es sei denn, daß erratisches Material auf dem Hauptdolomit liegt.
Als Beispiele der Alpinen Rendzina unter Seslerio-Semperviretum mögen die Profile Nr.149,41,15 und 35 dienen.

Profil Nr.149. 11.10.1967

Südwestlich Weißschrofenspitze, westlich der früher untersuchten Profile Nr.60-63, 2420 m N.N., Exposition SW, Neigung 55-60%. Muttergestein: Hauptdolomit-Hangschutt. Blöcke, Steine und Grus bedecken 30% der Bodenoberfläche, Vegetation 80%.

Der hier vorhandene Bestand des Seslerio-Semperviretum enthält als Entwicklungsrest noch *Carex firma*, was wohl dem verhältnismäßig skelettreichen Boden zuzuschreiben ist. Von den ohnehin nie reichlich vorhandenen Charakterarten der Assoziation wurden auf 10 m^2 um das Profil keine festgestellt, im übrigen aber war die Zusammensetzung sehr typisch:

Deckungsgrad (%)	80
Seslerietalia-Kennarten:	
Sesleria varia	1.2
Globularia nudicaulis	1.2
Anthyllis vulneraria	1.1
Androsace chamaejasme	1.1
Scabiosa lucida	+
Elyno-Seslerietea-Kennarten:	
Festuca pumila	+

Begleiter:	
Carex sempervirens	2.2
Ligusticum mutellina	1.1
Gentiana campestris	1.1
Poa alpina	+
Sukzessions-Relikt:	
Carex firma	2.2

Die Alpine Rendzina unter dieser Vegetation ist noch verhältnismäßig skelettreich und hat den folgenden Aufbau:

A_h (0-20 cm) stark skeletthaltiger steiniger Horizont mit stark humosem, schwarzem (bereits trocken 10YR 2/1) schluffigem Lehm, krümeliger Struktur, stark durchwurzelt. Skelett hinterläßt auf der Feinerde schwache helle Verwitterungsabdrücke;

A_hC_v (20-40 cm) skelettreicher Horizont mit (hell)braunem (10YR 6/3, feucht 10YR 5/3) sandigem Schluff, stark durchwurzelt. Unter 40 cm beginnt

C_v noch skelettreicher und stark durchwurzelt.

Die Reaktion mit der HCl 1:3 ist in der Feinerde des A_h-Horizontes verzögert und schwach aber lange andauernd, Im A_hC_v dagegen stark; das Skelett reagiert auf dem frischen Bruch mehr hörbar als sichtbar, auf den angewitterten Außenflächen dagegen etwas stärker.

Profil Nr.41. 26.8.1964

Widdersetin 242o m N.N., oberhalb des Pfades durch die Widderstein-Schlucht. Exposition E, Neigung 59%. Muttergestein: Hauptdolomit. Verwitterungsprodukte des Hauptdolomits sind in Form eines kleinen "Schuttfächers" (in welchem jedoch feines Material vorherrscht) über anstehendem Hauptdolomit-Fels angesammelt. Die untere Seite dieses Fächers ist etwa 5 m lang, die obere nur 0,5 m, die Breite beträgt 3 m. Diese Fläche ist gut bewachsen. Skelett bedeckt nur 5%, kahler Boden 2%. Der Rasen ist durch Betreten leicht zu Stufen geformt.

Das S e s l e r i o - S e m p e r v i r e t u m Br.-Bl.1926 (vgl.Bild 3) liegt hier in einer schneefeuchten Ausbildung mit *Gnaphalium supinum* vor und zeigt auf 5m² folgende, nicht gerade üppige Zusammensetzung:

Aufnahmefläche (m²):	5
Seslerietalia-Kennarten:	
Sesleria varia	1.2
Carduus defloratus ssp.rhaeticus	1.2
Androsace chamaejasme	1.2
Veronica aphylla	1.1
Phyteuma orbiculare	+.2
Elyno-Seslerietea-Kennart:	
Festuca pumila	2.2
Begleiter:	
Carex sempervirens	2.2
Homogyne alpina	1.2
D Gnaphalium supinum	1.2
D Soldanella pusilla	1.2
Gentiana campestris	1.1
Selaginella selagionides	1.1
Ligusticum mutellina	1.1
Salix retusa	+.2
Campanula scheuchzeri	+
Bartsia alpina	+

	Aster bellidiastrum	+
D	Ranunculus alpestris	+
D	Parnassia palustris	+
	Polygonum viviparum	+
	Euphrasia spec.	+
	Tortella tortuosa	+.2
	Plagiochila asplenioides	+.2
	Ctenidium molluscum	+

Unter nur schwach angdeuteten O_L- und O_H-Horizonten befindet sich (Bild 11):

A_{h1} (0-4 cm) dunkelgraubrauner (10YR 3/1,5, feucht 10YR 2/2), stark humoser, lehmiger Ton, krümeliger Struktur, schwach skeletthaltig, stark durchwurzelt. Im feuchten wie im trockenen Zustand ist die Reaktion der Feinerde mit HCl 1:3 stark verzögert, nicht sichtbar und nur schwach hörbar, aber lange andauernd; Skelett, meist Grus und kleine Steine, reagiert an den angewitterten Außenflächen sofort schwach positiv.

A_{h2} (5-22 cm) dunkelgraubrauner (10YR 3,5/2, feucht 10YR 2,5/2), humoser lehmiger Ton, krümeliger Struktur, mäßig skeletthaltig (Steine und Grus, alles Hauptdolomit). Helle Verwitterungsabdrücke des Skeletts an sonst dunklem Boden. Gute Durchwurzelung. Die Reaktion der Feinerde mit HCl 1:3 ist sehr schwach, meist nur hörbar (erst nach minutenlanger Verzögerung langsames Aufquellen!), deutlicher ist sie nur bei dolomitischen Grobsandkörnern und in hellen Verwitterungsabdrücken.

A_{h3} (25 -35 cm) dunkelgraubrauner (10YR 3,5/2, feucht 10YR 2,5/2), humoser, lehmiger Ton von krümeliger Struktur, etwas stärker skeletthaltig als A_{h2}-Horizont, aber sonst ähnliche Eigenschaften aufweisend. Deutlicher Übergang in

A_hC_v (35-45 cm) skelettreicher Horizont mit braunem (10YR 5/3, feucht 10YR 4/3), lehmigem Sand, der größtenteils aus Dolomit besteht. Einzelne Wurzeln. In der Feinerde ist die Reaktion mit HCl 1:3 nur schwach verzögert und deutlich positiv (auch sichtbar), besonders wenn der Boden trocken ist.

C_v (45-100 cm) ähnlicher Eigenschaften, geht bei 100 cm in

C_n $\pm$ horizontale, anstehende Dolomitfelsen über.

Profil Nr.15. 3.7.1964

Koblat im Nebelhorn-Massiv (Bild 1), 2180 m N.N., Exposition S, Neigung 78%- Muttergestein: Hauptdolomit, Verwitterungsschutt über anstehendem Fels. Die Bodenoberfläche ist stufenförmig gegliedert. Die Vegetation bedeckt etwa 60-70%, Skelett, meist auf den Verebnungen der Stufen liegender Steine, zum Teil aber auch anstehende Felsblöcke, 30%; kahler ungeschützter Boden etwa 10% der Fläche.

Auf 15 m^2 um das Profil wurden in der alpinen Blaugras-Horstseggen-Halde (S e s l e r i o - S e m p e r v i r e t u m Br.-Bl.1926) keine Charakterarten der Assoziation verzeichnet, wohl auch deswegen, weil die Vegetation noch nicht voll entwickelt war. Im übrigen war die Zusammensetzung:

Aufnahmefläche (m^2):	15
Seslerion variae-Kennarten:	
Gentiana clusii	+
Pedicularis cf-rostrato-capitata	+
Seslerietalia-Kennarten:	
Anthyllis vulneraria ssp.alpestris	2.2
Sesleria varia	1.2
Gentiana verna	1.1
Globularia nudicaulis	+
Arabis corymbifolia	+
Elyno-Seslerietea-Kennart:	
Festuca pumila	+.2
Begleiter:	
Carex sempervirens	3.2
Viola biflora	2.2
Salix retusa ssp.retusa	1.2
Aster bellidiastrum	1.1
Ranunculus montanus	1.1
Polygonum viviparum	+.2
Thesium alpinum	+.2
Bartsia alpina	+
Tortella tortuosa	+.2

Die alpine Rendzina unter dieser Vegetation zeigt folgenden Aufbau:

O_L (0-1 cm) lose, trockene Seggen- und Grasblätter;

A_h (1-20 cm) sehr dunkler (grau) brauner (10YR 3/1,5, feucht 10YR 2/2) stark humoser lehmiger Ton, skeletthaltig (Steine und Grus). Gefüge krümelig. Durchwurzelung stark. In der Feinerde brausen mit der HCl 1:3 nur einzelne Körner schwach auf; das Skelett reagiert auf frischem Bruch nur schwach, auf den verwitterten Außenflächen dagegen wesentlich intensiver. Zum Teil in einer Tiefe von 20 cm, zum Teil aber erst nach einem bis zu 20 cm mächtigen, etwas weniger humosen und helleren, aber immer noch dunkelbraunem Übergang, beginnt

C_v anstehender Dolomitfels mit nur wenig Dolomitsand.

Profil Nr.35. 21.8.1964

Walser Gaißhorn (Bild 4) Südseite, 3 m unter dem Gipfel, 2364 m N.N., Exposition SW, Neigung 46%. Muttergestein: Hauptdolomit. Lose und anstehende Steine bedecken 15%, kahler ungeschützter Boden 5% und die Vegetation 80%. Der Hang ist etwas in Treppen aufgelöst (Bild 12).
Die alpine Blaugras-Horst-Seggenhalde (S e s l e r i o - S e m p e r v i r e t u m Br.-Bl.1926 zeigt hier folgende Zusammensetzung:

Seslerion variae-Kennart:	
Pedicularis cf.rostrato-capitata	+
Seslerietalia variae-Kennarten:	
Sesleria varia	3.2
Androsace chamaejasme	1.2
Carex atrata ssp.atrata	1.2
Minuartia verna ssp.gerardi	+.2
Helianthemum nummularium ssp. grandiflorum	+.2
Anthyllis vulneraria ssp.alpestris	+
Galium anisophyllum	+

Elyno-Seslerietea-Kennarten:	
Gentiana nivalis	1.1
Festuca pumila	+.2
Begleiter:	
Carex sempervirens	2.2
Silene acaulis	1.3
Salix retusa ssp.serpyllifolia	1.3
Polygonum viviparum	1.2
Homogyne alpina	1.1
Euphrasia cf.picta	1.1
Gentiana campestris	+.2
Carex firma	+.2
Tofieldia calyculata	+
Myosotis alpestris	+
Campanula scheuchzeri	+
Leontodon hispidus	+
Nardus stricta	+°
Cetraria islandica	+
Cladonia cf.pyxidata	+
Moose	+.2

Der Boden ist eine reife, im oberen Teil des Profils bereits größtenteils entcarbonatisierte *Alpine Rendzina* mit folgender Horizontfolge:

O_L (0-1 cm) trockene Grasreste, nicht die ganze Bodenoberfläche deckend;

A_{h1} (1-4 cm) schwarzer (10YR 2/1) Humus-Horizont, sehr stark durchwurzelt, fast verfilzt. Nur wenig Skelett: Hauptdolomit-Grus. Kein Aufbrausen mit HCl 1:3.

A_{h2} (4-15 cm) sehr dunkler brauner (10YR 2/2, feucht: schwarz 10YR 2/1), sehr stark humoser lehmiger Ton krümeliger Struktur, gut durchwurzelt, stark skeletthaltig (ca.50% Hauptdolomit-Steine und Grus). Skelett weist weiße "Verwitterungsrinden" auf; in diesen ist die Reaktion mit HCl 1:3 stark. In der Feinerde ist sie meist negativ (bis auf die Dolomit-Körner), im Skelett auf dem frischen Bruch nur ganz schwach positiv. In einer Tiefe von 15-20 cm deutlicger Übergang in

A_hC_v (30-50 cm) absolut skeletthaltiger steiniger Horizont. Die Hauptdolomit-Steine sind ± plattig und vertikal gestellt, zum Teil lose, meist aber anstehend. In den Spalten befindet sich noch etwas graubrauner (10YR 4,5/2,5, feucht 3,5/3), humoser, lehmig-schluffiger Sand und viele Wurzeln. Die Reaktion mit HCl 1:3 ist in der Feinerde stark verzögert und trotz des hohen Carbonatgehaltes nur schwach positiv (dauert aber sehr lange), der Hauptdolomit reagiert auf frischem Bruch nur ganz schwach. Auf den Hauptdolomit-Steinen befinden sich weiße Verwitterungsrinden.

Diskussion der Ergebnisse

Ein Vergleich der Profil-Eigenschaften und der Labor-Analysen (Tab.2 und 3) der Böden unter den untersuchten Pflanzengesellschaften zeigt, daß im allgemeinen vom Thlaspietum rotundifolii über das Caricetum firmae zum Seslerio-Sempervіretum folgende Veränderungen der Boden-Eigenschaften zu verzeichnen sind:

1) Vergrößerung des Anteils der Feinerde, wenigstens in den obersten Horizonten;

Tabelle 2.
Korngrößenzusammensetzung einiger Böden auf dem Hauptdolomit aus der alpinen Rasenstufe der Allgäuer und Lechtaler Alpen

Profil	Horizont	Tiefe	Skelett in der Bodenprobe	Feinerde enthält in %				
				Ton 0,002	Schluff 0,002- -0,02	Staub 0,02- 0,062	Feinsand 0,062- 0,2	Grobsand 0,2-2,0
		cm	%	mm	mm	mm	mm	mm
Beweglicher Hauptdolomitschutt-Rohboden mit T h l a s p i e t u m r o t u n d i f o l i i								
60.Weißschrofenspitze	A_1	2-10	70,6	4,4	7,0	14,9	16,9	56,8
	C_{v2}	25-40	65,7	4,6	9,9	19,2	18,5	47,8
Initiale Alpine Polsterrendzina über anstehendem Hauptdolomit mit D r y a s - S t a d i u m des C a r i c e t u m f i r m a e								
62.Weißschrofenspitze	A_h	1-15	32,5	14,7	8,2	24,6	34,7	17,8
	A_hC_v	15-30	41,8	15,7	16,8	31,9	20,2	15,4
Initiale Alpine Polsterrendzina auf Hauptdolomit-Schutthalde mit Initialstadien des C a r i c e t u m f i r m a e								
61.Weißschrofenspitze	A_h	1-7	39,1	5,4	9,6	25,7	27,2	32,1
	A_hC_v	7-15	70,5	3,9	10,8	32,6	23,0	29,7
	C_{v1}	20-50	78,6	4,0	11,7	24,7	18,8	40,8
Alpine Rendzina unter C a r i c e t u m f i r m a e								
63.Weißschrofenspitze	A_h	0-20	33,2	10,1	8,6	11,3	53,6	16,4
	C_{v1}	30-50	63,3	2,7	9,2	24,4	29,8	33,9
	C_{v2}	50-80	53,7	2,8	10,7	22,5	24,0	40,0
38.Walser Gaißhorn	A_h	1-15	38,8	27,2	13,8	35,4	18,6	5,0
	A_hC_v	20-60	91,5	-	-	-	-	-
	C_v	60-90	64,2	4,7	9,1	37,2	20,0	29,0
42.Widderstein	A_{h1}	1-20	41,4	38,2	15,1	21,6	22,0	3,1
	A_{h2}	20-35	41,9	50,3	12,6	13,8	19,8	3,5
	A_{h3}	38-50	40,3	50,3	12,9	12,6	20,1	4,1
	A_hC_v	50-80	67,9	8,9	7,8	37,1	38,9	7,3
Alpine Rendzina unter S e s l e r i o - S e m p e r v i r e t u m								
149.Südwestlich Weißschrofensp.	A_h	0-20	67,6	17,1	18,0	37,6	21,6	5,7
	A_hC_v	20-40	82,6	6,3	18,1	47,6	23,2	4,8
41.Widderstein	A_{h1}	0-4	2,3	60,5	21,1	12,7	4,9	0,8
	A_{h2}	5-22	5,7	64,4	17,7	9,7	7,9	1,3
	A_{h3}	23-35	32,4	60,2	17,8	9,4	10,0	2,6
	A_hC_v	35-45	43,9	10,4	7,2	28,6	42,1	11,7
15.Koblat	A_h	1-20	11,2	62,3	26,5	8,5	2,0	0,7
35.Walser Gaißhorn	A_{h1}	1-4	14,0	-	-	-	-	-
	A_{h2}	4-15	23,7	54,1	30,0	9,3	5,8	0,8
	A_hC_v	20-50	44,4	-	-	-	-	-

Tabelle 3.

Einige chemische Eigenschaften der Böden auf Hauptdolomit der alpinen Rasenstufe der Allgäuer und Lechtaler Alpen

Profil	Horizont	Tiefe cm	pH in H_2O	pH in n/10 -KCl	Erdalkalicarbonate als $CaCO_3$ nach SCHEIBLER %	Ca (in HCl 1:2) %	Mg (in HCl 1:2) %	Ca:Mg (atomar)	Org. Substanz %	C %	N %	C:N %
Beweglicher Haupdolomitschutt Rohboden mit *Thlaspeetum rotundifolii*												
60. Weißschrofen-	A_i	2-10	8,2	8,0	105,4[1)]	21,16	11,82	1,09	0,7	0,4	0,02	20,0
spitze	C_{v2}	25-40	8,4	8,2	105,8	20,72	11,97	1,05	0,7	0,4	0,02	20,0
Initiale alpine Polster-Rendzina über anstehendem Hauptdolomit mit *Dryas* - Stadium des *Caricetum firmae*												
62. Weißschrofen-	A_h	1-15	7,6	7,3	78,6	16,11	8,88	1,10	7,2	4,2	0,27	15,4
spitze	A_hC_v	15-30	8,1	7,7	82,9	16,47	9,36	1,07	0,9	0,5	0,04	11,3
Initiale alpine Polster-Rendzina auf Hauptdolomit-Schutthalde mit Initialstadien des *Caricetum firmae*												
61. Weißschrofen-	A_h	1-7	7,6	7,3	96,4	19,56	11,58	1,02	5,0	2,9	0,13	22,3
Spitze	A_hC_v	7-15	8,1	7,9	101,7	20,64	11,99	1,04	1,7	1,0	0,05	20,0
	C_{v1}	20-50	8,4	8,2	103,2	21,24	12,40	1,04	1,2	0,7	0,02	-
Alpine Rendzina unter *Caricetum firmae*												
63. Weißschrofen-	A_h	0-20	7,6	7,4	90,0	18,71	10,82	1,05	7,6	4,4	0,49	9,0
Spitze	C_{v1}	30-50	8,3	8,2	104,3	-	-	-	0,9	0,5	0,06	8,3
	C_{v2}	50-80	8,4	8,3	105,0	-	-	-	0,5	0,3	0,04	-
38. Walser	A_h	1-15	7,3	7,0	54,7	-	-	-	16,4	9,5	0,94	10,1
Gaißhorn	A_hC_v	20-60	7,8	7,5	99,9	-	-	-	2,1	1,2	0,10	12,0
	C_v	60-90	8,1	7,6	99,8	-	-	-	0,5	0,3	0,03	10,0
42. Widderstein	A_{h1}	1-20	7,3	7,0	50,9	-	-	-	15,2	8,8	0,73	12,1
	A_{h2}	20-35	7,5	7,0	41,1	-	-	-	9,7	5,6	0,59	9,5
	A_{h3}	38-50	7,6	7,0	35,7	-	-	-	8,3	4,8	0,50	9,6
	A_hC_4	50-80	7,8	7,3	100,7	-	-	-	-	-	0,08	-

Alpine Rendzina unter S e s l e r i o - S e m p e r v i r e t u m

149.Südwestlich	A_h	0-20	7,3	7,1	50,9	-	-	-	24,7	14,3	1,40	10,2
Weißschrofensp.	A_hC_v	20-40	7,8	7,5	94,3	-	-	-	4,1	2,4	0,21	11,4
41.Widderstein	A_{h1}	0-4	6,9	6,3	8,5	-	-	-	18,1	10,5	0,91	11,5
	A_{h2}	5-22	7,5	6,9	16,1	-	-	-	11,2	6,5	0,64	10,2
	A_{h3}	23-35	7,5	6,9	26,3	-	-	-	8,1	4,7	0,53	8,9
	A_hC_4	35-45	7,8	7,4	94,3	-	-	-	-	-	0,07	-
15.Koblat	A_h	1-20	7,0	6,6	2,8	1,82	1,17	0,94	25,3	14,7	1,25	11,8
35.Walser	A_{h1}	1-4	7,6	6,4	1,8	2,00	0,65	1,87	53,1	30,8	2,44	12,6
Gaißhorn	A_{h2}	4-15	6,9	6,7	10,4	3,60	1,53	1,43	30,7	17,8	1,71	10,4
	A_hC_v	20-50	7,1	6,9	87,1	18,31	10,29	1,08	3,4	2,0	0,22	9,1

[1] Die Werte über 100% kommen dadurch zustande, daß bei der Bestimmung nach SCHEIBLER aus dem volumetrisch bestimmten CO_2 Gesamtcarbonat auf $CaCO_3$ umgerechnet wird. Da wegen des niedrigeren Atomgewichtes des Mg der Dolomit mehr Gewichtsprozente CO_2 enthält als $CaCO_3$, könnten theoretisch bei reinem Dolomit SCHEIBLER-Werte bis 108,5 erwartet werden.

2) Starke Zunahme des Ton- und Schluffgehaltes bei gleichzeitiger starker Verminderung des Grobsandanteils;
3) Bildung der stark humosen A_h-Horizonte;
4) starke Anreicherung des Stickstoffs im Boden. Die N-Gehalte lagen im beweglichen Hauptdolomit-Schutt-Rohboden mit Thlaspietum rotundifolii bei 0,02% und stiegen im A_h-Horizont der Initialen Alpinen Rendzina auf 0,13-0,27%, in der Alpinen Rendzina unter Caricetum firmae auf 0,49-0,94% und in der Alpinen Rendzina unter Seslerio-Semperviretum auf 0,91-2,44%.
5) Verminderung des Carbonatanteils, und zwar zum Teil durch die Verdünnung der mineralischen Substanz durch die neugebildeten organischen Stoffe und durch die Staubzufuhr, mehr aber noch durch die Lösung des Dolomits im Laufe der Verwitterung und Auswaschung der Lösungsprodukte, sowie durch die Anreicherung der nicht löslichen Anteile der mineralischen Substanz;
6) im Zusammenhang mit der Entcarbonatisierung und Anreicherung der organischen Substanz steht auch die Erniedrigung der pH-Werte vom schwach alkalischen (pH(H_2O) 8,0 im A_i-Horizont des Carbonat-Rohboden) zum schwach sauren Bereich (pH(H_2O) 6,6 im A_h-Horizont der am stärksten entcarbonatisierten Alpinen Rendzina).
Eine weitergehende Bodenversauerung als zum schwach sauren Bereich wurde auf Hauptdolomit in der alpinen Rasenstufe der Allgäuer- und Lechtaler-Alpen nicht festgestellt.
Die Veränderungen der Bodeneigenschaften laufen nicht ganz parallel zueinander und zu den Pflanzengesellschaften. So findet man zum Teil bereits unter Caricetum firmae den lehmigen Ton, während das Seslerio-Semperviretum stellenweise auf Böden von etwas gröberer mechanischer Zusammensetzung stockt. Offensichtlich besteht die Möglichkeit, daß sich bestimmte Bodeneigenschaften untereinander sowie durch die Lage im Gelände kompensieren.
Unter Caricetum firmae findet man auch humusreichere A_h-Horizonte, als sie in der Tabelle 3 dargestellt sind. Bezüglich des Gehaltes an organischer Substanz gelten auch für die Allgäuer und Lechtaler Alpen in etwa dieselben Variationsgrenzen wie sie J.BRAUN-BLANQUET & H.JENNY (1926,p.321) bei ihren Untersuchungen gefunden haben. Bei den hohen Gehalten an organischer Substanz handelt es sich aber meist um Standorte, an denen auf Grund der besonderen Standortsverhältnisse als Dauerstadium das

Caricetum firmae anzusehen ist.
Diese Schwankungen in den Bodeneigenschaften unter einzelnen Pflanzengesellschaften vermögen aber nicht die allgemeine Tendenz der Boden- und Vegetations-Entwicklung auf dem Hauptdolomit in der alpinen Rasenstufe zu verdecken. Da die Entwicklung allgemein von einfachen zu komplizierteren Formen fortschreitet, erscheint es dennoch berechtigt, die angeführten Formen als aufeinanderfolgende (z.T.parallele) Stadien einer charakteristischen Entwicklungsreihe der Böden und der Vegetation anzusehen, die sich durch folgendes Schema darstellen läßt:

Hauptdolomit Fels	⟶	Initiale Alpine Polster-rendzina mit Dryas-reichen Initial-stadien des Caricetum firmae	⟶	Alpine Rendzina unter Caricetum firmae	⟶	Reife und relativ mächtige Alpine Rendzina unter Seslerio-Sempervíretum
↓						
Beweglicher Hauptdolomit Schuttrohboden mit Thlaspietum rotundifolii						

Im Sinne von H.ELLENBERG (1967) handelt es sich um eine primär progressive, aber nicht rein autogene (nur durch Tätigkeit der Pflanzen bedingte), sondern eine auto-allogene Sukzession, bei der auch die Zufuhr von Flugstaub und gelegentlich auch die Ansammlung des Verwitterungsmaterials aus den höheren Lagen von Bedeutung sind. Sie ist allgemein klimatisch bedingt und führt zu Böden und einer Vegetation, die auf dem Hauptdolomit der alpinen Rasenstufe als zonal zu betrachten sind. Wie immer bei zonalen Böden und Pflanzengesellschaften, ist auch hier ihre Entwicklung nur auf gewissen Standorten "mittlerer Eigenschaften" möglich. So ist z.B. nicht zu erwarten, daß das Seslerio-Sempervíretum jemals die stark windausgesetzten oder extrem schneefeuchten Standorte besiedeln könnte.
Die Entwicklung, dargestellt im obigen Schema, ist langfristig, sowohl wenn sie vom Hauptdolomit-Fels, als auch wenn sie vom Hauptdolomit-Schutt ausgeht. Der Übergang vom Fels zum Schutt spielte sich im Laufe langer Zeiträume ab, die meisten derzeitigen Schutthalden dürften bereits im Spätglazial angelegt gewesen sein, die Oberfläche der meisten von ihnen wird aber durch die weitere Zufuhr des Schuttes aus den Felsen und durch die Umlagerung ständig erneuert, so daß die Bodenbildungen auf den Schutthalden wesentlich jünger sind als die Schutthalden selbst.

Auch wenn die heutige Bodenbildung vom anstehenden Dolomit ausgeht, kann ihre Geschwindigkeit sehr unterschiedlich sein je nachdem ob der Dolomit als fester Felsen oder als tektonisch stark beanspruchtes leicht verwitterbares Ausgangsmaterial vorliegt.

Bei der Bodenbildung auf Carbonatgesteinen wurde wiederholt die Frage gestellt, ob dieser auf dem Gestein oder aus dem Gestein erfolgt (vgl. z.B. I.SOLAR 1964,p.8), wobei die Antwort oft unterschiedlich ausfiel. SOLAR's Antwort bezüglich des Wetterstein-Kalkes auf der Raxalpe war "auf dem Gestein". Diese Antwort gilt aber nicht für die hier dargestellten Böden, denn der physikalische Zerfall des Dolomits, dargestellt durch den Carbonatgehalt, baut nicht nur bei Rohböden, sondern auch bei den fortgeschritteneren Entwicklungsstadien einen wesentlichen Teil der mineralischen Substanz im Boden auf. Natürlich wird die Entwicklung zum Seslerio-Sempervíretum durch jede Zufuhr der nichtcarbonatischen anorganischen Substanz oder durch eventuelle alte Bodenreste und Zwischenschichten, die reicher an dieser Substanz sind, beschleunigt; der Dolomit spielt aber keineswegs die Rolle einer passiven Unterlage, auf der sich die Bodenbildung durch Staubzufuhr und Pflanzentätigkeit vollzieht, wie dies von I.SOLAR (1964) für Wettersteinkalk dargestellt wurde.

Eine analoge Weiterentwicklung der reifen Alpinen Rendzina zu stark sauren Böden und zum Caricetum curvulae, wie sie von BRAUN-BLANQUET & JENNY (1926) aus dem Schweizerischen Nationalpark beschrieben wurde, habe ich auf dem Hauptdolomit in den Lechtaler und Allgäuer Alpen trotz der bedeutend höheren Niederschläge nicht feststellen können. Im unteren Teil der alpinen Rasenstufe konnte sich auf Hauptdolomit auch kein Nardetum entwickeln, es sei denn, daß Hauptdolomit von nichtcarbonatischem Material überlagert wurde.

Auch OBERDORFER (1950) fand in den Allgäuer Alpen kein Caricetum curvulae und auch das kleine Fragment einer Sesleria disticha-Gesellschaft (Caricion curvulae) fand er am Kreuzeck nicht auf dem Hauptdolomit, sondern über den Allgäuer Schichten ("Lias-Fleckenmergel").

Aus den Lechtaler Alpen ist der Krummseggen-Rasen bekannt. Ich selbst hatte Gelegenheit, ihn im Gebiet des Monzabon-Sattels zu studieren, nicht aber über dem Hauptdolomit, sondern auf einem stark sauren Boden, der aus Kreideschiefern hervorgegangen ist.

Es handelt sich wohl um ein Curvulo-Nardetum im Sinne von OBERDORFER (1959).

Das Fehlen des Krummseggen-Rasens über dem Hauptdolomit kann demnach, wenigstens in den Lechtaler Alpen, nicht auf pflanzengeographische Gründe zurückgeführt werden. Der Grund dafür liegt in der Beschaffenheit des Hauptdolomits, der als sehr reines Carbonatgestein, auch unter Berücksichtigung der Zufuhr durch Flugstaub, nicht genügend nicht-carbonatisches Material liefert, um die Entwicklung eines stark sauren Bodens zu ermöglichen. Die Vegetation allein kann in der alpinen Rasenstufe auf Hauptdolomit keine stark sauren organischen Auflagen zustande bringen (das C:N-Verhältnis von 9,0 bis 12,6 zeugt von guten Humifizierungsbedingungen unter Caricetum firmae und Seslerio-Semperviretum). Wenn BRAUN-BLANQUET & JENNY (1926) im Schweizerischen Nationalpark die Entwicklung zum Caricetum curvulae festgestellt haben, dann bedurfte es dafür keines silikatischen Moränenmaterials über dem dortigen Carbonatgestein wie in den Beispielen von ELLENBERG (1953,p.243-244) aus dem Fimbertal und auch keiner ungenauen geologischen Karte. Die Caricetum firmae-Caricetum curvulae-Serie (BRAUN-BLANQUET 1964,p.662) ist dort auf dem Murtèr-Sattel (bzw. der Murtèr-Alp) durch die Beschaffenheit des Muttergesteins bedingt, das bereits von BRAUN-BLANQUET & JENNY (1926,p.328)richtig als "Rätkalk und Mergel" bzw. in der geologischen Karte von W.HEGWEIN (1934) als rätischer Plattenkalk bezeichnet wird.

Der namengebende Kalk dieser Gesteinsserie enthält im Bereich des Murtèr-Sattels etwa 4-10 mal soviel nicht-carbonatische Bestandteile wie der meiste Hauptdolomit, außer ihm sind aber noch Mergel und Tonschiefer in dieser Serie enthalten, die noch mehr dem Boden die anorganische nicht-carbonatische Substanz liefern (eigene unveröffentliche Untersuchungen). Kein Wunder daher,daß dort die Boden- und Vegetationsentwicklung weiter fortschreiten konnte als auf dem Hauptdolomit.

Somit bestätigen diese Untersuchungen die Auffassung von ELLENBERG (1953,1963), daß die Boden- und Vegetationsentwicklung auf sehr reinen Carbonatgesteinen nicht zu sehr sauren Böden und zum Caricetum curvulae führen. Sie stehen aber auch keineswegs im Widerspruch zu den Ergebnissen von BRAUN-BLANQUET & JENNY (1926), und von BRAUN-BLANQUET (1964,p.666 ff.), die im Bereich der nicht ganz so reinen Carbonatgesteine eine Entwicklung zu stark sauren Böden und zum Caricetum curvulae festgestellt haben; diese Ergebnisse darf man nur nicht auf alle Carbonatgesteine übertragen.

ZUSAMMENFASSUNG

Die zonale Boden- und Vegetationsentwicklung auf dem Hauptdolomit führt in der alpinen Rasenstufe der Allgäuer und Lechtaler Alpen vom Hauptdolomit-Fels mit Fels- und Spaltenbewohnern oder vom Hauptdolomit-Schuttrohboden mit Thlaspietum rotundifolii über die Initiale Alpine Polsterrendzina mit *Dryas*-reichen Initialstadien des Caricetum firmae und Alpine Rendzina mit typischem Caricetum firmae zur reifen und relativ mächtigen Alpinen Rendzina unter Seslerio-Semperviretum.

Im Laufe dieser Bodenentwicklung vergrößert sich der Anteil der Feinerde, wenigstens in den obersten Horizonten; Ton- und Schluffgehalt nehmen bei gleichzeitiger starker Verminderung des Grobsand-Anteils zu; es werden stark humose A_h-Horizonte gebildet; Der Gehalt an Stickstoff nimmt zu, der an Carbonaten nimmt ab. Die pH-Werte sinken von ca pH(H_2O) 8,0 auf 6.6.

Zu einer weitergehenden Versauerung des Bodens und zur Entwicklung des Caricetum curvulae kommt es auf Hauptdolomit im untersuchten Gebiet nicht. Dazu reicht der Gehalt an nichtcarbonatischen Bestandteilen im Hauptdolomit nicht aus.

SUMMARY

Development of soil and vegetation on the Hauptdolomite in the belt of alpine meadows of the Allgäu and Lechtaler Alps

The zonal evolution of soil and vegetation on the Hauptdolomite in the belt of alpine meadows begins with Hauptdolomite-rock bearing some plants growing on rocks and Hasmophytes, or with Hauptdolomie-rubble raw soil with Thlaspietum rotundifolii, and comes then to the Initial Alpine Cushion-Rendzina with initial stages of Caricetum firmae rich with *Dryas*, and Alpine Rendzina with typical Caricetum firmae and at last to the mature and relatively mighty Alpine Rendzina under Seslerio-Semperviretum.

In course of this soil evolution the portion of fine earth enlarges, at least in the upper horizons; the contents of clay and silt increase with simultaneous considerable diminution of coarse sand (Table 2); A_h-horizons very rich of organic matter are formed; the contents of nitrogen increase, that of carbonates decrease. The pH changes from pH(H_2O) 8,0 to 6,6 (table 3). Stronger acidification of the soil and evolution of Caricetum curvulae on Hauptdolomite does not take place in the investigated region. The contents of non-carbonatical components in Hauptdolomite (table 1) is not sufficient for that.

LITERATUR

AMPFERER,O. -1932- Erläuterungen zu den geologischen Karten der Lechtaler Alpen i.M. 1:25.000.- Geologische Bundesanstalt. Wien.

BRAUN-BLANQUET,J- -1964- Pflanzensoziologie.- Wien, New York.

-- -- & JENNY,H. -1926- Vegetations-Entwicklung und Bodenbildung in der alpinen Stufe der Zentralalpen (Klimaxgebiet des Caricion curvulae). Mit besonderer Berücksichtigung der Verhältnisse im schweizerischen Nationalpark.- Denkschr.Schweiz.Naturf.Ges.63 (2): I-IX und 183-349. Ergebn.wiss.Unters.schweiz.Nationalpark 4. Zürich.

ELLENBERG,H. -1953- Führt die alpine Vegetations- und Bodenentwicklung auf reinen Karbonatgesteinen zum Krummseggen-Rasen (Caricetum curvulae)?- Ber.dtsch.bot.Ges.66. Stuttgart.

-- -- -1963- Vegetation Mitteleuropas mit den Alpen in kausaler,dynamischer und historischer Sicht.- Stuttgart.

-- -- -1967- Wichtige Bedriffe der Sukzessions-Forschung. Diskussions-Grundlage für das Symposion 1967 in Rinteln. Dieser Bericht p.

GAMS,H. -1961- Die Pflanzenwelt. In: K.Ilg (Edit.): Landes- und Volkskunde, Geschichte, Wirtschaft und Kunst Vorarlbergs 1. Landschaft und Natur.- Innsbruck.

GRACANIN,Z. -1968- Höhenlage, Aufbau und bodensystematische Stellung der Streifenböden in den Allgäuer und Lechtaler Alpen und Engadiner Dolomiten.- Mitt.dtsch.bodenkdl.Ges.8. Göttingen.

HAMANN,P.J. -1956- Die Geologie des Hochvogel-Gebietes in den Allgäuer Alpen.- Marburg/Lahn. (Diss.).

HEGWEIN,W. -1934- Geologische Karte der Quaterval-Gruppe im Schweizerischen Nationalpark 1:25 000. Ergebn.wiss.Unters.Schweiz.- Bern.

HUCKRIEDE,R. -1957- Die Geologie der Umgebung von Kaisers (Lechtaler Alpen).- Marburg/Lahn. (Diss.).

JACOBSHAGEN,V. -1957- Geologie des Hornbach-Gebietes und der Allgäuer Hauptmulde südöstl.Oberstdorf.- Marburg/Lahn. (Diss.).

JENNY,H. -1930- Hochgebirgsböden. In: Blanck,E.(Edit.): Handbuch der Bodenlehre 3.- Berlin.

OBERDORFER,E. -1950- Beitrag zur Vegetationskunde des Allgäu.- Beitr.naturkdl.Forsch.Südw.-Deutschl.9 (2). Karlsruhe.

-- -- -1959- Borstgras- und Krummseggen-Rasen in den Alpen. Ein Beitrag zur Abgrenzung der Nardo-Callunetea gegen die Caricetea curvulae.- Beitr.naturkdl.Forsch.Südw.-Deutschl.18: 117-143. Karlsruhe.

-- -- -1962- Pflanzensoziologische Exkursionsflora für Süddeutschland und angrenzende Gebiete. 2.Aufl.- Stuttgart.

PIWOWAR,A. -1903- Über Maximalböschungen trockener Schuttkegel und Schutthalden.- Vierteljahresschr.Naturforsch.Ges.48: 335-359. Zürich.

SCHÜLER,B. -1959- Geologie der Berge südlich von Mittelberg im Kleinen Walsertal.- Berlin. (Diss.Fr.Univ.Berlin).

SENDTNER,O. -1854- Die Vegetations-Verhältnisse Südbayerns nach den Grundsätzen der Pflanzengeographie und mit Bezugnahme auf Landeskultur. Literarisch-artistische Anstalt.- München.

SOLAR,I. -1964- Zur Kenntnis der Böden auf dem Raxplateau.- Mitt.Österr. bodenkdl.Ges.8: 1-72. Wien.

WESTRUP,J. -1966- Das Gerstrubener Fenster in den Allgäuer Kalkalpen und seine Umgebung.- Marburg/Lahn. (Diplom-Arbeit).

DISKUSSION

H.ELLENBERG

Herr Dr.GRAČANIN hat gerade mit den letzten Bildern gezeigt,wie sehr die primäre Verschiedenheit der Ausgangssituation für die Entstehung der Böden von Bedeutung ist. Die klassische Reihe von BRAUN-BLANQUET & JENNY ist einfach eine Reihe, die man als Zonierung bezeichnen muß, die zwar auf dem geologisch gleichen Substrat gefunden wurde, aber dieses geologisch-zeitlich Gleiche ist petrographisch von vornherein verschieden gewesen. Und man muß sich also sehr hüten, solche Zonierungen im Sinne der Entwicklungslehre zu deuten. Ich glaube, daß auch einiges von dem, was Herr GRAČANIN zeigte, von vornherein verschiedene Ausgangsmaterialien waren, auch auf dem Kalk. Und zwar ist die Wirkung der Eiszeit dabei zu beachten. Das Feinmaterial, das Herr Dr. GRAČANIN als von den Felsen herunterrieselnd erklärte,dieses erstaunlich viele Feinmaterial in den Schutthaldenböden, ist nicht, nach meiner Erfahrung, eine reine Schutthaldenbildung, sondern es sind Reste von Moränen. Unsere Alpen sind besonders in der vorletzten Eiszeit bis zu großer Höhe von Eis bedeckt gewesen, die Zentralalpen bis 3000 m Höhe, und nur wenige Felsrippen - man sah auf den Bildern deutlich die morphologische Grenze - schauten aus diesem allgemein sich bewegendem Eis oder aus den seitlichen Firnfeldern heraus. Es ist bekannt, daß die schürfende Wirkung des Eises eine sehr wirksame Verfeinerung des Materials bewirkt, eben diese Entstehung der mehr oder weniger lehmigen Moränenbestandteile, die die Feinerdekomponente in vielen dieser Profile darstellen. Die Seitenmoränen sind zwar nicht so scharf geschliffen, daß darin abgerundete Steine vorkämen, aber vieles von dem Feinmaterial, das sich in seinen Profilen zeigte, würde ich als geologisch verschieden, nämlich aus einer anderen eiszeitlichen Geschichte her,erklärbar halten, so daß selbst in dem Kalkbereich nicht immer Entwicklung, sondern auch Zonierung vorlag.

R.TÜXEN

Vor 40 Jahren lernten wir, in Nordwestdeutschland ist das F a g e t u m Klimax, auch im Flachlande, d.h.alle primären Gesteinsunterschiede werden unter der alles beherrschenden Wirkung des Allgemeinklimas soweit ausgeglichen, daß zum Schluß ein klimatisch bedingter einheitlicher Boden und eine zu diesem Boden gehörige, einheitliche und klimatisch bedingte Vegetation ist. Mit

DIEMONT plagten wir uns immer wieder mit Zweifeln, ob wirklich am Teutobuger Wald, oder im Harzvorland, oder hier im Wesergebirge, dort, wo der Kalk - und zwar zum Teil ein äußerst hochprozentiger Kalk etwa der Kreide - von glazialem Material überlagert wurde, eine Entwicklung vom Kalk-Fagetum - wir sprachen damals vom Fagetum calcareum - bis zum Querco-Betuletum vor sich gegangen sei. Wir haben in jahrelangem ernstem Ringen gegen die Lehre, der wir folgten, uns zu Begriffen durchgerungen, die wir damals 'Klimaxgruppe' und 'Paraklimax' nannten. D.h. wir haben gesehen, daß jedes Gestein, Kalk, Silikat und Quarzgestein, eine eigene parallele Entwicklung zu einem eigenen Endstadium durchmachten, die mit einander nichts zu tun haben. Es gewährt mir eine große Befriedigung zu sehen, wie jetzt diese von uns damals begonnene revolutionäre Bewegung sich immer noch fortsetzt.

R.CARBIENER
Wir haben in der Pollenanalyse auch ein sehr schönes Mittel, um syngenetische Studien im Gebirge zu machen, weil ja meistens die Hochgebirgsböden äußerst humusreich sind, und die Pollen sich hier gut erhalten. Da kann man darauf hinweisen, daß in diesen Fällen, wo eine Pollenanalyse von solchen alpinen Humusböden durchgeführt wurde, es sich herausstellte, daß von Anfang an die gleiche Gesellschaft auf diesen Böden jahrtausendelang gewachsen ist. In den Südsevennen, in den Vogesen, in den Monts d'Or sind solche Untersuchungen von LEMEE angestellt worden. Nur ganz am Anfang bestanden Pionierstadien von Moosen usw., die pollenanalytisch nicht erfaßbar sind. In der Nacheiszeit hat sich aber dann sehr schnell eine Dauergesellschaft eingestellt, die absolut stabil blieb. Und man kann annehmen, daß wir in der alpinen Stufe eben mit solchen Dauergesellschaften zu tun haben, die eine Anpassung an die so äußerst wechselvollen geomorphologischen Prozesse sind, die sich aber auch sehr komplex mit alten geomorphologischen Prozessen mischen, und daß dann also in den meisten Fällen keine wirkliche Entwicklung einer Gesellschaft zur anderen vorliegt, sondern daß die meisten Gesellschaften außerordentlich stabil sind.

Otti WILMANNS
Vor Jahren wurde mir gesagt, daß die Auffassungen von KUBIËNA (1953) ganz überholt seien. Wenn ich jetzt diesen Vortrag höre

und mit KUBIËNA vergleiche, dann sehe ich eigentlich keinen Unterschied. Können Sie mir sagen, wo der Unterschied vielleicht zu suchen ist?
Gibt es eine Sukzession vom Thlaspietum rotundifoliae zu Leontidetum montani, oder sind auch hier die edaphischen Unterschiede primär vom Gestein bedingt?

Z.GRAČANIN

Die Bodensystematik von KUBIËNA ist größtenteils noch gültig. Es gibt zwar einiges, was in der Bodensystematik weiterentwickelt wurde, für die alpine Rasenstufe haben wir aber noch nichts Besseres.
Für eine eventuelle Entwicklung von Thlaspietum zum Leontidetum habe ich keine Beobachtungen.
Dann zu der Bemerkung von Herrn Prof. ELLENBERG: Ich würde das Feinmaterial in den untersuchten Bodenprofilen nicht ganz oder vorwiegend auf die Eiszeit zurückführen. Bei den wenig entwickelten Bodenprofilen auf der Schutthalde spricht schon der hohe Carbonatgehalt in der Feinerde gegen eine Verfrachtung durch Eis; beim glazialen Transport müßte nicht-carbonatisches Fremdmaterial aus verschiedenen in der Umgebung anstehenden geologischen Formationen beigemischt sein, was nicht der Fall ist. Auf der Schutthalde konnte man dagegen junge Zufuhr von Hauptdolomit-Zersatz aus den Gipfel-Felsen des Weißschrofens beobachten. Andererseits stimme ich mit Ihnen überein, daß in den Alpen in sehr vielen Fällen eine Beimischung von Moränenmaterial oder Fremdmaterial vom Oberhang die Boden- und Vegetationsentwicklung auf Carbonatgesteinen beeinflußt, und zwar manchmal bis in die Gipfellagen. So liegt z.B. im Schweizerischen Nationalpark noch in einer Höhe von 2600 m ü.NN auf dem isolierten, aus Wettersteinkalk bestehenden Gipfelplateau des Munt la Schera erratisches nichtcarbonatisches Material aus dem Verrukano möglicherweise auch Buntsandstein, das als letzter Rest von Moränen der größten Vereisung gedeutet wird. Es war daher nicht leicht, Stellen zu finden, wo wie keine fremde Beimischung haben.

Bild 1. Nebelhorn, 2224 m, Allgäuer Alpen: Blick auf den Hauptdolomit-Kamm ö vom Gipfel. Auch die scheinbar kahlen Gipfelfelsen liegen ganz in der alpinen Rasenstufe. Matten: S e s l e r i o - S e m p e r v i r e - t u m Br.-Bl.1926 auf Alpiner Rendzina (vgl.Profil 15).

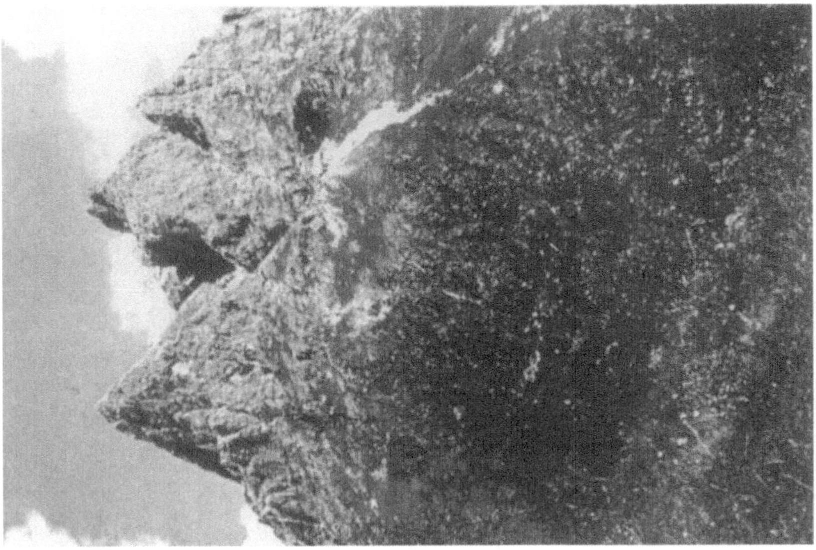

Bild 2. Widderstein, 2536 m, Allgäuer Alpen, Hauptdolomit, erscheint auf der Südseite von weitem ab ca 2130 m kahl; in dieser Höhe liegt aber nicht die obere klimatische Grenze der alpinen Rasenstufe. Das Aufhören der Alpenmatten ist geomorphologisch und petrographisch bedingt.

Bild 3. Widderstein: S e s l e r i o - S e m p e r v i r e t u m auf gut entwickelter Alpiner Rendzina (Profil 41) in 2420 m Höhe. Die geringere Neigung ermöglichte hier die Ansammlung des Verwitterungsmaterials.

Bild 4. Walser Gaißhorn, 2366 m, Allgäuer Alpen, vom Gemstelpaß (1971 m), größtenteils Hauptdolomit. Geschlossene Rasenflächen reichen bis zum Gipfel (Profil 35). Oberste Latschen bei ca 2050 m.

Bild 5. Bizarre Hauptdolomit-Grate und ausgedehnte Schutthalden unter dem Hochgehren (2253 m) und Schüsser (2258 m hinten) in den Allgäuer Alpen. Starke Schuttzufuhr besonders während und gleich nach der Schneeschmelze, wenn die Gefrornis im Fels auftaut.

Bild 6. Angewitterte Hauptdolomit-Felsen auf dem Südhang der Weißschrofenspitze (Lechtaler Alpen) bei 2440 m. Vorne noch Schnee. Trotz scheinbarer Kahlheit findet man hier und da Bewuchs, nicht nur Felsspaltbewohner, sondern auch Fragmente der Alpenmatten-Gesellschaften.

Bild 7. Hauptdolomitschutt-Rohboden (Profil 60) mit *Thlaspietum rotundifolii* Br.-Bl. auf Südhang der Weißschrofenspitze (Lechtaler Alpen) bei 2380 m. Die Bodenoberfläche ist fast völlig mit Steinen und Grus bedeckt.

Bild 8. Initiale Alpine Polsterrendzina über anstehendem Hauptdolomit-Fels (Profil 62) unter *Dryas*-Stadium des *Caricetum firmae* (Kerner) Br.-Bl. Die Mächtigkeit des A_h-Horizontes wird gegen den rechten Bildrand geringer, wenig weiter keilt der A_h-Horizont gänzlich aus. Südhang der Weißschrofenspitze bei 2440 m.

Bild 9. Initialstadium des Caricetum firmae (Kerner) Br.-Bl. nur 20% der Fläche deckend, auf Initialer Alpiner Polsterrendzina auf Hauptdolomit-Schutt. Südhang der Weißschrofenspitze (Lechtaler Alpen) bei 2380 m.

Bild 10. Alpine Rendzina (Profil 38) auf Hauptdolomit-Schutt unter Caricetum firmae (Kerner) Br.-Bl. Walser Gaißhorn 2292 m, Exposition W.

Bild 11. Alpine Rendzina (Profil 41) unter Seslerio-Semperviretum Br.-Bl. Widderstein (Allgäuer Alpen) bei 2420 m.

Bild 12. Seslerio-Semperviretum auf Alpiner Rendzina (Profil 35) unter dem Gipfel des Walser Gaißhorn (2366 m, Allgäuer Alpen).

Alle Auf.v.Verf.

DER UNBESTÄNDIGE KLIMAX UND SEINE KARTIERUNGSPROBLEME

A.W. K ü c h l e r

Einleitung.

Auf den weiten Ebenen zwischen Canada und Texas hat sich im Lauf der Zeit eine Graslandvegetation entwickelt, die besonders gut an die klimatischen Eigenheiten dieser Region angepaßt ist. Der wichtigste Zug des Klimas ist zweifellos der unzuverlässige Niederschlag. Die jährlichen und monatlichen Mittelwerte des Regenfalles sind zwar schon seit Jahrzehnten bekannt, sie verbergen aber den kritischsten Charakterzug, die Zuverlässigkeit. So kann ein Jahr zum Beispiel einen durchschnittlichen Regenfall verzeichnen und trotzdem an Dürre leiden, wenn nämlich der Regen in wenigen schweren Wolkenbrüchen herniedergeht, die rasch aufeinander folgen. WEAVER & ALBERTSON (1956), zwei der hervorragensten Autoritäten auf dem Gebiet der Prairieforschung, stimmen darin überein, daß periodische Dürren für das Klima charakteristisch sind. Solche Schwankungen zwischen trockenen und humiden Perioden haben einen höchst dynamischen Charakter der Vegetation zur Folge.

Das Phänomen der weiten klimatischen Schwankungen und die sich hieraus ergebenden Schwankungen der Vegetation ist keineswegs auf Nordamerika beschränkt. Es ist auch aus Argentinien, der Sowjet-Union, wie auch aus Indien, China und Australien und aus Afrika beiderseits des Äquators bekannt. Es ist also nicht gerade außergewöhnlich. In Wüsten ist der Regenfall am wenigsten zuverlässig. Die Klimaxvegetation bleibt sich aber im wesentlichen gleich, einerlei, ob es ungewöhnlich viel oder wenig regnet. Ähnliches trifft auf die humiden Klimazonen zu, wo die Vegetation gewöhnlich aus Wäldern besteht, die eine seltene Dürre natürlich überdauern. Selbst die krautreiche Vegetation des Waldbodens leidet relativ wenig, weil die schützenden Bäume die scharfe Zunahme an Temperatur und Wind abschwächen. Die Phytozönosen der Prärie genießen keinen solchen Schutz.

Dynamik und Klimax der Prärie

CLEMENTS wuchs in der Prärie auf, und man darf daher erwarten, daß seine Klimaxtheorie besonders an die Grasland-Vegetation Nordamerikas angepaßt ist. Die fundamentale Annahme ist, daß die Vegetation durch verschiedene Phasen der Sukzession das Klimax-stadium erreicht und so stabilisiert wird, weil sie dann in Harmonie mit dem Klima ist. Eine weitere Annahme ist, daß das Klima und andere Züge der Umwelt auch stabil sind. Ohne Zweifel trifft das meistens zu. Es ist aber interessant, daß ausgerechnet da, wo CLEMENTS seine Theorie entwickelte, das Klima großen und heftigen Schwankungen unterworfen ist, die sich ihrerseits getreu in den dramatischen Schwankungen der Vegetation widerspiegeln. Diese Schwankungen bedeuten, daß eine gegebene Stelle zu verschiedenen Zeiten von verschiedenen Phytozönosen eingenommen werden kann, und daß diese Phytozönosen verschiedene Phasen des lokalen Klimax sind.

In Kansas nehmen die Niederschläge von Osten nach Westen ab. Im Osten herrschen daher Hochgras-Prärien vor, im Westen Kurzgras-Prärien, und dazwischen liegt der breite Gürtel der gemischten Prärie. Die Veränderung von Osten nach Westen ist aber nicht einfach eine Abnahme der Niederschläge, sondern eine starke Zunahme in deren Unzuverlässigkeit. Es regnet oft oder selten, früh oder spät oder überhaupt nicht, und diese Unzuverlässigkeit liegt außerhalb der Toleranz der meisten östlichen Arten. Im Gegensatz hierzu können viele westliche Kurzgräser in eine Ruheperiode eintreten, wenn die Dürre zu intensiv wird, doch setzen sie ihr Wachstum fort, sobald genügend Regen fällt.

Die Hochgrasprärie wird vor allem von *Andropogon gerardii*, *A. scoparius*, *Sorghastrum nutans* und *Panicum virgatum* beherrscht. Viele andere Gräser und Kräuter sind häufig. In ihrem westlichen Teil geht die Hochgras-Prärie in die gemischte Prärie über. Während der großen Dürre der dreißiger Jahre wandelte sich ein 240 km breiter Streifen von Hochgras-Prärie in gemischte Prärie um, besonders in Kansas, Nebraska und South Dakota. Die Artenzusammensetzung der Prärie wurde manchmal völlig verändert, und sogar reine Phytozönosen der Kurzgras-Prärie waren nicht selten. Die Hochgräser starben, und westliche Gräser, besonders *Agropyron smithii* übernahmen schnell den freigewordenen Raum. Mit dem Wiedereinsetzen der Regen dehnten die Hochgräser ihr Gebiet wieder westwärts aus, aber sie mußten in ein dicht besetztes Gebiet einwandern. Ihre Rückkehr war daher langsam und ungleichmäßig. Selbst zwölf Jahre nach dem Ende der Trockenheit gab es immer noch viel gemischte Prärie im Gebiet der Hochgräser, und dann setzte die sehr intensive Dürre der fünfziger Jahre ein, die den Prozeß wieder umkehrte. Je nach den lokalen Verhältnissen brauchte die Hochgras-Prärie zwanzig Jahre, ehe sie wieder so war wie vor der großen Dürre der dreißiger Jahre. Andererseits fanden WEAVER & BRUNER (1945) im östlichen Nebraska an einer Stelle eine mehr oder minder normale Hochgras-Prärie schon nach acht Jahren, besonders nach drei sehr regenreichen Jahren.

Die gemischte Prärie wird vor allem von *Agropyron smithii* und *Aristida longiseta* beherrscht, desgleichen von *Bouteloua gracilis* und *B. curtipendula, Koeleria cristata, Sporobolus cryptandrus* und *Stipa comata*. Die Dominanz von *Bouteloua gracilis* deutet auf den westlichen Einschlag und auf die relativ einfache Struktur der gemischten Prärie mit einem kontinuierlichen Kurzgrasrasen und einer offenen Synusie von höheren Gräsern. *Andropogon gerardii* und *A. scoparius* kommen auch vor, wurden aber von CLEMENTS als post-climax betrachtet, eine für ihn typische aber schwer zu verteidigende Ansicht. Wesentlich aber ist die Erkenntnis, daß das Areal der gemischten Prärie dauernd wächst oder schrumpft, und daß eine einzelne Beschreibung ihr nicht gerecht werden kann.

Im Westen herrscht die Kurzgras-Prärie, dominiert von *Bouteloua gracilis* und *Buchloe dactyloides*. In der Übergangszone spielt die Topographie eine wichtige Rolle. Da bedecken die niedrigen Gräser die weiten Flächen der Hochebenen, aber in Depressionen, Talböden, buffalo wallows u.s.w. finden sich Hochgräser, weil (oder wenn) sich dort Wasser ansammelt. Hochgräser folgen auch den Flüssen nach Westen und Bächen talaufwärts weit in die trokkenen Ebenen, genau wie das die Galeriewälder in tropischen Savannen auch tun. Auf den Hängen spielen dann die mittleren Gräser eine wichtige Rolle.

In Dürreperioden verbreiten sich die Kurzgräser an den Hängen herab und vertreiben dort die höheren Gräser. So kann *Andropogon scoparius* völlig verschwinden, selbst wo es in großen Mengen auftritt. Andererseits breiten sich die mittleren Gräser in regenreichen Jahren nicht nur hangaufwärts, sondern auch weit über die Hochflächen aus, wo sie dann zusammen mit *Bouteloua gracilis* und *Buchloe dactyloides* wachsen. Gleichzeitig dehnen die Hochgräser in den tiefsten Stellen ihr Gebiet auf die unteren Hänge aus. Wie im Osten wird hier daher das Land bald von einer Phytozönose, bald von einer anderen besetzt. *Andropogon scoparius* verdrängt die Kurzgräser durch Beschattung, während die Kurzgräser durch ihre größere Dürreresistenz die höheren Gräser in Trocken-

perioden vertreiben können. Die tiefer gelegenen Stellen dienen daher den höheren Phytozönosen als Refugium in trockenen Jahren und als Verbreitungszentrum in feuchteren Zeiten.
Diese Beobachtungen bestätigen, daß die Vegetation in der zentralen Prärie sehr dynamisch ist und zwischen weiten Grenzen hin- und herschwankt infolge von Unregelmäßigkeiten in der zeitlichen und örtlichen Verteilung der Niederschläge. Die klimatischen Schwankungen sind hier fundamental,aber die edaphischen und anderen Eigenschaften der Biotope tragen das ihrige dazu bei, die Dynamik der Vegetation zu komplizieren. So kommt es vor, daß eine Gegend an akuter Dürre leidet, wenn unweit davon der Wasserbedarf reichlich gedeckt sein kann.

Kartierungsprobleme.

Die Kartierung der Vegetationsdynamik ist bereits an anderen Stellen besprochen worden (KÜCHLER 1961,1967). Es handelt sich dabei aber eigentlich meistens darum, Sukzession zu kartieren. Hier jedoch handelt es sich nicht um Sukzession, sondern ausschließlich um Klimaxvegetation. Nachdem die Natur der Vegetationsschwankungen nun erkannt ist, muß der nächste Schritt zur Entwicklung von Methoden führen, die es erlauben, die Klimaxdynamik zu kartieren.
Das fundamentale Problem besteht in der Tatsache, daß diese Klimaxdynamik aus zwei einander entgegengesetzten Bewegungen in der gleichen Gegend besteht.Da das Problem kompliziert ist, ist es vielleicht am besten, damit zu beginnen, daß man die Grenzen des betroffenen Gebietes feststellt.
Im Osten ist es immer feucht. Trockene Jahre kommen natürlich vor, aber die Dürreperioden sind weder häufig noch extrem, so daß sie die Klimaxvegetation nicht wesentlich beeinflussen. So führte die große Dürre der dreißiger Jahre zwar zu einer verminderten Produktion der Prärien von Iowa, Missouri und weiter östlich, aber die Arten blieben die gleichen wie in feuchten Jahren. Dies traf auch auf Prärien im östlichen Nebraska zu. Das bedeutet also, daß in solchen Prärien dieselben Dominanten hervortreten, gleichgültig ob es trocken oder feucht ist.
Weiter westlich ist das natürlich anders, und lange Trockenperioden können zu einer völligen Veränderung der Artenzusammensetzung der Prärie führen. Es muß also eine Grenze zwischen der Hochgras-Prärie, die trotz Klimaschwankungen gleich bleibt, und der Hochgras-Prärie geben, die sich infolge von Klimaschwankun-

gen mehr oder minder in eine gemischte Prärie verwandelt. Diese Linie sei hier die östliche Grenze genannt, weil sich die Diskussion auf das zentrale Prärie-Gebiet bezieht. Der östlichen Grenze entspricht natürlich eine westliche. Westlich von dieser wird die Kurzgras-Prärie vor allem von *Bouteloua gracilis* und *Buchloe dactyloides* beherrscht, gleichgültig ob trocken oder feucht. Östlich von dieser Grenze dagegen liegt das Gebiet, in das die Gräser der gemischten Prärie einwandern, wenn eine längere Regenperiode das ermöglicht, aus dem sie aber in Trockenjahren wieder verschwinden. Es ist also auch dies eine Grenze, die eine unveränderliche Kurzgras-Prärie von einer veränderlichen trennt.

Die östlichen und westlichen Grenzen schließen ein Gebiet ein, in welchem die Prärie Schwankungen unterliegt, im Gegensatz zu den Gebieten außerhalb der Grenzen, in welchen Schwankungen kaum Bedeutung erlangen. Das Ungewöhnliche an diesen Grenzen liegt darin, daß sie zeitweise die gleiche Vegetation auf beiden Seiten haben können, d.h. in solchen Zeiten scheinen sie keine ungleichen Phytozönosen von einander zu trennen. Mindestens zeitweise laufen diese Grenzen also durch einen Vegetationstyp anstatt drum herum, wie Grenzen das normalerweise tun. Dieser Fall tritt im Westen in trockenen Jahren ein, wenn sich die Kurzgras-Prärie nach Osten ausdehnt; der Fall tritt auch im Osten in feuchten Jahren ein, wenn die Hochgras-Prärie nach Westen wandert. Sind die Feuchtigkeitsverhältnisse aber umgekehrt, dann sind die Grenzen normal, d.h. sie liegen dann zwischen verschiedenen Vegetationstypen. Die Lage dieser Grenzen ist noch nicht klar erkannt. Es ist auch fraglich, ob sie je mit großer Genauigkeit festgestellt werden kann. Vielleicht findet man die Grenze am besten, wenn man die An- oder Abwesenheit einer Art benutzt, wie TÜXEN das für verschiedene Vegetationseinheiten schon vor Jahren empfahl.

Soll eine Vegetationskarte kleinen Maßstabs hergestellt werden, so kann man zunächst annehmen, daß der kleine Maßstab es nicht zuläßt, Schwankungen zu zeigen. Der Nachteil dieser Methode liegt darin, daß der Leser irregeführt wird. Die Karte macht nämlich den Eindruck, als sei die Vegetation im westlichen Kansas genau so stabil wie in Tennessee, und das ist sicher nicht der Fall.

Man kann Bewegung andeuten, indem man sowohl auf der Karte als auch in der Legende darauf hinweist, oder nur in der Legende. Die üblichen Methoden, Übergänge zu kartieren sind für diesen Zweck ungeeignet, da sie keine Bewegung, keine Richtung anzeigen. Bewegung und Richtung sind hier die Schlüsselworte, die irgendwie ihren Ausdruck finden müssen. Wenn Richtung gezeigt werden kann, dann bedeutet das bereits Bewegung, denn wo keine Bewegung ist, da ist auch keine Richtung.

Die einfachste Lösung ist der Gebrauch von drei Linien auf der Karte mit den entsprechenden Erklärungen in der Legende. Man könnte dies die "Drei Linien-Methode" nennen. In diesem Fall werden zwei Vegetationseinheiten durch eine Linie von einander getrennt. Ein Übergang wird nicht gezeigt, und die Trennungslinie erscheint in ihrer mittleren Lage. Sie heißt daher die "mittlere Grenze". Eine solche mittlere Grenze kann z.B. eine Kurzgras-Prärie, auf der Karte in gelb, von einer gemischten Prärie, auf der Karte in grün, trennen. Die mittlere Grenze zeigt man am besten als dünne schwarze oder graue Linie. Auf jeder Seite ist eine weitere Linie, die sogenante "äußere Grenze". Die äußere Grenze östlich der mittleren Grenze ist eine gelbe Linie; sie zeigt an, wie weit die Kurzgras-Prärie nach Osten vordringt,wenn die Dürre anhält. Entsprechend zeigt eine grüne Linie westlich der mittleren Grenze das maximale Vordringen der gemischten Prärie westwärts in regenreichen Jahren an. An der Ostgrenze der gemischten Prärie, wo diese auf die Hochgras-Prärie trifft,kann man die gleiche Methode anwenden. Der Leser muß jedoch verstehen, daß die äußeren Grenzen das maximale Vordringen anzeigen, was durchaus nicht immer der Fall zu sein braucht. Und selbst, wenn es so ist, heißt das doch noch nicht, daß ein Typ völlig verschwindet. Das kommt ganz auf die lokalen Verhältnisse an. Die "Drei Linien Methode" hat den großen Vorteil, daß man sie bei fast allen Maßstäben anwenden kann.

Vielleicht bedeutet es einen Fortschritt, wenn man die Typen in Fragmente auflöst. Diese können immer kleiner werden, weniger zahlreich sein, und so auf eine Bewegung hinweisen. Ein Vordringen ist gewöhnlich massiv an der Grenze, wo die Fragmente deshalb am größten sind; an der äußeren Grenze ist von einem Vordringen nur noch wenig zu spüren, daher sind die Fragmente hier ganz klein. Gleichzeitig könnte diese Darstellung einen Rückzug bedeuten; die wenigen kleinen Stellen an der äußeren Grenze sind dann Relikte. Man könnte das also als ein Hin- und Herfluten auslegen. Die Methode ist unzulänglich, weil sie den Eindruck macht, als sei ein Typ stabiler als der andere; auch ist die Bewegung nicht offenbar und man hält die Zone mit den Fragmenten für eine Übergangszone.

Eine andere Art der Darstellung ist die "Zickzack Methode", welche die Fragmentierung vermeidet. Die Grenze zwischen zwei Vegetationseinheiten läuft hin und her und macht den Eindruck einer Durchdringung. Imaginäre Linien, welche die Spitzen der Keile verbinden, entsprechen den äußeren Grenzen der "drei Linien Methode"; die mittlere Grenze fehlt hier. Die Winkel an den Keilspitzen sind alle gleich, und die entsprechenden Seiten sind einander parallel. Daher erscheinen kleine Keile viel schärfer, größere Keile scheinen eine breitere Basis zu haben. Obwohl die Keile in eine bestimmte Richtung weisen, bleibt der Gesamteindruck doch der eines Übergangs. Bewegung muß daher in der Legende erklärt werden.

Die wirkungsvollste Methode, eine Richtung anzudeuten, benutzt Pfeile. Pfeile können lang oder kurz, dick oder dünn sein; man hat eine große Auswahl. Kartierer können daher verschiedene Pfeile für verschiedene Zwecke anwenden, und so verschiedene Eindrükke erwecken.

Ein einfacher Gebrauch von Pfeilen beruht auf der "drei Linien Methode". Zwei Vegetationseinheiten, z.B. eine Kurzgras-Prärie und eine gemischte Prärie, werden von einander durch die mittlere Grenze getrennt. Dann werden sehr schlanke, parallele Pfeile darüber gedruckt, und zwar so weit, wie die äußeren Grenzen es erlauben. Nur werden die äußeren Grenzen natürlich nicht gezeigt. Die Pfeilspitzen sollten klein gehalten werden, damit sie sich nicht gegenseitig überdecken. Sie werden an beiden Enden angebracht, um so eine Bewegung in zwei entgegengesetzte Richtungen anzudeuten. Die Pfeile sind leicht zu konstruieren und werden in grau oder schwarz gedruckt; so stören sie die Vegetationsfarben nicht.

Es ist natürlich wirkungsvoller, die Pfeile farbig zu drucken. Man kann die Pfeile so anordnen, daß sie im gleichen Gebiet in zwei entgegengesetzte Richtungen weisen. Dies ist natürlich das Gebiet zwischen den beiden äußeren Grenzen. Je nach Richtung werden die Pfeile bald in der einen, bald in der anderen Farbe gedruckt,den beiden Vegetationstypen entsprechend. Die Pfeile beider Farben reichen von einer äußeren Grenze bis zur anderen,und sind daher ganz gemischt. Überall scheint es sowohl gelbe als auch grüne Pfeile zu geben (im Sinne der oben erwähnten Farben). Da die Pfeile gemischt sind und in zwei Farben erscheinen, muß ihr Hintergrund natürlich weiß bleiben. Das heißt also, daß zwischen dem Gelb der Kurzgras-Prärie und dem Grün der gemischten Prärie ein weißer Streifen liegt mit grünen und gelben Pfeilen. So ein weißer Streifen ist unerwünscht. Sind die Pfeile nicht sehr zahlreich, dann ist der weiße Streifen sehr prominent und stört. Versucht man aber, seinen Effekt dadurch zu mildern, daß man ihn mit mehr Pfeilen ausfüllt, dann entsteht ein so verwirrendes Bild, daß das auch störend wirkt. Pfeile, die in entgegengesetzte Richtungen weisen, sind sehr erwünscht, weil ja die Wanderung der Phytozönosen tatsächlich im gleichen Gebiet hin- und hergeht. Die daraus folgenden technischen Probleme sind einstweilen noch nicht befriedigend gelöst.

Man kann die Pfeile auch auf die Farbe der benachbarten Vegetation überdrucken, also z.B. gelbe Pfeile auf grün und grüne Pfeile auf gelb. Die Farben der Pfeile entsprechen dann der Farbe der sich ausdehnenden Phytozönosen; sie werden auf die Farbe der sich zurückziehenden Zönosen überdruckt. Dann sind sie aber nicht gemischt. Die Vegetationsänderung ist an der Grenze am größten, und hier benutzt man dann relativ große oder dicke Pfeile. Mit wachsender Entfernung werden die Pfeile kleiner, schlanker, weniger zahlreich und immer verstreuter, bis sie schließlich ganz aufhören. Man kann die Pfeile ganz schematisch auf die Karte setzen oder auch ganz unregelmäßig; beide Methoden können sehr wirkungsvoll sein.

Die Trennungslinie zwischen den Pfeilen ist die mittlere Grenze, an welcher sich die Grundfarben treffen, und von welcher die Pfeile fortweisen. Die Wirkung ist etwa die eines Spiegelbildes. Allerdings brauchen die beiden Streifen einander nicht gleich zu sein. So können die Pfeile auf einer Seite der mittleren Grenze viel weiter reichen als auf der anderen, und die Entfernung von der mittleren Grenze kann auf beiden Seiten ganz unregelmäßig verlaufen. In dieser Methode weisen die Pfeile also in zwei Richtungen, und kein weißer Streifen stört das Bild. Der Nachteil dieser Methode ist aber, daß die Pfeile nicht gemischt sind. Zwar scheint hier tatsächlich Bewegung in zwei Richtungen vorzuliegen. Aber es sieht so aus, als ob die Bewegung an irgend einem beliebigen Punkt doch in nur einer Richtung vor sich geht.

Zusammenfassung

Die Natur der nordamerikanischen Prärie mit den großen Schwankungen ihrer Klimaxphytozönosen verlangt besondere Kartierungsmethoden. Die fundamentale Schwierigkeit liegt in den Bewegungen, die im gleichen Gebiet in zwei entgegengesetzten Richtungen verlaufen. Die diskutierten Methoden sind einstweilen noch nicht zufriedenstellend für mindestens einen der folgenden Gründe: 1) das Gebiet, in dem die Bewegungen erfolgen, erscheint als Übergangsgebiet und als solches zu statisch; scheinbar bewegt sich da nichts. 2) Pfeile muten dynamisch an und können in zwei Richtungen weisen; auf weißem Untergrund stört das Fehlen der Farbe oder die große Zahl der Pfeile; 3) Beide Störungen können vermieden werden, aber scheinbar nur auf Kosten der beiderseitigen Bewegungen; an jedem Punkt scheint die Bewegung nur in eine Richtung zu gehen. Es ist offenbar, daß weitere Experimente nötig sind, um bessere Lösungen des Problems zu entdecken.

SUMMARY

It is recognized that the climax populations are not necessarily stable and that wide fluctuations may be the rule. Such major fluctuations occur primarily where the border between dry and humid climates passes through large plains with a predominantly herbaceous vegetation. The fluctuations of the climax vegetation result from major fluctuations of the precipitation. The demonstration is based on the conditions prevailing in the central prairie region of the United States. Fluctuations of vegetation is difficult to map because they are movements in two opposite directions within the same region. A number of methods to map fluctuating climax populations is discussed.

LITERATUR

KÜCHLER,A.W. -1961- Mapping the dynamic aspects of vegetation.- In: Gaussen,H.(ed.): Méthodes de la cartographie de la végétation.- C.N.R.S. 97e Internat. Toulouse.

-- -- -1967- Vegetations Mapping.- Ronald Press. New York.

WEAVER,J.E. & ALBERTSON,F.W. -1956- The grasslands of the Great Plains.- Johnsons Publishing Co. Lincoln,Nebr.

-- -- & BRUNER,W.E. -1945- A seven-year quantitative study of succession in grassland.- Ecol.Monogr.15: 297-319.

DISKUSSION

V.WESTHOFF
Es handelt sich hier hauptsächlich um ein Problem der kartographischen Darstellung. Dieses Problem ist in den Niederlanden gelöst worden, und zwar nicht von einem Biologen, sondern von einem Grafiker M.C.ESCHER. Dieser Grafiker hat in wunderbarer Weise das Problem des Überganges zwischen gegenseitiger Darstellung gelöst. Er läßt Vögel in Fische übergehen und Fische in Vögel, Elefanten in Pferde und Pferde in Elefanten, Land ins Meer und Meer in Land usw. Dieses Buch werde ich Ihnen gerne schicken.

D.MUELLER-DOMBOIS
Wie kann man sich das großräumige Vordringen der Kurzgräser während der Trockenjahre ökologisch erklären?

A.W.KÜCHLER
Die Ausdehnung in zwei Richtungen hat zwei verschiedene Gründe. Wenn es trocken wird, dann sterben die höheren Gräser ab, weil es zu trocken wird. Aber die niedrigeren Gräser, die Kurzgräser, haben eine so hohe Dürreresistenz, daß, wenn alle Gräser herum absterben, diese grün bleiben. Sie haben aber ein gewisses Ruhestadium, sie schlafen gewissermaßen ein. Sobald es regnet, wachsen sie wieder. Wird es aber feucht wird, ist es gerade umgekehrt, dann ist es ein Grad der Beschattung. Die höheren Gräser dringen dann zwar ein, beschatten die Sonnengräser, und ohne Licht, ohne Sonnenschein können diese nicht wachsen.

D.MUELLER-DOMBOIS
Meiner Ansicht nach müßten die Kurzgräser schon als Untergräser da sein, denn ein chronologisches Ausbreiten von Kurzgräsern auf trockenem Boden scheint mir ökologisch nicht wahrscheinlich.

A.W.KÜCHLER
Sie können ja sehr schnell hereinkommen. Der Boden ist nicht so trocken, daß die Kurzgräser sich nicht ausdehnen können. Wenn ich sage, es regnet überhaupt nicht, dann ist das vielleicht eine Übertreibung. Es kommt sehr selten vor, aber wenn es nur ein wenig regnet, dann ist es schon genug für die Kurzgräser, sie können dann wachsen.
Sie kennen alle die Erscheinung der Galeriewälder der tropischen Savannen, hier haben wir das gleiche. Wir haben Galeriehochgras-Prärie inmitten der der Kurzgras-Prärie. Hier haben Sie eine

Kurzgras-Prärie, und an den Bächen entlang eine Hochgras-Prärie.

W.G.BEEFTINK
Ich wußte nicht, daß die Prärie soviel mit den Gezeiten-Erscheinungen gemeinsam hat; denn wir haben dieselben Erscheinungen auch an den Küsten. Was Sie besprochen haben, ist eine durch Schwankungen in der Zeit gebildete instabile Situation. Wie Sie sagten, hat die Grenzsituation kaum eigene Arten. Gerade Entgegengesetztes hat man auch stabile Grenzsituationen genannt, welche in instabilen Situationen die Konzentration von Arealgrenzen gemeinsam haben. Die stabilen Grenzsituationen sind aber gekennzeichnet durch einen Reichtum an Endemismen. Das ist z.B. der Fall an einigen Stellen der Europäischen Küsten, wie Finnmark, der Bretagne, Westportugal.

A.W.KÜCHLER
Wir haben tatsächlich vier Arten, die auf die gemischte Prärie beschränkt sind. Das sind 2 Moose und 2 kleine Kräuter, kein Gras.

H.MERKER
Sie nannten als Niederschlag nur den Regen. Ich möchte fragen, ob auch der Tau in Frage käme?

A.W.KÜCHLER
Tau haben wir fast keinen. Die Luft ist gewöhnlich so trocken, daß Tau sich nie bildet. Wir haben auch im Winter keinen Rauhreif.

H.ELLENBERG
Das Vorschreiten der Kurzgräser kann man sich wohl unmöglich ökologisch so vorstellen, als schritten sie wirklich vor über hunderte von Kilometern. Das ist völlig unmöglich, eine Pflanze kann nicht so rasch wandern. Sie muß schon vorher da sein, und es handelt sich im Grunde doch nur um einen Mischbestand, innerhalb dessen physiognomisch und dominanzmäßig bald der eine Teil und bald der andere Teil zur Dominanz kommt. Sonst kann man sich unmöglich vorstellen, daß innerhalb von 3 oder 5 Jahren die Arten plötzlich erscheinen.

A.W.KÜCHLER
Ich habe darauf hingewiesen, daß wir uns hier im Herzen von Nordamerika befinden, auf großen Ebenen. Die Ebenen sind besonders groß, sie reichen vom Golf von Mexico bis zum Arktischen Ozean.

Infolgedessen sind sie, was die Atmospäre anbelangt, beinahe wie ein Ozean, d.h. die Luftmassen können frei über ungeheure Entfernungen hinweggehen, sehr große Geschwindigkeiten entwickeln. Außerdem ist die Luft relativ trocken, und infolgedessen können die Grasfrüchte, da sie ja sehr leicht sind, mit Leichtigkeit über hunderte von Kilometern geblasen werden. Wenn es sehr feucht ist, und die Hochgräser sehr dicht sind, dann haben sie natürlich keine Möglichkeit, aber sobald es trocken wird, dann kommen sie. Die Gräser sind nicht da, kommen aber sofort auf, wenn es trokken wird.

J.SCHMIHÜSEN

Auf der Erde haben wir in einem außerordentlich breiten Gürtel dasselbe Phänomen. Das ist nämlich der Grenzgürtel des nördlichen Waldes gegen die Tundra, wo wir in einer Breite von etwa 100 km in manchen Gebieten das Phänomen haben, daß, in ähnlicher Weise wie hier die beiden Typen der Steppe, die Tundra mit dem Wald wechselt. In dem Falle sind die Zeitspannen, um die es sich handelt, um das Vor- und Zurückgehen, durchaus bekannt und feststellbar durch die Altersbestimmungen bei den Bäumen. Der spielende Faktor dabei ist das Vorkommen der klimatischen Bedingungen für die Verjüngung. Die Tundrenwälder sind meistens sehr einheitlich in den Jahrgängen, und da kommen alle 50 oder 100 Jahre nur die Bedingungen für die Vermehrung zustande. Das ist die Erklärung für diese Durchdringung von Tundra und Wald an der Tundrengrenze, und wenn man das als Ganzes sieht, ist es genau dasselbe wie das, was uns hier auch vorgestellt wurde.

E.HÜBL

Zur Frage, ob die Gräser schon an der Stelle, wo sie dann infolge der Klimaschwankung aufkommen, schon vorhanden sind oder nicht, möchte ich einen Kompromißvorschlag unterbreiten. Man kann ja sicher annehmen, daß die Früchte auf jeden Fall schon anwesend sind, auch wenn sie nicht hingeblasen werden, weil sie ja fruktifizieren werden. Die Karyopsen haben ja einen Keimverzug, so daß sie mehrere Jahre keimfähig bleiben. Sie werden also sicherlich schon im Boden sein.
Wie weit sind diese Grenzverschiebungen auch menschlich bedingt, durch Feuer, oder Über- oder Unterbeweidung?

A.W.KÜCHLER

Der menschliche Einfluß ist gewöhnlich so, daß er die Wirkungen der natürlichen Schwankungen verstärkt, d.h., wenn eine Dürre eintritt, die einen gewissen Effekt hat, dann ist gewöhnlich der menschliche Eingriff so, daß das Unheil dadurch noch verstärkt wird. Besonders durch die schwere Beweidung und durch das Pflügen des Graslandes wird ein Milieu erzeugt, wo die Dürre viel akuter ist als es sonst der Fall wäre, und infolgedessen sind die Schwankungen dadurch tatsächlich verstärkt.

DIE MICROFLORA VAGINALIS ALS BEISPIEL DER SYNDYNAMIK BEI BAKTERIENGESELLSCHAFTEN

R. L i n d e r & Th. W u r c h

Neuerdings sind Bakterien die Hauptschauspieler der Forschung geworden. Sowohl in Genetik wie in Molekularbiologie hat sich herausgestellt, daß bei Microorganismen sich alles viel einfacher auslegt. Es geziemt sich also, daß der Phytosoziologe sich auch der Microflora zuwendet.

Die Aufnahmen sind jedoch schwierig anzufertigen, besonders bei Bodenflora. Ein günstiges Untersuchungsterrain bildet das geschlossene Ecosystem des Vaginalraumes der weiblichen Vertreter der Primaten, insbesondere bei *Homo sapiens*. Auf dem Glycogen ausscheidenden Epithelium wohnen hier endemisch folgende Bakterien: *Lactobacillus vaginalis* (L.v.Doederlein), *Haemophilus vaginalis*, *Leptothrix vaginalis* und das Flagellatentier *Trichomonas vaginalis*; oft leben sie gesellig mit Vertretern der Hautflora (*Staphylococcus*, *Candia albicans* etc.) und der Darmflora (*Enterococci*, *Escherichia coli* etc.).

Es zeigt sich, daß die Schwangerschaft die günstigsten fortwährenden ökologischen Bedingungen zur Ausbildung der Microflora vaginalis bietet. Die Aufnahmen von 100 wahllos sich vorstellenden Schwangerschaftsfällen entpuppen 4 Assoziationen:

1) Lactobacilletum vaginalis (71%; pH:4-4,5)	Subassoziation	typicum
	"	v.Enterococci
	"	v.Candida albicans
	"	v.Haemophilus vaginalis
2) Haemophiletum vaginalis (9%; pH:5-5,5)	Subassoziation	typicum
	"	v.Lactobacillus vagin.
3) Leptothricetum vaginalis (3%; pH:4-5,5)		
4) Trichomanadetum vaginalis (4%; pH: 6)	Subassoziation	v.Lactobacillus vagin.
	"	v.Candida albicans
	"	v.Haemophilus vaginalis

Tab.1.

Sous-association	type	à Cocci	à Candida	à Hæmophilus	type	à Lactobacillus		à Lac. / à Can. / à Hæm.
ASSOCIATION	LACTOBACILLETUM				HAEMOPHILETUM		LEPTOTHRICETUM	TRICHOMONADETUM
Nombre de relevés	21 33 2 2 1	1 3 1 2 2 1 1 1	1 1 1 1 1 1 1 1 1 1	1 1 1	1 1 1	1 1 1 1 1 1	1 1 1	1 1 1 1
Lactobacillus vaginalis	5 4 3 2 4	5 4 3 3 3 3 1 1	4 3 3 5 5 5 4 3 2 2	4 3 3		2 2 1 1 1 +	+1	4 +
Enterococci		1 + 1 1 + 1 5 2	+ + +		+			1 +
Staphylococcus		1 1	+					+
Candida albicans			3 1 1 2 1 + 2 3 3 1					3
Hæmophilus vaginalis		+		4 3 2	5 5 5	3 4 2 3 4 5	3 5 +	3 + 5
Leptothrix vaginalis	1						3 5 5	
Trichomonas vaginalis								4 5 5 4
Escherichia coli		2						
Fréquence	59 %	12 %	10 %	3 %	9 %		3 %	4 %
pH	4	4,5	3,8	4,7	5 - 5,5		4 - 5,5	6

Tab.1. Aufstellung von 100 Aufnahmen der Vaginalflora während der Schwangerschaft. Die Untersuchungen wurden wahllos an hundert in verschiedenen Stadien der Schwangerschaft stehenden Frauen angefertigt. Jede Aufnahme wird durch mikroskopische Beobachtung des Gram-gefärbten Vaginalabstrichs abgelesen. Jede Kolonne der Tabelle enthält ein Aufnahmebild; die beobachteten Arten sind durch ihren Abundanz-Dominanz-Wert verzeichnet, so wie ihn die Phytosoziologen verwenden: 5 = 75-100% der maximalen Abundanz-Dominanz die in diesem Falle die auf dem Objektträger dichteste Bakterienbedeckung darstellt, die überhaupt beobachtet werden kann: 4 = 50-75% der maximalen Bedeckung; 3 = 25-50%; 2 = weniger als 25%; 1 = bis 5%; + = nicht chiffrierbar; r = isoliert vorhanden. Identische Aufnahmen sind gruppiert: z.B. enthält die erste Kolonne 21 Aufnahmen mit *Lactobacillus* 5, die zweite Kolonne 33 Aufnahmen usw. Die dominante, charakteristische Art, welche die Assoziation bestimmt, ist durch vollen Strich umrandet; die Subassoziationen sind durch die Differentialarten gekennzeichnet (punktiert umrandet). Im untersten Teil der Tabelle sind Frequenz und mittlerer pH-Wert angegeben.

Hieraus ergibt sich, daß das L a c t o b a c i l l e t u m die vorherrschende Assoziation ist (84% der Fälle), dann folgt das H a e m o p h i l e t u m. Jede Assoziation entspricht einem bestimmten pH-Zwischenraum; unter pH = 5 entwickelt sich das L a c t o b a c i l l e t u m, über 5 das H a e m o p h i l e t u m. Man kann sogar behaupten, daß ein bestimmter pH-Wert jede Subassoziation charakterisiert; der sauerste pH-Wert ist jener der S u b a s s o z i a t i o n von C a n d i d a a l b i c a n s des L a c t o b a c i l l e t u m. Für die zwei übrigen Assoziationen, das L e p t o t h r i c e t u m und das T r i c h o m o n a d e t u m, reichen die

Angaben nicht, um sie näher zu bestimmen. Das L e p t o t h r i - c e t u m ist bei uns wenig verbreitet, es ist hauptsächlich in Mitteleuropa lokalisiert.

Tabelle 2.

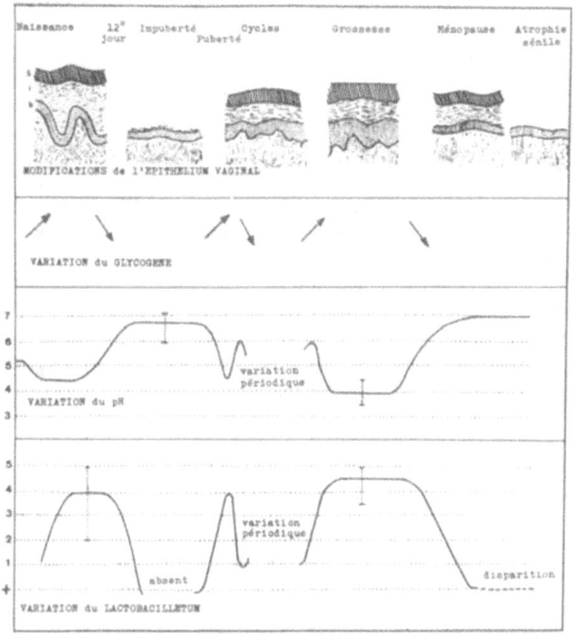

Tab.2. Syndynamische Fortentwicklung des vaginalen Milieus durch das ganze Leben der Frau hindurch: Vergleich der Veränderungen des L a c t o - b a c i l l e t u m (durch Abundanz-Dominanz-Koeffizienten dagestellt), des pH-Wertes und des Glycogens, im Zusammenhang mit den Epithelium-Abänderungen. Die Kurven entsprechen den Mittelwerten aus mehreren tausend Beobachtungen.Man bemerke das Verschwinden des L a c t o b a c i l - l e t u m , wenn das Epithelialsubstrat bis zu den Basalzellen (b) reduziert wird, Infolge Desquamierung der Oberflächenzellen (s) und der Intermidiärzellen (i). Das L a c t o b a c i l l e t u m variiert parallel zum Glycogengehalt und umgekehrt zum pH-Wert. Die vertikalen punktierten Linien zeigen die Amplitude der Variation an.

Verfolgen wir die Gesellschaftsdynamik der Vaginalflora von der Geburt bis ins hohe Alter der Frau, so stellt sich heraus, daß beim neugeborenen Mädchen das Vaginalmilieu steril ist, daß aber innerhalb 48 Stunden das L a c t o b a c i l l e t u m sich voll entwickeln kann, um am 12. Tag zu verschwinden (erste Genitalkrise). Dann bleibt der Raum praktisch vegetationslos - weil "bodenlos" - bis zur Pubertät (12-14 Jahr). Bei Eintreten der Menses beginnt eine periodische Wechselfolge des L a c t o b a c i l l e t u m bis

Tabelle 3.

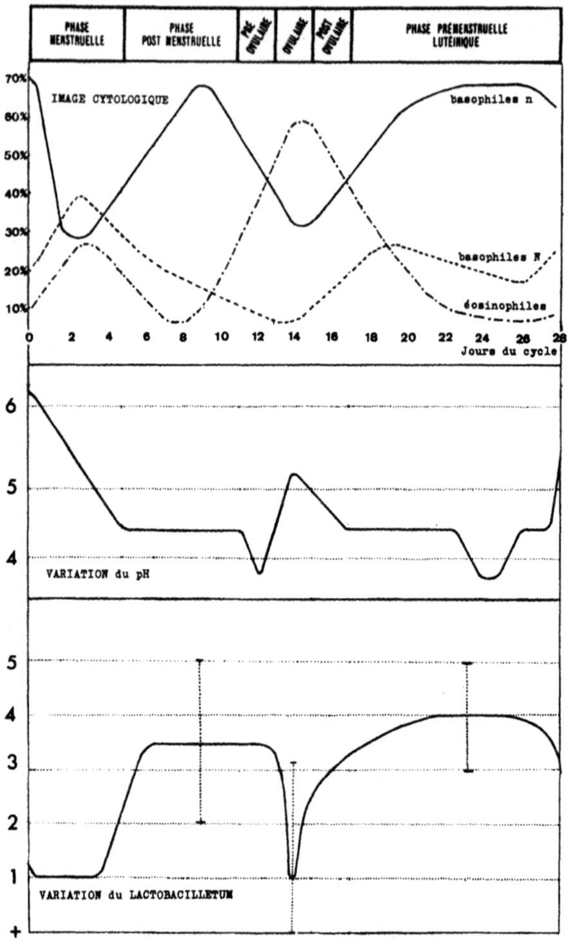

Tab.3. Graphischer Vergleich der syndynamischen Entwicklung des L e u c o - b a c i l l e t u m , des vaginalen pH-Wertes und des zytologischen Bildes während dse Monatszyklus. Das zytologische Bild ergibt sich aus den entsprechenden Prozenten der kleinkernigen basophilen Zellen (n), der großkernigen basophilen Zellen (N) und der eosinophilen Zellen, die im Vaginalabstrich ausgezählt wurden. Die n-basophilen-Kurve verläuft wie die L a c t o b a c i l l e t u m - Kurve, die durch den Abundanz-Dominanz-Koeffizienten des *Lactobacillus* dargestellt ist. Die pH-Kurve und die eosinophilen Kurven variieren korrelativ, aber in umgekehrter Richtung.

zur Schwangerschaft, die die Gesellschaft vom ersten Monat bis zur Entbindung stabil erhält. In den Wechseljahren regressiert das Epithelium - somit auch die Vaginalflora - bis zur senilen Atrophie, wo Pflanzendecke samt "Boden" verschwinden.

Syndynamisch interessant zeigt sich der Phasenwechsel während des Monatszyklus. Bei Einbruch der Menses (pH6) wird das L a c t o b a c i l l e t u m weggeräumt, steigt aber am 4.-5. Tag rasch an, bis zum Mittelwert 3-4 während der Postmenses (pH-4,5), stürzt bei der Ovulation (14.Tag) bis zum Wert +, und steigt wieder in der Lutealphase (16.-28.Tag) rasch zum Wert 4-5 an.

Somit läuft der Abundanzwert des L a c t o b a c i l l e t u m korrelativ sowohl mit der pH-Kurve, wie mit der Mengenkurve des Nährstoffs der Gesellschaft, des Glycogens. Beim H a e m o p h i l e t u m zeichnet sich diese Wechselfolge nicht so deutlich ab, weil hier der optimale pH-Wert bei 5,5 liegt.

Die überraschendste Beobachtung wird ersichtlich, wenn täglich die Microflora bei einem einzigen Individuum während des ganzen Zyklus verfolgt wird (Tab.4). Da zeigt sich, daß Störungen wie die Ovulation, so rasch und gewaltig das ökologische Gleichgewicht umwandeln können, daß innerhalb weniger Stunden, wenn nicht Minuten, ein L a c t o b a c i l l e t u m in ein H a e m o p h i l e t u m übergehen kann.

Tabelle 4.

	Jours		Lactobacillus	Enterococci	Haemophilus	
MENSES	1er	a		r		
		m		r		
		s		r		
	2.	a	r	1		
		m	r	1		
		s				
	3.	a	+	1		
		m	1	r		
		s				
PHASE FOLLICULINIQUE	4.	a				
		m	4			
		s	4			
	5.	a				
		m	3			
		s	3			
	6.	a	2			↑
		m	2			
		s				
	7.	a				
		m	4			
		s	2			
	8.	a	r	1		
		m	2			↑
		s	2	r		
	9.	a				
		m				
		s	2	r		
OVULATION	10.	a	4	r		↑
		m	2			
		s	1			↓
	11.	a	4	r		
		m				
		s				
	12.	a	2	r		
		m	r	+		
		s				
PHASE LUTEALE	13.	a	+		5	
		m				
		s	+		5	
	14.	a			5	
		m	+		5	
		s				
	15.	a			5	
		m	r		4	
		s	r		5	
	16.	a	+	+	5	
		m	+		5	
		s				
	17.	a	1		4	
		m	r		2	
		s	r	r	5	
	18.	a	r	r	5	
		m				
		s	2		3	↑
	19.	a		+	2	
		m		+	5	
		s	r	+	5	
	20.	a		r	1	↑
		m		r	5	
		s		+	5	
	21.	a		+	5	
		m				
		s				
MENSES	1.	a				

Tab.4. Laufende Aufnahmen der Vaginalflora durch einen kurzen, 21-tägigen Zyklus hindurch; es wurden 3 Aufnahmen pro Tag angefertigt (a = morgens, m = mittags, s = abends). Nach der Ovulation (10. Tag) geht das L a c t o b a c i l l e t u m in das H a e m o p h i l e t u m über. Ferner ist das Gleichgewicht der Gesellschaft gestört bei 8 m und 20 a durch Bakterien-Lyse, bei 10 a und 18 s durch Auftreten von Leucozyten, bei 6 a durch Erscheinen der Granulärform des *Lactobacillus*.

SCHLUSSFOLGERUNG

Microorganismen-Gesellschaften können in bestimmten, konstanten ökologischen Bedingungen charakterisiert werden, und zwar während der Schwangerschaft der Frau. Die Realität dieser Gesellschaften verwirklicht sich durch die Art und Weise womit sie die Probe der Milieu-Veränderung bestehen

1. während des Lebens der Frau,
2. während des Monatszyklus.

Die Ergebnisse dieser Syndynamik-Studie lassen erkennen, daß ganz enge Beziehungen zwischen Gesellschaft und Milieu existieren: das L a c t o b a c i l l e t u m wird jedesmal hergestellt, wenn die ihm geeigneten Umweltbedingungen auftreten; und zwar ist es der pH-Wert, der die Rolle des begrenzenden Faktors spielt. Der Gesellschaftsbegriff ergibt sich hier mit seiner zufriedenstellendsten Definition: einer biozönotischen Einheit, die mit einem bestimmten Milieu harmoniert.

Der Phytosoziologe hat also recht, wenn er sich auch den Microorganismen zuwendet. Diese Kategorie von Lebewesen erlaubt einen beschleunigten Beobachtungsrhytmus in hervorragend günstigen Bedingungen: genaue Analyse der mit wenig Arten auftretenden Gesellschaften, die Proben werden auf dem Objektträger endlos konserviert, das Milieu und seine Veränderungen können leicht zytologisch und chemisch untersucht werden. Ferner bietet das Vaginal-Milieu periodische Variationen, die genau bekannt sind; eine solche feine Umweltanalyse bleibt dem Soziologen bei höheren Pflanzengesellschaften meist unbekannt.

Die Definition der Assoziation stellt sich hier auf eine andere Stufe, sowohl vom Standpunkt der Raum-Statik, wie vom Standpunkt der zeitlichen Dynamik aus gesehen. Die phytosoziologische Methode, insofern sie in diesem Bereich angewandt wird, gewährt dem Soziologen die Satisfaktion einer besonders demonstrativen Syndynamik. Das Auftreten der Assoziation bei genau definierten Bedingungen wird ein vorhergesehenes Ereignis. Die Assoziation zeigt sich wohl als komplexes Produkt, welches das Gleichgewicht zwischen einem bestimmten Gesellschaftsbestand und einem bestimmten Milieu zum Ausdruck bringt.

Vom praktischen Standpunkt her gesehen, findet der Gynäkologe in der soziologischen Aufnahme ein wichtiges Behelfsmittel zur übersichtlichen Diagnose, nicht nur des Vaginalzustandes, sondern vielmehr des allgemeinen Gesundheitszustandes der Frau.

DIE ÄNDERUNG DER TEICHVEGETATION IN SÜDBÖHMEN NACH ZEHN JAHREN

R. N e u h ä u s l

Mittels indirekter Methoden kann man die Entwicklung der Pflanzengesellschaften nur sehr hypothetisch beurteilen, daher ist jede direkte Beobachtung für die Wissenschaft wertvoll. Zu diesem Zwecke kann man auch ältere Vegetationskarten mit gutem Erfolg auswerten, sofern nähere Angaben über Eingriffe in die Vegetation und die Umwelt feststellbar sind. Auch in unserem Falle wurde eine Kartierung der Teichvegetation aus dem Jahre 1956 als Grundlage zum Studium der syndynamischen Prozesse der Verlandungsgesellschaften herangezogen.

In den Jahren 1953-1956 wurden vom Verfasser die Röhricht- und M a g n o c a r i c e t e n- Bestände zweier südböhmischer Teiche eingehend untersucht und die gesamte Vegetation innerhalb der ursprünglichen Litorallinie im Sommer 1956 in eine Karte im Maßstab 1:6250 aufgenommen. Die Resultate dieser Untersuchungen wurden im Jahre 1965 publiziert (NEUHÄUSL 1965). Eine neue Kartierung eines der beiden Teiche wurde im Sommer 1966 durchgeführt; die gewonnenen Ergebnisse möchte ich hier bekanntgeben.

Vorerst jedoch einige wichtige Informationen über den untersuchten Teich. Der Teich "Nový Kanclíř" liegt im südöstlichen Teil des Wittinggauer Beckens in einer Höhe von 440 m NN. Er ist durch einen Zuflußkanal mit dem Bach "Chlumecký potok" verbunden, der verhältnismäßig ungedüngtes Wasser aus dem nadelwaldreichen Waldviertel zuführt. Der Teich nimmt eine Fläche von 31 ha ein, die durchschnittliche Tiefe schwankt in seinem mittleren Teil zwischen 180 und 220 cm. Die Unterlage bilden teilweise von jüngeren Anschwemmungen überdeckte Miozän-Sedimente, im Zuflußteil befinden sich ausgedehnte, aber nicht zu tiefe Schichten eines mesotrophen Flachmoortorfes. Die chemische Zusammensetzung des Wassers schwankt im Laufe des Jahres in Abhängigkeit von der Düngung von schwacher Oligotrophie bis zu einer mäßigen Eutrophie, am Zufluß des Teiches wurde während der Untersuchungen ein mäßig humoses, oligotrophes Wassermilieu festgestellt.

Der untersuchte Teich wurde durch die Bewirtschaftung nur sehr wenig beeinträchtigt. Als wichtigster Faktor kommt eine regelmäßige Kalk-, Phosphor- und Stickstoff-Düngung in Betracht. Die sublitoralen Verlandungsbestände werden nur ausnahmsweise gemäht oder ausgestochen.

Zustand der Vegetation im Jahre 1956

Bei der Kartierung der Teichvegetation im Jahre 1956 wurden 20 Einheiten unterschieden. Den Röhricht-Gesellschaften wurde besondere Aufmerksamkeit gewidmet; daher wurden sie auf Grund einer feineren Soziationsanalyse[1)] noch weiter gegliedert. Die niedrigsten floristisch-soziologischen Einheiten des Phragmition-Verbandes wurden auf Grund ihrer Struktur und unter Berücksichtigung der soziologischen Eigenschaften der Dominanten aller Schichten noch weiter geteilt, wie aus der Übersicht(p.251) hervorgeht. Es zeigte sich nämlich, daß auch die floristisch identischen, meist sehr artenarmen Röhrichtbestände in strukturell abweichende, dabei jedoch sehr stabile, den Standortsbedingungen entsprechende Mikro-Einheiten teilbar sind. Im Grunde genommen wurden 3 strukturelle Formen der niedrigsten floristisch-soziologischen Einheiten unterschieden:

1. Eine initiale Form, in der die Stengel einer einzigen Art nur durch das Wurzelsystem verbunden sind; die Einzelsprosse sind sehr locker voneinander räumlich ganz unabhängig; die Bestände üben praktisch keinen Einfluß auf das Milieu aus.
2. Eine Invasionsform, in der die Stengel einer oder mehrerer Arten schon deutlich ihr Milieu beeinflussen (Bodenbefestigung, Förna-Bildung, Beschleunigung der Wasserströmung usf.), jedoch die gegenseitige, sowohl direkte (mechanische) als auch indirekte Wirkung der Pflanzen von keiner entscheidenden Bedeutung ist. In die Invasionsform einer Röhricht-Gesellschaft können andere Arten eines gleichwertigen Sukzessionsstadiums (sublitorale Röhrichtarten) eindringen.
3. Eine Verlandungsform, in der die Stengel bzw. Blätter einer oder mehrerer Arten sehr dicht nebeneinander stehen, so daß eine starke mechanische Berührung und Beschattung entsteht. Die Röhricht-Gesellschaften der Verlandungsform sind floristisch und strukturell voll gesättigt. In die Verlandungsform können höchstens Arten fortgeschrittener Sukzessionsstadien

[1)] Durch die Soziationsanalyse kann man nicht nur die floristisch weiter unteilbaren Einheiten feiner gliedern, sondern auch im kleinen Areal zu abstrakten Einheiten (Soziationen) gelangen. Diese Methode gestattet, Soziationen auch auf Grund der Analyse in einzelnen Assoziationsindividuen (im Sinne von BRAUN-BLANQUET) zu erfassen - cf. DU RIETZ 1930, Vegetationsforschung auf soziationsanalytischer Grundlage, p.431: "Ob die Quadrate aus einer oder mehreren Siedlungen geholt sind, ist von keiner Bedeutung ... Die Forderung von BRAUN-BLANQUET u.a., daß jedes Quadrat aus einer neuen Siedlung geholt sein soll, muß deshalb abgelehnt werden".

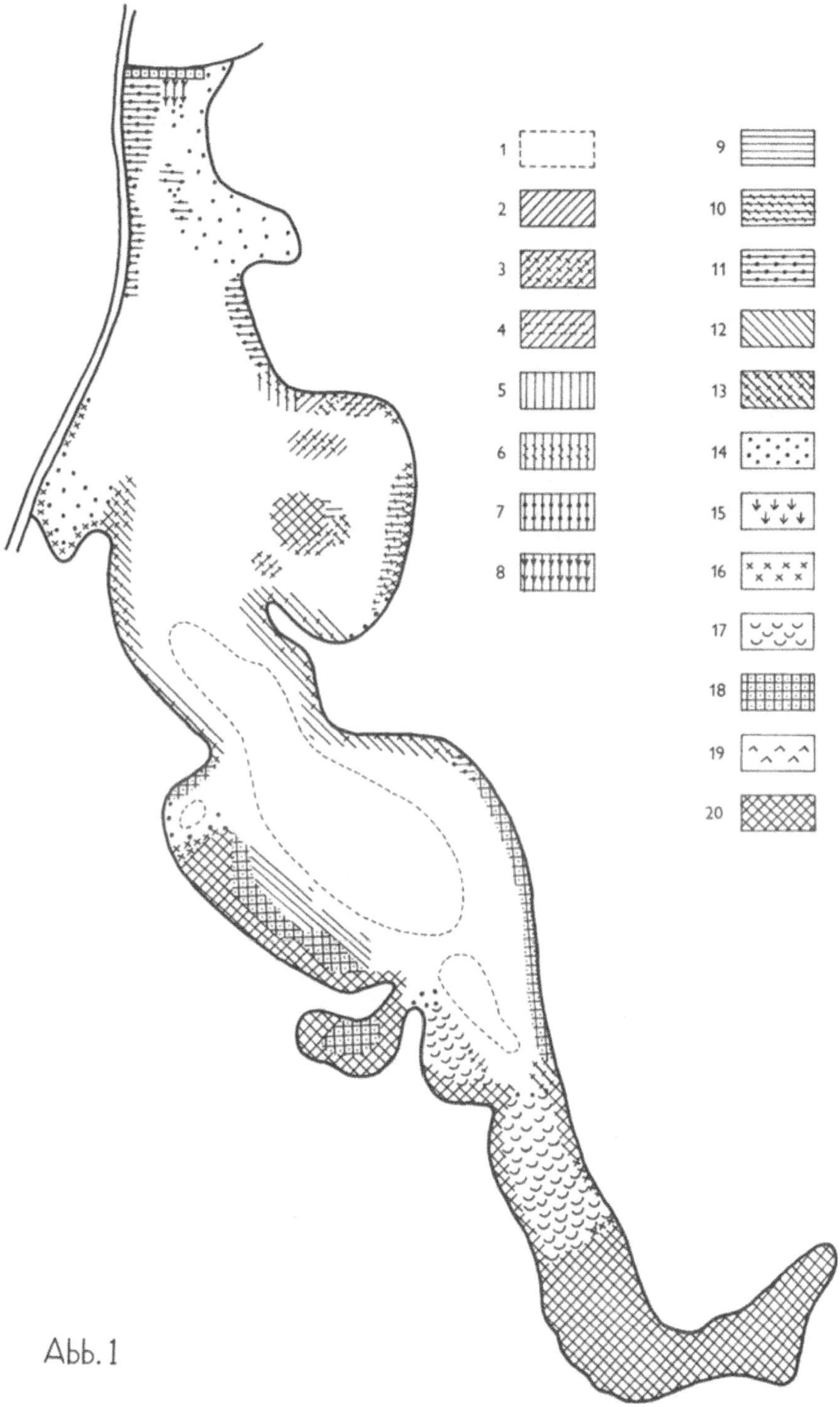

Abb. 1

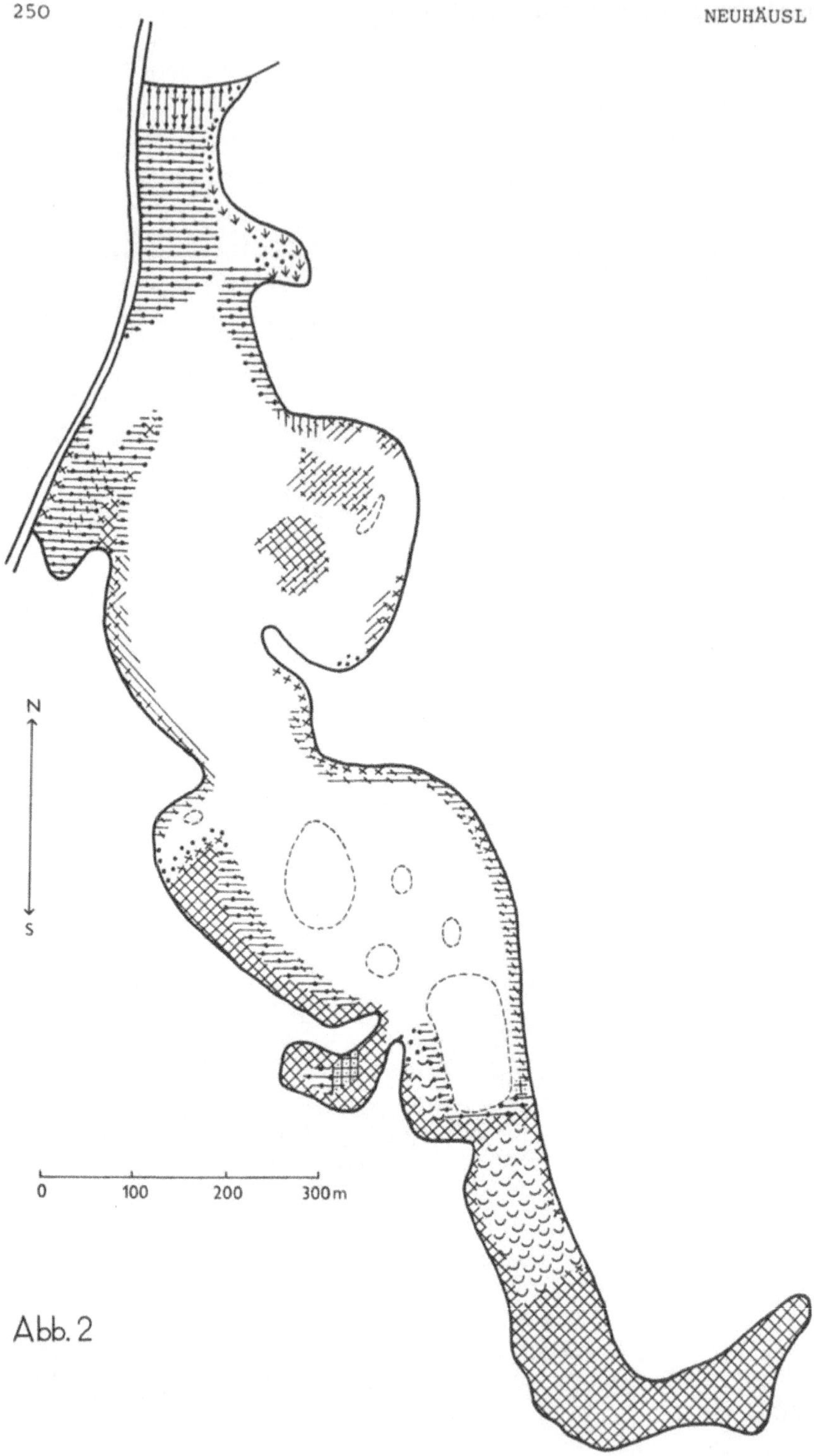

Abb. 2

(des Magnocaricion-Verbandes) eindringen.

Vom ökologischen Gesichtspunkt aus wurden weiter die eulitoralen (oberhalb der Litorallinie wachsenden) Bestände als selbstständige Kleineinheit unterschieden, da sie ebenfalls strukturelle Abweichungen, vor allem jedoch eine ziemlich abweichende Produktivität der Pflanzenmasse aufweisen. Sie wurden als eulitorale Form der betreffenden Röhrichtgesellschaften bezeichnet.

Zu diesen Kleineinheiten gelangten wir durch eine eingehende soziologische Analyse der Einzelbestände und durch eine induktive Synthese, wobei nicht nur floristische, sondern auch soziologische Eigenschaften der aufgenommenen Bestände (Abundanz, Deckungsgrad, Soziabilität, Repartition, Fertilität und Höhe der Arten) ausgewertet wurden.(Näheres über die Methodik in NEUHÄUSL 1956, 1965). Abstrakte Typen dieser Phytozönosen wurden als Soziationen bezeichnet, deren zugehöriger Praefix (Initial-, Invasion-, Verlandungs-, bzw. eulitorale Soziation) beigefügt wurden.

In Abb.1 sind die Vegetationsverhältnisse des untersuchten Teiches im Jahre 1956 aufgenommen. Folgende Einheiten wurden unterschieden (einschließlich einer Einheit, die erst im Jahre 1966 kartiert wurde):

1. Potamogeton natans - Nymphaea candida-Ass.

   Phragmition communis

Bezeichnung im florist.-soziol.System	Bezeichnung der Kleineinheiten
2. Scirpo-Phragmitetum typhosum-Initialstadium	Initial-Soziation Typha angustifolia
3. Scirpo-Phragmitetum typhosum	Invasions-Soziation Typha angustifolia
4. Scirpo-Phragmitetum lemnetosum, Fazies mit Typha angustifolia	Invasions-Soziation Typha angustifolia-Lemna minor
5. Scirpo-Phragmitetum phragmitosum-Initialstadium	Initial-Soziation Phragmites communis
6. Scirpo-Phragmitetum phragmitosum	Invasions-Soziation Phragmites communis
7. Scirpo-Phragmitetum phragmitosum	Verlandung-Soziation Phragmites communis
8. Scirpo-Phragmitetum phragmitosum-Abbaustadium	Eulitorale Verlandungs-Soziation Phragmites communis
9. Glycerietum aquaticae-Initialstadium	Initial-Soziation Glyceria aquatica
10.Glycerietum aquaticae	Invasions-Soziation Glyceria aquatica
11.Glycerietum aquaticae	Verlandungs-Soziation Glyceria aquatica
12.Acoretum calami-Initialstadium	Initial-Soziation Acorus calamus
13.Acoretum calami	Invasions-Soziation Acorus calamus

Magnocaricion elatae

14. Caricetum elatae
15. Caricetum rostratae (Caricetum inflato-vesicariae)
16. Caricetum gracilis
17. Acorus calamus - Carex rostrata-Ges. (Schwingrasen)
18. Mosaikartige Bestände des Sublitorals

Caricion canescenti-fuscae

19. Caricetum fuscae

Alnion glutinosae

20. Carici elongatae - Alnetum (incl. Entwicklungs- und Degradationsstadien)

In der Periode von 1956-1966 wurde der Teich auf normale Art und Weise bewirtschaftet. Der Wasserstand wurde praktisch auf gleichem Niveau gehalten; die Abfischung fand im Herbst eines jeden zweiten Jahres statt. Die Röhrichtbestände wurden fast überhaupt nicht gemäht. Nur in einer Bucht am Ostufer wurde ein schmaler Saum von Magnocaricetum ausgestochen. Das Teichwasser wurde alljährlich mit ca. 150 q Kalk, 200 q Kalziumkarbonat, 50 q Superphosphat, 20 q Salpeter und 500 q Mischdünger gedüngt.

Eine neue Kartierung nach zehn Jahren wurde aus folgenden Gründen vorgenommen:

1. Um die Stabilität der untersuchten Gesellschaften, vor allem der Kleineinheiten, zu beglaubigen;
2. die Empfindlichkeit der Gesellschaften gegen anthropogene Einflüsse und ihre Beziehungen zu den sich verändernden Standortsbedingungen festzustellen;
3. die Entwicklungstendenz, bzw. die syngenetischen Beziehungen der untersuchten Einheiten kennen zu lernen;
4. mit konkretem Material zur Kenntnis der Verlandungsprozesse beizutragen.

In Abb.2 sind die Resultate der Kartierung aus dem Jahre 1966 dargestellt. Sie dienten als Unterlage für die weiteren Deduktionen.

1. Stabilität der Gesellschaften in Bezug auf die Standortsbedingungen und anthropogene Einflüsse

Nach der syndynamischen Stabilität, bzw. Labilität, können wir im Bereich der Verlandungsserie die Kategorien der Einheiten wie folgt unterscheiden:

Labile Gesellschaften, welche meist an keine ausgeprägteren Standortsbedingungen gebunden sind, und die gegen Standortsänderungen oder direkte Eingriffe in die Phytozönose sehr empfindlich sind. Sie unterliegen auch während ihrer autonomen Entwicklung ziemlich schnellen Änderungen. In diese Gruppe gehören neben labilen mosaikartigen Beständen alle Initialformen der Röhrichtgesellschaften, weiter die zweisynusielle Soziation Typha angustifolia-Lemna minor) (die Lemna minor-Synusie entwickelt sich nur in der Zeit einer intensiven Stickstoffdüngung) und die Invasions-Soziation von Glyceria aquatica; die letztgenannte Einheit stellt daher nur eine Entwicklungsphase des Glycerietum aquaticae dar. Die Initialformen der Röhricht-Gesellschaften kann man teils als Entwicklungsphasen des Scirpo-Phragmitetum, teils jedoch nur als Entwicklungsstadien, aus denen mehrere Assoziationen entstehen können, werten.

b. Stabilisierte Gesellschaften, welche ein spezifisches Milieu für sich schon ausgebildet haben, bzw. die unter charakteristischen Umweltbedingungen vorkommen; infolge ihrer eigenen Lebenstätigkeit ändern sie ihre Umwelt nur sehr wenig und langsam (bei einer natürlichen Entwicklung sind größere Änderungen nur nach Dezennien feststellbar). Die stabilisierten Gesellschaften sind jedoch gegen Änderungen des Milieus relativ empfindlich, vor allem gegen diejenigen, die beim Alterungsprozeß (Eutrophierung) verlaufen. Zu dieser Gruppe gehören das sublitorale Caricetum elatae und die Nymphaea candida-Potamogeton natans-Ass. Sie reduzieren in künstlich gedüngten Teichen ihre Fläche und werden oft von anspruchsvolleren Gesellschaften zurückgedrängt. Als stabilisierte Gesellschaften erachten wir auch die Zönosen, die relativ stabil, jedoch gegen äußere, direkt auf die Bestände einwirkenden Eingriffe (Mähen) empfindlich sind. Hierher gehören die Invasionsformen mancher Röhrichtbestände (Invasions-Soziation von *Typha angustifolia*, *Phragmites communis*, *Acorus calamus*). Vom systematischen Gesichtspunkt aus stellen stabilisierte Gesellschaften entweder soziologisch ausgeprägte Einheiten (Assoziationen, bzw. niedrigere Einheiten) oder ihre relativ stabileren strukturellen Formen dar.

c. Stabile Gesellschaften zeichnen sich durch ihr eigenes Milieu aus, welches sich infolge der Lebenstätigkeit der Phytozönose nur sehr langsam ändert (s.b). Die stabilen Gesellschaften sind fähig, bestimmte äußere Eingriffe in ihre Umwelt (kurzdauernde

Änderungen des Wasserstandes) oder direkt in die Phytozönose (Mahd, Brand) auszugleichen, ohne ihre Struktur zu ändern. Sie können jedoch von konkurrenzfähigen oder höher organisierten Gesellschaften sukzessiv ersetzt werden (bestimmte Änderungen kann man schon nach zehn Jahren beobachten). Zu den stabilen Gesellschaften rechnen wir die Verlandungs-Soziationen von *Phragmites communis* und *Glyceria aquatica*, die Schwingrasen mit *Acorus calamus* und *Carex rostrata* und das C a r i c e t u m g r a c i l i s. Vom systematischen Gesichtspunkt aus handelt es sich um ausgeprägte soziologisch klassifizierbare Einheiten.

d. Dauergesellschaften, welche ihre klimatisch möglichen Endzustand bisher nicht erreicht haben, und welche noch eine größere Beständigkeit als stabile Gesellschaften aufweisen. Sie stellen die höchstorganisierte Einheit in einem unreifen Milieu (vom Gesichtspunkt der Klimax-Theorie) dar.In Dauergesellschaften wurden nach zehn Jahren keine soziologisch bedeuteren strukturellen oder räumlichen Änderungen, wie Verschiebung der Bestandesgrenzen, beobachtet. In diese Gruppe gehört von den untersuchten Einheiten nur das C a r i c i e l o n g a t a e - A l n e t u m.

2. Syngenetische Beziehungen der Gesellschaften in der Hydroserie

Ein Vergleich beider Vegetationskarten zeigt, daß die Bestände der Schwimmblattpflanzen und anderer Hydrophyten auf der freien Wasserfläche nicht als ein einziges Initialstadium der Sukzession betrachtet werden können. Die Verlandung im Bereich der Hydrophyten (im unteren Sublitoral) verläuft gewissermaßen selbständig und parallel mit der Verlandung durch Röhricht-Arten im litoralen Raum. In unteren Bodenschichten der seichten Uferpartien wurden unter den dünnen Flachmoor-Torfschichten keine Spuren von Hydrophyten stammenden Sedimenten gefunden. Während in unserem Falle der mittlere Teil des Teiches nur mit einer einzigen Assoziation verwächst, können als Erstbesiedler des litoralen Raumes Bestände verschiedener Röhricht-Arten (*Typha angustifolia, Glyceria aquatica, Phragmites communis, Acorus calamus*) aber auch der Steifen Segge (*Carex elata*) festgestellt werden. Diese Bestände gehören mehreren Assoziationen, bzw. vielen soziologischen Kleineinheiten an, die nicht nur strukturell, sondern auch ökologisch charakterisiert werden können. Den Charakter der ersten Sukzessionsstadien bestimmen einerseits die Anwesenheit von Diasporen gewisser Arten und die Bedingungen für ihr Aufkommen, andererseits vor allem die bestehenden Standortsbedingungen wie physikalische und chemische Eigenschaften des Bodens und des Wassers, Expositionsverhältnisse usw. Wie aus den Karten ersichtlich, können verschiedene strukturelle Formen von Röhrichtbeständen die litorale Hydro-Serie beginnen, welche ganz eindeutig von den bestehenden Standortsbedingungen abhängen. Bei der Weiterentwicklung haben wir folgende Entwicklungstendenz festgestellt:

a. Innerhalb der Bestände gleicher Artenzusammensetzung einen Übergang von einfacheren zu komplizierteren Strukturformen. Diese Änderung der Struktur beruht auf ökologischen Ursachen, welche entweder autonom (durch eigene Lebensprozesse der Phytozönose) oder durch Milieu-Änderung (Düngung usw.) entstehen.

b. Eine fortlaufende Verdrängung der anspruchlosen Gesellschaften durch anspruchsvollere. Dieser Prozeß wird meist durch anthropogene Einflüsse (Düngung, Zufuhr von Nährstoffen) beschleunigt; dabei vergehen das Caricetum elatae des Sublitorals, Acorus calamus - Gesellschaften und einfachere Formen von *Typha angustifolia*-Beständen. (Sie werden von Glycerietum maximae überwachsen).

c. Eine fortlaufende Verdrängung der für einen Wettbewerb weniger adaptierten Gesellschaften durch konkurrenzstärkere Bestände. Dieser Prozeß ist am besten im Nordteil des Teiches zu sehen. Einziger Sieger im Bereich der Röhricht-Gesellschaften eutropher Gewässer scheint *Phragmites communis* zu sein. Diese Art besitzt die größte Aussicht im sublitoralen Raume eutropher Gewässer andere Röhricht-Bestände zu ersetzen, was auch stratigraphische Analysen von Torflagern bestätigen.

d. Ein sukzessiver Abbau von Röhricht-Gesellschaften. Die Röhrichtbestände vermindern ihre Produktivität erst dann, wenn sie in eulitorale Bedingungen (oberhalb des mittleren Sommerwasserstandes) infolge von Verlandung gelangen. Hier verlieren sie auch ihre geschlossene Struktur und unterliegen weiteren floristischen Änderungen. Ihr Übergang in Großseggen-Gesellschaften oder Erlenbrüche hängt von den Bodenverhältnissen ab. In flachen Buchten mit stagnierendem hochstehendem Wasser verläuft eine schnelle Entwicklung über ein Praealnetum zu Alnion glutinosae- Gesellschaften, auf geneigten Ufern mit besserer Dränierung und nur periodischen Überschwemmungen entwickeln sich die Magnocaricion-Gesellschaften nur sehr langsam. Ähnlich verläuft auch die Sukzession der Schwingrasen, auf Uferteilen in Magnocaricion-, in zentralen Teilen in Alnion- Gesellschaften.

In Teichen muß man den Verlandungsprozeß als anthropogen sehr beeinflußt betrachten. Die natürliche Tendenz der Verlandung kann man infolgedessen zusammenfassend charakterisieren:
Im Bereich der freien Wasserfläche verläuft eine selbständige Verlandung durch sedimentäre Bestände, welche sich nicht immer als einziges Initialstadium der Hydro-Serie behaupten muß. Im seichten litoralen Raum verläuft eine Verlandung mit Röhricht- und Großseggen-Gesellschaften. Die ersten Stadien können verschiedenen soziologischen Einheiten und Strukturformen angehören,was von der Zufuhr von Diasporen bestimmter Arten, von den Bedingungen für ihre Entwicklung als auch von den Standortsbedingungen verschiedener Uferpartien abhängig ist. Der natürliche Sukzessionsprozeß ist durch eine autonome Eutrophierung des Milieus und Anhäufung organischer Masse bedingt und zeichnet sich durch das Entstehen komplizierter Strukturformen in gleichartigen Zönosen, Verdrängung anspruchsloser Gesellschaften durch anspruchsvolle und durch Zurücktreten konkurrenzschwächerer Gesellschaften aus. Alle diese Änderungen haben in erster Linie ökologische (standörtliche) Ursachen und können verschiedene Stufen im Sukzessionsprozeß darstellen. Ein weiteres Stadium folgt erst bei vorgeschrittener Verlandung (unter eulitoralen Bedingungen), wo wieder die bestehenden Standortsverhältnisse den Charakter der nachfolgenden Phytozönosen bestimmen.

ZUSAMMENFASSUNG

Eine vergleichende Kartierung der Teichvegetation im südöstlichen Teile des Wittingauer Beckens (Böhmen) wurde zur Untersuchung der syndynamischen Prozesse bei der Verlandung herangezogen. Besondere Aufmerksamkeit wurde den Röhricht-Gesellschaften gewidmet, welche nach den konstanten Dominanten als Soziationen klassifiziert und auf Grund ihrer Struktur und soziologischen Eigenschaften noch feiner in strukturelle Formen gegliedert wurden. Die einfachste Form wurde als Initial-, die höchstorganisierte als Verlandungs- und die dazwischen stehende als Invasions-Form bezeichnet.

Der Gesellschaftsvergleich nach 10 Jahren brachte folgende Resultate:

1. Die Stabilität der Gesellschaften und ihre Beziehungen zu den

sich ändernden Standortsbedingungen und anthropogenen Einflüssen wurden überprüft. Von diesem Gesichtspunkt aus kann man labile, stabilisierte, stabile und Dauergesellschaften unterscheiden und wie folgt charakterisieren:

Die l a b i l e n Gesellschaften unterliegen während ihrer autonomen Entwicklung ziemlich schnellen Änderungen und sind auch gegen winzige Standortsänderungen oder direkte Eingriffe sehr empfindlich. Die s t a b i l i s i e r t e n Gesellschaften besitzen schon ein spezifisches Milieu und ändern sich infolge ihrer eigenen Lebenstätigkeit nur sehr wenig. Sie sind gegen Änderungen des Milieus oder direkten Eingriffen gegenüber ziemlich empfindlich. Die s t a b i l e n Gesellschaften sind sehr gut fähig, äußere Eingriffe in ihre Umwelt oder direkt in die Phytozönose auszugleichen, sie können jedoch durch höher organisierte Gesellschaften sukzessiv ersetzt werden. Die D a u e r g e s e l l s c h a f t e n stellen die höchstorganisierten, sehr stabilen Einheiten im "unreifen" Milieu dar.

Was die syngenetischen Beziehungen der Gesellschaften in der Hydro-Serie betrifft, erhält man auf Grund unseres Vergleiches folgende Resultate:

Im Bereich der freien Wasserfläche verläuft eine selbständige Verlandung durch sedimentäre Bestände (P o t a m i o n , L e m n e t e a u.a.), welche sich nicht immer als einziges Initialstadium der Hydro-Serie behaupten muß. Die im litoralen Raum parallel verlaufende Verlandung durch Röhricht- und Großseggen-Gesellschaften kann mit verschiedenen soziologischen Einheiten und Strukturformen beginnen. Der Charakter der ersten Stadien hängt einerseits von der Zufuhr von Diasporen bestimmter Arten und von den Bedingungen für ihre Entwicklung, andererseits von den Standortsbedingungen der betreffenden Uferpartien ab.

Der natürliche, durch Eutrophierung und Anhäufung organischer Masse charakterisierte Sukzessionsprozeß ist durch das Entstehen komplizierter Strukturformen, Verdrängung anspruchsloser Gesellschaften durch anspruchsvolle und durch Zurücktreten konkurrenzschwächerer Zönosen gekennzeichnet. Das Endglied der Hydro-Serie ist eine Dauergesellschaft, deren Weiterentwicklung zur Klimax meist nur hypothetisch zulässig ist.

SUMMARY

A comparative recording of pond vegetations in the south-eastern part of Třeboň basin in Bohemia was undertaken to investigate syndynamic processes in filling of a reservoir (= Verlandung). Special attention was paid to reed communities which were classified as sociations according to the constant dominants, and, due to their structures and sociological characteristics, were divided into still more detailed structural forms. The simplest form was designated as an Initial, the most highly organized as Terrestrial and the intermediate one as an Invasion Form.

The comparison of the associations carried out after ten years brought about the following results:

1. The stability of the communities and their relationship to the changing habitat conditions after the lapse of ten years as well as the anthropogenic influences were first reviewed. On the base of this standpoint labile, stabilized, stable and permanent communities can be differentiated, and consequently, characterized. The labile communities during their autonomous succession undergo a rather rapid change and are fairly sensitive to even insignificant site changes or direct encroachment. The stabilized communities are found in specific environments and consequently, owing to their own mode of living, change very little. They are, however, rather sensitive to changes of environment or to direct encroachment. The stable communities are in a position to equalize the external encroachment on their environment or directly on the phytocoenosis, but can be successively replaced by higher organized communities. The permanent communities are represented by the most highly organized, very stable units in inmature environments. As to the syngenetic relationship of the communities in the hydroseries one arrives, on the ground of our comparison, to the following results:

Within the range of the open water surface there occurs an independent change to terrestrial conditions by means of sedimentary stands (Potamion, Lemnetea, and others) which need not always assert themselves as the only Initial stage of the hydrosere. The parallel development of terrestrial conditions in the litoral space by means of reed and high sedge communities may begin with various sociological units and structural forms. The character of the first stage depends on the one hand on the ingress of diaspores of certain species and on the conditions of their ecesis, on the other hand on the habitat conditions of the respective bank section. The natural succession characterized by eutrophication and accumulation of organic mass manifests itself by the rise of complicated structural forms, displacing unpretentious communities by the pretentious ones and by the retreat of less competitive coenosis. The last link on the process leading to terrestrial conditions is the permanent community whose succession to the climax can be admitted only in a hypothecical way.

DISKUSSION

J.BARKMAN

Ich halte diesen Übergang von A c o r u s c a l a m u s - Beständen, von dem A c o r e t u m zu diesem G l y c e r i e t u m unter Einfluß der Düngung für eine typische discessive Sukzession, wobei das Niveau der Lebensform das gleiche bleibt.

R.CARBIENER

Aus einer Erfahrung, die wir im Elsaß gemacht haben, möchte ich zeigen, wie stabil oder auch wie labil Hydrophyten-Gesellschaften sein können, wie schnell sie reagieren, wenn die Standortsverhältnisse sich ändern, hier am Beispiel von Fließgewässern. Wir hatten im Elsaß im Riet einige jetzt noch vorhandene Grundwasserbäche, das sind kalt-stenoterme oligotrophe Kalkgewässer, und viele dieser Gewässer wurden für eine gewisse Zeit trockengelegt. Eines dieser Gewässer war kolonisiert durch eine ziemlich seltene Gesellschaft mit *Potamogeton coloratus*, die ganz bezeichnend ist für diese Art von Gewässern im Elsaß. Das Gewässer war ein ziemlich schnellfließender Bach und wurde zwei Jahre vollständig trockengelegt. Im Bachbett wuchsen *Chenopodium*-Arten und andere Unkrautgesellschaften während dieser Zeit. Als die Arbeiten beendet waren, stieg der Grundwasserspiegel wieder, und ein Jahr nachher war der gesamte Bachlauf wieder fast schöner als vorher mit *Potamogeton coloratus* besetzt. Es ist mir sehr schwer zu erklären, wie so etwas möglich ist. Vielleicht kann man annehmen, daß die Rhizome überlebt haben, aber das scheint mir kaum möglich. Und wie, da es ein schnellströmender Bach ist, die *Potamogeton*-Art so schnell - es handelt sich um eine Strecke von 10 km - gegen den Strom diesen Standort wieder erobert hat, ist eine zweite Frage.

Das entgegengesetzte Beispiel aus einem gleichen Bach, der nicht trockengelegt wurde, aber Zufuhr bekam von etwas wärmerem, etwas sedimenthaltigem Wasser, auch durch dieselben Arbeiten. Da hatten wir eine Gesellschaft mit *Sium angustifolium* und *Mentha aquatica*; sie ist bestimmt schon beschrieben worden. Diese Gesellschaft ist charakteristisch für denselben Typ von Gewässern, die etwas wärmer sind und etwas langsamer fließen, aber immer ziemlich kalt-stenoterm und sehr gleichmäßig temperiert sind. Durch die Arbeiten wurden die Sommermaxima des Wassers ungefähr 3-4° erhöht, und der vorher rein grobkiesige Boden wurde durch Sedimentation etwas verschlammt. Und da wurde die *Sium angustifolium - Mentha aquatica*-Gesellschaft vollständig verdrängt durch eine Gesellschaft von *Callitriche obtusangula*, und das auch innerhalb von einem Jahr.

H. DIERSCHKE

Wäre es nicht besser, da Sie schon nach der BRAUN-BLANQUET-Methode gearbeitet haben, anstelle von Soziationen von Faziesbildungen innerhalb des Scirpo-Phragmitetum zu sprechen?

Haben Sie bei Ihren Untersuchungen Luftbilder verwendet? Wir konnten bei Geländearbeiten 1966 in einem Gebiet stark fortschreitender Verlandung anhand von etwa 10 Jahre älteren Luftbildern feststellen, wie schnell und in welchen großen räumlichen Ausmaßen die Entwicklung in dieser Zeit vorangegangen war. Gerade bei Verlandungsgesellschaften dürften Luftbilder, die in bestimmtem Zeitabstand geflogen werden, sehr wertvolle Grundlagen für Sukzessionen und überhaupt Änderungen liefern.

Können Sie mir sagen, welche Trennarten das Scirpo-Phragmitetum typhetosum, das 1942 von TÜXEN & PREISING aufgestellt wurde, im Gegensatz zum phalaridetosum der Flußröhrichte in Ihrem Gebiet aufweist? Bei einer Aufarbeitung des nordwestdeutschen Materials haben wir festgestellt, daß *Typha angustifolia* nicht in allen Teichröhrichten vorkommt, so daß hier wohl besser von einem *typicum* der Assoziation gesprochen werden kann.

Da *Acorus calamus* sowohl im Magnocaricion wie auch im Phragmition, also in zwei verschiedenen Verbänden, faziesbildend auftreten kann, möchte ich fragen, ob Sie in Ihrem Gebiet das Acoretum calami Schulz als eine eigene Assoziation aufrecht erhalten können?

R.TÜXEN

Ich möchte das hier noch weiter ausführen. Wir sind dabei, die Systematik unserer nw-deutschen Pflanzengesellschaften zu überprüfen, bzw. zu ergänzen. Wir hatten 1942 zusammen mit PREISING zwei Subassoziationen aufgestellt, das Teich- und Seeröhricht mit *Typha angustifolia* und noch einer anderen Trennart, und das Flußröhricht mit *Phalaris* und *Oenanthe aquatica*. Wir hatten damals schon gesehen, daß es Bestände gibt, die weder die eine noch die andere Gruppe von Differential-Arten haben. Bei der Überprüfung eines viel größeren Materials zeigte sich, daß es viele Bestände gibt, in denen weder die *Typha*- noch die *Phalaris*-Gruppe vorkommt, so daß wir geneigt sind, das typhetosum jetzt aufzugeben, während wir glauben, daß das phalaridetosum, also das an bewegtem, fließendem Wasser wachsende Röhricht, vielleicht eher als Subassoziation eine gewisse Berechtigung hat. Wir möchten aber ganz entschieden solche Subassoziationen ablehnen, die etwa konstruiert werden aus Scirpo-Phragmitetum- Arten, die in Mischung mit *Lemna*- oder mit *Potamogeton*-Arten oder mit *Stratiotes* oder ähnlichen vorkommen. Das halten wir, wie auch Herr NEUHÄUSL gezeigt hat, für Synusiengemische und nicht für eigentliche systematische Einheiten. Wir möchten auch nicht dynamische Entwicklungszustände unbedingt für Subassoziationen, also für systematische Einheiten, halten. Initialphasen sollte man besser vom dynamischen Standpunkt aus von den Degenerationsphasen unterscheiden, wobei natürlich eine gewisse Standortsveränderung so wirksam sein kann, daß auch eine systematische Einheit, besonders bei den Degenerationsphasen, aufgestellt werden könnte. Aber hier greift eine Reihe von Gesichtspunkten bei diesen sehr artenarmen Gesellschaften so stark ineinander, daß deren systematische Ordnung keineswegs zu den leichten systematischen Problemen gehört.

R.NEUHÄUSL

Luftbilder habe ich nicht benutzt. Diese Schwierigkeiten kennen wir auch sehr gut, und auch diese doppelte Kartierung leistete ich nur dort, wo es sich um primitiv bewirtschaftete Teiche handelt. Denn sonst ist es kaum möglich, alle Eingiffe, die bei der Bewirtschaftung vorkommen, zu kontrollieren. Ich habe also nur diesen Zeitabschnitt von 15 Jahren kontrolliert, und da es sich um einen wirklich primitiv erhaltenen Teich handelt, kann man diese zwei Etappen unter kontrollierten Bedingungen betrachten.

Was das Scirpo-Phragmitetum typhetosum betrifft, besteht da ein Irrtum. Ich sagte auch "typhosum", also eine Fazies mit *Typha angustifolia*. Es handelt sich nicht um typhetosum latifoliae.

Was das Acoretum calami Schulz betrifft, haben Sie, glaube ich, ganz recht, daß es keine Assoziation ist. Ich verzichtete auf diese systematischen Fragen, aber aus dem Vergleich war sehr gut zu sehen, daß es ganz verschwunden ist. Aber da ich mit einigen Einheiten arbeiten mußte, da wollte ich, weil es sich um mehrere Kleineinheiten handelte, möglichst viele Einheiten benutzen.

Warum habe ich nicht Fazies benutzt, sondern Soziationsanalyse gemacht? Ich habe das zu parallelisieren versucht. Mit der BRAUN-BLANQUET-Lehre kann ich bis zur Fazies kommen. Aber ich bekam innerhalb der gleichen Fazies verschiedene strukturelle Formen und, um im Phragmition ein einheitliches System einzuhalten, habe ich vom methodischen Standpunkt aus die feinere Soziationsanalyse benutzt, auch als eine methodische Überprüfung, mit der man ebenfalls gleichwertige Einheiten bekommen kann.

J.J.BARKMAN

Ich glaube, daß WESTHOFF schon vor 15 Jahren bei seiner Bearbeitung der Süßwassergesellschaften von Botshool in Holland darauf hingewiesen hat, daß man bei den Wasserpflanzengesellschaften manchmal besser mit Soziationen auskommt, und er sagt einfach: Versuchen wir, wo möglich Assoziationen und Subassoziationen und Varianten zu unterscheiden, und wo nicht möglich, wollen wir uns nicht dazu zwingen, Soziationen im skandinavischen Sinne unterscheiden zu wollen.

E.VAN DER MAAREL

Dieser Vortrag ist wieder ein Beispiel einer konkreten Sukzessionsforschung, wir haben hier wieder die Kombination von räumlicher und zeitlicher Variation. Herr VAN LEEUWEN und ich haben in unserem Vortrag etwas allgemeines darüber gesagt, und ich möchte darauf zurückkommen, indem ich Ihnen einige Fragen stelle:

1. Wie haben Sie die Stabilität bestimmt, quantitativ oder nur qualitativ? Es wäre außerdem sehr interessant, eine Karte der Stabilität zu machen, wie wir das gemacht haben.

2. Wie haben Sie die Struktur beschrieben, und haben Sie auch die Komplizierung der Struktur quantitativ bestimmt ?-Dann möchte ich noch auf eine Bemerkung zurückkommen, die ich selbst gemacht

habe. Dies ist ohne Zweifel eine junge, dynamische Landschaft, eine Ecotone-Landschaft, und wenn es Gradienten in einer solchen Landschaft gibt, könnte es sein, daß am Anfang und am Ende eines solchen Gradienten die relativ stabilen Gesellschaften vorherrschen. Nun habe ich nicht alle Gesellschaften behalten können, aber ich dachte doch, daß am unteren Ende Ihrer Karte der höchste Grad der Entropie herrschte, also das dürfte das eine Ende des Gradienten sein. Beginn und Ende sind beide extrem, darauf möchte ich einmal hinweisen.- Dann noch eine dritte Frage: haben Sie noch Informationen über den Gesamtartenreichtum dieses Gebietes?

R. NEUHÄUSL

Ich muß leider sagen, daß die Stabilitätsbestimmung nicht quantitativ gemacht worden ist, sondern nur durch diesen Vergleich einer prozentuellen Angabe von stabilen und instabilen Fällen der Einzelsegmente der Einheiten. Die Initialform von allen Gesellschaften wurde z.B. nur in einem einzigen Segment erhalten, alle anderen Fälle haben sich nach zehn Jahren geändert. In dem zweiten extremen Fall des A l n i o n g l u t i n o s a e - Verbandes blieben alle Segmente in ihrem Raum stabil. Die Strukturbestimmung ist eine kompliziertere Frage. Ich habe das schon in dem Buch über die Röhricht-Gesellschaften publiziert und in einer methodischen Studie über diese Schwingrasen. Es ist nicht ganz mathematisch belegt. Es basiert auf dem Prinzip des Vergleichs der strukturellen Merkmale der Pflanzen, d.h. Höhe, Dominanz u.a., analysiert in der Zeit. Aus dem Vergleich ist die Periode der maximalen Stabilität festgestellt, in der diese strukturellen Merkmale wie Abundanz, Höhe der Pflanzen, Dominanz, selbstverständlich für jede Art, auf Transekten verglichen sind. Wo es sich um strukturell einheitliche Gesellschaften handelt, da wurden alle diese Merkmale auf der einen Seite der Achse eingetragen, und die diakritischen Merkmale auf der anderen. Es ist also eine Vergleichsmethode. Die Stabilität der Landschaft ist in den letzten Jahren leider mehr und mehr zurückgegangen. Es handelt sich um den südlichsten Teil des Landes, ganz nahe der österreichischen Grenze mit kleiner Einwohnerzahl und extensiver Wirtschaft. Es sind die letzten Reste der ursprünglichen und relativ primitiv bewirtschafteten Teiche, bekannt seit dem 16. Jahrhundert, wo es nur sehr wenige Maßnahmen gibt, nur Wasserdüngung und in den größeren Teichen das Ausmähen der Bestände. Man kann also sagen, daß es im Vergleich zu der landwirtschaftlich oder industriell intensiv genutzten Landschaft, eine relativ stabile Landschaft ist. Was den Gesamtartenreichtum des Gebietes betrifft, ist es herzynisches Gebiet zwischen der kollinen und montanen Stufe, das besonders artenarm ist. Ich kann nicht auswendig sagen, wieviel Phanerogamen-Arten vorkommen, aber man kann es mit den artenärmsten Gebieten vergleichen, wo ein submontanes Klima herrscht, das diesen Floren-Reichtum beschränkt. Die miozänen Sedimente sind ein äußerst oligotrophes Substrat, und die Gneise im Waldviertel sind auch arm. Es ist auch ein Nadelwaldgebiet, jetzt aufgeforstet. Dort kommen mehrere Zwischenmoore vor, mit Rhynchosporion- Gesellschaften. Als Klimax der Landschaft kann man einen Eichen-Kiefernwald mit Tanne betrachten.

H.ELLENBERG

Herr Dr. NEUHÄUSL hat darauf hingewiesen, daß bei dieser Teichverlandung verschiedene Initialen festgestellt werden können. Das liegt daran, offenbar ökologisch bedingt, weil es bei diesen künstlichen Teichen von vornherein verschiedene Niveaus, also von vornherein verschiedene Bedingungen gab. Es ist aber einfach der Start unter verschiedenen ökologischen Bedingungen. Das kommt auch in Flußlandschaften vor. Wenn ein Fluß über die Ufer getreten ist, dann hinterläßt er entweder eine niedrige Sandbank, auf der gerade noch das Phalaridetum steht, oder er hat eine höhere Sandbank gemacht, dann findet sich das Phalaridetum und z.B. eine Weidenbusch-Gesellschaft, oder es ist zufällig noch eine höhere Bank entstanden, dann sind alle Stadien der sogenannten Flußsukzession gleichzeitig ansiedelbar.

S.SEGAL

Was hier unter Struktur verstanden wird, ist m.E. etwas anderes, als was sonst damit gemeint wird. Struktur hat physiognomische Elemente, wie Lebens- und Wuchsformen, eventuell Dimensionen und auch Elemente von Vertikal- und Horizontalverteilung, Schichtung usw.

Wenn *Acorus calamus* im Phragmition und im Magnocaricion Fazies bilden kann, dann glaube ich nicht, daß das ein

Argument dafür ist, daß es überhaupt keine Assoziation sein kann. Denn wir dürfen nicht vergessen, daß diese Einheiten, diese Verbände, Abstraktionen sind; es gibt allerhand Übergänge. Man kann z.B. auch fragen, ob das Cladietum marisci eine Einheit des Phragmition oder des Magnocaricion ist. Man kann beides verteidigen.

R.NEUHÄUSL

Bei der Hydro-Serie handelt es sich um die Entwicklung von einem gewissen Wassertyp zum Land. In der gleichen Tiefe habe ich die ersten Initialen der höheren Wasservegetation, drei verschiedene Verbände, festgestellt. Also in diesen oligotrophen Buchten beginnt es mit dem Magnocaricion, in den mesotrophen Teilen beginnt die Sukzession mit dem Phragmition und in den mittleren Teilen mit Potamion- Gesellschaften.

DIE BESIEDLUNG FEUCHTER BRACHEN IM BEREICH DES NEUSIEDLER SEES

E. H ü b l

Im Zuge einer noch laufenden pflanzensoziologischen Bearbeitung der Sumpfvegetation im Gebiet des Neusiedler Sees, die ich gemeinsam mit Dr. NIKLFELD (Graz) ausführte, ergab sich in dem abnorm nassen Jahr 1965 auch die Gelegenheit, die Vegetation tiefgelegener und in diesem Jahre zeitweise überschwemmter Äcker zu studieren und zwar hauptsächlich im Gemeindegebiet von St. Margarethen, westlich des Sees. Ich muß gleich betonen, daß die hier referierten Ergebnisse nur vorläufig sind. Die eingehende Auswertung des Aufnahmematerials steht noch aus. Wenn ich es trotzdem wage, schon hier darüber zu berichten, dann deshalb, weil ich bei der Anmeldung des Referats glaubte, mit der Auswertung weiter zu sein, und weil sich andererseits bezüglich des Verhaltens einzelner charakteristischer Arten gewisse Parallelen zu den Erfahrungen von Dr. HEJNY in der Südslowakei ergeben.Sein Buch "Ökologische Charakteristik der Wasser- und Sumpfpflanzen in den Slowakischen Tiefebenen" dürfte ja allgemein bekannt sein.

Das Klima des Untersuchungsgebietes kann als gemäßigt kontinental bezeichnet werden, mit mäßig kalten Wintern (Januarmittel zwischen -1° bis - 2° C) und relativ warmen Sommern (Julimittel über 19° C). Der Jahresniederschlag liegt etwas über 600 mm.

Das Haupt-Untersuchungsgebiet "Lange Wiesenäcker" südlich St. Margarethen, etwa 1 km von der ungarischen Grenze entfernt, liegt in einer natürlicherweise abflußlosen Senke am westlichen Fuße des Ruster Höhenzuges, der am Westufer des Neusiedler Sees entlangstreicht.
Der Untergrund wird aus kalkarmen, pleistozänen Schottern und darüberlagernden sandigen Sedimenten gebildet. Die Bodenart ist meist kalkarmer sandiger Lehm. Nur an quelligen Stellen wird der Boden durch karbonatreiches Wasser beeinflußt. Die tiefsten Senken, in denen der Wasserzug vom Ruster Höhenzug her nicht mehr wirksam ist, neigen zu einer mäßigen Versalzung, die aber nicht so stark ist, um den Ackerbau ernstlich zu stören.
Im Jahre 1965 verhinderte das steigende Grundwasser die Bestellung der Äcker in den Senken, so daß sich eine üppige Brachen-Vegetation entwickeln konnte.
So sehr oft auch die Artengarnitur von Parzelle zu Parzelle

schwankt, ergab sich doch ein gemeinsamer Grundstock von Arten. Die meisten Aufnahmen lassen sich der von H.WAGNER 1942 aufgestellten Veronica anagalloides-Lythrum hyssopifolia-Assoziation zuordnen. Die beiden namengebenden Arten scheinen gute Charakterarten der Gesellschaft von hoher Stetigkeit zu sein. *Veronica anagalloides* fehlt in fast keiner Aufnahme. Als weitere, zumindest lokal gültige Charakterart kommt vielleicht noch *Epilobium adnatum* hinzu. In den höher gelegenen Brachen wurde die genannte Gesellschaft meist von sehr artenreichen Windhalm-Beständen abgelöst, deren Zugehörigkeit zu den Aperetalia außer Zweifel steht.

In der Veronica anagalloides-Lythrum hyssopifolia-Gesellschaft waren die beiden namengebenden Arten zwar mit hoher Stetigkeit vertreten, aber nur *Veronica anagalloides* kam auch mehrmals zur Dominanz. Dominant traten ferner noch folgende Arten auf: *Agrostis alba* s.l. (wobei es sich im Untersuchungsgebiet sowohl um *Agrostis stolonifera* in der Subspezies *salina*, wie um *Agrostis gigantea* in der Subspezies *limosa* handeln kann. Beide Arten sind nicht leicht voneinander zu unterscheiden. Nach den wenigen, von A.NEUMANN genau bestimmten Belegen, dürfte auf den Brachen, mit Ausnahme der nassesten, *Agrostis gigantea* ssp. *limosa* überwiegen.), *Juncus articulatus*, *Bolboschoenus maritimus* ssp. *compactus* und *Phragmites communis*. Nicht direkt dominant, aber doch mit Deckungswerten bis zu 2, erschien in vielen Brachen *Echinochloa crus-galli*.

Mit höherer Stetigkeit bei meist niedrigen Deckungswerten kamen vor: *Typha angustifolia*, *Typha latifolia*, *Pulicaria dysenterica*, *Cirsium arvense*, *Ranunculus repens*, *Bidens tripartitus*, *Polygonum lapathifolium*, *Matricaria indora* und *Rumex stenophyllus*, den man vielleicht schon zur Gruppe der halophilen Arten rechnen kann.

Mehr oder weniger deutlich halophile Arten kamen in den meisten Brachen vor, aber niemals dominierend. Hier sind zu nennen: *Schoenoplectus tabernaemontani*, *Lotus tenuis*, *Atriplex hastata*, *Puccinellia distans*, *Chenopodium glaucum*, *Juncus gerardii*, *Aster tripolium* ssp. *pannonicus*. *Althaea officinalis*, *Crypsis aculeata* und *Trifolium fragiferum*. Vielleicht mit Ausnahme von *Spergularia marginata*, *Juncus gerardii* und der nur in einer Aufnahme vorkom-

menden *Crypsis aculeata* handelt es sich um fakultative, nicht um obligate Halophyten, die wie *Chenopodium glaucum* und *Atriplex hastata* als Ruderalpflanzen weit verbreitet sind. Die einzige dominant vorkommende Art, die man als halophil bezeichnen kann, ist *Bolboschoenus maritimus* ssp. *compactus*. Salztolerant sind wohl die meisten bisher angeführten Pflanzen, manche wie z.B. *Phragmites communis* bis zu einem sehr hohen Grad.
WAGNER hat die Veronica anagaloides-Lythrum hyssopifolia-Assoziation (mit fünf Aufnahmen belegt) von vernäßten Getreidefeldern des Wiener Beckens folgendermaßen beschrieben: "Der Boden ist durchwegs Lehm und die Gesellschaft bildet sich an Stellen aus, die im Frühjahr lange überschwemmt waren. Meist sind es kleine Mulden, in denen infolge der langandauernden Überschwemmung - infolge des dichten lehmigen Bodens ist ein Absickern nicht möglich - das Getreide überhaupt nicht keimt oder nur wenige cm hoch wird. Wohl überwiegen in der Tabelle die Arten des Secalinion und der Chenopodietalia beziehungsweise der Secalino-Chenopodietea, ich glaube aber trotzdem, daß es in diesem Falle ähnlich wie bei der Parallel-Gesellschaft des Centunculeto-Anthoceretum besser ist, die Gesellschaft dem Nanocyperion einzugliedern." Als Assoziations-Charakterarten führt WAGNER *Veronica anagaloides* und *Lythrum hyssopifolia*, als Verbands- und Ordnungs-Charakterarten *Plantago intermedia*, *Juncus bufonius*, *Potentilla supina* und *Limosella aquatica* an. Von diesen Arten fehlt bei St. Margarethen nur *Limosella aquatica*, dafür kommen die beiden Isoeto-Nanojuncetea-Klassen-Charakterarten *Cyperus fuscus* und *Centaurium pulchellum* dazu. Auch die Begleiter stimmen in beiden Gebieten weitgehend überein. Da wir wesentlich mehr Aufnahmen machen konnten als WAGNER, ist die Liste der Begleiter bei uns natürlich bedeutend länger. Bemerkenswert ist vielleicht, daß *Epilobium adnatum* bei WAGNER nicht erscheint, wogegen bei St. Margarethen *Sonchus paluster* fehlt. Interessant erscheint mir auch das Vorkommen der beiden halophilen Arten *Juncus gerardii* und *Bolboschoenus maritimus* an beiden Stellen.
Es sind zweifellos die meist einjährigen Isoeto-Nanojuncetea-Arten, die der Gesellschaft das Gepräge geben. *Veronica anagalloides* ist zwar ausdauernd, wirkt aber habituell als zarte, wenig verzweigte Pflanze ähnlich wie eine Annuelle.

HEJNY, der die Ökologie von *Lythrum hyssopifolia* und *Cyperus fuscus* in dem genannten Buch genau bespricht, führt interessanterweise *Veronia angalloides* für die Süd-Slowakei nicht an, obwohl man sie eigentlich auf den dortigen Reisfeldern erwarten sollte.

So wichtig die obige Artengruppe für die Einstufung in das Vegetationssystem ist, charakterisiert sie doch den gegenwärtigen Zustand des Wuchsortes, so wichtig sind für die Frage einer möglichen Weiterentwicklung die Pioniere hochwüchsiger ausdauernder Pflanzen, die meist dem Phragmition angehören und zum Teil schon sehr bald die Physiognomie der Gesellschaft bestimmen. Vielleicht wird manchem die Frage nach einer Weiterentwicklung müßig erscheinen. Denn natürlich werden in einem Normaljahr, wenn das Wasser zurückgeht, die Brachen wieder unter den Pflug genommen. Aber im Bereich des Neusiedler Sees herrscht in dem Grenzbereich, wo Ackerkultur gerade noch möglich ist, ein ständiger Kampf mit der Natur. Stellen, die in Trockenjahren umgepflügt wurden, müssen in feuchten Perioden wieder aufgegeben werden. Wir hatten in unserem Untersuchungsgebiet Flecken, die zweifellos einmal Acker waren, (oft sind die Parzellengrenzen und Abzugsgräben noch deutlich zu erkennen) die aber schon mehrere Jahre nicht mehr bestellt werden konnten. In der Vegetation stellt sich dann ein gewisser, sicher nur vorläufiger Gleichgewichtszustand ein. Die ausdauernden Phragmition-Arten bilden dann entweder allein, oder in Mischung miteinander Massenbestände, die wesentlich artenärmer sind, als die Brachen-Gesellschaften. Hier zwei willkürlich gewählte Beispiele:

Tabelle 1.

Bulboschoenus maritimus-Bestand. Deckungsgrad 60%, Wassertiefe 0 - 10 cm. Aufnahme 8.10.1965

4.5	Bolboschoenus maritimus	+	Cirsium brachycephalum
1.3	Mentha aquatica	+.2	Juncus articulatus
2.3	Agrostis stolonifera	+	Typha angustifolia
2.1	Veronica anagalloides	+	Calystegia sepium
1.3	Typhoides arundinacea	+.2	Batrachium trichophyllum
+	Lythrum salicaria	+	Alisma plantago-aquatica
+	Teucrium scordium	+	Carex riparia
+	Inula britannica	+	Rumex stenophyllus

\+ Schoenoplectus tabernaemontani

Tabelle 2.

Bestand von *Phragmites communis*, *Bolboschoenus maritimus* und *Schoenoplectus tabernaemontani*. Es handelt sich um einen tieferen, wahrscheinlich schon länger sich selbst überlassenen Bestand. Deckungsgrad 75%, Wassertiefe 30 cm. Aufnahme 19.9.1965

2.3	Bolboschoenus maritimus	1.2	Teucrium scordium
2.2	Schoenoplectus tabernaemontani	+	Typhoides arundinacea
2.1	Phragmites communis	+	Veronica anagalloides
3.4	Agrostis stolonifera	+	Mentha aquatica
1.1	Lythrum salicaria	(+)	Typha latifolia
1.1	Cirsium brachycephalum	(+)	Typha angustifolia
		4.4	Chara spec.

Zur Rolle der einzelnen Phragmition-Arten innerhalb der Vegetationsentwicklung ist folgendes zu sagen: die besten Ausbreitungsmöglichkeiten haben zweifellos die beiden *Typha*-Arten mit ihren sehr gut flugfähigen Früchten, die leicht keimen. HEJNY sagt von *Typha latifolia*, daß sie als einzige Röhrichtart in künstlichen Gewässern ephemer auftritt, sich rasch ausbreitet und bei Austrocknung und Verlandung auch schnell wieder verschwindet. Er schreibt: "Auf vernachlässigten (Reis-) Feldern bestimmt sie gemeinsam mit *Echinochloa* die Physiognomie. Dies ist um so auffallender, als ihre zönotische Bedeutung als Röhrichtkomponente in den natürlichen Gewässern der Umgebung minimal ist." Ich kann diese Erfahrung bestätigen. In der näheren Umgebung kommt *Typha latifolia* überhaupt nicht in der natürlichen Teichvegetation vor. Dies ist vielleicht damit zu erklären, daß ihre Optimal-Standorte in der natürlichen Vegetation nach HEJNY bei einer Wassertiefe von 20 bis 50 cm liegen und hauptsächlich dort, wo andere Röhrichtkomponenten degradiert sind. In den natürlichen kleinen Teichen des Untersuchungsgebietes sind aber entsprechende Standorte von anderen Pflanzen besetzt, so daß die konkurrenzschwächere Art nicht aufkommen kann.

Typha angustifolia trat stärker auf als *Typha latifolia*, wohl deshalb, weil es in den benachbarten Teichen natürliche Bestände gibt und zwar im tiefen Wasser, wo *Phragmites* nicht mehr aufkommen kann. Das stimmt wieder mit den Angaben von HEJNY überein, der eine optimale Tiefe von 80 bis 110 cm angibt. Auf den von ihm untersuchten Reisfeldern war die Art viel schwächer vertreten als *Typha latifolia*.

Juncus articulatus ist eine ausdauernde Pionierpflanze, die nicht dem Phragmition angehört, aber in den Vorstufen zu Phragmition-Gesellschaften eine wichtige Rolle spielt. In der Veronica anagalloides-Lythrum hyssopifolia-Gesellschaft wird er oft dominant. Nach HEJNY verträgt *Juncus articulatus* eine längere Überflutung durch eine 10 bis 15 cm hohe Wasserschicht, sowie ein starkes Austrocknen des Substrats und ist gegenüber der chemischen Beschaffenheit des Bodens indifferent. Dieser Anspruchslosigkeit verdankt er die Lebensmöglichkeit in verschiedenen Gesellschaften. OBERDORFER gibt in seiner Flora Scheuchzerio-Caricetea fuscae-Ges., Calthion, Agropyro-Rumicion und Nanocyperion an. Nur die Beschattung der hochwüchsigen Phragmition-Arten verträgt er nicht und verschwindet bald, wenn diese dichter zusammenschließen.

Eine typische Seichtwasserpflanze ist auch *Bolboschoenus maritimus* ssp. *compactus*. Nach HEJNY hängt sein häufiges Vorkommen mit der Verlängerung der litoralen Phase bis zum Hochsommer zusammen. Die Art wächst dort nicht mehr, wo ab Juni eine rasche Austrocknung erfolgt. In unserem Gebiet war die Art auf den Brachen häufig. Im Beobachtungsjahr waren die Lebensbedingungen durch die sommerliche Überschwemmung ausgezeichnet. Die Ausbreitung dürfte vorwiegend vegetativ erfolgt sein. Dies zeigt sich an hohen Soziabilitäts-Werten im Gegensatz etwa zu *Typha*, die meist gleichmäßig verteilt wuchs und auf den Brachen nirgends fruchtete, was für eine junge Ansiedlung spricht.

Noch stärker macht *Phragmites communis* den Eindruck geschlossenen Vorrückens von den Rändern her. Es trat in den jungen Brachen nur selten, dann aber gleich dominant auf. Die ökologische Amplitude der Art ist im pannonischen Gebiet sehr weit. Es gibt außer geschlossenem Wald kaum einen Standort, wo Schilf nicht auftreten könnte. Selbst dürre Schotterböden, Eisenbahn-Dämme und Ruinen werden manchmal besiedelt. Wahrscheinlich befähigt das relativ warme Klima, diese zwar nahezu kosmopolitisch verbreitete, aber doch wärmeliebende Pflanze zu dieser großen Ausbreitungsfähigkeit. HEJNY weist darauf hin, daß die ökologische Amplitude von *Phragmites communis* in der warmen Süd-Slowakei weiter ist als im kühlen Südböhmen. Auch der niederländische

Schilfspezialist VAN DER TOORN sagte mir bei einem Besuch am Neusiedler See im vergangenen Sommer, daß nach seiner Erfahrung *Phragmites* in den Niederlanden wesentlich mehr stenök ist. Zweifellos ist das Schilf, obwohl von allen Arten des Phragmition am wenigsten an das Wasser gebunden, im Röhricht die vitalste Art, welche die übrigen in die Randbereiche verdrängt. Eine menschlich bedingte Grenze wird dem Schilf in den Wiesen gesetzt, wo es durch Mahd so geschwächt wird, daß es keine geschlossenen Bestände bilden kann.

Zum Abschluß möchte ich darauf hinweisen, daß sehr schöne Phragmition-Gesellschaften in Südmähren, also in einem klimatisch ähnlichen Gebiet wie am Neusiedler See, im ehemaligen Kobyler See enwickelt sind. Ich konnte dank der liebenswürdigen Einladung der tschechoslowakischen Kollegen das Gebiet im vergangenen Jahre besuchen. Der See wurde im vorigen Jahrhundert abgelassen und der Boden bebaut. Nach dem Krieg füllte sich das Becken teilweise wieder durch Schäden im Entwässerungssystem und es entwickelten sich verschiedene Schlamm- und Seichtwasser-Gesellschaften, die zum Teil eine sehr große Ähnlichkeit mit den besprochenen Pflanzenbeständen haben. Ein wichtiger Unterschied scheinen mir die große Ausdehnung und Stabilität von *Typha latifolia*-Beständen zu sein.

ZUSAMMENFASSUNG

Im abnorm feuchten Jahr 1965 konnten die Bauern rund um den Neusiedler See ihre tiefgelegenen Felder nicht bestellen. Deshalb entwickelte sich eine üppige Brachen-Vegetation. Die bestuntersuchten Bestände können zur Veronica anagalloides-Lythrum hyssopifolia-Assoziation (Wagner 1942) gestellt werden. Als Dominante erscheinen: *Veronica anagalloides*, *Agrostis gigantea* ssp. *limosa*, *Agrostis stolonifera* ssp. *salina*, *Juncus articulatus*, *Bolboschoenus maritimus* ssp. *compactus* und *Phragmites communis*. Mit hoher Stetigkeit aber nicht mit hohem Deckungsgrad treten auf: *Typha angustifolia*, *Typha latifolia*, *Pulicaria dysenterica*, *Cirsium arvense*, *Ranunculus repens*, *Bidens tripartitus*, *Polygonum lapathifolium*, *Matricaria inodora* und *Rumex stenophyllus*. Eine Gruppe mehr oder weniger halophiler Arten bilden: *Schoenoplectus tabernaemontani*, *Lotus tenuis*, *Atriplex hastata*, *Puccinellia distans*, *Chenopodium glaucum*, *Juncus gerardii*, *Aster tripolium* ssp. *pannonicus*, *Althaea officinalis*, *Crypsis aculeata*, *Trifolium fragiferum*.

Von großer Bedeutung für die Entwicklung der Vegetation sind die hochwüchsigen ausdauernden Pflanzen (*Typha latifolia, Typha angustifolia, Juncus articulatus, Bolboschoenus maritimus* ssp.*compactus, Phragmites communis*); die meisten von ihnen gehören zum P h r a g m i t i o n. Sie bilden Bestände auf älteren Brachen. Die Standortsansprüche dieser Arten im Gebiet um den Neusiedler See werden diskutiert und mit den Ansprüchen in der Slowakei verglichen, die Dr. HEJNY untersucht hat.

SUMMARY

In the abnormally moist year 1965 in the territories around the Neusiedler See farmers could not cultivate their fields in low localities. Therefore, a luxuriant vegetation of fallow land developed. The communities investigated can be attached best to the V e r o n i c a a n a g a l l o i d e s - L y t h r u m h y s s o p i f o - l i a - A s s o c i a t i o n (WAGNER 1942). The dominants are: *Veronica anagalloides, Agrostis gigantea* ssp.*limosa, Agrostis stolonifera* ssp.*salina, Juncus articulatus, Bolboschoenus maritimus* ssp.*compactus* and *Phragmites communis*. Species with high constancy, but without a high degree of cover are: *Typha angustifolia, Typha latifolia, Pulicaria dysenterica, Cirsium arvense, Ranunculus repens, Bidens tripartitus, Polygonum lapathifolium, Matricaria inodora* and *Rumex stenophyllus*. A group of more or less halophilous species is: *Schoenoplectus tabernaemontani, Lotus tenuis, Atriplex hastata, Puccinellia distans, Chenopodium glaucum, Juncus gerardii, Aster tripolium* ssp.*pannonicus, Althaea officinalis, Crypsis aculeata, Trifolium fragiferum.*

The tall perennial plants (*Typha latifolia, Typha angustifolia, Juncus articulatus, Bolboschoenus maritimus* ssp.*compactus, Phragmites communis*) are of great importance for the development of vegetation. Most of them belong to the alliance of P h r a g m i - t i o n and form stands in older areas of fallow lands. The behaviour of these species in the territory around the Neusiedler See is discussed and compared with the behaviour in Slovakia investigated by Dr. HEJNY.

DISKUSSION

R. TÜXEN
Ich möchte anregen, daß Sie sich wegen *Puccinellia distans* mit Herrn Kollegen VICHEREK in Verbindung setzen, der weiß über diese kritische Art in Südosteuropa wahrscheinlich am besten Bescheid. Es ist vielleicht nicht das, was wir hier als *P. distans* bezeichnen.
Eine zweite Bemerkung: *Typha latifolia* ist bei uns in natürlichen Teichen sehr weit verbreitet, hier in Nordwestdeutschland und auch in Holland.

E. HÜBL
Ich muß gestehen, daß wir *Puccinellia distans* selbst bestimmt haben. Wir haben ja am Neusiedler See zwei mit *P. distans* sehr nahe verwandte Arten in den natürlichen Gesellschaften. Wir sind auf *P. distans* gekommen, obwohl wir die beiden anderen versucht haben, herauszubekommen. Aber vielleicht würde der Spezialist doch etwas S p e z i e l l e s finden.
Typha latifolia ist auch bei uns verbreitet, kam nur in unserem Untersuchungsgebiet nicht vor.

F. KÜHN
In Südmähren kommen auf Salzstellen in Feldern ähnliche Kombinationen von Arten alljährlich vor wie bei Ihnen. Die Kombinationen enthalten obligate und fakultative Halophyten, wie *Atriplex hastata f. incana*, *Potentilla anserina*, *Centaurium pulchellum*, *Chenopodium glaucum*, *Matricaria chamomilla f. bayeri*, *Potentilla supina*, *Melilotus dentatus*, *Lotus corniculatus* ssp. *tenuifolius*, *Lythrum hyssopifolia*, *Spergularia salina*, *Carex secalina*, *Atropis distans*, *Veronica anagallis*, *Trifolium fragiferum*, *Bolboschoenus maritimus*, *Inula britannica*, *Xanthium strumarium*, *Juncus gerardii*, *Heleochloa alopecuroides* und *H. schoenoides*, *Atriplex tatarica*, *Lepidium ruderale*, *Coronopus procumbens*, *Carex hordeistichos*, *Rumex stenophyllus*, *Ranunculus sardous*, *Chenopodium urbicum*, *Lythrum virgatum*, *Glaux maritima* und *Hibiscus trionum*.

E. HÜBL
Die Artengarnitur ist sehr ähnlich wie bei uns, nur kommt *Glaux maritima* am Neusiedler See nicht vor, das südöstlichste Vorkommen findet sich im nördlichen Niederösterreich.

E. BURRICHTER
Sie erwähnten, daß *Phragmites* im Vergleich zu *Typha* schlechter

keimt. Neben dieser Tatsache ist aber noch zusätzlich zu erwähnen, daß auch die Fruktifikation von *Phragmites* in den einzelnen Jahren sehr unterschiedlich sein kann. Nach unseren Erfahrungen können auf ein Jahr mit gutem Fruchtbesatz mehrere Jahre mit ganz geringen oder sogar ohne Karyopsen-Besatz der Rispen folgen, ähnlich wie wir das bei der Fruktifikation verschiedener Baumarten kennen. Dabei ist es möglich, daß sich selbst unmittelbar benachbarte Bestände in den einzelnen Jahren durchaus unterschiedlich verhalten.

I.S.ZONNEVELD
Die Ökologie von *Typha angustifolia* und *T.latifolia* ist ganz verschieden, wie Sie auch angaben. Sie haben das auf die Wassertiefe zurückgeführt. Es ist immer gefährlich, das auf eine Sache zu beschränken. Im allgemeinen kann man sagen, daß *T.latifolia* eine engere Amplitude hat als *T.angustifolia*. Die alte Einteilung bei TÜXEN & PREISING - *angustifolia* bis in die mesotrophe und eben oligotrophe Ecke zurückgedrängt - ist glücklicherweise aufgegeben worden, denn diese Art kommt ungefähr auf der ganzen Welt vor, in allen Klimaten, und geht weit bis in das Salz hinein. *Typha latifolia* bleibt immer in dem süßen Gebiet. In allen Flüssen, die in die Nordsee fließen, hat man eine Zone, wo *Typha latifolia* im süßen Gebiet bleibt,und *T.angustifolia* weit in das brackische Gebiet vordringt.

S. SEGAL
Typha latifolia ist in den Niederlanden sehr häufig und wächst dort oft an vom Menschen beeinflußten Stellen, besonders an Standorten, die eutrophiert worden sind; weiter an ruhigen Stellen, wo es entweder eine sehr schnelle Sapropel-Entwicklung gegeben hat, oder an Stellen mit bedeutender Streu-Entwicklung.
Die Untersuchungen von Professor BAKKER in Groningen haben gezeigt, daß *Phragmites communis* massenhaft keimen kann, wenn es über der Bodenoberfläche in der richtigen Zeit eine seichte Wasserschicht (bis zu 2 mm) gibt.

DIE AUSBILDUNG DER UNKRAUTVEGETATION UNTER VERSCHIEDENEN FELDFRÜCHTEN

F. K ü h n

Es wurde die Unkrautvegetation unter einer Fruchtfolge von zehn Versuchsflächen der Forschungsstation für Agrotechnik in Pohořelice in Südmähren verfolgt (nördliche geographische Breite 40°58, geographische Länge östlich 16°31', Meereshöhe 182 m, Jahrestemperatur 8,9° C, ebenes Gelände). Fruchtfolge: 1.Luzerne, 2. Luzerne, 3.Hafer, 4.Winterweizen, 5.Zuckerrübe, 6.Kartoffeln, 7.Winterweizen, 8.Winterroggen (für Korn), 9.Kornmais, 10.Sommergerste mit Untersaat. Im Jahre 1962 wurden anstatt des ausgefrorenen Winterweizens,Sommerweizen gesät, anstatt ausgefrorenem Winterrogen,Sommergerste. In günstigen Jahren wurden als Stoppelfrucht Bohnen mit weißem Senf gesät. Mit Kunstdünger wurden alle Früchte außer Gerste und Luzerne gedüngt, Zuckerrübe und Mais auch mit Stalldünger. In den Parzellen wurden vor jeder Ernte von 2-5 Quadratmetern die oberirdischen Teile der Unkräuter abgeschnitten, nach Arten gezählt, und die Trockenmasse je m^2 ermittelt. Es wurden 6 Jahrgänge erfaßt, ein Jahrgang wurde auch getrennt nach Unkrautarten gewogen. Die Unkrautvegetation der Lokalität kann man in die D e s c u r a i n i o - D e l p h i n i e t u m (SCHUBERT & MAHN 1959) einreihen. Die Versuchsfelder waren gut bearbeitet, die Verunkrautung gering. Die häufigsten Unkräuter waren *Taraxacum officinale*, *Chenopodium album*, *Echinochloa crus-galli*, *Anagallis arvensis* ssp.*phoenicea*, *Setaria viridis*. Der Homogenitätsgrad der Unkrautvegetation der Parzellen nach RAABE war 87.5, und daher recht hoch. Die Gemeinschafts-Koeffizienten nach JACCARD für die Unkrautvegetation der Parzellen, verglichen mit der charakteristischen Artenkombination nach RAABE, bewegten sich zwischen 67,7 und 84,8.

Die stärkste Verunkrautung dem Gewicht nach war in Luzerne,eine schwächere im Wintergetreide, noch weniger im Sommergetreide, weniger in Hackfrüchten, die schwächste in Stoppelfrüchten und auf geschälter Stoppel. Die Reihenfolge wurde oft durch den Witterungsverlauf beeinflußt. Nach Einführung von Herbiziden in Weizen und Hafer (Methylchlorphenoxyessigsäure-Präparat) war die Verunkrautung dem Gewicht nach in der ganzen Fruchtfolge geringer, außer unter Luzerne, Mais und Sommergerste mit Untersaat. Nach der Anzahl von Unkrautpflanzen war die Sommergerste mit Untersaat am dichtesten verunkrautet, dann folgten Luzerne, Hafer nach Luzerne, dann Wintergetreide, dann mit viel weniger Individuen Stoppelfrüchte, Mais, Kartoffeln, geschälte Stoppel, Zuckerrübe. Das höchste Gewicht je Pflanze war in Kartoffeln, das niedrigste in Sommergerste mit Untersaat. Die Unterschiede der Verunkrautung der Feldfrüchte waren nach Gewicht 2/5 statistisch gesichert, der Individuenzahl nach zu fast 1/2. Zwischen Verunkrautung und Ertrag der Kultur war eine negative Korrelation bei Luzerne im ersten Jahr, bei Winterweizen nach Kartoffeln, bei Winterroggen und Sommergerste, aber eine positive Korrelation bei Mais (vielleicht wegen geringer Schattenwirkung der Kulturpflanze). Im Jahre 1963 wurden die Gewichtsanteile der Unkrautarten festgestellt. In Luzerne waren im ersten Nutzungsjahr im Frühjahr am meisten *Taraxacum officinale*, im Sommer *Plantago lanceolata*, im Herbst *Polygonum aviculare* vertreten, im zweiten Nutzungsjahr im Frühjahr und Sommer *Taraxacum officinale*, im Herbst *Atriplex patula* vertreten, im darauffolgen Hafer *Taraxacum officinale*, im darauffolgenden Winterweizen, sowie in Zuckerrübe und Kartoffeln *Chenopodium album*, im nachfolgenden Winterweizen *Delphinium consolida*, in Winterroggen *Stellaria media*, in Mais *Echinochloa crus-galli*, in Sommergerste mit Untersaat *Anagallis arvensis* ssp.*phoenicea*. Weiter wurde die Individuenzahl der Unkrautarten je m^2 ausgewertet. In vergleichbaren Früchten wurde in Luzerne ein Zunehmen vom ersten zum zweiten Nutzungsjahr festgestellt bei *Stellaria media*, *Thlaspi arvense*, *Capsella bursa-pastoris*, *Taraxacum officinale*, und ein Abnehmen von *Amaranthus retroflexus*, *Chenopodium album*, *Sonchus asper*, *Conyza canadensis*. Im Winterweizen nach Hafer waren zahlreicher als im Winterweizen nach Kartoffeln: *Melandrium noctiflorum*, *Chenopodium album*, *Capsella bursa-pastoris*, *Medicago lupulina*, *Sonchus asper*, *Taraxacum officinale*, *Setaria glauca*, und weniger zahlreich *Papaver rhoeas*, *Delphinium consolida*, *Amaranthus retroflexus*, *Polygonum convolvulus*, *Sisymbrium sophia*, *Camelina microcarpa*, *Anagallis arvensis* ssp.*phoenicia*, *Euphorbia exigua*, *Plantago major* ssp.*vulgaris*, *Convolvulus arvensis* (die Getreideunkräuter waren im Material zahlreicher in Weizen nach Hackfrucht). In Stoppelfrüchten waren nach Winterroggen (frühe Bestellung) zahlreicher Unkräuter mit höherem Temperaturanspruch wie *Amaranthus retroflexus*, *Echinochloa crus-galli*,

Tab.1. Individuenzahl je 1 m^2 (vergleichbare Früchte der Fruchtfolge zusammengezogen)

Feldfrüchte	Luzerne-Frühjahr	Luzerne-Sommer	Luzerne-Herbst	Winter-roggen	Winter-weizen	Sommer-weizen	Sommer-gerste	Sommer-hafer	Zucker-rübe	Kartof-fel	Mais	Stoppel-früchte	Geschälte Stoppel
Medicago sativa	.	.			0,100			6,250					
Vicia villosa	.	.		0,133									
Avena sativa	.	.											+
Arrhenatheretum elatius	.	.						0,o55					
Festuca pratensis	.	.			0,033			4,417					
Festuca rubra	.	.						0,444					
Dactylis glomerata	.	.						4,666					
Poa pratensis	.	.			0,600			0,889					
Triticum aestivum	.	.										+	+
Secale cereale	.	.										+	+
Panicum miliaceum	.	.							0,250				
Algae	+	.			+								
Cyanthus vernicosus	.	.		+									
Coprinus sp.	.	.				+							
Octospora leucoloma	+	.		+									
Pottia intermedia	+	+	+	+									
Bryum argenteum	+	+	+	+	+								
Pleuridium alternifolium	.	+											
Ceratodon purpureus	+	+											
Delphinium consolida				1,000	2,550						0,050	0,167	
Papaver rhoes				0,067	0,092	0,250			0,050	0,567	0,100		0,095
Fumaria officinalis	0,033												
Arenaria serphyllifolia	0,200	1,125	0,274	1,400	0,850	0,250	0,762	0,111		0,067		1,500	0,190
Stellaria media	4,200	8,063	1,279	5,417	1,625	5,667	0,834	1,944	0,825	0,400	1,600	1,750	0,214
Melandrium noctiflorum	0,117	1,500	0,794	1,000	1,175	0,417	1,571	0,472	0,022	0,200	0,167	0,167	0,143
Melandrium album			0,o29										
Silene dichomata		0,063											
Amaranthus retroflexus	0,183	11,438	1,020	0,050	0,434	0,417	1,381	0,667	0,078	0,550	0,550	3,418	0,548
Amaranthus hybridus										0,033			
Chenopodium album	4,633	9,188	2,288	7,616	5,117	7,000	10,519	14,055	0,625	1,575	3,250	7,000	2,428
Chenopod.hybridum											0,050		
Atriplex patula	0,100	0,063	1,000				0,095	0,056					

Rumex obtusifolius			0,029										
Polygonum aviculare	3,733	1,813	1,335	4,667	4,775	1,167	1,738	1,222	0,025	0,075		0,167	0,143
Polygonum convolvulus	1,500	1,875	0,833	5,833	5,692	3,167	10,647	2,944	0,125	0,517	0,150	0,084	0,286
Polyg.lapathifolium				0,600			0,095			0,033			
Viola tricolor ssp. arvensis	1,100	0,438	0,392	2,817	0,517	2,417	1,478	0,444	0,275	0,033	0,050	0,833	
Sisymbrium sophia	0,250		0,029	0,917	2,050				0,150	0,067	0,300		0,048
Camelina microcarpa	0,050	0,063			0,692	0,250	0,048	0,056					
Thlaspi arvense	0,117			0,333	0,059		0,095	0,222	0,050	0,033			0,143
Capsella bursa-pastoris	0,600	0,125	0,705	0,266	0,150		0,095	0,278	0,025	0,600	0,182	0,083	0,143
Neslia paniculata		0,125					0,048						
Raphanus raphanistrum	0,050												
Anagallis arvensis ssp. phoenicea	2,767	2,750	3,579	5,300	4,658	2,167	21,618	2,388	0,450	0,275		0,333	0,143
Anagallis arvensis ssp. coerulea							0,048						
Euphorbia exigua	0,450	0,188	0,118	0,467	0,575	0,250	1,167	0,222	0,025				
Euphorbia falcata	0,033												
Euphorbia helioscopa							0,048	0,111		0,033			
Medicago lupulina	0,733	0,318	0,218	0,233	3,333	.	2,500	1,861	0,075	.	.	0,583	.
Vicia sativa ssp. angustifolia	.	0,063	.	0,067	.	.	.	.	.	.	.	.	.
Vicia hirsuta	.	.	.	.	0,050	.	.	.	.	.	.	.	.
Geranium pusillum	.	.	0,013	.	.	.	.	.	0,075	.	.	.	.
Erodium cicutarium	0,050	0,063	0,029	.	.	.	0,071	0,056	0,075	0,117	.	.	.
Daucus carota	.	.	.	.	.	.	0,048	.	.	.	.	.	.
Galium aparine	0,783	.	.	.	.	0,750	0,119	0,111	.	.	0,050	.	.
Convolvus arvensis	1,900	1,000	.	0,067	0,900	.	0,857	1,028	1,775	0,600	0,900	.	0,357
Lappula myosotis	.	.	.	.	.	.	.	0,056	.	.	.	0,083	.
Solanum nigrum	0,050	.	0,029	0,633	0,033	.	0,095	0,111	0,250	.	.	0,250	.
Veronica polita	1,650	0,188	0,196	0,133	.	.	0,048	0,389	.	0,117	.	.	.
Veronica persica	0,800	0,063	0,059	0,067	.	0,500	0,095	.	.	0,033	.	.	.
Veronica hederifolia	0,667	.	.	0,200	0,233	.	.	.	0,150	.	0,400	1,000	.
Veronica triphyllos	.	.	.	0,067	0,050	.	.	.	.	.	.	.	.
Plantago major	0,139	0,688	0,127		0,967		0,429			0,075			
Plantago lanceolata	0,084	0,188	0,088				0,238						
Plantago media	0,042		0,029										
Lamium amplexicaule	0,100						0,048	0,083		0,067	0,200	0,167	
Sonchus asper	1,867	3,625	1,441	0,533	0,825	0,500	1,881	1,306	0,025	0,292		0,500	0,476
Sonchus oleraceus		0,125		0,067									0,048
Crepis biennis										0,033			

Taraxacum officinale	13,600	26,250	20,915	0,950	0,450	0,634	1,356	3,083	0,150	2,158	0,200	0,250	0,143
Galinsoga parviflora		0,063	0,029	0,400		2,000		0,167	0,100	0,125	0,050	0,167	
Matricaria maritima ssp. inodora	0,050				0,033								
Cirsium arvense	0,100		0,196							0,208			
Carduus acanthoides				0,067								0,083	
Erigeron canadensis	0,350	0,188	0,078	0,067	0,033					0,067	0,050		
Avena fatua	0,050			0,200				1,500					
Agropyron repens	0,050							0,167					
Enchinochloa crus-galli	10,833	12,938	1,559	13,033	2,759	23,167	19,405	4,195	0,675	2,367	3,667	2,084	0,405
Setaria viridis	2,933	8,125	2,210	2,550	2,525	0,667	5,047	1,111	0,225	0,842	0,282	0,083	0,048
Setaria glauca	1,250	0,188	1,035	0,483	0,133	2,917	0,332	3,500	0,100	0,167	0,133		
Summe	57,734	92,938	41,894	57,633	43,225	54,250	82,500	44,028	6,400	12,325	12,867	12,800	6,000
Kulturpflanzen als Unkräuter				0,133	0,733			16,722				+	+
Anzahl Quadratmeter	23	16	46	17	29	5	20	17	40	27	18	12	11
Zahl der Unkrautarten	40	31	32	35	30	20	34	31	25	31	21	22	18

Ø = 28,645

und weniger zahlreich Unkräuter, welche bei niedrigeren Temperaturen keimen (*Veronica hederifolia, Arenaria serpyllifolia*). Es konnte nicht konstatiert werden, daß in stallmistgedüngten Beständen nitrophile Unkräuter häufiger wären; eher wäre das Gegenteil zu behaupten. Manche nitrophile Unkräuter waren nach Luzerne häufiger.
Die Artenzahl und die Individuenzahl der Unkrautarten je m^2 in den Feldfrüchten wurde nach Zusammenziehen der Zahlen in vergleichbaren Früchten ausgewertet (Durchschnittszahl von 2 Jahren Luzerne, 2 Winterweizen, 3 Stoppelfrüchten und Schälfurchen).Die höchste Zahl von Unkrautarten war in Luzerne (ganzjährige Vegetationszeit), dann folgten nach der Artenzahl Winterroggen, Sommergerste mit Untersaat, Hafer nach Luzerne, Kartoffel, Winterweizen, Zuckerrübe, Stoppelfrucht, Mais, geschälte Stoppel = durchschnittlich 1.Luzerne, 2.Getreide, 3.Hack- und Stoppelfrüchte. Die Unkrautarten, welche im Material eine niedrige Individuenzahl aufwiesen, fehlten zumeist in Früchten, welche eine niedrige Unkrautartenzahl aufwiesen. Die Homogenität nach RAABE für das Vorkommen von Unkrautarten in den Feldfrüchten der Fruchtfolge war 78,5, und die Differenzen zwischen der Unkrautvegetation der Feldfrüchte waren daher fast zweimal so groß wie zwischen der Unkrautvegetation der Parzellen im Material. Die Gemeinschaftskoeffizienten zwischen den Artenkombinationen in den Feldfrüchten und der charakteristischen Artenkombination nach RAABE waren recht hoch; am niedrigsten 38 (Luzerne:Schälfurche). Es war im untersuchten Material wegen der hohen Homogenität nicht möglich, die Unkrautvegetation der einzelnen Feldfrüchte verschiedenen Assoziationen anzureihen (ELLENBERG 1956). Nach den Artenkombinationen waren große Unterschiede zwischen Mais und Zuckerrübe, zwischen Winterroggen und Sommergerste. Die größte Individuenzahl hatten in Luzerne *Taraxacum officinale*, in Winterroggen *Echinochloa crus-galli*, in Winterweizen *Polygonum convolvulus*, in Sommergerste *Anagallis arvensis* ssp.*phoenicia*, in Sommerhafer *Chenopodium album*, in Zuckerrübe *Convolvulus arvensis*, in Kartoffeln und Mais *Echinochloa crus-galli*, in Stoppelfrüchten und auf Schälfurche *Chenopodium album*. Es wurde ein enger Zusammenhang zwischen Keimtemperaturen der Unkrautarten und deren Vorkommen in Feldfrüchten mit verschiedenen Saatterminen festgestellt.
Nach den Individuenzahlen je m^2 in den Feldfrüchten im Material

wurden die Unkrautarten in Gruppen eingeteilt. Es wurde dabei beachtet, ob die Art in der Frucht nicht nur keimend vorkommt, wie *Veronica hederifolia* in Kartoffeln/:

I Kulturpflanzen als Unkräuter:

A. Mit Saatgut und anderem Material ausgesäte Arten: *Vicia villosa* Roth (mit Roggensaat), *Panicum miliaceum* L. (mit Saatgut von Zuckerrübe oder mit Stallmist).

B. Arten aus Samen, welche bei der Ernte ausfallen und die Nachfrucht verunkrauten: *Avena sativa* L., *Triticum aestivum* L. em.Mansf., *Secale cereale* L.

C. Ausdauernde Arten, welche in der Nachfrucht austreiben: *Medicago sativa* L., *Arrhenatherum elatius* (L.) J.C.et C.B.Presl, *Festuca pratensis* W.Huds., *F.rubra* L., *Dactylis glomerata* L., *Poa pratensis* L.

II Kryptogame Unkräuter (makroskopisch wahrnehmbare Arten):

A. Auf verrottendem Mist: *Coprinus* sp.

B. Auf dem Erdboden wachsende Arten (im Material nur in Luzerne und Wintergetreide): Algen (verschiedene Arten), *Cyathus vernicosus* DC., *Octospora leucoloma* Hedw./Seav., *Pottia intermedia* Fürnr., *Bryum argenteum* L., *Pleuridium alternifolium*(Dicks.)Rabenh., *Ceratodon purpureus* Brid.

III Angiosperme Unkräuter:

A. Mit Saatgut ausgesäte Arten:

a. Mit Luzernesaat ausgesäte Arten: *Melandrium album* (P. Mill.)Garcke, *Rumex obtusifolius* L., *Daucus carota* L. ssp. *silvestris* (Mill.)Dom., *Plantago lanceolata* L., *P.media* L., *P.major* L. ssp.*vulgaris* (Hayne) Dost, (verbreitete sich nur vorwiegend mit Saatgut, kam auch wenig in nachfolgendem Hafer, Winterweizen und Kartoffeln vor, inkliniert zu B 3. In den Feldern kam er einjährig vor), *Atriplex patula* L. (im Material nur in Luzerne und ausklingend im folgenden Hafer, inkliniert zu B 2, wo er auch angeführt ist), *Medicago lupulina* L. (verbreitete sich teilweise mit Luzernesaat, war weniger häufig in anderen Früchten, inkliniert zu B 2,wo er auch angeführt ist).

b. Arten, welche mit anderem Saatgut ausgesät wurden: *Vicia hirsuta* (L.) S.F.Gray (im Material nur 1 Jahr in Winterweizen).

B. Unkrautarten, welche auf den Versuchsflächen dauernd

sich vermehrten:

a. Samenunkräuter:

1. Unkräuter, welche vorwiegend in Luzerne im Frühjahr und im Wintergetreide registriert wurden, in Hackfrüchten nur als Keimpflanzen. Sie haben niedrige Keimtemperaturen/TK 1 nach LAUER. In Luzerne regenerieren sie zumeist nicht nach der 1. Mahd: *Delphinium consolida* L. (wurde nicht in Luzerne registriert), *Sisymbrium sophia* L., *Vicia sativa* L. ssp.*angustifolia* (Grufb.) Gaud., *Veronica hederifolia* L., *V.triphyllos* L., *Erigeron canadensis*.
Alle angeführten Arten außer *Erigeron canadensis* waren in Wintergetreide zahlreicher als in Luzerne.

2. Unkräuter, die außer in Luzerne und Wintergetreide auch in Sommergetreide vorkamen. Die Arten haben etwas höhere Keimtemperaturen als Gruppe 1 /TK2 nach LAUER *Papaver rhoeas* L. (wurde nicht in Luzerne registriert), *Arenaria serpyllifolia* L.,*Atriplex patula* L., *Polygonum aviculare* L., *Camelina microcarpa* Andrz., *Neslia paniculata* (L.) Desv. (für kleine Individuenzahl unsichere Einreihung), *Anagallis arvensis* L. ssp.*phoenicia* (Scop.)Vollm., *Euphorbia exigua* L., *Medicago lupulina* L.,*Lappula myosotis* Moench. (für kleine Individuenzahl unsichere Einreihung),*Avena fatua* L.
Von den angeführten Arten war in Luzerne häufiger nur *Atriplex patula*, und in Wintergetreide waren seltener als in Sommergetreide nur *(Anagallis arvensis* ssp. *phoenicia)*,*(Medicago lupulina)* und *Avena fatua*.

3. Unkräuter, welche in allen Typen von Feldfrüchten vorkamen (welche auch in den Hackfrüchten häufig vorkamen). Die Arten keimen später als Gruppe 1 und 2, und keimen gewöhnlich bis in den Herbst hinein (TK3-5 nach LAUER; von TK2 nur einige, welche im Rahmen ihrer Gruppe höhere Optima und Minima haben, oder von den Unkrautarten solche, welche in Hackfrüchten verhältnismäßig selten waren):

3.1 Arten, welche in Luzerne und Getreide häufiger waren: *Stellaria media* (L.)Vill., *Melandrium noctiflorum* (L.) Fr., *Chenopodium album* L., *Polygonum convolvulus* L.

(inkliniert zu B a 2), *P. lapathifolium* L. (wenig Exemplare, Einreihung unsicher), *Viola tricolor* L.ssp. *arvensis* (Murr)Gaud., *Thlaspi arvense* L., *Euphorbia helioscopia* L., *Galium aparine* L. (inkliniert zu B a 2), *Veronica persica* Poir. (in Hackfrüchten nur wenig Exemplare), *Sonchus asper* (L.) Hill, *S. oleraceus* L., *Echinochloa crus-galli* (L.) Pal.-Beauv., *Setaria glauca* (L.) Pal.-Beauv., *S. viridis* (L.) Pal.-Beauv.

Von diesen Arten waren im Wintergetreide durchschnittlich weniger häufig als in Sommergetreide *Chenopodium album*, *Euphorbia helioscopia*, *(Galium aparine)*, *(Sonchus asper)*, *Echinochloa crus-galli*, *Setaria glauca*.

3.2 Arten, welche in Hackfrüchten häufiger waren: *Geranium pusillum* Burm.f. (wenig Exemplare, Einreihung unsicher), *Erodium cicutarium* (L.) Hérit. (in Hackfrüchten nur wenig häufiger).

3.3 Arten mit unregelmäßigem Vorkommen, ohne Überwiegen in Getreide oder Hackfrucht: *Amaranthus retroflexus* (in Wintergetreide weniger häufig als in Sommergetreide), *Capsella bursa-pastoris* (L.) Medik., *Solanum nigrum* L., *Veronica polita* Fries, *Lamium amplexicaule* L., *Galinsoga parviflora* Cav. (in Getreide mehr als in Luzerne und Hackfrucht).

In der ganzen Gruppe B a 3 waren in Luzerne häufiger als in anderen Feldfrüchten *(Stellaria media)*, *Amaranthus retroflexus*, *(Capsella bursa-pastoris)*, *Veronica polita*, *Sonchus asper*, *Setaria viridis*.

4. Arten, welche wegen geringer Anzahl von Exemplaren (unter 0,008-0,006 je m^2 im Material) nicht in Gruppen eingereiht wurden: *Fumaria officinalis* L., (wohl B a 1), *Chenopodium hybridum* L. (wohl B a 3), *Amaranthus hybridus* L. ssp. *chlorostachys* (Willd.) Hej. (wohl B a 3), *Raphanus raphanistrum* L. (wohl B a 2), *Anagallis arvensis* L. ssp. *coerulea* (Nath.)Hartm. (wohl B a 3), *Euphorbia falcata* L. (wohl B a 2), *Matricaria maritima* L. ssp. *inodora* (L.) Soó (wohl B a 3).

b. Mehrjährige und ausdauernde Unkräuter (im Material oft in Luzerne: *Convolvus arvensis* L., *Taraxacum officinale* F.Weber, *Cirsium arvense* L., *Agropyron repens* (L.) Pal.-Beauv.

C. Zufällig gekeimte Pflanzen, welche sich in Feldern nicht fortpflanzten: *Carduus acanthoides* L., *Crepis biennis* L.

ZUSAMMENFASSUNG

Es wurde die Unkrautvegetation unter einer Fruchtfolge in Pohorelice in Südmähren verfolgt: 1.Luzerne, 2.Luzerne, 3.Hafer, 4.Winterweizen, 5.Zuckerrübe, 6.Kartoffeln, 7.Winterweizen, 8.Winterroggen (für Korn), 9.Kornmais, 1o.Sommergerste mit Untersaat. In günstigen Jahren wurde Stoppelfrucht gesät.

Nach den Individuenzahlen je m^2 im Material wurden die Unkrautarten in Gruppen eingeteilt:

I. Kulturpflanzen als Unkräuter, z.B. *Panicum miliaceum, Medicago sativa, Dactylis glomerate, Poa pratensis..*

II. Kryptogame Unkräuter (makroskopisch wahrnehmbare Arten), z.B. *Pottia intermedia, Bryum argenteum.*

III. Angiosperme Unkräuter:

A Mit Saatgut ausgesäte Arten:

a Mit Luzernesaat: *Melandrium album, Rumex obtusifolius, Daucus carota, Plantago lanceolata, P.media, P.major (Atriplex patula),(Medicago lupulina)*

b Mit anderem Saatgut: *Vicia hirsuta* (im Material nur 1 Jahr)

B Unkrautarten, welche auf den Versuchsflächen dauernd wuchsen und sich vermehrten:

a Samenunkräuter:

1 Unkräuter, welche vorwiegend in Luzerne im Frühjahr und im Wintergetreide registriert wurden, in Hackfrüchten nur als Keimpflanzen: *Delphinium consolida, Sisymbrium sophia, Vicia sativa-angustifolia, Veronica hederifolia, V.triphyllos, Erigeron canadensis*

2 Unkräuter, welche außer in Luzerne und Wintergetreide auch in Sommergetreide vorkamen: *Papaver rhoeas, Arenaria serpyllifolia, Atriplex patula, Polygonum aviculare, Camelina microcarpa, Anagallis arvensis-phoenicia, Euphorbia exigua, Medicago lupulina, Avena fatua*

3 Unkräuter, welche in allen Typen von Feldfrüchten vorkamen:

1 Arten, welche in Luzerne und in Getreide häufiger waren: *Melandrium noctiflorum, Chenopodium album, Polygonum convolvulus, Viola tricolor arvensis, Thlaspi arvense, Euphorbia helioscopia, Galium aparine, Veronica persica, Sonchus asper, S.oleraceus, Echinochloa crus-galli, Setaria glauca, S.viridis*

2 Arten, welche in Hackfrüchten häufig waren: *Erodium cicutarium*

3 Arten ohne Überwiegen in Getreide oder Hackfrucht: *Amaranthus retroflexus, Capsella bursa-pastoris, Solanum nigrum, Veronica polita, Lamium amplexicaule, Galinsoga parviflora*

b Mehrjährige und ausdauernde Unkräuter: *Convolvulus arvensis, Taraxacum officinale, Cirsium arvense, Agropyron repens*

C Zufällig gekeimte Pflanzen, welche sich in Feldern nicht fortpflanzten: *Carduus acanthoides, Crepis biennis.*

SUMMARY

The results of six years of observations of weeds in a 10 strip croprotation in Pohorelice (South Moravia) are described. A classification of the weed species is given, derived from their abundance in the crops of the studied rotation area. There is a dependency between the occurence of weed species in crops differing in sowing date and the germination temperature of the weed species. The differences between the combinations of weed species in the several crops were relatively small, so that it has not been possible to segregate the weed vegetation of the different crops in different associations.

LITERATUR

ELLENBERG,H. -1956- Aufgaben und Methoden der Vegetationskunde.- In: Walter H.: Einführung in die Phytologie 4 (1). Stuttgart.

JACCARD,P. -1928- Die statistisch-floristische Methode als Grundlage in der Pflanzensoziologie.- In: Abderhalden,E. (Red.): Handbuch der biologischen Arbeitsmethoden. Abt.XI, Teil 5: 165-202. Berlin-Wien.

LAUER,E. -1953- Über die Keimtemperatur von Ackerunkräutern und deren Einfluß auf die Zusammensetzung von Unkrautgesellschaften.- Flora 140: 551-595. Stuttgart.

RAABE,E. -1953-1954- Über den "Affinitätswert" in der Pflanzensoziologie.- Vegetatio 4: 53-68. Den Haag.

SCHUBERT,R. & MAHN,E.G. -1959- Vegetationskundliche Untersuchungen in der Mitteldeutschen Ackerlandschaft I: Die Pflanzengesellschaften der Gemarkung Friedeburg/Saale.- Wissenschaftl.Z. Martin Luther-Universität Halle-Wittenberg. Math.-nat.8 (6): 965-1012. Halle/Saale.

DISKUSSION

J.TÜXEN

Unsere nw-deutschen Äcker sind seit etwa 10 Jahren durch Herbizide so sehr verarmt, daß die Artenzahl ganz außerordentlich vermindert ist, und viele Unkräuter kurz vor der Ausrottung stehen. Wie sieht das bei Ihnen auf normal bewirtschafteten Feldern aus? *Fumaria officinalis* wurde von Ihnen als seltene Halmfruchtart aufgeführt. Bei uns in NW - Deutschland ist das eine typische Hackfruchtart.

F.KÜHN

Die Versuchsfelder waren hier recht gut bearbeitet. In normalen Feldern wird die Unkrautvegetation reicher sein. Es wird genug von Herbiziden gebraucht, aber man kann noch ganz gute Aufnahmen machen, in denen mehr als 20 Arten vorkommen. Zumeist in höheren Lagen in Feldern, wo die Kulturpflanzen nicht gut vorankommen, und auf Salzstellen kann die Artenzahl größer werden.
Zur Frage über *Fumaria officinalis* wollte ich sagen, daß das nur die Resultate aus den Versuchsfeldern waren, und da gab es eine kleine Artenzahl. Aber in normalen Feldflächen ist *Fumaria officinalis* häufig genug; sie kommt sowohl im Getreide als auch in Hackfrüchten vor. Häufig ist sie auch in Gärten.

H.MERKER

In Schweden ist die Luzerne die Feldfrucht, welche die größte Unterdrückung auf die Ackerunkräuter ausübt. Es ist einfach so, daß es sich draußen, in der praktischen Landwirtschaft, gar nicht lohnt, ein Luzernefeld zu untersuchen, weil man überhaupt nicht mehr, weder artenmäßig noch zahlenmäßig, soviel Unkräuter erfassen kann, daß eine Bearbeitung möglich ist. Es scheint mir in dem Falle ein besonderer Ausschlag der Versuchsparzelle zu sein, daß die Luzerne so ungeheuer unkrautreich ist, oder welche Auffassung vertreten Sie selbst? Wieviel Stallmist wurde in den Parzellen verwendet und wie verhält sich das zu der praktischen Landwirtschaft? In Schweden verwenden wir Stallmist überhaupt nicht mehr, oder nur in ganz seltenen Fällen.
Über wieviele Jahre erstrecken sich die Untersuchungen auf den Versuchsparzellen? Sie sagten, daß Körnermais in die Fruchtfolge eingeschlossen war. Gerade im Zuckermais haben sich bei uns ganz spezielle Verhältnisse gezeigt, und ich möchte fragen, ob die auch bei Ihnen auftreten?

Haben Sie jemals, vielleicht auch nur untergeordnet, *Veronica praecox* zusammen mit *Veronica triphyllos* feststellen können?

F. KÜHN

Die Luzerne ist zumeist wenig verunkrautet. Die Unkräuter sind da meistens klein, aber die Individuenzahl ist oft groß, zumal wenn der Bestand locker ist. Wenn er gut steht, ist von den Unkräutern oft überhaupt nichts zu finden. Ich mache ungern Aufnahmen in Luzerne oder Klee, weil da wenig Arten zu finden sind. Aber im zweiten Jahr, wenn man das ganze Jahr aufnimmt, wie im Frühjahr als auch im Herbst, kommen dort auch Unkräuter vor,und die Skala der Arten ist oft breiter als in anderen Feldfrüchten.

Von Stallmist wurden 400 Zentner je Hektar gegeben, zweimal in der Fruchtfolge. In der Unkrautvegetation wirkte sich das eher negativ aus:die nitrophilen Unkräuter waren nicht häufiger in den gedüngten Feldfrüchten.

Die Untersuchung erstreckte sich über sechs Jahre.

Der Körnermais war ganz speziell verunkrautet. Es war ein großer Unterschied zwischen der normalen Artenkombination in der ganzen Fruchtfolge und dem Mais. Es muß vielleicht eine allelopathische Beeinflussung der Unkrautvegetation dabei wirksam sein, so daß die Unkrautvegetation hier anders ausgebildet ist. Zuckermais wird bei uns sehr wenig angebaut, darüber kann ich nichts sagen.

Veronica praecox ist bei uns in Feldern sehr wenig vertreten,gewöhnlich nur auf Feldern, die nicht normal bearbeitet werden können, z.B.ein steiniges Feld an einem Hang. Dann kann man dort in den oberen Partien manchmal *Veronica praecox* finden, aber sie ist in Feldern recht wenig vertreten. *Delphinium consolida* ist recht oft mit *Veronica triphyllos* anzutreffen. Letztere ist bei uns meist in niedrigen Lagen verbreitet.

J.FEISE

Ich freue mich, daß man sich wirklich um Unkräuter kümmert. Ich halte das für sehr günstig, man kann dabei viel lernen. Ich gehe allerdings etwas anders vor: ich sehe mir zunächst den Boden an, und welcher Pflanzenbestand oder Wald hier eigentlich stehen müßte. Und dann sehe ich mir an, welche Fruchtfolge durchgeführt wird. Unsere Fruchtfolgen sind meistens 3 bis 7 Früchte lang. Und wenn man die Fruchtfolge kennt und weiß, zu welchem Bodentyp das Feld gehört, dann kann man sich auch schon ein Urteil erlauben, ob die Fruchtfolge richtig gewählt ist. Ist sie falsch gewählt, dann kriege ich viel Unkräuter, und die Unkräuter sind meine Lehrmeister. Wenn Unkräuter auftreten, daß sie mich ärgern, dann habe ich Fehler gemacht. Unkräuter, die mich nicht ärgern, sind für mich keine Unkräuter.

Sie sagten z.B., nach Stoppelfrüchten haben Sie ganz wenig Unkräuter. Es ist die Frage, ob ich überhaupt Stoppelfrüchte baue, ob ich die Arbeit, die die Stoppelfrüchte leisten sollen, nicht durch die natürlichen Unkräuter leisten lassen kann. Unter Umständen liefern mir die Unkräuter, die nach dem Aufschlag kommen, genau das, nämlich die Bodenbedeckung, die Gare-Erhaltung usw; dann sind das gar keine Unkräuter mehr für mich, sondern ganz billige Stoppelfrüchte.

Und so geht es auch beim Mais. Der Mais wächst tatsächlich besser, wenn Unkräuter darin sind. Wenn man keine Unkräuter hat, kann man sich das auch etwas erleichtern dadaurch, daß man Mais mit Sojabohnen im "Unterholz" baut, daß man also keine offenen Flächen hat. sondern unter dem Mais etwas anderes steht, was u. U. gar nicht geerntet werden braucht, was aber verhindert, daß Unkräuter kommen. Jedenfalls ist das viel besser, als mit Simazin zu arbeiten, was bei uns oft empfohlen wird. Simazin ist ein solches Teufelszeug, daß ich im nächsten Jahr wieder nichts anderes anbauen kann als Mais. Jede andere deutsche Frucht kann dieses Zeug nicht vertragen, es hält sich 3-4, u.U. 5 Jahre. Und wenn ich 4 Jahre hintereinander Mais baue, dann mache ich mir den Acker dadurch schlecht.

Es ist gar nicht notwendig, daß die Unkräuter alle vernichtet werden. Als Landwirt interessiert mich der Ertrag und die Arbeit, die ich dafür geleistet habe. Ob mit oder ohne Unkraut ist mir egal, Hauptsache ist, ich kriege Geld und brauche soviel weniger hineinzustecken.

Wir haben Angst, daß wir jetzt Unkräuter kriegen, weil wir nicht mehr Hackfrüchte bauen, sondern Getreide nach Getreide. Und wir können dies Getreide nach Getreide laut Vorschrift der Betriebswirtschaftler dadurch bauen, daß wir das Stroh drauflassen, unterpflügen usw. Die Folge ist, daß wir bei dem vielen Stroh, daß jetzt untergepflügt wird, wieder mehr Verunkrautung haben; die Erträge sinken und - und das ist auch wieder eine Selbstreinigung der Natur - in diesem Jahr sind in diesem mit Stroh gedüngten Unkräutern sehr viel *Nematoden* und sehr viel *Tipula* aufgetreten, die machen die Sache dann wieder gut. Unkräuter und Schädlinge versuchen immer wieder das, was der Mensch gewissermaßen verseucht hat, zu einer normalen, natürlichen Vegetation zurechtzurücken, und wenn der Mensch aufpaßt, pflanzt er nicht erst solche Sachen, die da nicht hingehören.

R.CARBIENER
Wir haben im Elsaß eine interessante Pflanze, die eine ausgesprochene Vorliebe für Luzernefelder hat, *Senecio vernalis*. Während ungefähr 10 Jahren trat sie massenhaft, und zwar ausschließlich in Luzerne-Feldern auf. Seitdem ist sie wieder fast verschwunden, warum, wissen wir nicht genau.

J.BARKMAN
Sie findet sich auch in Wacholder-Gebüschen in Dänemark.

H.ELLENBERG
Ich bin als jemand zitiert worden, der über den JACCARD'schen Gemeinschaftskoeffizienten entscheiden sollte, was Assoziations-Rang hat. Die Originalstelle lautet: "Das könnte man vielleicht". Ich muß davor warnen, dies so einfach, besonders bei Unkraut-Gesellschaften als Mittel anzuwenden. Man muß hierbei immer zuvor pflanzensoziologisch entscheiden; das sollte immer ein Hilfsmittel sein.

J.BARKMAN
Sie haben darüber gesprochen, daß zwischen der Dichte des Unkrautbestandes und des Getreides, des Kulturgewächses, eine negative Korrelation zu finden war, ausgenommen beim Mais, wo sie positiv war. Der Mais steht nicht so dicht, haben Sie selber gesagt. Glauben Sie, das ist einfach, weil der Boden, wo er für Mais gut ist, auch gut ist für Unkräuter, ist es also keine direkte Korrelation, oder glauben Sie vielleicht, Mais und seine Unkräuter beeinflussen sich gegenseitig positiv durch Allelopathie?

F.KÜHN
Die Waldgesellschaft an der Stelle, wo die Versuchsflächen liegen, wäre rekonstruiert nach der tschechoslowakischen Karte, das Q u e r c o - C a r p i n e t u m . Der Boden war Schwarzerde, mittelschwer. Es war Alluvium auf Diluvium. Die Fruchtfolge war 10-jährig; das ist keine gewöhnliche Fruchtfolge, denn gewöhnlich wird eine 8-jährige praktiziert. Ich habe nur eine Frucht-

folge studiert, weil ich aus technischen Gründen nicht mehr schaffen konnte. Die Stoppelfrüchte waren in manchen Jahren sehr rein, in anderen Jahren waren sie recht verunkrautet, und es kam dann zu einer Samenentwicklung, z.B. *Chenopodium album* oder *Amaranthus retroflexus*, so daß in manchen Jahren die Stoppelfelder auch zu einer Verunkrautung beitragen können. Gewöhnlich waren in den Stoppelfrüchten mehr Unkräuter als in den ungeschälten Stoppeln, die nicht weiter bearbeitet wurden. Der Mais wird bei uns jetzt als Monokultur gesät. Früher war sehr oft auf kleinen Parzellen zu sehen, daß Kürbisse oder Fisolen(Bohnen) darunter waren.
Bei den speziellen Unkräutern der Luzerne wurde *Senecio vernalis* angeführt. Spezielle Unkräuter für Luzerne kenne ich bei uns recht viele, es sind Unkräuter, die mit dem Saatgut verbreitet werden, z.B. *Melandrium album* oder *Silene dichotoma*. Zum großen Teil war auch *Rumex obtusifolius* nur in Luzerne, und auch *Plantago lanceolata*. Es kommen auch oft andere Pflanzen vor, wie *Crepis nicaeensis*, *Crepis biennis* oder andere, die mit dem Saatgut verbreitet werden und die dazu beitragen, daß die Unkrautflora in der Luzerne oft artenreich ist.
Ich wollte nur ausdrücken, daß die Differenzen in dem JACCARD'schen Gemeinschaftskoeffizienten nur klein waren, da hatte ich in ELLENBERG einen Anhaltspunkt. Die größte Differenz war 38 zwischen den studierten Kulturpflanzen. Zwischen der Unkrautflora der Hackfrüchte in dieser Fruchtfolge waren z.B. große Differenzen zwischen Mais und Zuckerrübe. Zuckerrübe war dem Durchschnitt der charakteristischen Artenkombination sehr ähnlich, während Mais ganz anders verunkrautet war und eine andere Artenkombination aufwies. Bei den Getreiden war Gerste recht anders als Winterroggen. Zwischen manchen Getreiden und manchen Hackfrüchten gab es oft kleinere Differenzen.

Beim Mais war die Korrelation mit der Verunkrautung positiv; da habe ich als Bemerkung aufgeführt - aber das kann ich nicht beweisen - daß es vielleicht damit zusammenhängt, daß die Schattenwirkung im Mais gering ist, und daß es wohl möglich ist, daß, wenn es in manchen Jahren gute Witterungsbedingungen für den Mais gibt, diese auch günstig sind für die Entwicklung seiner Unkrautvegetation. Bei anderen Früchten war die Korrelation oft negativ, und oft war es nicht zu sagen, es blieb unentschieden.

SUCCESSION IN THE VEGETATION OF SAND DUNES OF GREEK COASTS AND THE ROLE OF MOST IMPORTANT DYNAMOGENETIC VALUE SPECIES.

G.J. Lavrentiades

Introduction

In the past fifty years many efforts have been made to the solution of problems related to the important role of high dynamogenetic value species and their constructiveness in the development of sand dune formation and vegetation. Marked progress has been also made in the study of the effects of ecological factors upon vegetation. Nevertheless, all these are concerned with the vegetation of the coasts of the Baltic Sea, the Northern Sea (ELLENBERG 1963), the Atlantic Ocean (TÜXEN 1950,1961; GEHU 1964a, 1964b; GEHU & PETIT 1965; HOCQUETTE, GEHU & FAUQUET 1965), or the Western Mediterranean (DUCELLIER 1911; KUHNHOLZ-LORDAT 1923; ARENES 1924,1929; BRAUN-BLANQUET 1933, FREI 1937; Braun-BLANQUET, ROUSSINE, NEGRE 1951; PIGNATTI 1959; MOLINIER 1964, 1965). Works related to the coastal sand dune vegetation of the Eastern Mediterranean area, and especially to those of Greece, are very few. However, they deal mostly with the relations of plant to habitat, whether the latter be natural or modified by humus or plant remains, or they are simple floristical plant lists. Moreover, these are not written from a sociological standpoint and therefore they can not be satisfactorily applied in solving such problems, as those regarding the dynamogenetic value of species upon the development of vegetation (TURRILL1929; RECHINGER 1943; RECHINGER & RECHINGER-MOSER 1951). However, an attempt has recently been made to meet these needs as far as possible (OBERDORFER 1952; LAVRENTIADES 1963,1964; KNAPP 1965).

Further studies allowed us to arrive at more satisfactory conclusions related to the mechanism of evolution occurring in the vegetation of sand dunes at greek coasts and especially the important contribution of causative species in the succession of vegetation. An outline of these conclusions, based upon the results of 250 relevations made all along the coastal line of Greece, and particularly upon a selection of vegetation types and habitats, are given in the following paragraphs.

The results may not appear to be completely convincing, as the present work deals only with the greek coasts, which, when compared with those of the whole mediterranean area, constitute only a small sample, and obviously can not be regarded as representative of the whole of the mediterranean area. However, I must emphasize that the purpose of this survey is to supply some knowledge of the greek vegetation to those who intend to study the vegetation of the mediterranean area in general, and therefore it is written in accord with present-day sociological progress.

General observations

Greece, in spite of its small area, is surrounded by a comparatively enormously long line of coasts of about 14.000 km. All these coasts are not suitable for dune formation. Some parts of them lie at about the same level as the neighbouring sea. This often happens, for example, at about the middle of the Western Coasts of Greece, near Mesolongi. Too often they are at the bases of hills which abrubtly sink into the sea. The best examples in Northern Greece occur at the three Peninsulas of Chalkidiki, and in Western Greece, at the coasts which extend from Albania to Mesolongi. In several cases, sea coasts consist of deposits, composed of loamy sand, which, of course, is not suitable for dune formation. Also, in small islands no sand dunes appear, because of their predominantly mountainous character. In such cases, associations of the C a k i l e t e a group, appearing in their initial or optimal phase may be present in some very restricted coastal areas of wind-protected small gulfs. Stands rich in ammophilous vegetation occur locally at the exposed to to the SW winds low land coasts of the Western Peloponnese (LAVRENTIADES 1964).

Our essential problem in this survey was to select, as far as possible, vegetational references from every type of sand dune sufficient in kind and number to represent adequately the total range of succession in the vegetation. The arranging of the data collected led to the conclusion that the psammosere which develops on the sand dunes of greek coasts includes at least one pioneer stage, several transition stages, and a terminal stage. These stages, according to their floristical composition and to their ecological factors may be devided into five groups. Within the limits of a group various stages of development can be distinguished and floristically and ecologically defined.

a) C a k i l e t e a group

The first group, including the phases of C a k i l e t e a associations, develops on low level dunes, which are more or less covered by humus and plant remains.

Its initial phase begins with the appearance of individuals of pioneer species *Euphorbia peplis* with its horizontally elongated root system or with the therophyte *Salsola kali* with its vertically elongated root system. It should be noted, that the initial phase of this group, representing the initial stage of the whole psammosere, is present in certain coastal areas of Greece, even down to sea level, and especially in those places where they are favoured by the prevalence of mountains or islands. Their pioneer species are seen as individuals all along the coasts of Greece, especially where the beach is very low and fool of pebbles and cobbles, without humus and plant remains, or they may be present on embryonic dunes. Examples from Northern Greece are the Gulf of Thessaloniki, the coasts of the Chalkidiki Peninsula, etc., while from Southern Greece, the coastal line which extends from Patrai to Corinth, where one can easily see individuals of *Salsola kali* and more rarely *Euphorbia peplis* or fragments of plant communities of the C a k i l e t e a m a r i t i m a class, at these places where humus is accumulated. One of the best examples occurs also at the Gulf of Kalamata. The beach is composed especially of pebbles and cobbles of about 2-15 cm in size mixed together with small quantities of fine sand. Species occurring on this beach are individuals of the therophyte *Salsola kali* , which presents the higher frequency and develops better to its full capacity than any other species at the Gulf, and the frequently found therophyte *Euphorbia peplis*. In several cases, this last species accompanies the isolated patches of *Salsola kali*, which abundantly grows on the coast (LAVRENTIADES 1964). It should be evident from what has been said above that both these pioneer species may be considered as of high dynamogenetic value, for their important role in the development of sand dune vegetation on the greek coasts.

On the places where the amount of pebbles and cobbles decreases and that of sand, humus, and plant remains increases, a new type of vegetation develops. It represents a transitional stage of evolution, characterized by the present of more psammophilous and nitrophilous species. Its aspect is given by a plant group where dominant species are *Salsola kali* and *Cakile maritima*. At this stage of development the frequent presence of *Cynodon dactylon* should be noted. There are some interesting plant groups of this type of vegetation in Northern Greece, as for example, at the Gulf of Thessaloniki, although the most important groups of this vegetation extend to any favouring part of the greek coast. In the southernmost part of Peloponnese, by Methoni, there is a plant group with therophytes *Salsola kali*, and *Euphorbia peplis* as dominants, and *Anthemis muenteriana*, *Agropyron litorale*, and *Vitex agnus-castus*, as companion species (LAVRENTIADES 1964).

As representative type of vegetation of the next transition stage of development may be considered the well known all over Greece, Salsola kali-Xanthium strumarium ass. (TÜXEN 1950; OBERDORFER 1952; LAVRENTIADES 1963,1964). From the data available, it seems evident that there are significant correlations between Salsola kali-Xanthium strumarium association and the height of dune, as there is a relationship between the association and concentration of humus and plant remains. Floristic lists and other data show that stands rich in this vegetation type occur locally at low level dunes of about 0.20 to 0.80 m in height, where there is more humus and plant remains. It seems likely that this is the most dominant association of the Cakiletea class and that species occurring in the plots are much more nitrophilous than those of the previously described vegetation. Locally in the Northern, Central and Southern Greece there are patches of a mixed community, in which the Salsola kali-Xanthium strumarium association share dominance. Most of the stands of the Salsola kali-Xanthium strumarium association are wind-protected areas in which humus and plant remains, derived especially from *Zostera marina* leaves, are accumulated in large quantities. Such places may actually be found almost anywhere in suitable habitats of Northern Greece, as for example, at Keramoti (LAVRENTIADES 1963), Porto-Lago, Gulf of Thessaloniki or at Southern Greece (LAVRENTIADES 1964).

As terminal stage of development of the Cakiletea group may be considered the few known associations, Anthemis-Salsola kali-ass. (LAVRENTIADES 1964), and Atriplex tatarica-Atriplex hastata-ass. (LAVRENTIADES 1963).

Both these constant and dominant associations together are found to be the most useful means of characterising the terminal phase of an interesting series of plant communities, in which the pioneer species *Salsola kali* and *Euphorbia peplis* represent their initial phase. This type of vegetation is confined to strongly humose sand soils and so occurs most frequently on the low level dunes, which are occupied by humus or plant remains, derived especially from *Zostera* leaves.

b) A g r o p y r e t u m group

The second group, including the phase of the A g r o p y r e t u m m e d i t e r r a n e u m , is distinguished by the occurrence of rhizomatous geophytes, *Agropyrum junceum* ssp.*mediterraneum*, *Sporobolus pungens*, and *Galilea mucronata*. Stands rich in A g r o - p y r e t u m m e d i t e r r a n e u m occur locally at high level dunes, of about 1-2 m height. They are often associated with absence of humus and plant remains and are seldom as extensive as the communities of C a k i l e t e a group.

Typical pioneer species of the initial phase of this type of vegetation are the rhizomatous geophytes, *Agropyrum junceum* ssp. *mediterraneum*, *Sporobolus pungens*, or *Galilea mucronata*. They are undoubtfully the main causative species of sand dune formation. That is probably due to their elongated root system, to which enables them to resist the wind. They are also more widely scattered than any other species of the A g r o p y r e t u m group and appear as individuals or in patches on high dunes, or mixed together with associations of C a k i l e t e a group. The best example of this last case occurs at the western parts of Peloponnese, near Gialova, where thanks to the presence of humus and plant remains, the associations of E u p h o r b i o n p e p l i s. All occupy the largest part of the area, while the initial phase of A g r o p y r e t u m m e d i t e r r a n e u m, represented by patches of the species *Agropyrum junceum* ssp.*mediterraneum*, appears on the places where are entirely or to some extent free of humus and plant remains (LAVRENTIADES 1964). There are also occurrences in flat places on coastal lands of North Greece as at Keramoti (LAVRENTIADES 1963), Porto-Lago etc., where the species *Agropyrum junceum* ssp.*mediterraneum* is a distinctive feature of the C a k i l e t e a stands. The type dominated by *Agropyrum junceum* ssp. *mediterraneum* is the most common and may be only representative of the initial phase,although our observations suggest that,in more rare cases, two of the above species may be present together at this type of vegetation, but very locally.

At higher level sand dunes, where habitats become more and more suitable, the initial phase of the A g r o p y r e t u m group is succeeded by its optimal phase. This is characterized partly by the nature of sand dune, which is to some extent or entirely free of humus and plant remains, but more particularly by the overwhelming dominance of rhizomatous geophytes, *Agropyrum junceum* ssp.*mediterraneum*, *Sporobolus pungens*, and *Galilea mucronata*. It seems that both these species, considered as of high dynamogenetic value, are the main causative species of the A g r o p y r e t u m dunes. They are also more vigorously grown plants, more widely scattered and the space between them is occupied by the usual species of the A g r o p y r e t u m zone. Locally the few known patches of this kind of vegetation occur near sea-shores, which are exposed to the SW winds, and such places are, for instance, the Western Coasts of Greece and the coasts of the Aegean Islands. A particularly sharp optimal phase of this type of vegetation can be seen on the area between Kaiafa and Zacharo, at the Western Peloponnese Coasts (LAVRENTIADES 1964), as well as on the islands Crete, Cephalinia and Rhodes. One of the best examples of this type of vegetation occurs at the coasts of Rhodes island, by Phaliraki. The A g r o p y r e t u m zone cover at almost the entirely of the sea-shore and develops mostly on high dunes of about 1-2 m high. Typical, most abundant and frequently found species on these dunes are the rhizomatous geophytes *Agropyrum junceum* ssp.*mediterraneum*, *Sporobolus pungens*, and *Galilea mucronata*. Scattered individuals of *Pancratium maritimum* are found locally over most of the places. The floristic differences on these places from the optimal phase of the A g r o p y r e t u m m e d i t e r r a n e u m show features intermediate between the present optimal phase and the terminal phase of the A g r o p y r e t u m m e d i t e r r a n e u m , in which *Pancratium maritimum* is one of the most characteristic species.

Locally in the Northern, Central, and Western Coasts of Greece, as well as in the Aegean Islands, there are patches of a mixed community in which *Agropyrum junceum* ssp.*mediterraneum*, *Galilea mucronata*, and *Sporobolus pungens* share dominance. This type of vegetation as a whole is distinguished from the individuals of *Diotis maritima*, *Convolvulus soldanella*, *Echinophora spinosa*, and *Pancratium maritimum*, which occupy more rarely the space between the above mentioned dominant species. Such plant community represents the terminal phase of the Agropyretum mediterraneum. That can be concluded from the fact that the species *Diotis maritima, Convolvulus soldanella, Echinophora spinosa* and *Pancratium maritimum* do not occur at embryonic dunes and are at almost absolutely absent from the transition phases of the Agropyretum mediterraneum. The presence therefore of such species must always be considered as a significant diagnostic element of the terminal phase of the Agropyretum mediterraneum. One of the best examples occurs at the Western Peloponnese Coasts, near Kaiafa. The stands of this association here are large enough and occupy an area of many hectares along the sea-shore. The height of sand dunes, ranging from 1.5 to 2 m, extend to about 34 m far from the sea. In this case, the Agropyretum mediterraneum element consists of 5 characteristic species, *Agropyrum junceum* ssp.*mediterraneum, Galilea mucronata, Sporobolus pungens, Diotis maritima,* and *Echinophora spinosa,* which dominate the area. It must be pointed out that the characteristic species of alliance, *Medicago marina* and *Pancratium maritimum,* as well as the characteristic species of the class, *Euphorbia paralias* and *Eryngium maritimum,* are also present. Each of these may attain dominance or co-dominance. The presence of the herbaceous chamaephyte *Diotis maritima,* the perennial hemicryptophyte *Echinophora spinosa* and the frequently abundant bulbous geophyte *Pancratium maritimum* is an evidence that the Agropyretum mediterraneum has reached its terminal phase (LAVRENTIADES 1964). Similar patches of this type of vegetation often occur in South Greece, as for instance in Crete and Rhodes islands.

c) Ammophiletum group

In a few places, the terminal phase of the Agropyretum mediterraneum merges with the initial phase of the interesting Ammophiletum arundinaceae, which occupies locally the higher sand dunes of greek coasts, giving a patchy occurrence of a mixed association. The association as a whole is distinguished from the terminal phase of the Agropyretum mediterraneum by the prominence of *Ammophila arenaria,* which may be considered as the starting point of the third group of plant communities. The change from the terminal phase of the Agropyretum mediterraneum to the initial phase of the Ammophiletum arundinaceae is mostly due to the occurrence of the rhizomatous geophyte *Ammophila arenaria*, which has a high dynamogenetic value and therefore greatly influences the conditions and consequently the variation of the vegetation (WEAVER & CLEMENTS 1938). More typical stands of this kind of vegetation have been recorded in North Greece near Litochoron and Porto-Lago. The optimal phase of this association is commonly marked by the abundance of the only characteristic species, *Ammophila arenaria* and *Cutandia maritima*, which attain dominance where the moving sand dunes are more than 2 m high and extend to a large area. In Southern Greece, such places for instance, are the coasts of the Western Peloponnese, near Petrochori, Voidokilia, and Zacharo (LAVRENTIADES 1964).

d) Crucianelletum group

In some cases, the moving sand dunes of the previously described vegetation types are occupied by a certain proportion of therophytes which together with some other species greatly influence the variation in vegetation. The dominance and constancy of these species modify the moving sand dunes into old stabilized sand dunes and determine a new vegetational physiognomy, known as Crucianelletum (BRAUN-BLANQUET 1933,1951). In the Western Mediterranean, the optimal phase of the Ammophiletum arundinaceae is often replaced by the well marked Crucianelletum maritimae, in which the characteristic species *Crucianella maritima* L., *Teucrium polium* L. v.*maritimum* Alb.et Jah., *Helichrysum stoechas* DC. v.*maritimum* Rouy, *Artemisia campestris* L. v.*glutinosa* Ten. and *Pancratium maritimum* L. share dominance (BRAUN-BLANQUET 1933,1951). In Northern Greece the usual representative association of this zonation is the Ephedra distachya-Silene subconica-Ass. It is clear that this zonation represents an actual succession and is one of the most widespread types of vegetation in North Greece. This characteristic association, for instance, is often extensive on the coasts of Keramoti and Porto-Lago (OBERDORFER 1952; LAVRENTIADES 1963). A similar development of this kind of vegetation in areas of stabilized dunes occurs frequently in Southern Greece and the Aegean Islands, where it is often well represented by the widespread Euphorbia terracina-Silene nicaeensis association, which may be replaced by ammophilous and climax vegetation on older stabilized ground. This type of vegetation, for example, is found very often on the stabilized dunes of the Western Peloponnese Coasts, near Zacharo, Katakolon, and Kaiafa (LAVRENTIADES 1964), on Rhodes and Crete islands,etc. It must be pointed out that the plant communities of this zonation consist of a series which belong to the fourth group.

e) Climax vegetation

Where the variations in surface soil development cause sharp changes in stability and soil texture, both these associations usually pass directly or indirectly into a climax vegetation, which shows similarities to the association Quercetum ilicis-galloprovinciale Br.-Bl.(1915) 1936, or it is closely related floristically to Oleo-Lentiscetum Br.-Bl.et Mol.1951. We can judge this from the evergreen sclerophyle shrubs, *Pistacia lentiscus* L., *Myrtus communis* L., *Spartium junceum* L., *Arbutus unedo* L., *Erica arborea* L., etc., which are commonly found and gradually increase to dominance. This vegetation type includes a number of plant communities which may classified into the fifth group. From the data available it seems evident that in this case changes in vegetational zonation may indeed be due to the fact that the psammophilous vegetation of the stabilized sand dunes of greek coasts is usually near to the Quercetea ilicis Br.-Bl.1947 zonation. The dominant pioneer species of the initial phase of climax vegetation is undoubtfully the widespread and abundantly grown all over the greek coasts, *Pistacia lentiscus*. The occurrence of dense patches of *Pistacia lentiscus* nearby at sea level in Northern Greece, as for example at the coasts of Chalkidiki Peninsula, is due not only to favourable conditions but also to the lack of sandy soils. In Southern Greece,as for instance at the Western Peloponnese Coasts, the climax vegetation element consists of the species *Pistacia lentiscus* L., *Myrtus communis* L., *Spartium junceum* L., *Arbutus unedo* L., *Erica arborea* L., etc., which may each attain dominance or co-dominance (LAVRENTIADES 1964).

A broad outline of the principal features of the psammosere,and the successional inter-relationships, occurring in the vegetation of sand dunes at greek coasts is given by the following representative schema.

Outline of the principal features of the psammosere, occurring in the vegetation on sand dunes of greek coasts.

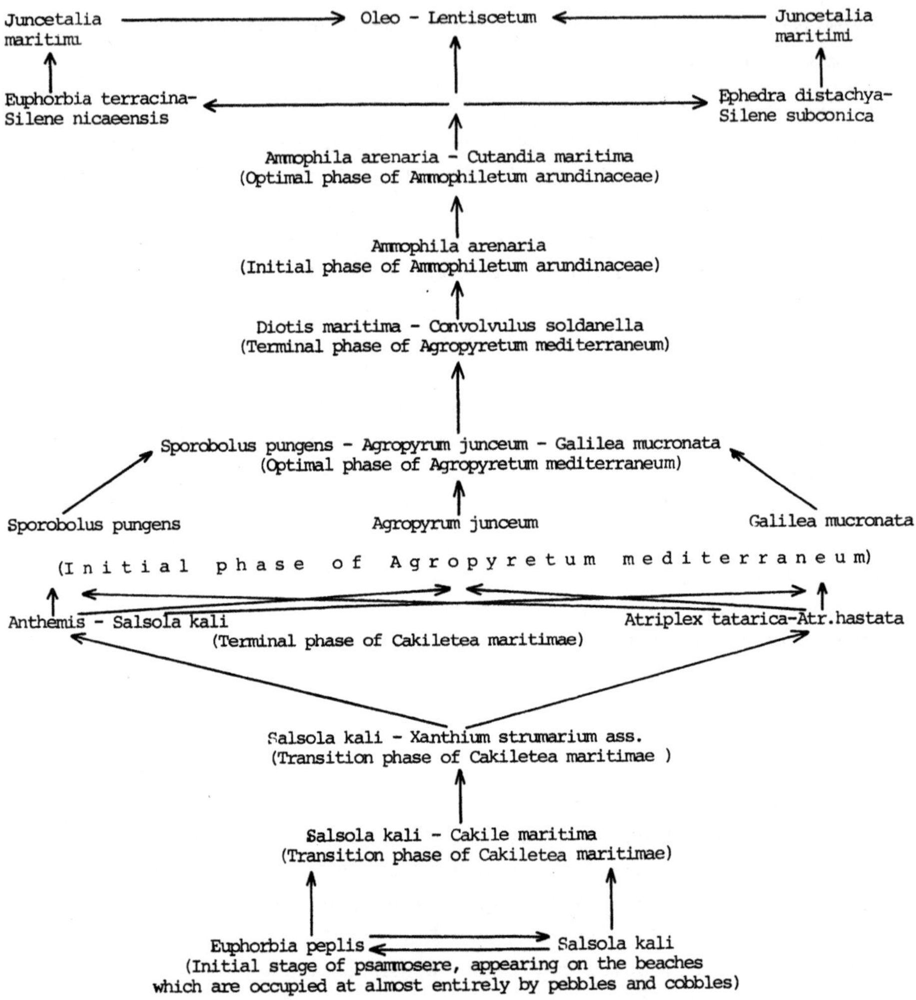

SUMMARY

The chief purpose of this paper is to give as far as possible a survey of the most important points of the succession, occurring in the vegetation of the sand dunes on greek coasts.

From the data available, it seems evident that the psammosere of sand dunes of greek coasts includes one pioneer stage, several transition stages, and a terminal stage which represents the climax vegetation. All these stages, according to their floristical composition and ecological factors, can be divided into the following five groups:

a) C a k i l e t e a group. It includes all phases of the C a k i l e t e a associations, and grows on low level dunes near beaches more or less occupied by humus or plant remains. Its initial phase, corresponding to the initial stage of the whole psammosere, is characterized by the appearance of individuals of the pioneer species *Euphorbia peplis* or *Salsola kali*, while its terminal stage is represented by the A n t h e m i s - S a l s o l a k a l i or A t r i p l e x t a t a r i c a - A t r i p l e x h a s t a t a associations.

b) A g r o p y r e t u m group. This group, including the phases of the A g r o p y r e t u m m e d i t e r r a n e u m is recognised especially by the presence of rhizomatous geophytes *Agropyrum junceum* ssp. *mediterraneum*, *Sporobolus pungens* and *Galilea mucronata*. It appears on dunes, of about 1-2 m above sea level, which are entirely or partly free from humus or plant remains. Its initial phase begins with the appearance of individuals of rhizomatous geophytes, *Agropyrum junceum* ssp. *mediterraneum*, *Sporobolus pungens* or *Galilea mucronata*, which possess higher dynamogenetic value, and its terminal phase can be distinguished by the presence of herbaceous chamaephytes, *Diotis maritima* and *Convolvulus soldanella*, the perennial hemicryptophyte *Echinophora spinosa*, and the bulbous geophyte *Pancratium maritimum*.

c) A m m o p h i l e t u m group. In this group are included all the phases of the A m m o p h i l e t u m a r u n d i n a c e a e, which appear especially on moving sand dunes more than 2 m high. The occurrence of the individuals of the rhizomatous geophyte with high dynamogenetic value *Ammophila arenaria* on these dunes represents the initial phase of this group, while the abundance of the characteristic species *Ammophila arenaria* and *Cutandia maritima* in the terminal phase.

d) C r u c i a n e l l e t u m group. Locally in a few places, where the moving sand dunes of A m m o p h i l e t u m group are modified by the presence of many therophytes in stabilized dunes, new vegetational types appear. The most characteristic associations of this zone are: in Western Mediterranean coasts C r u c i a n e l l e t u m m a r i t i m a e, while on greek coasts, E p h e d r a d i s t a c h y a - S i l e n e s u b c o n i c a and E u p h o r b i a t e r r a c i n a - S i l e n e n i c a e e n s i s associations.

e) Climax vegetation. As climax vegetation of the whole psammosere must be considered associations, which are closely related floristically to the association Q u e r c e t u m i l i c i s - g a l l o p r o v i n c i a l e , or most frequently to those of O l e o - L e n t i s c e t u m.

ZUSAMMENFASSUNG

In dieser Arbeit wird so weit wie möglich ein Überblick über die wichtigsten Punkte der Sukzession der Vegetationsstadien gegeben, die an den griechischen Küstendünen zu beobachten sind.
Die zur Verfügung stehenden Daten scheinen deutlich zu machen, daß die Pflanzenbesiedlung der griechischen Küstendünen (Sukzessionsserie) ein Anfangsstadium, mehrere Übergangs- und ein Endstadium durchläuft. Letzteres stellt die Vegetationsklimax dar. Diese Stadien lassen sich insgesamt nach ihrer floristischen Zusammensetzung und nach ökologischen Faktoren wie folgt in fünf Gruppen aufteilen:

a) C a k i l e t e a- Gruppe. In ihr treten alle Phasen der C a k i l e t e a - Assoziationen in Erscheinung. Sie gedeiht auf niedrigen Flachdünen in der Nähe von Küstenstränden. Die Dünen sind mehr oder weniger mit Humus und Pflanzenresten bedeckt. Die Initialphase ist - entsprechend dem Anfangsstadium der ganzen Sukzessionsserie - gekennzeichnet durch vereinzeltes Auftreten von Pionier-Arten wie *Euphorbia peplis* oder *Salsola kali*; während im Endstadium die A n t h e m i s - S a l s o l a k a l i - oder A t r i p l e x t a t a r i c a - A t r i p l e x h a s t a t a - Assoziationen auftreten (Schema).

b) A g r o p y r e t u m - Gruppe. Diese Gruppe umfaßt die Phasen des A g r o p y r e t u m m e d i t e r r a n e u m. Besonders ist sie gekennzeichnet durch das Auftreten von Rhizomgeophyten wie *Agropyrum junceum* ssp.*mediterraneum*, *Sporobolus pungens* und *Galilea mucronata*. Sie entwickelt sich auf Dünen von 1-2 m über dem Meeresspiegel, die völlig oder teilweise von Humus oder Pflanzenteilen frei sind. Ihre Initialphase beginnt mit dem Auftreten vereinzelter Rhizomgeophyten wie *Agropyrum junceum* ssp.*mediterraneum*, *Sporobolus pungens* oder *Galilea mucronata*, die starken dynamogenetischen Wert besitzen. Ihre Schlußphase ist ausgezeichnet durch das Auftreten von kriechenden ausdauernden Stauden (Chamaephyten), *Diotis maritima* und *Convolvulus soldanella*, der perennierende Hemikryptophyt *Echinophora spinosa* und den Knollengeophyten *Pancratium maritimum* (Schema).

c) A m m o p h i l e t u m - Gruppe. Diese Gruppe umfaßt alle Phasen des A m m o p h i l e t u m a r u n d i n a c e a e . Sie erscheint besonders auf Wanderdünen von mehr als 2 m Mächtigkeit. Vereinzeltes Vorkommen des Rhizomgeophyten *Ammophila arenaria* mit starkem dynamogenetischem Wirkungsgrad kennzeichnet die Initialphase dieser Gruppe, während überreichliches Auftreten der charakteristischen Arten *Ammophila arenaria* und *Cutandia maritima* die Schlußphase anzeigt (Schema).

d) C r u c i a n e l l e t u m - Gruppe. An wenigen Stellen, wo die Wanderdünen der A m m o p h i l e t u m -Gruppe durch das Auftreten von Therophyten stabilisiert werden, treten neue Vegetationstypen in Erscheinung. Folgende Assoziationen charakterisieren diese Vegetationszone: an westlichen Mittelmeerküsten das C r u c i a n e l l e t u m m a r i t i m a e , während an den griechischen Küsten die E p h e d r a d i s t a c h y a - S i l e n e s u b c o n i c a - und E u p h o r b i a t e r r a c i n a - S i l e n e n i c a e e n s i s - Assoziationen zu beobachten sind (Schema).

e) Klimaxvegetation

Als Klimaxvegetation der gesamten Sukzessionsserie ist das Auftreten von Assoziationen anzusehen, die floristisch mit dem Q u e r c e t u m i l i c i s g a l l o p r o v i n c i a l e oder häufiger mit dem O l e o - L e n t i s c e t u m verwandt sind (Schema).

LITERATUR

ARENES,J. -1924- Etude sur la zone halophile en Provence, Végétation des côtes basses.- Bull.Soc.Bot.Fr.71: 98-99. Paris.

-- -- -1929- Les associations végétales de la Basse-Provence.- Thèse. Paris.

BRAUN-BLANQUET,J. -1933- Prodrome des Groupements Végétaux (Ammophiletalia et Salicornietalia médit.), fasc.I. Montpellier.

-- -- ROUSSINE,N. & NÈGRE,R. -1951- Les Groupements Végétaux de la France Méditerranéenne. C.N.R.S. Montpellier.

DUCELLIER, L. -1911- Etude phytogéographique des dunes de la Baie d'Alger.- Rév.Gén.Bot. 23: 273-308, 321-340. Paris.

ELLENBERG,H. -1963- Vegetation Mitteleuropas mit den Alpen.- Stuttgart.

FREI,M. -1937- Studi fitosociologici su alcune associazioni littorali in Sicilia (Ammophiletalia e Salicornietalia).- Nuovo Giorn.Bot.Ital. n.s.44: 715-730. Firenze.

GEHU,J.-M. -1964a- Observations sur quelques greves a Euphorbia peplis dans le Nord-Ouest Français. Ecologie et Phytosociologie.- Bull.Soc.Bot. Nord Fr.17: 77-85. Lille.

-- -- -1964b- La végétation psammophile des îles de Houat et de Hoedic.- Bull.Soc.Bot.Nord Fr.17: 238-266. Lille.

-- -- & PETIT,M. -1965- Notes sur la végétation des dunes littorales de Charente et de Vendée.- Bull.Soc.Bot.Nord Fr.18: 69-88. Lille.

HOCQUETTE,M., GEHU,J.-M. & FAUQUET,M. -1965- Contribution a l'étude phyto-sociologique de l'estuaire de l'Authie.- Bull.Soc.Bot.Nord Fr.18: 114-133. Lille.

KNAPP,R. -1965- Die Vegetation von Kephallinia Griechenland.- Königstein.

KUHNHOLTZ-LORDAT,G. -1923- Les dunes du Golfe du Lion.- Thèse. Paris.

LAVRENTIADES,G. -1963- On the vegetation of the Keramoti coasts.- Boll. Ist.Bot.Univ.Catania, serie 3,4: 81-103. Catania.

-- -- -1964- The ammophilous vegetation of the Western Peloponnesos Coasts.- Vegetatio 12: 223-287. Den Haag.

MOLINIER,R. -1964- L'évolution du relief et les caractères de la végétation en Camargue.- Ann.de la S.S.N.A.T.V.: 52-79.

-- -- & TALLON,G. -1965- Etudes Botaniques en Camargue.- La Terre et la Vie, Rév.Hist.Nat. 1-2: I-192. Paris.

OBERDORFER,E. -1952- Beitrag zur Kenntnis der nordägäischen Küstenvegeta-tion.- Vegetatio 3: 329-349. Den Haag.

PIGNATTI,S. -1959- Développement du sol et de la végétation sur la plage de Carnon (Languedoc.- Delpinoa , n.s.I: 69-98. Napoli.

RECHINGER,K. (fil.) -1943- Flora Aegaea.- Denkschr.Akad.Wiss.Wien, math.-nat.-wiss.Kl.105, I.Halbbd. Wien.

-- -- & RECHINGER-MOSER,F. -1951- Phytogeographia Aegaea.-Denkschr.Akad. Wiss.Wien, math.-nat.-wiss.Kl.105, 2.Halbbd., 2.Abt. Wien.

TURILL,W. -1929- The plant-life of the Balkan Peninsula, p.121-125.- Oxford.

TÜXEN,R. -1950- Grundriß einer Systematik der nitrophilen Unkrautgesell-schaften in der Eurosibirischen Region Europas.- Mitt.Flor.-soz. Arbeitsgem.N.F.2: 94-175. Stolzenau/Weser.

-- -- -1961- Grundzüge der Entwicklung der natürlichen Dünenvegetation an den west-europäischen Küsten.- Angew.Pflanzensoz.17: 7. Stolzenau/ Weser.

WEAVER,J. & CLEMENTS,F. -1938- Plant Ecology. New York and London.

DISKUSSION

D.MUELLER-DOMBOIS

The different group sequences,a,b,c etc., are described as starting on environmentally quite different habitats, such as (a) on low dunes with some humus, (b) on 1-2 m high dunes that are practically free of humus, (c) on 2 m high moving sand and (d) on stabilized sand dunes. How then is it possible to merge all these into one successional sequence as presented in the paper? It is perhaps possible to see the developmental sequence from (c) to (d), but not from (a) to (b) to (c).
How do you interpret the appearance of annuals on the stabilized dunes? Do you consider this a cause or a symptom?

G.LAVRENTIADES

We can find all or a part of these successions in some special places of Greece, as the Western Peloponnese, which are quite vertical without these winds. But in some other places one or two stages of development may be found. It is very clear that the succession in the vegetation is due to some special species which have a high dynamogenetic value. So I tried to separate some groups and to define in every group special ecological factors.
Secondly you have told me that there is a correlation between the Crucianella maritima association and the Euphorbia-Ephedra distachya-Silene subconica ass. Both these associations closely belong to stabilized dunes. The correlation is due to the same ecological factors and the change from the moving sand dunes to the stabilized dunes is especially due to some therophytes which occupy is the place where the Ammophiletum arundinaceae.

R.TÜXEN

Ich habe in den letzten Jahren in zunehmenden Maße beobachtet, daß kaum eine andere Landschaft mehr zerstört wird als die Dünenlandschaft, durch den Tourismus und alles, was damit zusammenhängt. Meine Bitte ist die, daß unser Freund LAVRENTIADES in Griechenland dafür Sorge tragen möge, daß, wenn es noch möglich ist, rechtzeitig absolut geschützte Reservate geschaffen werden, um die schöne Dünenvegetation zu erhalten, wie es z.B. die Japaner gerade noch rechtzeitig gekonnt haben.
Pancratium und *Convolvulus soldanella* sind im Mediterrangebiet Ar-

ten der Vordünen, sie gehen bis an die Küste. Schon in NW-Spanien ist *Pancratium* in den hinteren Dünen, weit von der Küste entfernt, in dem Verband, den wir K o e l e r i o n a l b e s c e n t i s nennen und bei uns, in Holland oder in NW-Deutschland oder gar in Dänemark ist *Convolvulus soldanella* niemals mehr an der Küste, sondern weit in den Binnendünen zu finden. Je weiter wir nach Norden kommen, desto mehr gehen diese Arten vom Strand weg ins Innere. Ich glaube, aus Gründen der Wärme, weil es an der Küste bei uns zu kalt ist, und es in den Binnendünen wärmer wird.
Wir haben bei uns, in den westeuropäischen Dünen, 3 Klassen zu unterscheiden: die C a k i l e t e a auf den Spülsäumen, von diesen gibt es keine Sukzession zu den A m m o p h i l e t e a , sondern das sind 2 getrennte Zonen. Dann kommt die 2. Klasse, das sind die A m m o p h i l e t e a mit A g r o p y r e t u m und A m m o p h i l e t u m . Dann unterscheiden wir die - im Deutschen sagen wir - Kleingrasdünen mit K o e l e r i o n a l b e s c e n t i s , die wohl zur S e d o - S c l e r a n t h e t e a - Klasse zu stellen sind. Im Mediterrangebiet sind bisher, außer bei Ihnen, die A m m o p h i l e t e a und die dann folgenden, nicht K o e l e r i o n- sondern Zwergstrauch-Dünen mit *Ephedra* und vielen anderen Zwergsträuchern, nicht genügend unterschieden worden, sondern wie in den klassischen Arbeiten von BRAUN-BLANQUET ist das alles noch A m m o p h i l e t e a. Hier ist eine neue größere Einheit zu machen, die systematisch unbedingt von den A m m o p h i l e t e a abzutrennen ist.

G. LAVRENTIADES

Pancratium maritimum and *Convolvulus soldanella* are growing only in the moving sand dunes, in which the A m m o p h i l e t u m a r u n d i n a c e a e is developed.
About the C r u c i a n e l l e t u m m a r i t i m a e it is very important to notice here that there are no species of this association, like *Crucianella maritima*, *Helichrysum stoechas*, *Teucrium polium* var. *maritimum* growing on the coastal line of Greece, and I believe that these species are neither growing on the coastal line of Turkey and Israel. I would like very much to find this difference between the C r u c i a n e l l e t u m growing in the Western Mediterranean and that in the Eastern Mediterranean.
About the C a k i l e t e a m a r i t i m a e associations you told me that it is a very clear zone where the C a k i l e t e a associations grow. I have found some places in the C a k i l e t e a

zone where the Agropyretum mediterraneum is in its initial stages with *Agropyrum junceum* or *Sporobolus pungens* or *Galilea mucronata*. That means that this zone may be changed into the second zone, into the zone of Agropyretum mediterraneum. Another important fact is that I have not found associations of the Cakiletea class in the seashore of France. We have made excursions with Professor BRAUN-BLANQUET and I never found these associations of the Cakiletea group. Maybe they are not so extensive in the West mediterranean as in the East-Mediterranean.

R.TÜXEN
In Corsica very often.

A.H.FREIJZEN
Although the succession-scheme seems to have been constructed on the basis of fragments of the psammosere only, it is perhaps possible to mention the lenght of the period between the initial stage and the climax vegetation? The second question is related to this one. Which factor is the main cause of the succession? You have spoken about different habitat factors like the occurrence of humus etc., but I suppose there is a soil factor which changes during this succession, and I have got the impression that this succession as a whole is a very rapid and short one. Is the leaching out of the dune soils the cause of this succession or not?

G.LAVRENTIADES
It is not possible to find out how long the initial phase lasts, because we have collected vegetational types from different ecological types and we joined all these and made a hypothesis about the different stages of development.
As to the causes of the changes from one stage to the other, according to KUHNHOLTZ-LORDAT there is the "trinom dynamique" - the plant, the sand and the wind. These three factors are the causative agents of sand dune formation and consequently of the vegetation.

A.W.KÜCHLER
Have you vegetation maps prepared at different times, or have you permanent quadrats, or what method did you use in order to establish the sequence of plant communities in the succession?

G.LAVRENTIADES

I collect patches of every ecological type of vegetation of the beach, and measure according to the BRAUN-BLANQUET system. Then I make a hypothesis, because I have found a line from one stage of the psammosere to the terminal stage. I have found many scales. I studied these scales and came to the conclusion that the vegetation follows these scales from the initial stages of the psammosere to the terminal stages.

E.VAN DER MAAREL

In dutch dunes we have convincing evidence that zonation in dune vegetation should not be confused with succession. We have indications that the age of plant communities, also the most mature ones, is mostly much less than the age of the corresponding zone in the dune area. In general, the succession of the major communities is a secondary one and is bound to conditions which are quite different from the conditions suggested in the scheme.

R.CARBIENER

Wir haben hier einen typischen Fall einer Zonation, oder wollen wir sagen, einer allochthonen Sukzession. Es gibt immer wieder eine Hierarchie der ökologischen Faktoren, aber diese Hierarchie ist äußerst variabel. Hier ist es der edaphische Faktor, der dem klimatischen absolut überlegen ist. Deswegen gleichen sich auch die Dünen-Zonationen so sehr in klimatisch sehr verschiedenen Gebieten. Und deswegen bin ich auch in diesem Falle absolut einverstanden, daß man verschiedene Klassen von dieser Zonation macht, also C a k i l e t e a , A m m o p h i l e t e a und dann eventuell eine dritte Klasse für die stabilisierten hinteren Dünen, weil eben die Hierarchie der ökologischen Faktoren auch etwas sehr variabeles ist.

J.SCHMITHÜSEN

Anknüpfend an die Bemerkung von Herrn TÜXEN möchte ich hinzufügen, daß man die geschützten Gebiete auch studieren sollte! Eine der schönsten geschützten Dünen-Naturlandschaften ist das Welt-Naturschutzgebiet (Wild Life Conservation) an der Mündung des Guadalquivir. Soviel mir bekannt ist, hat dort bisher noch kein Pflanzensoziologe gearbeitet. Ich möchte den an den Dünen interessierten Herren empfehlen, sich gelegentlich an diesen Hinweis zu erinnern. Man muß nur den Eintritt vorbereiten; man kommt nur hinein, wenn man ein entsprechendes Papier hat. Es lohnt sich wirklich.

J.BARKMAN

As far as zonation is here to be extrapolated to succession, I think - if I understood you right - that your succession scheme, although of course more or less hypothetical, is not only based on zonation which you compiled from various fragments of the zonation you find on different parts of the Greek coast, and then you put them together, because they are overlapping, to one general scheme. But that was not your only argument. You also had the argument of the hight of the dunes. Of course in dune succession the height of the dunes increases, so that is another argument to say that a higher dune vegetation must follow a lower one. And the third factor you use is the behaviour of the species. I think it is not only zonation. You never must use zonation to extrapolate to succession! You have done more than only studying zonation, is that right?

G.LAVRENTIADES

It is clear that there is a zonation between the different types of vegetation. Sometimes we find the zonation clearly, but sometimes we can find in one zone species of the other zone. That means that zonation is the same succession. For example, in the Agropyretum zone we always found species of the Cakiletea maritimae zone, and these species are found in their last point of vitality. When the Agropyretum zone is in its terminal phase of development - and that we can know from such species as *Diotis maritima*, *Convolvulus soldanella*, *Echinophora spinosa* - than we can find in this zone *Ammophila arenaria* individuals. That means that this zone is succeeded by the initial phase of the Ammophiletum arundinaceae zone. We can conclude from these facts that zonation and succession sometimes is the same.

ZONATION UND ENTWICKLUNG DER VEGETATION ZENTRALCHILENISCHER KÜSTENDÜNEN

A. K o h l e r

Während durch die Arbeiten von OBERDORFER (1960) und SCHMITHÜSEN (1954, 1956) die Wälder und ihre Ersatzgesellschaften des mittleren Chile schon gut bekannt sind, fehlten bisher Untersuchungen über die Vegetation der Küstendünen. Nur bei REICHE (1895 u.1907) finden wir einige ältere Angaben zur Vegetation von Dünengebieten Chiles, die insofern von großem Interesse sind, als sie uns Veränderungen der Pflanzengesellschaften seit der Jahrhundertwende zu erschließen gestatten. So ist für das Verständnis der heutigen Vergesellschaftungen die Tatsache von entscheidender Bedeutung, daß Neophyten, insbesondere die nord-amerikanisch-pazifische Komposite *Ambrosia chamissonis*, die Gesellschaften der Sandstrände und meernahen Dünen überlagert - und entscheidend umgestaltet haben (KOHLER 1966). Auf einer Küstenlänge von ca. 1500 km zwischen dem Randgebiet der Atacama-Wüste im Norden und dem südchilenischen Lorbeerwald im Süden bildet *Ambrosia* heute auf diesen Standorten das beherrschende Element von größtem soziologischen Bauwert.

Im folgenden sollen Teilergebnisse von Untersuchungen mitgeteilt werden, die im Bereich des sandigen Litorals entlang der Küste von 1965 bis 1967 durchgeführt wurden (KOHLER 1970). Die in diesem Zusammenhang zu besprechenden Probleme der Strand- und Dünenvegetation beziehen sich nur auf die zentralchilenische Küste. Sie tangiert das mittelchilenische Hartlaubgebiet mit Gesellschaften der L i t h r a e o - C r y p t o c a r y e t e a (OBERDORFER 1960). Das Klima hat mediterranen Charakter mit Sommerdürre und Winterregen und ist nach LAUER (1960) durch jährlich 5 bis 8 aride Monate gekennzeichnet.

Die chilenische Pazifikküste bietet einen Wechsel von steilen Felsen und flachen sandigen Buchten mit stellenweise beachtlichen Flugsand-Dünen.

Komplette Zonationsreihen, die das gesamte Inventar der Strand- und Dünen-Gesellschaften enthalten, sind im Bereich Mittelchiles recht selten. Verhältnismäßig gut ausgebildet ist die Sequenz im Dünengebiet bei El Tabo (Provinz Santiago), das als Ausgangspunkt unserer Untersuchungen diente (vgl. Zonations - Schema).

Zunächst soll die typische Abfolge von Vegetationseinheiten des psammischen Litoralbereiches der mittelchilenischen Küste kurz geschildert, und dann die Frage erörtert werden, inwieweit das räumliche Nebeneinander Ausdruck eines zeitlichen Nacheinanders, also der Sukzession sein kann.
Auf Grund der pflanzensoziologischen Analyse läßt sich folgende Reihe von Vegetationstypen unterscheiden:

1. Nolanetum paradoxae ass.nov.
2. Poo-Ambrosietum ass.nov.
3. Chorizanthe vaginata-Margyricarpus se-

4. Baccharis concava-Buschdünen
5. Mittelchilenischer Hartlaubwald (Cryptocaryion-Verb.).

In der Regel wird die Reihe durch das Nolanetum paradoxae eröffnet, eine Gesellschaft mit strenger Bindung an die unmittelbare Meeresnähe. Die Entwicklung dieser Einheit geht meist vom flachen Sandstrand und von Spülsäumen aus, wo *Nolana paradoxa* und *Ambrosia chamissonis*, zuweilen begleitet von dem stark blattsukkulenten *Carpobrotus chilensis*, die wichtigsten Pioniere darstellen. An Standorten, die eine gewisse Zeit von der Zerstörung durch die Brandung verschont werden, bauen diese Arten niedrige Vordünen auf, wie bei El Tabo während des zweijährigen Untersuchungszeitraumes deutlich zu beobachten war. In der weiteren Entwicklung der Dünen übernimmt dann aber *Ambrosia chamissonis* die Führung: Ihr vom Sand verschüttetes, sich reichlich verzweigendes Sproßsystem zeigt Dickenwachstum, verholzt und bildet das stabile Grundgerüst der meernahen Dünen.
Der ephemere Geophyt *Nolana paradoxa*, der meist nur vom Winter bis zum Frühsommer (Juli bis Dezember) oberirdische Organe trägt, vermag zwar mit Hilfe einzelner Rhizome mit *Ambrosia* den Sand zu durchwachsen, spielt aber beim Aufbau der *Ambrosia*-Dünen lediglich eine untergeordnete Rolle. Meist macht diese einzige Kennart der Assoziation *Nolana* auf den vordersten *Ambrosia*-Dünen halt und dringt nicht weiter landeinwärts vor.
Ökologisch ist das Nolanetum gekennzeichnet durch Sandbewegungen, eine regelmäßig erfolgende (Zer)störung durch Meerwasser-Einbrüche, sowie durch während der humiden Zeit austretendes süßes Grundwasser, das auch die Periodizität der Gesellschaft maßgeblich bestimmt.
Ist ein breiteres Feld von Dünen vorhanden, in denen mehr oder

weniger starke Sandbewegungen stattfinden, so schließt sich landeinwärts eine Gesellschaft an, in der *Nolana paradoxa* stets fehlt, *Ambrosia chamissonis*, gefolgt von *Carpobrotus chilensis*, aber eine beherrschende Rolle spielt: das Poo-Ambrosietum.

Je nach der Intensität der Sandbewegungen und nach der Stärke des Meereswasser-Einflusses lassen sich verschiedene Ausbildungen der Gesellschaft unterscheiden. Den extremsten Flügel der Assoziation nehmen artenarme *Ambrosia*-Bestände ein, die streckenweise nur aus dieser einen Art bestehen; wo der Salzeinfluß geringer ist, kann sich eine taxonomisch noch nicht näher geklärte diözische, xeromorphe *Poa*-Art (*P. aff. lanuginosa*) ansiedeln, die für die Gesellschaft sehr bezeichnend ist. Mit weiterem Nachlassen der extremen Umweltbedingungen, vor allem der Intensität der Sandbewegungen, wird die Assoziation durch neu hinzutretende Arten bereichert: *Hypochoeris toltensis*, *Astragalus valparadisiensis*, *Alstroemeria ligtu* var., *Euphorbis portulacoides* u.a.m. Der artenreichste Flügel der Gesellschaft, in dem kaum mehr Sandumlagerungen stattfinden, enthält u.a. frühlingsephemere Therophyten, wie *Leuceria*-Arten und *Chamissonia dentata*; er leitet bereits zu den Zwergstrauch-Dünen über.

Oftmals besiedelt das Poo-Ambrosietum ganze Scharen von kuppigen Dünenhügeln, deren Zwischenräume im artenarmen Flügel oft vegetationsfrei sind. Schon in der vordersten Zone befindet sich auf den Dünenkuppen regelmäßig der sklerophylle immergrüne Strauch *Baccharis concava*, wenn auch in niedrigen, der Düne angedrückten Polsterexemplaren, der in den meerfernen Teilen die bestimmende Art der Buschdünen darstellt. Die artenarmen meernahen, stark kuppigen Bestände gehen meist allmählich in die artenreicheren Ausbildungen der Gesellschaft über, wobei auch das Dünenrelief sanfter wird.

Die Vitalität von *Ambrosia chamissonis* läßt dort deutlich nach, wo die Sandzufuhr nahezu aufhört. Hier gewinnen neben frühlingsephemeren Therophyten die Zwergsträucher *Margyricarpus setosus* und *Chorizanthe vaginata* stärker an Raum. In den einzelnen Dünengebieten der Küste herrscht bald die eine, bald die andere der beiden Zwergstraucharten vor, eine Tatsache, die vorerst noch nicht eindeutig zu erklären ist.

In Zentralchile (aber auch im ariden Kleinen Norden) besiedeln die Zwergstrauch-Gesellschaften meist muldenartige Standorte im Schutz meerwärts vorgelagerter höherer Dünen. Ökologisch ist dieser Vegetationstyp durch extreme mikroklimatische Verhältnisse und kaum mehr stattfindende Sandumlagerungen gekennzeichnet.

Der wichtigste Pioniergebüsch-Typ wird von dem sklerophyllen Strauch *Baccharis concava* gebildet, dem sich zwei weitere Hartlaubsträucher beigesellen: *Schinus polygamus* und *Lycium chilense*. Diese Dünen-Gebüsche finden sich an verschiedenen Stellen als kleine Gruppen; nur selten sind größere geschlossene Strauchbestände anzutreffen. *Baccharis* vermag offenbar nur dort ihre normale Höhe (ca. 1 m) zu erreichen, wo sie nicht allzu großer Salzsprühwirkung ausgesetzt ist.
Eng verflochten mit den *Baccharis concava*-Gebüschen sind Hartlaubbusch-Wälder auf den Dünen. Den wichtigsten Pionier dieser Dünengehölzbestände, *Azara celastrina*, kann man oft in verschiedenen Altersstadien im Schutz eines *Baccharis*-Busches antreffen.

Auch der Hartlaubbaum *Lithraea caustica* ist häufig als Pionier an der Waldentwicklung auf Dünen beteiligt. Zu *Azara* und *Lithraea* gesellt sich noch ein ganzer Schwarm von Elementen der Charakterklasse des Hartlaubgebietes, der L i t h r a e o - C r y p t o - c a r y e t e a , wie *Cryptocarya*, *Eupatorium glechonophyllum*, *E. salvia*, *Cestrum parqui*, *Schinus latifolius*, *Maytenus boaria*, *Villarezia mucronata*, *Peumus boldus* u.a.m. Im allgemeinen sind die Dünen-Buschwälder wahrscheinlich durch Salzeinwirkung stark verformt und strukturell von den *Baccharis*-Dünengebüschen kaum zu differenzieren.
Die hier skizzierte Abfolge von Strand-Dünengesellschaften an der zentralchilenischen Küste dürfte wohl zunächst mit Recht als Ausdruck ökologischer Gradienten betrachtet werden, von denen Meerwasser-Einbrüche, Salzsprühwirkungen, Windintensität und Sandumlagerungen sowie gewisse Bodeneigenschaften als differenzierende Faktoren in Betracht kommen. Innerhalb der Sequenz ließ sich bezüglich edaphischer Faktoren eine rasche Entsalzung, kaum eine Entkalkung und somit auch keine Versauerung der Dünenböden (P_H in KCl: 8,3 - 7,1), sowie eine Reihe vom Rohboden zur Pararendzina feststellen.
In gewisser Beziehung sind Abfolge und floristische Zusammensetzung der Dünen-Gesellschaften auch von zoo-anthropogenen Faktoren beeinflußt; fast überall werden die Küstendünen u.a. durch Rinder und Pferde extensiv beweidet. Eine Anzahl von Dünenpflanzen wie *Ambrosia chamissonis*, *Scirpus nodosus*, *Baccharis concava* u. a. werden von den Weidetieren abgefressen. Aufgrund der hohen

Regenerationsfähigkeit dieser Arten tritt jedoch kaum eine entscheidende Störung des Vegetationsgefüges ein. Dagegen ist eine gelegentliche Eutrophierung durch Weidetiere festzustellen, die sich vor allem in weniger bewegten Dünen im Erscheinen holarktischer, nitrophiler Pflanzen wie *Hordeum murinum*, *Sonchus asper* und *Medicago hispida* äußert.
Auch auf die Artenzusammensetzung der Buschwälder dürfte die Beweidung einen gewissen Einfluß ausüben, zumal manche Gehölze von den Rindern stark heimgesucht (wie *Villarezia mucronata* und *Maytenus boaria*), andere stark aromatische Arten (*Peumus boldus* und *Cryptocarya alba*) offenbar aber gemieden werden.
Glücklicherweise verursacht der Erholungstrieb kaum eine Störung der meernahen Dünen, was einmal auf die hohe Resistenz von *Ambrosia* gegenüber mechanischen Beschädigungen, zum anderen darauf zurückzuführen ist, daß während der Badesaison die *Ambrosia*-Dünen wegen der empfindlich stechenden Früchte von den Badenden gemieden werden.
Nachdem wir die Reihe vom Nolanetum paradoxae bis zum Hartlaub-Buschwald als Ausdruck ökologischer Gradienten bezeichnet haben, ist nun die Frage zu stellen, ob diese Reihe auch das Abbild zeitlich aufeinanderfolgender Vegetationstypen sein kann. Hierbei scheint es wichtig, zu unterscheiden, ob sich aus dem räumlichen Nebeneinander eine Entwicklungstendenz, bzw. eine in der Zukunft folgende "potentielle" Sukzession ableiten läßt, oder ob diese Reihe Ausdruck einer tatsächlich abgelaufenen Sukzession ist.
Wenn wir unter einer Sukzession die Abfolge von verschieden zusammengesetzten Pflanzenbeständen auf *demselben* Wuchsort verstehen, so dürfen wir für die vorliegenden Verhältnisse folgendes feststellen: Echte Sukzessionsschritte lassen sich zwischen einzelnen Gliedern der Reihe erkennen: so zwischen Nolanetum und Poo-Ambrosietum oder zwischen *Baccharis*-Gebüsch und Hartlaubwald. In der ganzen Reihe von Einheiten aber eine Entwicklungstendenz zu sehen, erscheint jedoch problematisch. Eine Sukzession vom Nolanetum über die einzelnen skizzierten Stadien zum Endstadium des Hartlaubwaldes an ein und demselben Ort dürfte nur dort erfolgen, wo eine gleichbleibende Sandförderung oder (und) wo eine stetige Hebung des Litorals stattfindet. Unter derartigen Bedingungen könnten in Strandnähe immer wieder neue Dünen entstehen, wodurch ältere Sandmassen der Ein-

wirkung von Brandung und Wind allmählich entzogen würden. Damit wären ökologische Voraussetzungen für eine zeitliche Abfolge von verschiedenen Vegetationstypen an einem Ort gegeben. Wir wissen aber, daß die Küste Chiles in verschiedenen Gebieten durch tektonische Vorgänge bewegt wird. Außerdem wird von geologischer Seite eine aktuelle Dünen-Neubildung weitgehend verneint. In Mittelchile werden nach PASKOFF & HERM (mündl.) heute kaum mehr Dünensande aus dem Meer aufgeweht; vielmehr werden die bei früheren Meeresrückzügen geförderten Sandmassen z.T. nur umgebildet und weiterverfrachtet.

Daß die aufgezeigte Reihe nicht Abbild einer bereits erfolgten Sukzession sein kann, dürfte als erwiesen gelten. Denn die geobotanischen Befunde zeigen, daß die meernahen Gesellschaften des Nolanetum paradoxae und des Poo-Ambrosietum in der heute vorliegenden Zusammensetzung und Struktur den Dünen-Hartlaubwäldern zeitlich nicht vorausgegangen sein können. Die durch *Ambrosia chamissonis* beherrschten Artenverbindungen stellen ganz junge Vergesellschaftungen dar (der Neophyt *Ambrosia* hat sich, wie erwähnt, erst seit der Jahrhundertwende in Chile ausgebreitet) und dürfen somit nicht als Vorläufer der oft viel älteren Dünenwälder betrachtet werden.

Einiges spricht sogar dafür, daß *Ambrosia* nicht nur die vorhandenen offenen Dünen überlagert, sondern auch neue Standorte besiedelt hat. Die im Vergleich mit den autochthonen Dünenpflanzen gegen mechanische Einflüsse, aber auch gegen Meerwasser viel robustere Art hat möglicherweise in manchen Fällen meerwärts *vor* den vorhandenen Gesellschaften eine neue Dünenzone geschaffen. Dafür spricht, daß in Mittelchile die nach REICHE (1895 u.1907) für die unmittelbare Meeresnähe bezeichnenden Dünenpflanzen heute hauptsächlich in den meerfernen Ausbildungen des Poo-Ambrosietum erscheinen.

Mit dem Nachweis, daß die *Ambrosia*-Dünen junge Bildungen sind, gilt als gesichert, daß noch heute eine Sandförderung aus dem Meer stattfindet, auch wenn diese als geologischer Prozeß unerheblich ist.

Die aufgeworfenen Fragen sollen zeigen, daß es problematisch ist, aus der ökologischen Reihe der Dünenvegetation eine potentielle Sukzession abzuleiten und es schließlich völlig verfehlt wäre, in ihr den Ausdruck einer bereits erfolgten Sukzession sehen zu wollen.

Zonationsschema von zentralchilenischen Küstendünen

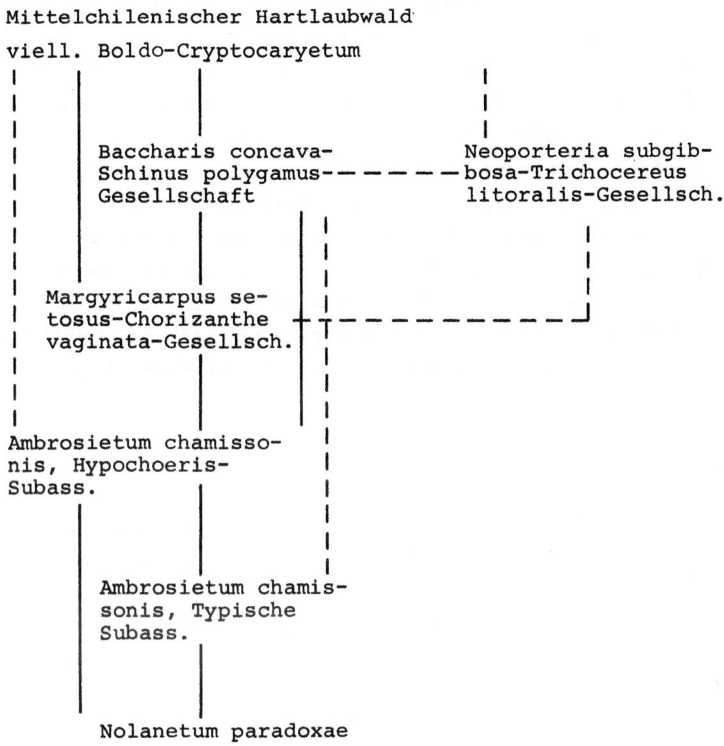

ZUSAMMENFASSUNG

Im Bereich der Sandstrände und Dünen der zentralchilenischen Küste läßt sich die folgende charakteristische Reihe von Vegetationstypen ermitteln: Nolanetum paradoxae (auf Strand und Vordünen) - Poo-Ambrosietum (meist auf Sekundärdünen) - *Chorizanthe vaginata-Margyricarpus setosus*-Zwergstrauchdünen - *Baccharis concava*-Buschdünen - Mittelchilenischer Hartlaubwald (Cryptocaryion). Die Sequenz dieser Vegetationseinheiten wird als durch ökologische Gradienten bedingte Standortreihe aufgefaßt. Aus der Reihe den Gang der künftigen Sukzession abzuleiten, erscheint problematisch u.a. deswegen, weil eine dauernde starke Sandförderung nicht vorliegt. Die Sequenz kann auch nicht das Abbild einer bereits erfolgten Vegetationsentwicklung darstellen, da die meernahen Dünen-Gesellschaften sehr junge Bildungen sind und nicht den älteren Hartlaub-Waldbeständen zeitlich vorausgegangen sein können. Sukzessionsschritte zwischen einzelnen Gliedern der Standortreihe dürfen jedoch angenommen werden.

SUMMARY

Within the area of the sand beaches and coastal dunes of Central Chili the following series of carasteristic plant communities can be found: Nolanetum paradoxae (on beaches and primary dunes) - Poo-Ambrosietum (mainly on secondary dunes) - *Chorizanthe vaginata-Margyricarpus setosus*-community (heathland plant formation) - *Baccharis concava* (dune shrub community) - sclerophyll wood (Cryptocaryion).

The series of these plant communities is considered to be affected by some ecological gradients. It seems to be problematic to deduce from it the potential future succession, mainly because there is no permanent fresh supply with sand from the sea. Neither this series can represent the past chronological succession of plant communities, since dune associations close to the beach are very young neophytic formations and cannot be predecessors of the older indigenous sclerophyll wood. However, single succession steps from one community to the other can be assumed.

LITERATUR

KOHLER,A. -1966- Ambrosia chamissonis (Less.) Greene, ein Neophyt der chilenischen Pazifikküste.- Ber.deutsch.bot.Ges.79: 313-323. Stuttgart.

-- -- -1970- Geobotanische Untersuchungen an Küstendünen Chiles - zwischen 27 und 42 Grad südl.Breite.- Bot.Jb.90: 55-200. Stuttgart.

LAUER,W. -1960- Klimadiagramme - Gedanken und Bemerkungen über die Verwendung von Klimadiagrammen für die Typisierung und den Vergleich von Klimaten.- Erdkunde, Arch.wiss.Geogr.14: 232-242. Berlin.

OBERDORFER,E. -1960- Pflanzensoziologische Studien in Chile.- Ein Vergleich mit Europa.- Flora et Vegetatia Mundi 2: 208. Weinheim.

REICHE,K. -1895- Die Vegetationsverhältnisse am Unterlauf des Rio Maule (Chile).- Engl.Bot.Jb.21 (1896): 1-52. Leipzig.

-- -- -1907- Grundzüge der Pflanzenverbreitung in Chile.- Die Vegetation der Erde 8: 374. Leipzig.

SCHMITHÜSEN,J. -1954- Waldgesellschaften des nördlichen Mittelchile.- Vegetatio 5/6: 497-486. Den Haag.

-- -- -1956- Die räumliche Ordnung der chilenischen Vegetation.- Bonner Geogr.Abh.17: 1-86. Bonn.

SPEZIFISCHER EINFLUSS VON MAHD UND WEIDEGANG AUF DIE BESTANDESENTWICKLUNG VON GRÜNLAND-GESELLSCHAFTEN

B. S p e i d e l

Die fortschreitende Intensivierung der Grünlandbewirtschaftung bringt es mit sich, daß immer mehr trittfeste Grünland-Flächen wenigstens zeitweilig unter Weidenutzung genommen werden. Dadurch wird auf der herkömmlichen 2 bis 3-schürigen Wiese, dem A r r h e n a t h e r e t u m bzw. T r i s e t e t u m ein Umschichtungs-Prozeß zu einer Ersatzgesellschaft 2. Grades eingeleitet, welcher schließlich zur Ausbildung des C y n o s u r e t u m führt. Allerdings verläuft dieser Prozeß nur sehr langsam. Je nach der Beweidungs-Intensität ergeben sich zunächst mannigfaltige Übergangsstufen, und es kann sogar Jahrzehnte dauern, bis er abgeschlossen ist. Das wird begreiflich, wenn man sich den spezifischen, weitgehend antagonistischen Einfluß des Mähens und des Weidens auf die Grasnarbe veranschaulicht, und wenn man außerdem noch berücksichtigt, daß in der Praxis die beiden Nutzungs-Arten in vielfältiger Weise kombiniert werden können. In jedem Falle aber ist der Umschichtungs-Prozeß die Folge der nutzungsbedingten Änderung der anthropo-zoogenen Standortsfaktoren.
Nachfolgend sollen die wichtigsten Bewirtschaftungs-Merkmale bei Wiese und Weide in ihrer Wirkung auf die Vegetation besprochen werden.

1. Das augenscheinlichste Merkmal ist die unterschiedliche Zahl der Nutzungen. Die Wiese wird 2 bis höchstens 3 mal pro Jahr genutzt, die Weide hingegen je nach Intensität bis 6 mal (und mehr). Häufigere Nutzung bei konstanten übrigen Bedingungen bewirkt immer eine Schwächung von Arten, welche ihre Speicherstoffe nur sehr langsam einzulagern vermögen und daher besonders nach einer frühen Mahd noch nicht genügend Reservestoffe besitzen, um anschließend sofort wieder nachwachsen zu können. Betroffen sind hiervon vor allem die ertragsstarken Obergräser und hochwachsende Kräuter. Einjährige Arten (z.B. *Bromus mollis*) können auch dadurch verdrängt werden, daß sie bei frühem Schnitt nicht mehr zur Samenreife und zum Fruchtfall kommen.

   Begünstigt und dadurch indirekt gefördert werden hingegen weniger empfindliche, aber meist ertragsschwache Untergräser und Bodenkräuter.

Mit dieser Umschichtung ist deshalb fast immer eine Ertrags-Depression verbunden.

2. Die Intensivweide wird stärker gedüngt als die Mähwiese. Verstärkte Düngung, vor allem mit N, vermag aber die unter 1 genannten Erscheinungen aufzuhalten und sie u.U. sogar ins Gegenteil umzukehren. Durch N-Düngung gefördert werden vor allem die anspruchsvollen Obergräser, die *Poa*-Arten und hochwachsende Kräuter. Benachteiligt (durch Konkurrenz und Beschattung) werden anspruchslose Untergräser, die Leguminosen und eine Reihe von Bodenkräutern. Zusätzliche N-Düngung in Verbindung mit Vielschnitt verhindert daher in der Regel nicht nur einen Ertrags-Abfall, sondern bewirkt meistens sogar einen Ertrags-Anstieg.
Tabelle 1 zeigt das Verhalten landwirtschaftlich wichtiger Arten am Beispiel von 38 einheitlich angelegten Versuchen mit gestaffelter Schnittzahl und Düngung. Die angegebenen Mengenanteile sind jeweils die Mittelwerte aus den Versuchen mit Vorkommen der betreffenden Art.

3. Auf der Wiese wird durch den Schnitt und durch die von Fahrspuren herrührende Bodenverdichtung zwar eine gewisse Selektion ausgeübt, sie ist aber relativ gering gegenüber der direkten und indirekten Selektion durch das Weidevieh. Aus Gründen der Schmackhaftigkeit bevorzugt das Vieh bestimmte Arten, die es immer wieder verbeißt, während andere, weniger schmackhafte Arten nur ungern gefressen, oft sogar ganz verschmäht und nicht einmal zertreten werden. Bei schlecht geregelter Weidenutzung können sich diese daher ungestört ausbreiten,während die bevorzugten Arten zurückgedrängt werden. Als Beispiel hierfür sei an die mit *Deschampsia caespitosa, Cirsium arvense, Urtica*- und *Rumex*-Arten verseuchten Weideflächen erinnert.
Zu der direkten Selektion kommt durch den Tritt der Tiere noch eine meist stärker wirkende indirekte. Alles, was nicht trittfest ist, wird mit zunehmender Intensivierung der Weidenutzung mehr und mehr verschwinden, um trittfesten Arten Platz zu machen, und das ist letzten Endes das ausschlaggebende Moment für die Umschichtung einer Wiesen- zu einer Weide-Gesellschaft. Man kann eine direkte Abhängigkeit der Bestandsentwicklung von der Intensität und Dauer des Tritteinflusses feststellen. Im Extrem bildet sich ein reiner Trittrasen heraus, der überhaupt nicht mehr unter den Begriff des Wirtschafts-

Tabelle 1

Einfluß von Schnittzahl und Düngung

(Auswertung 3-4-jähriger Versuche)

Mengenanteile in %

Schnittzahl		2	3	3	3	
Düngung		PK	PK	N_1PK	N_2PK	
Bromus mollis	5 Vers.	7,9	2,5	2,5	2,7	Schnitt schwächt
Achillea millefolium	13 "	3,3	2,3	2,3	2,6	
Arrhenatherum elatius	14 "	7,0	4,1	6,0	8,2	
Trisetum flavescens	27 "	6,5	4,6	5,3	5,0	Schnitt schwächt
Dactylis glomerata	32 "	3,9	2,9	5,7	6,6	N-Düngung fördert
Alopecurus pratensis	37 "	14,5	11,1	20,8	23,6	
Festuca rubra	37 "	14,4	18,8	11,7	9,2	Schnitt fördert
Trifolium repens	37 "	7,1	9,2	1,1	0,5	N-Düngung schwächt
Lolium perenne	8 "	6,1	7,3	5,5	3,9	
Trifolium pratense	34 "	5,4	5,5	0,5	0,2	N-Düngung schwächt
Plantago lanceolata	24 "	3,8	4,0	0,9	0,5	
Festuca pratensis	38 "	7,8	8,9	10,5	10,2	
Poa pratensis	35 "	6,6	7,0	9,3	11,9	
Poa trivalis	15 "	2,9	3,0	5,5	6,6	N-Düngung fördert
Heracleum sphondylium	23 "	4,3	4,6	6,4	6,5	
Anthriscus silvestris	13 "	5,6	5,3	9,6	9,2	
Rumex acetosa	11 "	0,9	0,8	1,6	1,9	
Holcus lanatus	36 "	7,4	8,6	8,6	8,3	
Taraxacum officinale	33 "	4,9	4,5	4,9	4,6	± indifferent
Sanguisorba officinalis	9 "	3,1	2,5	2,9	2,6	
Ertrag dz/Trockenmasse (Mittel aus allen Versuchen)		62,5	57,3	74,5	83,2	

Düngung P_2O_5 = 120 kg/ha K_2O = 160 kg/ha
N_1 = 100 kg/ha N_2 = 160 kg/ha

grünlandes fällt. Eine nach modernen Gesichtspunkten (d.h.mit kurzen Freßzeiten und langen Ruhepausen für die Vegetation) beweidete Wiesen-Narbe wird sich daher nur sehr langsam umstellen, andererseits bietet die veraltete sogenannte Standweide, welche fast während der gesamten Vegetationszeit betreten wird,und auf welcher es so gut wie keine Ruhepausen gibt, die besten Voraussetzungen für die Ausbilding des L o l i o - C y n o s u r e t u m. Die folgenden Tabellen mögen das veranschaulichen.

Tabelle 2 gibt einen Ausschnitt aus einem kombinierten Nutzungsversuch. Versuchsglied 1 wurde als normale 2-schürige Wiese genutzt. Bei Versuchsglied 5 wurde zwischen zwei Weide-Auftrieben eine Schnitt-Nutzung eingeschaltet. Versuchsglied 6 diente als Umtriebsweide mit drei bis vier Auftrieben und Versuchsglied 8 schließlich als Portionsweide mit maximal sechs Auftrieben. Es muß hierzu erwähnt werden, daß die einzelnen Parzellen so klein gewählt wurden, daß sie jeweils an einem Tag abgefressen waren. Das Weidevieh war also pro Auftrieb jeweils nur etwa 8 Stunden auf den Koppeln, auf der Portionsweide noch weniger. Ein Vergleich der Bestandeszusammensetzung der einzelnen Versuchsglieder zeigt, daß selbst nach sieben Jahren noch keine Umschichtung erfolgt ist. Die kurzen Auftriebszeiten reichten für eine Trittwirkung nicht aus, und eine durch die häufigere Nutzung an sich mögliche Bestandesverschiebung wurde durch die gleichlaufende Erhöhung der Düngung kompensiert.

In Tabelle 3 ist die verkürzte Artenliste von zwei Beständen aufgeführt, die ursprünglich mit der gleichen Saatmischung angesät und dann eine zeitlang einheitlich als Wiese genutzt wurden. Während Fläche A auch heute noch Wiese (mit spätem 1. Schnitt) ist, wurde Fläche B vor 15 Jahren zur Umtriebsweide umgestellt, auf welcher das Vieh bei jedem Auftrieb jeweils mehrere Tage auf derselben Koppel bleibt, Fläche A trägt ein, wenn auch nicht sehr typisches A r r h e n a t h e r e t u m ,Fläche B hingegen zeigt schon deutliche Anklänge an das L o l i o - C y n o s u r e t u m . Auf Fläche B sind alle nicht weidefesten Arten, besonders *Bromus mollis*, *Alopecurus pratensis* und *Achillea millefolium*, zurückgegangen, während andererseits weidefeste Arten, besonders *Festuca pratensis*, zugenommen haben, bzw. neu hinzugekommen sind.

Tabelle 4 vermittelt die Artenzusammensetzung von fünf verschieden bewirtschafteten Grünland-Flächen eines arrondierten Großbetriebes. Die von A bis E zunehmend intensivere Nutzung bedingt

Tabelle 2

Kombinierter Nutzungsversuch Eichhof
(7. Versuchsjahr)

Mengenanteile in %

	P_2O_5	K_2O	N	Weidetage pro Jahr
Vers.Glied 1: 2 Schnitte	90	120	-	-
5: Weide-Schnitt-Weide	90	120	80	2
6: Umtriebsweide	120	120	120	3-4
8: Portionsweide	180	180	180	4-6

Versuchsglied	1	5	6	8
Gräser	90,0	91,7	89,3	91,3
Leguminosen	0,5	0,3	+	+
Kräuter	9,5	8,0	10,7	8,7
Alopecurus pratensis	47,5	47,5	48,7	48,0
Festuca pratensis	14,0	17,0	10,5	12,3
Poa prat.+ triv.	24,5	24,0	26,7	27,7
Festuca rubra	2,8	1,0	0,5	0,3
Agrostis alba	+	1,3	1,3	0,5
Holcus lanatus	+	0,3	1,0	+
Deschampsia caespitosa	1,3	0,3	+	1,7
Lolium perenne	+	0,5	0,5	+
Trifolium repens	0,5	0,3	+	+
Taraxacum officinale	3,8	3,3	1,8	1,5
Heracleum sphondylium	1,0	+	+	+
Ranunculus repens	4,0	4,5	4,3	7,3

Tabelle 3
Lorsch/Bergstraße

A = alte Wiese
B = alte Wiese, seit 15 Jahren Mähweide

	Mengenteile in % A	B
Arrhenatherum elatius	+	
Galium mollugo	+	
Galium verum		
Colchicum autumnale	+	
Sanguisorba officinalis	+	
Trifolium pratense	+	
Vicia cracca	+	
Lathyrus pratensis	1,5	
Chrysanthemum leuc.	0,5	
Bromus mollis	18,0	+
Holcus lanatus	2,0	+
Achillea millefolium	6,0	+
Agrostis tenuis	7,5	3,8
Alopecurus pratensis	12,5	4,1
Poa pratensis + trivialis	25,0	24,0
Dactylis glomerata	17,5	18,8
Festuca rubra	4,0	5,3
Festuca pratensis	3,0	32,3
Taraxacum officinale	+	3,0
Agropyron repens	+	4,2
Lolium perenne		0,9
Trifolium repens		0,5
Ranunculus repens		1,3
Phleum pratense		+
Poa annua		+
Plantago major		+
Leontodon autumnalis		+
Rumex crispus		+
Raphanus raphanistrum		+
Lolium multiflorum		+

Tabelle 4
Holzkirchen/Allgäu

A = Zweischürige Wiese, Spätschnitt
B = Mehrschürige Wiese, Frühschnitt
C = Mähweide, Schwerpunkt Schnittnutzung
D = Mähweide, Schwerpunkt Weidenutzung
E = Dauerweide

	Ertragsanteile in %				
	A	B	C	D	E
Artenzahl	40	31	24	25	20
Gräser	53,0	64,4	84,0	92,2	97,0
Leguminosen	3,6	7,6	2,3	6,4	1,5
Kräuter	43,4	28,0	13,7	2,4	1,5
Trifolium repens	3,6	7,6	2,3	6,4	1,5
Alchemilla vulgaris	0,1	+	+	0,2	0,5
Trisetum flavescens	3,7	1,8	1,3	(+)	
Bellis perennis	+	+	+	+	
Carum carvi	0,1	2,0	2,0		
Heracleum sphondylium	14,8	(+)			
Crepis biennis	+				
Pimpinella major	(+)				
Tragopogon pratensis	(+)				
Anthriscus silvestris	(+)				
Lolium perenne	(+)	(+)	1,0	29,0	27,5
Agropyron repens	(+)	1,6		4,4	2,0
Poa annua	(+)	(+)	+	+	+
Veronica serpyllifolia	(+)	(+)		+	+
Plantago major		(+)	(+)	+	+
Poa pratensis + triv.	27,6	52,0	47,0	52,8	37,5
Dactylis glomerata	18,6	5,0	30,7	2,2	15,0
Festuca pratensis	3,0	4,0	4,0	2,6	15,0
Ranunculus acer	3,7	4,8	1.0	0,6	0,5
Plantago lanceolata	0,1	+	+	+	+
Rumex acetosa	2,8	0,6	+		
Trifolium pratense	+	+	(+)		
Centaurea jacea	(+)	+			
Sanguisorba officinalis	0,2				
Lathyrus pratensis	(+)				
Taraxacum officinale	18,2	16,8	10,4	1,0	0,5
Rumex obtusifolius	3,4	1,6	0,3	0,4	+
Achillea millefolium	+	1,4	+	(+)	
Vicia sepium	+				

(+) bedeutet, daß die betreffende Art nur sporadisch gefunden wurde.

nicht nur eine entsprechende Verringerung der Gesamt-Artenzahl, sondern auch eine Verschiebung der Mengenanteile einzelner Arten, bzw. Artengruppen. Besonders deutlich ist die Verschiebung der Gräser- und Klee-Anteile.
Aus Tabelle 5 ist schließlich der Einfluß der Trittwirkung auf den Pflanzenbestand am deutlichsten zu erkennen. Unittelbar am Weidetor, also dort, wo die Tiere bei jedem Betreten und Verlassen der Fläche auf engstem Raum zusammenkommen, findet sich ein sehr artenarmer und lückiger Trittrasen. Mit zunehmender Entfernung vom Weidetor verringert sich der Tritteinfluß langsam, dementsprechend können mehr und mehr auch etwas empfindlichere Arten aufkommen und die Narbe schließt sich. Etwa 3 Meter vom Weidetor entfernt beginnen die Arten des L o l i o - C y n o s u r e t u m zunehmend an Bedeutung zu gewinnen. Bei noch größerem Abstand vom Tor folgt dann nach und nach eine Reihe weiterer A r r h e n a t h e r e t a l i a - bzw. M o l i n i o - A r r h e n a t h e r e t e a - Arten, aber erst in einer Entfernung von etwa 20 Metern entspricht der Bestand der tatsächlichen Wirtschafts-Intensität.
4. Ein weiterer Unterschied zwischen Wiese und Weide besteht darin, daß der hohe Wiesenbestand immer zu einem bestimmten Grad beschattend und dadurch hemmend auf die Bodenpflanzen wirkt, wodurch die Narbe sich lockert und etwas lückig wird. Auf der immer niedrig gehaltenen Weide hingegen kommen alle Pflanzen etwa gleichmäßig in den Lichtgenuß. Das Licht als Konkurrenzfaktor ist also weitgehend ausgeschaltet, und die Bodenpflanzen können sich ebenso entfalten, wie die hochwachsenden. In Verbindung mit einer ebenfalls vom Weidegang herrührenden stärkeren Bestockung der einzelnen Pflanzen entwickelt sich daher auf der Weide gewöhnlich eine dichtere Narbe, als auf der Wiese. Erst dann, wenn der Tritt allein zum bestimmenden und begrenzenden Faktor wird, also auf dem reinen Trittrasen, kommt es dann wieder zu einer Narben-Auflockerung und u.U.zu einer starken Lückenbildung (wie aus Tabelle 5 ersichtlich wird).
5. Ein letzter Faktor soll noch erwähnt werden: Auf der Wiese erfolgt bei jeder Nutzung eine schlagartige Entfernung fast der ganzen assimilationsfähigen organischen Substanz. Zurück bleibt eine meist etiolierte oder von abgestorbenen Blattscheiden umgebene assimilations-unfähige Stoppel. Im Gegensatz hierzu wird auf der Weide die Photosynthese-Tätigkeit nie ganz unterbrochen.Die

Tabelle 5

Einfluß der Trittwirkung
Typische Weidelgrasweide des Niederrheins
(nach Graf BOTHMER, gekürzt)[1]

Ertragsanteile in %

	Entfernung vom Weidetor in Metern						
	0-1	1-2	2-3	3-4	4-5	9-10	19-20
Poa annua	62,0	39,6	23,4	15,4	11,4	0,2	0,2
Plantago major	11,0	7,2	8,8	3,6	1,6	0,2	0,2
Polygonum aviculare	9,2	20,4	10,0	5,2	2,0	-	-
Lolium perenne	16,0	24,0	42,0	59,0	57,0	45,8	33,2
Trifolium repens	1,4	2,0	3,4	3,4	4,8	4,2	4,0
Taraxacum officinale	0,4	0,4	0,6	1,6	2,0	3,2	4,4
Agrostis alba		4,0	6,6	5,0	6,0	11,8	7,8
Agropyron repens		0,8	0,4	0,8	0,8	1,0	0,8
Ranunculus repens		+	0,4	0,2	0,6	3,4	2,6
Poa pratensis		1,6	3,0	3,2	5,2	4,8	4,4
Ranunculus acer		+	+	+	+	0,6	2,2
Alopecurus pratensis			0,6	1,0	4,0	8,0	7,2
Poa trivialis			0,4	0,8	1,2	1,6	1,4
Leontodon autumnalis			+	0,2	0,2	0,4	1,0
Dactylis glomerata			0,2	0,6	0,2	0,6	1,4
Cirsium arvense			+	+	1,0	5,2	6,8
Phleum pratense					0,4	1,8	3,0
Festuca pratensis					0,6	2,4	7,0
Agrostis tenuis					0,6	2,6	4,0
Trifolium pratense						0,2	0,8
Holcus lanatus						1,4	3,4
Festuca rubra						1,0	2,6
Cynosurus cristatus							0,8
durchschn.Artenzahl	4,2	6,4	7,6	9,6	12,0	15,6	20,o
durchschn.Lücken %	59,0	14,0	4,0	1,0	0,4		

[1] Z.Acker- u.Pflanzenbau 96 (1953)

Pflanze auf der Wiese muß also in jedem Falle die in der Wurzel oder in der Stoppel gelagerten Reservestoffe angreifen und mobilisieren, um wieder assimilationsfähige Blattorgane zu bilden. Die Weidepflanze hingegen kann ohne Unterbrechung, wenn zunächst auch in etwas vermindertem Maße, mit den ihr verbliebenen Blattresten neue organische Substanz aufbauen.

Diese Aufzählung der auf Wiese und Weide in verschiedener Weise zur Wirkung kommenden Standortsfaktoren erhebt nicht den Anspruch der Vollständigkeit. In Wirklichkeit kann auch kaum ein Faktor für sich allein erfaßt und betrachtet werden, sondern es müssen durch eine kritische Betrachtung der Komplexwirkung aller Faktoren jeweils Rückschlüsse auf den Wirkungsgrad eines einzelnen Faktors gezogen werden.

ZUSAMMENFASSUNG

Die Bestandesentwicklung auf der Mähwiese einerseits und auf der Weide andererseits ist in erster Linie bedingt durch die (mit der Nutzungsart verbundenen) unterschiedlichen anthropogenen und zoogenen Faktoren.

Wiese	Weide
1. höchstens 2 bis 3 Nutzungen pro Jahr	bis zu 6 Nutzungen pro Jahr und mehr (je nach Intensität)
2. relativ schwache Düngung	starke Düngung, vor allem mit N, dazu Exkremente der Tiere
3. nur schwache Selektion und Bodenverdichtung (durch Mahd bzw. Fahrspuren)	starke Selektion und Bodenverdichtung durch Verbiß und Tritt der Weidetiere
4. Narbenauflockerung	Narbenverdichtung
5. durch den Schnitt zunächst schlagartige Unterbrechung der Photosynthese	beim Verbiß bleiben immer Reste assimilationsfähiger Blätter zurück.

SUMMARY

The development of the species composition of the meadow on the one hand and of the pasture on the other hand is mainly caused by the different anthropogenic and zoogenic factors connected with the land use system.

Meadow	Pasture
1. at most 2 or 3 cuttings per year	up to 6 utilizations per year and more (dependent on intensity)
2. relatively poor fertilizing	heavy fertilizing, particularly with N, in addition excrements of animals
3. only slight selection and soil compaction (caused by mowing resp. tracks)	heavy selection and soil compaction caused by grazing
4. loosening of the turf	compaction of the turf
5. sudden interruption of the photosynthesis caused by cutting	during grazing there always remain rests of leaves which are capable of assimilation.

DISKUSSION

H.DIERSCHKE

In den Feuchtwiesen NW-Deutschlands, Bayerns und wohl auch in anderen Gebieten läßt sich eine starke Mengenzunahme von *Holcus lanatus* feststellen. Man glaubt, daß diese Tatsache auf starke Stickstoffdüngung zurückgeht. Ich hatte deshalb die Art in Ihrer 1. Tabelle in der Gruppe der stickstoffgeförderten Arten erwartet, sie erscheint aber bei den indifferenten Arten. Verhält sich *Holcus lanatus* bei Ihnen anders als in Molinietalia-Gesellschaften?

B.SPEIDEL

Holcus lanatus gehört eigentlich in die letzte Gruppe, bei der die Reaktion nicht einheitlich ist. In unseren Versuchen waren Flächen, in denen *Holcus lanatus* durch die N-Düngung gefördert wurde, es waren aber auch Flächen dabei, wo die Reaktion genau umgekehrt verlief. Woran das liegt, ist mir nicht klar.

R.TÜXEN

Holcus lanatus dürfte auf feuchten Böden auf Stickstoff positiv reagieren. So scheint mein qualitativer Eindruck.
Eine Bemerkung zu der Tabelle 4, wo Sie die Abnahme der Artenzahl von 40 auf 20 zeigten von der Wiese zur Weide. Und ganz unten standen die Wertzahlen. Die Wertzahlen nach KLAPP stiegen gewaltig an,wenn ich mich nicht irre, von 6 auf 7,7. Wenn man das liest, fragt man sich, warum werden nicht alle unsere Mähweiden auf 20 Arten reduziert, wenn eine solche Steigerung des Wertes damit verbunden ist? Aber was sagt die Kuh?

B.SPEIDEL

Zweierlei! Sie frißt sicherlich auf einer guten Weidelgrasweide lieber als auf einer schlechten Wiese. Aber auf der anderen Seite kann man immer wieder beobachten, daß der Apfel aus des Nachbars Garten besser schmeckt als der im eigenen Garten. Beobachten Sie einmal, daß Weidekühe in einer Koppel unter dem Zaun hindurch einen halben Meter vom Nachbarn abfressen, wenn auch in der eigenen Koppel das Futter noch steht. (Zwischenruf: *Molinia* fressen sie, die armen Viecher.) Sie fressen auch *Calluna*. Zwischenruf: Ja, aus Not.) Nein, nein! Es gibt in unseren Mittelgebirgen Gegenden, wo das Heu von *Nardus*-Wiesen säuberlich nach Hause gebracht wird, getrennt gelagert wird, als Medizinalheu. Warum? Weil es sehr krautreich, weil es sehr abwechslungsreich ist. Es hat sicherlich weniger Nährstoffe als das Heu von einem guten Arrhenatheretum oder von einem Lolio-Cynoseretum, aber es ist abwechslungsreicher und schmackhafter.

J.FEISE

Das Wohlbefinden der Kuh messen wir nicht an der Literzahl, die wir melken, sondern an dem Fettgehalt. Wenn eine Kuh gut zuwege ist, dann hat sie einen hohen Fettgehalt in der Milch. Wir haben es häufig so gemacht, wenn die Kühe nicht recht wollen, mähen wir Heide und geben Heide zu, weil in dieser Heide Gerbsäure,und Kieselsäurebestandteile und schwer verdauliche Kupfersalze enthalten sind, Spurenelemente, die die Kuh so nötig hat, wie wir das Salz.

In der 1. Tabelle stand der Gesamtwert Doppelzentner je Hektar Trockenmasse. Das wollen wir ja gar nicht, wir wollen gesunde Milch verkaufen von Kühen, die sich wohl fühlen.

Wir machen viel Fehler damit, daß wir die Kühe auf diesen lächerlichen Fettgehalt treiben.Es wäre uns ja viel mehr damit gedient, ständig Kälber und keine Unfruchtbarkeit zu haben, auf Milch zu verzichten, die wir doch nicht loswerden, sondern wir wollen gesunde Tiere, die lange leben.

Noch eins zur Ehrenrettung der Zahlen von KLAPP. KLAPPs Zahlen sind sehr gut und besser als gar keine. Aber sie werden manchmal auch falsch angewendet. Eine Weide mit 100% *Lolium perenne*, so wird von manchen errechnet, hätte die Wertzahl 8,0. Das stimmt aber nicht. KLAPP hatte eine Fußnote, wenigstens in den späteren Ausgaben, die oft übersehen wird: Je mehr von einer Art darin ist, desto größer sind die Abzüge, die man machen muß. Eine 100% Lolium perenne - Wiese hat höchstens den Futterwert von 5,5. *Lolium perenne*,ist fast ein Zeichen einer Überdüngung mit Kali und Stickstoff,kann ein Zeichen einer falschen Beweidung, einer Mißhandlung sein, und es ist die Frage, ob man die Fläche nicht schonen muß, ob man nicht versuchen muß, *Cichorium intybus* anzubauen, um das zu unterdrücken. Was wirtschaftlich richtig ist, das kann man nicht so ohne weiteres sagen, ich glaube, es ist jedenfalls unwirtschaftlich, was wir in den letzten 20 Jahren getan haben!

H.ELLENBERG

Ich möchte auf die Frage nach *Holcus lanatus* zurückkommen, und überhaupt auf die Frage der Förderung oder Nicht-Förderung von bestimmten Arten durch bestimmte Behandlungen oder Düngungen. Durch Stickstoffdüngung werden alle Grünlandpflanzenarten gefördert, wenn man sie in Reinkultur zieht. Das gilt auch für *Holcus lanatus*, für die Leguminosen und alle anderen sog. "Stickstoffmangel-Zeiger". Erst durch Konkurrenz anderer Arten, die relativ mehr gefördert werden, gehen die "Mangelzeiger" zurück, weil sie überschattet werden. Man muß sich also darüber klar sein, daß grundsätzlich jede Art durch Düngung gefördert wird und daß nur die Konkurrenz, das gegenseitige Verhalten der Arten, dann über den Endeffekt entscheidet. Dabei spielen dann auch die anderen Faktoren, z.B. die Selektion durch das Vieh usw. eine Rolle.

D.MUELLER-DOMBOIS

In diesem Zusammenhang dürfte es vielleicht interessieren, daß *Holcus lanatus* sich momentan auf der südlichen Hawaiischen Insel, Hawaii, zwischen 1200-2000 m auf der Ostflanke Mauna Loa's sehr schnell ausbreitet, und zwar auf Stellen, die von Wildschweinen bevorzugt werden. Dort verdrängt *Holcus lanatus* die endemische *Deschampsia umbigena*, ein Zeichen verhältnismäßig großer Agressivität von *Holcus lanatus*.

A.W.KÜCHLER

Ihre Tabelle 2 zeigt, daß sich die Produktion der verschiedenen Gräser nicht ändert, trotz verschiedener Nutzungsmethoden von Doppelmahd bis zur Dauerweide. Gleichzeitig zeigt die Tabelle eine Zunahme des Düngerverbrauchs. Da die Produktion immer teurer wird, frage ich, warum wird nicht alles Grünland gemäht?

B.SPEIDEL

Das beruht auf einem Irrtum. Der Ertrag stieg selbstverständlich an, ist aber in der Tabelle gar nicht erschienen. Das ist der prozentuale Mengenanteil der Arten am Gesamtbestand. In allen 4 Versuchswiesen haben Sie etwa 45-50% *Alopecurus pratensis* und Sie haben überall 10-15% *Festuca pratensis* und etwa 25% *Poa pratensis* + *P.trivialis*. Vom Ertrag habe ich nichts gesagt, der steigt selbstverständlich von links nach rechts an, das ist ja der Sinn der Düngung.

Dann noch eine Ergänzung zu den Ausführungen von Herrn Professor ELLENBERG. Selbstverständlich gelten alle diese Zahlen nur für den Bestand, nicht für die einzelne Art. Und man kann sogar, um auf den Klee zurückzukommen, beobachten, wenn wir auf einer Mäh-

wiese immer nur P-K düngen, kein N, daß die Leguminosen zunächst gefördert werden. Das geht einige Jahre. Plötzlich können wir aber dann sehen, daß die Leguminosen zurückgehen, obwohl wir keinen Stickstoff gedüngt haben. Warum? Die Leguminosen haben in diesen paar Jahren soviel N gesammelt, der Boden hat sich mit N angereichert, jetzt kommen die Gräser, machen sich diesen Stickstoff zu eigen und wachsen über die Leguminosen hinaus, beschatten sie, und die Leguminosen gehen zurück.
Ich glaube doch, ich muß bei den Wertzahlen die Ehre von Herrn Professor KLAPP verteidigen. Herr Dr. FEISE hat recht, daß in dem Rezept von KLAPP bei gewissen Arten, wenn sie in höheren Mengen auftreten, Abzüge gemacht werden. Das sind aber einmal *Deschampsia cespitosa*, dann *Sanguisorba officinalis*, dann sind es die *Juncus*-Arten usw., es ist aber nicht *Lolium perenne*. Aber ich stimme mit ihm insofern überein, daß ich einer Wiese mit 100% *Lolium* auch nicht die Wertzahl 8 geben würde, weil sie zu monoton ist. Aber sonst finden wir mit steigendem Anteil an *Lolium perenne* auch einen Qualitätsanstieg.

R.TÜXEN
Warum werden Kühe steril?

B.SPEIDEL
Ja, das ist etwas ganz anderes. Das kann - aber das ist meine ganz persönliche Ansicht - mit dem Dünger zusammenhängen. Wir haben früher ziemlich niedrig konzentrierten Dünger gehabt, das heißt, wir haben viel Ballaststoffe mit unserem Dünger ausgestreut. Wir haben Kainit gestreut, mit 20% Kali, das andere war Natrium, Magnesium usw. Heute haben wir hochprozentige Dünger, 50% Kali, da ist der Ballast zum großen Teil weggefallen. Und jetzt kommt die Kehrseite, jetzt müssen wir das, was früher automatisch mit dem Ballast gegeben wurde, zusätzlich als Spurenelemente dazu geben.

SOZIOLOGISCHE VERÄNDERUNGEN IN ZWEI DAUERQUADRATEN EINER WESER-WIESE BEI STOLZENAU (KRS.NIENBURG) VON 1945-1978[1)]

R. Tüxen

Arbeiten aus der Arbeitsstelle für Theoretische und Angewandte Pflanzensoziologie/Rinteln Nr.187

Dem verehrten Lehrer
JOSIAS BRAUN-BLANQUET

nach mehr als 50jähriger Freundschaft
zur Vollendung seines 94ten Lebensjahres
in Dankbarkeit gewidmet

Wenn die Weser Hochwasser führt und irgendwo über die Ufer strömt, werden in der Aue zuerst die Flutrinnen überschwemmt, die oft am Fuß der Terrassen liegen. Dort bleibt das zunächst strömende Wasser schließlich am längsten stehen. Es setzt dabei Schlick ab. Auch Schmelzwasser sammelt sich hier nach schneereichen Wintern. In manchen Jahren werden diese Rinnen gar nicht, in anderen kürzere oder längere Zeit überflutet.

In einer solchen flachen Flutrinne bei Stolzenau haben wir im Jahre 1945 zwei aneinandergrenzende Probeflächen angelegt, die eine (A) 16, die andere (B) 12 m^2 groß und beide alljährlich (mit Ausnahme weniger Jahre)pflanzensoziologisch aufgenommen.In der Regel geschah das im April oder Mai, wenn die Gräser spannhoch waren, ausnahmsweise auch im März oder August. Der frühe Zeitpunkt war gewählt worden, um uns alle, die wir mit Grünland-Gesellschaften zu tun hatten, in der Erkennung der Gräser nach dem "langen Winterschlaf" in sterilem Zustand wieder zu üben.Zugleich konnte zu dieser Zeit *Ranunculus ficaria* sicher gefunden werden.

Wir hatten 1967, als dieser Bericht während des Symposion erstattet wurde, einen Zeitraum von 22 Jahren verfolgt, den wir jetzt (1978) auf 33 Jahre erweitern konnten (Tab.A,B).(In den Jahren 1951, 1964-1966, 1969 und 1971 sind keine Aufnahmen gemacht worden). Bis 1961 machten alle wissenschaftlichen Mitar-

[1)] Das Referat von 1967 wurde nach Auswertung weiterer Aufnahmen bis 1978 überarbeitet und erheblich ergänzt. Die Diskussion konnte darauf natürlich nicht eingehen. Ihre grundsätzlichen Bemerkungen behalten aber ihre Gültigkeit.

beiter der Bundesanstalt für Vegetationskartierung in Stolzenau - jeder für sich - diese Aufnahmen, die zu einer Tabelle vereinigt und anschließend an den Probeflächen diskutiert wurden. Über die Ergebnisse des Vergleichs dieser Aufnahmen von verschiedenen Bearbeitern auf den jeweils gleichen Flächen habe ich seinerzeit hier berichtet (R.TÜXEN 1972).

Die kritische Überprüfung der einzelnen Aufnahmen ergab, daß sie nicht nur in den Mengenschätzungen, sondern auch in der Zahl der gefundenen Arten voneinander abweichen. Kein Bearbeiter hat alle in dem untersuchten Quadrat wachsenden Pflanzen gefunden.

Unter Berücksichtigung der Artenkenntnis der einzelnen und der Bestätigung einer Art durch mehrere Bearbeiter wurden, soweit nötig, die Artenlisten und ebenso Unterschiede in der Mengenschätzung bereinigt.

Meine späteren Aufnahmen wurden meist in Begleitung von fachkundigen Kollegen gemacht. Sie dürften in der Vollständigkeit und Mengenschätzung nicht stärker von den früheren abweichen, zumal die Vergleichstabellen stets mitgeführt und nicht mehr auftretende Arten besonders sorgfältig gesucht wurden. Es ist dennoch nicht auszuschließen, daß auch in meinen Aufnahmen die eine oder andere Art trotz sorgfältigen Suchens an Hand der Vergleichstabelle übersehen worden ist, wenn sie in sehr geringer Menge und schlecht entwickelt vorkam.

Dieser Befund verdient Beachtung, zeigt er doch, daß selbst der Geübte, (meine damaligen Mitarbeiter waren fast alle seit vielen Jahren mit sterilen Gräsern und allen anderen Grünland-Pflanzen bestens vertraut), die eine oder andere in kleiner Menge vorkommende Art übersehen oder ihre Menge, (selbst nach der BRAUN-BLANQUET-Skala, um von %-Schätzungen ganz zu schweigen), ungleich einschätzen kann. Wer aber glaubt, die Genauigkeit unserer Aufnahmen gering achten zu müssen, möge zunächst vor der eigenen Tür kehren. Zwei oder drei gute Artenkenner würden auf der gleichen Aufnahme-Fläche vielleicht doch auch die eine oder die andere übersehene Art entdecken. Das wäre ganz normal!

Das standörtliche Schicksal der Dauerflächen war im Laufe der Beobachtungszeit recht wechselhaft: Bis etwa 1970 wurde die Flutrinne meist im Spätwinter kürzere oder längere Zeit fast alljährlich von der Weser oder vom Schmelzwasser der Schneedecke überflutet (vgl.Tab.A u.B). Mehrfach war sie auch von einer Eisdecke von wechselnder Dicke bedeckt. Schlick wurde nach Weserhochwasser

Additional material from Gesellschaftsentwicklung (Syndynamik),
ISBN 978-94-011-7594-4 (978-94-011-7594-4_OSFO2),
is available at http://extras.springer.com

EXTRA
MATERIALS
extras.springer.com

besonders 1956(!), 1957, 1961,1962 und 1967 manchmal in erheblicher Stärke (1956 13-15 cm!) abgelagert, so daß die Grasnarbe sich erst verspätet hindurch arbeiten, dann aber dank der natürlichen Düngung besonders üppig entwickeln konnte. Als Folge der Überschlickung siedelte sich vorübergehend die Kugelalge *Botrydium granulatum* auf dem feuchten Schlick an. Im Sommer 1956 fand eine ungewöhnliche Überflutung des Wesertales statt, die starke Schäden in den Wiesen hinterließ. In den letzten Jahren fanden, mit Ausnahme von 1976, keine Überflutungen durch die Weser statt.

Düngung erfolgte zunächst auf der Fläche A mit Jauche oder auch mit Kompost, während die Fläche B ohne Düngung blieb, weil die Ziegen des Eigentümers kein Futter fraßen, das von einer gedüngten Wiese stammte (1947). Ab 1973 wurden beide Flächen nach einem Besitzwechsel und einer Bodenuntersuchung mit Thomasmehl und Kali (2-3 Ztr. pro Morgen) und Kupfer-Gaben gedüngt.
Die Mahd erfolgte zunächst nicht gleichzeitig auf beiden Flächen, jedoch mehrmals im Sommer. Im Herbst wurde schwach beweidet. In den letzten Jahren sind beide Flächen mit Unkraut-Ex gespritzt worden.
Auf diese natürlichen und menschlichen Wirkungen, die bei wechselndem Wetter der verschiedenen Jahre, besonders der früher oder später beginnenden Frühjahrserwärmung stattfanden, stellten sich unmittelbar sowohl die Arten-Verbindungen als noch deutlicher die Mengen einzelner Arten ein. Dem Pendeln der Standortsbedingungen folgte eine entsprechende Fluktuation in der Zusammensetzung der Pflanzendecke in Arten und Mengenanteilen (Tab.A,B, Fig.1-6).[1),2)]
In den letzten 15 Jahren wird eine deutliche bis jetzt bleibende starke Abnahme der Artenzahlen in beiden Flächen erkennbar (Fig. 1,2).
Im einzelnen erfolgt der Wechsel der Artenzahlen und ihre Abnahme nicht übereinstimmend in den beiden Flächen, wie der Vergleich von Tab.A und B zeigt. Während in Fläche A zunächst Arten wie *Trifolium dubium, Daucus carota, Festuca rubra, Tripleurospermum*

1) Für die freundliche Mitteilung seiner Erkundigungen über die Bewirtschaftung der Flächen in den letzten Jahren danke ich Herrn A. v.HÜBSCHMANN herzlich.
2) Für die gewissenhafte Anfertigung der Tabellen-Reinschriften und der Kurven danke ich Herrn YUKITO NAKAMURA herzlich.

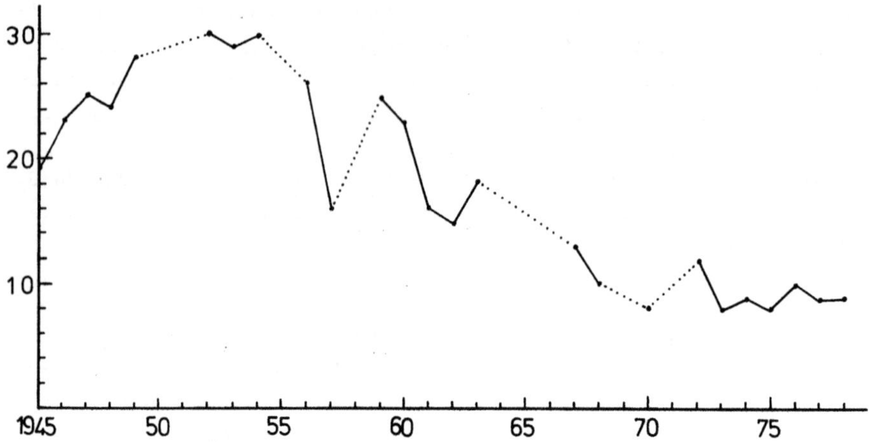

Fig. 1. Artenzahlen der Aufnahmen in der Fläche A von 1945-1978.

inodorum, Chrysanthemum leucanthemum, dann auch *Plantago lanceolata, Achillea millefolium, Trifolium repens* und *Phleum pratense* ausfallen, denen Anfang der 60er Jahre *Leontodon autumnalis, Bellis perennis, Festuca pratensis, Cerastium holosteoides* und *Poa pratensis* sowie *Lysimachia nummularia* und am Ende dieses Jahrzehnts auch *Agrostis stolonifera* folgen, vollzieht sich der Wechsel und die Abnahme der Arten in der einige cm höher liegenden Fläche B anders. Zunächst verschwinden hier Feuchtezeiger wie *Silaum silaus* und *Symphytum officinale,* sowie auch *Vicia cracca* (1950). Dafür treten *Alopecurus geniculatus, Festuca rubra, Glechoma hederacea, Cerastium holosteoides, Bromus mollis* und etwas später *Agrostis tenuis, Brachythecium rutabulum, Bellis perennis* und *Plantago major* und einige Moose auf, was leicht durch das Lückigwerden der Grasnarbe verständlich wird. Dann aber fallen bald nach allen genannten Arten, wieder mit dem Ende der sechziger Jahre *Agrostis stolonifera, Lysimachia nummularia* und schon etwas früher *Potentilla reptans* und *Leontodon autumnalis* aus.

Diese beiden Arten, die in beiden Flächen etwa gleichzeitig verschwinden, dürften mit den bis dahin häufigen, jetzt aber seltener und kürzer oder gar nicht mehr auftretenden winterlichen Überflutungen von den dagegen empfindlichen Gräsern *Alopecurus pratensis* und *Poa trivialis* überwachsen und verdrängt worden sein. In der Fläche B beteiligt sich auch die Quecke (*Agropyron repens*) durch zunehmende Menge an diesem Wandel. Zuletzt stellt sich *Anthriscus sylvestris* seit 1972 bleibend ein, und auch *Rumex crispus* weicht nicht mehr.

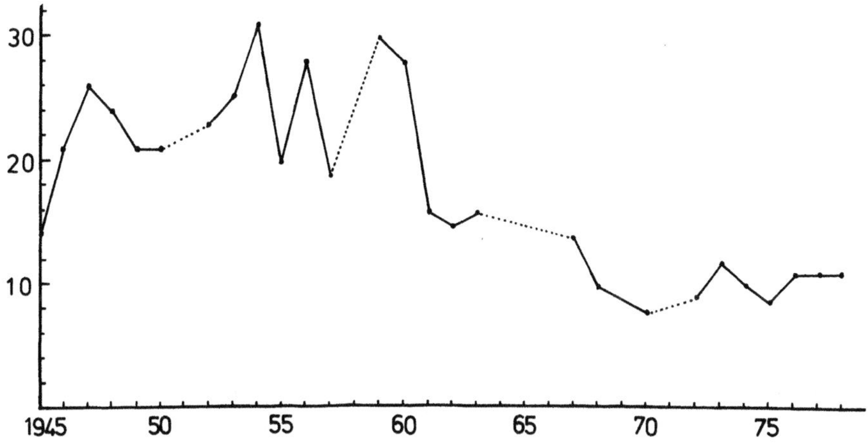

Fig.2. Artenzahlen der Aufnahmen in der Fläche B in den Jahren 1945-1978.

Bis dahin zeigten beide Dauerflächen im Laufe der Jahre erhebliche Schwankungen der Arten-Zahlen, eine Erscheinung, die wir besonders im Hinblick auf die noch stärkeren und rasch verlaufenden Änderungen der Mengen einzelner Arten (z.B. *Agrostis stolonifera, Ranunculus repens, Poa trivialis, Agropyron repens, Alopecurus pratensis* als "Harmonika-Sukzession" bezeichnet haben (TÜXEN 1950, p.149).

Es handelt sich dabei allerdings nicht um eine einseitig gerichtete, den Bestand endgültig verändernde Entwicklung von einer in eine andere Gesellschaft. Vielmehr ist dieser Wechsel wohl eine Folge der von Jahr zu Jahr verschieden langen oder ganz fehlenden Überflutungen und deren unterschiedlichen Schlick-Absätzen, und vielleicht der verschieden häufigen Mahten und möglicherweise auch von Beweidung und Düngung. Darum sollte man besser von Fluktuation sprechen, die um einen mittleren Zustand pendelt.

Der Vergleich von Fig.1 und 2 zeigt, daß diese Schwankungen auf den beiden aneinander grenzenden Flächen keineswegs gleich verlaufen, sondern erheblich voneinander abweichen. Diese Unterschiede dürften auf die etwas verschiedene Höhenlage und auf unterschiedliche Nutzung (Mahd) u.a. Wirtschaftseinflüsse (Düngung, Beweidung) zurückgeführt werden.

Beiden Kurven (Fig.1,2) aber ist die starke Abnahme der Artenzahlen etwa seit den Jahren 1963 gemeinsam.

In der Fläche A (Fig.1) betrug das Mittel der Artenzahlen von 1946-1960 24, von 1968-1978 10. In Fläche B (Fig.2) waren die entsprechenden Werte 25 und 9; die Abnahme war hier also noch stärker.

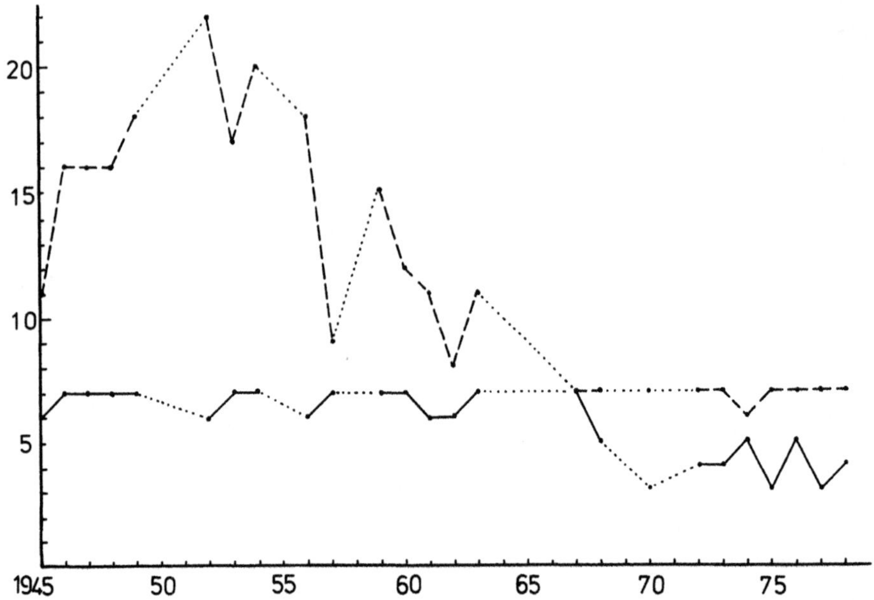

Fig.3. Fläche A. Anzahl der A g r o p y r o - R u m i c i o n - (——) und der übrigen M o l i n i o - A r r h e n a t h e r e t e a - Arten (----).

Die Zerlegung der Arten in syntaxonomische Gruppen: A g r o p y - r o - R u m i c i o n - und übrige M o l i n i o - A r r h e n a t h e - r e t e a - Arten läßt die starken Schwankungen der Artenzahlen vor allem in der letzten Gruppe deutlich werden (Fig.3 u.4). Sie sind in Fläche B aber noch auffallender als in Fläche A. Erst in den letzten 10 Jahren bleibt ihre Zahl in Fläche A mit 7 nahezu konstant, während sie in B zwischen 6 und 8 wechselt. Die A g r o p y r o - R u m i c i o n - Arten verhalten sich ausgeglichener. Sie pendeln in A in den ersten beiden Jahrzehnten zwischen 6 und 7, um im letzten stark abzufallen, ohne jedoch konstant zu bleiben (Fig.3). In B (Fig.4) sind ihre Schwankungen stärker: bis 1967 zwischen 4 und 9, später zwischen 3 und 5.

Die eigentlichen strukturellen Veränderungen der Bestände werden deutlicher, wenn außer der Anzahl der Arten auch ihre Menge berücksichtigt wird, was wir (nach TÜXEN und ELLENBERG 1937) mit Hilfe der Gruppenmenge getan haben. Für die Errechnung der Gruppenmengen (Fig.5,6) und der Futterwerte (Fig.7,8) haben wir die Mittelwerte der BRAUN-BLANQUET-Skala der Mengenschätzung und die Wertziffern von KLAPP et al.(1953) zu Grunde gelegt. Da immer der gleiche Fehler oder die gleiche Ungenauigkeit hinge-

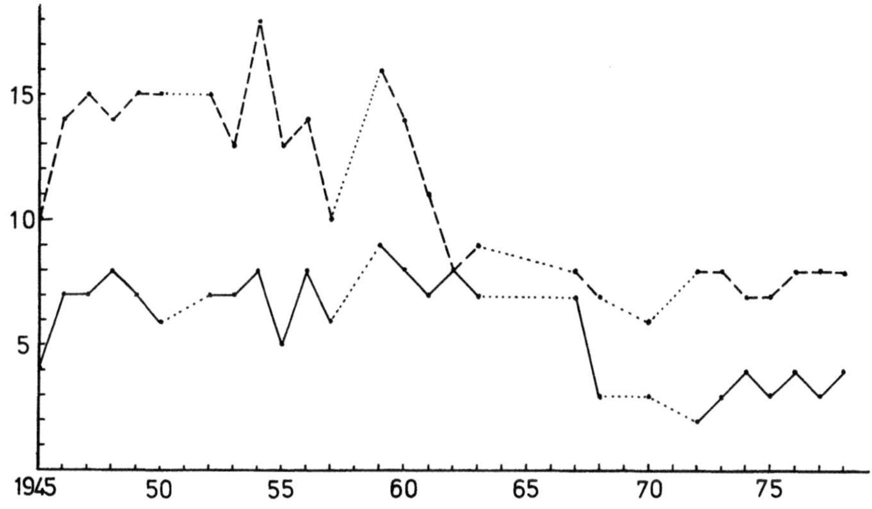

Fig.4. Fläche B. Anzahl der Agropyro-Rumicion- (——) und der übrigen Molinio-Arrhenatheretea-Arten (----).

nommen werden muß, erhält man zwar keine genauen absoluten Werte, aber die Unterschiede zwischen einzelnen Jahren und Jahr-Gruppen sind so stark, daß die Größenordnung der Schwankungen deutlich wird.

Die Gruppenmenge der Agropyro-Rumicion-Arten fällt in beiden Flächen zunächst sehr stark ab (Fig.5,6), wobei die Fläche A nachhinkt. Bis etwa 1962 bleibt die Gruppenmenge der Agropyro-Rumicion-Arten niedrig, sie wechselt in A stärker als in B. Dann steigt sie in beiden Teilen erheblich an, ohne aber wesentlich über die Hälfte des Anfangswertes zu kommen. Im letzten Jahrzehnt zeigt sie mäßige Schwankungen. Alle diese Unterschiede werden durch den Wechsel in der Dominanz von *Agrostis stolonifera*, *Ranunculus repens* und (schwächer) von *Agropyron repens* erzeugt.

Die Gruppenmenge der übrigen Molinio-Arrhenatheretea-Arten wechselt auch nicht unerheblich zwischen Höhepunkten zu Beginn und kurz vor dem Ende (1975/1976) der Beobachtungszeit (Fig.5,6). Im einzelnen unterscheiden sich die gleichen Jahre in beiden Flächen meist nicht sehr erheblich, wenn man von einigen Abweichungen absieht.

Diese Befunde, welche für die syntaxonomische Beurteilung der Syndynamik aufschlußreich sind, spielen für den Ertrag der Flä-

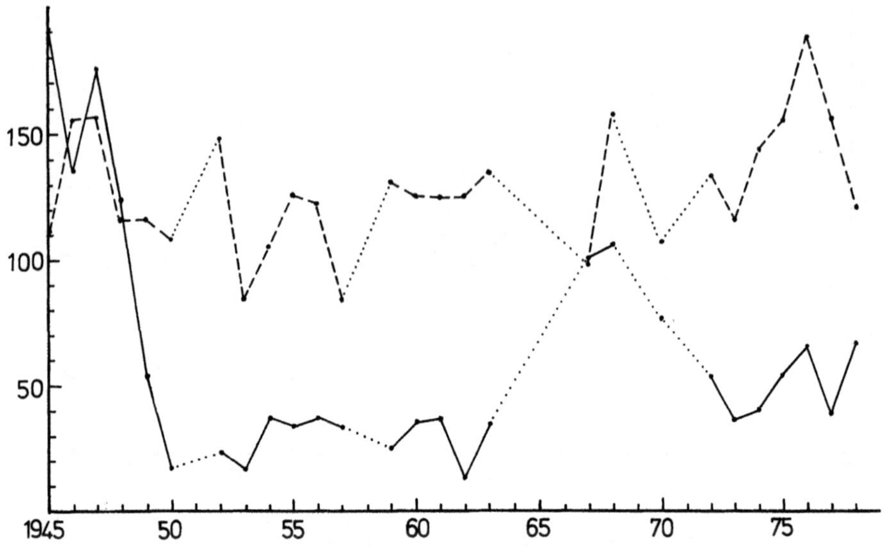

Fig.5. Fläche A: Gruppenmengen der Agropyro-Rumicion- (——) und der übrigen Molinio-Arrhenatheretea-Arten (----).

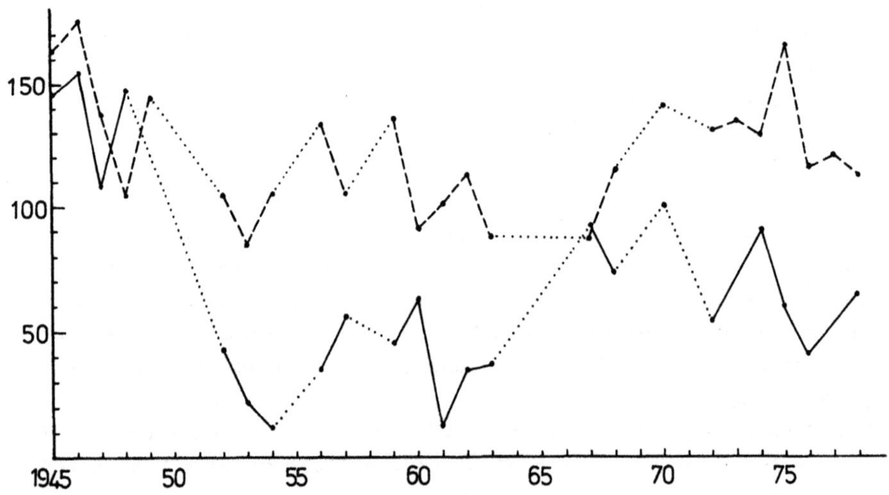

Fig.6. Fläche B: Gruppenmengen der Agropyro-Rumicion- (——) und der übrigen Molinio-Arrhenathareteа-Arten (----).

chen keine Rolle. Dieser hängt vom Futterwert der einzelnen Arten und deren Menge ab. Obwohl wir die Masse der Arten nicht prozentual an der Gesamtmasse geschätzt haben, wollen wir eine gewisse Vorstellung des "Futterwertes" der beiden Bestände und dessen jährlicher Veränderung zu gewinnen versuchen, indem wir die geschätzte mittlere Menge der einzelnen Zahlen mit den Wertzahlen von KLAPP et al. (1953) multiplizieren und die Summen aller Arten miteinander vergleichen (Fig.7,8).

Beide Flächen weichen in ihrem so errechneten "Futterwert" mit Ausnahme der Jahre 1967/1968 und 1976-1978 nur unbedeutend von einander ab. Beide zeigen eine starke Abnahme des "Futterwertes" in den ersten 10 Jahren, der dann wieder (besonders Fläche B) deutlich ansteigt. Die letzten 10 Jahre lassen jedoch einen erneuten Abfall erkennen. Offenbar hängt der Wert des Futters jetzt im wesentlichen vom Mengenverhältnis *Ranunculus repens* und *Agropyron repens* ab. *Poa trivialis* und *Alopecurus pratensis* beeinflussen ihn weniger, da sie ziemlich gleiche Mengen hervorbringen.

Verknüpft man das wohl vor allem durch chemische Mittel erzeugte Absinken der Artenzahl, d.h. der Diversität des Futters mit diesem Befund, so wird es wahrscheinlich, daß der wirkliche Nährwert der Bestände trotz des höheren Pflege-Aufwandes nach dem Ausfall der natürlichen Düngewirkung durch die Weser-Überflutungen nicht erhalten geblieben ist,oder genauer gesagt, nach einem vorübergehenden Tiefstand nicht wieder erreicht und gehalten werden konnte.

Das Schicksal unserer Probeflächen ist besonders im Hinblick auf die Verarmung an Arten bei gleichzeitiger Dominanz weniger Gräser und von *Ranunculus repens* wohl als ein bezeichnendes Beispiel für die Wandlungen nordwestdeutscher Grünlandgesellschaften in den letzten Jahrzehnten zu werten.

Nachdem das Ranunculo repentis-Agropyretum repentis Tx.1977 erkannt worden war, ist es nicht schwierig, unsere Bestände der Subass. von Alopecurus pratensis dieser Assoziation zuzuordnen (Tab.C). Die heutige Ausbildung (seit etwa 10 Jahren) stellt allerdings nur noch ein Fragment dieser Gesellschaft dar, in dem *Poa trivialis*, *Agropyron repens*, *Alopecurus pratensis* und *Ranunculus repens* abwechselnd dominieren.

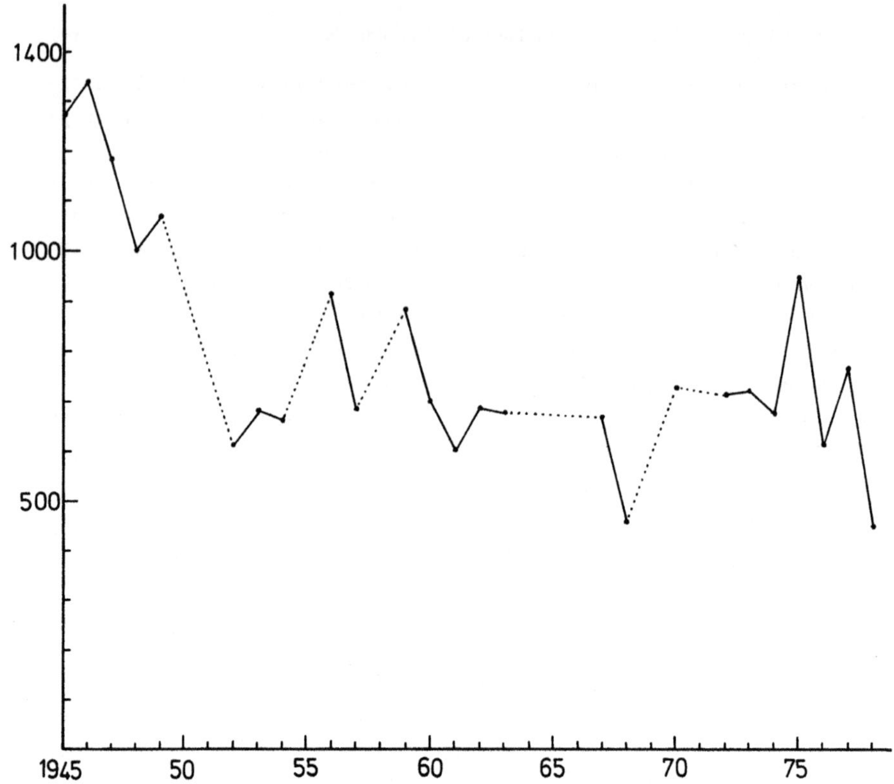

Fig.7. Fläche A: Schwankungen des "Futterwertes" von 1945-1978.

Die Fläche B zeigt durch die bleibende Ansiedlung von *Anthriscus sylvestris* eine leichte Annäherung an die Subass. von Anthriscus des Ranunculo-Agropyretum.

Die Stetigkeitsangaben der Tabelle C sind nicht gleichwertig: In A und B geben sie die Stetigkeiten der Arten je eines Bestandes über die Zeit von 1945-1978, in C die Stetigkeit aus 29 verschiedenen Beständen wieder. Abweichungen in den Stetigkeits-Klassen haben daher keine syntaxonmische Bedeutung, (z.B. *Cardamine pratensis, Rumex acetosa, Ranunculus ficaria*).
Die Dominanz von *Alopecurus pratensis* in unserer Gesellschaft lenkt die Aufmerksamkeit auf eine von manchen Autoren als Alopecuretum pratensis bezeichnete Gesellschaft. Dieser Name dürfte in erster Linie der Dominanz von *Alopecurus pratensis*, nicht aber der Artenkombination Rechnung tragen. Diese sagt aber synökologisch und syntaxonomisch weit mehr aus als die Dominanz. So kann *Alopecurus pratensis* sowohl im Arrhenatheretum, in Calthion-Gesellschaften und hier im Ranunculo repentis-Agropyretum repentis (Agropyro-Rumicion) zur Dominanz kommen. Das 'Alopecure-

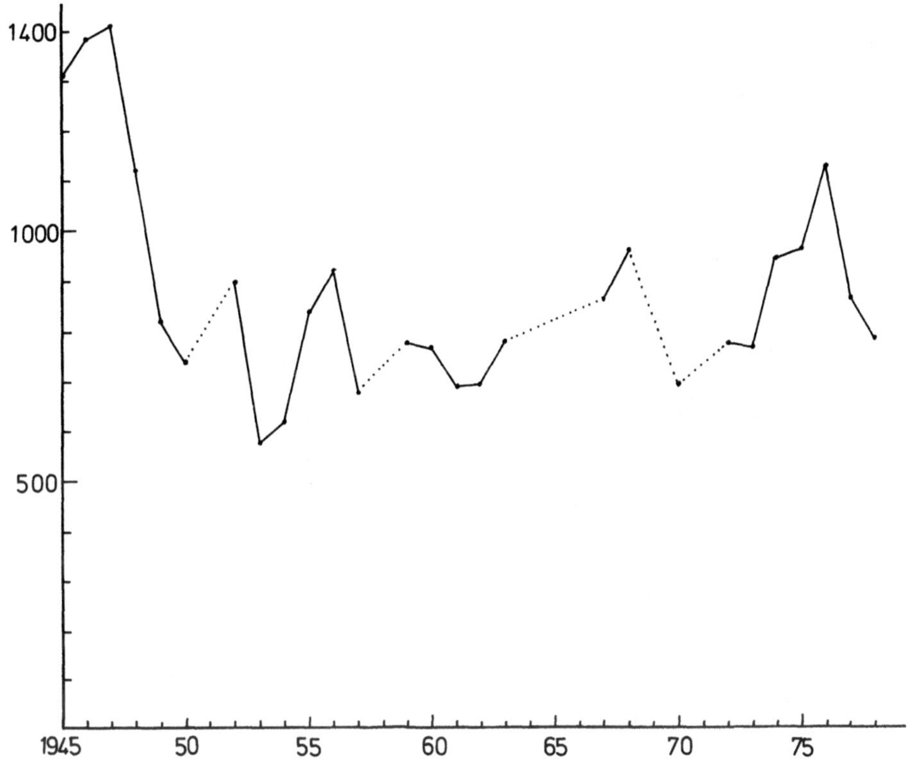

Fig.8. Fläche B: Schwankungen des "Futterwertes" von 1945-1978.

t u m p r a t e n s i s ' ist also keine auf einheitlicher Arten-Verbindung gegründete Gesellschaft und sollte nicht in unserem System verwendet werden.

Tab.C. Vergleich der Dauerflächen A.u.B. mit der Stetigkeitstabelle des Ranunculo repentis-Agropyretum repentis Tx.1977 (c).

	A	B	C
Kennart			
Agropyron repens	V^{+-4}	V^{+-4}	V^{+-5}
Trennarten gegen Rumici-Alopecuretum geniculati			
Taraxacum officinale	V^{+-2}	V^{+-2}	IV^{+-2}
Dactylis glomerata	II^{+-1}	I^{+}	$+^{+}$
Trennarten der Subass.			
Alopecurus pratensis	V^{1-4}	V^{1-4}	IV^{+-4}
Lysimachia nummularia	IV^{+-2}	IV^{+-3}	III^{+-2}
Leontodon autumnalis	III^{+-2}	III^{+-3}	III^{+-2}
Potentilla reptans	III^{+-2}	III^{+-1}	III^{+-3}
Poa pratensis	II^{+-1}	II^{+-1}	III^{+-3}
Verbands-u.Ordnungskennarten			
Ranunculus repens	V^{1-4}	V^{+-3}	V^{+-4}
Agrostis stolonifera	IV^{1-4}	IV^{+-4}	III^{+-3}
Lolium perenne	IV^{+-2}	II^{+-1}	II^{+-1}
Rumex crispus	II^{+-1}	III^{r-1}	IV^{r-2}
Plantago major	$+^{+}$	I^{+}	II^{i-+}
Alopecurus geniculatus	I^{+}	II^{+-1}	r^{+}
Veronica serpyllifolia	I^{+-1}	.	.
Festuca arundinacea	.	I^{+}	.
Klassenkennarten			
Poa trivialis	V^{1-4}	V^{2-4}	V^{+-4}
Rumex acetosa	V^{+-1}	V^{+-2}	I^{+-1}
Cardamine pratensis	IV^{+-2}	V^{+-2}	II^{+-1}
Trifolium repens	II^{r-3}	III^{+-3}	III^{+-2}
Festuca pratensis	II^{+-1}	III^{+-2}	I^{+-1}
Festuca rubra	II^{+-1}	II^{+-2}	I^{+-1}
Phleum pratense	II^{+-1}	II^{+-2}	$+^{+-1}$
Cerastium holosteoides	II^{+-1}	II^{r-+}	$+^{+}$
Plantago lanceolata	II^{+-1}	II^{+}	$+^{+}$
Daucus carota	II^{+}	I^{r-+}	I^{+}
Bellis perennis	II^{+-1}	I^{+}	I^{+-1}
Vicia cracca	I^{+-1}	I^{+}	I^{+-1}
Holcus lanatus	I^{+}	I^{+-1}	r^{+}
Anthriscus sylvestris	r^{r}	II^{+-1}	r^{1}
Achillea millefolium	I^{+}	r^{+}	I^{1-2}
Bromus mollis	II^{+-1}	I^{+-2}	.
Symphytum officinale	$+^{+-1}$	I^{r-+}	.
Galium mollugo	I^{+}	.	r^{+}
Crepis biennis	r^{+}	.	r^{+}
Trifolium pratense	.	r^{+}	$+^{+-2}$
Trifolium dubium	II^{+}	.	.
Chrysanthemum leucanthemum	I^{i-+}	.	.
Silaum silaus	.	II^{+-1}	.
Begl.			
Ranunculus ficaria	IV^{+-2}	V^{+-2}	r^{+}
Tripleurospermum inodorum	I^{i-+}	$+^{+}$	II^{+-1}
Poa annua	I^{+}	I^{+}	II^{+-2}
Polygonum aviculare agg.	r^{+}	r^{+}	II^{r-+}
Brachythecium rutabulum	II^{+-2}	II^{r-+}	.
Glechoma hederacea	I^{+-1}	II^{+}	.

u.a.

A. Tab.A, B. Tab.B., C. Ranunculo repentis-Agropyretum repentis, Subass. v. Alopecurus pratensis, 17 Aufn., vgl. Tüxen.R. 1977, p. 220.

ZUSAMMENFASSUNG

In der Zeit von 1945-1978 wurden (mit wenigen Ausnahmen) alljährlich genaue pflanzensoziologische Aufnahmen von zwei nebeneinander liegenden 12 und 16 m^2 großen Dauerflächen des Ranunculo repentis-Agropyretum repentis (Tab.C) in einer flachen Flutrinne des Wesertales bei Stolzenau gemacht (Tab.A,B). Die Flächen wurden zunächst fast alljährlich von der Weser oder von Schnee-Schmelzwasser meist am Ende des Winters überflutet und dabei oft (von Weser-Wasser) überschlickt. Sie wurden z.T. mit Jauche und Kompost und seit 1973 beide mit Thomasmehl und Kali, sowie mit Kupfer gedüngt. Sie wurden mehrfach im Jahr gemäht und im Herbst kurze Zeit beweidet.
Im Laufe der ersten beiden Jahrzehnte war die Artenzahl beider Bestände mehr als doppelt so hoch als im dritten Jahrzehnt, wo sie offenbar durch Unkraut-Ex erniedrigt wurde (Fig.1,2).
Die Dominanz der Gräser *Agrostis stolonifera, Agropyron repens, Poa trivialis* und *Alopecurus pratensis*, sowie von *Ranunculus repens* wechselte stark.
Artenzahlen und Gruppenmengen der Agropyro-Rumicion und der sonstigen Molinio-Arrhenatheretea-Arten zeigen gleichsinnige Fluktuationen und seit 10 Jahren einen starken Abfall (Fig.3-6). Der "Futterwert" fiel zunächst stark ab, um zuletzt wieder vorübergehend zu steigen (Fig.7,8). Bei der starken Arten-Verarmung dürfte die einseitige Zusammensetzung des Futters trotz hoher Masse dessen Bekömmlichkeit herabmindern. Der Wert des Futters schwankt jetzt vor allem mit der wechselnden Dominanz von *Agropyron repens* und *Ranunculus repens*.

SUMMARY

Nearly every year from 1945 to 1978 phytosociological relevés have been made in two permanent plots (12 resp. 16 m^2) of the Ranunculo repentis-Agropyretum repentis (table C); they are situated in a shallow relief river bed in the Weser valley near Stolzenau (table A,B). The areas were flooded by the Weser river or by water of snow-melting in late winter and were thereby covered with mud. This took place nearly every year till 1970, thereafter only rarely. Both areas received liquid manure ("Jauche") and compost and were manured since 1973 with phosphate (Thomasmehl), potassium and copper salts. They were mown several times a year and grazed for a

short time in autumn.
During the first twenty years the number of species in both stands was more than double the number as in the third decade when a decrease was probably due to the use of herbicides (Unkraut-Ex) (fig.1,2).
The dominance of the grasses *Agrostis stolonifera, Agropyron repens, Poa trivialis* and *Alopecurus pratensis* as well as *Ranunculus repens* alternated considerably.
Numbers and "Gruppenwert" of the species of the Agropyro-Rumicion- and Molinio-Arrhenatheretea show the same fluctuations and a heavy decline since 10 years (fig.3-6). The value of the forage crop ("Futterwert") at first decreased strongly and after a temporary increase went down again (fig.7,8). Due to the heavy diminishment of species and the consequent one-sided composition of the forage crop, the over-all nutritional value was decreased. The value of the forage crop varies especially with the changing dominance of *Agropyron repens* and *Ranunculus repens*.

LITERATUR

Bundesanstalt für Vegetationskartierung Stolzenau (1961) - Vegetationskarten deutscher Flußtäler. Mittlere Weser bei Stolzenau. 1:10 000.- Stolzenau/Weser.

KLAPP,E.et al. -1953- Wertzahlen der Grünlandpflanzen.- Das Grünland 2 (5): 38-40. Hannover.

TÜXEN,R. -1950- Grundriß einer Systematik der nitrophilen Unkrautgesellschaften in der Eurosibirischen Region Europas.- Mitt.Flor.-soz. Arbeitsgem.N.F.2: 94-175. Stolzenau/Weser.

-- -- -1972- Kritische Bemerkungen zur Interpretation pflanzensoziologischer Tabellen.- In: Tüxen,R. (Ed.): Grundfragen und Methoden in der Pflanzensoziologie.- Ber.Intern.Sympos.Rinteln/Weser 1970.

-- -- -1977- Das Ranunculo repentis-Agropyretum repentis, eine neu entstandene Flutrasen-Gesellschaft an der Weser und anderen Flüssen.- Mitt.Flor.-soz.Arbeitsgem.N.F.19/20: 219-224. Todenmann-Göttingen.

-- -- & ELLENBERG,H. -1937- Der systematische und ökologische Gruppenwert.- Mitt.Flor.-soz.Arbeitsgem.3: 171-184. Hannover.

DISKUSSION

J.J.MOORE

Man kann mit diesen Diagrammen anscheinend sehr genaue Zahlen bekommen. Man muß aber damit rechnen, daß bei Gelände-Arbeiten immer Fehler gemacht werden. Und deshalb ist es nur Schein,wenn man von dem quantitativen Vorgehen so exakte Zahlen bekommt.

D.MUELLER-DOMBOIS

Zu Herrn Dr.MOORE's Bemerkung möchte ich sagen, daß die quantitativen Methoden auch ihre Berechtigung haben. Es kommt immer auf den Zweck an, über den man etwas aussagen möchte, und welche Methoden man anwendet. Deshalb braucht gar kein Widerstreit zwischen qualitativen und quantitativen Methoden zu sein. Sie verfolgen verschiedene Ziele.

Zu Herrn TÜXEN's Referat möchte ich fragen, ob aus der Interpretation des fluktuativen Verhaltens der Arten zu Überschwemmungen auch eine direktive Veränderung herausgelesen werden kann?

R.TÜXEN

Wenn in normalem Zustande die Weser ein Jahr überschwemmte, die nächsten 2 Jahre nicht, dann wieder, paßte sich die Vegetation in überfluteten Rinnen und ebenso am Ufer an die Zahl, die Zeit und die Dauer der Überschwemmungen an. Das geschieht zwar nicht in der Weise, daß sich die Grenzen der Gesellschaft wesentlich verschieben,sondern indem sich die soziologischen, oder wenn Sie wollen, die ökologischen Artengruppen in Zahl und Menge der Arten ändern, wie die Kurven zeigen. Als ich noch glaubte, dies sei eine Sukzession, habe ich das eine Harmonika-Sukzession genannt. Genau genommen sind diese Änderungen in der Tat aber eher Fluktuationen, die durch reversible Einflüsse verursacht werden.

Eine Frage von Prof. HÜBL nach der syntaxonomischen Einstufung der untersuchten Bestände konnte damals nicht beantwortet werden. Sie ist inzwischen geklärt (vgl.p.

J.J.BARKMAN

Es freut mich sehr, daß hier die Frage der Fluktuationen erörtert wurde, ohne die man die Frage der Sukzession nicht verstehen kann.

Drei kurze Bemerkungen:

1. Ist es möglich, daß die Fluktuationen in der Dichte der Arten nicht nur eine Folge jährlicher Unterschiede in der Überflutungs-

dauer ist, sondern auch eine Folge jährlicher Unterschiede in der Winter- und Vorfrühlingswärme? Die Aufnahmen wurden ja sehr früh gemacht (März und April), so daß nach einem kalten Winter einige Arten vielleicht nur noch unterirdisch leben? Es würde dann bei den Beobachtungen ein Misch-Resultat von Fluktuationen und Periodizität vorliegen.

2. Inwieweit hat die seit vielen Jahren jährlich durchgeführte Untersuchung derselben Probefläche durch den Tritt vieler Pflanzenzoziologen Einfluß auf die Artenzusammensetzung und Vegetationsstruktur ausgeübt?

Und meine 3. brennende Frage: Warum finden Sie, daß dies eine Fluktuation ist? Ist sie das, weil sie reversibel ist, oder ist sie das, weil die Gesellschaft in allen ihren Formen die Jahre hindurch, nach Ihrer Ansicht, immer zur gleichen Agropyro-Rumicion-Assoziation gehört, wenn diese auch noch nicht feststeht? Und würden Sie, wenn es eine Fluktuation sein würde, den Wechsel von einer reinen Alopecurus geniculatus-Ranunculus repens-Assoziation zu einem reinen Arrhenatheretum und wieder zurück, dann auch noch eine Sukzession nennen? Wo liegt das Kriterium?

R.TÜXEN

Die Veränderungen in der Arten-Verbindung und ihren Mengen mögen durchaus neben der Überflutungs-Zeit und -Dauer auch durch Wärme-Unterschiede verschiedener Jahre bedingt sein. Manchmal lag lange Eis auf der Fläche, in anderen Jahren war sie gar nicht vereist, wohl unter Wasser oder auch ganz trocken.

Tritt-Zeiger haben wir auf den Flächen nicht bemerkt, so stark war unser Einfluß keineswegs, daß wir solche Arten begünstigt hätten. Wohl sind einige Einjährige wie *Stellaria media*, *Capsella bursa-pastoris*, *Atriplex hastata*, *Tripleurospermum inodorum* u.a. vorübergehend aufgetreten, besonders dann, wenn die Flächen stark überschlickt oder mit Kompost gedüngt worden waren. Das vereinzelte Vorkommen von *Plantago major* ist in Agropyro-Rumicion-Gesellschaften normal und nicht auf Tritt zurückzuführen.

Der Übergang von Fluktuation zur echten einseitigen Sukzession ist in unserem Falle weder vorauszusagen, noch leicht zu begrenzen. Vielleicht ist es wenig fruchtbar, darüber lange zu streiten. Die Tatsache, daß sich die Artenverbindung und ihre Mengen-Verhältnisse verändern, gilt zunächst vor allem festgehalten zu werden. Wie man das Geschehen nennt, ist vielleicht Nebensache.

J.J.BARKMAN

Ja, doch nicht ganz; denn wenn man nämlich sagt, es ist reversibel und durch Überschwemmungsschwankungen entstanden, dann würde man sich auf den Standpunkt stellen, es sind immer Schwankungen oder Fluktuationen. Das hat auch praktische Bedeutung für die Synsystematik. Dann kann man auch nicht die Gesellschaften der Äcker zu verschiedener Jahreszeit auf dem gleichen Acker als verschiedene Assoziationen betrachten; man müßte sie dann konsequent als Fluktuation innerhalb des Jahres betrachten. Die Frage ist nicht nur akademisch.

R.TÜXEN

Acker-Assoziationen der Sommer- und Winterfrüchte kommen in Gebieten, wo sie nicht durch Rotation der Wirtschaft immer wieder durcheinander gemengt werden, sauber getrennt vor. In Weinbau-Gebieten, wo nie Winterfrucht gebaut wird, gibt es keine Aperetalia; in reinen Ackerbau-Gebieten, wo z.B. nur Weizen kultiviert wird, findet man keine Chenopodietalia albi. Erst auf der Stufe der Klasse fassen wir als Stellarietalia mediae beide Ordnungen zusammen. In unserem Falle handelt es sich hier um das, was WESTHOFF & VAN LEEUWEN eine zeitlich instabile Grenze nennen.

J.J.BARKMAN

Wenn diese Fluktuation aus einem reinen Agropyro-Rumicion in einem Jahr in ein reines Arrhenatheretum übergehen würde, dann würde man das doch nicht als eine Fluktuation betrachten können?

R.TÜXEN

Nein, das glaube ich auch nicht. Ein ähnlicher Fall wäre ein Kahlschlag, wo sich das Atropetum belladonnae einstellt, bis nach einiger Zeit das Fagetum wiederkommt. Das ist eine sekundäre progressive Sukzession. Aber soweit geht die Veränderung hier lange nicht. Unsere Gesellschaft bleibt Agropyro-Rumicion bis eine entscheidende irreversible Standortsänderung eine ebenso starke Wandlung in ihrem Arten-Gefüge hervorrufen würde, das nun erneut um einen mittleren Zustand schwanken würde.

D.MUELLER-DOMBOIS

Begriffe, die einmal aufgestellt sind, sollte man nicht als Ersatz für eine Beschreibung ansehen. Denn eine Beschreibung macht so etwas erst ganz klar. Ich finde auch der Ausdruck Sukzession muß einmal nach dem Maßstab der geographischen Größe beschrieben werden, bzw.danach, was da vor sich geht, sowie auch nach dem zeitlichen Ablauf. Zum Beispiel kann man so etwas meiner Ansicht nach ruhig als eine kurzfristige Sukzession in sehr kleinem Maßstab bezeichnen. Man muß ja nicht immer bei Sukzession nur in Vegetationstypen denken.

D.J.BELLAMY

I was studying swamps in India in the monsoon, and there were two types of basin in the ground nearby. One has steep sides and the other is like you described. In the dry season there are no plants. The monsoon drown begins and the one basin fills very quickly. The other basin fills very slowly. At the height of the rains, when the water is deepest, we have a zonation with perennial plants, rooted plants and swimming plants. This is a zonation but not a succession because next year the water then disappears and all the plant material has disappeared. How would you go about to describe and classify this vegetation?

R.TÜXEN

I think the classification of plant communities is in first line not an ecological problem. I should make a number of relevés and then make a table and so I should find in the one basin some different combinations of species, that means different plant communities, and in the other basin, we should probably also find a mosaic. Only if we find the same communities in the other basin and elsewhere too, we can call them vegetation types.

J.SCHMITHÜSEN

Ich möchte zu der Frage von Herrn BARKMAN einerseits und zu diesem letzten Beispiel andererseits ein weiteres hinzufügen, das diese Kompliziertheit der Anwendung von Begriffen noch einmal zeigt, aber auch in demselben Sinne beantworten, daß es im Grunde nicht darauf ankommt, wie man das nennt. Ein ausgezeichnetes, rein natürliches Beispiel ist die sogenannte "Blühende Wüste". Ich hatte 1952 das große Glück, dieses Phänomen in Nordchile zu erleben. In Gebieten, wo es oft jahrzehntelang nicht regnet und absolut keine Pflanze an der Oberfläche sichtbar ist, dort war für im ganzen etwa 8-9 Wochen lang eine erstaunlich üppige Vegetation entstanden. Ich habe rund 100 Arten gesammelt. Man mußte aber ungefähr jede Woche dort hingehen, weil nämlich jede Woche etwas anderes aufblühte. In der ersten Woche kamen ganz winzige annuelle Pflänzchen, die kaum 1 cm groß waren, und die aus dem Samen ihre Entfaltungskraft schöpften. Nach 2 oder 3 Wochen waren hauptsächlich Pflanzen da, die zunächst das Wasser in sukkulenten Blättern speicherten und sich dann mehr Zeit bis zur Samenreife nahmen, die also etliche Wochen zur Vollendung ihres Lebens brauchten. Am Schluß, nach etwa 8 oder 9 Wochen, war ein ganz anderer Artenbestand vorhanden als etwa in den ersten Wochen. Alle 10 bis 30 Jahre kommt so etwas vor, d.h. die Fläche ist durch 10, 20 oder 30 Jahre pflanzenleer, und dann ist sie in einem Jahr, innerhalb einer ganz kurzen Zeit, durch nur wenige Wochen dauernde, unterschiedliche Pflanzenbestände besetzt. Man erfaßt also gar nicht den ganzen Pflanzenbestand mit einer Aufnahme, wenn man das Ganze eine Gesellschaft nennen wollte. Am Anfang ermittelt man nur einen Bruchteil, und am Ende erhält man einen anderen Bruchteil und in der Mitte wieder etwas anderes. Die erste Frage wäre also: ist das Ganze eine Gesellschaft? Zweite Frage: ist diese Abfolge in den 8 Wochen eine Sukzession? Das Phänomen kann man sehr schön beschreiben, es ist ein ausgesprochener Grenzfall. Ich pflege meinen Studenten zu sagen, mit Grenzfällen kann man alles beweisen.

H.ELLENBERG

Ich freue mich, daß Herr SCHMITHÜSEN diese Frage angeschnitten hat. Ich hatte auch das Glück, dieses Phänomen in Gegenden zu sehen, in denen nicht so viel Regen gefallen war, und hier die gleichen Standorte zu studieren. Man findet pro Quadratmeter hunderte von Samen, die nachweislich - im Experiment kann man das zeigen - lebendig sind. Deshalb möchte ich behaupten, hier ist nicht plötzlich eine Gesellschaft neu entstanden, sondern die Gesellschaft ist immer da. Sie hat nur längere Ruhezeiten, so wie auch eine Ackerunkraut-Gesellschaft auf dem gepflügten Acker nicht sichtbar ist. So ist das auch hier, nur daß die Bedingungen eben etwas wechselnder sind als beim Pflügen. Und wenn wir etwa einen Frühlingswald ansehen, wo jetzt etwa *Corydalis* blüht, die man nachher oberflächlich nicht mehr findet, so hat niemand Zweifel daran, daß diese *Corydalis* zu derselben Gesellschaft gehört, wie meinetwegen *Elymus europaeus* oder andere Arten, die später da sind. Ich würde also das Ganze als Aspekte, jahreszeitliche oder in dem Falle in der Entwicklungsperiode liegende Aspekte ein und derselben ephemeren Gesellschaft ansehen. Da ist also kein grundsätzlicher, sondern nur ein größenordnungsmäßiger Unterschied gegenüber dem, was wir von anderen Gesellschaften kennen. Das ist also auch eine Assoziation. Wahrscheinlich sind es so gar mehrere Assoziationen, jedoch nicht an derselben Stelle, sondern nach meinen Beobachtungen gibt es große, kaum bemerkbare Dellen im Gelände, wo das Wasser zusammenläuft und dadurch längere Zeit zur Verfügung steht. Dort gibt es eine andere Gesellschaft als an anderen Orten, wo das Wasser leichter abläuft. Es gibt also eine Fülle von Gesellschaften in der "Blühenden Wüste".

J.SCHMITHÜSEN
Manche von diesen Gesellschaften enthalten nur Dauerpflanzen und in anderen leben nur Annuellen. Insofern ist es keine Schwierigkeit, wenn man das auf die Spitze treibt in der Auffassung, daß man das als eine durchgehende Gesellschaft durch das dauernde Vorhandensein der Samen auffaßt. Die Samen kommen in den Gebieten, die ich meine, alljährlich mit dem Südwind. Wir wissen aber nicht, wie lange sie keimfähig bleiben, so daß eine ständige Samenverbreitung dort nicht unbedingt durch die ganze Zeit gegeben ist.

H.ELLENBERG
In Peru ist es ausgeschlossen, daß sie alljährlich kommen; sie sind ständig am Ort, und man kann sie zu hunderten pro Quadratmeter auszählen. Ich möchte das als den normaleren Fall ansehen, als das sie ständig wieder einwandern.
Darf ich noch eine Frage zum Referat stellen? Warum sind die Aufnahmen so früh im Jahr gemacht worden und nicht später? Sind nicht durch diese früh liegenden Aufnahmen die Unterschiede überbetont worden, die vielleicht im Laufe der späteren Jahreszeit irgendwie anders gestaltet wären? Gibt es für die gleichen Flächen Aufnahmen aus dem August?

R.TÜXEN
In den Tabellen A und B sind zwei August- mit zahlreichen März-, April- und Mai-Aufnahmen vereinigt. Für unsere Betrachtung ergeben sich aus der zeitlichen Lage der Aufnahmen keine merklichen Abweichungen.

In der zweiten August-Aufnahme (1947/(3) kommen als einmal in der Tabelle auftretende Arten: (+) *Polygonum aviculare*, (+) *Sisymbrium officinale* und (1 St.) *Sonchus asper* vor. In diesem Jahr war die Fläche mit Kompost gedüngt worden! In den Mengen weicht keine Art in diesen Aufnahmen merklich von den Frühlings-Aufnahmen ab. Der Grund für den jahreszeitlich frühen Zeitpunkt der Aufnahmen war das Bedürfnis, so früh wie möglich wieder enge Fühlung mit sterilen Gräsern zu bekommen. Wenn *Arrhenatherum* blüht, sind alle seine Begleitarten leicht zu erkennen. Im frühen Frühling muß man sich aber stets von neuem in die noch sterilen eben austreibenden Gräser einsehen, und sich deren Merkmale wieder neu einprägen.

H.ELLENBERG

Ist dann wirklich Aufnahme mit Aufnahme vergleichbar? Das, was in den Kurven dargestellt wurde, ist zum Teil das Ergebnis jahreszeitlicher Aspekte.

R.TÜXEN

Das könnte bestenfalls für Aufn. 1 und 3 zutreffen, welche die einzigen August-Aufnahmen sind. Die anderen liegen zwischen (März),April und Mai nur wenige Wochen auseinander; phänologisch gesprochen liegen sie - die Jahre sind ja nicht ganz gleich - in der gleichen Phase und sind daher sicher vergleichbar. Endlich zeigt das Schicksal der Bestände über einen Zeitraum von mehr als drei Jahrzehnten, daß nicht phänologische, sondern ökologische Ursachen dafür bestimmend waren.

ZUR DYNAMIK DER ARTMÄCHTIGKEIT INNERHALB SÜDMÄHRISCHER CNIDION VENOSI-AUENWIESEN

Emilie B a l á t o v á - T u l á č k o v á

In diesem Referat möchte ich auf Änderungen innerhalb der Assoziationen der alljährlich überschwemmten Wiesengesellschaften des C n i d i o n - Verbandes hinweisen, welche durch die sich von Jahr zu Jahr wechselnden Standortsgesellschaften hervorgerufen werden. Das Faktum, daß die Pflanzenzusammensetzung die Dynamik der direkten ökologischen Faktoren widerspiegelt, kommt hier vornehmlich im Wasserfaktor zum Ausdruck. Es spielen dabei die Wasserverhältnisse am Anfang der Vegetationsentwicklung, besonders die Tiefe und Dauer der Überschwemmung, eine entscheidende Rolle.

Was die Struktur unserer C n i d i o n - Gesellschaften aus der Umgebung von Lanžhot betrifft, beteiligen sich an deren Zusammensetzung konstant jene Arten, die nicht nur starke Austrocknung des Bodenprofils in der Trockenperiode ertragen, sondern auch fähig sind, langandauernde Überschwemmungen zu überleben. Diejenigen Arten, die den zeitweise vorkommenden Katastrophen-Überschwemmungen nicht angepaßt sind, kommen nur in den trockeneren Ausbildungsformen der Gesellschaften vor. Von diesen sind zu nennen: *Festuca pratensis*, *Poa pratensis*, *Lathyrus pratensis*, *Galium boreale* und *Silaum silaus*. Durchgehend sind in den von uns untersuchten C n i d i o n - Gesellschaften folgende Artengruppen vertreten:

a) Von den M o l i n i o - A r r h e n a t h e r e t e a - (bzw. A r r h e n a t h e r e t a l i a -) Arten: *Alopecurus pratensis*, *Cardamine pratensis*, *Taraxacum officinale*, weniger *Ranunculus acer* und *Prunella vulgaris*.

b) Von den M o l i n i e t a l i a - Arten: *Serratula tinctoria*, *Symphytum officinale*, *(Ranunculus auricomus)*, *Sanguisorba officinalis*, *Senecio erraticus* ssp. *barbaraeifolius*, *Lythrum salicaria* und *Filipendula ulmaria*.

c) Von den C n i d i o n - Arten: *Gratiola officinalis*, *Cnidium venosum*, *Scutellaria hastifolia*, *Lythrum virgatum*, weniger die Assoziationsdifferentialart *Carex praecox-suzae*.

d) Von den übergreifenden A g r o p y r o - R u m i c i o n - Arten: *Lysimachia nummularia*, *Ranunculus repens*, *Rumex crispus*, *Potentilla reptans*, *Agrostis alba* und *Trifolium hybridum*.

e) Von den übergreifenden C a r i c i o n g r a c i l i s - Arten: *Poa palustris, Carex vulpina, Eleocharis palustris, Galium palustre* und *Phalaris arundinacea.*

Von anderen Begleitern spielen nur *Mentha arvensis, Stellaria palustris* und *Polygonum amphibium* eine wichtigere Rolle.

Auch vom Standpunkt der vollen Entwicklungszeit der generativen Sprosse gibt es hier mehrere Gruppen, wobei die meisten C n i d i o n v e n o s i - als auch einige A g r o p y r o - R u m i c i o n - Arten erst in der Zeit vor dem zweiten Schnitt (in den Sommermonaten) zu blühen beginnen. Dasselbe gilt auch für einige M o l i n i e t a l i a - Arten wie *Serratula tinctoria, Sanguisorba officinalis, Lythrium salicaria.* Die meisten Arten mit breiterer ökologischer Amplitude bilden Knospen und Blüten in den Frühjahrsmonaten (z.B. *Alopecurus pratensis*). Am Anfang der Vegetationsentwicklung können von diesen solche eine gewisse Rolle spielen,die eine raschere Entwicklung der vegetativen Sprosse vollziehen. Werden diese durch für sie günstige Verhältnisse gefördert (z.B. durch Überschwemmung von bestimmter Tiefe und Dauer), dann können sie in solchem Maß herrschen, daß die sich später entwickelnden Arten nicht mehr zum Optimum gelangen können (geschwächte Bildung der Assimilationsorgane infolge der Beschattung, Wurzelkonkurrenz usw.).

Nicht weniger aufschlußreich ist die Art und Weise, auf welche die im C n i d i o n v e n o s i vorkommenden Pflanzen fähig sind, die langdauernden Überschwemmungen zu überleben. Von diesem Blickpunkt aus kann man in den C n i d i o n - Wiesen folgende Artengruppen unterscheiden:

1) Arten, die von den langdauernden Überschwemmungen nur die seichten vertragen. Bei diesen Arten bleiben oft nur die höheren jungen Stengelblätter erhalten, die übrigen verfaulen - z.B. bei *Galium boreale* und *Silaum silaus.* Die meisten Arten der Gruppe 2) dagegen überleben eine seichte Überschwemmung ohne Beschädigung.

2) Arten, die nicht nur den zeitweise vorkommenden tieferen und langdauernden Überschwemmungen, sondern auch deren eventueller Strömung angepaßt sind. Die in diese Gruppe gehörenden Pflanzen können in mehrere Untergruppen geordnet werden. Man kann unterscheiden

   a) Arten, die ihre Assimilationsfähigkeit auch unter Wasser behalten können, z.B. *Lysimachia nummularia, Cardamine pra-*

tensis, Ranunculus flammula, Eleocharis palustris.

b) Arten, deren oberirdische, assimilationsfähige Organe nach gewisser Zeit den Fäulnisprozessen unterliegen und bei denen es nach Rückgang der Überschwemmungen zur raschen Regeneration der Pflanze kommt, z.B. *Alopecurus pratensis, Potentilla reptans,*

c) Arten, die den Überschwemmungen durch die Fähigkeit der Sproßverlängerung angepaßt sind: die Pflanze folgt dem steigenden Wasserspiegel in der Richtung des strömenden Überschwemmungswassers, wobei die Assimilationsfähigkeit der jüngeren Blätter erhalten bleibt (die älteren sterben ab, (Abb.1). Solche Arten, deren Stengelgewebe reich an Sklerenchym ist, bleiben nach dem Sinken des Überschwemmungswassers aufrecht, die anderen legen sich um. Die ersten werden z.B.durch *Cnidium venosum, Serratula tinctoria,* die anderen durch *Carex disticha* und *Caltha palustris* vertreten. Manche Arten mit sklerenchymreichen Stengeln können unter Umständen auch der folgenden Untergruppe angehörende Formen bilden wie z.B. *Lythrum*-Arten und *Phalaris arundinacea.*

d) Wie oben (sub c), die umgelegten Arten aber treiben nach dem Sinken des Überschwemmungswassers am alten Stengel sowohl neue adventive Wurzeln als auch Nebensprosse aus.Die Hauptachse stirbt später ab, die neugebildeten Sprosse dagegen schlagen neue Wurzeln und werden zu neuen Pflanzen (vegetative Vermehrung). Hierher gehören z.B. *Gratiola officinalis, Mentha arvensis, Poa palustris, Galium palustre, Rorippa silvestris,* ferner auch *Lythrum salicaria, Lythrum virgatum* und andere Arten mit Sklerenchym-reichen Achsen, falls sie dem Einfluß des stärker strömenden Überschwemmungswassers ausgesetzt sind (Abb.2 und 3).

Wie die hier angeführten Momente in den Änderungen der Dominanz der pflanzensoziologisch wichtigen Gruppen und Arten innerhalb der untersuchten Cnidion venosi- Wiesen zum Ausdruck kommen können, wird weiter an einigen Beispielen illustriert. Das Material stammt aus den Jahren 1961-1966, wo es zu Verschiebungen in der Artmächtigkeit einer ganzen Reihe von Arten nicht nur von Jahr zu Jahr, sondern auch vom ersten zum zweiten Schnitt kam. Wie schon angeführt, kommen hier enge Koinzidenzbeziehungen mit der Dynamik des Wasserregimes, besonders mit der Unregelmäßigkeit der Zeit, Dauer und Tiefe der Überschwemmungen zum Aus-

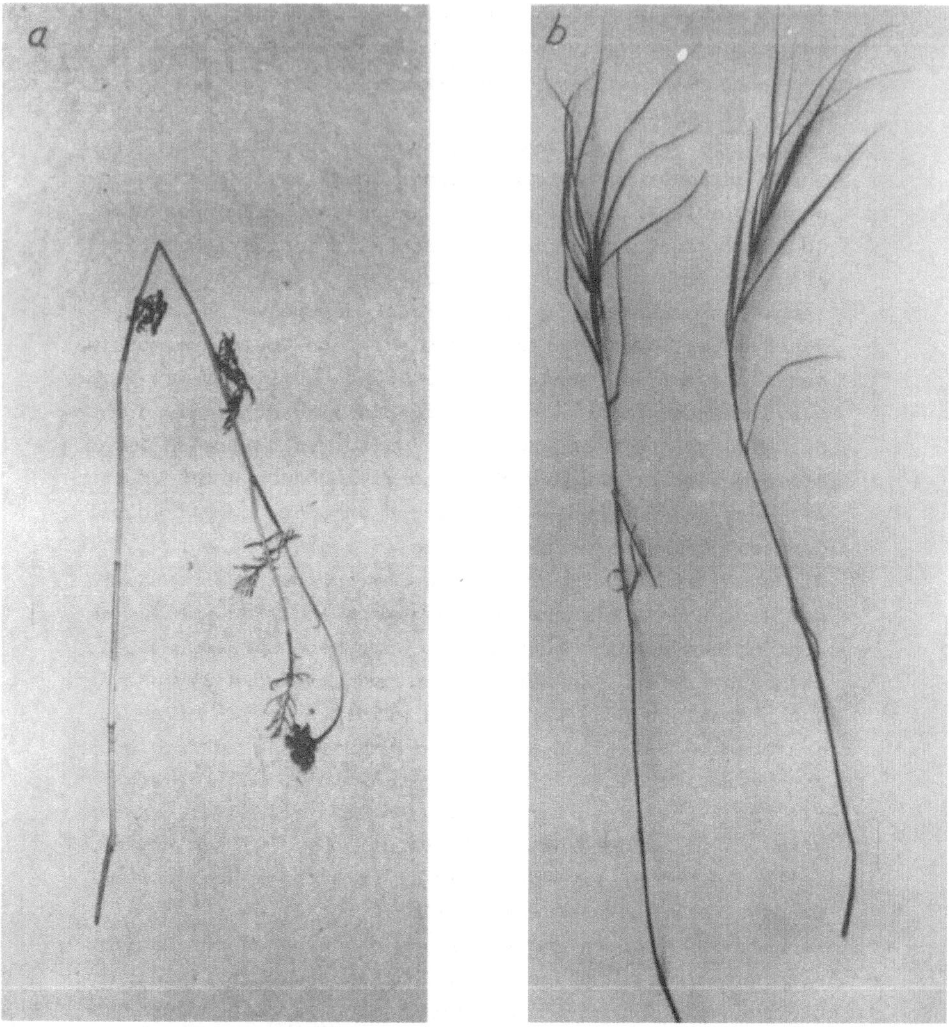

Abb.1. *Cnidium venosum* (a) und *Carex disticha* (b) gehören zu den Arten, welche Überschwemmungen durch die Sproßverlängerung und gleichzeitige Bildung neuer Blätter angepaßt sind.

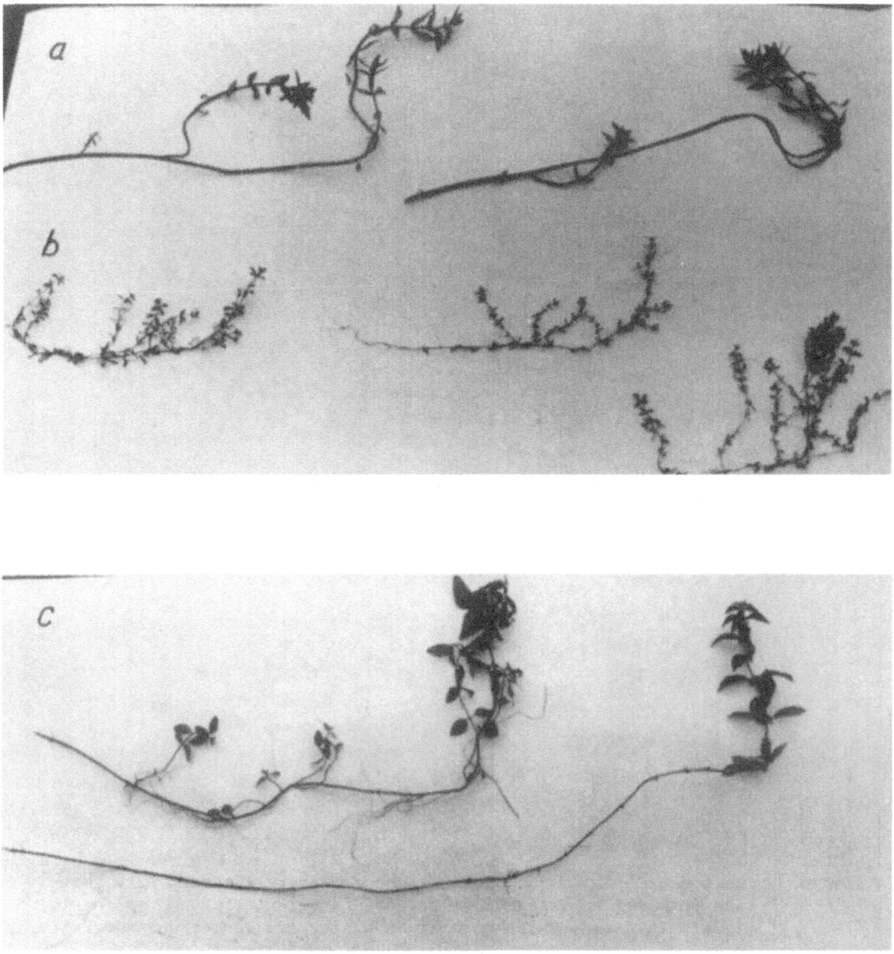

Abb.2. Gruppe der den Überschwemmungen durch den Verlängerungsprozeß angepaßten Pflanzen, an deren schlaffen Stengeln adventive Wurzeln und neue Nebensprosse nach dem Überschwemmungsabfall austreiben: a) *Gratiola officinalis*, b) *Galium palustre*, c) *Mentha arvensis*.

Abb. 3. Falls die Strömung des Überschwemmungswassers zu stark ist, legt sich auch der an Sklerenchym reiche Stengel von *Lythrum salicaria* nieder. In diesem Falle erfolgt auch hier die Vermehrung vegetativ.

druck. Infolgedessen geraten auch die bestandsbildenden Arten in verschiedene gegenseitige Konkurrenzbeziehungen, was auch zu einer Änderung des soziologischen Wertes der betreffenden soziologischen Einheit führen kann. Bei den meisten unserer C n i d i o n v e n o s i - Wiesen, mit deren Studium in den trockenen Jahren 1961/1962 begonnen wurde, kam es zu bestimmten Verschiebungen nur im Rahmen der Assoziationen (siehe unten). Eine Ausnahme bildeten jene Ausbildungsformen der C n i d i o n v e n o s i - Assoziationen, welche an der Grenze des C n i d i o n und A g r o p y r o - R u m i c i o n standen und noch der G r a t i o l a o f f i c i n a l i s - C a r e x p r a e c o x - s u z a e - A s s o z i a t i o n , S u b a s s . von R o r i p p a s i l v e s t r i s zugerechnet werden. Hier kam es in den an Überschwemmungen reichen Jahren 1965 und 1966 zu einer wesentlicheren Ausbreitung der A g r o p y r o - R u m i c i o n - oder C a r i c i o n g r a c i l i s - Arten, so daß man jetzt kaum von einer M o l i n i e t a l i a-Wiese sprechen kann. Als Beispiel kann die Untersuchungsfläche 27 dienen (Tab.1).

Nicht weniger lehrreich war die Untersuchungsfläche 19, wo 1961 *Festuca pratensis* dominierte. In den Jahren 1962 und 1963 wurde an derselben Stelle - durch Frühjahrsüberschwemmungen gefördert - *Alopecurus pratensis* bestandsbildend, in den nachfolgenden abnormal nassen Jahren 1965 und 1966 dagegen überwog *Phalaris arundinacea* (Katastrophen-Überschwemmungen!). Diese Tatsache ist auch aus Tabelle 1 ersichtlich.

In C n i d i o n - Wiesen, die am Anfang der Untersuchungen typisch entwickelt waren (größerer Antel der C n i d i o n - Charakterarten), kam es zwar auch zu mehr oder weniger größeren Verschiebungen zwischen den bestimmenden Arten, die Einheit blieb aber wenigstens im Rahmen der Assoziation. Als Beispiel sind zwei verschiedene C n i d i o n v e n o s i - Wiesen gewählt: die eine trockener, im Kontakt zum M o l i n i o n , die andere feucht-naß mit Beziehungen zu den eutrophen Großseggen-Gesellschaften. Die Ergebnisse der Untersuchungen sind graphisch dargestellt. In den Abbildungen, welche die jährlichen Änderungen in der Artmächtigkeit (A+D) der Pflanzen im Zeitabschnitt 1963-1966 betreffen, werden die Verhältnisse innerhalb der soziologisch wichtigen Artengruppen unter Benutzung der Deckungskoeffizienten berücksichtigt. Aus den Grundwasserganglinien ist ersichtlich, wie weit das Wasserregime der Untersuchungsflächen in den

Tab.1 Änderung der Artmächtigkeit der pflanzensoziologisch wichtigen Gruppen innerhalb der zwei untypisch entwickelten Ausbildungsformen der G r a t i o l a o f f i c i n a l i s - C a r e x p r a e c o x - s u z a e - Assoziation.

Untersuchungsfläche	Charakterarten von	Mai, bei 19.August 1962		Mai 1964		Mai 1966	
		Artenzahl	Artmächtigkeit1)	Artenzahl	Artmächtigkeit1)	Artenzahl	Artmächtigkeit 1)
27. (Subass.von Rorippa silvestris)	Cnidion venosi 2)	2	0,6	2	0,6	1	2
	Molinietalia	5	1,8	4	3,0	4	5,2
	Molinio-Arrhenatheretea	3	6,4	4	26,6	4	1,1
	Agropyro-Rumicion	6	39,1	6	33,0	8	35,8
	Caricion gracilis	6	8,4	6	33,9	8	35,5
19.	Cnidion venosi 2)	2	2,2	2	6,2	2	0,4
	Molinietalia	3	4,0	5	13,4	4	8,8
	Molinio-Arrhenatheretea	10	32,4	9	50,2	5	18,8
	Agropyro-Rumicion	11	4,5	7	31,8	6	3,0
	Caricion gracilis	4	3,2	3	6,6	3	42,2

1) Die Artmächtigkeit ist durch die Summe der Deckungskoeffizienten der einzelnen Arten ausgedrückt, wobei + ... 0,4, 1 ... 2, 2 ... 10, 3 ... 20, 4 ... 30, 5 ... 40 ist.

2) incl. Carex praecox-suzae

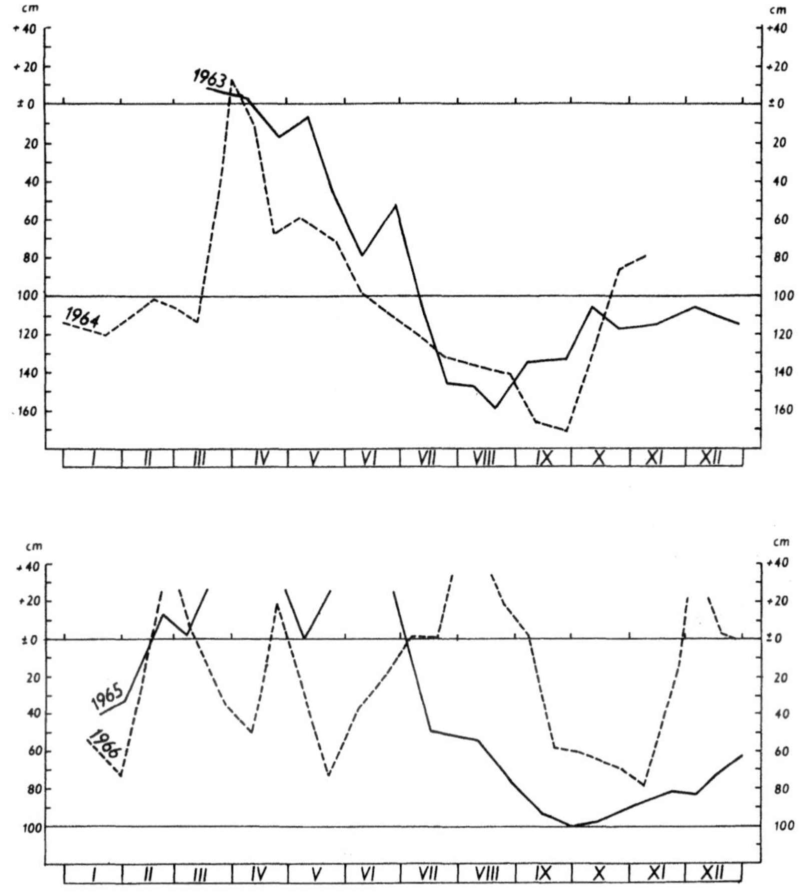

Fig.1. Grundwasserganglinien der G r a t i o l a o f f i c i n a l i s - C a r e x p r a e c o x - s u z a e - Assoziation in den Jahren 1963-1966 (U.-Fläche 25).

Untersuchungsjahren Schwankungen unterlag.

a) Die Situation bei einer trockeneren Cnidion venosi-Gesellschaft. Es handelt sich um die Untersuchungsfläche 25,die Gratiola officinalis-Carex praecox-suzae-Ass., welche am Anfang der Untersuchungen (im Jahre 1961) eine typisch entwickelte Subassoziation von Galium boreale mit dominierender *Festuca pratensis* darstellte (lokale Subassoziations-Differentialarten: *Galium boreale, Silaum silaus, Centaurea jacea* ssp. *angustifolia, Chrysanthemum leucanthemum*). Die Grundwasserganglinien aus den Untersuchungen der Jahre 1963-1966 sind aus Fig.1 ersichtlich. Die ersten zwei Jahre, 1963 und 1964, waren relativ trocken, vornehmlich das Jahr 1964, wo die Überschwemmungen nur kurzfristig waren, so daß es zu einer größeren Durchfeuchtung des Bodens nicht kommen konnte. Die übrigen zwei Untersuchungsjahre, 1966 und besonders das Jahr 1965, standen dagegen unter dem Einfluß der Katastrophen-Überschwemmungen, wobei die größte Tiefe des Überschwemmungswassers nur ausnahmsweise den Wert von +40 cm überschritten hat. Die nachfolgende Figur 2 zeigt Änderungen in der Artmächtigkeit der pflanzensoziologisch wichtigen Arten. Am Anfang der Untersuchungen im Jahre 1963 überwogen an der Untersuchungsfläche 25 die Molinio-Arrhenatheretea- Arten (insgesamt 12 Arten) nicht nur qualitativ, sondern auch quantitativ (Zusammenhang mit der nicht lange dauernden seichten Überschwemmung). Von diesen sind zu nennen: *Alopecurus pratensis* (dom.), *Poa pratensis, Ranunculus acer, Festuca pratensis, Lathyrus pratensis* und *Vicia cracca*. Die drei letztgenannten Arten zeigen dabei die größte Empfindlichkeit gegen die Katastrophen-Überschwemmung: in den Jahren 1965 und 1966 war ihr Deckungsgrad belanglos, dagegen im trockenen Sommer des Jahres 1964 erreichten sie den höchsten Deckungsgrad. *Alopecurus pratensis* hingegen zog sich im selben Zeitabschnitt stark zurück, während die Ausbreitungstendenz der Art während der Überschwemmungen des Jahres 1965 und auch des Frühjahrs 1966 nicht beeinflußt wurde. Zusammenfassend bewertet war die Artmächtigkeit der Molinio-Arrhenatheretea sinkend, nicht aber die der Cnidion- und Molinietalia- Arten. Die Cnidion-Arten blieben bis zur Sommer-Überschwemmung etwa gleichmäßig vertreten, wobei in den trockeneren Jahren *Carex praecox-suzae*, in den feuchteren (besonders nach der Sommer-Überschwemmung 1966) *Gratiola officinalis* zur Geltung kam. Die Molinietalia- Arten zeigten im Durchschnitt ab Sommer 1964 eine

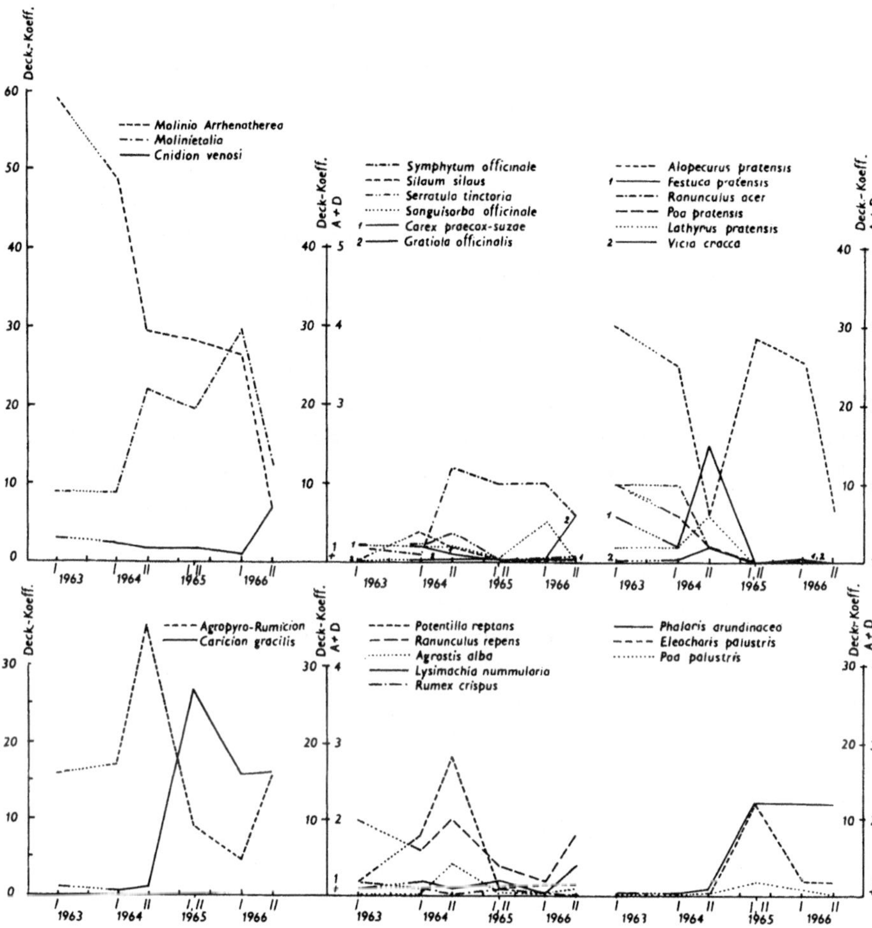

Fig.2. Änderungen in der Artmächtigkeit der wichtigen pflanzensoziologischen Artengruppen und Arten der G r a t i o l a o f f i c i n a l i s - C a r e x p r a e c o x - s u z a e - Assoziation in den Jahren 1963-1966 (U.-Fläche 25).

steigende Tendenz; ein wesentliches Sinken wurde erst nach der Sommer-Überschwemmung 1966 beobachtet. Daran beteiligten sich: *Symphytum officinale*, im trockenen Sommer 1964 auch *Serratula tinctoria* und im Mai 1966 auch *Sanguisorba officinalis*. Für die Subassoziations-Differentialart *Silaum silaus* war nur der etwas trockenere Frühling 1964 günstig. Von den Begleitern spielten in den trockenen Jahren (besonders im trockenen Sommer 1964) die A g r o p y r o - R u m i c i o n - Arten eine wichtige Rolle. Von diesen wiesen eine ausgeprägte Dynamik auf: *Ranunculus repens* mit Höchstentwicklung in der Zeit der zweiten Mahd, wo *Alopecurus pratensis* im Minimum stand, dann *Potentilla repans* und *Agrostis alba*, beide mit höchster Dominanz im Trockensommer 1964; *Lysimachia nummularia* wurde erst vor der Sommer-Überschwemmung 1966 etwas gefördert. Mit den Katastrophen-Überschwemmungen 1965 und 1966 ist eine starke Zunahme der C a r i c i o n g r a c i l i s - Arten ab 1965 eng verknüpft. Daran beteiligten sich hauptsächlich *Phalaris arundinacea* und *Eleocharis palustris*, weniger *Poa palustris*. Während sich der Anteil der beiden letztgenannten Pflanzen im nachkommenden Jahr verminderte, blieb *Phalaris arundinacea* auf derselben Höhe erhalten.

b) Die Situation bei einer feucht-nassen Gesellschaft der C n i d i o n v e n o s i - Wiesen, der L a t h y r u s p a l u s t e r - G r a t i o l a o f f i c i n a l i s - Assoziation. Der Gang der Grundwasserganglinien verlief in den Beobachtungsjahren 1963-1966 analog wie bei der vorigen Assoziation, die Kurven wiesen aber eine größere Schwankungstendenz auf (Fig.3): in den trockenen Jahren 1963-1964 sank das Grundwasser bis zur undurchlässigen Tonschicht (fast in die Tiefe von 200 cm), in den nassen Jahren dagegen erreichte das Überschwemmungswasser eine Maximalhöhe von ca. +100 cm. Die großen Unterschiede im Wasserregime der einzelnen Jahre kamen natürlich auch in der Struktur der untersuchten Pflanzengesellschaft stärker zum Ausdruck als bei der vorigen Assoziation (Fig.4). Dies betrifft nur die Änderungen in der quantitativen Vertretung; ein Neuauftreten oder voller Rückzug der Arten während der Überschwemmungen 1965 und 1966 war belanglos. Am Anfang der Untersuchungen beteiligten sich an der Pflanzenzusammensetzung wesentlich die M o l i n i e t a l i a -, A g r o p y r o - R u m i c i o n - und C a r i c i o n g r a c i l i s - Arten. Die M o l i n i e t a l i a - sowie C n i d i o n - Arten, vertreten insgesamt durch 14 Pflanzen, erreichten den höchsten Deckungsgrad im

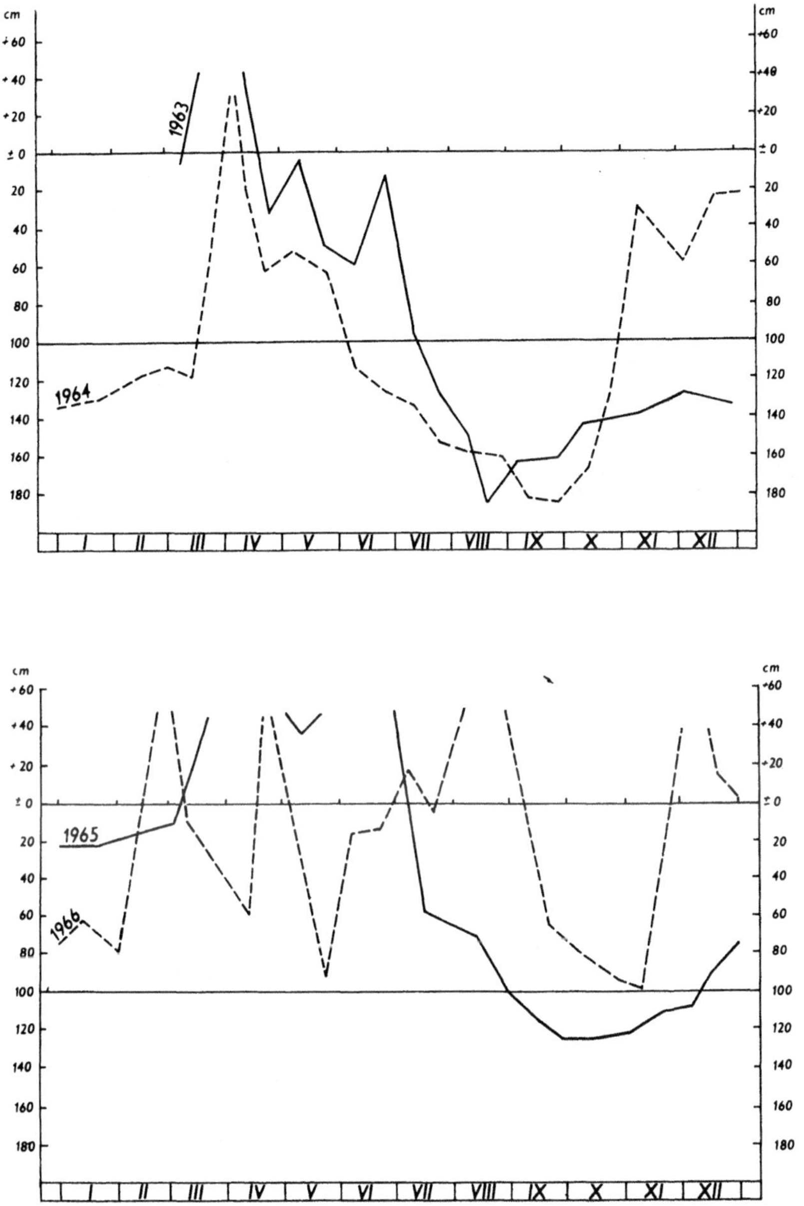

Fig.3. Grundwasserganglinien der L a t h y r u s p a l u s t e r - G r a - t i o l a o f f i c i n a l i s - Assoziation in den Jahren 1963-1966.

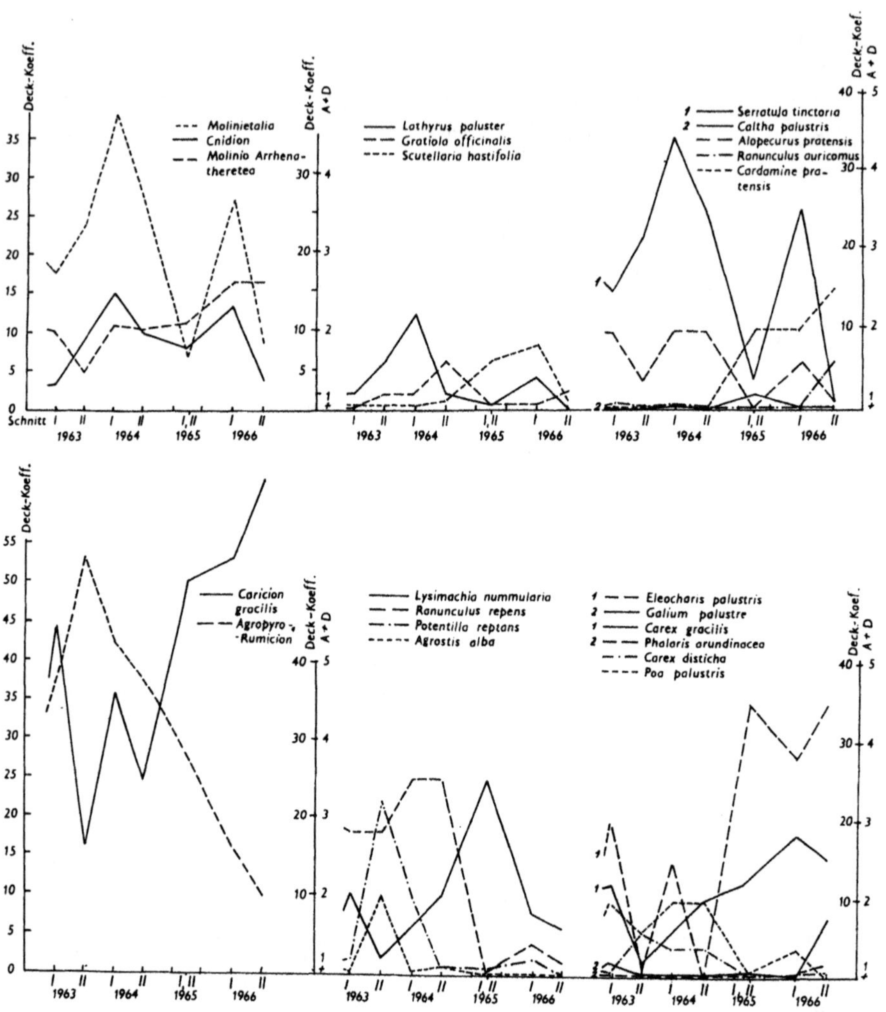

Fig.4. Änderungen in der Artmächtigkeit der wichtigen pflanzensoziologischen Artengruppen und Arten der Lathyrus paluster-Gratiola officinalis-Assoziation in den Jahren 1963-1966.

Mai 1964, einen zweiten im Jahr 1966 vor dem ersten Schnitt. Daran beteiligten sich hauptsächlich *Serratula tinctoria* und *Lathyrus paluster*, und im nassen Frühling 1966 auch *Scutellaria hastifolia*. Die bis Anfang Juni anhaltende Katastrophen-Überschwemmung 1965 und ebenso die Sommer-Überschwemmung 1966 wirkten dagegen auf *Serratula tinctoria* und *Lathyrus paluster* zurückdrängend. Eine analoge Tendenz war auch bei *Alopecurus pratensis* zu sehen. Bei einer anderen Klassencharakterart, *Cardamine pratensis*, war dagegen eine Ausbreitung nach den Katastrophen-Überschwemmungen auffallend. Ebenso positiv wirkte die Sommer-Überschwemmung 1966 auf *Ranunculus auricomus*. Von den übrigen Arten ist eine bestimmte Ausbreitungstendenz bei *Gratiola officinalis* (Cnidion-Art) in der Zeit des zweiten Schnittes 1964 bemerkenswert, ebenso eine mäßige Ausbreitung von *Caltha palustris* nach der Katastrophen-Überschwemmung 1965. Bei den Begleitern ist die gegenläufige Ausbreitungstendenz der zwei wichtigsten Gruppen , der Agropyro-Rumicion crispi- und Caricion gracilis-Arten, höchst bemerkenswert (je 7 Arten). Indem die erste die größte Ausbreitung im trockenen Sommer 1964 erreichte (dank der lichtliebenden *Potentilla reptans* und *Agrostis alba*), wies die andere ihr Optimum im Jahr 1965 (Katastrophen-Überschwemmung!) sowie im nachfolgenden ebenso nassen Jahr 1966 auf (mehrere Überschwemmungen, auch Sommer-Überschwemmung!).Von den Agropyro-Rumicion-Arten wurde auch *Ranunculus repens* durch die abnormen Verhältnisse des Jahres 1965 gehemmt; die ebenso lichtliebende *Lysimachia nummularia* dagegen fand in demselben Jahr ihr Optimum (Fähigkeit des Wachsens und der vollen Assimilation unter Wasser). Von den Caricion gracilis-Arten zeigte *Eleocharis palustris* die größte Empfindlichkeit gegen die Änderungen des Wasserregimes. Auffallend ist ihr Zurücktreten in den Trockenperioden während der Jahre 1963 und 1964 und ihre Vollentwicklung in der Überschwemmungszeit 1965 und 1966. Auch *Carex gracilis* wurde auf unserer Untersuchungsfläche durch sämtliche beobachteten Überschwemmungen gefördert, *Galium palustre* nur durch die Sommerüberschwemmung 1966. *Poa palustris* und gewissermaßen auch *Carex disticha* wurden dagegen durch die anomalen Wasserverhältnisse der Jahre 1965 und 1966 gehemmt.

Schlußfolgerungen

Für die Überschwemmungswiesen des Cnidion venosi-Verbandes wird das von Überschwemmungen bestimmte Wasserregime zum entscheidenden ökologischen Faktor. Weil das Vorkommen, die Dauer sowie die Tiefe des Überschwemmungswassers von Jahr zu Jahr wechseln, kommt es auch innerhalb der Cnidion venosi - Gesellschaften zu großen Verschiebungen in der Artmächtigkeit der bestandsbildenden Klassen-Kennarten, namentlich *Alopecurus pratensis*, *Festuca pratensis* ferner von *Serratula tinctoria*, *Symphytum officinale* sowie einiger Agropyro-Rumicion- (z.B. *Rorippa silvestris*, *Ranunculus repens*, *Agrostis alba*) und Caricion gracilis - Arten (wie *Phalaris arundinacea*, *Poa palustris* und *Eleocharis palustris*). Während der Artenentwicklung am Anfang der Vegetationsperiode und auch während der Assoziationsregeneration nach dem Rücktritt der zerstörenden Spätfrühlings- oder Sommer-Überschwemmungen kommt es hier zu deren gegenseitiger Beeinflussung. Diese Wirkung kann zur Ausbreitung einmal der einen, andermal der anderen Art führen. Man kann hier über Fluktuation oder über "associations pivots" im Sinne von DUVIGNEAUD sprechen. Die wissenschaftliche Systematik unserer mesophytischen Überschwemmungswiesen auf der Dominanz der Arten mit breiterer ökologischer Amplitude zu gründen, wäre daher unsicher. Auf diese Weise hätten wir im Laufe einiger nacheinanderfolgender Jahre verschiedene Assoziationen, nicht selten aus verschiedenen höheren systematischen Einheiten, zu beschreiben. Wenn wir z.B. für die oben angeführten Ausbildungsformen die Benennung Festucetum pratensis, Alopecuretum pratensis bzw. Phalaridetum arundinaceae wählen würden, dann würden wir uns nur über die momentanen Feuchtigkeitsverhältnisse, nicht aber über die langjährige Wirkung der Überschwemmungen auf den Pflanzenbestand, der für unsere Wiesen typisch ist, äußern.

ZUSAMMENFASSUNG

Im vorliegenden Referat werden Resultate der Untersuchungen über die Änderungen in der Pflanzenzusammensetzung der mesophytischen Auenwiesen aus dem trockenen Gebiet Süd-Mährens (Umgebung von Lanžhot), deren Wasserregime durch Überschwemmungen bestimmt ist, zusammengefaßt. Weil sich die Zeit der Überschwemmungen, sowie deren Dauer und Tiefe von Jahr zu Jahr ändern, unterliegen auch die untersuchten Wiesen-Gesellschaften der Variabilität. Auf Grund der Untersuchungen aus den Jahren 1961-1966 wurde festgestellt, daß es bei den meisten typisch entwickelten Assoziationen des Cnidion venosi- Verbandes zu grundsätzlichen Änderungen in der Pflanzenzusammensetzung nicht kommt (wenn doch, dann sind sie belanglos). Die quantitative Vertretung (Artmächtigkeit) einer ganzen Reihe der Arten innerhalb der Assoziation und damit auch der Anteil der Kennarten der höheren systematischen Einheiten schwankt dagegen nicht nur von Jahr zu Jahr, sondern auch im Verlauf desselben Jahres (Tab.1, Fig.2,4).

SUMMARY

The results of a study of the changes in the cover values of the vegetation of flooded meadows from the dry area of southern Moravia (in the neighbourhood of the village of Lanžhot) are summarized in this report. On the basis of investigations covering the period 1961-1966, it was concluded that qualitative changes of species structure of well developed communities of the alliance Cnidion venosi, are negligible. However, marked changes in the quantitative representations (cover and abundance) of a number of the species are evident, not only between years, but in the same year (Fig.2,4 and Table 1).

The hydrological conditions prevailing within the habitat (the date and duration of flooding, and the depth of the flood waters (Figs.1,3), appear to be the controlling factor.

In the light of these results, difficulties arising from the use of quantitative data in the delimitation of the higher phytocoenologic systematic units are discussed.

LITERATUR

BALÁTOVÁ-TULÁČKOVÁ,Emilie -1966- Synökologische Charakteristik der südmährischen Überschwemmungswiesen.- Rozpravy ČSAV 76 (1). Praha.

RAABE,E.W. -1960- Über die Regeneration überschwemmter Grünländereien in der Treene-Niederung.- Schr.naturw.Ver.Schlesw.-Holst.31: 25-55. Kiel.

ROBOTNOV,T.A. -1955- Raznogodičnaja izmenčivost' lugov. Bjuleten Mosk. ispytat.prirody, Otd.biolog.60 (3): 9-30. Moskva.

DISKUSSION

R.TÜXEN

Ich glaube, daß es zu den charakteristischen Eigenschaften bestimmter Pflanzengesellschaften gehört, im Laufe ihres Daseins nicht konstant zu sein, sondern, weil sie eben unter fluktuierenden Einflüssen stehen, auch vor allem in ihren Mengenverhältnissen zu schwanken.

J.BARKMAN

Die Form der Periodizitätskurven von Arten und Artengruppen hängt weitgehend davon ab, wie man die Symbole der kombinierten Schätzung in Zahlen "übersetzt". Ich habe festgestellt, daß Sie eine "TÜXEN-SCHWICKERATH-Modifikation" gebraucht haben, nämlich eine Modifikation, die intermediär ist zwischen der Gruppenmenge nach TÜXEN und ELLENBERG und der Artmächtigkeit nach SCHWICKERATH. Die Übersetzungsskala von Frau BALÁTOVÁ (+=0,4, 1=2, 2=10, 3=20, 4=30, 5=40) scheint mir ein weniger glückliches hybrides Verfahren zu sein, das sich auf beide Prinzipien zugleich stützt.

Frau BALÁTOVÁ-TULÁČKOVÁ

Als ich 1951 in einem Futterinstitut arbeitete, da hatte ich mehrere Aufgaben, die mit der Praxis zusammenhingen. Damals kannte ich leider keine der anderen Skalen, über die jetzt gesprochen wurde. Bei uns existierte damals die Skala von VÁLEK;die schien mir aber nicht entsprechend für die höheren Prozente. Die habe ich dann abgeändert, um einen Vergleich mit meinen älteren Arbeiten zu haben.

I.S.ZONNEVELD

Wir haben unter anderem über den sklerenchymatischen Bau von *Lythrum salicaria* im Zusammenhang mit der Überflutungsintensität gesprochen. Es wäre wertvoll, die Skleromorphie, also die sklerenchymatischen Unterschiede in ein Lebensformensystem von Sklerotypen zu fassen, und so das Milieu zu indizieren, etwa wie ich das vor 10 Jahren in Stolzenau vorgeschlagen habe. Besonders für die Trennung von allogenen und autogenen Tendenzen wäre das wertvoll.

J.J.MOORE

Man könnte auch mit der DOMIN-Skala arbeiten, die mit der von BRAUN-BLANQUET leicht zu verwechseln ist. Vor etwa 6 Jahren stand im Journal of Ecology eine Umrechnung, die man verwerten kann.

VEGETATIONSZYKLEN BEI NORDWESTDEUTSCHEN PFLANZENGESELLSCHAFTEN

F. Runge

Mit Hilfe von Dauerquadraten lassen sich Vegetationszyklen finden. Hier sollen drei aufgezeigt werden, die relativ gesichert erscheinen. Ob es sich in allen Fällen um echte Vegetationszyklen oder um ähnliche Vorgänge handelt, bleibe dahingestellt.

Einen eindrucksvollen Vegetationszyklus gibt es in flachen Heideweihern Nordwestdeutschlands. Ich fand ihn mit Hilfe von zwei Dauerquadraten,die ich im Heideweiher des Naturschutzgebietes "Heiliges Meer" im Kreise Tecklenburg vor 8 Jahren anlegte.

Diese oligotrophen Gewässer zieren in manchen Jahren die Polster des seltenen Sumpfhartheus *(Hypericum elodes)*. Es gehört zur Vielstengelsimsen-Gesellschaft, dem Eleocharetum multicaulis. Neben dem Sumpfharteu gedeihen darin *Eleocharis multicaulis*, oft auch *Litorella uniflora* und die seltene *Deschampsia setacea*.

In der Nachbarschaft der Vielstengelsimsen-Gesellschaft wachsen im wenige Zentimeter tieferen Wasser *Lobelia dortmanna*, recht oft zusammen mit *Littorella uniflora* sowie *Characeen*. Diese Arten bilden die Lobelien-Gesellschaft, das Isoeto-Lobelietum.

Sowohl das Eleocharetum multicaulis wie das Isoeto-Lobelietum wurzeln im flachen, nährstoffarmen Wasser in Sand.

Haben wir nun im Herbst desselben Jahres, im Winter oder im Frühjahr des folgenden Jahres einen hohen, vielleicht auch normalen Wasserstand, so ändert sich die Vegetation recht schnell: Es breitet sich eine Torfmoos *(Sphagnum)*-Decke aus, in meinen Dauerquadraten eine solche aus *Sphagnum auriculatum (Sph. subsecundum* var. *auriculatum)*.

Unter dieser Torfmoos-Decke verschwinden sowohl das Isoeto-Lobelietum wie das Eleocharetum multicaulis und mit ihnen alle die seltenen Arten.

So lag beispielsweise 1949 ein hellblauer Schleier aus Tausenden, wenn nicht Millionen, *Lobelia*-Blüten über dem Heideweiher des Naturschutzgebietes "Heiliges Meer". 1950, also im Jahre darauf, war kaum noch eine *Lobelie* zu sehen, und böse Zungen behaupteten, ein Dozent hätte mit seinen Studenten die *Lobelien* ausgerottet. Das war bei der Riesen-Menge natürlich unmöglich.

Bei hohem oder normalem Wasserstand wird die Torfmoos-Decke immer dichter. Sie bildet eine fest verwobene Schlamm-Torfschicht, die sich leicht vom Sanduntergrund abheben läßt.
Herr Dr. F.KOPPE, Bielefeld, und ich überlegten 1954, ob wir angesichts der dichten, abhebbaren Schlamm-Torfdecke im Heideweiher des NSG "Heiliges Meer" diese Schicht auf eine Strecke hin entfernen sollten, um die selbständige Wiederansiedlung der *Lobelia* und anderen Kostbarkeiten zu ermöglichen, sahen aber von diesem unnatürlichen Eingriff ab. Die seltenen Pflanzen schienen unwiederbringlich verloren zu sein.
Herr C.ALTEHAGE,Osnabrück, erkannte als erster die entwicklungsfeindliche Wirkung der *Sphagnum*-Decke und ließ im "Ahlder Pool" (Kreis Lingen) die Schlamm-Torfdecke auf weite Strecken hin abräumen.
Die Schlamm-Torfdecke nimmt bei hohem oder normalem Wasserstand an Dicke von Jahr zu Jahr zusehends zu. Ich maß nach 3 Jahren bis zu 6 cm.
Folgt nun ein Dürrejahr wie das Jahr 1959, dann trocknet der flache Heideweiher ganz oder weitgehend aus. Das Torfmoos verblaßt und stirbt ab. Die Schlamm-Torfdecke trocknet ein, sie zerreißt in polyedrische Schollen und zerbröckelt dann, und der reine Sand des Untergrundes erscheint. Bald besteht die Schlamm-Torfdecke nur noch aus stecknadelkopfgroßen Partikelchen und schließlich aus staubförmigem Pulver. Diesen entscheidenden Vorgang habe ich 1959 im Naturschutzgebiet "Heiliges Meer" sowie in anderen Heideweihern Nordwestdeutschlands genauestens verfolgt.

Jetzt können sich auf dem Sand Arten des trockenen Sandbodens einfinden. Beobachtet habe ich Birken *(Betula)*-Keimlinge, auch *Polygonum* und *Sagina*, wohl *S.procumbens*.
Steigt der Wasserspiegel im Herbst oder Winter wieder, so findet sich auf dem nunmehr reinen Sandboden sowohl das Eleocharetum multicaulis wie das Isoeto-Lobelietum mit *Littorella uniflora*, *Eleocharis multicaulis*, *Hypericum elodes* und *Lobelia dortmanna* wieder ein.
Diese Gesellschaften kehren aber nur zögernd zurück, die ersten Pflanzen im nächsten Jahr, die meisten erst nach 2-3 Jahren.
Nun läßt sich auch diese Massenentwicklung von *Lobelien* des Jahres 1949 im Heideweiher leicht erklären, denn 1947 war ein Dürrejahr.

Im Dürrejahr 1959, im Jahre des Zerfalls der Schlamm-Torfdecke und des Wiedererscheinens des Sandes konnte ich das Wiederauftreten der *Lobelien* und anderer seltener Arten für 1960 und 1961 mit Sicherheit voraussagen. Sie kehrten tatsächlich zurück.
Nun können wir auch erklären, warum *Lobelia dortmanna* so selten erscheint, und warum man immer wieder hört: "Hier wuchs einstmals die *Lobelie*". Sie kommt sicherlich wieder!
Dieser Kreislauf scheint mir sicher zu sein. Einzelheiten möge man einer Veröffentlichung an anderer Stelle entnehmen.
Von diesem Kreislauf abweichende Ausnahmen:
a) An steilen Ufern, beispielsweise im "Erdfallsee" des NSG "Heiliges Meer", kommt es nicht zur Schlamm-Torfbildung. Hier erscheinen das Eleocharetum und das Isoeto-Lobelietum in allen Jahren.
b) In Heideweihern, in denen viel gebadet wird, unterbinden die Badenden die Schlamm-Torfbildung, und Littorella- Rasen bedecken in allen Jahren den Sanduntergrund, beispielsweise im "Schnippenpohl" in der Brechte bei Ochtrup (Kreis Steinfurt).

Der Vegetationszyklus in Heideweihern wird also durch extreme Dürreperioden ausgelöst.
Vegetationszyklen beobachten wir auch bei tiefgreifenden Heidebränden. Hier löst im allgemeinen der Mensch den Kreislauf aus; denn Heidebrände infolge von Blitzeinschlägen sind äußerst selten.
Die Wirkung eines Heidebrandes und die nachfolgende Besiedlung untersuchte ich mit Hilfe eines 1960 angelegten Dauerquadrates in einer abgebrannten Bergheide oder Besenginsterheide Calluno-Sarothamnetum im Sauerland im Kreise Meschede in 317 m Meereshöhe. Darüber berichtete ich bereits 1963 im Archiv für Naturschutz und Landschaftsforschung H.2: 173. 1963 war aber der Zyklus noch nicht geschlossen, wie ich seinerzeit vermutete.
Im Jahre 1959, also im Jahre vor dem Brande, hatte ich in derselben Bergheide eine soziologische Aufnahme gemacht. Im Frühjahr des nächsten Jahres 1960 brach der Brand aus. Ein Bauer hatte in der Heide Papiersäcke verbrannt und so den Heidebrand entfacht. In der kohlschwarzen Fläche legte ich 1960 das Dauerquadrat an.

Ob nun zunächst eine Gesellschaft aus grünen Algen auftrat, vermag ich nicht zu beurteilen, weil ich das Dauerquadrat in den ersten Monaten zu wenig kontrollierte.

Im Jahre darauf folgte eine Moosgesellschaft, anscheinend das von HÜBSCHMANN 1957 beschriebene Funarietum hygrometricae.
Drei bis fünf Jahre nach dem Brande löste ein Trockenrasen (Festuco-Sedetalia) die Moosgesellschaften ab, der aus *Festuca ovina, Agrostis tenuis* und *Deschampsia flexuosa* bestand.

Erst 1966, also nach sechs Jahren, war die Bergheide wieder so charakteristisch wie vor dem Brande ausgebildet.
Der Kreislauf scheint gesichert, zumal er bei der nachfolgend beschriebenen Brandstelle in einer anderen Heide ähnlich verlief.
Im Calluno-Genistetum typicum des NSG "Heiliges Meer" (Kreis Tecklenburg) legte ich gleich nach einem tiefgreifenden Brande auf einer kohlschwarzen Fläche ein Dauerquadrat an (Tabelle). Man hatte an dieser Stelle am 1.2.1962 einen Holzstoß verbrannt, also nicht etwa die Heide nur leicht abgeflämmt, wie es sonst geschieht. Die Hitze drang also tief in den Boden und schädigte oder zerstörte das Wurzelwerk von *Calluna vulgaris.*
Das Dauerquadrat nahm ich Jahr für Jahr mehrmals auf. (Die Tabelle enthält aber nur wenige Aufnahmen).
Nach etwa 1/2 Jahr besiedelte eine Gesellschaft aus grünen Algen das Dauerquadrat. Im Herbst desselben Jahres, also 3/4 Jahr nach dem Brande, stellte sich die Brandstellen-Pilz-Gesellschaft, das Geopyxidetum carbonariae ein, hier bestehend aus dem Kohlen-Graublatt *Lyophyllum cabonarium.*
Nach ein bis drei Jahren erschien das Funarietum hygrometricae.
Vier Jahre später besiedelte ein Trockenrasen (Festuco-Sedetalia) die frühere Brandstelle.
Ob demnächst die trockene Heide wiederkehrt, oder sich ein Birkengestrüpp auf der Brandstelle entwickelt, müssen wir abwarten. (Die Birken des Dauerquadrates wurden 1965 und 1966 geschlagen.)

Der Kreislauf entspricht etwa dem von HEGI (V (3),p.1696) nach Untersuchungen von FRITSCH,PARKER & SALISBURY in Großbritannien gefundenen Zyklus, doch faßt HEGI Algen und Pilze, sowie Moose Gräser und Kräuter zusammen und beschreibt ein dann nochmals folgendes Algen-Flechten-Stadium. Dieses habe ich aber bisher nicht angetroffen, obwohl ich sehr darauf achtete. Auch K.HANSEN hat

Dauerquadrat "Brandstelle in trockener Heide"
im NSG "Heiliges Meer", Kreis Tecklenburg
Brand am 1.2.1962

Jahr	1962				63	64	65	66
Tag	3. 8.	30. 8.	15. 10.	23. 11.	30. 8.	7. 9.	1. 8.	2. 8.
Gesamtbedeckung %	7	9	8	16	25	95	99	100
grüne Algen, Bed.%	7	7	.	.	.	.	.	.
Kohlen-Graublatt, Lyophyllum carb., Zahl	.	11	91	39	.	.	.	.
Kohlen-Schüppling, Pholiota carb., Zahl	.	.	4	1	.	.	.	.
Drehmoos, Funaria hygrometrica, Bed.%	.	2	7	12	25	25	25	10
Weißbirke, Betula verrucosa, Bed.%	.	1	1	1	2	45	(35)	(45)
Silbermoos, Bryum argenteum, Bed.%	.	.	.	.	1	10	1	.
Löwenzahn, Taraxacum officin., Zahl	.	.	.	.	1°	1°	2	.
Waldweidenröschen, Epilobium ang., Zahl	.	.	.	.	2	70	29	16
Waldweidenröschen, Epilob. ang., Bed.%	.	.	.	.	2	25	10	5
Kleiner Ampfer, Rumex cf. acetosella, Zahl	.	.	.	.	1°	4	4	6°
Weiße Melde, Chenopodium album, Zahl	.	.	.	.	.	1°	.	.
Rotes Straußgras, Agrostis tenuis %	.	.	.	.	.	2	20	25
Feinschwingel, Festuca tenuifolia, %	.	.	.	.	.	1	15	25
Rotstengelmoos, Entodon schreberi, %	.	.	.	.	.	1	1	1
Besenmoos, Dicranum scoparium, Bed.%	.	.	.	.	.	.	1	3
Hornkraut, Cerastium caespit., Bed.%	.	.	.	.	.	.	10	20
Ruchgras, Anthoxanthum odor., Bed.%	.	.	.	.	.	.	1°	5
Vielblüt. Simse, Luzula multifl., Zahl	.	.	.	.	.	.	1	1
Hornzahnmoos, Ceratodon purpur., %	.	.	.	.	.	.	1	20
Gefalt. Nabeling, Omphalia umbell., Zahl	.	.	.	.	.	.	.	1

Im Herbst 1960 aber erschien eine Brandstellen-Pilzgesellschaft mit *Rhizina inflata*. Sie gehört wohl zum G e o p y x i d e t u m c a r b o n a r i a e Ebert 1958.

in West-Jütland ähnliche Beobachtungen gemacht.
Diese 3 Zyklen erscheinen mir hinreichend gesichert, wenn auch einzelne Stadien noch mit Fragezeichen versehen und Feinheiten später herausgeschält werden müssen.

ZUSAMMENFASSUNG

Mit Hilfe von Dauerquadraten konnten mehrere Vegetationszyklen gefunden werden, unter ihnen Zyklen in Heideweihern und auf Brandstellen in Zwergstrauchheiden.
In flachen Heideweihern wachsen das Eleocharetum multicaulis und das Isoeto-Lobelietum auf Sand im oligotrophen Wasser. Bei hohem oder normalem Wasserstand verschwinden diese Gesellschaften unter einer sich bildenden *Sphagnum*-Decke (in den Dauerquadraten *Sph. auriculatum*). Diese erzeugt Schlamm bzw. Torf. Die Schlamm-Torfdecke wird von Jahr zu Jahr dicker. Kommt ein Dürrejahr (1947, 1959), so trocknet der Heideweiher weitgehend aus, das *Sphagnum* stirbt, die Schlamm-Torfschicht trocknet ein, zerreißt und zerfällt. Auf dem wieder erscheinenden, noch trockenen Sand können Pflanzen des trockenen Bodens (z.B. *Betula*-Keimlinge) Fuß fassen. Steigt der Wasserspiegel wieder, so kehren Eleocharetum multicaulis und Isoeto-Lobelietum innerhalb von 1-3 Jahren zurück.
In einem Calluno-Sarothamnetum des Sauerlandes trat nach einem Brande folgender Zyklus ein: Gesellschaft aus grünen Algen (?) - Geopyxidetum carbonariae mit *Rhizina inflata* - Moosgesellschaft (Funarietum hygrometricae) - Trockenrasen mit *Agrostis tenuis* und *Festuca* cf. *ovina* - Calluno-Sarothamnetum.
In einem Calluno-Genistetum des Naturschutzgebietes "Heiliges Meer", Kreis Tecklenburg, vollzog sich nach einem tiefgreifenden Brande folgende Sukzession: Gesellschaft aus grünen Algen - Geopyxidetum carbonariae - Funarietum hygrometricae - Trockenrasen (s.Tabelle). Der Zyklus war 1967 noch nicht geschlossen.

SUMMARY

With help of permanent quadrats several vegetation cycles could be found, among them cycles in heath ponds and its sites of heath fires.
In shallow heath ponds the Eleocharetum multicau-

l i s and the I s o e t o - L o b e l i e t u m grow on sand in the oligotrophic water. If the water level is high or normal these associations disappear under a developing cover of *Sphagnum* (in the permanent quadrats *S.auriculatum*). The carpet produces mud or peat. The mud-peat cover grows thicker from year to year. When is a very dry year (1947,1959) much of the heath pond dries up, *Sphagna* die, the cover of mud or peat shrivels up, disintegrates and decays. Plants of dry soil (seedlings of *Betula*) can now colonise the dried sand. If the water level rises E l e o - c h a r e t u m m u l t i c a u l i s and I s o e t o - L o b e l i e - t u m return within 1-3 years.
After a fire in a C a l l u n o - S a r o t h a m n e t u m of Sauerland the following succession occurred: community of green algae (?) - G e o p y x i d e t u m c a r b o n a r i a e with *Rhizina inflata* - moss community (F u n a r i e t u m h y g r o m e t r i c a e) - dry grassland with *Agrostis tenuis* and *Festuca* cf. *ovina* - C a l l u n o - S a r o t h a m n e t u m .
In a C a l l u n o - G e n i s t e t u m t y p i c u m of the Nature Conservation area "Heiliges Meer", Kreis Tecklenburg, the following succession appeared after a severe fire: community of green algae - G e o p y x i d e t u m c a r b o n a r i a e - F u n a r i e - t u m h y g r o m e t r i c a e - dry grassland (see table). The succession was not yet completed in 1967.

DISKUSSION

J.J.MOORE
Die Engländer haben auch über diese Fragen gearbeitet. WATT hat vor etwa 20 Jahren in seinem berühmten "Pattern and Process in the Plant Community" in Heiden ungefähr dieselben Zyklen gezeigt.

J.PFADENHAUER
Was geschieht mit dem zu Staub zerfallenen Torf? Wird er vom Wind weggeweht oder durch Humuszersetzung aufgelöst?
Wie geschieht die Neubesiedlung mit *Lobelia* und *Eleocharis multicaulis*? Bleiben Samen oder Pflanzenteile, die sich regenerieren können, zwischen Torf und Sandboden erhalten?

F.RUNGE
Zu Pater MOORE möchte ich sagen, die Engländer haben darüber gearbeitet, aber die Definition des Vegetationszyklus habe ich nicht gefunden.

Die Auflösung der Torfdecke, ob durch Humuszersetzung oder Wind kenne ich nicht. Ich habe diese Torfteilchen durch den Wind nicht wegblasen gesehen. Ich hatte vermutet, daß das Wasser diese winzigen Teilchen, die man kaum noch sehen kann, zur Weihermitte hin wegschwemmen würde; das ist aber nicht der Fall. Ob der Regen irgendwie mitgewirkt hat, weiß ich nicht.
Wie nun *Lobelia, Eleocharis multicaulis, Littorella, Deschampsia setacea* usw. wieder dorthin kommen, weiß ich auch nicht. Denn, wenn man die Torfdecke abhebt - das haben wir stellenweise getan - sieht man darunter nichts.

H.ELLENBERG

Haben Sie einmal gegraben? Da würde man bestimmt Reste finden.

C.ALTEHAGE

Im Naturschutzgebiet "Keienvenn" in Emsbüren,Krs,Lingen, hat die Trockenheit des Jahres 1959 die gesamte Wasserfläche austrocknen lassen, wobei nur an einer kleinen, tieferen Stelle etwas Wasser blieb, das allerdings durch Einfließen von Fäkalien aus dem Stall eines benachbarten Bauernhauses verunreinigt wurde. Das noch 1959 stark entwickelte Lobelietum wurde dadurch zerstört und ist seitdem nicht wieder beobachtet worden.
In einem zweiten Naturschutzgebiet,dem "Ahlder Pool", gleichfalls im Krs. Lingen, wurde durch Austrocknung der Fläche ebenfalls das Lobelietum vernichtet. Hier ist später versucht worden, die Entwicklungsmöglichkeit des *Lobelia*-Bestandes wieder zu sichern. Einzelne Pflanzen wurden im Jahre 1966 wieder beobachtet. Worauf das zurückzuführen ist, weiß ich nicht.

F.RUNGE

Doch, jetzt wissen wir es. Ich sprach soeben von diesem Ahlder Pool, wo Herr ALTEHAGE diese Torfdecke hat wegräumen lassen.Also *Lobelien* sind wiedergekommen, wie man das ja auch erwarten durfte. Und in das Keienvenn,einen anderen Heideweiher im Kreise Lingen,das sagte Herr ALTEHAGE eben schon, mündet ein Graben hinein mit eutrophem Wasser. Dort beobachtet man eine leichte Eutrophierung. Dann sieht die Geschichte ganz anders aus.
aus.

R.TÜXEN

Eine kurze Bemerkung, zu der mich der Kreislauf der Heideregeneration angeregt hat. Wir kennen die nw-deutsche *Calluna*-Heide seit 40 Jahren, aber wir wissen nicht, wie sie sich erhält und regeneriert. Wir haben angenommen, man müsse unsere Heide entweder abplaggen, oder abmähen, oder man müsse sie brennen, oder man müsse sie beweiden, und wenn man das nicht tut, dann würde sie zerstört. So war lange Zeit unsere Annahme. Und diese Annahme ist falsch! Die Heide kann sich allein erhalten. Die Heide stirbt zwar in ungünstigen Jahren, und wenn sie alt genug ist, dann stirbt sie, und dann regeneriert sie sich ganz von selbst. Ich habe Dauerquadrate, wo die toten Heidestengel herumliegen, dazwischen wachsen *Cladonia impexa*, *Cladonia mitis*, *Cornicularia aculeata* und andere *Cladonien*, und überall sieht man, ohne daß geplaggt oder gebrannt worden ist, oder geweidet wurde, junge Heide-Keimlinge aufkommen. Nach drei Jahren ist die Heide so dicht wie sie war, bevor sie starb. Bäume kommen nicht! Ich kenne viele Stellen, seit 40 Jahren, wo nicht eine Birke gekommen ist, wo die Heide sich seit 40 Jahren erhalten hat. Das ist der eine Fall, der andere ist dieser: Zu Beginn des Krieges hatte ich eine Heide, am Blanken Flat Krs.Neustadt zum letzten Mal als Heide gesehen, eine wunderbar blühende *Calluna*-Heide. Und dann konnte ich fünf oder sechs Jahre diese Heide nicht besuchen. Ein neuseeländischen Kollege besuchte mich eines Tages nach dem Kriege mit dem Auto, und wollte eine *Calluna*-Heide sehen. Wir fuhren zum Blanken Flat. Ich wollte ihm jene *Calluna*-Heide zeigen, und was war dort? Keine Bäume! Aber ein Rasen von 100% *Festuca ovina*. Und heute (1967) ist dort wieder eine *Calluna*-Heide. Ich habe alle Leute gefragt, von denen ich glaubte, daß sie etwas davon verständen und habe nur eine Antwort bekommen von einem polnischen Kollegen Prof. RZEDOWSKI. Er hat mir gesagt, wahrscheinlich ist diese Heide früher - und das ist wohl richtig - sehr intensiv von Schafen beweidet worden, und dadurch ist sie eutrophiert worden, und darum kam nicht in diesem Falle nach dem Absterben der Heide, als sie alt genug war, gleich *Calluna* wieder, sondern darum kam die etwas eutrophere, eutraphentere *Festuca ovina*, und erst dann kam *Calluna*. (Allerdings kann ein Brand dieser Fläche nicht ausgeschlossen werden.)

F.RUNGE

Herr WESTHOFF hat das auch beschrieben, kann er etwas darüber sagen?

V.WESTHOFF

Wie Herr Dr.MOORE schon gesagt hat, sind zyklische Sukzessionen viel weiter bekannt und erforscht, als Dr.RUNGE meint. Nach dem

Vorbild von Dr. A.S.WATT (Pattern and Process in the Plant Community, J.Ecol.35: 1-22, 1947) sind zyklische Sukzessionen in Großbritannien und den Niederlanden mehrfach erforscht worden. Der Vegetationszyklus der niederländischen Heide ist zuerst eingehend von P.STOUJESDIJK untersucht worden (Heaths and Inland Dunes on the Veluwe, Wentia 1959, Thesis, Utrecht). Die Sukzession auf Heidedauerquadraten ist nachher vor allem von J.Th. DE SMID verfolgt worden, der diese Ergebnisse hier hätte darstellen wollen, doch leider durch Krankheit verhindert war, an diesem Symposium teilzunehmen. Es ist nun so, daß ich völlig mit Herrn Professor TÜXEN übereinstimme, daß man nicht im allgemeinen sagen kann, daß für die Erhaltung der Heide entweder Brand oder Plaggen oder Mähen notwendig sind. Das Umgekehrte kann man aber auch nicht sagen, das hängt vom Fall ab. Wir haben die rein natürlichen Küstenheiden in unseres westfriesischen Inseln verfolgt. Diese Heide ist soweit natürlich, daß sie nicht aus einem Wald hervorgegangen ist. Sie ist immer Heide gewesen und vorher war hier eine Xero-Serie der Dünensukzession. Diese Heide hat nur eine *Calluna*-Decke, wenn sie beweidet und gebrannt wird. Dann wird *Calluna* dominant. Wenn diese Heide aber sich selbst überlassen wird, stirbt zwar die *Calluna* völlig ab, aber das ist nur eine Zeiterscheinung. Es regeneriert dann ein Mosaikkomplex, der auch ein Regenerationskomplex ist, also eine zyklische Sukzession, wo man *Empetrum*, *Calluna*, *Erica*, *Carex arenaria* , *Salix repens*-Bestände abwechselnd in mosaikartiger Form vorfindet.Und dieses Mosaik erneuert sich ständig und ist eigentlich das natürliche Bild der Heide. Wenn wir uns eine Heide vorstellen, es sei eine Dominanz von *Calluna*, dann ist das ein falsches Bild, so ungefähr wie ein Kartoffelacker, den haben wir auch künstlich geschaffen. Eine natürliche Heide ist nicht ein dominanter *Calluna vulgaris*-Bestand, sondern sie besteht aus verschiedenen Arten, die jede für sich dominieren können, und in welchen die *Calluna* eine bestimmte Rolle spielt.

Bei den Heiden im Inland sind ganz andere Fälle da, die ich in meiner Arbeit "Het beheer van heidereservaten", Natuur en Landschap 14(4) 1961 beschrieben habe. Es gibt Heiden die, wenn sie sich selbst überlassen werden, in Wald übergehen, das hängt vom Bodenprofil ab. Es gibt auch Heiden, die das nicht tun, und in denen eine derartige zyklische Sukzession zu verfolgen ist. Auch in diesem Fall ist, soweit wir wissen, Dominanz von *Calluna* nicht eine natürliche Erscheinung.

D.J.BELLAMY

I would like to develop the cycle in *Calluna* a little farther if I can. When a *Calluna* plant starts to grow, we have a bush largely getting larger and then at a certain stage, the bush of *Calluna* begins to open out and new species come in of the bare ground, *Cladonia* and grasses and many other plants. So if the system is left, then we will have a cycle of plant communities, pure heather and heather mixed with other plants. We have measured the rate of this process, because we have here the green biomass supplying the brown biomass with material. When the "Green to Brown" ratio gets to a certain level, then there is not enough photosynthesis, there is not enough energy to supply the respiration of the whole heather plant and the heather plant begins to break down. In Northern England, in the Pennines, we have measured the rate of this process, it takes 25 years, and the rate of productivity of the heather is 177 g/m^2. So with this rate of production we get to the critical stage in about 25 years. Where Dr.WATT has studied his "Pattern and Process", in SE-England, it reaches this stage in only 12 years and here the rate of production of the heather is 240 g/m^2.

So here we have a natural cycle which is governed by the rate of productivity of the heather, the rate of energy storage. Now we have been given three examples of cyclical succession. In all these are given some factors which removed the stored energy: we have the fire, we have the lowering of the water level and in this case it is the removal of the stored energy by the ecosystem itself, by the respiration.

I.S.ZONNEVELD

In the southwest of the Netherlands (Ossendrecht, Kalthout) I have found a local cyclical alternation of forest and heather. During the Holocene, starting about 8000 years ago, at least 6 or 7 times the forest - oak forest a kind of *Querco-Betuletum* - was replaced by heather and vice versa. In the heather stage, in the very poor stage, even during the warm and moist Atlanticum, there were windblown soils. Partly man may have given rise to the forest devastation, by fire. But could your explanation of the short time relation between the energy balance and the cyclical succession also be appropriate for these long term successions of about 1000 years for one cycle?

D.J.BELLAMY

I think that we can have such a cycle in forest but I could only guess at the mechanism of control. In the back of my mind there is a statement that in tropical rain forest, in high forest in the tropics, all the mineral matter is in the standing crop. If this could be true for the deciduous forest of the temperate zone - which I do not know and I know of no measurement to proof this - then we could have a cycle in which all the mineral matter moved into the standing crop and was gradually cycled between the trees: one tree would die, this mineral matter would go back into the ground and another tree could grow in its place. But if anything came along to upset this cycle, then we could have a break-down of the whole system and it would take a long time for the forest ecosystem to develop again. This is exactly what happens in the tropics when we have the forest there cut down and it is turned over the shifting agriculture, it takes a certain amount of time to start again. But this is pure hypothesis!

J.BARKMAN

Ich möchte betonen, wie wichtig es ist, daß Herr Dr. RUNGE hier die zyklischen Sukzessionen erörtert hat, denn ich habe das bis jetzt sehr vermißt, und doch kommen sie ganz allgemein vor. Nicht nur WATT, auch andere Engländer haben viele Fälle beschrieben und in KERSHAW "Quantitative and Dynamic Ecology" kann man viele Beispiele auch aus rezenten Publikationen über alle möglichen Zyklen in Vegetationen finden.

Zu den Bemerkungen von Professor WESTHOFF über den Heidezyklus in den Niederlanden möchte ich noch folgendes ergänzen: Nach STOUTJESDIJK wird er dadurch ausgelöst, daß *Calluna* in einem gewissen Alter spontan stirbt, während DE SMIDT dafür die unregelmäßig, aber meistens einmal in 4-5 Jahren auftretenden Epidemien des Heidekäfers (*Lochmaea suturalis*) verantwortlich hält. Allerdings greift *Lochmaea* nur alte Heidepflanzen an, so daß eine vor wenigen Jahren abgebrannte Heide auch in solchen Unfalljahren nicht von dieser Plage befallen wird. Ein flächenhaftes Absterben erfolgt nur, wenn alle Heidepflanzen gleich alt sind. Die Heide kann dann großflächig in eine *Festuca*-Gesellschaft verwandelt werden. Unter "natürlichen" Umständen - wenn nicht regelmäßig alle 10-15 Jahre großflächig abgebrannt wird - entsteht ein Mosaik von 1.*Calluna*-Heide, 2.*Festuca-Cladonia*-Gesellschaft (C l a d o n i e t u m m i t i s) und 3. C o r y n e p h o r e t u m . Phase 3 kann fehlen und tritt nur dann auf, wenn durch intensiven Tritt die lockeren *Cladonia*-Teppiche, die ja keine Wurzel haben und bei trockenener Witterung sehr spröde sind, vernichtet werden. Jetzt hat STOUTJESDIJK auf der Veluwe den relativen Flächenanteil bestimmt, den diese 3 Stadien einnehmen. Aus dem Anteil der Gesamtoberfläche, der von jedem dieser drei Stadien eingenommen wird, errechnet er, wie lange jedes Stadium dauert in Bezug zum Gesamtzyklus, der etwa 40 Jahre ist. Das ist ein Hilfsmittel; Herr ELLENBERG wird wahrscheinlich skeptisch sein, denn es ist wieder eine indirekte Methode. Wenn man aber mit Dauerquadraten den Zyklus wirklich beobachtet hat, dann kann man an Hand dieser relativen Oberfläche berechnen, wie lange jedes dieser 3 Stadien dauert. Eine Mosaikstruktur verhindert nicht nur, daß eine Heide abstirbt, sondern sogar alte *Calluna*-Pflanzen werden oft geschont, auch wenn sie als "Insel" nur durch wenige Meter offenes Gelände von geschlossenen Beständen alter Heide getrennt sind; dies scheint die Migration der Käfer offenbar doch stark zu behindern.

F. RUNGE

Ich glaube auf Grund eines Dauerquadrates, das ich seit 12 Jahren im Naturschutzgebiet "Heiliges Meer" in der trockenen Heide liegen habe, und über das zwei Veröffentlichungen in Vegetatio vorliegen, folgendes sagen zu können: Erstens möchte ich Herrn Professor TÜXEN (und Herrn RZEDOWSKI) recht geben, die diese starke Bildung des Rasens, *Festuca* und auch *Agrostis tenuis*, auf die starke Schaf-Beweidung zurückführen. Das haben wir ebenfalls. Zweitens kann es, wie Herr BARKMAN eben sagte, auf einen Schädling - vielleicht ist es auch noch ein anderer - zurückgeführt werden. Und drittens kann Trampelei die Ursache sein. KNAPP hat auch ein Dauerquadrat gehabt; dort ist sehr viel in der Heide herumgefahren und herumgetrampelt worden, die Heide geht dann sehr schnell ein. Sie nimmt ferner, wenn kein Schnee liegt, bei sibirischer Kälte, sehr schnell ab zu Gunsten der *Cladonien*; *Empetrum* vermehrt sich dann auch. *Calluna* kann ferner in Dürrejahren sehr stark abnehmen, darauf hat BRAUN-BLANQUET für Irland hingewiesen. Das sind aber alles Fluktuationen. Einen Vegetationszyklus in einer Heide, wie ihn eben Herr BELLAMY beschrieb, habe ich bis jetzt noch nicht bemerkt; das entspricht ungefähr dem, was Herr WESTHOFF über das Vegetationsmosaik in der Heide darlegte.

R.CARBIENER
Für die Heidezyklen gibt es ganz rezente Arbeiten englischer Bodenkundler über die Mykorrhizawirkungen der *Calluna*. Vor etwa 3 Jahren haben diese festgestellt, daß die *Calluna*-Mykorrhizen eine sehr starke antibiotische Wirkung haben im Gegensatz zu Mykorrhizen von *Pinus* und *Betula*. Gewöhnlich sind es dieselben Arten. Dabei ist festgestellt worden, daß die Mykorrhizen der *Calluna* eben auch mitverantwortlich sind an dieser verjüngungshemmenden Wirkung der *Calluna*-Heiden.

H.ELLENBERG
Die früheste Definition des Vegetationszyklus, die ich gefunden habe, ist von GAMS, und von dem habe ich sie übernommen.

E.FURRER (Zürich) weist (nachträglich) auf Henry C.COWLES hin, der bereits 1911 den Ausdruck Vegetationszyklus verwendet hat. (The causes of vegetative cycles in Bot.Gaz.51.)

SUKZESSIONEN BEI EINIGEN TORFBILDENDEN PFLANZENGESELLSCHAFTEN (NACH ERGEBNISSEN VON GROSSREST-UNTERSUCHUNGEN AN TORFEN)

G. G r o ß e - B r a u c k m a n n

1) Einleitung

Da wir Torfe als fossil gewordene Reste torfbildender Pflanzengesellschaften ansehen können, stellt jedes Moorprofil gleichsam einen in seiner Gesamtheit nach und nach fossil gewordenen und damit konservierten und "registrierten" Sukzessionsablauf dar. Der Vegetationscharakter der einzelnen Sukzessionsphasen läßt sich dabei durch Großrest-Untersuchung (Makrofossil-Analalyse) der Torfe weitgehend rekonstruieren, während eine Rekonstruktion des zeitlichen Ablaufs der Sukzessionsschritte durch pollenanalytische oder durch ^{14}C-Datierungen möglich ist.

Es ist somit bei den torfbildenden Pflanzengesellschaften möglich, gut fundierte und verhältnismäßig exakte Aussagen über die an einem Wuchsort während mehrerer Jahrhunderte oder Jahrtausende abgelaufenden Sukzessionen zu machen. Ganz allgemeine Aussagen dieser Art sind übrigens bereits in jeder Moorprofil-Beschreibung enthalten, in der die Aufeinanderfoge der durch kennzeichnende Pflanzenreste definierten Torftypen wiedergegeben wird.In pflanzensoziologischer Hinsicht differenziertere Ergebnisse,die sich auch auf die gut umschriebenen Einheiten der heutigen Vegetation beziehen lassen, sind allerdings nicht ohne eingehende Laboratoriumsuntersuchungen möglich, denn sie erfordern die Berücksichtigung einer möglichst großen Zahl von Pflanzenresten aus nicht zu kleinen, in möglichst engem Abstand entnommenen Torfproben. Über Ergebnisse derartiger Untersuchungen[1)] soll im folgenden berichtet werden.

2) Zur Arbeits- und Darstellungsweise

In der bereits früher beschriebenen Weise (z.B.GROSSE-BRAUCKMANN 1963) wurden Torfproben von etwa 50 cm^3 bei meist 1 cm

[1)]Die Arbeiten wurden durch eine Sachbeihilfe der Deutschen Forschungsgemeinschaft wirksam gefördert. Dank schuldet der Verfasser weiterhin Herrn Dr. F.KOPPE, Bielefeld, der freundlicherweise wieder eine große Zahl kritischer Moosreste durchsah, sowie Herrn Dr.W.TRAUTMANN, Bad Godesberg, der zahlreiche wertvolle Anregungen und Hinweise zu Fragen der Darstellung und Deutung der Untersuchungsbefunde gab.

Mächtigkeit (gelegentlich mehr , so bei mit dem Bohrer entnommenem Material) nach Aufbereitung mit KOH auf die in ihnen enthaltenen Gewebereste untersucht und deren Anteil zunächst in Prozent der Gesamtmenge der strukturierenden Pflanzenreste geschätzt, außerdem wurden die vorhandenen Früchte und Samen ausgezählt.Mit Rücksicht auf eine übersichtliche Darstellungsweise wurden diese quantitativen Befunde dann in eine einfache Skala "übersetzt", bei der die Buchstaben s, m, h und H steigende Mengen von Früchten und Samen bedeuten, während - in Analogie zur BRAUN-BLANQUETschen Skala der "Artmächtigkeit" - die Symbole + und 1 bis 5 für steigende Mengen von Geweberesten verwendet wurden, gleichgültig, ob daneben noch Früchte und Samen vorkamen. Für die Definition der einzelnen Stufen der Skalen muß hier auf die schon genannte frühere Veröffentlichung hingewiesen werden.
Da im folgenden der Sukzessionsablauf, also die einzelnen Sukzessionsphasen und ihre Übergänge im Mittelpunkt der Betrachtung stehen, wurde für die Ergebnisse eine tabellarische Darstellungsweise gewählt, die die Beziehungen der Makrofossil-Befunde zu den soziologisch bekannten Einheiten der "rezenten" Vegetation besonders gut hervortreten läßt. Im einzelnen wird die Art der Darstellung am besten an Hand einer Tabelle erläutert.

3) Profil Giehler Moor (Topogr.Karte 1:25000; Blatt 2619) [2)]

In Tabelle 1 sind die Befunde von einem Profil des Giehler Moores wiedergegeben, einem etwa 30 km nördlich von Bremen am Westrand der Hammeniederung ("Teufelsmoor") gelegenen Hochmoor. Das Profil entstammt einem randlichen Bereich des (heute entwässerten) Moores; es wurde etwa 150 m von der Grenze des Hochmoores gegen ein von Niedermoortorfen erfülltes Tal entnommen.
Die Tabelle enthält in den senkrechten Reihen die Untersuchungsbefunde der einzelnen Torfproben in ihrer natürlichen Aufeinanderfolge, die ältesten (untersten) Proben, links, die jüngsten (obersten) rechts. Wo stärkere floristische Kontraste zwischen zwei benachbarten Proben zu erkennen waren, wurde in der Tabelle ein etwas vergrößerter Kolonnenabstand eingeschaltet, um dadurch die Gliederung des gesamten Sukzessionsablaufes in eine Anzahl in sich einigermaßen einheitlicher Phasen besonders zu

[2)] Siehe die inzwischen erschienene,ausführlichere Darstellung der Ergebnisse (GROSSE-BRAUCKMANN 1969); die hier in der Tabelle 1 wiedergegebenen Befunde entstammen dem 1969 als E bezeichneten Profil.

Additional material from Gesellschaftsentwicklung (Syndynamik),
ISBN 978-94-011-7594-4 (978-94-011-7594-4_OSFO3),
is available at http://extras.springer.com

EXTRA
MATERIALS
extras.springer.com

unterstreichen.
Die Anordnung der Arten geschah so, daß diese zu Gruppen von im wesentlichen gleichsinnigem Verhalten zusammengestellt wurden. Dabei ergaben sich, wie zu erwarten, zugleich auch Gruppierungen von ökologisch (und teilweise auch soziologisch) in vieler Hinsicht einheitlichem Charakter, wie das aus der Tabelle im einzelnen zu erkennen ist.[4)]
Lediglich bei den Gehölzen wurde abweichend verfahren: Ihre Reste sind in den meisten Torfen mehr oder weniger akzessorische Elemente, indem mit ihrem Auftreten im allgemeinen keine sonstigen kennzeichnenden Änderungen der Artenzusammensetzung parallel gehen. Daher werden die Gehölze gleichsam als Begleiter am Schluß der Tabelle als gesonderte Gruppe aufgeführt.
Eine Gruppierung unter ausschließlicher Berücksichtigung ausgesprochen soziologischer (Charakter- oder Differential-) Artengruppen war im vorliegenden Fall nicht sinnvoll, und zwar aus verschiedenen Gründen:

a) In der Tabelle müssen Vegetationseinheiten sehr verschiedener floristischer Zusammensetzung vereinigt werden, zwischen denen es im übrigen keineswegs scharfe Zäsuren sondern die mannigfachsten Übergänge gibt. Die von Phase zu Phase übergreifenden Arten tragen dabei teilweise den Charakter von Pionieren (z.B. *Eriophorum vaginatum*), teilweise den von Sukzessionsrelikten (z.B. *Cladium mariscus*). Eine scharfe, rein soziologische Gliederung müßte solche Arten demnach während ihres optimalen und "typischen" Auftretens als Charakterarten stark herausheben, während sie als Relikte und Pioniere lediglich wie Begleiter oder allenfalls Differentialarten an untergeordneter Stelle einzureihen wären.
b) Es gibt einige Arten, deren Vorkommen (und auch Fehlen) in mancher Hinsicht kennzeichnend für bestimmte Teilabschnitte des Sukzessionsablaufes ist, obwohl es sich hierbei im soziologischen Sinn um mehr oder weniger gesellschaftsvage Arten handelt (z.B. *Sphagnum palustre*).
c) Eine Anzahl Arten tritt zwar in den Torfen wie eine echte soziologische Gruppe auf, berücksichtigt man aber die Gesamtheit ihrer Vorkommen, auch außerhalb der Moore, so erweisen sie sich als soziologisch außerordentlich heterogen (z.B. *Calluna* - *Aulacomnium* - *Sphagnum magellanicum*).

4) Zur näheren Kennzeichnung der Gruppen siehe GROSSE-BRAUCKMANN 1968.

Innerhalb der jeweils nach einer oder zwei kennzeichnenden Arten benannten Gruppen wurden die Arten nicht nach ihrer Gesamtstetigkeit geordnet, sondern im allgemeinen so, daß die zuerst aufgetretenen am Ende, die zuletzt aufgetretenen (oder die sich am längsten haltenden) an den Anfang ihrer Gruppe gestellt wurden. Dadurch ließen sich etwaige Differenzierungen innerhalb der Gruppen bzw. die ihnen in der Moorentwicklung entsprechenden kleineren Wandlungen mitunter besonders deutlich machen - ebenso, wie die allgemeinen Tendenzen des gesamten Sukzessionsablaufes und der damit parallel laufenden standörtlichen Änderungen durch das Zunehmen und Abnehmen der Gruppen in ihrer Gesamtheit sehr klar zum Ausdruck kommen.

Das Profil läßt sich an Hand der Tabelle - und unter Verwendung pollenanalytischer Befunde - folgendermaßen beschreiben:

Im untersten, 2 m mächtigen Teil, (a-c) der noch in die Vorwärmezeit hineinreicht und die gesamte Frühe und Mittlere Wärmezeit umfaßt, bleibt die floristische Zusammensetzung weitgehend unverändert: Am Beginn der Moorbildung stehen moosreiche Gesellschaften mit beigemischter Weide und Birke, die bald in eine Vegetation mit Schilf und Birke sowie Moosen in wechselnden Mengen übergehen. Eine gewisse Zäsur setzt hier noch das zusätzliche Auftreten von *Cladium* und *Pinus* (Phase c), die sich beide erst im Lauf der Mittleren Wärmezeit einfinden. Eine schärfere Vegetationsumstellung ist dann mit Einsetzen der Phase d, etwa zu Beginn der Späten Wärmezeit, zu erkennen; denn hier erscheinen in großer Zahl die Arten der *Sphagnum teres-Menyanthes*-Gruppe, bei denen es sich im wesentlichen um Vertreter aus den Kleinseggen-Riedern handelt. Begleitet sind sie von *Myrica gale*, anfänglich auch noch von *Betula* und *Cladium*, die einem späteren Teilabschnitt der Phase d jedoch fehlen (die reichlichen Birkenfrüchte dürften aus der Nähe eingeweht sein). Das Verschwinden der letzten Arten der *Sphagnum teres-Menyanthes*-Gruppe beim Übergang in die Phase e (noch Späte Wärmezeit) ist gekoppelt mit dem ersten stärkeren Hervortreten von *Sphagnum palustre* und *Sphagnen* der Sektion *Cuspidata* sowie von *Eriophorum vaginatum*; dadurch kündigt sich das späte Erscheinen des eigentlichen (ombrotraphenten) Hochmoores (Phase f, nachwärmezeitlich) bereits an. Die Phase e selbst hat jedoch noch unter (schwach) minerotrophen Bedingungen gestanden (*Sphagnum palustre*).

An Hand der Tabelle läßt sich die allmähliche Verwandlung der "euminerotraphenten" Pflanzengesellschaften in das ombrogene Hochmoor gut im einzelnen verfolgen. Obwohl dieser Prozeß,aufs Ganze gesehen, einen einigermaßen kontinuierlichen Eindruck macht, lassen sich doch einige verhältnismäßig gut gegeneinander abgesetzte Stadien erkennen:

1) die an Röhricht-Pflanzen reichen Phasen (von denen es wegen der vorhandenen Gehölze und der zeitweilig reichlichen Moose[5] allerdings fraglich erscheint, wie weit sie soziologisch noch zu den eigentlichen Röhrichten gerechnet werden können),
2) die folgende Phase des Kleinseggenrieds mit *Myrica* (bzw. des *Myrica*-Busches mit einer entsprechenden Bodenvegetation), die nur gegen Ende schwache Zeichen einer gewissen Nährstoffverarmung erkennen läßt.
3) die Phase, die durch eine Reihe ombro- neben schwach minerotraphenten Arten gekennzeichnet ist und daher am zutreffendsten als hochmoorartig bezeichnet wird, und schließlich
4) die Hochmoorphase, die durch das Fehlen aller Minerotraphenten ausgezeichnet ist.

Die Übergangsphasen d und e sind hier verhältnismäßig langgezogen, stärker als es bei einem Profil mit nicht so ausgesprochener Lage am Hochmoorrand der Fall wäre. Den *Myrica*-Busch bzw. das Kleinseggenried wird man als die Vegetation des Laggs aufzufassen haben, während die hochmoorartigen Gesellschaften vom Randgehänge stammen dürften, das hier wohl nur flach geneigt war und daher keinen Standort für Bäume bot. Zur Zeit der beiden randlichen Übergangsphasen war weiter mooreinwärts das ombrotraphente, stark zersetzten Torf bildende Hochmoor bereits in kräftigem Wachstum begriffen; der stark zersetzte, sogenannte ältere Hochmoortorf ("Schwarztorf") ist in dem wiedergegebenen Profil daher nicht zur Ausbildung gekommen.

Abgesehen vom Fehlen des "Schwarztorfes" und der sehr starken Ausbildung der den Übergang zum Hochmoor bildenden Torfe - beides Folgen der Lage am Hochmoorrand - entspricht das in Tabelle 1 wiedergegebene Profil einem in Nordwestdeutschland sehr verbreiteten Typhus: dem des Versumpfungsmoores, das das Hochmoorstadium erreicht hat. Im folgenden seien nun noch die Untersu-

[5] Eine "rezente" Entsprechung hierzu beschreibt GROSSER 1963 (p.144) als S p h a g n u m r e c u r v u m - P h r a g m i t e s - G e s e l l - s c h a f t .

chungsbefunde von einem weiteren Versumpfungsmoor-Profil wiedergegeben, bei dem die Entwicklung nach Ausweis der Torfzusammensetzung viel weniger "geradlinig" vor sich gegangen ist als im ersten Beispiel:

4) Profil Poggenpohlsmoor (Topogr.Karte 1:25000; Blatt 3016)
Am östlichen Rand der Hunte-Aue, nördlich des Dorfes Dötlingen (gut 30 km westsüdwestlich von Bremen) liegt das etwa 40 ha große, wegen zahlreicher floristischer Besonderheiten seit 1939 unter Naturschutz stehende Poggenpohlsmoor; von seinem östlichen Rand, der etwa 30 m von der ziemlich steil ansteigenden Geest entfernt ist, stammt das in Tabelle 2 wiedergegebene Profil.
In der Tabelle ist im wesentlichen die gleiche Gruppen-Gliederung wie bei Tabelle 1 vorgenommen worden[6], lediglich mit der Abweichung, daß die *Sphagnum palustre*- und *Cuspidata*-Artengruppe zusammengefaßt wurden und daß als eine weitere Einheit die *Potentilla-Rubus*-Gruppe erscheint. Bei dieser handelt es sich offenbar um Arten, aus denen sich Hinweise auf ein wenigstens stellenweise verlangsamtes Moorwachstum entnehmen lassen; dabei stehen allerdings *Potentilla erecta* und *Hypnum cupressiforme* (einschl. *H. ericetorum*) als stark säureertragende und wenig nährstoffanspruchsvolle, auch an offenen Standorten vorkommende Arten in einem gewissen Gegensatz zu den drei anderen Arten, die als Wald- (oder Waldlichtungs-)Pflanzen und zugleich (wenigstens z.T.) als Zeiger günstiger Ernährungsbedingungen und einer gewissen Nitrat-Freisetzung gelten (OBERDORFER 1962).
Vegetationsgeschichtlich weist das Profil - nach den pollenanalytischen Befunden, die hier nicht im einzelnen wiedergegeben werden können - insofern eine Besonderheit auf, als es offenbar gegen Ende der Entwicklungsphase d (bzw. zwischen d und e) zu einer völligen oder weitgehenden Stagnation des Moorwachstums gekommen ist, die den größten Teil der Mittleren und Späten Wärmezeit umfaßt; diesem Hiatus entspricht auch ein verhältnismäßig scharfer Kontrast in der Artenzusammensetzung der Torfe, der bei allen Artengruppen zum Ausdruck kommt.
An Hand der Tabelle lassen sich die folgenden Moorentwicklungsphasen feststellen:

a) moosreiches Kleinseggenried, anfänglich mit *Empetrum*, ferner

[6] Obwohl *Sphagnum magellanicum* fehlt, wurde die Gruppe der ombrotraphenten Arten wie in Tabelle 1 als *Sph. magellanicum*-Gruppe bezeichnet.

zeitweilig mit *Salix* und *Betula nana* (überwiegend vorwärmezeitlich),

b) Zwergbirkenbestand mit moosreicher Bodenvegetation vom Charakter eines Kleinseggenrieds, durchsetzt sowohl mit Arten der (anspruchsvolleren) *Dryopteris*- wie der (anspruchsloseren) *Sphagnum magellanicum*- und *Sph.palustre*-Gruppe,

c) kurze "hochmoorartige" Phase,

d) *Phragmites-Cladium*-Röhricht mit einzelnen Moosen aus der *Sphagnum magellanicum*- und *Sph.palustre*- sowie der *Sphagnum teres-Menyanthes*-Gruppe. Das ebenfalls vorhandene Erlenholz ist hier wohl wenigstens teilweise sekundär eingewachsen und stammt dann aus der Zeit des weitgehenden Moorwachstums-Stillstandes.

e) Moosreiches Kleinseggenried, in dem aber zugleich zahlreiche Vertreter der *Sphagnum magellanicum*-Gruppe (besonders *Sph.imbricatum* und *Erica*) vorkommen, und dazu schließlich auch noch die sämtlichen Arten der *Potentilla-Rubus*-Gruppe.

Die sonderbare Artenkombination während der jüngsten Entwicklungsphase kann wohl nur als Folge eines engräumig mosaikartigen Wechsels zwischen zwei soziologisch und ökologisch sehr ungleichen Vegetationseinheiten aufgefaßt werden, wobei die Arten der *Sphagnum teres-Menyanthes*-Gruppe in sehr nassen Schlenken gelebt haben (*Scorpidium, Characeen!*), während die Vertreter der *Sphagnum magellanicum-, Sph.palustre-* und *Potentilla-Rubus*-Gruppe auf Bulten vorgekommen sein müssen, bei so engräumigen Übergängen, daß die beiden Vegetationseinheiten in den abgelagerten Torfen nicht mehr voneinander getrennt werden können (Vegetationskomplex im Sinne von GROSSER 1965). Dafür spricht auch, daß noch heute auf einer kleinen Fläche des Poggenpohlsmoores ein in Bulte und Schlenken gegliederter Vegetationskomplex vorkommt, in dem *Sphagnum imbricatum* und *Sph.rubellum, Erica, Calluna* u.a. neben *Scorpidium, Drepanocladus intermedius, Calliergon giganteum* u.a.[7] in unmittelbarer Nachbarschaft nebeneinander (aber nicht durcheinander) leben. Damit handelt es sich hier um eines der wenigen Moore, die wenigstens auf einem Reststück noch von der natürlichen (und ursprünglichen) torfbildenden Vegetation bedeckt werden, so daß

[7] Unter diesen auch *Cinclidium stygium* und *Leiocolea rutheana* (nach freundlicher Bestimmung durch Herr Dr.F.KOPPE); von beiden sind bislang sonst keine Fundorte in Niedersachsen bekannt (vgl.KOPPE 1964).

die sie zusammensetzenden Arten im eigentlichen, engsten Sinne hier als Relikte angesprochen werde können.
Das Profil ist in noch anderer Hinsicht bemerkenswert: Es läßt nämlich keine einheitlich durchgehende Entwicklungstendenz vom Eu- zum Oligotrophen erkennen, vielmehr haben die frühwärmezeitlichen Phasen b und c und die nachwärmezeitliche Phase e trotz deutlicher floristischer Unterschiede manche miteinander gemeinsame Züge durch das Neben- und (scheinbar) Durcheinander von eutraphenteren und oligotraphenteren Arten und Artengruppen, unter denen die Laubmoose und *Sphagnen* zu beiden Zeiten eine sehr große Rolle gespielt haben. Dieser gewissen Konstanz seiner Pflanzendecke müssen ebenso konstante Standortsbedingungen zugrunde liegen, besonders hinsichtlich der Hydrologie, die hier offenbar von Anfang an bestimmt gewesen ist durch das am Auenrand austretende und mehr oder weniger hochgedrückte, wohl relativ kalk- und nährstoffreiche Wasser der angrenzenden, wesentlich höher liegenden Geesthöhen.
Den beiden Versumpfungsmoor-Profilen vom Neu-Kuhstedter und Poggenpohlsmoor seien im folgenden nun noch die Befunde aus einem Verlandungsmoor gegenübergestellt:

5) Dreckmoor am Steinhuder Meer (Topogr.Karte 1:25000; Blatt 3521)
Das Steinhuder Meer, Niedersachsens größter Binnensee (etwa 30 km nordwestlich von Hannover gelegen), wird an seinem westlichen Rand von ausgedehnten Mooren umsäumt, die zum größten Teil auf den organischen Sedimenten des Sees, einer mehr oder weniger konsolidierten Mudde, aufgewachsen sind[8]. Es handelt sich also um den sehr bekannten, wenn auch in den altpleistozänen Gebieten Nordwestdeutschlands nicht eben häufigen Typus des Verlandungsmoores. Der allgemein geläufigen Vorstellung von der Aufeinanderfolge der Pflanzengesellschaften beim Verlandungsprozeß entspricht auch die Vegetationszonierung des am nordwestlichen Seeufer gelegenen Dreckmoores; denn vom offenen Wasser her folgen einander hier, teilweise mit ziemlich scharfen Grenzen, ein Schilfröhricht, ein Erlenbruchwald und ein (allerdings großenteils in Grünland umgewandelter) Birkenbruchwald.
Aus dem Bereich des Birkenbruchwaldes, etwa 500 m vom offenen Wasser entfernt, stammt das Profil, dessen pflanzliche Zusammen-

[8] Hierzu siehe die inzwischen erschienenen Arbeiten von GRAHLE 1968, sowie MÜLLER 1968 und 1969.

setzung in der Tabelle 3 wiedergegeben ist[9]. Die Gruppeneinteilung der Arten entspricht in den Grundzügen wieder derjenigen der Tabellen 1 und 2, lediglich mit den folgenden Abweichungen:

1) *Drepanocladus fluitans* wurde zur *Sphagnum cuspidata*-Gruppe gestellt,
2) es wurde eine Gruppe nur in der Mudde vorkommender Arten ausgeschieden,
3) *Agrostis canina, Carex echinata, Juncus articulatus* und *Rubus idaeus*, die nur vereinzelt in der Tabelle vorhanden sind, wurden zu einer (*"Rubus*- und *Juncus-")*Gruppe vereinigt, da sie in anderen Profilen desselben Moores, in denen sie häufiger auftraten, eine gewisse Zusammengehörigkeit erkennen ließen; im einzelnen wird darauf noch zurückzukommen zu sein.

Die Torfschichten über der Mudde lassen eine Viergliederung erkennen, wobei die erste Phase (a) durch die eutraphente *Dryopteris*-Gruppe und eine Reihe(ebenfalls relativ eutraphente) Vertreter der *Sphagnum teres-Menyanthes*-Gruppe sowie durch weitgehendes Fehlen der Vertreter der oligo- bzw. ombrotraphenten *Sphagnum magellanicum*-Gruppe sehr scharf gegen die weiteren Torfschichten abgesetzt ist. In der letzten Phase (d) sind ausschließlich ombrotraphente Arten vorhanden (zu denen hier offenbar auch *Drepanocladus fluitans* zu rechnen ist), so daß die Torfe also als eigentliche Hochmoortorfe angesprochen werden können. In den übrigen beiden Phasen (b und c) dominieren ebenfalls die Ombrotraphenten; da daneben aber (fast) regelmäßig die Arten der *Sphagnum palustre*-Gruppe vorhanden sind, kann hier nur von hochmoorartigen Torfen bzw. Pflanzengesellschaften die Rede sein. Dabei ist die Phase b durch das Fehlen, die Phase c durch das Vorhandensein von *Carex echinata* und *Agrostis canina* ausgezeichnet.

Eine Reihe seewärts anschließender Profile wie einen im wesentlich gleichen, lediglich im Detail modifizierten Aufbau auf;das wird durch die Skizze des Profilschnittes (Abb.1) verdeutlicht. Der Weg, der zu dieser Skizze führte, ähnelt dem beim Entwurf einer pflanzensoziologischen Karte üblichen Weg: An Hand zahlreicher, aus mehreren Profilen stammender Torfuntersuchungs-Befunde (pflanzensoziologischen Aufnahmen analog) wurden nach der

[9] Befunde von zwei weiteren Profilen, auch mit Angaben zur Datierung, siehe in GROSSE-BRAUCKMANN 1968.

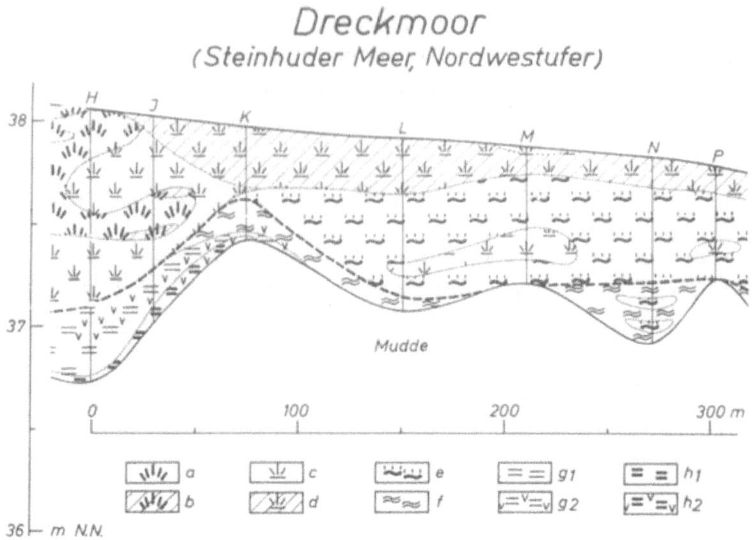

Abb. 1. Verteilung pflanzensoziologisch definierter Torfarten in einem nord-südlich verlaufenden Schnitt durch das Dreckmoor am Steinhuder Meer (Südrichtung rechts, = Richtung zum Schwingrasenrand). Die verschiedenen Signaturen entsprechen den in den Stetigkeitstabellen 4 und 5 ausgeschiedenen Torf-Ausbildungsformen; die Buchstabenbezeichnungen sind dieselben (siehe hierzu auch die Fußnoten 10 und 11).

bei der pflanzensoziologischen Tabellenarbeit üblichen Arbeitsweise zunächst floristisch definierte, durch eine charakteristische Artenverbindung ausgezeichnete Torftypen ausgeschieden und diese in der Skizze in ihrer räumlichen Verteilung dargestellt. Die unterschiedenen Typen gehen aus den beiden Stetigkeitstabellen 4 und 5 hervor[10], die mit Buchstaben bezeichneten Ausbildungsformen entsprechen den in der Skizze dargestellten, pflanzensoziologisch unterschiedenen Torflagen[11].

Hiernach stellt sich der Sukzessionsablauf in dem dargestellten Teilstück, das nach pollenanalytischen Befunden im Bereich der Profile H und J eine Zeitspanne von etwa einem Jahrtausend, im Bereich der Profile M bis P dagegen nur einige Jahrhunderte vor der Gegenwart umfaßt, folgendermaßen dar: Nach der Ablagerung der Mudde fanden sich zunächst e n t w e d e r (Prof. H-K) eu-

[10] Ihnen liegen allerdings noch Untersuchungsbefunde aus einer Reihe weiterer, in Abb. 1 nicht wiedergegebener Profile desselben Moores zugrunde.

[11] Nicht dagegen alle durch Zahlenindices bezeichneten; diese wurden übrigens teilweise nach mehr quantitativen Unterschieden im Vorkommen der Artengruppen voneinander abgesetzt (e_1/e_2), teilweise aber auch nach dem Vorkommen bestimmter "Differential"-Gruppen.

traphente Pflanzengesellschaften, etwa vom Charakter der C i c u t a v i r o s a - C a r e x p s e u d o c y p e r u s - A s s o z i a t i o n (Ausbildungsform h; *Carex pseudocyperus*-Gruppe!) ein, o d e r (bei Profil L und N) die erste torfbildende Vegetation trug insgesamt mesotraphenten Charakter durch (gleichzeitig oder eng mosaikartiges oder zeitlich rasch wechselndes) Vorkommen von anspruchslosen und anspruchsvollen Arten (Ausbildungsform f), oder schließlich (bei Profil M und P) die erste Pflanzengesellschaft war bereits ein oligo-mesotraphentes S c h e u c h z e r i e t u m , allerdings in einer durch Vertreter der *Menyanthes*-Gruppe ausgezeichneten, nicht extrem nährstoffarmen Variante (Ausbildungsform e). Dieses S c h e u c h z e r i e t u m breitete sich bald großflächig aus und blieb geraume Zeit erhalten. Erst in jüngerer Zeit verschwanden hier *Scheuchzeria* und *Carex limosa* und es blieb eine Vegetation (Ausbildungsformen c und d), die zwar soziologisch als eine O x y c o c c o - S p h a g n e t e a - Gesellschaft anzusehen ist, aber doch nicht rein ombrotraphent (*Sph.magellanicum*-Gruppe!) sondern leicht minerotroph beeinflußt ist (*Sph.palustre*-Gruppe, teilweise auch *Menyanthes*-Gruppe!). Diese "hochmoorartige" Vegetation ist im Bereich der Profile H und J bereits frühzeitiger aufgetreten, und hier hat sich aus ihr zeitweilig und stellenweise auch eine rein ombrotraphente Vegetation (Ausbildungsform a und b) entwickelt. Im Bereich K bis P ist die hochmoorartige Vegetation noch besonders dadurch ausgezeichnet, daß dort zusätzlich die Arten der *Juncus*-Gruppe auftreten (Ausbildungsform d). Es ist sehr wahrscheinlich, daß es sich hier um Symptome von Störungen der Moorvegetation handelt: einerseits um die Wirkung einer geringfügigen Entwässerung (darauf scheint die *Potentilla-Rubus*-Gruppe hinzudeuten; Ausbildung d_1), zum andern aber, wenigstens zeitweilig und stellenweise, auch um die Wirkung einer schwachen Durchweidung (darauf deutet wohl die insgesamt etwas eutraphentere *Carex canescens*-Gruppe; Ausbildung d_3), mit der man bei dem geringen Alter dieser Torfe mit einiger Sicherheit wird rechnen dürfen. Gelegentlich fanden sich übrigens die "Störungszeiger" auch in den echten Hochmoortorfen (Ausbildungsform b).

Auffällig ist an der Entwicklung des hier geschilderten Verlandungsmoores das starke Hervortreten meso- bis oligo-mesotraphenter Pflanzengesellschaften, mit denen - im Gegensatz zu dem,was man nach der heutigen Vegetationszonierung annehmen möchte -

stellenweise sogar die Moorbildung beginnt, obwohl das Steinhuder Meer ein eindeutig eutropher See mit einer entsprechenden Muddeablagerung ist. Diese Erscheinung hängt offenbar ursächlich damit zusammen, daß nach Ausweis der Pollenanalysen die Torfbildung von K bis P nahezu gleichzeitig oder sehr rasch zum See hin fortschreitend, also sicherlich nach einer sehr starken Muddeablagerung begonnen hat, und unter solchen Voraussetzungen sind offenbar die Bedingungen für eine sehr rasche Verarmung, sei es durch Erschöpfung der vorhandenen oder mangelnde Zufuhr neuer Nährstoffvorräte besonders leicht gegeben. Im einzelnen kann diese (von KULCZYNSKI 1949 eingehend erörterte, von GROSSER 1963 (S.131) gelegentlich berührte) Frage hier jedoch nicht weiter diskutiert werden.

6) Folgerungen

Aus den gebrachten Beispielen und einigen weiteren Untersuchungsfunden lassen sich - über die Rekonstruktion der jeweiligen Moorentwicklung hinaus - die folgenden allgemeineren Aussagen über Sukzessionsvorgänge torfbildender Pflanzengesellschaften ableiten:

1) Einerseits gibt es offensichtlich eine Art autonomer Sukzessionen im Laufe des Moorwachstums. Sie führen von nährstoffanspruchsvolleren zu nährstoff-anspruchsloseren Pflanzengesellschaften, im Extrem zum ombrogenen Hochmoor (erstmals von WEBER erkannt und 1907 beschrieben).
2) Andererseits beharren viele Pflanzengesellschaften der Moore an ihrem Wuchsort ohne nennenswerte Veränderungen während langer Zeiträume, u.U. viele Jahrhunderte hindurch. Daraus läßt sich folgern, daß in diesen Fällen die Wirkung konstanter Standortsbedingungen entscheidend war und dadurch irgendwelche "autonomen" Entwicklungstendenzen nicht zur Auswirkung kommen konnten.
3) Wenn es nach langdauernder pflanzensoziologischer Konstanz zu mehr oder weniger plötzlichen Wandlungen des Vegetationscharakters kommt - unter Umständen (aber nicht unbedingt) im Sinne der Tendenz vom Eu- zum Oligotraphenten -, so muß daraus auf gleichzeitige, durch äußere Umstände bewirkte Änderungen der Standortsbedingungen geschlossen werden. Und ebenso sind (stand-) örtliche Verschiedenheiten selbstverständlich die Ursache für alle Variationen im Sukzessionsverlauf vom Eu- zum Oligotraphenten, gleichgültig, ob sie mehr im Quantitativen liegen (verschiedene Dauer, u.U.sogar

Ausfall einzelner Entwicklungsphasen) oder mehr im Qualitativen (verschiedenartige Gesellschaften während einander sonst entsprechender Entwicklungsphasen).

4) Die gegenwärtige Zonierung torfbildender Pflanzengesellschaften muß nicht den wirklich abgelaufenen, aus den Torfen abzulesenden Sukzessionen entsprechen. Das gilt im Besonderen für die Verlandung der Seen, zumal hier die Sukzessionen der torfbildenden Pflanzengesellschaften außerordentlich von der Art und Geschwindigkeit der voraufgehenden Mudde-Sedimentation abhängen.

Außerdem sollte durch die gebrachten Beispiele gezeigt werden, daß botanische Großrest-Untersuchungen an Torfen zur Vermehrung unserer Kenntnisse von den Sukzessionen torfbildenden Pflanzengesellschaften wesentlich beitragen können,und der bei der Auswertung und Darstellung beschrittene Weg sollte hiermit zur Diskussion gestellt werden.

SUMMARY

Macrofossil analyses of peat samples give an idea of the local successions of the peat-forming vegetation of the past. This is demonstrated by the analysis of three examples of peat profiles from different sites of North Western Germany.

Some general results are as follows:

(1) The present zonation of peat-forming plant communities can but does not necessarily correspond to the successions having actually taken place.

(2) The successional direction from ± eutrophic to ± oligotrophic types seems to be an endogenous tendency of all peat-forming vegetations (cf.C.A.WEBER 1907), but the real process of such successions is highly controlled by environmental, especially by water conditions (cf. KULCZYNSKI 1949).

(3) In most cases successional phases have lasted for more than a hundred, sometimes for some thousands of years without any essential floristic changes.

The following results are of more special interest:

(4) When a lake and its organic sediments are eutrophic, the first peat formation in the terrestrialisation process does not necessarily have to be eutrophic, too. In contrast, in the eutrophic lake investigated here, partly oligo-mesotrophic peats derived from oligo- mesotrophic plant communities were

found. Several factors contributing to an explanation of this phenomen are discussed.

(5) Plant communities forming oligo-mesotrophic peats under certain circumstances undergo specific floristic changes when they are disturbed by grazing or by small range drainage (e.g. by shallow and distant trenches). This is clearly expressed by the floristic composition of the peat.

(6) In the case of a particular raised bog it was possible to reconstruct the original character of vegetation of the marginal zones (lagg and marginal slope) today covered by the uncharacteristic secondary vegetation known of drained bogs.

Methodically the paper demonstrates the application of some phytosociological concepts to the findings of botanical peat investigations and the possibilities of interpretation of the results with respect to composition, ecology and succession of peat forming plant communities.

ZUSAMMENFASSUNG

Großrestuntersuchungen an Torfen, die an Proben ausreichenden Volumens und nicht zu großer Dicke vorgenommen werden, ermöglichen im allgemeinen eine Rekonstruktion der ursprünglichen torfbildenden Vegetation. Demgemäß geben solche Untersuchungen,wenn man sie an einer genügend engen Probenfolge eines ganzen Moorprofils durchführt, Aufschluß über die dort abgelaufenen Sukzessionen. Eine übersichtliche Darstellung solcher Befunde ist unter anderem mit Hilfe von Tabellen bei geeigneter Artenanordnung möglich.

Untersucht wurden ein Verlandungsmoor und zwei Versumpfungsmoore, von denen eines das Hochmoorstadium erreicht hatte. Dabei ergaben sich die folgenden allgemeineren Ergebnisse:

In den Mooren gibt es eine autonome Entwicklungstendenz von anspruchsvolleren zu anspruchsloseren Pflanzengesellschaften, im Extrem zum Gesellschaftskomplex des Hochmoores, die von einer zunehmenden Verarmung des Standorts bedingt wird. Ob und wann diese Entwicklungstendenz aber zur Auswirkung kommt, und welchen soziologischen Charakter und welche Dauer die dabei durchlaufenden Entwicklungsphasen haben, das ist von Moor zu Moor äußerst verschieden und scheint aufs engste von den jeweiligen, in manchen Fällen sich plötzlich ändernden Standortsbedingungen (besonders den hydrologischen) gesteuert zu werden. Die Dauer

der in sich ziemlich konstanten Entwicklungsphasen der Moore kann dabei beträchtlich sein und viele Jahrhunderte, u.U.selbst einige Jahrtausende betragen. Die tatsächlich abgelaufenen Sukzessionen müssen nicht der gegenwärtigen Zonierung der torfbildenden Pflanzengesellschaften entsprechen.
Weitere speziellere Ergebnisse betreffen den Charakter der Vegetation am Randgehänge und im Randsumpf eines nordwestdeutschen Hochmoores, die Auswirkung von Störungen (Durchweidung, auch leichte Entwässerung) auf die Zusammensetzung hochmoorartiger Pflanzengesellschaften und die Anfangsphase der Moorbildung bei der Seenverlandung, wobei auch im Fall eines eutrophen Sees als erste torfbildenden Vegetation unter bestimmten Umständen eine nur oligo-mesotraphente Pflanzengesellschaft auftreten kann.

LITERATUR

GRAHLE,H.-O. -1968- Limnogeologische Probleme in Nordwestdeutschland.- (Geologische Untersuchuchungen an niedersächsischen Binnengewässern IV). Z.Deutsche Geol.Ges. (Jahrg.1965) 117 (2-3): 727-737. Hannover.
GROSSE-BRAUCKMANN,G. -1963- Über die Artenzusammensetzung von Torfen aus dem nordwestdeutschen Marschen-Randgebiet (Eine pflanzensoziologische Auswertung von Großrestuntersuchungen).- Vegetatio 11: 325-341.Den Haag.
-- -- -1968- Einige Ergebnisse einer vegetationskundlichen Auswertung botanischer Torfuntersuchungen, besonders im Hinblick auf Sukzessionsfragen.- Acta Bot.Neerl.17: 59-69. Leiden.
GROSSER,K.H. -1968- Die Moor- und Bruchgesellschaften.-In Scamoni: 122-167.
-- -- -1965- Vegetationskomplexe und Komplexgesellschaften in Mooren und Sümpfen.- Feddes Repert.Beih.142: 208-216. Berlin.
KOPPE,F. -1964- Die Moose des Niedersächsischen Tieflandes.- Abh.naturw. Ver.Bremen 36: 237-424. Bremen.
KULCZYNSKI,S. -1949- Peat bogs of Polesie.- Mém.Acad.Polon.Sci.Lettr.,Cl. Sci.Mat.Nat.,Sér.B.Sci.Nat.15: 356. Cracovie.
MÜLLER,H. -1968- Zur Entstehung und Entwicklung des Steinhuder Meeres. Geologische Untersuchungen an niedersächsischen Binnengewässern VI.- Gas- u.Wasserfach 109: 538-541. Hamburg.
-- -- -1969- Diskordanzen und Umlagerungserscheinungen in holozänen Sedimenten flacher Seen Nordwestdeutschlands.- Mitt.Intern.Ver.Limnologie 17: 211-218. Stuttgart.
OBERDORFER,E. -1962- Pflanzensoziologische Exkursionsflora für Süddeutschland und die angrenzenden Gebiete.- Stuttgart. 987 pp.
SCAMONI,A. -1963- Natur, Entwicklung und Wirtschaft einer jungpleistozänen Landschaft, dargestellt am Gebiet des Meßtischblattes Thurow (Krs.Neustrelitz). Teil I.- Wiss.Abh.Deutsch.Akad.Landwirtschaftswiss.Berlin Sekt.Landeskultur u.Naturschutz. Berlin. 340 pp.
WEBER,C.A. -1907- Aufbau und Vegetation der Moore Norddeutschlands.- Englers bot.Jahrb.40, Beibl.90: 19-34. Leipzig.

Nachtrag (September 1978): Über die Untersuchungen die diesem 1967 gehaltenen Vortrag zugrundelagen, wurde inzwischen in mehreren Publikationen berichtet, die neben ausführlicherem Tabellenmaterial auch chemische, pollenanalytische und ^{14}C-Befunde im Detail enthalten:

GROSSE-BRAUCKMANN,G. -1969- Zur Zonierung und Sukzession im Randgebiet eines Hochmoores (nach Torfuntersuchungen im Teufelsmoor bei Bremen).- Vegetatio 17: 33-49. Den Haag.

-- -- & DIERSSEN,K. -1973- Zur historischen und aktuellen Vegetation im Poggenpohlsmoor bei Dötlingen (Oldenburg).- Mitt.Flor.-soz.Arbeitsgem.N.F.15/16: 1+9-145. Todenmann u.Göttingen.

-- -- -- -- -1975- Über Beziehungen zwischen chemischen Merkmalen von Torfen und ihrem botanischen Charakter.- Moor und Torf in Wissenschaft und Wirtschaft: 145-155. Bad Zwischenahn.

-- -- -- -- -1974 u.1976- Zum Verlauf der Verlandung bei einem eutrophen Flachsee (nach quartärbotanischen Untersuchungen am Steinhuder Meer) I u.II.- Flora 163: 179-229 u. 165: 415-455. Jena.

DISKUSSION

D.MUELLER-DOMBOIS

Die am Poggenpohlsmoor dargestellte Überlagerung in der Moorbildung von eutrophischen über oligotrophische Elemente ist sehr interessant, weil es dem allgemein bekannten Bild nicht entspricht, und wohl auch einen weniger verbreiteten Fall darstellt. Im südöstlichen Manitoba haben wir am Fuße einer Endmoräne eine ähnliche Folge gefunden, wo eutrophische, mosaikartige Überlagerungen wohl von ungleichmäßigen Hangwasserzuschüssen herrührten. Der Fall zeigt, daß man bei derartigen Mooruntersuchungen immer auch die umliegenden physiographischen Landformen berücksichtigen sollte.

R.TÜXEN

Wir sahen eine sehr heterogene Gruppe in Ihren Tabellen, die *Rubus*-Gruppe. Da stand *Rubus fructicosus* und *R.idaeus*.Könnte es da nicht möglich sein, daß auch *Rubus chamaemorus* darin wäre, der heute noch an mehreren Stellen in NW-Deutschland vorkommt und von dem wir nicht wissen, seit wann er hier ist?

G.GROSSE-BRAUCKMANN

Die geomorphologischen Lageverhältnisse jeden Moores sind sicher von entscheidender Bedeutung für den Gesamtcharakter der Moore und der von ihnen durchlaufenen Sukzessionen. Darauf ist übrigens in jüngster Zeit besonders von polnischen Kollegen mehrfach hingewiesen worden, die gerade jetzt versuchen, eine geomorphologische Klassifikation der Moore aufzustellen.
Die Steinkerne von *Rubus chamaemorus* sind mit Sicherheit von den *R.fructicosus-R.idaeus*-Früchten, die morphologisch häufig etwas ineinander übergehen, zu unterscheiden. *Rubus chamaemorus*-Steinkerne wurden in nw-deutschen Mooren überhaupt noch nicht angetroffen.

H.ELLENBERG

Ich beglückwünsche Herrn GROSSE-BRAUCKMANN zu seinen Ergebnissen, die endlich genauen Aufschluß über den Verlauf langdauernder Verlandungssukzessionen geben. Großflächige, plötzlich einsetzende oligotraphente Vermoorungen sind z.B. aus Estland bekannt und hingen dort mit tektonisch bedingten Verlagerungen der Wasserscheide zusammen, was dazu führte, daß die Mineralbodenwasserzufuhr, im Sinne von DU RIETZ, nun nicht mehr stattfand.

Könnten ähnliche tektonische Ursachen auch am Steinhuder Meer zutreffen?

V.WESTHOFF

Diese Methodik der Erforschung der Sukzession durch Analyse des Substrates ist in den Niederlanden schon früher angewandt worden und zwar von Frau Dr.W.A.E.VAN DONSELAAR-TEN BROKKEL HUININK in der niederländischen Zeitschrift Wentia (1959, The Ecology of old river beds) bei der Erforschung von Altwässern der Flüsse.

Wie sind die Termini "anspruchsvoll" und "anspruchslos" hier zu verstehen? Die stenöken Arten des ombrotrophen Hochmoores stellen doch die höchsten Ansprüche an die Umwelt, und die mehr vagen (euryöken) Arten wie *Phragmites* sind vielmehr als anspruchslos zu bezeichnen.

D.J.BELLAMY

One classic case of this reverse succession to be seen in England due to agricultural practices is Wybunbury Moss (WALKER & POORE 1956). On the small catchment area they are putting lime and other agricultural chemicals and now we have Niedermoor where before these was a Hochmoor. But one very interesting thing in the study of these peat profiles and succession: The hydrologic and chemical peat types are fossilized with the plant and animal remains and one can analyse the profile chemically and see all these changes. And this can help greatly to understand how the changes have come about. If one classifies peat types or vegetation types from peat lands on the ground of hydrology there appears to be seven types. These are seven types which were first described by HUGO SJÖRS for Scandinavia and these seem to be almost universal throughout Europe and North America. I think by analysing chemically the profiles one can get a much clearer picture of the successions.

F.RUNGE

Die Untersuchungen von Herrn GROSSE-BRAUCKMANN sind schon deshalb zu begrüßen, weil wir jetzt besser darüber Bescheid wissen, wie lange bestimmte Pflanzen wie *Potentilla erecta, Lychnis flos-cuculi, Galium palustre, Cicuta virosa* bei uns einheimisch sind.

Es wäre erwünscht, wenn dort, wo der Hochmoorrand noch erhalten ist, einmal die Großreste im Torf, ausgehend von den heutigen Pflanzengesellschaften des Moorrandes, in Richtung Moormitte untersucht würden. Es wäre doch sehr interessant, wenn nun einmal die Pflanzengesellschaften nicht nur in der Vertikalen, sondern auch einmal in der Horizontalen untersucht würden.

G.LAVRENTIADES
Kann man genau wissen, ob alle die Pflanzengruppen ursprünglich sind, oder ob sie auf menschlichen Einfluß zurückzuführen sind?

G.GROSSE-BRAUCKMANN
Zu der Frage von Herrn Professor ELLENBERG möchte ich sagen: Mehr oder weniger plötzliche hydrologische Änderungen, wie das Entstehen einer "Wasserscheide-Situation", sind sicher immer die Ursache dafür, wenn eutrophe Ablagerungen, bzw. eutraphente Pflanzengesellschaften mit einer ziemlich scharfen Zäsur in oligotrophe bzw. oligotraphente übergehen, worauf ja für die Moore KULCZYNSKI besonders eindringlich hingewiesen hat. Im Falle des Steinhuder Meeres ist eine Spiegelabsenkung, durch die auch größere Muddeflächen trockengefallen sein könnten, gut denkbar; allerdings schafft wohl auch ohne Seespiegelsenkung schon eine "überstürzte" Muddeausschwemmung und -ablagerung in einem größeren Bereich des Sees ähnliche hydrologische Bedingungen.
Zu der Frage von Herrn WESTHOFF: Mit "anspruchsvoll" war durchweg "nährstoffanspruchsvoll" im Sinne von mehr oder weniger eutraphent gemeint.
Chemische Untersuchungen, wie sie Herr BELLAMY vorschlug, der botanisch bearbeiteten Moorprofile können natürlich eine wertvolle Ergänzung liefern; zu den vorgeführten Profilen liegen übrigens auch einige chemische Untersuchungsbefunde vor. Es muß aber bei chemischen Untersuchungen an Torfen immer bedacht werden, daß durch die hohe Kationen-Umtauschkapazität der Torfe auch sekundäre Veränderungen im Elektrolytgehalt der Torfe denkbar sind; daher muß nicht in jedem Fall der Chemismus eines Torfes auch ein getreues Abbild der chemischen Standortsverhältnisse der ursprünglichen, torfbildenden Pflanzengesellschaften sein.

Eine "paläosoziologische Kartierung" in der die Horizontalen, auf die Herr Dr.RUNGE hinwies, wäre äußerst interessant, erfordert aber einen beträchtlichen Aufwand an Untersuchungsarbeit. Hinweise in dieser Richtung lassen sich jedoch auch schon aus Profil- R e i h e n entnehmen. So ergaben weitere, nach dem Moorrand hin, an das gezeigte Teufelsmoor-Profil anschließende Profile, daß die Hochmoortorfe bald ganz auskeilen, während die Übergangsbildungen (mit *Myrica*) zunächst an Mächtigkeit zunehmen, schließlich aber brechen auch diese ab, und es schließt sich ein Niedermoorgebiet an - es handelt sich um ein kleines Bachtal - in dem die Torfbildung erstaunlicherweise schon bald nach Beginn der Mittleren Wärmezeit zum Stillstand gekommen ist. Im Randbereich des Hochmoores läßt sich also eine deutliche, wenn auch begrenzte, Transgression feststellen, während im Zentrum, bzw. im Hauptteil des Moores die Hochmoorbildung mehr oder weniger gleichzeitig, und auch schon viel früher als am Rand, begonnen hat.

Was die Frage von Prof.LAVRENTIADES betrifft: das Auftreten der *Juncus*- und *Carex canescens*-Gruppe dürfte zum größten Teil anthropogen bedingt sein (schwache Entwässerung, leichte Eutrophierung durch Beweidung); die übrigen Gruppen gehören zur Vegetation der natürlichen Moor-Entwicklungsstadien.

H.ELLENBERG

Ich bin Herrn WESTHOFF dankbar, daß er diese grundsätzliche Frage angeschnitten hat und möchte doch sagen, daß wir Pflanzensoziologen und -Ökologen uns doch angewöhnen sollten, eine exaktere Ausdrucksweise hinsichtlich der sogenannten Ansprüche der Arten zu gebrauchen. Niemand wird "anspruchsvoll" in diesem vermenschlichten Sinne gebrauchen. Aber was wissen wir überhaupt über die Ansprüche der physiologischen Notwendigkeiten der Pflanzen? In allem, was wir experimentell geprüft haben, haben wir gesehen, daß die Amplituden der Arten, wenn man sie ohne Konkurrenten kultiviert, häufig sehr viel breiter sind, auch in Richtungen hinein, die man in der Natur gar nicht ahnen würde. *Bromus erectus* ist gar keine trockenheitsliebende Pflanze, und *Luzula luzuloides* wächst bei pH 8,5 ebenso gut wie bei pH 3. Wir sollten deshalb mit Ausdrücken, die irgendwelche Ansprüche oder Vorlieben enthalten, also acidophil usw., wirklich Schluß machen und nur den Befund, den wir draußen im Gelände finden, ansprechen: also: "vorwiegend auf saurem Medium vorkommend", "vorwiegend auf nährstoffreichem Boden vorkommend", usw. Also z.B. "säure-stet" oder"nährstoff-stet" gebrauchen, im Fremdwort "acido-konstant" oder "acidotrop" usw., und nicht"nährstoffliebend", "säureliebend" usw. Dann werden auch solche Mißverständnisse vermieden!

ZUR ENTSTEHUNG UND ENTWICKLUNG DER TERRAINBEDECKENDEN MOORE WESTIRLANDS

J.J. M o o r e

Anlaß dieser Mitteilung war eine vegetationsgeschichtliche Untersuchung gemeinsam mit Archäologen, die ein vom Torf überdecktes neolithisches Megalithgrab freilegten. Die Grabstelle war auf Mineralboden errichtet worden.

Im nord-westlichen Mayo, wo neolithische Gräber in großer Zahl anzutreffen sind, kommt das terrainbedeckende Moor auf basenarmem Gestein in großflächiger Ausbildung vor; nicht dagegen (oder kaum) auf Moränen oder an steilen Hängen. Die Vegetation dieser Moore wird trotz ihrer soziologischen Verwandtschaft zu den Heidemooren am besten zu den S p h a g n e t a l i a gestellt.

Diese große Moorlandschaft sieht sehr natürlich aus, pollenanalytische und Großrest-Untersuchungen dagegen beweisen das Gegenteil.

Ausgewählte Fragen zum Thema: Syndynamik

1) Ist die derzeit anzutreffende Vegetation spontan (ombrosoligen) oder anthropogen bedingt? Zur Beantwortung dieser Frage wurden in der Nähe der Megalithgräber folgende Entwicklungsphasen festgestellt (s.Abb.1):

a. P i n u s - Wald mit *Betula*, *Sorbus aucuparia*, *Juniperus*, (*Quercus*), *Ilex aquifolium*.
b. Rodungsphase und dadurch bedingte Massenentwicklung der Gräser; Getreidearten und *Plantago lanceolata*.
c. Regeneration der früheren Waldgesellschaft.
d. Rodung (Brand?).
e. Massenentwicklung der Gräser (wohl M o l i n i e t a l i a - Gesellschaften mit vornehmlich *Succisa* und *Juncus* spec.div.).
f. Verheidung des Grünlandes.
g. Abermalige Dominanz der Gräser.
h. Großflächige Vermoorung. Fragmentarisch ausgebildete Waldbestände finden sich nur auf den sporadisch vorkommenden Moränen oder Bächen.

2) Zur Entstehung des charakteristischen Artengefüges der Moore.

a. *Sphagnum* spec.div. sind im *Pinus*-Wald vorhanden; Großreste gehen in unter 1. aufgezählter Reihenfolge zurück. In der Moorphase sind die Torfmoose dominant.

b. Samen von *Erica tetralix* finden sich nur in der Moorphase.

c. Heidekomponente: *Potentilla erecta* findet sich sowohl in den Heide-, Grünland- und Moorstadien. Sicherlich besteht eine Kontinuität zwischen der derzeitigen und heutigen *Potentilla*-Population.

d. Pollen von *Schoenus nigricans* sind nur in rezentem Torf beobachtet worden.

SUMMARY

Three pollen profiles from the blanket bog of north Co.Mayo have been studied along with their associated macrofossils. Two of these sites come from an area of intense Neolithic activity where Court Cairn type Megalithic tombs have been burried by the growing peat. The third Profile is a deeper one taken 19 km to the west in an area of less Neolithic activity (Glenamoy). Results suggest that the vigorous development of blanket peat under bog vegetation (Sphagnetalia) is due mainly to human interference with the former pine wood ecosystem (Abb.). There is a continous record of *Potentilla erecta* (type) pollen from the time of grassland and heath that followed the final destruction and the other differential species of the western blanket bog show a genetical continuity with species growing within earlier vegetation. A selection of bog-viable ecotypes would have occurred over the years. *Schoenus*-pollen has not, however, been detected any earlier than the beginning of the bog phase.

DISKUSSION

J.PFADENHAUER

Wie kann sich eine Torfdecke von zirka 1 m auf dem Mineraluntergrund bilden, wenn sich nur *Pinus*-Wälder mit *Lonicera* angesiedelt haben?

Sie sprachen davon, daß diese Moore durch menschliche Einflüsse etwa 3000 v.Chr. entstanden seien. Haben Sie irgendwelche Theorien entwickelt, wie das geschehen konnte? Es muß doch eine Vernässung eingetreten sein, daß sich Molinietalia-Arten ansiedeln können?

Konnten Sie aus Ihren Pollendiagrammen Schlüsse ziehen über die Krautvegetation der zweiten *Pinus*-Phase?

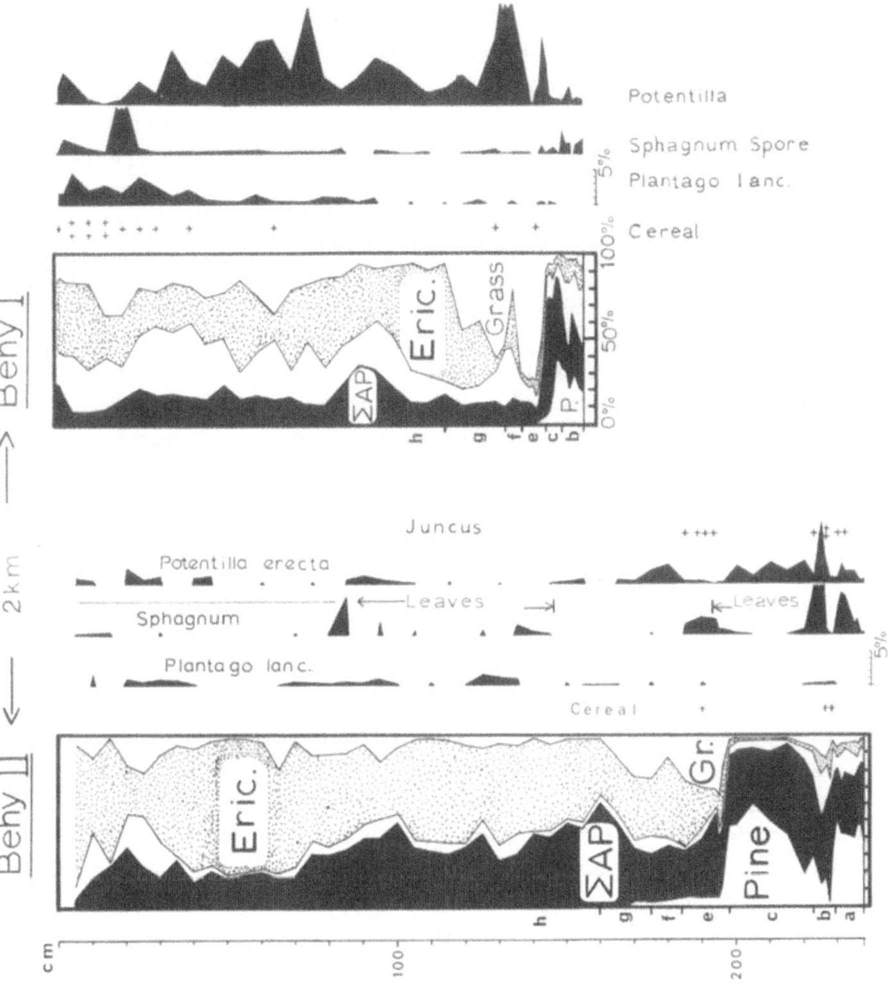

Vereinfachte Pollendiagramme aus Schwarztorf (Blanket bog) Ballycastle, Mayo, Irland.
Behy I ca 5 km vom Moorrand bei einem Megalithgrab (150 m ü.M.);
Behy II 2 km weiter auf flachem Land etwa 20 m ü.M.

E.HÜBL

Soviel ich weiß, verläuft die heutige natürliche *Pinus*-Grenze durch die Mitte von Schottland. Irland liegt vollkommen außerhalb derselben. Sollte der Mensch, ohne Mitwirkung des Klimas, eine Baumart wirklich so vollkommen ausgerottet haben?
Hängt der Übergang von *Schoenus nigricans* auf mehr oligotrophe Standorte in Irland nicht mit der allgemeinen Erscheinung zusammen, daß Moorpflanzen von Nordost auf Südwest auf immer nährstoffärmere Standorte übergehen im Zusammenhang mit dem günstigeren Klima (längere Vegetationsperiode)?

D.J.BELLAMY

I must agree with Father MOORE that I believe there is some theoretical basis to this change. I would like to demonstrate exactly the same change in another part of the world. On the Shillong Plateau in Assam, we have the wettest place in the world where in one year they have had 20 m of rain. It is well documented that 400 years ago all the Shillong Plateau was covered with climax forest of *Quercus* and *Erica arborea* and a lot of hepatics and mosses. Then man came into the region and he has cut the primary forest and used it for agriculture. This then was left and it went over to *Pinus insularis* and then this developed into a good secondary forest with trees 150 years old and again it was cut and now it is a blanket bog. In this stage *Sphagnum* was very rare, in small patches in the forest or on the steep sides of the mountains. As man has changed the habitat, so we have had an explosion of *Sphagnum* into the newly created blanket peats and also of many other species, and it is beautiful to see because here are the same types of plants as we find in the blanket peat of Western Ireland, there is *Parnassia* and others, I think that before the blanket peat period in Ireland, *Schoenus* was perhaps in small colonies on rocks, and along the sides of streams and the woodland was cut and it exploded into its new habitat. This cycle, I am sure, is governed by minerals. We have much mineral material in the standing crop, it is cut and it is exchanged for the climax woodland and then again cut, and all the material is leached away and finally only the blanket peat can survive.
If we study succession in a basin where peat is developing, we go through the classic way of succession from Niedermoor to Übergangsmoor to Hochmoor. The work of HUGO SJÖRS and my own work have shown that there are 7 to 10 stages and these stages are

controlled by hydrology. Here we have water all the time flowing and slowly the peat grows, the energy is stored, the water flow slowes, the water is held back: as KULCZINSKI says; the retention of system increases. So here we have, with stored energy, growth changes, major changes in the habitat template. But in each of these stages we may also have succession from open communities to woodland. The one process is controlled by one type of energy storage, the other is controlled by another and I think here, in Hochmoor, we are dealing with a special case.

V.WESTHOFF
Das schon längst bekannte merkwürdige Auftreten von *Schoenus nigricans* in den ombrotrophen blanket bogs Westirlands ist ökologisch ungeklärt. Dr.MOORE und Dr.BELLAMY haben diese Erscheinung historisch zu deuten versucht. Ohne etwas dagegen einzuwenden, sei dennoch bemerkt, daß damit das ökologische Problem nicht gelöst ist, wie *Schoenus* in diesem ombrotrophen Bereich wachsen kann und sich zu erhalten vermag. Wenn ich mich recht entsinne, haben OSVALD oder SJÖRS die Hypothese geäußert, daß die Ombrotrophie in den westlichen blanket bogs nur relativ sei, indem der Niederschlag an der atlantischen Küste relativ nährstoffreich sei (wegen der Beimischung von Meeressalzen), und daß daher ein gewisser Mineralgehalt des dennoch ombrotrophen blanket bogs vorliegen würde, woraus sich das Auftreten von *Schoenus* erklären ließe. In einem rezenten Beitrag im Journal of Ecology ist diese Hypothese von einem irischen Botaniker bodenchemisch nachgeprüft worden, jedoch mit negativem Ergebnis. Ich möchte daher Herrn MOORE fragen. wie seine Meinung über dieses Problem jetzt ist.

R.TÜXEN
Ich möchte das, was Herr BELLAMY dargestellt hat, noch ein klein wenig erweitern. Er zeigte die Explosion der S p h a g n e t e n nach der Vernichtung des Waldes. Das ist im extrem-atlantischen Klima möglich. Bei uns, im subatlantischen, haben wir die gleiche Explosion, nach der Zerstörung des Q u e r c o - B e t u l e t u m , also der acidophilen *Quercus*-Wälder mit *Calluna*. Und im subkontinentalen Klima haben wir die Zerstörung des Waldes und bekommen Grasland, das wir als Steppenheide bezeichnen (F e s t u c o - B r o m e t e a). Ich glaube, hier liegt eine ganz allgmeine Regel, die klimatisch gesteuert wird.

D.MUELLER-DOMBOIS

Vielleicht wird eine kleine Beobachtung über den Einfluß der Waldbedeckung aus den südöstlichen Manitoba hier von Interesse sein. Dort bildet sich auf Flachmooren *Sphagnum* im Zusammenhang mit der Waldbestockung von *Larix laricina* und *Picea mariana*. Wo der *Picea mariana*-Moorwald besonders dicht im Kronendach ist, wird *Sphagnum magellanicum* durch Braunmoose verdrängt, wie *Pleurozium schreberi* und *Hylocomium splendens*. Wenn dann Lücken im Kronendach durch Ausfall von Schwarzfichten (*P.mariana*) entstehen, dann kommt oftmals wieder eine Überlagerung mit *Sphagnum* vor. Man kann also einen Wechsel von Braunmoosen und *Sphagnen* beobachten, der im Zusammenhang mit der Kronendachbedeckung steht.

J.J.MOORE

Es ist hier in Deutschland schwer zu verstehen, wie *Pinus* auf einem Meter Torf wachsen kann. In unserem Klima wächst *Pinus* auf Moor. In Nord-Schottland kommt *Pinus* sogar auf sehr tiefem Moor vor.

Wann kam das Moor und damit eine Vernässung? Man muß daran erinnern, daß die Niederschläge in West-Irland sehr hoch sind.Sie sind immer höher als die Evaporation und die Transpiration. Und wenn sich ein Podsol entwickelt und die natürliche Entwässerung des Bodens schlecht ist, kann sehr schnell eine Moor-Entwicklung stattfinden. Auch folgendes ist merkwürdig: Man behauptet immer, daß der Westen mit seinen hohen Niederschlagsmengen auch viel Torfmoose besitzt; in den blanket bogs sind *Sphagnen* weniger vertreten als in den Hochmooren, den raised bogs. Nur im Schwarz-Torf finden sich Fragmente von *Sphagnum*-Blättern und wenn überhaupt, sind solche nur mikroskopisch feststellbar. *Sphagnum*-Arten sind hier nicht die Torfbildner, sondern höhere Pflanzen wie *Schoenus*, *Eriophorum vaginatum* etc.

Ich glaube, daß die These von der Ausrottung von *Pinus* ein Märchen ist! Aus den Pollendiagrammen geht hervor, nicht nur aus den meinigen, sondern auch aus anderen, daß die Kiefer immer vorhanden war. Damals bildete sie, wie aus den dünnen Kurven der Pollendiagramme hervorgeht, nicht geschlossene Wald-Komplexe, sondern siedelte sporadisch an ihr zusagenden Stellen.

Über die *Schoenus*-Frage wurden sehr viele Hypothesen aufgestellt. Natürlich haben wir den Mineralanteil des Regenwassers gemessen! Er ist ziemlich hoch, eben so hoch wie in West-Schottland; dort

kommt *Schoenus* auf den ombrotrophen Flächen nicht vor, obgleich diese Art in Schottland nicht fehlt. Deshalb ist in Irland etwas besonderes geschehen, aber wir wissen nicht, was es ist. Ich glaube, daß es etwas Genetisches ist.Der hohe Aluminium-Gehalt des Regenwassers kann nur ein untergeordneter Faktor sein.
The illustration you gave, Dr.BELLAMY, of almost the exact same process was really very interesting. The only slight difference I noticed: your first stage was a *Quercus* stage. *Quercus* was not so common in our first woodland stage, it became a little more common in the second woodland stage and when the pine was gone, there was a fair amount of *Quercus* in the curves and I think not on the developing bog surface obviously but along by the streams where there was some mineral ground to be had.
Also gab es keine explosionsartige Entwicklung von *Sphagnum;* man findet weder Sporen noch Blattfragmente. Natürlich kommen sie im *Pinus*-Wald vor, aber nicht in der Grün- oder Grasland-Phase. Nur in Heiden kommen sie wieder auf, in Verbindung mit der Ausbreitung von *Cyperaceae* und einer beginnenden Moor-Entwicklung.

VEGETATIONSENTWICKLUNG AUF WALDBRANDFLÄCHEN IN SILIKATGEBIETEN DER SLOWENISCHEN SÜDOSTALPEN

M. W r a b e r †

Zwei seit langen Zeiten eingebürgerte Wirtschafts-Systeme haben in der Vergangenheit entscheidend in die Waldungen der Urgesteinsmassive in den slowenischen Südost-Alpen und deren Voralpenwelt eingegriffen, das Waldgefüge wesentlich geändert und die in regelmäßigen Zeitabständen wiederkehrenden Sukzessions-Erscheinungen hervorgerufen. Es waren dies der vor etwa zwei Jahrhunderten in unserem Lande eingeführte und planmäßig ausgeübte Groß-Kahlschlag und die seit uralten Zeiten übliche Praxis der Gereutwirtschaft. Beide Bewirtschaftungsformen wurden unabtrennbar mit Schlagbrennen verbunden und stehen nun seit 1953 unter Verbot; sie können nur ausnahmsweise mit behördlicher Bewilligung noch vorgenommen werden.

Beide Bewirtschaftungssysteme haben gewisse gemeinsame Züge und auch grundsätzliche Unterschiede. Beide haben den Gleichgewichtszustand der Waldnatur für längere oder kürzere Zeit ins Schwanken gebracht und somit eine starke Dynamik der Vegetationsentwicklung ausgelöst. Von den tiefgreifenden Änderungen der Waldstandorte und des Waldbodens wollen wir an dieser Stelle absehen und unser Augenmerk nur auf die Sukzessionserscheinungen in der Vegetationsdecke richten. Es muß noch betont werden, daß in beiden genannten wirtschaftlichen Systemen geplante, unter Kontrolle stehende Brandführung ausgeübt wurde. Es handelt sich also nicht um Waldbrände, die durch die Fahrlässigkeit des Menschen hervorgerufen wurde.

Das Feuer spielt in beiden Fällen die Rolle eines wirtschaftlichen Regulators, und sein Einflußbereich auf die Wald-Vegetation wird derart gesteuert,daß ein Überspringen auf benachbarte Komplexe innerhalb der Vegetation nicht vorkommt.

Der Groß-Kahlschlag

Die völlige Wald-Vernichtung durch Kahlschlag und darauffolgenden Brand rufen eine intensive Vegetations-Dynamik hervor, die sich in mehreren, je nach Standorts- und Vegetationsverhältnissen unterschiedlich verlaufende Sukzessionsreihen äußern. Die natürliche Neubesiedlung, oder künstliche Aufforstung der kahlen Flächen durchläuft viele Stadien und stellt zu einem Teil regressive und zum anderen progressive Vegetationsentwicklung dar. Wenn nach dem Kahlschlag noch das Brennen eingesetzt wird und die abgebrannten Flächen weiter ackerbaulich oder als Weide genutzt werden, wird der Degradations- bzw. Regenerations-Prozeß indifferenter. Auffallendste Erscheinung ist dann, daß der Abbau der Vegetationsdecke rascher vonstatten geht als deren Wiederaufbau.

Aus der Vielfaltigkeit dieser Entwicklungsvorgänge möchte ich nur ein Beispiel herausgreifen, das sich auf die auf Urgestein stockende, klimatozonale Gesellschaft der Fichte und der Großen Hainsimse (Luzulo silvaticae-Piceetum M.Wraber 1963) bezieht. Diese weit ausgedehnten Wälder bedecken die abgerundeten Kuppen, flachen Rücken und sanften Flanken des Pohorje-Gebirges und anderer Urgesteinsmassive oberhalb von 1200 bis 1300 m NN. Der Wald hat hier seine besondere Geschichte: Er war zum größten Teil Eigentum der Groß-Grundbesitzer, die ihn seit Jahrhunderten plünderten und rodeten, um Weiden zu gewinnen, oder die Schmelzhütten, Glasfabriken und Sägewerke mit ungeheuren Mengen Holz und Kohle zu versorgen. Nachdem sich allmählich ein Holzmangel fühlbar machte, oder doch zu befürchten war, ging man zwar zu einer formal geregelten, aber im wesentlichen zu einer in ein starres System gefesselten, räumlich und zeitlich geordneten Bewirtschaftung über, die einer Raubbauwirtschaft sehr nahe kommt. Infolge dieses widernatürlichen Wirtschaftssystems wurde selbstverständlich die ursprüngliche Vegetation stark beeinflußt und besonders in ihrer Struktur und floristischen Zusammensetzung sehr verändert. Tanne, Buche und Bergahorn gingen stark zurück, in manchen Waldstrecken verschwanden sie ganz. Ihren Platz übernahm die Fichte, wodurch der Bestandesaufbau floristisch sehr verarmte. Die mehr oder minder reinen, gleichaltrigen, streifen- oder schachartig angeordneten Fichtenbestände des Luzulo-Piceetum sind also sekundäre, halbnatürliche Gebilde. Wie bereits gesagt, gehört die Bewirtschaftungsmethode mit Großkahlschlag und Brand bereits der Geschichte an, aber ihre ungünstigen Folgen werden noch in mehreren Waldgenerationen zu erkennen sein.

Nach jedem Kahlschlag blieben ungeheuer große Mengen von Schlagabfällen (Äste, Wipfelstücke, schadhafte Klotzabschnitte etc.) liegen und bedeckten die Schlagfläche in großen Haufen (Abb.1) Es ist Gewohnheit geworden, die Holzreste gleichmäßig über die gesamte Fläche zu verteilen und sie nach dem Trocknen bei günstiger Witterung (Windstille) nach einem besonderen Verfahren zu verbrennen. Es würde nämlich zu viel Zeit beanspruchen, bis das knorrige und harzreiche Abfallholz verrotten würde. Das Schlagbrennen (Frattenbrennen) ging planmäßig und streng überwacht vor sich, damit die gesamte organische Masse vollständig verascht wurde und auch die dicken Stammreste möglichst gut verkohlten.

Das Schlag-Brennen hatte noch einen weiteren spekulativen Sinn, nämlich die Gewinnung von Ackerland, zumindest für kurze Zeit, um den in den Hochlagen siedelnden Bauern und Holzarbeitern Land zu verschaffen. Es wurden auf den aschereichen Böden vorwiegend Kartoffeln, in günstigen Lagen auch Roggen und Hafer angebaut, die sehr gute Ernteerträge brachten. Im zweiten oder dritten Anbaujahr wurde dem Saatgut ein Gemisch von Fichten- und Lärchensamen beigemischt, die nach dem Schnitt gut aufliefen. Leider setzte gleichzeitig auch die ungeregelte Weide ein, wodurch der progressive Entwicklungsprozeß stark gehemmt, aufgehalten oder gar rückgängig gemacht wurde. In späteren Zeiten wurde die Aussaat von Baumsamen durch das Setzen von Jungpflanzen abgelöst; dadurch kam es zu einer rascheren Verdichtung, besonders wenn ein mehrfaches Nachsetzen auf freier Fläche vorgenommen wurde.

Der Kahlschlag mit Brand, landwirtschaftlicher Bodennutzung und Beweidung unterlag regelmäßigen, zeitlichen Intervallen; in den höchsten Lagen betrugen diese Zeitabschnitte 100 - 120 Jahre,in den mittleren 80 - 100. Dadurch, daß der Gesamt-Waldbestand für das Schlag-Brennen parzelliert wurde, überrollte ihn binnen etwa eines Jahrhunderts diese Wirtschaftsmethode. Noch heute liefern Waldstreifen unterschiedlichen Alters (Kahlflächen, Jungbestände, ausgewachsene Fichten-Abschnitte und hiebreife Wälder) hierzu das typische Bild; eine für die Sukzessions-Forschung wertvolle Gegebenheit (Abb.2).

Das Sukzessionsschema nimmt auf das eingangs bereits erwähnte L u z u l o - P i c e e t u m Bezug, das als Klimax-Wald betrachtet werden kann. Die sich auf Kahlschlag- und Brandflächen abspielenden Vegetationsänderungen waren vielfach Gegenstand vorübergehender oder eingehender Studien in allen Weltteilen, in den Balkan-Ländern dagegen nur in geringem Ausmaß. Die einschlägige Literatur bezieht sich bei uns größtenteils nur auf Karbonat-Gebirge (S.TREGUBOV 1941, M.GLIŠIĆ 1951, E.VURIČEVIĆ 1965 u.a.m.), während wir über die bezüglichen Vorgänge auf Urgesteinsmassiven wenig unterrichtet sind. Es sei dies ein kleiner Beitrag hierzu.

Ohne mich in die Feinheiten des sehr komplizierten, in fast unzähligen Formen und Gängen auftretenden Entwicklungs-Prozesses auf Kahlschlag, Brand- und Weideflächen einzulassen, möchte ich hier nur in groben Zügen dieses Naturgeschehen vor Augen führen. Es ist festzustellen, daß sowohl die regressiven, besonders noch die progressiven Großstadien, d.h. die länger andauernden Hauptphasen des Sukzessionsprozesses, einen recht einheitlichen Verlauf aufweisen, während die kleinen und kleinsten, sehr kurzfristigen Abänderungen, viel größere Unterschiede aufweisen. Natura in minimis maxima, in maxima unica!

Das Sukzessionsschema des L u z u l o - P i c e e t u m umfaßt folgende Hauptserien mit entsprechenden Groß-Stadien (Schema im Anhang):

Serie 1 (Kahlschlag und Brand, gewöhnlich auch Beweidung)

Auf der abgebrannten Schlagfläche wird der aschenreiche Boden umgebrochen und landwirtschaftlich genutzt, und zwar etwa zwei bis drei Jahre. Während dieser Zeit herrscht die sog. Schlagruhe; waldbauliche Maßnahmen werden nicht vorgenommen.

1. Großstadium

Bereits während der ackerbaulichen Nutzung stellt sich kurzfristig ein schwach ausgeprägtes Ruderalstadium ein, das sich aus einer schwach nitrophilen Ackerunkraut-Vegetation zusammensetzt; Therophyten überwiegen.

2. Großstadium

Schon im ersten Stadium lassen sich einzelne Exemplare des Brandzeigers *Chamaenerion angustifolium* finden; aber erst nach drei bis vier Jahren schließt sich *Chamaenerion* zu dichten, hochwüchsigen Herden zusammen und liefert im Sommer auf den Kahlschlägen eine Blütenpracht. Dies ist das erste, besser ausgeprägte und länger andauernde Stadium (fünf bis acht Jahre), das biologisch aber noch immer eine Regression darstellt. Im dichten Wuchs dieser Hochstaude gehen die meisten Therophyten unter. Hier bietet sich anthrakophilen Gehölzen die erste günstige Siedlungsmöglichkeit, die sich jedoch gegenüber dem nachfolgenden Grasstadium nur wenig oder gar nicht zu behaupten vermögen. Wenn allerdings den *Chamaenerion*-Stauden Fichten-Setzlinge zugefügt werden, finden letztere ausreichenden Schutz und kommen auf.

3. Großstadium

Als nächstes kommt nun für die Gehölze ein äußerst entwicklungsfeindliches Grasstadium auf, das mit seiner Hochwüchsigkeit und dichter Rasenbildung die *Chamaenerion*-Herden ablöst und eine Dauer von 15 bis 20 Jahren besitzt. Dieses waldentwicklungsfeindliche Stadium verhindert weitgehend das Aufkommen von Sträuchern und Bäumen (Abb.3). Es handelt sich hierbei um *Calamagrostis arundinacea* (niedere Lagen, Skelettboden), *C.epigeios* (feuchter oder sogar oberflächig versumpfter Boden), und *C.villosa*(sehr saurer, mäßig tiefer und frischer oder nasser Boden in höheren Lagen). Es kommen gruppenweise noch dazu *Deschampsia caespitosa, D.flexuosa, Luzula silvatica* ssp.*silvatica, Agrostis alba, Carex leporina, Cirsium palustre* u.a.m. Im Herbst machen die vergilbten, grasbewachsenen Kahlschlagflächen den Eindruck eines schnittreifen Getreidefeldes. Der erste Schnee drückt die langhalmigen Gräser zu Boden, sie verrotten jedoch nicht, und nach ihrem Ausaspern bleibt noch lange eine Schicht von halbverfaulten Grasresten liegen. Darunter ersticken die meisten Strauch- und Baumsämlinge. Um die Fichtensetzlinge vor einem derartigen Verderben zu schützen, muß der Grasbestand gleich einem Roggenfeld auf die Stoppel gesetzt werden und zwar vor dem ersten Schnee. Das besprochene Grasstadium muß zur Serie des Regressionsprozesses gestellt werden.

4. Großstadium

Sehr langsam stellen sich die ersten Pioniersträucher ein, dehnen sich aber nur allmählich aus. Es sind dies vor allem *Rubus idaeus* und *R.hirtus*, die sich in heftigem Konkurrenzkampf mit den Gräsern behaupten können. In diesem *Rubus*-Stadium kommt auch *Sambucus racemosa* hoch. Dank seines vegetativen Fortpflanzungsvermögens kann sich besonders *Rubus hirtus* stark ausbreiten und durch Beschattung die Gräser zurückdrängen. Mit dem Aufkommen der Sträucher beginnt eigentlich erst die progressive Waldentwicklung. Die Eroberung des dicht mit Gras bewachsenen Kahlschlagbodens geht vom Waldrand her und rückt langsam gegen die offene Grasflur vor. Dies benötigt allerdings sehr viel Zeit (10 - 20 Jahre), bis der Grasteppich von einer mehr oder minder geschlossenen Strauchschicht abgelöst worden ist. Damit steht der Wiederbewaldung die Tür offen. Der aus Saat, Pflanzung oder aus natürlichem Samenanflug hervorgegangene Fichten-Jungwuchs ist nun lebenskräftig genug, um sich ziemlich ungehindert entwickeln zu können.

An quelligen und bodenwassernahen Stellen kommt es nie zu einer oben erwähnten Grasfazies, sondern schon sehr früh kommt eine Hochstaudenflur mit *Senecio fuchsii, Eupatorium cannabium, Doronicum austriacum, Adenostyles alliariae, Crepis paludosa, Deschampsia caespitosa, Cirsium palustre* u.a.m. zur Entwicklung, die für die Wiederbewaldung weniger entwicklungshemmend ist als der Grasbestand.

5. Großstadium

Bereits im Strauchstadium kommen vereinzelt Vertreter der Baumschicht hoch und überwachsen die Strauchschicht. So finden sich die Pioniergehölze *Populus tremula, Betula pendula, Salix caprea* und die später hinzukommende *Sorbus aucuparia.* So wächst aus der Strauchschicht allmählich eine schüttere Baumschicht hervor, die Vorstufe eines Waldes.

Man ist geneigt, dieses Aspen-Birkenstadium als eigene Assoziation (Populo tremulae-Betuletum) aufzufassen (M. GLIŠIĆ 1951: 124-129); es stellt dagegen als Initialphase nur den Vorläufer einer kommenden Waldvegetation dar. Die schütteren Aspen-Birkenbestände besitzen zwar eine eigentümliche Morphologie und Ökologie, sind in ihrem Aufbau aber gar nicht einheitlich. Je nach standörtlichen Gegebenheiten kommt es zu Aspen- und Birken-Reinbeständen, wobei die mesophilere und ökologisch anspruchsvollere Aspe die besseren Wuchsorte einnimmt, und der ökologisch bescheideneren Birke die mageren und trockenen Böden zufallen. Auch die Weiterentwicklung verläuft in beiden Fällen etwas anders. Die Birke beeinflußt mit ihren, vom Wind leicht zu bewegenden Zweigen die Jungtriebe und Knospen der Fichte mechanisch, so daß eine Verzögerung in der Sproßbildung auftritt, wogegen die Aspe das Aufkommen der Fichte nicht behindert,

Es hängt natürlich von der Pflege des Fichten-Jungwuchses ab, wie lange das Strauch-Baumstadium mit *Populus* und *Betula* andauert. Der sich selbst regulierende natürliche Entwicklungsgang ist langsamer und biozoenologisch stabiler und lebenskräftiger als die durch waldpflegliche Hilfe beschleunigten mehr oder minder reinen Fichtenbestände.

6. Großstadium

Es kommt zu einer Fichtendeckung, in der die 30 bis 50 Jahre alten Fichten einen recht geschlossenen, drei bis auch sechs Meter hohen Bestand bilden. Unter dem dichten Bestandesschluß geht jede andere Vegetation zugrunde; der Boden ist absolut kahl, und nur vereinzelte Überhälter von *Sorbus aucuparia, Populus tremula,*

Fagus silvatica oder *Acer pseudoplatanus* überragen die Fichten.

7. Großstadium

Die Weiterentwicklung zum Fichten-Stangenholz geht verhältnismäßig rasch vor sich. Unter Konkurrenz-Druck sterben viele in ihrer Entwicklung zurückgebliebene Gehölze ab, oder werden von Menschenhand eliminiert. Wenn auch noch immer nur wenig Licht in das Bestandesinnere dringt, bildet sich dennoch allmählich eine Moosschicht aus, in der *Pleurozium schreberi, Hylocomium splendens, Rhytidiadelphus triquetrus* und *Polytrichum formosum* dominieren. Von der natürlichen Auslese oder forstlichen Maßnahmen ist es abhängig, wann und wie sich die endgültige Waldgesellschaft, das Luzulo silvaticae-Piceetum entwickeln wird.

Der gesamte Entwicklungsprozeß von der gebrannten Schlagfläche bis zum ausgewachsenen (sekundären) Fichten-Wald dauert 100 bis 120 Jahre, wenn nicht Jahrhunderte. Die Entwicklungsdauer hängt letztlich von waldbaulichen Maßnahmen ab, die einzelne Stadien, wie die von *Chamaenerion* oder *Calamagrostis*, verkürzen können. Auch hier ist der kürzere (wirtschaftlichere) Weg nicht der biologisch günstigste. Die natürliche Entwicklung nimmt gewöhnlich "Umwege" und einen gewissen "Zeitverlust" mit in Kauf. Die biologisch eingestellten Forstleute sind sich dessen wohl bewußt und berücksichtigen bei ihren Planungen und Kalkulationen die von der Natur gegebenen "Wegweiser".

Serie II (Kahlschlag ohne Brand, mit oder ohne Beweidung)

Die Entwicklungsreihe des Hainsimsen-Fichtenwaldes stellt sich ein, wenn nach der Rodung der Brand vollends entfällt und der Weidegang nur untergeordnet eine Rolle spielt. Das ursprüngliche Waldgefüge kann sich dann am raschesten wieder aufbauen. Vor allem das *Chamaenerion*-Stadium entfällt vollends, und es kommt gleich zu einem Gras-Stadium, das bei völligem Ausfall einer Beweidung rasch in das Stadium der Sträucher und Pionier-Gehölze übergeht (Abb.4).

Eine andere, regressive Entwicklungsrichtung finden wir auf den flachmuldigen, wellig geformten Hochebenen und den abgerundeten Rücken des Pohorje-Gebirges, wenn das Luzulo-Piceetum sphagnetosum geschlagen, gebrannt und beweidet wird. Es entwickelt sich allmählich das Arnico-Nardetum, das in ein Calluno-Vaccinietum übergeht. Nur sehr langsam entwickelt sich aus der letztgenannten Gesellschaft ein Fichten-Wald, der reich an Torfmoosen und *Vaccinium uliginosum* ist. Brand und ungeregelter Weidegang führen zu einer Verdichtung des Bodens und einem Aufkommen einer Moorvegetation mit einzelnen verkrüppelten Fichten (Sphagnetum piceetosum) oder einem vermoorten Latschen-Bestand (Sphagnetum mugetosum); die so entstandenen Gesellschaften stellen in der Sukzessionsreihe jeweils das Endglied dar.

Die Gereut-Wirtschaft

Mit der anfangs genannten Wirtschaftsform nicht zu verwechseln ist das Gereutbrennen oder Schwenden. Diese althergebrachte Methode wird heute kaum noch angewandt und wenn, dann nur unerlaubt.

Gereut-Wirtschaft und Groß-Kahlschlag unterscheiden sich in folgendem: Der Grund für das Gereutbrennen liegt nicht im Bedürfnis, große Mengen Holz zu gewinnen, sondern an mangelnder Acker-Fläche oder fehlendem Weideland. Die Umtriebszeit war bei der Gereutwirtschaft unverhältnismäßig kürzer als beim Groß-Kahlschlag (ca 10-20 Jahre); gleichermaßen waren die gerodeten Wald-Flächen (0,5-1 ha im Gegensatz zu 100 ha) kleiner.
Das Gereutbrennen ging wie folgt vor sich: Das aus verschiedenen Baumarten (*Picea, Alnus, Betula, Populus* u.a.m.) bestehende Jungholz auf Wald- und Hutflächen wurde alle 10-20 Jahre abgetrieben, das minderwertige Prügelholz beseitigt und das übrige Abfallholz gleichmäßig über die Fläche verbreitet und verbrannt (Abb.5), der so gewonnene, aschegedüngte Boden mit Roggen, Weizen, Hafer, Gerste, Buchweizen oder Raps bestellt. Die anfänglich sehr hohen Ernteerträge gingen schon nach dem zweiten bis dritten Anbaujahr zurück.
Nach der Ackerwirtschaft wurde der Gereutboden einer unkontrollierten Weidenutzung überlassen und nie aufgeforstet. Die nicht genutzten Abschnitte nahm *Alnus incana* als erste in Besitz; überhalb von 900 bis 1000 Metern war es *Alnus viridis* (Abb.6). Unter dem Schutz der Erlen konnten angewehte Fichtensamen gut auflaufen und die Nadelgehölze gut zur Entwicklung kommen. Dazu kam es aber nur,wenn nicht zu früh gerodet wurde. Dieser Zyklus wiederholte sich ständig, so daß jedes Jahr ein mehr oder weniger zugewachsenes Weideland vernichtet wurde. Eine standortsbedingte Vegetation konnte sich nur noch auf für Mensch und Vieh unzugänglichen Steilhängen halten, aus denen sich die potentielle natürliche Vegetation ablesen läßt. Hierzu gehören folgende Wald-

Gesellschaften: Q u e r c o - C a r p i n e t u m (l u z u l e t o s u m), L u z u l o - F a g e t u m , G a l i o r o t u n d i f o l i i - A b i e t e t u m , B a z z a n i o - A b i e t e t u m , M y r t i l l o - P i n e t u m a u s t r o a l p i n u m .

Auf den aufgelassenen Gereutflächen gedeihen 30-80 jährige Fichten-Reinbestände, die biologisch gesund sind und gute Wuchsleistungen zeigen. Sie stocken auf nährstoffreichen, frischen und biologisch aktiven Böden, die sich unter dem stickstoff-bindenden Erlengebüsch durch die zyklische Wiederholung der Gereut-Wirtschaft gebildet haben.

Die Gereut-Wirtschaft hat sich zu einer Kulturform entwickelt, die seit der Nachkriegszeit im Verschwinden begriffen ist. Leider wurde bislang dieser interessanten Betriebsform wissenschaftlich nur wenig Aufmerksamkeit geschenkt. Man hat diese Kulturform wohl mehrfach vom ethnologischen, kulturhistorischen und wirtschaftlichen Standpunkt (A.ŠIVIĆ 1954) gewürdigt, während vegetationskundliche Studien nur in Manuskripten vorliegen (M. WRABER 1958, 1959, 1960, 1961). Es sei abschließend noch bemerkt, daß die Gereut-Wirtschaft eine sehr lebhafte Vegetationsdynamik hervorgerufen hat. So hat, vom Grau- bzw. Grünerlen-Stadium abgesehen, dem ein Fichten-Stadium folgte, durch die frühzeitigen Rodungen nur selten ein Entwicklungsstadium voll ausreifen können. Die genannte Betriebsform hat nur ausnahmsweise eine regressive Entwicklungsrichtung eingeleitet, außer an sehr steilen Hängen, wo es durch Regengüsse zu Boden-Erosionen oder gar Erdrutschen gekommen ist, sonst aber eine progressive Entwicklung genommen. Durch das säkulare Gereutbrennen ist vor allem die biologisch und ökologisch empfindliche Tanne aus ganzen Landstrichen verdrängt worden, und es ist bislang noch nicht zu erkennen, wann sie wieder ihre ursprünglichen Siedlungsorte einnehmen kann und wird. Das gilt auch für die autochthonen Laubgehölze, jedoch gewinnen sie ihren Platz leichter wieder.

ZUSAMMENFASSUNG

In den Slowenischen Südost-Alpen haben in der Vergangenheit zwei Wirtschaftsformen die natürliche Wald-Vegetation wesentlich beeinflußt und das Waldbild stark verändert. Es waren dies der mit Brand verknüpfte Groß-Kahlschlag und die uralte, ebenfalls auf Schlag und Brennen beruhende Praxis der Gereutwirtschaft.

A. Es werden zunächst die tiefgründigen Auswirkungen des Groß-Kahlschlages und des Brandes am Beispiel einer abgeleiteten Fichtenwald-Klimax-Gesellschaft der höchsten Lagen des Pohorje-Gebirges und anderer Urgesteinsmassive, des Luzulo silvaticae-Piceetum, anhand mehrerer Entwicklungs-Serien näher beschrieben.

Die beigefügte Skizze erläutert die Hauptsukzessions-Reihen des durch Kahlschlag, mit oder ohne Brand, mit oder ohne Beweidung, hervorgerufenen Entwicklungsganges. Der unter natürlichen Bedingungen sich abspielende Entwicklungsprozeß wird durch waldbauliche Eingriffe mannigfach gestört, wodurch gewisse Entwicklungsstadien abgekürzt, oder sogar übersprungen werden.

Nach Schlag und Brand wird der ehemalige Waldboden zwei bis drei Jahre ackerbaulich genutzt, worauf ein brandbedingtes, fünf bis acht Jahre andauerndes *Chamaenerion angustifolium*-Stadium folgt, worin die "Fichtensaat" zwar aufkommt, aber bald dem nächstfolgenden, langandauernden (15-20 Jahre und darüber) Grasstadium (*Calamagrostis* u.a.) zum Opfer fällt. Erst mit dem Aufkommen einiger Sträucher (*Rubus*, *Sambucus*) setzt eine progressive Entwicklung mit Andeutung zu einem Fichtenjungwuchs ein. Den Sträuchern gesellen sich bald einige Pioniergehölze zu (Aspe, Birke, Vogelbeere und Salweide), die eigentliche Vorstufe des aufkommenden Waldes. GLIŠIĆ bezeichnet dieses Entwicklungsstadium als Populo tremulae-Betuletum, eine morphologisch und ökologisch heterogene Gesellschaft. Unter dem Lichtschutz der Laubgehölze kommt es endlich zu einem ± dichten Jung-Fichtenbestand, der unter seinem dichten Schirm keine Krautschicht mehr aufkommen läßt.

Mit rasch zunehmendem Längenwachstum kommt es zum sog. Fichten-Stangenholz mit einer zunehmend sich entwickelnden dichten Moosdecke (z.B. *Pleurozium schreberi*). Als endgültige Klimaxgesellschaft stellt sich schließlich das Luzulo-Piceetum ein. Bleibt nach Kahlschlag der Brand aus, und setzt auch eine Beweidung nicht ein, so führt die Entwicklung unter Ausfall des

Chamaenerion-Stadiums unter Umgehung des Gras-Stadiums auf kürzestem Wege zur Wiederherstellung der vorhergehende Fichtenwald-Gesellschaft.

Die nasse Ausbildung des Fichtenwaldes (Luzulo-Piceetum sphagnetosum) tritt in flachmuldigen Hochplateau-Lagen auf. Hier ist der Verlauf der Vegetations-Entwicklung nach Kahlschlag, Brand und Beweidung ausgesprochen regressiv und führt über ein feuchtes Arnico-Nardetum zum Calluno-Vaccinietum, das unter natürlichen Bedingungen langsam dem Fichtenwald zustrebt, sonst aber in einem Moor mit verkrüppelten Fichten (Sphagnetum piceetosum) endet.

B. Die Gereutwirtschaft ist gegenüber dem Großkahlschlag-Verfahren sehr verschieden, da ihr Zweck nicht die Gewinnung von Holz, sondern notwendigen Ackerbodens und zusätzlichen Weidelandes ist. Diese Wirtschaftsform ist stark zurückgegangen und gehört schon der Geschichte an. Sie wurde in Silikatgebieten vielfach ausgeübt, wo der bäuerliche Großgrundbesitz herrschte, und es für die Nahrungsgewinnung zu wenig Eigenbesitz gab. Die Umtriebszeit war hier unvergleichbar kürzer (10-20 Jahre) als beim Großkahlschlag (100-120 Jahre); auch die gerodete Waldfläche war hier viel kleiner (bis 1 ha) als dort (bis 100 ha). Der kurzfristig und kleinflächig wiederkehrende Rodungsprozeß wurde nur in Höhenlagen bis 1000/1200 m ausgeübt. Das gefällte wertlose Prügelholz wurde abgetrieben, die reichlichen Schlagabfälle wurden über die Fläche gleichmäßig verteilt, verbrannt, und der aschegedüngte Boden zwei bis drei Jahre ackerbaulich genutzt. Der Ertrag war anfangs sehr groß (zehnfach) ging dann aber rasch zurück.

Der aufgelassene Gereutboden wurde einem ungeregelten Weidegang und der natürlichen Entwicklung überlassen und nie aufgeforstet. Den Neuboden besiedelten als erste *Alnus incana* und über 1000 m *Alnus viridis*. Die Grauerle wuchs ungemein rasch heran und bildete binnen weniger Jahre ein Dickicht, worin der Fichtensame ausgezeichnet keimte, und der Jungwuchs schnell in die Höhe schoß. In 20 bis 30 Jahren würde der Fichtennachwuchs die Erle überholen, es kam aber gewöhnlich schon vorher zur periodisch wiederkehrenden Rodung. Fast jedes Jahr fiel ein anderes Waldstück der Gereutwirtschaft zum Opfer, und so gab es kaum noch eine verschont gebliebene Waldfläche mit einigermaßen natürlicher Vegetation.

Auf den aufgelassenen Gereutflächen geseihen nun 30 bis 80-jährige Fichten-Reinbestände, die biologisch gesund sind und gute Wuchsleistungen zeigen. Sehr allmählich stellen sich darin Baumarten aus dem ursprünglichen Mischwald (Querco-Carpinetum, Luzulo-Fagetum, Galio-Abietetum, Myrtillo-Pinetum austroalpinum) ein. Durch das säkulare Gereutbrennen wurde insbesondere die ökologisch empfindliche Tanne vertrieben.

SUMMARY

Two main features of forest management have deeply influenced the natural forest vegetation in the Slovene part of the Southern Alps. These features were the wood-cutting and burning of slash debris. The same method, wood-cutting on small spots, burning, and subsequent grazing has been used for centuries.

First the effects of wood-cutting and burning of the climax spruce forests, Luzulo silvaticae-Piceetum), on the topmost parts of the Pohorje mountains will be desribed. For illustration several series of development are represented in Tab.1. (wood-cutting with or without burning, and with or without grazing).It should be stressed, however that the natural process of reforestation has been often deeply influenced by the measures of forestry, owing to that some steps of development were shortened or even passed over.
After clearing and burning the forest soil is exploited for 2 or 3 years for farming purposes. When farming is given up, then 5-8 years of *Chamaenerion angustifolium* stage sets in as a consequence of the burning. The spruce seedling grow therein, it is true, but very soon they are suppressed by a 15-20 year period of high-grass stage (*Calamagrostis* and others), possessing a very high concurrence power. Only when the pioneer shrub vegetation (*Rubus, Sambucus*) appears, by which the grass stage is suppressed, a new remarkable advance of spruce sets in. The pioneer trees (aspen, birch, mountain ash, osier) join the bush vegetation. The appearance of these trees is a significant step towards the reforestation and may be regarded as a true preliminary step of the forest. This stage, built up of pioneer trees with bush undergrowth is considered as a special association (Populo tremulae-Betuletum Glišić 1951), however, that is neither ecologically nor morphologically uniform. Only at this stage the young spruce growth joins in the dispersed heliophile broad-leaved trees and soon forms a thicket; deprived

of light all shrub vegetation dies out and the soil underneath becomes throughout bare. The next stage, pole spruce forest (Fichtenstangenholz) is characterized by a vigorous growth of young trees in height; simultaneously a thick moss growth starts covering the soil (*Pleurozium schreberi* and others).
Finally, L u z u l o - P i c e e t u m , the end of the stage is attained as a secondary climax association.
In the case when the clearing is not followed by burning and grazing is omitted, the *Chamaenerion* stage does not appear, grass stage is remarkably shortened and the development of a new forest proceeds directly to the primary spruce forest association.

On the flat or slightly concave plateaus, where primarily wet type of spruce forests prevails, the clearing, burning and grazing leads into a regressive stage. A wet A r n i c o - N a r d e t u m turf is the first stage that later on passes into a wet C a l l u n o - V a c c i n i e t u m heath. The latter shows a weak trend to pass under natural conditions into a spruce forest, otherwise it develops into a moor with crippled trees (S p h a g n e t u m p i c e e t o s u m).
The new land management on the contrary is very different from the previous methods. In this case the main purpose is not the lumbering but acquiring till able soil for farming and additional pasturage. This form of management is very seldom to be met with nowadays, it belongs history. It was used in the mountains, on areas of igneous rocks, where large estates were held by peasants. The nutritional needs compelled them obtain new land for farming. The rotation on the fresh land soil was much shorter, only 10-20 years of duration. Meanwhile, in the case of extensive clearing that period lasted 100-120 years. The surface cleared for the fresh land method was also much smaller (only 1 ha), than above mentioned (100 ha).
Such fresh land management of short duration on small surfaces, accompanied with burning was used only at the altitudes of 1000-1200 m. The pole-wood, that was cut down, was removed, but the remains of branches have been scattered all over the surface and dried to be later burnt. In such a way fertilized soil was then ploughed up and used for 2 or 3 years for crops (potatoes, rye, wheat, barly, buckwheat, oats, turnip). In the beginning the production was enormously rich, but later on it decreased quickly. When farming was given up on the fresh land, it was used as

pasture from time to time but never reforested again.
So the first tree settled here was grey alder (*Alnus incana*) and at altitudes over 900/1000, *Alnus viridis*. The grey alder grew up in a short time so that after a few years there was already a thicket in which the spruce seeds germinated exceptionally well and soon broke through. After 20-30 years the spruce would have predominated if a new cutting cycle had not started.
Such management was going on from year to year, step by step. Thus there is hardly any area (of greater importance) untouched and possessing natural vegetation.
At present, on these areas the new pure spruce forests (30-80 years old) show a vigorous growth. Only the trees of the original mixed woods (Querco-Carpinetum, Luzulo-Fagetum, Galio-Abietum, Myrtillo-Pinetum austroalpinum) join these spruce forests gradually.
Through the centuries the repeated wood-cuttings led to a consequence that the ecologically very delicate fir (*Abies*) has been expelled from large areas. The same could be said also for some broad-leaved trees, which, owing to these measures have been weakened in vigour and in number and they reoccupy an area only step by step.

LITERATUR

BRAUN-BLANQUET,J. -1964- Pflanzensoziologie 3.Aufl.- Wien.
GLIŠIĆ,M. -1951- Fitocenoloski pogledi na posumljavanje sumskih pozarista (Phytocoenologische Ansichten über die Aufforstung der Waldbrandflächen).- Godisnjak Biol.inst.u.Sarajevu,III/1-2 (1950). Sarajevu.
SIVIĆ,A. -1954- Poziganje gozdnih frat (Das Frattenbrennen/Schlagbrennen).- Slov.etnograf VI/VII: 73-82.
TREGUBOV,S. -1941- Les forêts vierges montagnardes des Alpes Dinariques. Massif de Klekovatcha-Guermetsch.- Comm.SIGMA 78. Montpellier.
VUKIČEVIĆ,E. -1965- Sukcesija vegetacije i prirodno obnavljanje suma na sumskim pozaristima u Srbiji (Succession of vegetation and the natural regeneration of forests on burns in peoples republic of Serbia).- Glasnik Sumar.fak.Univ.u Beogradu 29. Belgrad.
WRABER,M. -1958- Tipoloska podoba gozdne vegetacije v okolici Slovenjega Gradca (Typologisches Bild der Waldvegetation in der Umgebung von Slovenj Gradec).- Mskr.verviel. Ljubljana. 56 pp.
-- -- -1959- Pregled in opis glavnih gozdnih zdruzb na podrocju Mozirja in Gornjega grada (Übersicht und Schilderung der Haupt-Waldgesellschaften im Gebiet von Mozirje und Gornji grad).- Mskr.vervielf. Ljubljana. 45 pp.
-- -- -1961- Gozdna vegetacija na Solcavskem in v okoloci Sostanja (Waldvegetation im Solvaca-Gebiet und in der Umgebung von Sostanj).- Mskr.vervielf. Ljubljana. 71 pp.
-- -- -1960- Gozdne zdruzbe v Zgornji Savinjski dolini (Waldgesellschaften im oberen Savinja-Tal).- Mskr.vervielf. Ljubljana. 58 pp-
-- -- -1963- Gozdna zdruzba smreke in godzne bekice v slovenskih Vzhodnih Alpah (Die Waldgesellschaft der Fichte und der Waldhainsimse in den Slowenischen Ostalpen, Luzulo silvaticae-Piceetum Wraber 1953).- Razprave IV. razr.SAZU 7: 79-175.

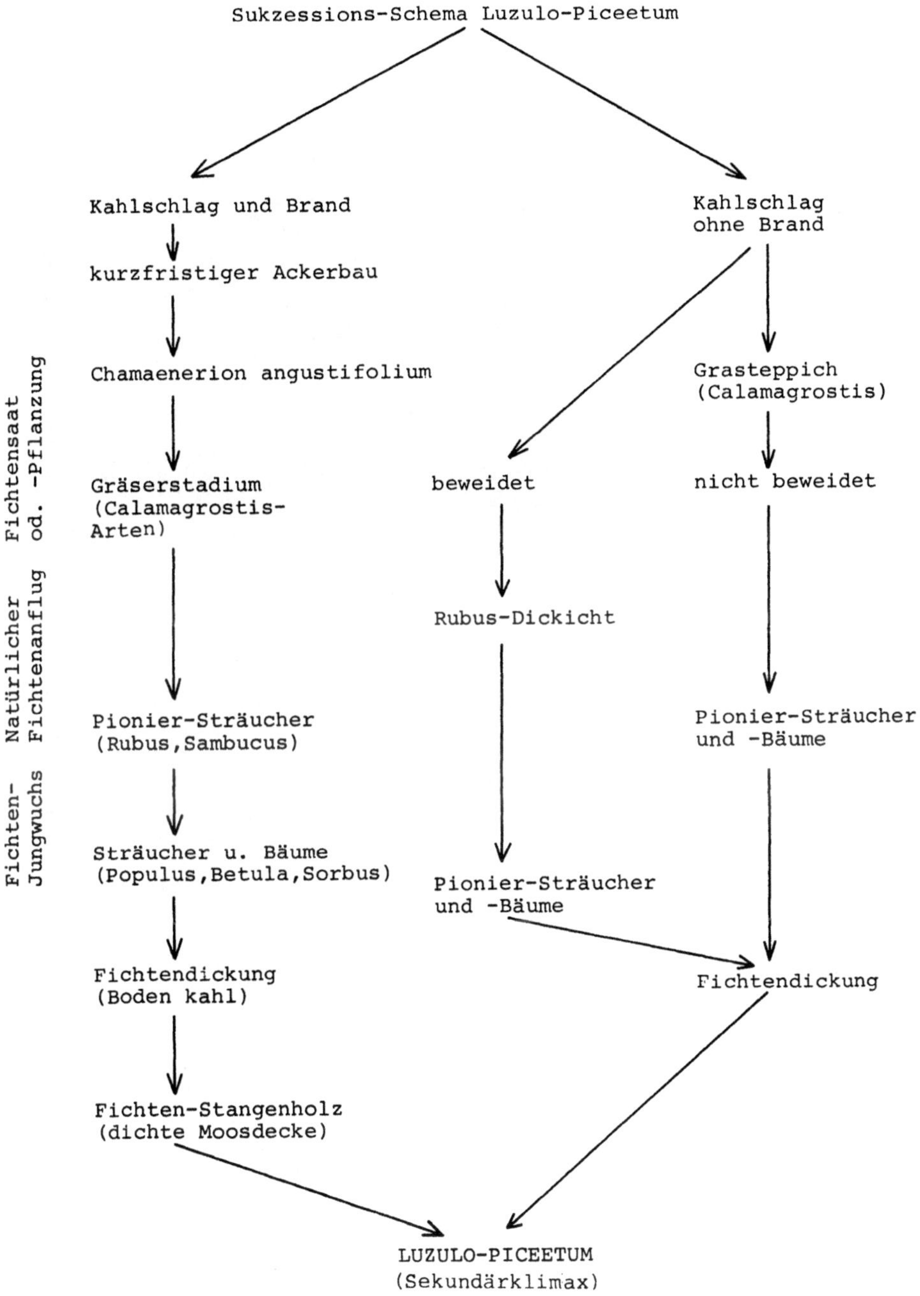
Sukzessions-Schema Luzulo-Piceetum
Kahlschlag und Brand
kurzfristiger Ackerbau
Chamaenerion angustifolium
Gräserstadium (Calamagrostis-Arten)
Pionier-Sträucher (Rubus, Sambucus)
Sträucher u. Bäume (Populus, Betula, Sorbus)
Fichtendickung (Boden kahl)
Fichten-Stangenholz (dichte Moosdecke)
Fichtensaat od. -Pflanzung
Natürlicher Fichtenanflug
Fichten-Jungwuchs
Kahlschlag ohne Brand
beweidet
Rubus-Dickicht
Pionier-Sträucher und -Bäume
Grasteppich (Calamagrostis)
nicht beweidet
Pionier-Sträucher und -Bäume
Fichtendickung
LUZULO-PICEETUM (Sekundärklimax)

Abb.1. Waldrevier Rakovec am Pohorje-Gebirge (ca 1460 m): Nach erfolgtem Kahlschlag überdecken die Holzabfälle den Boden in dicker Lage.

Abb.2. Waldrevier Rakovec am Pohorje-Gebirge (ca 1300 m): Die auf Kahlschlagflächen heranwachsenden künstlichen Fichtenkulturen sind in breiten Gürteln angeordnet.

Abb.3. Waldrevier Rakovec am Phorje-Gebirge (ca 1450 m): Die Kahlschlagfläche und den aufgelockerten reinen Fichtenbestand überwuchert *Calamagrostis villosa* in dichtem Rasen.

Abb.4. Südhang von Kašni vrh in den Savinja-Alpen (ca 1150 m): Durch Weidegang aufgelockerte Fichten- und Lärchenkulturen am Steilhang.

Abb.5. Mrzli vrh auf dem Ter-Berg in den Savinja-Alpen (ca 1080 m): Das Gereutbrennen ist vorrüber, es raucht aber noch auf der Brandfläche.

Abb.6. Bela peč in den Savinja-Alpen (1300-1350 m): Die aufgelassenen Gereut- und Weideflächen besiedelt dicht geschlossenes Grünerlen-Gebüsch *(Alnus viridis)*.

Alle Aufnahmen: M.WRABER.

DISKUSSION

F.K.HARTMANN †

Sie haben angeführt, daß im Luzulo silvaticae-Piceetum gewisse Grasstadien aufkommen, entweder unter Zwischenschaltung eines *Epilobium*-Stadiums nach Brand oder eines direkten Grasstadiums ohne Brand. Wenn wir das mit unseren Mittelgebirgs-Waldgesellschaften künstlicher Prägungen vergleichen, etwa Fichtenforst-Gesellschaften in dem Mittelgebirgsraum nördlich der Alpen, dann können wir sagen, wir haben immer ein *Epilobium*-Stadium zwischengeschaltet, auch ohne Brand. Ich möchte sagen, sogar noch ein üppigeres. Das kann nur durch Wildäsung unterbrochen werden, da *Epilobium* immer sehr schnell abgeäst wird. Und das, glaube ich, ist wohl darauf zurückzuführen, daß in den mittleren Gebirgslagen eben mehr Streu produziert und schneller umgesetzt wird durch die besseren Vegetations- und Temperaturverhältnisse, und daß aus diesem Grunde eine Nitrifizierung stattfindet, die wir oben in den Hochlagen vielleicht nicht haben.
Eine andere Ausgangslage wäre vielleicht die, daß in den Hochlagen-Piceeten schon vor dem Kahlschlag ein gewisses Verlichtungsstadium eingetreten ist, das zu einer Vergrasung führt, wenn man nicht brennt. Und wenn man brennt, dann würde eben durch das Verbrennen des Reisigs eine bessere Nitrifizierung oder eine bessere Nitratlieferung, die das *Epilobium*-Stadium schafft, gewährleistet sein.
Eine 2. Frage. Sie haben gewisse Initialstadien mit Grünerle und auch Grauerle geschildert. Das trifft wahrscheinlich die feuchteren Hänge und die quelligeren Lagen?
Eine 3. Frage ist die: Sie haben *Calamagrostis villosa = hallerina*, *C.arundinacea* und *C.epigeios* sozusagen in einem Atemzug genannt. Bei uns ist eine sehr deutliche höhenzonale Gliederung. Wir haben oben in den echten Piceeten unserer Mittelgebirge - soweit echte Piceeten am Rande des böhmischen Kessels in unseren Mittelgebirgen bis zum Harz vorhanden sind - *C.villosa* als charakteristische Art. Und wenn wir *C.arundinacea* schon etwas dazwischen haben, dann haben wir den Verdacht, daß da bereits eine Verzahnungsstufe zur Fichten-Buchenwaldstufe oder zur Fichten-Buchen-Tannenwaldstufe eintritt. So haben wir z.B. im Harz festgestellt, daß auf dem Wurmberg in 950 m Höhe, wo er nach der Südseite überkippt, es werden dort Ausgrabungen gemacht, noch einzelne Horste von *C.arundinacea* in einer großen Wüste von *C.*

villosa auf Kahlschlägen vorkommen, und wenn man die Waldgeschichte dazu näher untersucht, dann sind dort einzelne Buchen sozusagen als hochgeschobene Verzahnungsstufe von Buchen-Fichtenwald tatsächlich festgestellt worden. Man würde also bei uns eine Trennung schaffen müssen, wenn man soziologisch scharf gliedern will zwischen *C.villosa*-Stufe mit echtem Fichtenwald und *C.arundinacea* mit kleinflächiger *C.villosa* als Verzahnungsstufe zwischen Fichtenwald mit Buchenwald und *C.epigeios* ist also Ubiquist der von der Eichenmischwald-Stufe bis in die Niederungen des norddeutschen Flachlandes hin und auch bis in die mittleren montanen Stufen übergeht.

M.WRABER †

Es handelt sich bei uns um ein ausgeprägtes *Chamaenerion*-Stadium nur auf Brandflächen. Da wird stark gebrannt. Man achtet darauf, daß möglichst alle Abfälle verbrannt werden, sogar die Holzstümpfe. Man legt auch Abfallholz auf die Baumstümpfe, damit die Stümpfe möglichst verbrennen. Und auf solchen stark gebrannten Flächen kommt es zur Ausbildung eines ausgeprägten *Chamaenerion*-Stadiums. *Chamaenerion* kommt ja auch auf anderen Kahlschlagflächen vor, aber nicht als eigenes Stadium, nur vereinzelt oder in kleinen Gruppen.

Was die Grauerlen und Grünerlen betrifft, muß man bedenken, daß in unseren Alpen die Niederschläge außerordentlich hoch sind, ständig über 1500-2000-3000 mm Jahresdurchschnitt, und sehr frische Böden bedingen, wo sich die Erlen überall einstellen können.

Es ist ganz berechtigt, die verschiedenen *Calamagrostis*-Stadien auseinanderzuhalten. Ich habe es vielleicht zu wenig betont,daß bei uns *C.villosa* in den höheren Lagen und auf etwas sauerem, frischerem Boden wächst, und *C.arundinacea* tiefere Lagen einnimmt und einen etwas ärmeren und skelettreicheren Boden bewächst, während *C.epigeios* durchgehend ist und nur etwas feuchtere Stellen einnimmt.

H.SUKOPP

Mit den Bemerkungen über den Rückgang der Tanne ist die Frage angeschnitten worden nach der Bedeutung der Änderung des Florenbestandes für die Vegetationsentwicklung. Ist die Tanne in einem größeren Gebiet ganz verschwunden und wenn ja, kann sie in einem solchen Falle noch als ein Element der potentiellen natürlichen Vegetation betrachtet werden, und wie steht es in diesem

Falle mit der Definition: "menschlicher Einfluß schlagartig ausgeschaltet"?

R.TÜXEN

Ich möchte zu der Bemerkung von Herrn Professor HARTMANN nur hinzufügen, daß wir hier im Westen, also von unserem Gebiet an weiter nach Westen, auf Fichtenforsten folgend nicht mehr - weil wir zu niedrig liegen - *Calamagrostis villosa* kennen, auch nicht mehr - weil wir zu westlich liegen - *C.arundinacea*, sondern nur noch *C.epigeios* haben, und daß sich *Epilobium angustifolium* bei uns regelmäßig bereits *Digitalis purpurea* beimengt. Das ist der Unterschied gegen den Osten. Lohnt es sich wirklich, meine Herren Forstleute, solche *Calamagrostis epigeios*-Bestände hier bei uns zu erzeugen, die dann auch so aussehen, wie jene Aufforstungen, in denen alle 20 Meter einmal eine Fichte sich hochquält? Ich habe vor einigen Tagen erst welche gesehen, und der Eigentümer sagte mir, er wisse auch nicht, was er jetzt mit diesem *C. epigeios* eigentlich machen solle. Viele Arbeitskräfte gibt es nicht, und Gift will er nicht gerne. Aber bei uns hier im Gebirge wird die Fläche, aus der Rinteln sein schönes Trinkwasser bezieht, jetzt mit Gift behandelt.

M.WRABER †

Ich möchte noch auf die Frage von Herrn SUKOPP antworten. Die Tanne ist vom Gebiet nicht verschwunden, sie ist flächen- und mengenmäßig stark zurückgegangen, aber auch oben, ganz auf dem Kamm, in der Höhe von 1450, 1500, 1550 m, findet man ab und zu noch einige Tannen, besonders aber in den unteren Hanglagen gegen die Erosionsgräben. Da stehen immer noch einige Tannen,aber sonst war die Tanne, wie man vermuten kann, früher viel verbreiteter, besonders weil die Tannengesellschaft, das sogenannte G a l i o r o t u n d i f o l i a e - A b i e t e t u m , ja bis zur Fichtengrenze hinaufreicht, nämlich bis zur Höhe von 1150-1200, sogar bis 1300 m. Diese Gesellschaft wird ja von der Tanne gebildet, mit Fichte und Buche vermengt. Da hat Herr Professor HARTMANN recht, daß es sich da, wo *Calamagrostis arundinacea* auftritt, um eine Verzahnungszone zwischen einem Buchen-Tannenwald, also einem G a l i o - A b i e t e t u m , ab und zu auch einem A b i e t i - F a g e t u m a u s t r o a l p i n u m und einem Fichtenwald handelt.

STATISTISCHE UNTERSUCHUNGEN ÜBER DIE WIEDERBEWALDUNG AUF DEM TRIESTER KARST

D. L a u s i - S. P i g n a t t i - L. P o l d i n i

Die natürliche Wiederbewaldung der Karstweiden im Gebiet um Triest bietet ein interessantes Feld für Untersuchungen über die Dynamik der Vegetation.
Das Untersuchungsgebiet liegt in unmittelbarer Nähe der Stadt und stellt ein ausgedehntes Karstplateau dar, das auf meistens kompakter Kalkunterlage eine dünne Schicht tertiärer "Terra rossa" zeigt. Die Bodenschicht beträgt heute 10 bis maximal 50 cm, in Erosions-Trichtern oder -Spalten erreicht die "Terra rossa" sogar einige Meter Tiefe.
Das Klima ist trotz der Nähe des Meeres ziemlich kontinental getönt mit mittlerer Jahrestemperatur von etwa 11° C (Triest 14°C) und einem durchschnittlichen jährlichen Niederschlag von 1200 mm (Triest 1078 mm).
Die natürliche Schlußvegetation dieses Gebietes scheint ein azidophiler Eichenmischwald zu sein. Durch Kahlschlag, Feuer und Beweidung wurde diese ursprüngliche Vegetation zu einer kargen Weide degradiert, die noch vor wenigen Jahrzehnten den größten Teil des Gebietes bedeckte. Das Ende der Weidewirtschaft ermöglichte durch Aufhebung der degradierenden Wirkung im ganzen Gebiete den Beginn der Wiederbewaldung. Dieser Prozeß, der vor etwa 20 Jahren begonnen hat, wird in dieser Arbeit erstmals untersucht.
Der Ausgangspunkt der Vegetationsserie für unsere Studien ist das C a r i c i - C e n t a u r e e t u m r u p e s t r i s , eine karge Weide, die etwa mit den mitteleuropäischen X e r o b r o m e t e n zu vergleichen wäre, die aber den illyrischen S c o r z o n e r o - C h r y s o p o g o n e t a l i a zuzuschreiben ist.
Die nächstfolgende Assoziation ist schon eine Waldgesellschaft, das S e s l e r i o - O s t r y e t u m c a r p i n i f o l i a e , ein illyrischer Buschwald mit *Ostrya carpinifolia*, *Fraxinus ornus*, *Quercus pubescens* und *Quercus petraea* in der Baum- und *Sesleria autumnalis* in der Krautschicht.
Die Erfassung der Zwischenstadien mit den gewöhnlichen pflanzensoziologischen Aufnahmen ist unseres Erachtens nicht angebracht, da die Änderungen der Vegetation mit dieser Methode nicht ermit-

telt werden. Es handelt sich nämlich um ein grundsätzlich statisches Verfahren, an das wir alle gewöhnt sind und das für die Beschreibung von Assoziationen wertvoll ist. Wir lenkten hingegen unser besonderes Augenmerk gerade auf solche Vegetationsmosaike, die nach der üblichen Methode beiseite gelassen werden, indem wir uns in diesem Falle gerade auf kleinste Assoziationsfragmente beschränken mußten.

Die Wiederherstellung des Waldes erfolgt durch Aufwachsen von Straucharten in der Weide; die Sträucher wachsen in die Höhe und in die Breite, beschatten die Bodenoberfläche, bilden kleine ökologische Nischen und bringen allmählich immer größere Änderungen in der floristischen Zusammensetzung der Weide, bis diese in einen lichten Wald umgewandelt wird.

Wir konzentrierten also unsere Aufmerksamkeit auf das Studium der einzelnen Sträucher, die auf diesen Flächen vorkommen.

Es wurde daher eine besondere Aufnahmekarte erarbeitet, auf der die Art des Strauches und außerdem Höhe, Durchmesser, Form,Stammeshöhe und -Durchmesser desselben notiert wurden und noch dazu die Entfernung der nächsten Strauchgruppen (und deren Höhe) und die Arten aus der Krautschicht, die unmittelbar um den Stamm wachsen.

Die Namen aller Arten wurden kodifiziert und durch Zahlen ersetzt, was eine große Vereinfachung darstellt, größere Geschwindigkeit in der Geländearbeit ermöglicht und einen unumgänglichen Schritt für weitere Bearbeitung durch Computer darstellt. Die Bearbeitung der aufgenommenen Angaben ermöglichte uns, weitere Informationen über die Ökologie und den Aussagewert der wichtigsten Straucharten zu gewinnen.

Die Untersuchung ist noch nicht abgeschlossen und wurde bisher auf Grund von 273 Karten durchgeführt; in jeder Karte sind 14 quantitative und wenigstens 6 (meistens aber 10-15) floristische Angaben. Die bisher gesammelten Informationen gehen also in die Zehntausende.

Eine erste Frage ist es: Welche Straucharten dringen in das C a r i c i - C e n t a u r e e t u m ein? Es sei dabei bemerkt, daß diese Assoziation in reinen Beständen praktisch keine Strauchart besitzt.

Wir fanden auf 15 Stationen von je 100 m^2 folgende nach abnehmender Häufigkeit geordnete Sträucher:

Juniperus communis	98 Individuen	Cerasus mahaleb	26 Individuen
Fraxinus ornus	74 "	Pinus nigra	19 "
Ostrya carpinifolia	40 "	Cotinus coggygria	17 "
Quercus pubescens	26 "	Rhamnus rupestris	16 "
	Weitere Arten 30 Individuen		

Eine zweite Frage ist, ob das Erscheinen dieser Sträucher irgendeiner Gesetzmäßigkeit unterliegt.
Als erste Bearbeitung wurden die Sträucher nach ihrer Höhe geordnet, weil die Höhe, abgesehen von einigen Arten mit niedrigem Wuchs, vom Alter abhängt.

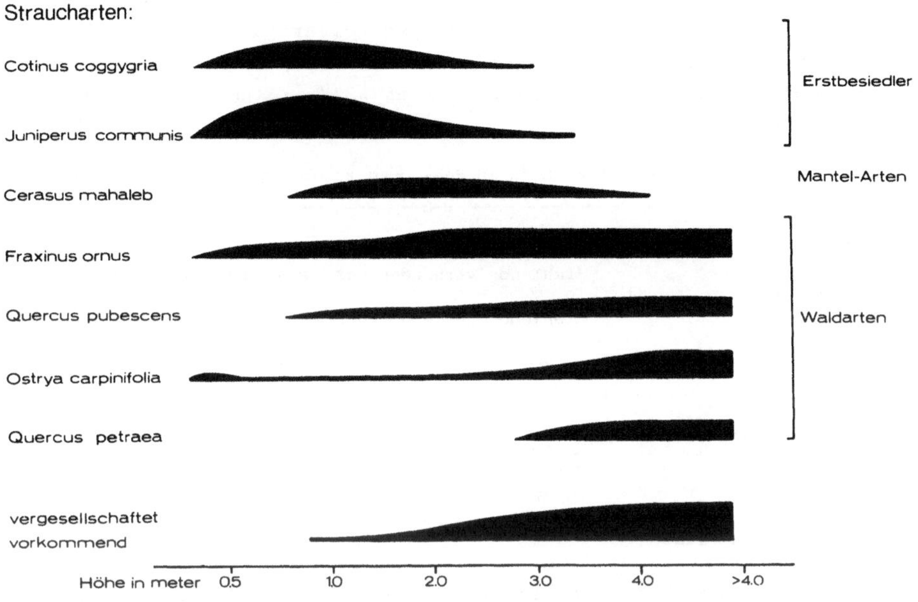

Fig. 1. Frequenz der einzelnen Straucharten auf Grund ihrer Höhe.

Die Resultate dieser Bearbeitung sind in Fig. 1 zusammengefaßt. In den sieben ersten Diagrammen sind die Frequenzen (auf Grund ihrer Höhe) der einzelnen häufigsten Straucharten dargestellt, im letzten Diagramm hingegen die Häufigkeit der Fälle, in denen diese Sträucher im selben Gebüsch vergesellschaftet vorkommen. Die Korrelation zwischen den ersten Diagrammen und dem letzten gibt uns nun eine Information über das Auftreten der einzelnen Straucharten in der Zeit.
Aus den Angaben kann man den Schluß ziehen, daß sich *Juniperus communis* als Erstbesiedler in der strauchlosen Weide verhält; auch *Cotinus coggygria* ist auf die erste Phase der Bewaldung beschränkt, vor allem auf steinigem Boden. *Cerasus mahaleb* verhält sich als ausgesprochene Mantelpflanze. Erst in der darauffolgenden Phase, d.h. in ausgereiften und zusammenfließenden Mantelabschnitten, erscheinen *Quercus pubescens*, *Ostrya carpinifolia*,

Quercus petraea in oben genannter Reihenfolge.
Eine Sonderstellung nimmt *Fraxinus ornus* ein, die gleich am Anfang erscheint und bis in die beginnende Waldphase hinein immer mehr an Bedeutung gewinnt. Wie aus anderen Angaben bekannt ist, verbleibt diese Art weiterhin in der Baumschicht des Waldes.
Weitere Gesetzmäßigkeiten werden durch die Korrelation zwischen Sträuchern und Krautarten ermittelt. In Tab.1 werden die prozentuellen Präsenzwerte der verschiedenen Sträucher nach dem Vorhandensein oder Fehlen von zwei wichtigen Leitarten angegeben. Die Leitarten sind: *Brachypodium pinnatum* in der Form *rupestre* (eine Art des Waldmantels) und *Sesleria autumnalis* (eine ausgesprochene Art des Waldes).

Tab.1. Präsenzwerte von Sträuchern in Beziehung zu ökologischen Leitarten

Leitarten					
Brachypodium pinnatum	fehlend	vorhanden	vorhanden	fehlend	fehlend
Sesleria autumnalis	fehlend	fehlend	vorhanden	vorhanden	dominant
Sträucher					
Cotinus coggygria	12	5	2	4	-
Juniperus communis	28	28	9	7	7
Cerasus mahaleb	9	10	5	8	5
Fraxinus ornus	32	35	36	36	38
Quercus pubescens	14	15	29	25	21
Ostrya carpinifolia	5	7	15	20	29

Die Angaben der Tab.1 bestätigen die allgemeinen Schlüsse über das Verhalten von *Juniperus communis*, *Cotinus coggygria*, *Quercus pubescens* und *Ostrya carpinifolia*. Es ist aber besonders hervorzuheben, daß auch in dieser Hinsicht *Fraxinus ornus* eine Sonderstellung einnimmt, da diese Art in allen Kombinationsklassen der Tab.1 völlig gleiche Werte aufweist. Das bedeutet, daß sie die einzige Art ist, die während des ganzen Wiederbewaldungsprozesses die gleiche Häufigkeit zeigt. Man scheint daraus schließen zu können, daß *Fraxinus ornus* wegen seiner breiten ökologischen Amplitude die wichtigste Rolle sowohl bei der Wiederbewaldung als auch im Walde selbst spielen kann.
Bisher wurden nur die kolonisierenden Sträucher untersucht. Nun wollen wir zu erfassen versuchen, welche Änderungen in der Vegetation das Auftreten der Sträucher bewirken kann. Das geschieht auf Grund der Beziehungen zwischen Höhe der dominierenden Strauch-

arten und Zusammensetzung der Krautschicht. Die Resultate sind in Fig.2 zusammengefaßt. Die Höhen der einzelnen Sträucher werden auf sechs Klassen verteilt, d.h. von 1,2,3,4,5 und 5 Meter Höhe.

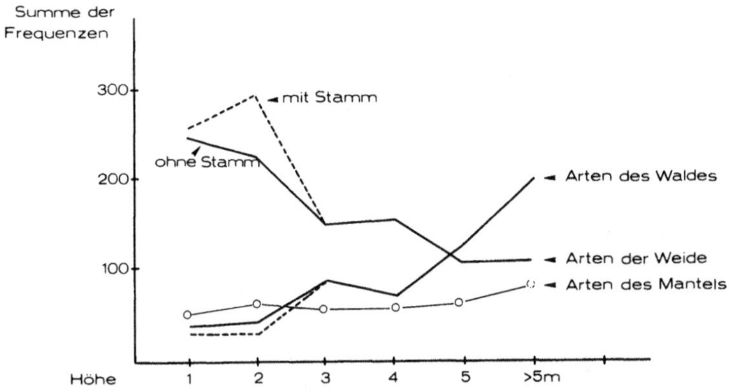

Fig.2. Frequenzen der wichtigsten Kennarten der Weide und des Waldes in Beziehung zur Höhe der Sträucher.
(Die Frequenzen werden als Summe der % jeder Art angegeben)

Es war nötig, solche Sträucher zu unterscheiden, die nur einen Stamm besitzen, da in diesem Falle die beschattende Wirkung über die Krautschicht völlig anders ist. Das wurde aber nur für die beiden untersten Klassen ausgeführt, bei denen die Unterschiede sehr deutlich sind (vgl.Fig.2), während die höheren Klassen nur unwesentliche Differenzierungen aufweisen, so daß die Angaben dennoch einheitlich dargestellt werden können.

Eine globale Schätzung der Änderungen in der floristischen Zusammensetzung wurde bisher wegen des zeitraubenden Vergleiches zwischen etwa 150 Arten noch nicht unternommen, ist aber für eine weitere Bearbeitung durch Computer vorgesehen. Bisher wurden nur die häufigsten Arten bearbeitet, und zwar folgende drei Gruppen:

Arten der Weide: *Carex humilis, Potentilla tommasiniana, Satureja montana, Festuca sulcata, Bromus erectus.*

Arten des Mantels: *Brachypodium rupestre, Teucrium chamaedrys, Dactylis glomerata.*

Arten des Waldes: *Sesleria autumnalis, Hedera helix, Helleborus* aus der *viridis*-Verwandtschaft *(H. istriacus, H.odorus, H.multifidus), Cyclamen europaeum, Mercurialis ovata, Festuca heterophylla, Hieracium barbatum, Quercus petraea* (Keimlinge).

In Fig.2 werden also die Beziehungen zwischen den wichtigsten Kennarten der Weide und des Waldes dargestellt. Aus dieser Bearbeitung lassen sich folgende Schlüsse ziehen: Die Bedeutung der Form des Strauches ist am größten, wenn die Strauchhöhe 1-2 Meter beträgt. In diesen Höhenklassen, unter mehr oder weniger kugelförmigen Sträuchern (d.h. ohne Stamm), erreichen die Arten der Weide den Wert 228, unter bäumchenförmigen (d.h. mit Stamm) aber den Wert 294, d.h. um 28% mehr. Die Weidearten, die an stärkere Lichtintensität angepaßt sind, kommen wesentlich häufiger unter diesen bäumchenförmigen Sträuchern vor, während die kugelförmigen Individuen schon einen wesentlichen Abbau der Weide bewirken. Dementsprechend nehmen die Waldarten unter bäumchenförmigen, auch unter *Fraxinus ornus* mit und ohne Stamm ab (Fig.3).

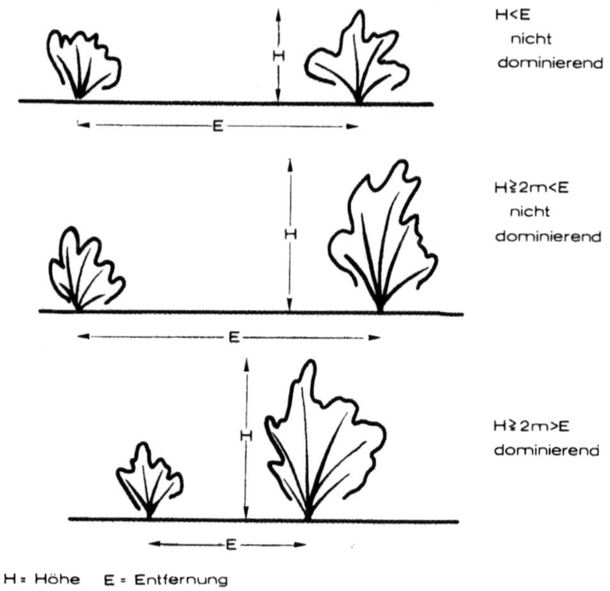

Fig.3. Beziehungen zwischen Nachbarstrauch (links) und dominierendem Strauch (rechts). Weitere Erklärungen im Text.

Als Gesamtergebnis läßt sich sagen, daß die Weidearten relativ häufig unter Sträuchern bis 2 Meter Höhe vorkommen; mit weiterem Aufwachsen der Sträucher aber sind ihre Frequenzen wesentlich geringer, was schon auf ein Beginnen der Waldphase hindeutet. Dementsprechend steigen die Frequenzen der Waldarten allmählich mit zunehmender Höhe der Sträucher an und erst, wenn diese 4 Meter und mehr erreicht haben, überwiegen sie.

Die Mantelarten zeigen fast konstante Frequenzwerte bis zu einer Strauchhöhe von etwa 4 Metern, dann zeigen sie verhältnismäßig hohe Werte in der letzten Klasse (diese Klasse ist freilich noch nicht der endgültige Wald, sondern muß noch zum Waldmantel gerechnet werden).

Zusammenfassend können wir beobachten, daß das Auftreten von Sträuchern in der sonst strauchlosen Weide eine allgemeine Verschiebung der floristischen Zusammensetzung bewirkt. Die Arten des Waldsaumes treten allmählich auf, und wenn die durchschnittliche Höhe der Sträucher 4 Meter überschreitet, breiten sich auch die Waldpflanzen aus.

Die Form der Sträucher ist maßgebend für die Geschwindigkeit dieser Ausbreitung während der ersten Stadien.

Weitere Informationen über den Prozeß der Wiederbewaldung können wir auch gewinnen, wenn wir die Entfernungen zwischen den einzelnen Strauchgruppen in Betracht ziehen. Das wesentliche Merkmal scheint in diesem Falle das Vorhandensein von einem "dominierenden Strauch" zu sein. Darunter verstehen wir einen ziemlich großen Strauch, der wenigstens zwei Meter Höhe erreicht hat und dessen Entfernung vom beherrschten,sagen wir, "Nachbarstrauch" die eigene Höhe nicht, oder höchstens unwesentlich, überschreitet (Fig.3).

Die Wahl der beiden kritischen Werte (Höhe über 2 Meter, Entfernung kleiner als die Höhe) wurde empirisch ausgesucht. Die Höhe von zwei Metern steht aber in gutem Einklang mit den Angaben vom Diagramm in Fig.2. Der Wert für die oben angegebene Entfernung zwischen den beiden Sträuchern scheint von großer Bedeutung zu sein. Die Ergebnisse dieser weiteren Untersuchung sind in Tab.2 zusammengefaßt.

Tabelle 2. Beziehungen zwischen Zusammensetzung der Krautschicht und Vorhandensein eines dominierenden Strauches.

Zusammensetzung der Krautschicht		Untersuchte Sträucher		
		Total	mit dominierendem Strauch	
Brachyp.rupestre	Sesleria autumnalis	N.	N.	%
fehlt	fehlt	62	34	54
vorhanden	fehlt	32	18	57
vorhanden	vorhanden	45	38	84
fehlt	vorhanden	56	51	91
fehlt	dominierend	36	35	97

Das Auftreten der beiden Mantel- und Waldgräser scheint vom Vorhandensein eines "dominierenden Strauches abhängig zu sein.

Das Problem scheint hingegen für den ersten Fall (s.Tab.2) offen zu bleiben. Wir können uns nämlich fragen, warum in der Hälfte

der Probeflächen doch ein dominierender Strauch vorhanden ist. Die Aufnahmekarten des ersten Falles wurden deshalb weiter untersucht. Fast alle diese Karten gehörten zu solchen Sträuchern, die entweder von geringer Höhe (unter 2 Meter) waren, oder, obwohl von Höhen über 2 Meter, eine ungünstige Gestalt (bäumchenförmig, verkehrtpyramidenförmig oder zylindrisch mit Durchmesser unter einem Meter) aufwiesen, so daß deswegen die Wirkung des dominierenden Strauches aufgehoben wurde. Andere Karten zeigten das Auftreten von weiteren Waldpflanzen (*Hedera helix* und *Helleborus viridis*). Nur in wenigen (insgesamt 3%) Fällen ist es uns bisher nicht möglich, eine logische Erklärung für das Fehlen der Mantel- und Waldarten zu geben.
Der Einfluß des "Nachbarstrauches" ist in Fig.4 für vier typische Fälle ersichtlich.

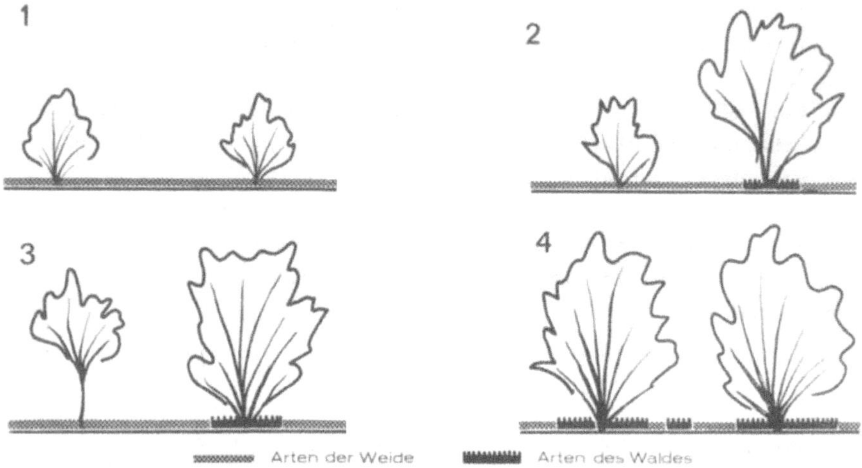

Fig.4. Einfluß des Nachbarstrauches auf die floristische Zusammensetzung der Krautschicht
1 - kein Einfluß, weil beide Sträucher zu klein sind.
2 - Kein Einfluß, weil der linke Strauch zu klein ist.
3 - kein Einfluß, weil der linke Strauch bäumchenförmig verzweigt ist.
4 - Auftreten der Waldart *Sesleria autumnalis*.

Die Verschiebungen in der Zusammensetzung der Krautschicht unter verschieden großen und verschieden vergesellschafteten Sträuchern wird größtenteils in dieser ersten Phase durch Änderungen des Standort-Lichtklimas bedingt. Es wurden an klaren Tagen, sowohl im Winter (6.2.1967), als auch im Sommer (18.9.1967), direkte Lichtmessungen am Standort durch Lange-Luxmeter unternommen.Die Ergebnisse werden in Fig.5 zusammengefaßt. Die Korrelation zwischen Abnahme der Lichtintensität und Zunahme der Häufigkeit waldsteter Arten kommt in Tab.3 besonders klar hervor. Es sei dabei bemerkt, daß die Lichtintensität allmählich abnimmt, während die Änderungen der Vegetation sprunghaft sind (Fig.5).

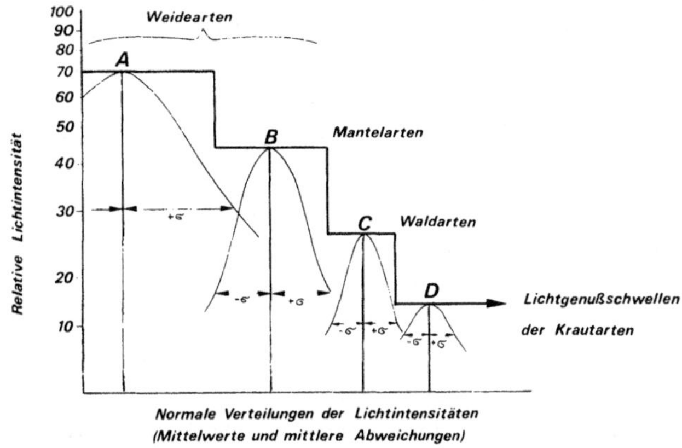

A- In der Weide (Bodennahelichtintens.)

	Mittlere Höhe m		Mittlerer Durchmesser des Laubwerkes m	
B- Unter Sträuchern von	0,7	u.	0,7	(Präsenz nur von Weidearten in der Krautschicht)
C- " " "	2,0	u.	1,8	(" von Mantelarten " ")
D- " " "	3,6	u.	2,0	(" von Waldarten " ")

Fig.5.

Aus dem oben angegebenen Sachverhalt und anderen Teiluntersuchungen, die hier der Kürze halber nicht wiedergegeben werden können wir schließen, daß die Wiederbewaldung des Carici-Centaureetum rupestris von folgenden Bedingungen abhängt:

1. Nur bestimmte Straucharten sind imstande, die Wiederbewaldung einzuleiten; die Kolonisierung der Weide durch *Pinus nigra*, *Rosa* sp. pl. und *Crataegus monogyna* bleibt meistens erfolglos; bessere Pioniere sind *Fraxinus ornus*, *Cerasus mahaleb*, *Quercus pubescens*, *Ostrya carpinifolia*.
2. Die betreffenden besiedelnden Sträucher müssen gewisse Dimensionen erreichen (zur Vereinfachung wurde von uns die Höhe berücksichtigt, die 2 m überschreitet).
3. Die kolonisierenden Sträucher müssen Wuchsformen ohne Hauptstamm haben, und in der Breite wenigstens einen Meter Durchmesser erreichen.

Die Dynamik der Wiederbewaldung wird durch die Bildung von Her-

den einzelner Sträucher beschleunigt, die eine dichtere Besiedlung erreichen. Als Schwelle der Massenwirkung haben wir das Vorhandensein eines dominierenden Strauches angenommen.
Es scheint, daß man von diesem Moment an nicht von unbestimmten Herden sprechen kann, die von getrennten einzelnen Individuen gebildet sind, sondern, daß hier die gegenseitige Beeinflussung der einzelnen Individuen beginnt. Diese erstreckt sich auf die pedologischen und mikroklimatischen Eigenschaften, aber auch auf die floristische Zusammensetzung der Assoziationsfragmente. Mit weiterer Entwicklung schmelzen die einzelnen Individuen zusammen, so daß wir meistens Gruppen beobachten, die aus mehreren Arten zusammengesetzt sind. Hier können wir den entscheidenden Augenblick der Vergesellschaftung feststellen.
Bei diesen sprunghaften Vegetationsveränderungen können wir zwei bedeutende Phasen unterscheiden und zwar:

1. Besiedlung durch Einzelindividuen ohne wirkliche gegenseitige Beeinflussung.
2. Bewaldung, wenn die Wirkung der Vergesellschaftung anfängt. Diese beginnt, wenn bereits zwei Sträucher nebeneinander wachsen und sofern einer derselben die passende Höhe und Form erreicht hat, und die Entfernung zwischen beiden unter einem gewissen Wert liegt.

Aus diesen Untersuchungen können wir einige allgemeine Schlüsse ziehen:
Das Studium der feineren "Mechanik der Änderungen" in der Vegetation läßt sich am besten durch die Untersuchung kleinster Flächen und Vegetationsfragmente verfolgen; dabei ist es notwendig, mit umfangreichen samplings und Bearbeitungen auf statistischer Grundlage vorzugehen.
Es scheint, daß die Entwicklung von dem Massenwirkungsgesetz aus und von den beschränkenden Faktoren gleichzeitig geregelt wird, so daß die Wirkung gewisser ökologischer Verhältnisse latent bleiben kann, soweit einer der unentbehrlichen Faktoren im Minimum liegt. Wenn dieser Faktor eine gewisse Schwelle überschreitet, hat man eine spontane Variation, so daß die Vegetation zu einem weiteren Gleichgewichtszustand kommt, der von einem anderen Minimalfaktor bestimmt wird (das ist für den Lichtfaktor aus Fig.5 ersichtlich). Aus dem oben gesagten ergibt sich, daß der Übergang von einer Assoziation zu einer anderen nicht ein linearer und kontinuierlicher Vorgang, sondern vielmehr ein sprunghafter mit mehreren sich ablösenden Stadien ist.

Tab.3. Abnahme der Lichtintensität unter Sträuchern zunehmender Dichte[1)]

Sträucher	Höhe	Laubwerk (Durchmesser)	Dominierender Strauch	Absolute u. relative Lichtintens.		Krautschicht unter den Sträuchern Weide-Arten	Mantel-Arten	Wald-Arten
	m	m		Klux	%			
WEIDE (Carici-Cent.rup.)				60,0	100			
ERSTBESIEDLER								
Juniperus communis	0,5	0,3	-	24,0	40			
Cotinus coggygria	0,8	3,0	Ostrya carp.	18,0	30			
Cotinus coggygria	1,0	2,5	-	8,0	14			
MANTELARTEN								
Cerasus mahaleb	0,7	0,8	-	39,5	66			
Cerasus mahaleb	1,0	1,0	-	21,0	35			
Crataegus monogyna	0,8	3,0	-	22,0	37			
WALDARTEN								
Fraxinus ornus	0,3	0,4	-	35,0	58			
Fraxinus ornus (mit Stamm)	1,0	1,0	-	13,0	22			
Pinus nigra	4,0	2,5	Ostrya carp.	1,5	2			
Quercus petraea	4,5	2,0	Quercus petr.	26,0	43			
Quercus petraea	2,5	1,5	Ostrya carp.	17,0	28			
Fraxinus ornus (ohne Stamm)	1,5	0,5	-	13,0	22			
Fraxinus ornus	2,5	1,0	Ostrya carp.	7,5	13			
Fraxinus ornus	3,0	1,5	Ostrya carp.	5,0	12			
Quercus petraea	3,0	1,5	Ostrya carp.	15,0	25			
Quercus petraea	5,0	3,5	Ostrya carp.	5,0	12			
Ostrya carpinifolia	1,8	1,5	Quercus petr.	4,5	7			
Ostrya carpinifolia	5,5	1,5	Ostrya carp.	4,5	7			
Fraxinus ornus	2,0	2,0	Quercus petr.	3,6	6			

[1)] Lichtmessungen vom 18.9.1967, 13.30^{h} - 14^{h}. Völlig klarer Tag.

ZUSAMMENFASSUNG

Die Autoren haben eine statistische Methode erarbeitet, um die Dynamik der Wiederbewaldung der Karstweiden im Gebiet um Triest zu untersuchen. Das Ende der Weidewirtschaft im Gebiet ermöglichte den Beginn der natürlichen Wiederbewaldung, die vor etwa 20 Jahren begonnen hat. Dieser dynamische Prozeß scheint nicht ein kontinuierlicher Vorgang zu sein, sondern sprunghaft vor sich zu gehen. Mehrere Stadien wurden unterschieden, die von dem Massenwirkungsgesetz und von beschränkenden Faktoren gleichzeitig geregelt werden.

SUMMARY

The authors have examinated the possibilities of a statistical method to study the natural reafforestation of the karstic grazing grounds near Trieste. These pastures, having been left for about twenty years, are becoming wooded again by a dynamic process that seems to evolve in an interrupted or abrupt manner with several different equilibrium levels. This process appears to be regulated by the combined action of the mass effect of invading bushes and the generalised law of limiting factors. This action creates new ecological niches, so that the initial forest phase can start.

DISKUSSION

Z.GRAČANIN

Aus meinen Untersuchungen im kroatischen Karst ist mir bekannt, daß auf Karstweiden lebende Stöcke vieler Holzarten vorhanden sind. Handelt es sich nun bei den Sträuchern, die sich auf Ihren Flächen entwickeln, um Pflanzen aus Samen oder um Stockausschläge? Falls es sich um Sämlinge handelt, wo kommt dann *Quercus pubescens* her? Eicheln sind bekanntlich schwer und können nicht sehr weit vom Wind transportiert werden.

Gibt es auf Ihren Flächen neben *Juniperus communis* auch *J.oxycedrus* oder Bastarde *J.communis* x *oxycedrus*?

Haben Sie bei der Bestimmung der dominierenden Sträucher neben der Entfernung und Höhe der Sträucher auch die Himmelsrichtung, in welcher sie zueinander stehen, berücksichtigt? Ich kann mir vorstellen, daß ein großer Strauch einen kleineren Strauch, der nördlich davor steht, viel stärker in der Entwicklung beeinflußt als umgekehrt.

L.POLDINI

Zu den Fragen kann ich sagen, daß diese Arten sowohl als Folge von Samenanflug als auch von Stockausschlägen vorkommen.

In bezug auf *Quercus pubescens* haben wir feststellen können,daß die Keimlinge und nachher die aufkommenden Pflanzen immer von Sträuchern umgeben werden, die dornig und sehr reich an Bitter- und Gerbstoffen sind.

Nur *Juniperus communis* kommt vor. Wir haben in der Nähe der Stadt nur eine Stelle, wo *J.oxycedrus* wächst. Nur in Istrien gibt es sowohl *J.oxycedrus* als auch die Hybride.

Was die Himmelsrichtung betrifft, haben wir immer alle Eigenschaften des Standortes notiert.

H.ELLENBERG

Zu dem dominierenden Nachbarstrauch ist mir nicht ganz klar geworden, wie Sie die Bestimmung vornehmen. Sie zeichneten eine gestrichelte Linie mit Winkel; wodurch wird dieser Winkel bestimmt?

Aber ich möchte noch etwas nachtragen, angeregt durch eine Bemerkung von Herrn POLDINI. Ich habe gesagt, wenn man von einzelnen Arten spricht, dann darf man nichts über ihre Ansprüche nur auf Grund ihres Vorkommens aussagen. Das gilt nicht für Pflanzengesellschaften. Pflanzengesellschaften, Kombinationen von Arten, kommen ja unter ganz bestimmten Bedingungen vor, und dann kann man diese Ausdrücke durchaus verwenden und z.B. sagen, das Luzulo-Fagetum sei azidophil.

DIE WALDENTWICKLUNGSDYNAMIK IM SÜDOST-ALPENRAUM IN PALYNOLOGISCHER SICHT

A. Š e r c e l j

Der südostalpine Raum wurde einst als ein Refugium der mitteleuropäischen Floren angesehen. FIRBAS (1923,1949) vertrat die Ansicht, daß wenigstens die Fichte in diesen Gegenden die letzte Eiszeit überdauert hätte. LÄMMERMAYR (1926) hat die Nordgrenze des Buchenareals während der Eiszeit durch Slowenien gezogen. HORVAT (1959), beeinflußt von dem Artenreichtum des *Q u e r c o - C a r p i n e t u m*, war fest überzeugt, daß diese Waldgemeinschaft tertiären Ursprungs wäre, und das ganze Pleistozän in dem Gebiet existent gewesen wäre. BORHIDI (1963) glaubt, die Ansicht, daß das *F a g i o n i l l y r i c u m* seit dem Tertiär ununterbrochen in diesem Raum bestanden hätte, durch seine Üppigkeit und seinen Arten-Reichtum beweisen zu können. Viele andere Forscher sind zwar in ihren Äußerungen vorsichtiger, sprechen jedoch gern über "tertiäre Relikte".
Den Waldgesellschaften pflegt man desto höheres Alter beizumessen, je üppiger sie entwickelt sind und je reicher sie an Pflanzenarten sind. Dem Faktor "Zeit" wird also eine große Bedeutung zugeschrieben, die fördernd wie hemmend wirkenden klimatischen und edaphischen Faktoren dagegen werden eher vernachlässigt. Darum versetzt man gern die Entstehung und Entwicklung der heutigen Floren möglichst weit in die Vergangenheit, ja sogar ins Tertiär zurück, ungeachtet der katastrophalen Zäsuren des Pleistozäns.

Es gab bis jetzt wenig Anlaß und nur unzuverlässige Beweise,sich dieser allgemein herrschenden Meinung zu widersetzen. Man hatte ja sehr unklare Vorstellungen über das Pleistozän, ja sogar heute noch ist unser Wissen über die pleistozänen Ereignisse strittig. Nicht nur unter den Biologen, sondern auch unter den Geologen und Geographen und besonders unter den Palynologen herrschen darüber starke Auseinandersetzungen. Solche Verwirrungen mußten auch bei den Vegetationsgeschichtlern zum Ausdruck kommen. Die meisten sind sich zwar über die pleistozänen Ereignisse im klaren, ihre Auswirkung auf die jeweilige Pflanzenwelt haben sie aber nicht in allen weittragenden Folgen erfaßt.
Wenn man nun das rein hypothetische Argument "hohes Alter" näher betrachtet, dann ist folgendes dazu zu sagen: Jede Biozöno-

nose kann als ein Organismus höheren Ranges angesehen werden. Und überall in der Natur werden wir sehen, daß die charakteristische Lebenskurve eines jeden Organismus ihren Höchstpunkt schon im mittleren Alter erreicht, während sie im höheren Alter allmählich zurückgeht. Und man weiß auch, daß in den letzten 10.000 Jahren im europäischen Raum mehrere Waldgesellschaften in einer bestimmten Reihenfolge aufeinander gefolgt sind. Warum sollten dann zur Entwicklung einiger Pflanzengesellschaften Jahrmillionen notwendig sein?

Auch die Vielgestaltigkeit der Phytozönosen ist kein überzeugender Beweis für tertiäre Herkunft. Es sind zwar aus dem Tertiär artenreiche und ziemlich differenzierte Floren bekannt, die aber unter andersartigen Bedingungen entstanden und inzwischen vernichtet worden sind. Das Pliozän war eine ruhige und klimatisch ausgeglichene, warme Jahrmillionen dauernde Zeitperiode, die weite Verbreitung der Pflanzen ermöglichte, sowie durch Anpassung und Differenzieren das Entstehen neuer Formen begünstigte. POTONIÉ (1952) nennt das Tertiär die "Amplifikationssphase". Darunter versteht er eine Anreicherungszeit der Floren an Arten. Zu dieser Amplifikation zählt er auch die parallel zur Florenanreicherung verlaufende Verbreitung der parasitischen und epiphytischen Pflanzen. So ist diese "gesättigte" Pflanzenwelt in eine Art "Bioananke" oder "Bioklimax" geraten, aus dem es keinen Ausweg zur Normalisierung gab. Darum spricht man mehr über tertiäre Taxa als von veralterten, archaischen Elementen. Und tatsächlich waren manche von ihnen nicht den neuen Lebensbedingungen gewachsen und verschwanden aus dem europäischen Raum. In Nordamerika dagegen haben sich sogar viele den tertiären Floren beiderseits des Atlantik gemeinsamen Elemente bis ins Holozän erhalten. So ist die rezente Flora eines Teiles von Nordamerika wenigstens zum Teil aus den "tertiären" Elementen aufgebaut. Und dennoch spricht man da nicht von tertiären und archaischen Gesellschaften. Vielmehr sieht man in ihnen eine "moderne" quartäre Vegetation, nicht aber eine direkte Fortsetzung der pliozänen Floren, obwohl sie aus denselben Elementen besteht.

Mit dem Auftreten der Eiszeiten ist das biotische Gleichgewicht gefährlich gestört worden, und die Gesellschaftsbindungen wurden immer lockerer. Endlich sind die kaum noch zusammenhaltenden Gesellschaften zerfallen, so daß die einzelnen Elemente in loser Folge die Migration nach den Refugien angetreten haben. Genauso

verlief die Wiederbewaldung: Es siedelten sich sukzessiv einzelne Bestandteile (Anaphase der Pollendiagramme) an und schlossen sich an Ort und Stelle zu einer Gemeinschaft zusammen.
Stark wechselnde Lebensbedingungen wirken ausgesprochen selektiv auf alle Lebewesen. So sind manche tertiäre Elemente in den jeweils neu auftretenden Floren ausgefallen. Das Pleistozän hat sich somit als ein wirksames Sieb für die zu empfindliche oder überalterte pliozäne Taxa erwiesen.
Den Entwicklungsgang der anthropogenen ruderalen Gesellschaften kann man wohl im Laufe einer menschlichen Generation vom Initial- bis zum Endstadium verfolgen. Bei der Waldentwicklung ist das aber nur indirekt möglich. Man kann zwar theoretisch verschiedene, örtlich getrennte Stadien der Waldentwicklung zusammenstellen und so eine hypothetische Sukzessionsfolge konstruieren.Das wäre aber immer noch kein direkter Beweis, weil man das relative Alter der einzelnen Stadien nicht kennt. Außerdem werden Waldsukzessionen seit Jahrtausenden vom Menschen stark beeinflußt und willkürlich gelenkt.

Bis zu einem gewissen Grad ist das Werden einer Waldgesellschaft durch genaue Untersuchungen und Beobachtungen an Rodungs- und Brandflächen oder auf Bergschutthalden und Flußschottern "meßbar". H.MAYER (1967) hat mit mehreren Mitarbeitern komplexe Untersuchungen an Bergschutthalden im Wimbachgriestal unternommen, in einer absoluten Meereshöhe von 800 bis 1300 m. Mit Hilfe phytosoziologischer, palynologischer, pedologischer und Radiokarbon-Methoden wurde eindeutig bewiesen, daß in verschiedenen Zeitabschnitten, vom 6.Jahrhundert bis zur Gegenwart eine auf den Bergschutthalden sich ansiedelnde Vegetation bis zu einem P i c e e t u m s u b a l p i n u m , stellenweise mit *Abies*, 400 bis 500 Jahre braucht. Damit ist schon ein direkter Beweis erbracht, daß sich unter heutigen Bedingungen eine ganze Waldsukzession im Laufe von 500 Jahren abspielen kann.
Die Pollenanalyse kann in dieser Streitfrage Entscheidendes beitragen. Leider sind aber die gegenseitigen Beziehungen beider Disziplinen noch immer zu locker, als daß sie einen regen Gedankenaustausch begünstigten. Es gibt zwar eine Menge Pollendiagramme aus holozänen Schichten, die aber nicht alle ins Spätglazial reichen. Die Anzahl derer, welche die Vegetationsgeschichte älterer, zusammenhängender Perioden darstellen, ist schon wieder erheblich geringer. Es gibt bisher nur wenige Pollendiagramme, die in einem durchgehenden Profil mehrere Glaziale und Intergla-

ziale umfassen. Und gerade solche sind für einen überzeugenden Beweis maßgebend!

Für das Gebiet der jugoslawischen Ost-Alpen liegen bereits Pollendiagramme aus allen Interglazialen, angefangen mit dem Pliozän, vor. Es werden nur die führenden Wald-Gesellschaften angeführt, daneben wird uns aber auch der Stand des angeblich aus dem Tertiär stammenden Querco-Carpinetum und Fagion interessieren.

Das Pollendiagramm aus dem obersten Pliozän von Velenje (SERCELJ 1968) weist noch eine Anzahl tertiärer Elemente auf, von denen aber keines eine ausgesprochene Vorherrschaft erreicht. *Fagus* erlangt hier zum letzten Mal 30% des Pollenwertes.

Die Vegetation des Interglazials Donau-Günz bei Ilirska Bistrica (ŠERCELJ 1965) war ein Caryo-Quercetum mit *Cedrus*; *Fagus* ist nur noch untergeordnet.

Im Günz-Mindel Interglazial von Zalog bei Novo mesto (ŠERCELJ 1961) tritt das Caryo-Quercetum noch stärker hervor, während *Fagus* nur sporadisch erscheint. *Carpinus* erreicht hier 5 bis 10%.

Die Waldgeschichte vom ersten Mindelstadial bis heute wird aus dem beigefügten Pollendiagramm von Ljubljansko barje (Laibacher Moor) deutlich (ŠERCELJ 1966).

Für das Mindel-Riß Interglazial war *Tsuga* der bedeutendste Waldbaum, nebst noch einigen tertiären Elementen in geringer Anzahl. *Fagus* fehlt in diesem Abschnitt vollkommen.

Mit Beginn des Riß-Würm Interglazials tritt ein echtes, üppiges Querco-Carpinetum mit *Carpinus*-Werten bis 43% auf, *Fagus* erscheint neuerdings, aber nur sporadisch.

Im ersten Würm-Interstadial findet sich *Fagus*, zwar mit niedrigen Werten, aber in geschlossener Kurve im Pollen-Diagramm.

Erst das Holozän weist ein kräftiges Fagion auf, mit *Fagus*-Werten bis 60%. Die Buche ist in dieser Gegend, im Gegensatz zum nördlichen Alpenraum, um etwa 3000 bis 4000 Jahre vorausgegangen. Aus dem vorliegenden Diagramm geht eindeutig hervor, daß das Fagion die jüngste quartäre Waldgesellschaft ist und deshalb üppig und artenreich auftritt.

Aus den folgenden Überlegungen und den Pollendiagrammen können wir schließen:

1. Die pleistozänen Refugien der jeweiligen Interglazialen Floren oder Florenelemente sind nicht hier, wie bislang vermutet,

Additional material from Gesellschaftsentwicklung (Syndynamik),
ISBN 978-94-011-7594-4 (978-94-011-7594-4_OSFO4),
is available at http://extras.springer.com

EXTRA
MATERIALS
extras.springer.com

sondern weiter im Süden der Balkan- und Appenin-Halbinsel zu suchen. Der Südost-Alpenraum war daher in jeder Kaltzeit frei von thermophiler und mesophiler Waldvegetation. Die eigentliche stationäre Vegetation muß hier eine Art Taiga oder Tundra gewesen sein.

2. Durch stark wechselnde klimatische Extreme wurde das Pleistozän in getrennte biotische Phasen geteilt, und ein jeder dieser längeren oder kürzeren Zeitabschnitte wirkte sich als selbständige florogenetische Phase für thermo- oder frigoriphile Phytozönosen aus.

3. Daher ist es ausgeschlossen, daß irgendeine mesophile Vegetation im Gebiet das gesamte Pleistozän hindurch überdauert hat, oder während der Eiszeiten hier ein Refugium gehabt haben könnte, mit Ausnahme vereinzelter mesophiler Baumgruppen, die möglicherweise an besonders günstigen Standorten,Nunataks, ausdauern konnten, worauf sporadische Pollenfunde dieser Arten hinweisen.

4. Weder Querco-Carpinetum noch Fagion sind tertiären Ursprungs oder tertiäre Relikte, sondern jüngerer Herkunft; der Eichen-Hainbuchen-Wald enstammt dem letzten Interglazial, während das Fagion dem Holozän angehört. Gleichermaßen sind die heutigen *Pinus*-Wälder keine glazialen Relikte, sondern sind nur als Pionier-Vegetation auf degradierten oder mageren Böden anzusehen.

5. Die Pollendiagramme, die in einem Profil mehrere Interglaziale erfassen, weisen eindeutig darauf hin, daß die Zeitdauer eines Interglazials oder Interstadials zur geschlossenen Entwicklung einer Vegetationsfolge ausgereicht hat, und deshalb langdauernde geologische Perioden nicht zur Hilfe gerufen werden müssen.

6. Der Begriff "Relikt" ist zu unklar definiert und muß näher abgegrenzt werden.

SUMMARY

The SE Alps have long been considered as the refuge for the Middle European floras during the Pleistocene glaciations.FIRBAS (1923),1949) had supposed *Picea* should have persisted on this territory during the last glaciation. LÄMMERMAYR (1926) has drawn an exact line through Slovenia to which *Fagus* should have reached during the last glaciation. According to HORVAT (1959) Querco-Carpinetum should have persisted on favourable spots since Tertiary. The aggressiveness of the actual Fagion illyricum led BORHIDI (1963) to the supposition, it should have existed here uninterruptedly since Tertiary until the present times. Several authors like to speak about "Tertiary" or

"Pleistocene" relics, thus avoiding a direct statement of persistency of certain plant species or communities.
Palynological investigations during recent years, however, brought evidence that Pleistocene glaciations should have had the same disastrous effect upon these floras as they have had on those in Middle Europe.
Every Interglacial plant community has been destroyed by the following glaciations and every subsequent Interglacial or Interstadial brought back the forest elements, one after another, to form on the site new communities, not quite identical with the former ones. There is also enough palynological evidence to confirm that in every succeeding interglacial period "Tertiary" elements have been decreasing in species and frequency, while the Quaternary ones were gaining more and more importance.
The formation of these floras was therefore not a process of a sudden change in vegetation as a consequence of climate deterioration, but it took its course as a gradual supplanting in each warm period of Tertiary elements by Quaternary ones.
Ruderal, anthropogenic societies colonize mostly over shorter periods and can be followed and controlled in course of a human life. The forest communities however, take much more time for their development and cannot be controlled or followed in course of a human generation. With this fact in mind some botanists tended to postulate thousands or even millions of years for a regional development and spreading out of an actual climax forest society.
Yet, it is difficult to agree with the idea that the modern aggressive and vigourous plant societies should be the oldest ones.

A new approach to this problem has been made by H.MAYER and his collaborators (1967). Their investigations in Wimbachgries showed that under the existing climatic conditions a Piceetum subalpinum needs approximately 300 to 500 years for its complete development from the initial vegetational stages on the mountain gravels.
A direct proof is yielded by adequate pollen diagrams comprising several sequential Pleistocene periods. The palynological evidence permits conclusions for these territories as follows:
1. The interglacial and interstadial floras did not have their refuges in this region; they must have been scattered more southward and southeastwards, along the Appenine and Balkan Peninsulas. So the southeastern Alps lacked the thermophile and mesophi-

le vegetation during every Pleistocene stadial. Except maybe some scattered mesophilic tree groups in nunataks, as sporadic finds of their pollen suggest.

2. Because of the strong climate extremes, Pleistocene was divided into several biotic phases. Thus each of these periods of longer or shorter duration is to be regarded as a separate and independent florogenetic phase as well for thermophilic as for frigoriphilic phytocoenoses.

3. The Pollen diagrams that include several subsequent sections of Pleistocene demonstrate that a period of interglacial or interstadial suffices and must have sufficed for a complete development of the vegetation from initial stages to a certain climax. There is therefore no need to call for long geological periods for their development.

4. Neither Querco-Carpinetum nor Fagion illyricum are of Tertiary origin or Tertiary relics, as often supposed. Both most important forest communities of this area are of much younger age, the first from the last interglacial and the latter being of Holocene provenience. Similarly also the recent *Pinus* forests cannot be regarded as glacial relics; they are a pioneer vegetation responding to unfavourable ecological conditions caused either by nature or by man.

5. The term "relic" is evidently much too vague, so it should be defined more precisely and used only for well-documented cases of the relic state for a given taxon or vegetation unit.

LITERATUR

BORHIDI,A. -1963- Die Zönologie des Verbandes Fagion illyricum.- Acta bot. Acad.Sci.Hung.3-4. Budapest.

FIRBAS,F. -1923- Pollenanalytische Untersuchungen einiger Moore der Ostalpen.- Lotos 71. Prag.

-- -- -1949- Spät- und nacheiszeitliche Waldgeschichte Mitteleuropas nördlich der Alpen.- Jena.

HORVAT,I. -1959- Die Pflanzenwelt Südosteuropas als Ausdruck der erd- und vegetationsgeschichtlichen Vorgänge.- Acta Soc.bot.Poloniae 28 (3). Kraków.

LÄMMERMAYR,L. -1926- Fagus silvatica. Pflanzenareale, R.I (1-2).

MAYER,H., SCHLESINGER,B., Thiele,K. -1967- Dynamik der Waldentstehung und Waldzerstörung auf den Dolomit-Schuttflächen im Wimbachgries (Berchtesgadener Kalkalpen). Jb.Ver.z.Schutze Alpenpfl.32. München.

POTONIE,R. -1952- Gesichtspunkte zu einer paläobotanischen Gesellschaftsgeschichte (Soziogenese).- Beih.Geol.Jb.45. Hannover.

ŠERCELJ,A. -1938- Palinologija - zgodovina gozdov (Palynologie - die Waldgeschichte).- Gozd.vestn.15 (6-7). Ljubljana.

-- -- -1961- Staropleistocenska vegetacija v Zalogu pri Novem mestu (The Lower Pleistocene Vegetation from Zalog near Novo mesto).- Diss. IV.Cl.Acad.Sci.Slov.6. Ljubljana.

ŠERCELJ,A. -1965- Staropleistocenska flora in Bukovice pro Ilirski Bistrici (Altpleistozäne Flora su Bukovica bei Ilirska Bistrica.- Diss.IV. Cl.Acad.Sci.Slov.8. Ljubljana.

-- -- -1966- Pelodne analize pleistocenskih in holocenski sedimentov Ljubljanskega barja (Pollenanalytische Untersuchungen der pleistozänen und holozänen Ablagerungen von Ljubljansko barje).- Diss.IV.Cl.Acad.Sci.Slov.9. Ljubljana.

-- -- -1968- Pollenanalytische Untersuchungen des Hangenden von Velenje mit Funden von Mastodonten.- Diss.IV.Acad.Sci.Slov.11. Ljubljana.

DISKUSSION

M.WRABER †

Die palynologischen Forschungsergebnisse dieses Referates haben die allgemeine Auffassung, das Querco-Carpinetum croaticum sei tertiären Ursprungs, - eine Auffassung, die auch der verstorbene Professor HORVAT vertrat - für uns etwas enttäuschend, aber überraschend widerlegen können.

H.ELLENBERG

Ich habe lange auf eine solche Arbeit gewartet und begrüße den exakten Nachweis, daß sogenannte "Relikt"-Pflanzengesellschaften tatsächlich keine Relikte sind, sondern nach der Eiszeit in relativ kurzer Zeit aus ihren Komponenten entstanden sind. Es handelt sich bei den sog. Reliktgesellschaften, wie sie uns von GAMS und von E.SCHMID insbesondere genannt wurden, um Refugien von konkurrenzschwachen Arten, die sich nur auf relativ trockenen, nährstoffarmen oder in anderer Hinsicht extremen Standorten zu halten vermögen, weil hier raschwüchsigere, aber anspruchsvolle Arten ausgeschlossen bleiben. Es ist also dieser Relikt-Föhrenwald, oder was wir sonst im Auge haben, grundsätzlich nichts anderes als jede andere Pflanzengesellschaft, ein gesetzmäßig unter der Wirkung der Umwelt zustandegekommenes Gefüge von Arten, wobei die Konkurrenz eine mehr oder weniger große Rolle spielt. Ich glaube, daß in dieser Hinsicht die Arbeit von Herrn SERCELJ geradezu umwälzende Bedeutung hat.

A.W.KÜCHLER

Ich möchte gern auf die Terminologie zurückkommen. Wenn der Ausdruck tertiär auf Arten wie *Carya* und *Liriodendron* angewandt wird, dann muß das auch mit *Picea* und *Fagus* usw. geschehen, die sind genau so tertiär.

A.ŠERCELJ

Ich möchte noch etwas ergänzen. Ich sagte nämlich nicht, Arten seien "tertiäre", sondern für Gesellschaften treffe dies nicht zu. An dem Alter der Arten kann man nicht zweifeln, aber die Gesellschaften, die sie bilden, sind als solche nicht als tertiär anzusehen, sondern sie sind "modern" quartär.

J.SCHMITHÜSEN

Zu der Bemerkung von Herrn Professor ELLENBERG möchte ich zur Ehrenrettung von E.SCHMID nur bemerken, daß dieser seine Reliktföhrenwälder niemals als voreiszeitliche Relikte bezeichnet hat. Das ist ihm später unterstellt worden. Er meint tatsächlich so etwas Ähnliches wie Relikte, die sich an konkurrenzlosen Standorten gehalten haben.

FLORENWANDEL UND VEGETATIONSVERÄNDERUNGEN IN MITTELEUROPA WÄHREND DER LETZTEN JAHRHUNDERTE

H. Sukopp

Jede Pflanzengesellschaft hat sich im Laufe der Erdgeschichte mit der sich wandelnden Flora zu ihrer heutigen Zusammensetzung entwickelt und unterliegt auch weiterhin Veränderungen im Zusammenhang mit dem Wechsel des Florenbestandes. Dieser Wechsel geht - vom Standpunkt des Menschen aus betrachtet - i.a. sehr langsam vor sich. Jedoch ist er in neuerer Zeit durch den wirtschaftenden Menschen, speziell im industriellen Zeitalter, stark beschleunigt worden. Dieser Wandel zeigt sich sowohl im Rückgang und Aussterben, wie auch in der Einführung und Einbürgerung von Sippen. Die "naturalisierten Arten" nehmen in unserer Flora und Vegetation einen festen Platz ein. Sie verändern die Zusammensetzung natürlicher und anthropogener Pflanzengesellschaften.In Ackerunkraut- und Ruderalgesellschaften ist die Dynamik besonders augenfällig. Es fällt schwer, sich die Ruderalvegetation Berlin ohne die heute sehr häufigen Arten *Atriplex nitens*, *Corispermum leptopterum*, *Sisymbrium altissimum*, *S. loeselii*, *Diplotaxis muralis*, *Solidago canadensis* und *Bidens frondosa* vorzustellen - alles Arten, die noch nicht hundert Jahre Heimatrecht in Berlin besitzen. Über die interessantesten Neubürger, nämlich diejenigen, die in natürlichen Pflanzengesellschaften Mitteleuropas Fuß fassen konnten, wurde auf dem Symposion über "Anthropogene Vegetation" berichtet (SUKOPP 1966a). Von diesen Neophyten (sensu THELLUNG 1912) läßt sich für Mitteleuropa feststellen, daß von etwa 100 hierher zu rechnenden Arten etwa die Hälfte regionales Vorkommen zeigt; die anderen sind lokal eingebürgert. Die meisten dieser Neophyten werden in Pionier- oder Dauergesellschaften auf offenen oder zeitweilig offenen Standorten gefunden (vgl. These 16 des Vortrages von SEGAL). Mit wenigen Ausnahmen fehlen Neophyten in naturnahen Wäldern und Gebüschen. Einige Arten haben sich in freie ökologische Nischen eingepaßt; andere konkurrenzfähige Neophyten verdrängen Arten der alten Gesellschaft. Einigen Neophyten mit hohem soziologischen Bauwert gelang es sogar, eigene Pflanzengesellschaften aufzubauen, wie das Spartinetum townsendii.

Den Neubürgern steht eine große Zahl von ausgestorbenen,verschollenen und zurückgehenden Arten gegenüber, über die heute berichtet werden soll. Die Aufzeichnung des raschen Wandels von Flora und Vegetation in aller Welt ist eine wesentliche Aufgabe der botanischen Forschung in diesem Jahrhundert. Da die Arten die Grundeinheiten der meisten geobotanischen Arbeiten darstellen, ist eine Erörterung des Florenwandels von Bedeutung für unser Thema: Vegetationsentwicklung. Bemerkungen über Florenwandel klangen mehrfach während des Symposions an. Die folgenden Beispiele sind auf Gefäßpflanzen beschränkt.

In der Flora von Deutschland - einer "Neulandflora", in der kleinräumig verbreitete Endemiten kaum eine Rolle spielen, betrifft ein Rückgang immer nur Teilareale von Sippen. Nach unserer heutigen Kenntnis sind folgende einheimische Pflanzenarten in Deutschland ausgestorben oder verschollen[1]: *Carex capitata, C.microglochin* (SCHULTZE-MOTEL 1968), *Orchis spitzelii* (KRAUSS 1925), *Minuartia stricta* (FRIEDRICH 1962), *Herniaria incana* (EHRENDORFER 1967), *Veronica paniculata* (HARTL 1966), *Carpesium cernuum* (WAGENITZ 1966) und *Artemisia laciniata* (WENDELBERGER 1960).Dabei handelt es sich um seltene Arten der Moore (Glazialrelikte) und um einige südliche und südöstliche Arten, die Deutschland nur mit wenigen Fundorten erreicht haben.

Bezogen auf die Flora von Deutschland mit 2572 urwüchsigen und eingebürgerten Arten stellen die genannten Arten nur 0,3% des Bestandes dar. Für die Schwäbische Alb und den Schwarzwald wird ein Prozentsatz von 0,5 bzw. 1% ausgestorbener und verschollener Sippen angegeben (LITZELMANN 1957). In Sachsen und in Schleswig-Holstein gelten nach den Floren zwischen 2 und 3% der Arten als ausgestorben. Von den rund 1300 urwüchsigen und eingebürgerten Gefäßpflanzen der Niederlande sind nach WESTHOFF (1956) 4,2% ausgestorben und verschollen, 33% zurückgegangen, bedroht und ausgestorben. Als "in Zukunft" bedroht werden sogar 54% aller Arten der Flora der Niederlande betrachtet (VAN LEEUWEN & WESTHOFF 1961).

Einen starken Rückgang zeigt die Flora in der Umgebung von Großstädten und in Industriegebieten. Wegen der verschiedenen Flächengröße, der verschiedenen Abgrenzung, die häufig politisch oder kommunal bedingt ist, und wegen der Verschiedenartigkeit der berücksichtigten Arten (urwüchsige, eingebürgerte oder Teilgruppen davon) sind die Gebiete nur beschränkt vergleichbar.So

[1]Nomenklatur nach ROTHMALER 1958,1963.

kann die Tabelle nur ein ungefähres Bild vom Verlust an Pflanzenarten (zwischen 6 und 13% des Bestandes vor etwa 100 Jahren vermitteln.

Betrachten wir das Gebiet von Berlin genauer. Seit 1859 sind 124 Arten ausgestorben und verschollen. Das sind rund 13% des Bestandes an einheimischen und archäophytischen Arten (965) Berlins vor 100 Jahren. Starke Verluste haben die *Potamogetonaceae* (mit 41% des Bestandes vor 100 Jahren) und die *Orchidaceae* (39%) aufzuweisen; sowie von kleineren Familien z.B. die *Elatinaceae* (3 Arten vor 100 Jahren: 1 Art heute), *Gentianaceae* (7:4) und *Alismataceae* (7:4).

Welche Aufschlüsse über Vegetationsveränderungen können wir aus dieser Pflanzenliste gewinnen? Betrachten wir zu diesem Zweck eine Aufgliederung der ausgestorbenen und verschollenen Arten nach Vegetationstypen (Tab.1).

Die meisten ausgestorbenen und verschollenen Arten wuchsen demnach in Ödlandrasen, auf Äckern und in Wäldern. Jedoch geben diese Zahlen (Tab.1,Sp.1 und 2) ein falsches Bild, wenn wir sie nicht mit dem gesamten Artenbestand eines jeden Vegetationstyps vergleichen. Der Versuch einer Zuordnung aller Arten zu den Vegetationstypen (Sp.3) birgt manche Fehlermöglichkeiten, zumal manche Arten in verschiedenen Vegetationstypen vorkommen, hier aber nur einmal gezählt wurden. Es wird aber durch diesen Vergleich (Sp.4) deutlich, daß der Rückgang in der Wasser- und Teichrandvegetation, auf Mooren, feuchtem Grünland und Äckern weit größer ist, als die erste Aufstellung vermuten ließ.

Den stärksten Rückgang zeigt bei dieser Betrachtung die Vegetation der Gewässer, der Ödlandrasen und Äcker, die zwischen der Hälfte und einem Fünftel ihres Artenbestandes eingebüßt hat.

Erst danach folgen Moore, feuchtes Grünland und Wälder. In Schleswig-Holstein gehören nach Angaben von RAABE (1955) Ödlandrasen und Moore ebenfalls zu den Vegetationstypen mit vielen gefährdeten Arten. Das Aussterben von Arten der Gewässer ist besonders hervorzuheben. Das Ausmaß des Rückganges der Gewässervegetation wird noch deutlicher, wenn wir bedenken, daß die Flächengröße der Gewässer (mit Ausnahme der Teiche, vgl.Sp.5) sich in dem betrachteten Zeitraum nicht verändert hat und dennoch ein so starker Rückgang an Arten auftritt. Auch auf den Äckern ist wahrscheinlich der Rückgang an Fläche nicht allein ausschlaggebend. Die Ruderalvegetation schließlich zeigt bei starker Zunahme der

Tab. 1. Einheimische und archäophytische Gefäßpflanzen Berlins nach Vegetationstypen (aus SUKOPP 1966b)

Standort	Ausgestorbene und verschollene Arten		Einheimische und archäophytische Arten		Veränderung der Flächengröße während der letzten hundert Jahre (+ Zunahme − Abnahme)
	Anzahl	in% (n=124)	Anzahl	davon ausgestorben und verschollen in %	
	1	2	3	4	5
Wälder	18	16,5	171	10,5	-
Gebüsche, Waldränder Säume	6	4,8	65	9,2	-
Ödlandrasen	26	20,9	148	17,6	- -
Wasser	15	12,1	56	26,8	keine
Teichränder	8	6,5	18	44,4	- -
Ufer von Seen und Fließgewässern	3	2,4	102	2,9	- -
Moore	10	8,1	61	16,4	- -
feuchtes Grünland	12	9,7	91	13,2	- -
Fettwiesen und -weiden	0	0	42	0,0	-
Äcker	18	14,5	90	20,0	- -
Ruderalstellen, Mauern	8	6,5	121	6,6	+ +
Gebiet von Berlin	124	100,0	965	-	

Tab.2. Ursachen des Rückgangs höherer Pflanzen

1. Rückgang durch natürliche Veränderungen der Umweltbedingungen (MEYER 1952), besonders durch Änderungen (Schwankungen) des Klimas (vgl. RAABE 1955, PERRING 1965)

2. Rückgang durch den Menschen bedingt
   a) durch direkte Einwirkung auf Pflanzen
      - Rückgang durch Sammeln und Pflücken (vgl.SCHEERER 1960,p.13; LANDOLT 1961,p.67)
      - Rückgang durch mechanische Schäden infolge Erholungsbetriebes: Sportboote, Badebetrieb
      - Rückgang durch Bekämpfungsmaßnahmen: Saatgutreinigung (vgl.ARNDT 1955)
        Bekämpfung von Schmarotzern (wie *Cuscuta* u.a.)
        Verwendung von Herbiciden (vgl.WESTHOFF & ZONDERWIJK 1961)
        Reinigung von Gräbern
      - Rückgang durch Konkurrenz eingeführter Arten (vgl.GRAEBNER 1934 über *Allium paradoxum*; MOOR 1958, p.263 über die nordamerikanischen Flußufer-Compositen)
      - Rückgang durch Einführung von Pflanzenkrankheiten oder Insekten (vgl.FOSBERG 1959)
      - Rückgang durch Veränderungen im Anbau von Kulturpflanzen (Rückgang des Anbaus von Flachs, von *Fragaria moschata*

   b) durch Veränderungen der Umweltbedingungen

      Änderung der Standorte
      - Rückgang durch Änderungen des Wasserhaushaltes: Grundwassersenkung
        Verstädterung der Dörfer (vgl.SCHOLZ 1960)
      - Rückgang durch Änderung der chemischen Faktoren: Eutrophierung, Düngung, auch durch verwehten Düngerstaub, Verschmutzung des Oberflächenwassers (WEBER 1967)
        Verunreinigung der Luft (vgl.BOWEN 1965)
      - Rückgang durch Änderungen in der Form der Landnutzung: Veränderungen des Verhältnisses Wald zu Freiland,
        Veränderungen der Holzartenzusammensetzung, speziell Rückgang der Bodenvegetation unter dem Einfluß von Nadelhölzern (NOIRFALISE 1967), bzw. Änderungen in der Waldbewirtschaftung (z.B. Rückgang von *Taxus baccata* durch Übergang zum Dunkelwald (WILLERDING 1968)
        Aufgeben der Brachlandwirtschaft (vgl.ARNDT 1955), Aufforsten von Heiden und Sandfeldern (vgl.BERGER-LANGEFELDT & SUKOPP 1965)

      Zerstörung von Standorten
      - Rückgang und Verluste durch Bebauen, Zu- und Aufschütten, Abtragen, Flurbereinigung, Bau betonierter Wirtschaftswege, Camping, Restaurieren von Mauern u.a.

Fläche - heute fast die Hälfte des Stadtgebietes - dennoch einen merklichen Rückgang an einheimischen und archäophytischen Arten! Freilich wird dieser Verlust bei weitem durch den Zugang an neophytischen Ruderalpflanzen (SCHOLZ 1960) wettgemacht.

Ursachen des Aussterbens

In Tab.2 sind wichtige Faktoren zusammengestellt, die einen Rückgang von Pflanzen bedingen können, wobei zu einigen speziellen Fällen auf neuere Literatur verwiesen wird. In intensiv vom Menschen umgestalteten Gebieten kennen wir allerdings keinen Fall, in dem eine Pflanze mit Sicherheit durch natürliche Veränderungen ausgestorben wäre. Unter den anthropogenen Ursachen des Rückganges haben die Veränderungen der Standorte (b) viel weiterreichende Folgen gehabt als die direkten Einwirkungen auf Pflanzenvorkommen (a). Etwas überspitzt hat STRICKER (1962,p.466) diese Bedeutung der Standortsveränderungen hervorgehoben, indem er schreibt, daß sich nicht eigentlich die Pflanzen im Rückgang befänden, sondern viel eher oder nur die Standorte, an denen sie zu wachsen vermögen. Wichtige Faktoren, auf die der Rückgang zahlreicher Arten von Gefäßpflanzen zurückzuführen ist, sind Grundwasserabsenkung, Eutrophierung und Gewässerverschmutzung. Danach dürften Änderungen der mechanischen Faktoren wie Bebauen, Auf- und Abtragen, sowie Aufgeben der Brache folgen. In Zukunft wird die Verwendung von Herbiciden immer größere Bedeutung erlangen. Ein interessantes Beispiel für den Einfluß der Luftverunreinigung auf höhere Pflanzen bieten die Abb.1 und 2 aus der Arbeit von BOWEN (1965).

Die Ursachen der Veränderungen lassen sich auch unter einem weiteren Aspekt zusammenfassen. Die eine Gruppe von Veränderungen hatte bereits vor hundert Jahren zu wirken begonnen. Sie umfaßt alle Veränderungen der land- und forstwirtschaftlichen Nutzung, die sich durch die Umstellung von extensiver auf intensive Wirtschaft ergeben. Hierzu gehören lokale Grundwassersenkungen, das Aufforsten oder Beackern von Heiden und Sandfeldern, das Aufgeben der Brache, Einführung von Düngung und Saatgutreinigung. BENGT PETTERSSON(1949) hat diese Veränderungen, die die gesamte land- und forstwirtschaftlich genutzte Kulturlandschaft betreffen, zusammenfassend als *"Inversion"* (Umkehrung) bezeichnet. Erst später setzten bestimmte Veränderungen der Landschaft ein, die durch Verstädterung und Industrialisierung i.e.S. bewirkt werden. Hierzu kann man besonders großflächige Bebauung, Luftverunreinigung, Gewässerverschmutzung, großräumige Änderungen des Grundwasserhaushalts und weitreichende Anwendung von Herbiciden zur Unkrautbekämpfung rechnen. WESTHOFF & VAN LEEUWEN sprechen von einem "allgemeinen verpestenden Einfluß". Ein Ergebnis dieses Vorganges ist die "Ruderalisation der Landschaft" (SCHOLZ 1960). FREYER (1966) hat jüngst den Vorgang der durch die Industrialisierung bewirkten Veränderungen der Landschaft als *"Abstraktion"* bezeichnet. Schon jetzt greifen die Auswirkungen dieses Vorganges über den Ausgangspunkt, die engere Umgebung von Großstädten und Industriegebieten, hinaus. Für beide Phänomene wurden die in der Literatur vorhandenen Termini beibehalten, auch wenn sie recht allgemein sind und auch in anderen Zusammenhängen benutzt werden. Das Prägen neuer Ausdrücke sollte jedoch vermieden werden.

Für beide Fälle wollen wir je ein Beispiel betrachten. Der starke Rückgang von Ackerunkräutern dürfte im wesentlichen durch "Inversion" bedingt sein. Damit steht die Tatsache im Einklang, daß Ackerunkräuter bei Änderung der Landnutzung in vielen Gebieten zurückgehen, wie dies z.B. WESTHOFF (1956) für die Niederlande erwähnt. Als Beispiel für die zweite Gruppe wird meist der Rückgang epiphytischer Flechten infolge Luftverunreinigung angeführt (für Berlin vgl. NATHO 1964). Unter den Phanerogamen mögen der auffällige Rückgang an Wasserpflanzen in verschmutzten Gewässern sowie die allgemeine "Ruderalisation der Landschaft" genannt werden.

Bei allen Vergleichen früherer Vorkommen mit dem heutigen Pflanzenbestand stellt sich jedoch auch immer wieder heraus, daß die Florenkonstanz bei ungestörten Biotopen mindestens ebenso eindrucksvoll ist wie die Florenänderung bei Biotopwandel.

Bedeutung des Florenwandels für die Vegetation

Bevor wir die Bedeutung des Florenwandels für die Vegetationsveränderungen im einzelnen betrachten, seien einige allgemeine Bemerkungen über die Bedeutung des Verlustes von Arten vorangestellt. Wenn man den Extremfall, völliges Aussterben einer Art auf der Erde betrachtet, läßt sich die genetische und phylogenetische Bedeutung klar erkennen. Mit dem Aussterben einer Sippe ist nicht nur ein bestimmter unwiederbringlicher Typ, Ergebnis einer langen Entwicklung verschwunden, sondern es sind zugleich alle zukünftigen Entwicklungsmöglichkeiten abgeschnitten, die von dieser Sippe hätten ausgehen können. Eine ausgelöschte Art kann nicht wieder entstehen. Dagegen können verlorene Teilarea-

le, wenn die Art noch vorhanden ist, wieder besiedelt werden.
Aus der Tierwelt sind seit langem zahlreiche Beispiele von ausgerotteten+) Säugetieren und Vögeln bekannt. Eine neue Zusammenstellung nennt mehr als 200 Vogel- und Säugetierformen, die in den letzten 300 Jahren auf der Erde ausgerottet wurden (ZISWILER 1965). Über das Ausmaß der Ausrottung von Pflanzen fehlt uns eine Übersicht. Eine ganze Reihe von Insel-Endemiten ist ausgestorben. Als Beispiel sei die Flora von Hawaii erwähnt, aus der u.a. verschiedene *Malvaceen*, *Lobelioideen* und *Compositen* - zumindest als Wildpflanzen - verschwunden sind (PORSCH 1930, DEGENER 1945). Daß aber nicht nur Bewohner von Inseln besonders bedroht sind, zeigt z.B. der Fall von *Astragalus physocalix* aus Südbulgarien (TURRILL 1951).
Über die Folgen des Aussterbens von Arten hat man sich schon frühzeitig Gedanken gemacht. Der berühmte Naturforscher Carl v. LINNÉ hat 1760 (Politia naturae, zit. nach REMANE 1950, p.405) - vor dem Aufkommen der Abstammungslehre - formuliert: "Ja, ich wüßte nicht einmal, wie die Welt ohne Schaden bestehen sollte, wenn auch nur eine einzige Tierart auf ihr fehlen würde". Inzwischen wissen wir, daß Organismen sowohl "natürlich aussterben" als auch durch den Menschen ausgerottet werden. Das Gleichgewicht in der Biozönose mag sich dann verschieben, die gesamte Natur jedenfalls bricht nicht zusammen. In historischer Zeit sind meistens solche Arten ausgestorben, die bereits am Rande des Überlebens standen wie viele Relikt-Endemiten. Das Eingreifen des Menschen hat in diesen Fällen nur die bereits vorhandene Tendenz beschleunigt, aber keine grundsätzlich neue Situation geschaffen.
Eine andere Bedeutung hat das Aussterben von Arten in kleinen Gebieten, wenn dadurch nur ein Teil des Areals der betreffenden Sippe ausfällt. Als Beispiel einer solchen "Arealschrumpfung" seien erwähnt der Rückgang von *Ledum palustre* an dessen westlicher Arealgrenze (Karte bei MEUSEL 1943, p.92). Auch in diesen Fällen wird es manchmal nicht mehr möglich sein, einzelne Merkmale von Lokalrasen oder das Verhalten einer Sippe an der Arealgrenze festzustellen. Bei der Betrachtung kleiner Gebiete tritt jedoch der phylogenetische Gesichtspunkt an Bedeutung zurück.
In solchen Gebieten wird die Aufmerksamkeit überwiegend auf die

+) ZISWILER (1965, p.56-58) unterscheidet zwischen dem "natürlichen Aussterben" von Tierformen und der durch den Menschen verschuldeten Ausrottung.

Ursachen der Verluste gerichtet. Ebenso wie der Totalverlust von Arten ist auch der Rückgang der Arten in den meisten Fällen durch Veränderungen der Standorte bedingt. Wir benutzen den Verlust an Arten als ein Anzeichen für Vegetations- und Standortsveränderungen. In der Kulturlandschaft, in der die meisten Änderungen der standörtlichen Differenzierung durch den Menschen verursacht sind, bietet uns daher der Verlust an Arten ein einfaches Maß für die Auswirkungen des menschlichen Einflusses.

Listen der ausgestorbenen oder zurückgehenden Arten lenken zugleich unser Augenmerk auf bedrohte oder gefährdete Pflanzengesellschaften. Aus den Verlusten der Berliner Flora kann man schließen, daß verschiedene Nanocyperion- und Littorellion-Gesellschaften, das Najadetum marinae und andere Kleinlaichkraut-Gesellschaften, das Ranunculetum fluitantis, das Adonido-Brachypodietum und andere Orchideen-reiche Trockenrasen, bestimmte Typen des Alnetum betuletosum mit *Huperzia selago*, *Blechnum spicant* und *Corallorhiza trifida*, sowie bestimmte Pinion-Gesellschaften mit *Pulsatilla patens*, *P. vernalis* und *Arctostaphylos uva-ursi* in den letzten hundert Jahren verschwunden sind. Stark im Rückgang sind Molinion- und Eriophorion latifoliae-Gesellschaften, Rhynchosporetum, Scheuchzerietum u.a. In vielen Fällen, besonders im heute bebauten Stadtgebiet, liefern die floristischen Angaben die einzigen direkten Hinweise auf die frühere Vegetation. Die für Berlin genannten Pflanzengesellschaften oder regional entsprechende Typen sind in den meisten Gebieten Mitteleuropas gefährdet (vgl. TÜXEN 1955).

In den eben betrachteten Fällen handelt es sich in erster Linie um einen Rückgang der Standorte, an denen bestimmte Pflanzen und Pflanzengesellschaften zu wachsen vermögen. Besonderes Interesse verdienen noch die Fälle, in denen Vegetation und Standort erhalten bleiben, aber einzelne Arten ausfallen. Bei den in Deutschland ausgestorbenen Arten ist über die Bedeutung ihres Ausfallens für den Vegetationsaufbau nichts bekannt; sie dürfte aber nicht groß sein. Für Mitteleuropa ist jedoch auf folgende vier Fälle zurückgehender Arten, die größere Bedeutung haben könnten, hinzuweisen. *Taxus baccata*, die Eibe (vgl. WILLERDING 1968) war noch im 17. und 18. Jahrhundert viel verbreiteter als heute. In Brandenburg, Posen und Schleswig-Holstein z.B. kommt sie urwüchsig nicht mehr vor. In den Gegenden, in denen die Art seit mehreren Jahrhunderten ausgerottet ist,wird sie mit Recht nicht mehr als Element der potentiellen natürlichen Vegetation betrachtet. Immerhin ergeben sich bei der allmählichen Zurückdrängung einer Art alle Übergänge bis zu solchen Fällen, in denen etwa die Buche heute gebietsweise praktisch fehlt, wo dennoch als potentielle natürliche Vegetation ein buchenreicher Wald kartiert wird. *Abies alba*, die Weißtanne, befindet sich in den Randgebieten ihres Areals in ständigem Rückzug. An ihrer nördlichen Arealgrenze im Thüringer Wald ist u.U. mit ihrem völligen Aussterben zu rechnen. Der Ausfall der Tanne ist sowohl durch vom Menschen geschaffene als auch durch natürliche Ursachen bedingt. Die durch diesen Ausfall hervorgerufenen "Vegetationsveränderungen sind derart einschneidend, daß die Rekonstruktion des natürlichen Vegetationsbildes oder gar die Konstruktion der potentiellen natürlichen Vegetation ... auf größte Schwierigkeiten stößt",schreibt SCHLÜTER (1965,p.57). Der dritte Fall betrifft *Ulmus carpinifolia*, die Feldulme, bei der das Ulmensterben - sonst von Straßenbäumen bekannt - in den letzten zwei Jahren in Mitteldeutschland auch auf naturnahe Pflanzengesellschaften der Flußauen und Restgehölze im Bereich der Ackerflächen übergegriffen haben soll (freundliche Mitteilung von Prof.Dr.R.SCHUBERT). Besonders in den Jahren 1931-1934 sind die Seegraswiesen aus *Zostera marina* in den europäischen Gewässern und an der nordamerikanischen Atlantikküste fast vollständig durch eine Krankheit vernichtet worden, deren Erreger nach RENN (1936) wahrscheinlich ein Myxomycet, *Labyrinthula macrocystis* ist. Eine Wiederbesiedlung erfolgte nur langsam (vgl.z.B. DETLEFSEN 1965). An einigen Orten nahmen andere *Zostera*-Arten, die nicht von der Krankheit befallen wurden, bei verminderter Konkurrenz von *Z.marina* zu (SCULTHORPE 1967).In allen diesen Fällen sind wir über die soziologischen Auswirkungen des Ausfalles dieser Arten nicht genau orientiert. Vergleichsweise können wir nur auf zahlreiche Arbeiten hinweisen, die die rezenten Veränderungen der Eichen-Kastanien-Mischwälder des östlichen Nordamerika nach dem Absterben von *Castanea dentata* durch Befall mit einem um 1900 aus Asien eingeschleppten Ascomyceten, *Endothia parasitica* beschrieben haben. *Castanea dentata* war früher zu mehr als 40% in der Baumschicht der Klimax-Gesellschaft des Gebietes vertreten. 30-40 Jahre nach dem Rückgang der Art auf einen ganz unbedeutenden Anteil (etwa 1% der früher vorhandenen Pflanzen leben noch, GLEASON & CRONQUIST 1964,p.123) hat sich ein neues Gleichgewicht noch nicht völlig eingestellt. Es ist zu einer Zunahme der Co-Dominanten gekommen. *Castanea dentata* ist nicht länger ein bedeutender Bestandteil des einst von ihr mitbeherrschten Vegetationstyps. In diesem Falle sind wahrscheinlich nicht nur die Anteile verschiedener Arten verschoben, sondern die gesamte Struktur der Vegetation ist hochgradig verändert (FOSBERG 1959). Inzwischen hat das Edelkastanien-Sterben auf Europa übergegriffen. Im Tessin waren nach BAZZIGHER (1964) stellenweise 45 bzw. 85% der *Castanea sativa*-Bäume befallen. Selbst im Falle von *Castanea dentata*, der sich unter ständiger Beobachtung in diesem Jahrhundert vollzogen hat, wissen wir nur ungefähr über die Bedeutung des Florenwandels für die Vegetation Bescheid.

SUMMARY

Changing flora and vegetation appear both in retreat or extinction, and in the introduction and naturalisation of taxa. Naturalized species occupy a certain position in our flora and vegetation and alter the composition of natural and man-made plant communities. Interesting newcomers are those plants which are established at natural or semi-natural habitats. Coastal areas and river valleys seem to be their favorite places. With a few exceptions they will not be found in climax vegetation.

In contrast to these naturalized or established aliens there are many extinct, lost or retreating species. In Germany we know 8 species of higher plants which are extinct (= 0,3% of the flora with 2572 native and naturalized species). In smaller areas,e.g. Saxony and Schleswig-Holstein 2-3% have disappeared. A strong retreat is seen in the flora of great cities and industrial regions (6-8% of the species). In Berlin 124 species (= 13% of the flora) have disappeared since 1859. Especially the vegetation of open water, extensive grasslands and fields show the greatest retreat. Change of habitats (drainage and eutrophication) has a much more far-reaching effect than a more direct influence on plants (seed cleaning, use of herbicides). One may say that it is not the plants which are retreating, but their habitats. That is why we use the loss of species as an indication of change in vegetation and habitat. In the cultivated landscape where most habitat changes are caused by man this loss gives us a simple indication of human influence.

LITERATUR

ARNDT,A. -1955- Wandlungen der Ackerunkrautflora in der westlichen Niederlausitz. Beitr.z.Flora u.Veget.Brandenburgs 1.- Wiss.Z.Pädagog.Hochsch. Potsdam. Math.-nat.R.1 (2): 149-151. Potsdam.
BAZZIGHER,G. -1964- Die Ausbreitung der Endothia-Seuche im Kanton Tessin.- Schweiz.Z.Forstw.115: 320-330. Zürich.
BERGER-LANDEFELDT,U.& SUKOPP,H. -1965- Zur Synökologie der Sandtrockenrasen, insbesondere der Silbergrasflur.- Verh.Bot.Ver.Brandenburg 102: 41-98. Berlin-Dahlem.
BOWEN,H.J.M. -1965- Sulphur and the distribution of British plants.- Watsonia 6: 114-119. Arbroath.
BURCK,O. -1925- Veränderungen in der Flora Frankfurts seit 100 Jahren.- Natur u.Museum 55: 423-433. Frankfurt/M.
DEGENER,O. -1945- Plants of Hawaii National Park.- Ann Arbor.
DETLEFSEN,N. -1965- Das Seegras kommt wieder.- Die Heimat 72: 339. Neumünster.
EHRENDORFER,F. -1967- Liste der Gefäßpflanzen Mitteleuropas.- Graz,254 pp.
FOSBERG,F.R. -1959- Man as a dispersal agent.- Southwestern Naturalist 3: 1-6.
FREYER,H. -1966- Landschaft und Geschichte.- In: Mensch und Landschaft im technischen Zeitalter. (39-70). Gestalt und Gedanke, 10.Folge.München.
FRIEDRICH,H.-Ch. -1962- Caryophyllaceae (2.Teil).- In: Hegi,Illustrierte Flora von Mitteleuropa III, 2.Lfg.5. 2.Aufl. München.
GLEASON,H.A.& CRONQUIST,A. -1964- The Natural Geography of plants.- New York and London,420 pp.
GRAEBNER,P. -1934- Allium paradoxum Don. Sonderbarer Lauch.- In: Kirchner-Loew-Schroeter: Lebensgeschichte der Blütenpflanzen Mitteleuropas. Bd.I, Abt.3: 454-457.
-- -- -1964- Die Pflanzenwelt des Paderborner Raumes.- Schr.Reihe Paderborner Heimatver.2. Paderborn.
HARTL,D. -1966- Scrophulariaceen. 2.Teil.- In: Hegi, Illustrierte Flora von Mitteleuropa VI (1), Lfg.2. 2.Aufl. München.
KRAUSS,H.A. -1925- Das Kommen und Gehen einer Orchis-Art in Württemberg.- Veröff.Staatl.Stelle f.Naturschutz Württ.Landesamt f.DenkmalPfl.2: 115-119. Stuttgart.
KREH,W. -1951- Verlust und Gewinn der Stuttgarter Flora im letzten Jahrhundert.- Jh.Ver.vaterl.Naturk.Württemb.106: 69-124. Stuttgart.
LANDOLT,E. -1961- Schutz der einheimischen Pflanzenwelt.- Schweizer Naturschutz 27: 66-69.
LEEUWEN,C.G.van & WESTHOFF,V. -1961- De nivellering van flora en vegetatie.- Natura 58: 132-140. Den Helder.
LITZELMANN,E. -1957- Pflanzengeographie und Geschichte in ihren Wechselbeziehungen.- Alemannisches Jb.1957: 166-196.
MEUSEL,H. -1943- Vergleichende Arealkunde 1.- Berlin-Zehlendorf.
MEYER,D.E. -1952- Untersuchungen über Bastardierung in der Gattung Asplenium.- Bibliotheca botanica 123. Stuttgart.
MOOR,M. -1958- Pflanzengesellschaften schweizerischer Flußauen.- Mitt. schweiz.Anst.forstl.Versuchsw.34: 221-360. Zürich.
NATHO,G. -1964- Die Verbreitung der epixylen Flechten und Algen im Demokratischen Berlin.- Wiss.Z.Humboldt-Univ.Berlin Math.-nat.R.13: 53-75. Berlin.
NOIRFALISE,A. -1967- Conséquences écologiques de la monoculture des conifères dans la zone de feuillus de l'Europe tempérée.- IUCN.Publ.new ser.9: 61-71.
PERRING,F.H. -1965- The Advance and Retreat of the British Flora.- In: C.G.Johnson & L.P.Smith (ed.): The Biological Significance of Climatic Changes in Britain (51-59). London and New York
PETTERSSON,B. -1949- Gotländska vegetationsproblem. Preliminärt meddelande.- Bot.Notiser 1949: 377-394. Lund.

PORSCH,O. -1930- Kritische Quellenstudien über Blumenbesuch durch Vögel. V.- Biologia Generalis 6: 133-246. Vienna/New York.
RAABE,E.W. -1955- Über die Verarmung der Landschaft.- Schr.Naturw.Ver. Schlesw.-Holst.27(2): 171-189.Kiel.
REMANE,A. --1950- Ordnungsformen der lebenden Natur.- Studium Generale 3: 404-410. Berlin.
RENN,C.E. -1936- The wasting disease of Zostera marina. I.A phytological investigation of the diseased plant.- Biol.Bull.mar.biol.Lab., Woods Hole 70: 148-158. Lancaster,Pa.
ROTHMALER,W. -1958- Exkursionsflora von Deutschland. Gefäßpflanzen.- Berlin. 502 pp.
-- -- -1963- Exkursionsflora von Deutschland.Kritischer Ergänzungsband. Gefäßpflanzen. Berlin 622 pp.
SAVIDGE,J.P., HEYWOOD,V.H. & GORDON,V. -1963- Travis's Flora of South Lancashire.- Liverpool. 400 pp.
SCHLÜTER,H. -1965- Vegetationskundliche Untersuchungen an Fichtenforsten im Mittleren Thüringer Wald.- Die Kulturpflanze 13: 55-99.
SCHEERER,H. -1960- Gefährdung und Schutz unserer Wildpflanzen.- Kosmos-Bibliothek 226: 66. Stuttgart.
SCHOLZ,H. -1960- Die Veränderungen in der Ruderalflora Berlins. Ein Beitrag zur jüngsten Florengeschichte.- Willdenowia 2 (3): 379-397. Berlin.
SCHULTZE-MOTEL,W. -1968- Cyperaceae.- In: Hegi, Illustrierte Flora von Mitteleuropa.Bd.II,1. Lfg.2. 3.Aufl.- München.
SCULTHORPE,C.D. -1967- The Biology of Aquatic Vascular Plants.- London. 610 pp.
STRICKER,W. -1962- Das Leipziger Hafengelände-Einwanderungstor seltener und fremder Pflanzenarten.- Sächs.Heimatbl.6: 464-473. Radeberg.
SUKOPP,H. -1966a- Neophyten in natürlichen Pflanzengesellschaften Mitteleuropas (275-284). In: Tüxen,R. (Ed.): Ber.intern.Sympos.Stolzenau/Weser 1961, Anthropogene Vegetation. Den Haag.
-- -- -1966b- Verluste der Berliner Flora während der letzten hundert Jahre.- Sber.Ges.naturf.Freunde N.F.6: 126-136.
THELLUNG,A. -1912- La flore adventice de Montpellier.- Mém.Soc.nat.Sci. nat.-Math.Cherbourg 38.
TÜXEN,R. -1955- Aufruf zur soziologischen Aufnahme gefährdeter Pflanzengesellschaften.- Mitt.Flor.-soz.Arbeitsgem.N.F.5: 203. Stolzenau/Weser.
TURILL,W.B. -1951- Some problems of plant range and distribution.- J.Ecol. 39: 205-227. Cambridge.
WAGENITZ,G. -1966- Compositae II.Teil.- In: Hegi, Illustrierte Flora von Mitteleuropa. Bd.VI/3, Lfg.3. 2.Aufl. München.
WEBER,D.W. -1967- Über die Wasserpflanzenflora Ostfalens.- Braunschweigische Heimat 53: 11-15. Braunschweig.
WENDELBERGER,G. -1960- Die mitteleuropäischen Reliktvorkommen der Artemisia-Arten aus der Sektion Heterophyllae.- Verh.zool.-bot.Ges.Wien 98 u. 99: 57-95 1959,(1960). Wien.
WESTHOFF,V. -1956- De verarming van flora en vegetatie.- In: "Vijftig jaar Natuurbescherming in Nederland": 151-186.
-- -- & VAN LEEUWEN,C.G. -1959- De zwarte adem.- Levende Nat.62: 106-114. Arnhem.
-- -- & ZONDERWIJK,P. -1961- The effects of herbicides on the wild flora and vegetation in the Netherlands.- IUCN Symposium Warszawa 15-24. VII.1960: 69-78. Leiden.
WILLERDING,U. -1968- Beiträge zur Geschichte der Eibe (Taxus baccate L.)- Plesse-Archiv 3: 96-155. Göttingen.
ZISWILER,V. -1965- Bedrohte und ausgerottete Tiere. Eine Biologie des Aussterbens und des Überlebens.- Verständliche Wissenschaft 86: 134. Berlin-Heidelberg-New York.

DISKUSSION

F.KÜHN

Bei der rezenten Florenentwicklung natürlicher Standorte geht es nicht immer um eine Verarmung. Bei Šlapanice bei Brno wurde von G.NIESSL 1865 eine botanische Studie der Talhänge ausgearbeitet. H.ILTIS veröffentlichte 1914 eine ausführliche Studie desselben Tales. Darin konstatierte er, daß 325 Arten gefunden wurden; die Anzahl mitteleuropäischer Arten war ungefähr gleich geblieben; es wurden etwa 90 Arten neu gefunden, zum größten Teil xerotherme Arten. Ich habe dieselbe Lokalität 100 Jahre nach NIESSL und 50 Jahre nach ILTIS untersucht: Manche Arten, die ILTIS nicht wiederfand, wurden erst jetzt wiedergefunden, und etwa 120 Arten wurden neu registriert, zumeist xerotherme Arten. Das Flächenausmaß der Lokalität hat sich in den letzten 100 Jahren verringert.

M.WRABER

Das sind m.E. gegenseitige Verschiebungen von Pflanzengruppen, mehr oder weniger lokal. Bei Herrn Dr.SUKOPP handelte es sich wirklich um absolute Verluste und absoluten Zuwachs in der Flora.

F.RUNGE

Im allgemeinen verarmt unsere Flora sicher, die Arten werden weniger. Eine Ausnahme bilden wohl unsere Ostfriesischen Inseln, wo die Flora in den letzten 150 Jahren reicher geworden ist.Das liegt aber daran, daß diese Dünen auf diesen Inseln sich entwikkeln und altern.

R.TÜXEN

Herr SUKOPP hat uns gezeigt, wie eine gewisse Abnahme der Artenzahl und andererseits eine Zunahme der Artenzahl erfolgt. Er hat die Bilanz gezogen, daß bei uns in Mitteleuropa die Abnahme im Augenblick stärker als die Zunahme ist. Er hat auch von den Veränderungen der Pflanzengesellschaften gesprochen und hat darauf hingewiesen, daß manche Pflanzengesellschaften, oder sagen wir, viele Bestände bestimmter Pflanzengesellschaften, verschwunden sind, so daß diese Pflanzengesellschaften vielleicht hier und da, wenn auch nur fragmentarisch, noch bestehen mögen. Aber ist das wirklich alles? Ich dachte während seines Vortrages oft an ein Wort, das mich sehr getroffen hat. Es stammt von Herrn MERKER. Er spricht von der Devitalisierung des menschlichen Lebensraumes. Das ist ein schlimmes Wort. Wenn ich an die letzten 40 Jahre zurückdenke, die ich hier in NW-Deutschland ziemlich intensiv gearbeitet habe mit Flora und Pflanzengesellschaften, und wenn ich an die Tabellen erinnern darf, die uns Herr SPEIDEL zeigte, wo durch die Zunahme der Beweidung die Mähwiesen vernichtet wurden - auch darüber hat ELLENBERG sich hier vor Jahren einmal geäußert, er wollte ein A r r h e n a t h e r e t u m zum Naturschutzgebiet machen - da muß ich sagen, es ist eine von Herrn SUKOPP versteckt, oder doch nicht ganz deutlich ausgesprochene, ungeheure Verarmung, um nicht zu sagen, Verödung unserer Vegetation nicht nur eingetreten, sondern in vollem Gange. Denken Sie an den Verlust unserer Unkrautgesellschaften, von denen nur noch die Arten da sind, die jeder Sextaner kennt, die anderen gibt es ja gar nicht mehr, oder denken Sie an den Reichtum unserer Blumenwiesen. Als ich Student war, da waren die süddeutschen M o l i n i e t a l i a - oder A r r h e n a t h e r e t a l i a - Wiesen Blumengärten, und als wir im vorigen Jahr in NO-Bayern waren, das hat Freund DIERSCHKE vorhin gesagt, da sind wir einen ganzen Tag lang durch *Holcus lanatus*-Wiesen gefahren und sonst haben wir fast nichts gesehen. Das ist keine Devitalisierung, aber der Reichtum an Vielseitigkeit einer ungeheuer vielgestaltigen Landschaft ist nivelliert worden durch eine Eutrophierung, d.h. Anreicherung, und dadurch ungeheuer verarmt. Das wollte ich noch dazu gesagt haben, um auf die ganze Tragik eigentlich, oder gar Gefahr dieses Vorganges - denken Sie an die Sterilität der Rinder - noch einmal hinzuweisen.

H.ELLENBERG
Im Prinzip bin ich durchaus dieser Meinung. Aber wir sehen jetzt die Veränderung der Flora und Vegetation in einer immer dichter besiedelten Landschaft, zu der ja Mitteleuropa gehört. Wenn wir jetzt einmal in die Zeit zurückgehen, als das noch gar nicht dicht besiedelt war, als es Urwald war, da war sowohl die Flora als auch die Vegetation sehr viel ärmer, als wir sie in unseren Jugendzeiten kennengelernt haben. Denn da gab es noch gar keine *Arrhenathereten*, die sind vielleicht 1000 Jahre alt. Jedenfalls gibt es sehr viele Pflanzengesellschaften und auch viele Florenelemente, die erst durch die Tätigkeit des Menschen in unsere Gegend gekommen sind und sich hier gebildet haben, das ist ein allmählicher Prozeß.

Der Urwald ist sozusagen das Langweiligste, was man sich vorstellen kann. Ich habe das Glück gehabt, viele Wochen in einem wirklichen Urwald zu leben, in den Amazonas-Urwäldern. Man kommt sich

vor, als wenn man an dem Grunde eines Meeres läuft, aus dem man nie richtig herauskommt, außer am Ufer eines Flusses.Es ist furchtbar, es ist dunkel darin und tierarm, der wirklich unberührte Wald ist etwas Monotones. Und man ist geradezu erlöst, wenn man in eine besiedelte Landschaft kommt, die wieder einen etwas lebendigeren Wechsel aufweist.
Das, was wir so schön finden an unserer Landschaft, hat großenteils der Mensch gemacht. Da muß man also dem Menschen gegenüber gerecht sein. Dieses ist jetzt in der letzten Zeit eine Entwicklungsphase, die sicher, da stimme ich völlig überein, zu einer zum Teil ganz erstaunlichen und vor allen Dingen leider auch gedankenlosen Verarmung führt. Da müssen sich die Verantwortlichen über diese Folgen Rechenschaft geben, die auch sogar für die Gesundheit existieren. Wenn man es nun weltweit betrachtet, ist die Gefahr allerdings nicht so groß, denn das ist eigentlich noch immer auf verhältnismäßig kleine Teile dieses Erdballs beschränkt. Heute sind auch schon Gegenreaktionen im Gange. Das Bewußtwerden der Gefahr ist deutlich, da bin ich einfach Optimist.

D.MUELLER-DOMBOIS

Dieses Thema ist das erste, das über die floristischen Verhältnisse in bezug auf Sukzession etwas aussagt, denn wenn wir Vergleiche aus dem Nebeneinander anstellen, um auf das zeitliche Nacheinander hinzuführen, dann spielt gerade hier das Weggehen einiger Arten und das Hinzukommen anderer eine große Rolle.Denn floristische Verschiebungen können dadurch sehr leicht entstehen, und das macht diese Basis, um aus dem räumlichen Nebeneinander zu einem zeitlichen Nacheinander zu kommen, nämlich etwas unsicher. Das ist ein Punkt, der diese Unsicherheit aufdeckt.
Der zweite Punkt ist der, daß wir den Einfluß des Menschen immer von der aktiven Seite her betrachten. In Indien ist es z.B. genau umgekehrt, da hat der Mensch eine passive Rolle gespielt, und die Natur ist zumindest um die Städte herum, stärker degradiert als in Gebieten, wo der Mensch eine aktive Rolle gespielt hat. Und zwar ist das darauf zurückzuführen, daß der Mensch, der ja sowieso ein sehr unspezialisierter Organismus ist, sich an alle möglichen Situationen anpassen kann, sich infolgedessen kolossal vermehrt hat, aber auf Kosten anderer Organismen. Wenn man heute in eine Stadt wie Benares kommt, dann findet man dort keine Singvögel mehr, sondern nur noch einige kleine Populationen von Raubvögeln und großen Drosseln. Alles andere ist ausgestorben. Genau

so ist es mit der Vegetation, dort findet man nur noch "Unkräuter" ,und eine Monotonie der Flora, die auf Konkurrenzdruck einiger Populationen zurückzuführen ist; bedingt durch das passive Verhalten des Menschen konnten sich diese kolossal ausbreiten und zur Degradation der anderen von Natur aus nicht so konkurrenzfähigen beitragen.

H.SUKOPP

Ich habe versucht, die Florenveränderungen als Verlust und Gewinn darzustellen, und viele Autoren haben ja eben die Worte Verarmung für Rückgang und Bereicherung oder im schlimmen Falle Florenfälschung für Zugang benutzt. Dabei wird unterstellt, daß der Eingriff des Menschen in unsere Flora diese in ihrer natürlichen Zusammensetzung "verändere". Es träte eben eine Verarmung auf im Vergleich zu sozusagen "reichen Natur". Dabei wird vergessen,daß die Tätigkeit des Menschen ohnehin die Landschaft und auch die natürliche Vegetation gründlich verändert hat und ständig verändert. Und zu der reichen Flora der Vergangenheit, dazu würde ich das Beispiel aus Brno mit extensiver Wirtschaft rechnen, dazu das Beispiel der friesischen Inseln, das keine Ausnahme, sondern durchaus typisch ist. Diese Flora war eben früher reich, und die genannten Beispiele wurden vom Menschen durch extensive Nutzung von Wald, Grünland und Acker gebildet. Die entsprechenden Zahlenvergleiche hat ELLENBERG in dem Buch über Wiesen und Weiden bereits einmal gezogen.

Ein Beispiel für die Flora von Finnland liefert KALLIOLA insoweit, daß dort 42% der Arten unter dem Einfluß des Menschen häufiger geworden sind als in der Naturlandschaft, und nur 22% der ursprünglichen Arten in der Flora Gebietsverluste erlitten haben. Mindestens also die Mengenverhältnisse waren eben auch früher nicht nur durch die natürlichen Standortsverhältnisse bedingt, sondern durch die Tätigkeit des Menschen. All diese Feststellungen betreffen letztlich also eigentlich nur Schwankungen in der Flora und der Vegetation in der Kulturlandschaft. Aber auch ein solcher Vergleich ist eben wohl notwendig, er zeigt den Einfluß verschiedener Formen der Landnutzung. Man darf eben nur nicht davon ausgehen, Rezentes mit Progression vergleichen zu wollen. Denn wie käme es sonst gerade zu einem Rückgang auf kulturbedingten Plätzen wie Ödlandrasen und Äckern? Es ist durchaus richtig, daß sich der Einfluß des Menschen in Verarmung und Bereicherung zeigen kann. KREH hat die Rechnung für Stuttgart aufgestellt und gesagt, 6% Arten verloren, 17% Arten dazu; die Flora ist bereichert. Das erklärt sich daraus, daß er "naturalisierte" Arten genommen hat und Siedler, von denen nicht klar ist, ob sie im Gebiet bleiben werden. Wenn ich das nicht tue, und wirklich nur naturalisierte Arten rechne, dann ergibt es für alle intensiv beeinflußten Gebiete, speziell für das von mir genauer untersuchte Berlin, einen Rückgang.

J.SCHMITHÜSEN

Um auf die Diskussion TÜXEN-ELLENBERG noch einmal zurückzukommen. Ihnen allen ist bekannt, daß wir in Mitteleuropa so etwas wie Wallfahrtsorte für Botaniker haben, d.h., wenn man etwas Bestimmtes sehen will, viele Arten und spezielle Gesellschaften, da gibt es nur noch wenige Punkte, und man muß sich von Spezialisten führen lassen, um manche Arten in Mitteleuropa überhaupt noch sehen zu können. Dies gilt nicht nur für Mitteleuropa, sondern es gilt für viel größere Teile der Erde, als man sich normalerweise vorstellt.

Als ich 1952 zum ersten Male nach Chile kam, und die Chilenen mich nach 3 Monaten fragten, was mich am meisten beeindruckt hätte, da habe ich gesagt, daß es in Mittelchile keine chilenischen Pflanzen gibt. Der ganze Raum besteht praktisch, von einzelnen Lokalitäten abgesehen, die man genau so aufsuchen muß wie hier, aus einer immigrierten Flora. OBERDORFER hat sie studiert, der größte Prozentsatz der Fläche wird dort von solchen Fremdfloren eingenommen. Und man muß dieses räumlich-flächenhaft Quantitative vielleicht doch einmal stärker sehen als bloß das Zahlenmäßige von den Arten aus. Als wir das zweite Mal dann gemeinsam mit OBERDORFER und KÜHLWEIN nach Chile gingen, um von jedem Vegetationstypus einen echten Urwaldbestand zu studieren, da hatten wir größte Mühe und brauchten mehrere Wochen, um dann notdürftigerweise 3 oder 4 Bestände von dem laubwerfenden *Nothofagus obliqua*- Wald zu finden. Da gibt es keine Urwälder mehr, sondern es gibt nur noch einige wenige ganz kleine Reste, ähnlich als wenn Sie hier in Europa noch einen Urwald suchen wollten.

Als wir dann nach Chiloe gingen, also nach Westpatagonien, wo nach all unseren Karten noch immer unendlich viel Wald sein soll, und wir den nordpatagonischen Regenwald auf dem nördlichen Drittel von Chiloe studieren wollten - ich war da 1952, also 5 1/2 Jahre vorher noch in großen geschlossenen Urwaldgebieten gewesen - da sind wir 3-4 Tage herumgereist. Es war inzwischen alles

abgebrannt.
In Neuseeland ist durch die Bewirtschaftung auf schätzungsweise 2/3 der gesamten Landesoberfläche ebenfalls nur noch eine Flora zu beobachten, die mit der ursprünglichen Flora des Landes überhaupt nichts zu tun hat. Man kann hunderte von Kilometern durch das Land fahren und braucht, wenn man die neuseeländische Vegetation studieren will, gar nicht auszusteigen, denn es sind alles Immigranten, die dort die Weideflächen bilden, und auch in den Gebieten, wo die Wälder abgebrannt sind. Als Gehölze finden wir *Pinus insignis* und *Cupressus macrocarpus* und andere heute international vor allem in den Subtropen verbreitete Hölzer,und es sind die internationalen Wiesengräser, meistens Europäer,die dort die Weideflächen bilden. Selbst die größeren Nationalparke, die wunderschön sind, und die insofern noch Urwälder sind, daß dort die Urwaldbäume noch stehen. Dort haben Sie größte Mühe, irgendwelche Bestände zu finden, von denen man wirklich noch sagen kann, es ist Urwald, weil nämlich die seit COOK eingeführten Hirsche und Rehe und Wildschweine inzwischen, auch in den Gebieten, die der Mensch nie betreten hat, die Bodenflora völlig vernichtet haben; auch dort sind die Fremdlinge eingedrungen.
Ich erzähle diese Beispiele hier einmal, um die Dimensionen zu zeigen.Das bevorstehende Aussterben von endemischen Baumarten im Süden Madagascars durch die Vorgänge der modernen Intensivierung der Nutzung in diesen Gebieten ist ein anderes Beispiel. In Japan ist die Situation erstaunlicherweise ganz anders, und das ist ein dichtbevölkertes Industrieland. Der entscheidende Grund ist dort, daß es in Japan bis zur Entwicklung des modernen Industriestaates keine Weidewirtschaft gegeben hat, keine Viehzucht, und daher haben sich in den Gebirgen die Wälder und damit viele andere Dinge erhalten, und die Japaner haben sie rechtzeitig schützen können. Das ist also in den einzelnen Ländern unterschiedlich.
Aber die Beispiele habe ich hier auch noch aus einem anderen, grundsätzlichen Punkt gewählt, der mir in allen diesen Diskussionen noch immer viel zu wenig zum Ausdruck kommt, weil wir bei der Betrachtung eigentlich meistens uns nur in unserem holarktischen Florenbereich bewegen. Das ist die Konkurrenz der Floren. Es gibt konkurrenzschwache und konkurrenzstarke Floren, es gibt ganze Länder, in denen die einheimische Flora so konkurrenzschwach ist, daß sie schlagartig in wenigen Jahren oder Jahrzehn-

ten auf Riesenflächen, die 99% des Landes ausmachen können, vor den einwandernden internationalen Wiesen- und Weidepflanzen verschwinden, meistens holarktische oder europäische Arten.

D.MUELLER-DOMBOIS

Für meine Begriffe hat Herr SCHMITHÜSEN die Begriffe Vegetation und Flora in der Anwendung durcheinander gebracht. Wenn wir über diese Situation klagen, dann müssen wir auch dafür sorgen, daß wir etwas dafür tun können. Wenn wir über Vegetation sprechen, dann müssen wir uns auch um die Konservierung von Ecosystemen, d.h. die Vegetation mit dem Land dazu, kümmern. Wenn wir aber über sehr seltene Pflanzen sprechen, die langsam verschwinden, dann können wir das auch auf eine andere Weise tun, nämlich diese Pflanzen in einen botanischen Garten bringen. So haben es die Japaner jetzt z.B. gemacht mit ihren alpinen Pflanzen, die am Aussterben sind. Genau so machen wir es jetzt in Hawaii,auch mit einigen Arten, für die die Gefahr besteht, auszusterben.Die kommen in einen kleinen botanischen Garten in ihrem natürlichen Habitat, wenn möglich, und auf diese Weise werden dann die vorhandenen Gene und auch die Arten selbst konserviert.

NACHWORT im September 1978

Untersuchungen der letzten Jahre haben gezeigt, daß bei den meisten Pflanzengruppen zwischen 40 und 50% der Arten in ihrem Vorkommen gefährdet sind, mit Ausnahme der Moose, die regional weniger bedroht sind (SUKOPP & MÜLLER 1976).

Bei allen Organismengruppen überwiegt in Mitteleuropa der Rückgang einheimischer Arten zahlenmäßig die Einbürgerung neuer Arten.

Bei den Farn- und Blütenpflanzen sind in den Flächenstaaten der Bundesrepublik Deutschland 30-40% aller Arten gefährdet, in der Umgebung von Großstädten mehr als 50% der Arten.

Eine zahlenmäßige Aufstellung der ausgestorbenen und gefährdeten Farn- und Blütenpflanzen zeigt die nachfolgende Tabelle (aus KORNECK, LOHMEYER, SUKOPP & TRAUTMANN 1977).

Tabelle 2. Gefährdete und ausgestorbene Farn- und Blütenpflanzen der Bundesrepublik Deutschland

Gefährdungsgrad	Zahl der Arten
1.1 ausgestorben oder verschollen	58
1.2 vom Aussterben bedroht	161
2 stark gefährdet	175
3 gefährdet	213
4 potentiell gefährdet	215
insgesamt	822

Fast 31% der einheimischen und eingebürgerten Arten von Farn- und Blütenpflanzen der Bundesrepublik Deutschland und Berlin (West) sind ausgestorben oder gefährdet.
Ausgestorben in phylogenetischer Bedeutung, d.h. als Sippe auf der Erde überhaupt verschwunden, ist von den Blütenpflanzen der Bundesrepublik Deutschland nur der Amphibische Steinbrech *(Saxifraga oppositifoloia* ssp. *amphibia)*. Alle anderen in der Bundesrepublik Deutschland nicht mehr vorkommenden Arten existieren noch in anderen europäischen Ländern. Das Maß der Eingriffe läßt sich am leichtesten durch den Artenrückgang demonstrieren; jedoch wird dadurch die Reichweite der Veränderungen nicht deutlich. Bei einer flächenhaften Darstellung ergibt sich ein weit ungünstigeres Bild, als es sich nach den Artenzahlen allein darstellt.
Unter den anthropogenen Einwirkungen haben Veränderungen der Standorte viel weitreichendere Folgen gehabt als direkte Einwirkungen auf Pflanzenbestände (mindestens 10 : 1).
Charakteristika bedrohter Pflanzenarten sind Stenözie, beschränktes Verbreitungsgebiet und spezialisierte Fortpflanzungsbedingungen (SUKOPP 1972). Weil die Populationen bedrohter Arten durch die zunehmende Landnutzung fortschreitend stärker isoliert werden, ist die Verbreitungsökologie dieser Arten von großer Bedeutung.

LITERATUR

KORNECK,D., LOHMEYER,W., SUKOPP,H.% TRAUTMANN,W. -1977- Rote Liste der Farn- und Blütenpflanzen (Pteridophyta et Spermatophyta). 2.Fassung.- In: Rote Liste der gefährdeten Tiere und Pflanzen in der Bundesrepublik Deutschland.- Naturschutz aktuell (1): 45-58. Greven.

SUKOPP,H. -1972- Grundzüge eines Programmes für den Schutz von Pflanzenarten in der Bundesrepublik Deutschland.- Schriftenr.LandschPfl. Naturschutz 7: 67-79. Bonn-Bad Godesberg.

-- -- & MÜLLER,P. -1976- Symposium über Veränderungen von Flora und Fauna in der Bundesrepublik Deutschland.- Ergebnisse und Konsequenzen.- Schriftenr.Vegetationskde.10: 401-409. Bonn-Bad Godesberg.

ASPEKTE DER SUKZESSIONSFORSCHUNG AUF DER INSEL HAWAII

D. Mueller-Dombois

Die Insel Hawaii, die südlichste und größte der Hawaiischen Inselgruppe, ist zugleich auch die geologisch jüngste. Sie hat noch vier aktive Vulkane, und die zwei aktivsten, Mauna Loa (4150 m) und Kilauea (1200 m) befinden sich im National Park. Datierte Lava-Substrate des Mauna Loa geben die Möglichkeit, die Vorgänge der Pflanzeninvasion und die folgende Sukzession auf gleichaltrigen und gleichartigen Substraten über Höhenunterschiede vom Meeresspiegel bis über 4000 m zu verfolgen. Zum anderen sind Vergleiche zwischen den Vorgängen im humiden und semi-ariden Klima bei gleichen Temperaturverhältnissen möglich. Der fast alljährlich einmal ausbrechende Kilauea ermöglicht Sukzessions-Vergleiche auf gleichaltrigen Substraten verschiedener physikalischer Beschaffenheit im gleichen Makroklima. Dauerquadrate im tropisch-humiden Klima können schon in 5 bis 7 Jahren aufschlußreiche Resultate bringen. Diese günstige Situation der Insel Hawaii für Sukzessionsforschung ist bedingt einmal durch ihre Lage im tropischen Bereich, in dem die Wachstumsvorgänge relativ schnell ablaufen, zum anderen durch die regelmäßigen Nord- und Passatwinde, die orographischen Regenfall hervorrufen und damit ein am jeweiligen Standort recht gleichmäßiges, aber über sehr kurze Entfernungen wechselndes Makroklima hervorrufen. Die häufigen Lava-Ausbrüche sind niemals so großflächig, daß sie ganze Vegetationsbedeckungstypen vernichten würden. Das bedeutet, daß man viele primäre Vegetationsentwicklungsstadien vorfinden kann, vom ersten Pionierstadium bis zu Stadien, die mit dem Makroklima weitgehend im Gleichgewicht stehen. Das kann allerdings nur in wenigen Fällen vom Boden gesagt werden.

Trotz der idealen Situation Hawaiis für Sukzessionsstudien, ist bisher noch wenig auf diesem Gebiet gearbeitet worden. Erst jetzt war es uns möglich, durch einen National Science Foundation Grant dort ein intensiveres ökologisches Programm zur Sukzessionsforschung zu beginnen. Deshalb können in diesem Vortrag nur gewisse Richtlinien aufgezeigt werden, die aus bisherigen Arbeiten hervorgehen.

Die Schwierigkeiten, die in der Interpretation des Gleichgewichtszustandes der Vegetation mit ihren Standortsfaktoren liegen, haben zu mindestens drei verschiedenen Klimax-Theorien geführt: führt:

1. die Monoklimax-Theorie
2. die Polyklimax-Theorie
3. die Theorie der Klimax-Populationsschwärme und ihrer Struktur

Ich will in diesem Zusammenhang nicht weiter auf diese eingehen, sondern nur andeuten, daß ein Teil dieser konzeptionellen Schwierigkeiten wohl darauf zurückzuführen ist:

1. Daß es außerordentlich schwierig ist, den Zeitfaktor in der Vegetationsentwicklung einigermaßen genau zu bestimmen, da es nur wenige Gebiete auf der Welt gibt, in denen d a t i e r t e Substrate verschiedenen Alters im gleichen Makroklima nebeneinander vorkommen.
2. Durch eine allgemeine Tendenz in der Vegetationsforschung dieser Art (= rezente Entwicklungsforschung), aus dem geographischen "Nebeneinander" von verschiedenen Vegetationstypen auf ein zeitliches "Hintereinander" zu schließen, ohne eine genaue Untersuchung der Umweltsfaktoren mit in Angriff zu nehmen oder kritisch zu interpretieren.

Diesem zweiten Aspekt haben wir in Hawaii durch unabhängige Erstellung einer Vegetationsbedeckungstypen-Karte (1:12000) und anschließende Analysierung des Klimas und der Substrate vorgebeugt, so daß wir einigermaßen sicher sind, welche Vegetationstypen zeitlich in eine Reihe gesetzt werden können und welche nicht.

Der erste Punkt ist in Hawaii, wie in kaum einem zweiten Gebiet der Erde, gegeben. Es wird deshalb möglich sein, über die wichtige Frage - die Dauer verschiedener Entwicklungsstadien und Sukzessionsreihen - eine etwas bessere Vorstellung zu bekommen, als das bisher der Fall war.

Wie sieht nun die primäre Sukzession in Hawaii aus?

Zu dieser Frage sind bis jetzt nur vier publizierte Untersuchungen vorhanden.[1)]

FORBES (1912) beschränkte seine Untersuchungen auf das semi-aride Gebiet auf der Lee-Seite des Mauna Loa. Er machte Pflanzenaufnahmen auf nebeneinanderliegenden Lavasubstraten verschiedenen Alters und kam zu dem Schluß, daß die Vegetationsentwicklung sich in den folgenden fünf Stadien vollzieht:

1. Zuerst erschienen die Kryptogamen (er meinte damit wohl *Ste-*

reocaulon vulcani, die auf der blockartigen, aus lose zusammenliegenden Steinen bestehenden, AA-Lava dominant ist.

2. Dann kann man ein Farn-Stadium erkennen, in dem besonders *Polypodium peleucidum* und *Sadleria cyatheoides* auftreten, aber gleichzeitig findet sich auch schon die Baumart *Metrosideros collina* ssp.,*polymorpha*.
3. Weitere Stadien erscheinen fließend und lassen eine Angleichung an die benachbarte Flora im Unterwuchs des *Metrosideros*-Waldbestandes erkennen.
4. Dann beginnt die Einstellung der stabilen, natürlichen (Gleichgewichts-) Vegetation durch den Übergang zum *Acacia koa*-Wald, wo immer der Lavastrom den *Acacia koa* Vegetationsgürtel durchquert hat.
5. Weitere Veränderungen werden durch Einwanderung der "naturalisierten" Flora herausgebracht, ein Einfluß, der durch den Menschen verursacht ist.

In einer späteren Arbeit stellten ROBYNS & LAMB (1939) nur drei Stadien fest:

1. Die Risse in der neuen Lava (gemeint ist hier die flach ausgelagerte, felsenflächige Pahoehoe-Lava) besiedeln sich gleich mit Angiospermen und Farnen der benachbarten Bestände. Diese Besiedlung wird unterstützt durch Kryptogamen (die also kein ausgeprägtes Stadium zeigen).
2. Eine Verdichtung der Vegetation setzt mit der Ausfüllung der Lava-Oberfläche ein. Ein Busch-Stadium kann erkannt werden, in dem *Dodonaea viscosa (Sapindaceae, Styphelia tameiameiae (Epacridaceae)* und *Metrosideros collina* besonders hervortreten.
3. Danach beginnt die floristische Angleichung an die Vegetation der Umgebung.

Beide Arbeiten sind sehr generell, und die chronologischen Stadien wurden von Beobachtungen aus dem "Nebeneinander" erschlossen. Jedoch waren diese Arbeiten richtungweisend. Zum Beispiel stellten sie die Frage des Unterschiedes in der Besiedlung der beiden hauptsächlichen Laven, der AA- und Pahoehoe-Lava. SKOTTSBERG(1941) legte im Jahre 1926 sechs Quadrate auf zwei datierten Lavasubstraten an. Er kehrte 1938 zur Wiederaufnahme dieser Quadrate zurück und fand u.a., daß in einem Vergleich die Vegetationsentwicklung schneller auf der AA-Lava fortgeschritten war und in einem anderen Vergleich schneller auf der Pahoehoe-Lava. Er führt aus, daß feinere Substratunterschiede als AA und Pahoehoe, und ebenfalls, daß besonders die Entfernung zu intakten Beständen von Bedeutung seien. Außerdem führt er aus, daß ein Kryptogamen-Stadium durchaus erkennbar auftritt (wie schon FORBES anzeigte, im Unterschied zu ROBYNS & LAMB).

DOTY (1961) richtete eine Reihe von Dauerquadraten auf neuer Lava ein, die 1955 in 300 m Höhe an der Südost-Ecke von Hawaii ausbrach und von dort ins Meer floß. Das Klima ist hier tropisch-warm und humid. Genaue, wiederholte Beobachtungen der Lava ergaben hier das Erkennen der Initialstadien als:

1. Algen (*Scytonema hofmanii* hauptsächlich).
2. Moose (verschiedene *Campylopus*-Arten, *C.densiflorus* am häufigsten).
3. Farne (*Nephrolepis*-Arten) zusammen mit höheren Pflanzen (*Metrosideros*-Sämlinge) und nicht einheimische Kräuter (*Arundena* und *Spathoglottis*, beide *Orchidaceen*), und
4. Flechten (*Stereocaulon vulcani* als die letzte Gruppe unter den Pionierpflanzen, die sich über ein schwärzliches Netz der Algengattung *Stigonema* auf der Lavaoberfläche (nicht in den Rissen) erheben.

Diese Anfangs-Sukzession vollzog sich bei einem jährlichen Niederschlag von 2 000 mm mit fast gleichmäßiger Jahresverteilung und nur etwas Nachlaß im Sommer in 6 bis 7 Jahren. Die ersten Algen wurden nach einem Jahr entdeckt. Dikotyledonen-Sämlinge fanden sich schon vor den Algen ein, aber nur für sehr kurze Zeit. Verschiedene Ansiedlungen von sehr ephemerer Dauer wurden von der Composite *Erechtites valerianaefolia* beobachtet. Diese wurden aber nicht als Pionierpflanzen angesehen, da sie sich nicht auf dem Standort fest ansiedelten und zur Fortpflanzung gelangten, sondern durch von außen kommende gelegentliche Einsamungen unterstützt, nur ein vorübergehendes Dasein in feuchteren Perioden fristeten.

Zur unterschiedlichen Besiedlung von AA- und Pahoehoe-Lava stellte DOTY fest, daß kein floristischer, sondern nur ein Unterschied in der Geschwindigkeit des Populationswechsels zu bemerken sei. Eigene, vorläufige Messungen auf DOTY's Beobachtungsflächen ergaben 10 Jahre nach der Lava-Aufschüttung (Kamaile):

Auf Pahoehoe durchschnittlich auf einem Quadratmeter:

22 *Nephrolepis*-Büschel,
verschiedene Individuen von *Metrosideros* und *Arundena*, und 120 Tausend *Stereocaulon*-Stämmchen.

Auf AA ebenfalls auf einem Quadratmeter im Schnitt:

1 *Nephrolepis*-Büschel,
kein *Metrosideros*-Sämling und
keine *Arundena*-Pflanze, aber
170 Tausend *Stereocaulon*-Stämmchen.

Die genannten Zahlenergebnisse dürfen nur als vorläufig angesehen werden, da eine gründlichere statistische Auswertung noch vorgenommen werden muß, und ebenfalls die Beobachtung feinerer Standortsunterschiede, wie z.B. Grad der Asche-Überlagerung über der Lava, und Nähe und Art der benachbarten intakten Bestände noch näher ausgewertet werden müssen. Die Zahlen unterstreichen jedoch DOTY's Beobachtungen der unterschiedlichen Geschwindigkeit der Besiedlung. Zweifellos haben sich die Farne und höheren Pflanzen viel dichter auf der Pahoehoe angesiedelt (im gleichen Zeitraum), während die Flechten etwas dichter auf der AA zu stehen scheinen, eine Beobachtung, die die von FORBES eingangs gemachte unterstützt.

Während nach DOTY's Feststellung die Populationsfolge der primären Anfangs-Sukzession auf Lava sich wie die klassische Evolutionsentwicklung verhält, ist dieses nach vorläufigen Beobachtungen auf neuer vulkanischer Asche nicht der Fall. Auf der Kapoho-Asche von 1960 wurde als einwandfrei etablierte Anfangspopulation Gruppen-Kolonien von einer unkrautartigen, perennierenden Cyperacee, *Cyperus polystachyos*, beobachtet. Hinzu treten hier vom Rande her kommend die nichteinheimischen Phanerogamen, *Casuarina equisetifolia* und *Pluchea odorata*.

Auf einem großen Aschenfeld ungefähr gleichen Alters (1959 Kilauea Iki Devastation Area) in 1200 m Höhenlage, zeigen seit 1961 laufende Beobachtungen u.a., daß die nicht-einheimischen Phanerogamen, wie *Buddleja asiatica* und *Rubus penetrans*, anfangs sehr dominant auftraten, jetzt aber von einheimischen Pflanzen ähnlicher Lebensform verdrängt werden, wie *Railliardia scabra* und *Coprosma ernodeoides*. Auch hier ist die anfängliche floristische Folge nicht im Sinne der klassischen Evolutionsfolge wie auf Lava, sondern gleicht mehr der Lebensformenfolge höherer Pflanzen, wie in bekannten Beispielen der zweiten Sukzessionsfolge auf schon verwitterten Böden des humiden Klimas.

Das gleiche große Aschenfeld gab auch Gelegenheit zur Beobachtung von graduellen Zerstörungs- und Wiederausschlag-Effekten der von der Asche überschütteten Vegetation. Hier zeigte es sich, daß die wiederausschlagenden *Metrosideros*-Bäume ihre apikalische Dominanz verloren hatten und einen buschigen Wuchs herunter bis zur neuen Substrat-Oberfläche annahmen, eine Änderung in der Lebensform,die scheinbar irreversibler Art ist. Außerdem produzierten viele dieser Bäume Luftwurzeln, die inzwischen zum Teil im neuen Substrat Fuß faßten, zum anderen Teil aber in Ballenform an den Zweigen hängend verblieben, ohne weiter nach unten zu wachsen.

Unter den Sträuchern zeigte besonders das endemische *Vaccinium reticulatum* eine starke Wiederausschlagsfähigkeit und beantwortete die Ascheüberschüttung in diesen Fällen mit weitaus vitalerem Wuchs als das vor dem Ascheregen der Fall war. Diese und weitere Beobachtungen und die teilweisen Gründe für das Verhalten in diesem Gebiet werden momentan in einer Dissertation behandelt, die ungefähr in 2-3 Jahren fertig sein wird.[2)]

SUMMARY

The island of Hawaii is exceptionally well suited for studies of primary succession on recent volcanic substrates. This convenience is provided by the frequent volcanic outbreaks of Mauna Loa and Kilauea, both of which are located in the National Park. Therefore, the upcoming new vegetation is kept relatively free of human influences. Lava substrates of the same age are present in a variety of different macroclimates. Moreover, volcanic substrates of the same age, but of different physical characteristics, are in several areas exposed to the same macroclimate.

The results of previous succession research on Hawaii up to the present are briefly outlined and supplemented by own observations. It has been found that the floristic sequence on lava surfaces parallels in essence the classic populations sequence of plant evolution, while this does not seem to hold true for pyroclastic surfaces. Differing rates of population invasion, but not of the floristic sequence, were observed on the two, basically different lava types, AA and Pahoehoe.

The rate of change of successional stages is primarely dependant upon the amount and distribution of rainfall. In humid-tropical areas, initial stages of primary succession to *Metrosideros* tree stands (in sapling stage) can already be observed within 5-7 years in permanent quadrats. In climates with summer-drought

[2)] cf.p.500

the lava remains free of higher plants within the same time span.

The questions of the duration of the different developmental stages and of the equilibrium stage of certain vegetation cover types can be answered by further sudies on dated Hawaiian volcanic material with greater clarity than has been the case in the past and current climax theories. For this reason, future succession studies from Hawaii may result in important contributions to the central ecological concepts of succession and climax.

LITERATUR

DOTY,M.S. -1961- Lava flow pioneer populations.- In: "Guide to Excursion III. Tenth Pacific Science Congress" (by F.R.Fosberg): 139-144.

FORBES,C.N. -1912- Preliminary observations concerning the plant invasion on some of the lava flows of Mauna Loa, Hawaii.- Benice P. Bishop Museum, Occasional Papers 5 (1): 15-23. Honolulu.

ROBYNS,W. & LAMB,S.H. -1939- Preliminary ecological survey of the island of Hawaii.- Bull.Jard.Bot.Brux.15: 241-293. Bruxelles.

SKOTTSBERG,C. -1941- Plant succession on recent lava flows in the island of Hawaii.- Gotborgs Kungl.Vetenskap- och Vitterhets-samhalles Handlinger Sjatte fojden, Ser.B 1 (8): 1-32. Göteborg.

Der größte Teil der Informationen im vorstehenden Text basiert auf:

DOTY,M.S. & MUELLER-DOMBOIS,D. -1966- Atlas für bioecology studies in Hawaii Volcanos National Park.- University of Hawaii, Hawaii Botanical Science Paper 2. Honolulu. 507 pp.

DISKUSSION

A.W.KÜCHLER

Diese Ausführungen haben bewiesen, daß wir hier ein großzügiges natürliches Laboratorium haben, in dem wir die Vegetation in ihrer ganzen Sukzession studieren können. Herr MUELLER-DOMBOIS hat zwar allerlei Dauerquadrate angelegt, es ist aber meines Erachtens unerläßlich, daß die Vegetation in so einem Lande wiederholt kartiert wird. Ich sehe natürlich die unglaublichen Komplikationen,die dabei entstehen, aber ein Quadrat verschwindet, ein Quadrat ist nur ein Punkt. Wenn man eine Karte hat, dann geht das besser.

R.LINDER

Was kommt zuerst auf der Lava als Phanerogame? Es zeigt sich doch meistens, daß sich auf diesem Substrat hauptsächlich "physiologische" Endemiten einstellen. In allen Gebieten, wo Vulkane sind, gibt es Arten, die man im Garten nicht kultivieren kann, die man nirgendwo anders kultivieren kann als an der Stelle, wo sie eben wohnen. Und dazu möchte ich noch sagen, daß es auch bestimmte Gattungen sind, in denen sich die Endemiten bilden, z.B. die Gattungen *Viola* oder *Silene*. Haben Sie so etwas Ähnliches auf Hawaii?

D.MUELLER-DOMBOIS

An und für sich sind nur wenige Arten endemischer Natur, die sich an diesen Prozessen der Initialphase beteiligen. Das kann allerdings nicht ganz verallgemeinert werden, z.B. *Metrosideros collina* ist eigentlich keine endemische Art, wir haben aber endemische Subarten und Varietäten. Es ist gut möglich, daß *Metrosideros*, die man in Neuseeland auf den Laven findet, sehr große Ähnlichkeit hat. Vergleiche könnte man sehr leicht anstellen. Es ist ganz erstaunlich, was dieser Baum für eine morphologische Flexibilität hat. Es stimmt nicht ganz, daß er nur auf junger Lava wächst, er wächst auch auf alten Substraten und zwar ist er an fast alle Situationen sehr anpassungsfähig. Er braucht bloß Licht. Wenn diese Bestände ins reife Altersstadium kommen, dann brechen die Bäume zum Teil herunter, sie verjüngen sich dann aber auf höheren Punkten; z.B. wachsen sie gern auf den Baumfarnen, senken ihre Wurzeln herunter und fassen dann immer neu Fuß. Das zeigt die morphologische Flexibilität.
Eine andere Baumgruppe, die auf den Cameroon Mountains in Afrika auch in einem sehr humiden Gebiet auftritt, sind die Feigen, die vielleicht eine noch etwas größere morphologische Flexibilität besitzen.

H.MERKER

Ich möchte Sie fragen, ob Sie irgendwelche Bestimmungen gemacht haben, welche Nährstoffe von der Luft immer wieder auf die Pflanzengesellschaften herunterfallen? Manche Vulkane brechen jährlich aus und da müssen doch enorme Mengen anfallen.
Könnte man nicht die Entstehung von Luftwurzeln bei *Metrosideros* so erklären, daß bei einer neuerlichen Überschichtung des Wurzelsystems der Baum ein neues Atmungsorgan entwickeln muß, so daß die Luftwurzeln hauptsächlich Atmungsorgane sind, um die abgeschnittene Gaszufuhr der Wurzeln zu kompensieren?

D.MUELLER-DOMBOIS

Die Nährstoffzufuhr haben wir noch nicht direkt untersucht. Sie wird allgemein als Problem der jungen Laven angesehen: Wo kommt der Stickstoff her? Alle anderen Elemente sind nämlich vorhanden. Das würde bei uns in Hawaii kein zu großes Problem sein, weil unsere Laven meistens wie Handtücher herunterkommen und nicht so großflächig sind. So besteht die Möglichkeit, daß die Blätter von der Seite her auf die Lava gestreut werden und in die Ritzen fallen, und dann ist organisches Material da und damit auch die Komponente Stickstoff. Die Lava ist basaltisch und sehr reich an Nährstoffen, und alles ist vorhanden, außer Stickstoff. Eine Antwort auf Ihre zweite Frage versuchen wir mit Experimenten herauszubekommen.

J.SCHMITHÜSEN

Zu der Frage *Metrosideros* möchte ich noch etwas hinzufügen, und zwar scheint mir das Stichwort "morphologische Flexibilität" wichtig zu sein. Die Gattung *Metrosideros* gehört zu der relativ jungen Flora im pazifischen Raum, die ihre Verbreitung hat etwa von Neu-Guinea - Neu-Caledonien - Neuseeland und Hawaii in diesem großen Dreieck. D.h., sie hat sich erst entwickelt und verbreitet, nachdem die Trennung nach den anderen Richtungen schon vollzogen war, während die übrigen Florenelemente darübergreifen. *Metrosideros* hat ein sekundäres Entfaltungszentrum auf Hawaii; sein Schwergewicht aber auf Neu-Caledonien und auf Neuseeland. Die Gattung ist insofern hochinteressant, da sie von den Wuchsformen aus, wenn man das Wort in diesem Zusammenhang wiederholen darf, morphologisch äußerst flexibel ist. Ich glaube, es gibt in Neuseeland ungefähr ein Dutzend Arten, davon sind etwa die Hälfte Lianen, dann sind einige strauchartige Pflanzen, die auf den Bäumen als Epiphyten wachen, und dann sind einige richtige Bäume wie z.B. der Cristmas tree von Neuseeland, und andere sind Baumwürger, die auf den Bäumen als epiphytische Sträucher zunächst entstehen, dann Luftwurzeln entsenden und zwar in so dichter Folge, daß man sie mit COCKAYNE Wurzelstämme nennen kann. Das gibt also Riesen-Urwaldstämme mit 2 Meter Durchmesser, deren Stämme aus Wurzeln bestehen. Und da liegt also die Parallele.Es gibt also diese *M.collina* nicht in Neuseeland, aber von da aus sieht man leicht, daß das in der Gattung liegt, daß sie Luftwurzeln bildet. Und vielleicht mag eine solche Anlage dann gelegentlich bei einer Art in einer besonderen Situation noch irgendwie wieder aktiviert werden. Ich wundere mich gar nicht über die Tatsache, daß Ihr *Metrosideros*-Baum nun plötzlich Luftwurzeln bildet, weil es eben *Metrosideros*-Bäume gibt, die nur aus Luftwurzeln bestehen.

R.TÜXEN
Zu dem 2. oder 3. Dauerquadrat, das Sie zeigten, mit den hochwüchsigen Farnen und der flachen Lava mit *Stereocaulon*, möchte ich noch bemerken, daß darin zwei ganz deutlich verschiedene Gesellschaften: eine Spalten- und eine Flechtengesellschaft waren.

[1] Anm.zu p.492
Weitere Arbeiten werden berücksichtigt in MUELLER-DOMBOIS & SMATHERS (1975) Sukzession nach einem Vulkanausbruch auf der Insel Hawaii. In: Ber.Intern. Sympos.Intern.Vereinigung für Vegetationskunde, Band Sukzessionsforschung, Redaktion W.Schmidt, p.159-188.

[2] Anm.z.p.496
Inzwischen publiziert unter dem Titel: Invasion and recovery of vegetation after a volcanic eruption in Hawaii, Smathers & Mueller-Dombois. National-Park Service Scientific Monograph Series No.5. 1974. 129 pp. Washington,D.C.

SCHLUSSWORTE

M.WRABER †

Wenn ich abschießend noch einige Worte sagen darf - wir müssen uns ja Rechenschaft abgeben über unser Symposium - so glaube ich, daß ich im Namen aller spreche, wenn ich sage, daß wir mit diesem Symposium einen großen Erfolg erreicht haben. Wir können zufrieden sein. Ich danke allen Referenten und allen Diskussionsrednern und sage auch den Nichtdiskutanten herzlichen Dank fürdie Zusammen- und Mitarbeit. Vor allem aber gilt unser Dank unserem Herrn Professor TÜXEN. Wir scheiden nun in der Hoffnung, uns das nächste Jahr hier wiederzutreffen um das 12. Symposium dann ebenso zahlreich und erfolgreich mitmachen zu können, und ich bitte Herr Professor TÜXEN, dieses Symposium nun zu schließen.

R.TÜXEN

Meine lieben Freunde!

Ich bin ganz mit Herrn WRABER einverstanden, daß unser diesjähriges Symposium ein voller Erfolg war, und zwar wissenschaftlich wie auch menschlich. Wir haben, mit 36 Vorträgen in diesen 4 Tagen, beinahe ein ganzes Semester abgeleistet. Wir haben eine ganze Menge lernen können, und auch gelernt, aus den verschiedensten Bereichen und geographischen Gebieten. Ich glaube auch,daß wir nicht nur wissenschaftlichen Gewinn davongetragen haben,sondern daß wir auch manches für den Naturschutz und Einsichten für die Erhaltung des menschlichen Lebens gewinnen konnten. Daß sich viele freundschaftliche Bindungen gefestigt haben und neue angeknüpft wurden, und daß wir uns als Menschen aus so außerordentlich verschiedenen Lebensbereichen, wie wir hier versammelt waren, jetzt viel besser kennen und verstehen gelernt haben als etwa aus Zeitungen, aus dem Fernsehen oder aus dem Radio, das bedarf ja wohl keiner weiteren Begründung. Ich glaube, wir sollten weiter solche menschlichen Beziehungen pflegen und durchführen.

Dann habe ich an erster Stelle den Herren Präsidenten sehr herzlich zu danken für die sehr mühevolle Aufgabe, der sie sich unterzogen haben. Ich möchte weiter meinen Dank aussprechen Herrn Oberstudiendirektor ROTH, daß er uns diesen Saal zur Verfügung gestellt hat. Dann gilt mein Dank meinen Mitarbeitern, vor allem meinem Freund HARTMUT DIERSCHKE. Vor allem aber habe ich den Teilnehmern zu danken, den jungen und den mittleren und den älteren, am meisten aber unseren vier Gästen aus der UdSSR, die zum ersten Mal in unserem Kreise mit uns gearbeitet, gedacht und diskutiert haben, und die jetzt einen persönlichen Eindruck von uns allen mitnehmen können.

Dank allen, Dank und auf Wiedersehen im nächsten Jahre, so fröhlich und so freundlich wie in diesem. Herzlichen Dank, und damit darf ich dieses schöne 11. Symposium schließen.

GPSR Compliance

The European Union's (EU) General Product Safety Regulation (GPSR) is a set of rules that requires consumer products to be safe and our obligations to ensure this.

If you have any concerns about our products, you can contact us on ProductSafety@springernature.com

In case Publisher is established outside the EU, the EU authorized representative is:

Springer Nature Customer Service Center GmbH
Europaplatz 3
69115 Heidelberg, Germany

Batch number: 10682411

Printed by Printforce, the Netherlands

Springer Aerospace Technology

The *Springer Aerospace Technology* series is devoted to the technology of aircraft and spacecraft including design, construction, control and the science. The books present the fundamentals and applications in all fields related to aerospace engineering. The topics include aircraft, missiles, space vehicles, aircraft engines, propulsion units and related subjects.

More information about this series at http://www.springer.com/series/8613

Shaoming He · Chang-Hun Lee ·
Hyo-Sang Shin · Antonios Tsourdos

Optimal Guidance and Its Applications in Missiles and UAVs

Springer

Shaoming He
School of Aerospace Engineering
Beijing Institute of Technology
Beijing, China

Hyo-Sang Shin
Aerospace, Transport and Manufacturing
Cranfield University
Cranfield, UK

Chang-Hun Lee
Department of Aerospace Engineering
Korea Advanced Institute of Science and Technology
Daejeon, Korea (Republic of)

Antonios Tsourdos
Aerospace, Transport and Manufacturing
Cranfield University
Cranfield, UK

ISSN 1869-1730 ISSN 1869-1749 (electronic)
Springer Aerospace Technology
ISBN 978-3-030-47347-1 ISBN 978-3-030-47348-8 (eBook)
https://doi.org/10.1007/978-3-030-47348-8

This Springer imprint is published by the registered company Springer Nature Switzerland AG
The registered company address is: Gewerbestrasse 11, 6330 Cham, Switzerland

Preface

Autonomous guidance has gained increasing attention in the aerospace industry due to its significant importance in various military and civil applications. Optimal control theory provides engineers a systematic way to establish new guidance laws to simultaneously satisfy various constraints and optimize a meaningful performance index. This book explores the possibility on how optimal guidance law can be utilized to achieve different mission objectives for missiles and UAVs. The key feature of this book is the analysis of physical meaning and working principles of different new optimal guidance laws. In practice, this information is important in ensuring confidence in the performance and reliability of the guidance law when implementing it in a real system, especially in aerospace engineering where the reliability is the first priority.

The main contents of this book are the summary and extension of the authors' recent works in the field of guidance for missiles and UAVs. We start from the introduction of a general optimal error dynamics for missile guidance law design and then focus on the development of various new optimal guidance laws for different mission objectives, e.g., target observability enhancement, reducing sensitivity against unknown target maneuvers and active gravity utilization. In the second part of this book, the problem of optimal waypoint-following guidance law design for fixed-wing UAVs is studied and various constraints, e.g., arrival angle constraint, autopilot dynamics and obstacle avoidance, are considered in an integrated manner. The comprehensive analysis of physical meaning and systematic treatment of practical issues in optimal guidance law is one of the major features of the book, which is particularly suited for readers who are interested to learn practical solutions in guidance law design. The book can also benefit researchers, engineers and graduate students in the field of Guidance, Navigation and Control, applied optimal control, etc. It is our hope for this book to serve as a useful step for permitting further advances in the field of guidance.

The authors have carefully reviewed the content of this book before the printing stage. However, it does not mean that this book is completely free from any possible errors. Consequently, the authors would be grateful to readers who will call out attention on mistakes as they might discover.

Finally, the authors would like to thank colleagues from the Institute of Autonomous UAV Control, Beijing Institute of Technology and the Center of Cyber-Physical Systems, Cranfield University, for providing valuable and constructive comments. Without their support, the writing of the book would not have been a success.

Beijing, China — Shaoming He
Daejeon, Republic of Korea — Chang-Hun Lee
Cranfield, UK — Hyo-Sang Shin
Cranfield, UK — Antonios Tsourdos
April 2020

Contents

Acronyms

AIM	Adaptive intermittent maneuver
APNG	Augmented proportional navigation guidance
BPNG	Biased proportional navigation guidance
FIM	Fisher information matrix
FOV	Field-of-view
G2C	Guidance-to-collision
IACG	Interception angle control guidance
LOS	Line-of-sight
LQR	Linear quadratic regulation
MPC	Model predictive control
OGL	Optimal guidance law
OWFGL-0	Lag-free optimal waypoint-following guidance law
OWFGL-1	First-order lag optimal waypoint-following guidance law
PI	Proportional-integral
PIP	Predicted interception point
PNG	Proportional navigation guidance
POPOGL-1	Point-to-point optimal guidance law with first-order autopilot dynamics
RMSE	Root mean square error
SGWL	Synthetic waypoint guidance law
SMC	Sliding mode control
TPG	Time-to-go polynomial guidance
TSG	Trajectory shaping guidance
TSWGL	Trajectory shaping waypoint guidance law
TWOG2CIAG	Time-to-go weighted optimal guidance-to-collision impact angle guidance
TWOIAG	Time-to-go weighted optimal impact angle guidance
UAV	Unmanned aerial vehicle
UKF	Unscented Kalman filter
ZEA	Zero-effort angle
ZEM	Zero-effort-miss

Chapter 1
Introduction of Optimal Guidance

Abstract Guidance is one of the key parts of modern aerospace vehicles since it enables improved autonomy by guiding the vehicle to approach a specific stationary/dynamic position. Optimal control theory is known as a valuable and systematic tool in the design and development of guidance algorithms. We explore the characteristics of classical proportional navigation guidance and present the motivation of the development of new optimal guidance laws in this chapter. The overall aim of this book is then summarized and the organization of the entire book, together with some highlights of each chapter, are presented.

1.1 Background and Motivation

In the last few decades, numerous achievements and massive efforts have been witnessed to improve the performance of autonomous aerospace vehicles. Nowadays, aerospace science and technology has brought various changes in not only military but also civil engineering applications. It is known that the guidance technology plays an important role in modern aerospace vehicles since it enables improved autonomy, thus having attracted a remarkable level of attention in research and development. Guidance is the process of guiding a vehicle to reach a specific stationary/dynamic target, often by generating acceleration/attitude command to its autopilot or motion controller based on vehicle's states and target information. Figure 1.1 presents the general flowchart of the guidance process.

Classical proportional navigation guidance (PNG) law has been widely used in the aerospace industry since 1960s and is still a benchmark for modern guidance laws. The main concept of PNG is to generate a lateral acceleration command to change the vehicle's heading such that the vehicle can converge to a constant collision course to approach the target. A major advantage of PNG, contributing to its longevity as a favored guidance algorithm over the last few decades, is its efficacy and simplicity of implementation. As elaborated in Chap. 2, PNG is also an optimal guidance law in the sense of minimizing terminal miss distance. Although PNG is originally developed

S. He et al., *Optimal Guidance and Its Applications in Missiles and UAVs*, Springer Aerospace Technology, https://doi.org/10.1007/978-3-030-47348-8_1

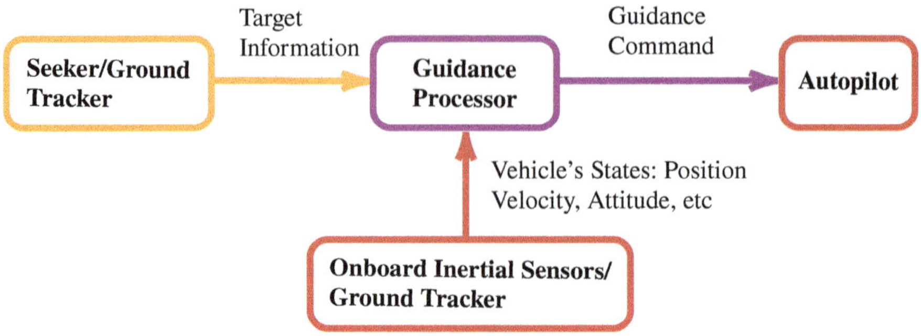

Fig. 1.1 The general flowchart of the guidance process

for missile guidance [1, 2], it has been utilized in obstacle avoidance [3, 4], path following [5], wall following [6], formation flight [7] of unmanned air vehicles (UAVs).

With the increasing complexity of application scenarios, however, real-world guidance problems in autonomous aerospace systems will be characterized by numerous practical constraints to achieve different mission objectives, indicating that PNG might be ineffective [8]. For example, anti-tank and anti-ship missiles are usually required to intercept the target with a specific impact angle to attack the vulnerable point. When passing one waypoint, the UAV might need to constrain its flight path angle as a desired value to satisfy the requirement of radar detection avoidance or terrain following. For UAV rendezvous missions, e.g., aerial refuelling, the UAV needs to align its velocity vector with the target aircraft to cater for the mission constraint [9]. During the midcourse guidance phase, maximizing terminal velocity of tactical missiles enables high survivability and kill probability in the following terminal guidance phase. In order to counteract with the close-in weapon systems of modern warships, the concept of simultaneous interception is introduced to improve the kill probability and therefore the terminal impact time becomes an additional constraint. Typically, there is an explicit assumption that the autopilot dynamic response is fast enough in guidance law design. In practice, guidance laws with lag-free assumption could inevitably experience performance degradation such as a significant miss distance, resulted from an autopilot lag in the guidance loop. The performance degradation associated with the issue of autopilot dynamics could be significantly exacerbated when the vehicle's moving velocity is fast. The requirement for satisfying multiple constraints and better performance, therefore, continues to push guidance law development.

1.2 Optimal Guidance Problem

Optimal control theory is known as a valuable and systematic tool in the design and development of guidance algorithms with multiple constraints. Optimal guidance brings in the philosophy of trajectory optimization that minimizes a meaningful performance index while satisfies certain terminal constraints. The general performance index in optimal guidance problems is usually defined as a scalar function as

$$J\left[x(t), u(t)\right] = \phi\left[x\left(t_f\right)\right] + \int_t^{t_f} \mathcal{L}\left[x(\tau), u(\tau)\right] d\tau \tag{1.1}$$

where $x(t)$ is the system state vector and $u(t)$ is the guidance command. Note that the system state vector $x(t)$ of the guidance problem is usually defined by the kinematics model. The function $\phi\left[x\left(t_f\right)\right]$ stands for the terminal constraint with $x\left(t_f\right)$ being the terminal state and $\mathcal{L}\left[x(t), u(t)\right]$ is the running cost that can be utilized to quantify the transient effect.

In aerospace applications, energy consumption is an important criterion in guidance law evaluation and hence minimizing energy consumption is a worthy goal in guidance law design. Basically, there are two different types of energy performance indices

$$\mathcal{L}_1 \text{ norm: } \mathcal{L} = |u(t)| \tag{1.2}$$

$$\mathcal{L}_2 \text{ norm: } \mathcal{L} = u(t)^2 \tag{1.3}$$

To illustrate the difference between these two different energy performance indices, we consider the following simple optimization problem

$$\mathcal{L}_1 \text{ norm: } \min_u J = \int_0^1 |u(t)|\, dt \tag{1.4}$$

$$\mathcal{L}_2 \text{ norm: } \min_u J = \int_0^1 u(t)^2 dt \tag{1.5}$$

$$\begin{aligned} &\dot{x}_1 = x_2 \\ &\dot{x}_2 = u \\ \text{Subject to: } &(x_1(0), x_2(0)) = (0, 0) \\ &(x_1(1), x_2(1)) = (1, 0) \\ &|u| \le 6 \end{aligned} \tag{1.6}$$

The comparison results of different energy performance indices are presented in Fig. 1.2. From this figure, it can be clearly noted that minimizing the $\mathcal{L}_1$ norm energy performance index generates sparse control input and hence it is suitable for on/off actuators, e.g., spacecraft, satellite. In contrast, leveraging the $\mathcal{L}_2$ norm energy consumption as a cost function provides continuous control. Since the induced

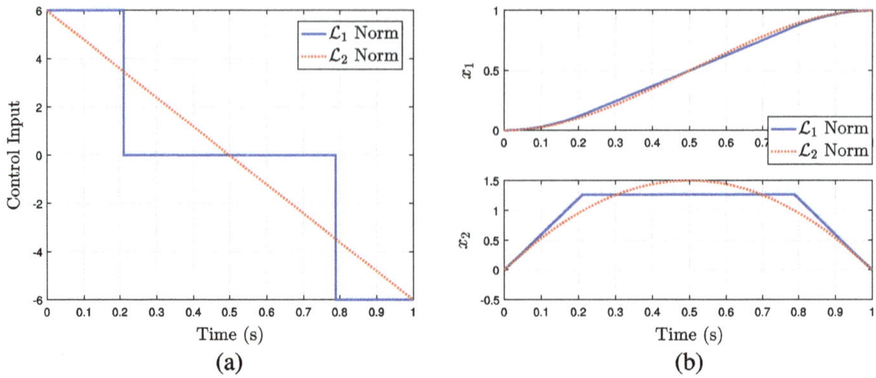

Fig. 1.2 Comparison results of different energy performance indices. **a** Control input; and **b** system states

drag of aerodynamically-controlled vehicles is proportional to the quadratic form of the lateral acceleration, the $\mathcal{L}_2$ norm energy performance index is suitable for endo-atmospheric applications. This means that minimizing the $\mathcal{L}_2$ norm energy consumption is helpful to reduce the velocity loss for aerodynamically-controlled vehicles. Another advantage of minimizing the $\mathcal{L}_2$ norm type cost function is that it permits closed-form solutions in a feedback form and hence is suitable for online applications. For this reason, we will consider the $\mathcal{L}_2$ norm type objective functions in the following chapters.

In some practical applications, we might need to shape the guidance command to achieve different mission objectives. For example, it is desirable to reduce the magnitude of the guidance command at the beginning of the terminal homing phase to mitigate the sensitivity against the handover process. Since the air density exponentially decreases as the altitude increases, it is better to gradually limit the maneuverability of surface-to-air missiles. A general cost function for command shaping in guidance law design is given by a weighted energy form as

$$J = \int_t^{t_f} W(\tau) u(\tau)^2 dt \tag{1.7}$$

where $W(t)$ is weighting function.

By properly designing the weighting function $W(t)$, we can shape the guidance command as a desired pattern for various mission objectives. Once the performance index is defined, the guidance command can then be derived by solving the optimization problem formulated. Note that the kinematics model is usually a nonlinear dynamics model and hence deriving the analytic solutions directly from the nonlinear model is generally intractable except for some special cases. For this reason, optimal guidance laws are typically derived by using linearized kinematics models based on small angle assumptions. Consequently, analytic optimal guidance laws are usually derived based on the linear quadratic optimization theory.

1.3 Aim and Organization

The overall aim of this book is to advance the optimal guidance laws and provide readers a deep understanding of how optimal guidance law can be utilized to achieve different mission objectives for missiles and UAVs. The key part of this book is the analysis of physical meaning and working principles of different new optimal guidance laws. In practice, this information is important in ensuring confidence in the performance and reliability of the guidance law when implementing it in a real system, especially in aerospace engineering where the reliability is the first priority. This book presents a comprehensive summary of the authors' recent works in this field and the main contents of this book consists of two parts: (1) optimal guidance in missile applications; and (2) optimal guidance in UAV applications.

In Part I, we first introduce a general optimal error dynamics for missile guidance law design and analyze the physical meaning of various optimal guidance laws by using the proposed error dynamics in Chap. 2. This approach also provides a unified way in establishing new optimal guidance laws to achieve various mission objectives. As an application of the optimal error dynamics proposed in Chap. 2, a new optimal trajectory shaping guidance law is suggested in Chap. 3 to cater for the requirement of impact time and impact angle control. The look angle constraint is also considered to support practical applications when the seeker's field-of-view is limited. In order to improve target estimation performance in passive guidance, Chap. 4 proposes a new linear optimal guidance law for observability enhancement. The working principle and closed-form solution of the proposed guidance law are also theoretically analyzed to provide better understandings of the physical meaning of observability-enhancement guidance. In Chap. 5, we demonstrate how the classical proportional-integral (PI) concept can be leveraged in optimal guidance law design to mitigate the effect of target maneuver on the guidance performance. Through an inverse approach, the working principle of the proposed optimal PI guidance law is theoretically analyzed and the reason why the PI concept is helpful in reducing the sensitivity to unknown target maneuvers is illustrated. In Chap. 6, a new concept by utilizing the gravitational effect in missile guidance law design is introduced. This concept enables energy saving and guarantees zero final guidance command to improve the operational margins to cope with undesired external disturbances. Based on the geometric conditions, we also analyze the relationships of the optimal gravity-turn-assisted guidance law with classical PNG and guidance-to-collision law. To maintain advantageous homing engagement geometry against a target, we extend the gravity-turn-assisted concept to optimal intercept angle guidance in Chap. 7.

In Part II, a minimum-effort waypoint following guidance with an arbitrary number of waypoints and arrival angle constraints is proposed in Chap. 8 to improve the endurance of a fixed-wing UAV. Based on the closed-form solutions, the main characteristics of the proposed guidance law is also analyzed and the results reveal that classical point-to-point optimal guidance laws are special cases of the proposed formulation. We also reveal that the proposed guidance law is helpful in reducing the transition effect when passing one waypoint. In order to mitigate the performance degradation resulted from autopilot lag, we then extend the results of Chap. 8 by incorporating a general autopilot dynamics in Chap. 9. Finally, the problem of integrated waypoint-following and obstacle avoidance is investigated in Chap. 10. This new guidance law is an extension of Chap. 9 by considering additional obstacle avoidance constraints.

References

1. Zarchan P (2012) Tactical and strategic missile guidance. American Institute of Aeronautics and Astronautics, Reston
2. Murtaugh SA, Criel HE (1966) Fundamentals of proportional navigation. IEEE Spectr 3(12):75–85
3. Han S-C, Bang H, Yoo C-S (2009) Proportional navigation-based collision avoidance for UAVs. Int J Control Autom Syst 7(4):553–565
4. Clark M, Kern Z, Prazenica RJ (2015) A vision-based proportional navigation guidance law for UAS sense and avoid. In: AIAA guidance, navigation, and control conference
5. Medagoda EDB, Gibbens PW (2010) Synthetic-waypoint guidance algorithm for following a desired flight trajectory. J Guid Control Dyn 33(2):601–606
6. Suresh S, Ratnoo A (2020) Guidance using multiple sequential line-of-sight information. In: AIAA Scitech 2020 forum
7. Smith AL (2008) Proportional navigation with adaptive terminal guidance for aircraft rendezvous. J Guid Control Dyn 31(6):1832–1836
8. Palumbo NF, Blauwkamp RA, Lloyd JM (2010) Modern homing missile guidance theory and techniques. Johns Hopkins APL Techn Dig 29(1):42–59
9. Tsukerman A, Weiss M, Shima TY, Löbl D, Holzapfel F (2017) Trajectory shaping autopilot-guidance design for civil autonomous aerial refueling. In: AIAA guidance, navigation, and control conference

Part I
Optimal Guidance in Missile Applications

Chapter 2
Optimal Error Dynamics in Missile Guidance

Abstract This chapter investigates the optimal convergence pattern of the tracking error that frequently appears in missile guidance problems and proposes an optimal error dynamics for guidance law design to achieve various operational objectives. The proposed optimal error dynamics is derived by solving a linear quadratic optimal control problem through Schwarz's inequality approach. The properties of the proposed optimal error dynamics are discussed. The significant contribution of the proposed result lies in that it can provide a link between existing nonlinear guidance laws and optimal guidance laws for missile systems. Therefore, the advantages of both techniques can be fully exploited by using the proposed approach: existing nonlinear guidance laws can be converted to their optimal forms and the physical meaning of them can then be easily explained. Four illustration examples, including zero zero-effort-miss (ZEM) guidance, impact angle guidance, impact time control, impact angle control as well as impact angle and impact time control, are provided to show how the proposed results can be applied to missile guidance law design. The performance of the new guidance laws is demonstrated by numerical simulation.

2.1 Introduction

In essence, missile guidance law design is a kind of finite-time tracking problem depending on the operational objective. If only target interception is considered, the tracking error is defined as the ZEM distance [1–4]. Nullifying the ZEM results in perfect interception with zero miss distance. For some types of tactical missiles, constraining the final impact angle or intercept angle is helpful to enhance the kill probability of the warhead or maintain advantageous homing engagement [5–7]. To satisfy this requirement, the terminal impact angle error is considered as the tracking error in guidance law design [8–12]. In order to enhance the survivability of anti-ship missiles against advanced close-in weapon system of battleships, the concept of salvo attack [13] is introduced to achieve simultaneous attack among all interceptors. One typical implementation of salvo attack is the impact time guidance, in which the

S. He et al., *Optimal Guidance and Its Applications in Missiles and UAVs*, Springer Aerospace Technology, https://doi.org/10.1007/978-3-030-47348-8_2

missile is forced to intercept the target at a desired time instance and hence the final impact time error is considered as the tracking error in guidance law design [13–17].

After defining the tracking error for a specific missile guidance problem, many systematic nonlinear control theories, e.g., sliding mode control (SMC) [1, 2, 18–20], H_∞ control [21], Lyapunov function theory [22], predictive control [23] and feedback linearization control [24, 25], can then be utilized to drive the tracking error to zero asymptotically or in finite time. More specifically, the guidance command is derived to force the system trajectory to follow the desired error dynamics, which provides desired convergence pattern of the tracking error. This is a general procedure to establish new missile guidance laws using nonlinear control methodologies. However, most previous studies only focus on how to force the tracking error to converge to zero and disregard what is the optimal error dynamics in terms of a meaningful performance index.

Inspired by the preceding observations, this chapter attempts to examine the optimal convergence pattern of the tracking error and propose an optimal error dynamics with a meaningful cost function for missile guidance problems. To realize this, we solve the linear quadratic optimal control problem for a generalized tracking problem that is often observed in missile guidance law design using Schwarz's inequality approach. The potential significance of the proposed results is clear:

(1) The existing nonlinear guidance laws can be further improved by converting them to their optimal version and the physical meanings of existing nonlinear guidance laws can be easily explained;
(2) Various new missile guidance laws can be simply developed by using the proposed optimal error dynamics through a systematic prediction-correction approach.

For illustration, we present four examples to show how to apply the proposed optimal error dynamics to missile guidance law design. The presented examples include homing guidance, impact angle guidance, impact time guidance as well as impact time and angle guidance. By utilizing the proposed optimal error dynamics, the physical meaning of these guidance laws developed is theoretically analyzed and the cost functions provide useful guidelines on how to choose proper gains to guarantee bounded terminal guidance command. Theoretical analysis also reveals that some previous optimal guidance laws are special cases of the proposed optimal solutions. Hence, the proposed optimal error dynamics provides a unique and generalized way in establishing optimal guidance laws.

2.2 Preliminaries and Motivations

This section introduces some preliminary backgrounds on the missile-target relative kinematics model and also presents the motivations of suggesting a general optimal error dynamics for missile guidance law design. Before presenting the results, we make the following reasonable assumptions to facilitate the analysis in the following sections.

Assumption 2.1 Both the missile and the target are assumed as ideal point-mass models.

Assumption 2.2 The engagement occurs in a 2D plane.

Assumption 2.3 Both the missile and the target are flying with constant velocity.

Note that these assumptions are widely accepted in guidance law design for tactical missiles: (Assumption 2.1) Typical philosophy treats the guidance and control loops separately by placing the guidance kinematics in an outer-loop, generating guidance commands tracked by an inner dynamic control loop, also known as autopilot. (Assumption 2.2) It is assumed that the missile employs an ideal attitude control system that provides roll stabilization such that the guidance problem can be treated in two separate channels. Accordingly, homing engagement can be treated as a 2D problem. (Assumption 2.3) The vehicle's velocity is generally slowly varying and hence can be assumed as piece-wise constant.

2.2.1 Missile-Target Relative Kinematics

The planar homing engagement geometry is shown in Fig. 2.1, where M and T denote the missile and target, respectively. The notation of (X_I, Y_I) represents the inertial frame. The variables of σ and γ stand for the line-of-sight (LOS) angle and flight path angle, respectively. The symbol r denotes the relative distance between the target and the missile. a_M and a_T are the missile and target accelerations normal to the velocity vectors, respectively. Based on the principles of kinematics, the differential equations describing the engagement geometry, depicted in Fig. 2.1, are formulated as

$$\begin{aligned}
\dot{r} &= V_T \cos(\gamma_T - \sigma) - V_M \cos(\gamma_M - \sigma) \\
r\dot{\sigma} &= V_T \sin(\gamma_T - \sigma) - V_M \sin(\gamma_M - \sigma) \\
\dot{\gamma}_M &= \frac{a_M}{V_M} \\
\dot{\gamma}_T &= \frac{a_T}{V_T}
\end{aligned} \tag{2.1}$$

Note that optimal guidance law is usually derived by using linearized kinematics. For this reason, a new reference frame (X_R, Y_R) is defined by rotating from the inertial frame with a constant reference angle σ_0, as shown in Fig. 2.1. In the reference frame, the differential equations describing the engagement kinematics can be expressed as

$$\begin{aligned}
\dot{y} &= v \\
\dot{v} &= a_{T_\sigma} \cos(\sigma - \sigma_0) - a_{M_\sigma} \cos(\sigma - \sigma_0)
\end{aligned} \tag{2.2}$$

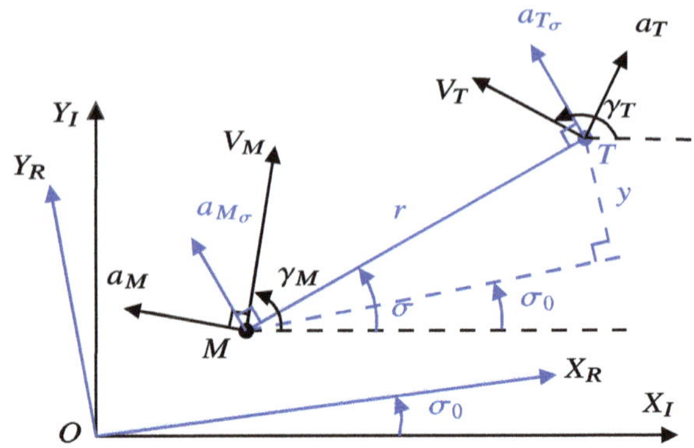

Fig. 2.1 The homing engagement geometry and parameter definitions

where y is the relative distance between the target and the missile perpendicular to the X_R-direction. The variables of a_{M_σ} and a_{T_σ} denote the missile and target accelerations normal to the LOS direction, respectively.

The complementary equation that describing the relationship between a_{M_σ} and a_M is given by

$$a_{M_\sigma} = a_M \cos(\gamma_M - \sigma) \tag{2.3}$$

In practical flight, the difference between σ and σ_0 can be very small, i.e., $\sigma - \sigma_0$ can be a small angle, by properly choosing the reference angle. In light of the fact that the terminal homing guidance phase generally follows a precise guidance handover from a proper mid-course guidance law. The reference angle can be chosen as the initial LOS angle to enable the validation of the small angle assumption. With this in mind, the nonlinear engagement kinematics can be linearized as

$$\begin{aligned} \dot{y} &= v \\ \dot{v} &= a_{T_\sigma} - a_{M_\sigma} \end{aligned} \tag{2.4}$$

Under the assumption that angle $\sigma - \sigma_0$ is small, Fig. 2.1 reveals that

$$\sigma - \sigma_0 = \frac{y}{r} \tag{2.5}$$

Taking the time derivative of Eq. (2.5) results in

$$\begin{aligned} \dot{\sigma} &= \frac{vr - y\dot{r}}{r^2} \\ &= \frac{yV_c + vr}{r^2} \end{aligned} \tag{2.6}$$

where $V_c = \dot{r}$ denotes the closing velocity.

For constant-moving vehicles, the relative range can be approximated as

$$r \approx V_c t_{go} \tag{2.7}$$

where $t_{go} \triangleq t_f - t$ denotes the remaining flight time, or the so-called time-to-go, with t_f being the final time.

Substituting Eq. (2.7) into Eq. (2.6) gives the LOS rate as

$$\dot{\sigma} = \frac{y + v t_{go}}{V_c t_{go}^2}. \tag{2.8}$$

2.2.2 Motivations

As stated earlier, missile guidance law design is a kind of finite-time tracking problem and the objective is to regulate a specific mission-dependent tracking error to zero in finite time The general form of the tracking problem that is often observed in missile guidance law design is

$$\dot{\varepsilon}(t) = g(t)\, u(t) \tag{2.9}$$

where $\varepsilon(t)$ represents the tracking error; $g(t)$ stands for a known time-varying function; and $u(t)$ denotes the control input.

According to a specific guidance problem under consideration, the tracking error can be ZEM, impact angle error, impact time error, heading error, etc. The function of $g(t)$ also depends on the guidance problem and it is invertible since the tracking problem is controllable, i.e., $g(t) \neq 0$. In order to regulate the tracking error to be zero, the desired error dynamics is first selected and various control theories such as SMC, Lyapunov function, and feedback linearization, can then be applied to system (2.9) to follow a desired error dynamics. In many previous works, one widely-accepted error dynamics for guidance law design is given by

$$\dot{\varepsilon}(t) + k\varepsilon(t) = 0 \tag{2.10}$$

where $k > 0$ is the guidance gain to regulate the convergence rate of tracking error.

It is easy to verify that the closed-form solution of differential equation (2.10) is determined as

$$\varepsilon(t) = \varepsilon(t_0)\, e^{-kt} \tag{2.11}$$

where $\varepsilon(t_0)$ denotes the initial tracking error.

The preceding equation reveals that the tracking error converges to zero asymptotically with an exponential rate governed by the guidance gain k. Accordingly, this desired error dynamics has two major drawbacks:

(1) The finite-time convergence is not strictly guaranteed;
(2) It only focuses on how to drive the tracking error to zero and never considers what is the optimal error dynamics in terms of a meaningful performance index.

Although the first issue can be addressed by using the SMC approaches [10, 11, 17], the optimal convergence pattern of the tracking error cannot be taken in to account. The resultant guidance law is also formulated in a complex form and hence it is difficult to analyze the physical meaning. Motivated by these observations, this chapter aims to investigate the optimal convergence pattern of the tracking error and propose an optimal error dynamics that achieves this optimal pattern with guaranteed finite-time convergence for missile guidance law design.

2.2.3 Preliminaries

This subsection introduces the well-known Schwarz's inequality that will be utilized in the following section.

Lemma 2.1 *Let $\eta_1(\tau)$ and $\eta_2(\tau)$ be any two real integrable functions in range $[t_1, t_2]$, then the Schwarz's inequality is given by*

$$\left[\int_{t_1}^{t_2} \eta_1(\tau)\eta_2(\tau)d\tau\right]^2 \leq \int_{t_1}^{t_2} \eta_1^2(\tau)d\tau \int_{t_1}^{t_2} \eta_2^2(\tau)d\tau \tag{2.12}$$

where the equality holds if and only if $\eta_1(\tau)$ and $\eta_2(\tau)$ have linear relationship as

$$\eta_1(\tau) = \Gamma\eta_2(\tau) \tag{2.13}$$

with Γ being a constant.

2.3 Optimal Error Dynamics

In this section, we first derive the proposed optimal error dynamics by using Schwarz's inequality and then analyze the properties as well as the potential significance of the proposed approach.

2.3.1 Derivation of the Proposed Optimal Error Dynamics

In this subsection, we discuss the optimal error dynamics for missile guidance law design. The main results are presented in Theorem 2.1.

Theorem 2.1 *Suppose that the system equation is given by Eq. (2.9) and the desired error dynamics is chosen as*

$$\dot{\varepsilon}(t) + \frac{\Gamma(t)}{t_{go}} \varepsilon(t) = 0 \tag{2.14}$$

where

$$\Gamma(t) = \frac{t_{go} R^{-1}(t) g^2(t)}{\int_t^{t_f} R^{-1}(\tau) g^2(\tau) d\tau} \tag{2.15}$$

with $R(t) > 0$ being an arbitrary weighting function.

Then, the resulting control input minimizes the performance index

$$J = \frac{1}{2} \int_t^{t_f} R(\tau) u^2(\tau) d\tau \tag{2.16}$$

Proof The control input that enables system (2.9) to follow error dynamics (2.14) is determined by substituting Eq. (2.14) into Eq. (2.9) as

$$u(t) = -\frac{\Gamma(t)}{g(t) t_{go}} \varepsilon(t) = -\frac{R^{-1}(t) g(t)}{\int_t^{t_f} R^{-1}(\tau) g^2(\tau) d\tau} \varepsilon(t) \tag{2.17}$$

Next, we seek to prove that control input (2.17) is the optimal solution of the following optimization problem

$$\min_u \ J = \frac{1}{2} \int_t^{t_f} R(\tau) u^2(\tau) d\tau \tag{2.18}$$

subject to

$$\dot{\varepsilon}(t) = g(t) u(t), \quad \varepsilon(t_f) = 0 \tag{2.19}$$

Integrating from t to t_f on both sides of Eq. (2.9) gives

$$\varepsilon(t_f) - \varepsilon(t) = \int_t^{t_f} g(\tau) u(\tau) d\tau \tag{2.20}$$

Imposing terminal constraint $\varepsilon(t_f) = 0$ on Eq. (2.20) gives

$$-\varepsilon(t) = \int_t^{t_f} g(\tau) u(\tau) d\tau \tag{2.21}$$

Introducing a slack variable $R(t)$ renders Eq. (2.21) to

$$-\varepsilon(t) = \int_t^{t_f} g(\tau) R^{-1/2}(\tau) R^{1/2}(\tau) u(\tau) d\tau \tag{2.22}$$

Applying Lemma 2.1 to the preceding equation yields

$$[-\varepsilon(t)]^2 \le \left[\int_t^{t_f} R^{-1}(\tau)\, g^2(\tau)\, d\tau\right]\left[\int_t^{t_f} R(\tau)\, u^2(\tau)\, d\tau\right] \tag{2.23}$$

Rewriting inequality (2.23) as

$$\frac{1}{2}\int_t^{t_f} R(\tau)\, u^2(\tau)\, d\tau \ge \frac{[-\varepsilon(t)]^2}{2\left[\int_t^{t_f} R^{-1}(\tau)\, g^2(\tau)\, d\tau\right]} \tag{2.24}$$

which gives a lower bound of the performance index.

According to Lemma 2.1, the equality of Eq. (2.24) holds if and only if there exists a constant C such that

$$u(t) = CR^{-1}(t)\, g(t) \tag{2.25}$$

Substitution of Eq. (2.25) into Eq. (2.21) results in

$$-\varepsilon(t) = C\int_t^{t_f} R^{-1}(\tau)\, g^2(\tau)\, d\tau \tag{2.26}$$

Solving Eq. (2.26) for C gives

$$C = \frac{-\varepsilon(t)}{\int_t^{t_f} R^{-1}(\tau)\, g^2(\tau)\, d\tau} \tag{2.27}$$

Substituting Eq. (2.27) into Eq. (2.25) gives the optimal control input as

$$u(t) = -\frac{R^{-1}(t)\, g(t)}{\int_t^{t_f} R^{-1}(\tau)\, g^2(\tau)\, d\tau}\varepsilon(t) \tag{2.28}$$

which is identical with Eq. (2.17). □

2.3.2 Discussion of the Proposed Optimal Error Dynamics

One of the interesting points of the proposed optimal error dynamics is that it guarantees finite-time convergence since it is obtained by directly solving the finite-time optimal tracking problem. Also, the optimal error dynamics (2.14) is given in a similar form as the existing desired error dynamics (2.10). The only difference lies in the proportional gain: a constant term k in the existing method and a time-varying term $\Gamma(t)/t_{go}$ in the proposed method. Unlike the usual one, the proposed proportional gain changes from an initial small value to a final infinite value as the time-to-go goes to zero, that is,

$$\lim_{t_{go}\to 0} \frac{\Gamma(t)}{t_{go}} = \infty \tag{2.29}$$

According to the value of $\Gamma(t)$, the evolving pattern of the proposed proportional gain further changes. Hereafter, let us discuss the characteristics of $\Gamma(t)$. From Eq. (2.15), $\Gamma(t)$ can be rearranged as

$$\Gamma(t) = \frac{\phi(t)}{\left(\int_t^{t_f} \phi(\tau)\, d\tau / t_{go}\right)} \tag{2.30}$$

where $\phi(t) = R^{-1}(t)\, g^2(t)$ is defined for notational convenience.

The function of $g(t)$ is given by the missile guidance problem under consideration and the weighting function $R(t)$ is the design parameter. According to different selections of $R(t)$, the time-varying term $\Gamma(t)$ changes differently. In the following, we reveal that $\Gamma(t) > 0$, $\forall t > 0$ during the homing engagement in Proposition 2.1. Proposition 2.2 provides the limiting values of $\Gamma(t)$ for different $\phi(t)$.

Proposition 2.1 *For given $g(t)$ and $R(t)$, $\Gamma(t)$ is always greater than zero, i.e., $\Gamma(t) > 0$.*

Proof By definition, we have $R(t) > 0$ and $g(t) \neq 0$. From Eq. (2.30), it is obvious that the numerator is positive since $\phi(t) = R^{-1}(t)\, g^2(t) > 0$. Also, the denominator is positive since the integration of positive function gives positive value. Accordingly, we have $\Gamma(t) > 0$. □

Proposition 2.2 *When function $\phi(t)$ keeps constant, the equality $\Gamma(t) = 1$ holds. If $\phi(t)$ decreases as $t \to t_f$, then $\Gamma(t) > 1$. If $\phi(t)$ increases as $t \to t_f$, we have $\Gamma(t) < 1$.*

Proof In Eq. (2.30), the denominator can be considered as the average value of $\phi(t)$, denoted by $\bar{\phi}(t)$, during the remaining time of interception, as shown in Fig. 2.2. Accordingly, the time-varying term $\Gamma(t)$ is the ratio of the current value $\phi(t)$ to the average value $\bar{\phi}(t)$. If the term of $\phi(t)$ is given by a constant value, then the $\Gamma(t)$ is unity as $\Gamma(t) = 1$ since $\phi(t) = \bar{\phi}(t)$. If the term $\phi(t)$ decreases as $t \to t_f$, $\Gamma(t)$ is greater than unity due to $\phi(t) > \bar{\phi}(t)$. Conversely, if the term of $\phi(t)$ has an increasing pattern, then $\Gamma(t)$ is less than unity because of $\phi(t) < \bar{\phi}(t)$. □

Remark 2.1 It follows from Eq. (2.14) that $\Gamma(t) \geq 1$ is desirable to ensure a stable and a fast convergence rate at the initial time. Accordingly, Proposition 2.2 reveals that it is desirable to impose a constant value or a decreasing pattern on the function $\phi(t)$.

Remark 2.2 As shown in Eq. (2.15), the computation of $\Gamma(t)$ contains of the integration of $\phi(t)$, which is given as a function of $g(t)$ and $R(t)$. If $\phi(t)$ is given by a closed-form function, $\Gamma(t)$ can be obtained analytically or numerically. Even

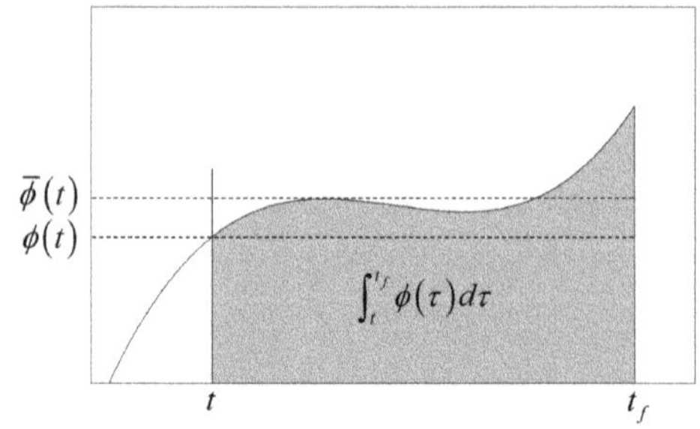

Fig. 2.2 The graphical interpretation of $\Gamma(t)$

though $g(t)$ is not given by a closed-form function due to the nature of the guidance problem, we can find closed-form solution of $\phi(t)$ through appropriate choice of the weighting function $R(t)$. Under this condition, the chosen optimal error dynamics only considers the minimization of a specific performance index since the physical meaning of the performance index shown in Eq. (2.16) changes according to choice of $R(t)$.

For a specific mission, the function $g(t)$ is fixed. Thus, by properly choosing the weighting function $R(t)$ for different objectives, the error dynamics is determined by Theorem 2.1. Here, we provide two special cases. In the case of $R(t) = 1$, the term $\Gamma(t)$ is given by

$$\Gamma(t) = \frac{t_{go} g^2(t)}{\int_t^{t_f} g^2(\tau)\, d\tau} \tag{2.31}$$

The optimal error dynamics with $\Gamma(t)$ shown in Eq. (2.31) minimizes the performance index

$$J = \frac{1}{2}\int_t^{t_f} u^2(\tau)\, d\tau \tag{2.32}$$

which provides an energy optimal guidance law.

In addition, if we choose the weighting function as $R(t) = g^2(t)/t_{go}^{K-1}$ with $K \geq 1$, then $\phi(t)$ is given by a function of time-to-go, regardless of $g(t)$, as

$$\phi(t) = t_{go}^{K-1} \tag{2.33}$$

In this case, $\Gamma(t)$ is a constant as

$$\Gamma(t) = K \tag{2.34}$$

Then, the desired error dynamics is given by

$$\dot{\varepsilon}(t) + \frac{K}{t_{go}}\varepsilon(t) = 0 \tag{2.35}$$

which is a simple Cauchy–Euler differential equation.

Solving Eq. (2.35) gives the closed-form of the tracking error as

$$\varepsilon(t) = \varepsilon(t_0)\left(\frac{t_{go}}{t_f}\right)^K \tag{2.36}$$

In practice, this optimal error dynamics, shown in Eq. (2.35), is useful due to the fact that the decreasing pattern of the tracking error is predictable as a function of time-to-go and the desired error dynamics is given by a simplified form. Therefore, in the next section, we will apply this desired error dynamics to various missile guidance problems for simplicity and practicality.

Following Theorem 2.1, the desired error dynamics (2.35) minimizes the performance index

$$J = \frac{1}{2}\int_t^{t_f} \frac{g^2(\tau)}{(t_f - \tau)^{K-1}} u^2(\tau)\, d\tau \tag{2.37}$$

It is obvious that performance index (2.37) with $g(t) = 1$ and $K = 1$ becomes the energy minimization case.

2.3.3 Potential Significance of the Proposed Optimal Error Dynamics

In this subsection, we discuss the potential significance of the findings in this chapter. First, the proposed results provide a way to improve existing missile guidance laws based on nonlinear control methods. For example, existing missile guidance laws using SMC are generally composed of two terms as

$$a_M = a_M^{eq} + a_M^{dis} \tag{2.38}$$

where a_M^{eq} and a_M^{dis} are the equivalent control part for disturbance-free dynamics and the SMC control part for compensating external uncertainties, respectively.

Using Eq. (2.38), some existing missile guidance laws based on SMC can be improved by converting the equivalent control parts to their optimal forms using the proposed optimal error dynamics and compensating the undesired disturbances by an add-on SMC control part.

Second, the proposed results present a theoretical background to provide a meaningful performance index of existing nonlinear missile guidance laws, which has not been clearly explained so far. For example, suppose that there is a missile guidance law using desired error dynamics (2.10). We can easily verify that the existing error dynamics (2.10) is a special case of the proposed optimal error dynamics (2.14) with $\Gamma(t) = kt_{go}$. Then, imposing this condition on Eq. (2.15) gives

$$k = \frac{R^{-1}(t)\, g^2(t)}{\int_t^{t_f} R^{-1}(\tau)\, g^2(\tau)\, d\tau} \tag{2.39}$$

For $t_f \to \infty$, we can obtain that the weighting function $R(t)$ satisfies

$$R(t) = g^2(t)\, e^{kt} \tag{2.40}$$

Substituting Eq. (2.40) into Eq. (2.16) with $t_f \to \infty$ gives the performance index of existing missile nonlinear guidance laws using Eq. (2.10) as

$$J = \frac{1}{2} \int_t^{\infty} g^2(\tau)\, e^{k\tau} u^2(\tau)\, d\tau \tag{2.41}$$

In a similar way, we can also examine the performance index of existing missile nonlinear guidance laws based on different desired error dynamics. In addition, if a nonlinear missile guidance law is designed by using the proposed optimal error dynamics, the performance index of the designed guidance law is given by Eq. (2.17) and its physical meaning is clear.

Finally, provided that the tracking error is suitably defined, the proposed results can be applied to different missile guidance problems for various operational objectives since we suggest a unified approach of guidance law design for a generalized tracking problem. Therefore, the proposed error dynamics provides the possibility of developing various new missile guidance laws for different practical problems.

2.3.4 *General Approach for Guidance Law Design*

This subsection presents the details of a general approach of how to utilize the proposed optimal error dynamics in guidance law design. To this end, let q be the interested variable that we would like to control for a specific mission objective. Suppose that the dynamics of q is given by a general nonlinear differential equation as

$$\dot{q}(t) = f(t) + b(t)u(t) \tag{2.42}$$

where $f(t)$, $b(t) \neq 0$ are time-varying functions and $u(t)$ denotes the command input.

By formulating the guidance command as a composite form as

$$u(t) = u_0(t) + u_b(t) \tag{2.43}$$

where $u_0(t)$ denotes the nominal control part and $u_b(t)$ stands for a biased term that can be designed by using the proposed error dynamics.

Based on the general command (2.43), the systematic way to utilize the proposed optimal error dynamics in guidance law design is to follow a prediction-correction manner:

(1) Predict the terminal error of the variable that we would like to control for a specific mission objective with a nominal trajectory, i.e., only controlled by u_0, as

$$q\left(t_f\right) = q(t) + \int_t^{t_f} f(\tau) + b(\tau)u_0(\tau)d\tau \tag{2.44}$$

(2) Define $\varepsilon\left(t\right) = q\left(t_f\right) - q_d$ as the tracking error with q_d being the desired state. Since the nominal control part u_0 imposes no effect on $q\left(t_f\right)$, the dynamics of tracking error $\varepsilon\left(t\right)$ under composite command (2.43) can then be readily determined as

$$\dot{\varepsilon}\left(t\right) = -b(t)u_b(t) \tag{2.45}$$

(3) Correct the tracking error by using a biased term u_b based on the proposed error dynamics.

2.4 Illustrative Examples

In this section, we provide some examples of how to apply the proposed optimal error dynamics in missile guidance law design. As illustrative examples, we consider guidance law design for homing, impact angle control, impact time control, and impact time and angle control.

2.4.1 Homing Guidance

The main objective of homing guidance is to guide the missile to intercept the target with zero miss distance. Following the general prediction-correction concept, we first need to predict the terminal miss distance based on a nominal trajectory. Ideally, it is desirable to intercept the target with zero control effort for energy minimization. Hence, the uncontrolled trajectory is chosen as the nominal trajectory and therefore the concept of ZEM is leveraged in this subsection for homing guidance law design. The ZEM is defined as the terminal miss distance between the missile and the target if the missile performs no maneuver from the current time onwards and hence is the predicted terminal miss distance of the zero-control-effort trajectory. Let z denote the ZEM, then, one can derive the dynamics of ZEM in the linearized engagement kinematics as

$$\begin{aligned} \text{Non-maneuvering target: } z &= y + v t_{go} \\ \text{Maneuvering target: } z &= y + v t_{go} + \frac{1}{2} a_{T\sigma} t_{go}^2 \end{aligned} \tag{2.46}$$

Differentiating Eq. (2.46) with respect to time yields

$$\dot{z} = -t_{go} a_{M\sigma} \tag{2.47}$$

In order to nullify the ZEM, it is natural to choose the tracking error as $\varepsilon_z = z$. In this case, we have $g(t) = -t_{go}$ and $u(t) = a_{M\sigma}$. We select the optimal error dynamics with respect to ε_z as

$$\dot{\varepsilon}_z + \frac{N}{t_{go}} \varepsilon_z = 0 \tag{2.48}$$

where N is a positive constant.

It follows from Eq. (2.16) that the corresponding performance index is obtained as

$$J = \frac{1}{2} \int_t^{t_f} \frac{1}{\left(t_f - \tau\right)^{N-3}} u^2(\tau)\, d\tau \tag{2.49}$$

The preceding index can be viewed as the energy minimization weighted by $1/t_{go}^{N-3}$. Accordingly, if the value of N is chosen as 3, the resulting optimal error dynamics ensures energy minimization. Additionally, we can safely predict that the designed guidance law using Eq. (2.48) with $N > 3$ guarantees zero terminal acceleration command since the weighting function with $N > 3$ becomes infinite as $t_{go} \to 0$. This property provides the missile with guaranteed operational margins to cope with undesired disturbances when the missile approaches the target.

Substituting Eq. (2.48) into Eq. (2.47) gives the optimal solution as

$$a_{M\sigma} = N \frac{z}{t_{go}^2} \tag{2.50}$$

Based on Eqs. (2.8) and (2.46), one can derive the final guidance command as

$$\begin{aligned} \text{Non-maneuvering target: } a_{M\sigma} &= N V_c \dot{\sigma} \\ \text{Maneuvering target: } a_{M\sigma} &= N V_c \dot{\sigma} + \frac{N}{2} a_{T\sigma} \end{aligned} \tag{2.51}$$

The obtained guidance command shown in Eq. (2.51) is the well-known PNG/augmented PNG (APNG) [26]. From the derivation presented in this subsection, the physical meaning of PNG is clear: the PNG is a predictor-corrector guidance law in the sense that it first predicts the terminal miss distance without any maneuver and then corrects the terminal miss distance by using desired error dynamics (2.48), thereby forcing the interceptor to maintain on the collision triangle.

Remark 2.3 For the considered example, the time-to-go power weighting function is utilized and the resulting guidance gain N is constant. Obviously, one can choose other weighting functions to accomplish different mission objectives. For example, in order to reduce the sensitivity with respect to the initial heading error to avoid an abrupt change during handover, a weighting function that provides a larger initial value can be designed as

$$R(t) = \frac{1}{t_{go}} + bt_{go}, \quad b > 0 \tag{2.52}$$

It is clear that the term bt_{go} dominate over $1/t_{go}$ during the initial time and thus can help to alleviate the transition effect, i.e., reduce the sensitivity to the initial heading error. Following Theorem 2.1, the resulting optimal error dynamics in this case is given by

$$\dot{\varepsilon}_z + \frac{N_1(t)}{t_{go}} \varepsilon_z = 0 \tag{2.53}$$

where the time-varying guidance gain $N_1(t)$ is determined by

$$N_1(t) = \frac{2b^2 t_{go}^4}{\left(1 + bt_{go}^2\right)\left[bt_{go}^2 - \ln\left(1 + bt_{go}^2\right)\right]} \tag{2.54}$$

For tactical missiles, the aerodynamic maneuverability exponentially decreases with the increasing of the flight altitude due to air density. In order to shape the guidance command against the loss of maneuverability, the weighting function can then be designed as

$$R(t) = \frac{1}{e^{mt_{go}} + n}, \quad m \geq 0, \quad n \geq -1 \tag{2.55}$$

According to Theorem 2.1, the resulting optimal error dynamics in this case in formulated as

$$\dot{\varepsilon}_z + \frac{N_2(t)}{t_{go}} \varepsilon_z = 0 \tag{2.56}$$

where the time-varying guidance gain $N_2(t)$ is determined by

$$N_2(t) = \frac{3m^3 t_{go}^3 \left(e^{mt_{go}} + n\right)}{\left(3m^2 t_{go}^2 - 6mt_{go} + 6\right) e^{mt_{go}} + m^3 n t_{go}^3 - 6}. \tag{2.57}$$

2.4.2 Impact Time Control

The ability to constrain the final impact time of anti-ship missiles is often desirable for increasing the survivability since this strategy enables penetrating ship-board self-defense systems by using multiple missiles to intercept a target simultaneously [13, 27–29]. One typical solution to realize simultaneous interception is the impact time control guidance, in which the missile is required to intercept the target at a specified time instant. In this subsection, we will show how to utilize the proposed optimal error dynamics to design impact time control guidance law. For simplicity, we only consider stationary target interception in this subsection.

Following the systematic prediction-correction guideline, we formulate the guidance command as

$$a_M = a_{M_0} + a_{IT} \tag{2.58}$$

where a_{M_0} denotes the nominal command to guarantee zero ZEM for target interception and a_{IT} stands for the biased correction part to nullify the impact time error.

For the purpose of illustration, we leverage the PNG as the nominal guidance command, i.e.,

$$a_{M_0} = N V_M \dot{\sigma} \tag{2.59}$$

As derived in [13], the estimated final interception time under PNG is given by

$$t_f = t + \frac{r}{V_M}\left(1 + \frac{\sin^2\theta}{2(2N-1)}\right) \tag{2.60}$$

where $\theta = \gamma_M - \sigma$ is the heading error.

Let t_d be the desired impact time. Then, the impact time error can be defined as $\varepsilon_t = t_d - t_f$. In order to regulate the impact time error, consider ε_t as the tracking error, which gives the error dynamics as

$$\begin{aligned}
\dot{\varepsilon}_t &= -\dot{t}_f \\
&= -\frac{\dot{r}}{V_M}\left(1 + \frac{\sin^2\theta}{2(2N-1)}\right) - \frac{r\sin\theta\cos\theta\dot{\theta}}{(2N-1)V_M} - 1 \\
&= \cos\theta\left(1 + \frac{\sin^2\theta}{2(2N-1)}\right) - \frac{r\sin\theta\cos\theta\left(\frac{a_{M_0}+a_{IT}}{V_M} - \dot{\sigma}\right)}{(2N-1)V_M} - 1 \\
&= \cos\theta\left(1 + \frac{\sin^2\theta}{2(2N-1)}\right) + \frac{(N-1)\sin^2\theta\cos\theta}{2N-1} - \frac{r\sin\theta\cos\theta}{(2N-1)V^2}a_{IT} - 1
\end{aligned} \tag{2.61}$$

In practical flight, the lead angle $\theta = \gamma_M - \sigma$ is small and thus $\sin\theta \approx \theta, \cos\theta \approx 1 - \theta^2/2$. Using these approximations and neglecting the higher order term of θ, Eq. (2.61) can be reduced to

$$\dot{\varepsilon}_t = -\frac{r \sin\theta}{(2N-1)\,V_M^2} a_{IT} \tag{2.62}$$

In this example, we have $g(t) = -r\sin\theta/(2N-1)\,V_M^2$ and $u(t) = a_{IT}$. The optimal error dynamics with respect to the impact time error ε_t is selected as

$$\dot{\varepsilon}_t + \frac{K}{t_{go}}\varepsilon_t = 0 \tag{2.63}$$

Substituting Eq. (2.63) into Eq. (2.62) gives the guidance command to nullify the impact time error as

$$a_{IT} = \frac{K\,(2N-1)\,V_M^2}{r\sin\theta t_{go}}\varepsilon_t \tag{2.64}$$

Combining Eq. (2.59) with Eq. (2.64) leads to the impact time control guidance law as

$$a_M = NV_M\dot{\sigma} + \frac{K\,(2N-1)\,V_M^2}{r\sin\theta t_{go}}\varepsilon_t \tag{2.65}$$

Note that if we choose $K = 4$ and $N = 3$, then guidance command (2.65) is identical to the impact time guidance law [13]. Also, in the case of $K = N + 1$, the obtained impact time guidance law becomes the guidance law proposed in [15]. Therefore, the obtained result is a generalized form of previous impact time guidance laws [13, 15]. Table 2.1 summarizes the relationship between the proposed approach and existing optimal impact time control guidance laws.

In addition, it follows from Theorem 2.1 that the corresponding performance index of dynamics (2.63) is given by

$$J = \frac{1}{2}\int_t^{t_f} \frac{r^2 \sin\theta^2}{(2N-1)^2 V_M^4 \left(t_f-\tau\right)^{K-1}} u^2(\tau)\,d\tau \tag{2.66}$$

By neglecting the constant terms in the performance index, Eq. (2.66) further reduces to

$$J = \frac{1}{2}\int_t^{t_f} \frac{r^2 \sin\theta^2}{\left(t_f-\tau\right)^{K-1}} u^2(\tau)\,d\tau \tag{2.67}$$

The performance index, given in Eq. (2.67), gives us general insights into the command pattern of the proposed guidance law and the behavior of guidance laws under the circumstance of impact time control. From Eq. (2.67), we can readily

Table 2.1 Summary of existing optimal impact time guidance laws

Guidance law	Guidance gain
Guidance law [13]	$N = 3$ and $K = 4$
Guidance law [15]	$K = N + 1$

observe that the magnitude of the weighting function decreases as r and θ decrease for a given time-to-go. Therefore, the resultant guidance command of impact time control tends to increase as r and θ decrease. This is a general phenomenon of impact-time-control guidance laws. For a stationary target, the parameter θ can be considered as a heading error. Therefore, decreasing r and θ means that the missile approaches a target, converging to a collision course. For the impact time control, the missile needs to make a detour away from the desired collision triangle, hence adjusting the flight time as desired. By the geometric rule, we readily know that changing the flight trajectory is getting easy as the missile deviates from the collision course. Namely, once the missile makes the collision course to a target, more control effort is required to correct the flight time as desired. This is a general characteristic of impact time control guidance laws, and performance index (2.67) implies this fact.

Based on performance index (2.67), we can also determine the command pattern of the proposed guidance law. It follows from Eq. (2.67) that the weighting function $r^2\sin^2\theta/t_{go}^{K-1}$ gradually decreases with the decrease of r and θ. For this reason, it is recommended to choose relatively large guidance gain K such that t_{go}^{K-1} has faster decreasing rate than $r^2\sin^2\theta$ to compensate for the decreasing of the weighting function. Notice that the decreasing rates of both relative range and velocity lead angle are governed by the PNG term. Since $r \approx V_M t_{go}$ when the interceptor approaches the target, the decreasing rate of r is proportional to t_{go}. In the case of PNG with constant navigation gain N, the closed-form solution of the velocity lead angle is $\theta = Ct_{go}^{N-1}$ with C being a constant determined by the initial condition [30]. Considering the fact that θ is small, the decreasing rate of $r^2\sin^2\theta$ is, therefore, proportional to t_{go}^{2N}. With this in mind, a suitable choice of K that guarantees a zero final guidance command is $K - 1 > 2N$.

2.4.3 Impact Angle Control

Constraining the impact angle is often desirable in terms of increasing the warhead effectiveness as well as inflicting the maximum damage of the target for both anti-ship and anti-tank missiles since it enables exploiting the structural weakness of the target. This subsection will show the details of how to utilize the proposed optimal error dynamics in trajectory shaping for impact angle control. For simplicity, we consider a stationary target interception scenario in this subsection.

Following the guideline of the general prediction-correction concept, the guidance command is formulated in a composite form as

$$a_M = a_{M_0} + a_{IA} \tag{2.68}$$

where a_{M_0} denotes the nominal guidance command that provides zero ZEM and a_{IA} represents the biased command that is utilized to control the impact angle.

Similarly, we choose the optimal PNG as the nominal guidance command, i.e.,

$$a_{M_0} = N V_M \dot{\sigma} \tag{2.69}$$

According to [9], the final flight path angle governed by PNG is given by

$$\gamma_{M_f} = \frac{N}{N-1}\sigma - \frac{1}{N-1}\gamma_M \tag{2.70}$$

Let γ_f be the desired final flight path angle. Then, the impact angle error can be defined as $\varepsilon_\gamma = \gamma_f - \gamma_{M_f}$. In order to nullify the impact angle error, consider ε_γ as the tracking error, which gives the error dynamics as

$$\begin{aligned}\dot{\varepsilon}_\gamma &= -\dot{\gamma}_{M_f} \\ &= -\frac{N}{N-1}\dot{\sigma} + \frac{1}{N-1}\dot{\gamma}_M \\ &= -\frac{N}{N-1}\dot{\sigma} + \frac{1}{N-1}\frac{a_M}{V_M}\end{aligned} \tag{2.71}$$

Substituting $a_M = a_{M_0} + a_{IA}$ into Eq. (2.71) gives

$$\dot{\varepsilon}_\gamma = \frac{a_{IA}}{(N-1)\,V_M} \tag{2.72}$$

In this example, the control input $u\,(t)$ and the function $g\,(t)$ match with a_{IA} and $1/\,(N-1)\,V_M$, respectively. For the impact angle error ε_γ, the optimal error dynamics is selected in a similar way as

$$\dot{\varepsilon}_\gamma + \frac{K}{t_{go}}\varepsilon_\gamma = 0 \tag{2.73}$$

where $K \geq 1$ is the guidance gain to be designed.

The corresponding performance index is obtained as

$$J = \frac{1}{2}\int_t^{t_f} \frac{1}{(N-1)^2 V_M^2 \left(t_f - \tau\right)^{K-1}} u^2\,(\tau)\,d\tau \tag{2.74}$$

Since the constant terms in the performance index has no effect on the optimal pattern, performance index (2.74) is identical to

$$J = \frac{1}{2}\int_t^{t_f} \frac{1}{\left(t_f - \tau\right)^{K-1}} u^2\,(\tau)\,d\tau \tag{2.75}$$

We can clearly observe from the preceding performance index that the weighting function with $K > 1$ becomes infinite as $t_{go} \to 0$. Therefore, one can imply that the

Table 2.2 Summary of existing optimal impact angle guidance laws

Guidance law	Guidance gain
OGL [31]	$N = 3$ and $K = 1$
IACG [9]	$N \geq 3$ and $K = 1$
TWOIAG [32, 33]	$N = K + 2$ and $K \geq 1$
TPG [34]	$N > K + 1$ and $K \geq 1$

optimal error dynamics with $K > 1$ guarantees zero impact angle guidance command at the final time. Additionally, if $K = 1$, the above performance index coincides with the energy optimal case.

From Eqs. (2.72) and (2.73), the optimal solution of a_{IA} is easily obtained as

$$a_{IA} = -\frac{K(N-1)V_M}{t_{go}}\varepsilon_\gamma \tag{2.76}$$

Combining Eq. (2.68) with Eq. (2.76), one can derive the final guidance command for impact angle control as

$$a_M = NV_M\dot{\sigma} - \frac{K(N-1)V_M}{t_{go}}\varepsilon_\gamma \tag{2.77}$$

If we choose $N = 3$ and $K = 1$, guidance law (2.77) becomes optimal guidance law (OGL) for impact angle control [31]. If the guidance gains satisfy $N \geq 3$ and $K = 1$, guidance law (2.77) reduces to interception angle control guidance (IACG) law [9]. If one enforces $N = K + 2$ and $K \geq 1$, guidance law (2.77) is identical with time-to-go weighted optimal impact angle guidance (TWOIAG) law [32, 33]. Finally, if one selects $N > K + 1$ and $K \geq 1$, guidance law (2.77) turns out to be time-to-go polynomial guidance (TPG) law [34]. Consequently, using the proposed approach generates a generalized optimal impact angle control guidance law that covers several previous results. Table 2.2 summarizes the relationship between the proposed approach and existing optimal impact angle control guidance laws.

From guidance command (2.77), the optimal impact angle guidance law can also be interpreted as a predictor-corrector guidance law: the optimal impact angle guidance law first predicts the terminal flight path angle by using the optimal PNG as its control input and corrects the terminal flight path angle error by using desired error dynamics (2.73).

2.4.4 Impact Angle and Impact Time Control

In this subsection, we extend the results of the previous two subsections to both impact angle and impact time control. Similarly, the guidance command is composed of two different parts as

$$a_M = a_{IAC} + a_{IT} \tag{2.78}$$

where a_{IAC} is the baseline guidance law utilized for impact angle control and the biased command a_{IT} is leveraged to nullify the impact time error.

Without loss of generality, the command a_{IAC} is defined as the well-known TWOIAG law [32] as

$$a_{IAC} = -\frac{V_M^2}{r}\left[2(m+2)\theta + (m+1)(m+2)\theta_f\right] \tag{2.79}$$

where $\theta_f = \gamma_f - \sigma$ and the integer $m \geq 0$ is a design parameter.

The advantage of utilizing TWOIAG lies in that it guarantees theoretical zero final guidance command when $m \geq 1$, thus providing enough operational margins to cope with undesired external disturbances. As derived in [32], the predicted impact time under guidance law (2.79) is determined as

$$t_f = \frac{r}{V_M}\left[1 + \frac{(m+2)\theta^2 + (m+1)^2(m+2)\theta_f^2 - (m+1)\theta\theta_f}{2(2m+3)(2m+5)}\right] \tag{2.80}$$

Define $\varepsilon_t = t_d - t_f$ as the impact time error. Differentiating ε_t with respect to time yields

$$\begin{aligned}\dot{\varepsilon}_t =& \cos\theta\left[1 + \frac{(m+2)\theta^2 + (m+1)^2(m+2)\theta_f^2 - (m+1)\theta\theta_f}{2(2m+3)(2m+5)}\right]\\ &- \frac{r}{2(2m+3)(2m+5)V_M}\left[(2m+4)\theta - (m+1)\theta_f\right]\dot{\gamma}_M\\ &- \frac{1}{2(2m+3)(2m+5)V_M}\left[(2m+3)\theta + \left(2m^2+6m+3\right)\theta_f\right]\sin\theta - 1\end{aligned} \tag{2.81}$$

Similar to the previous subsection, we assume that the heading error θ is small. Then, Eq. (2.81) can be approximated as

$$\begin{aligned}\dot{\varepsilon}_t =& \left(1 - \frac{1}{2}\theta^2\right)\left[1 + \frac{(m+2)\theta^2 + (m+1)^2(m+2)\theta_f^2 - (m+1)\theta\theta_f}{2(2m+3)(2m+5)}\right]\\ &- \frac{r}{2(2m+3)(2m+5)V_M}\left[(2m+4)\theta - (m+1)\theta_f\right]\frac{a_{IAC} + a_{IT}}{V_M}\\ &- \frac{1}{2(2m+3)(2m+5)V_M}\left[(2m+3)\theta + \left(2m^2+6m+3\right)\theta_f\right]\theta - 1\\ =& -\frac{r\left[(2m+4)\theta - (m+1)\theta_f\right]}{2(2m+3)(2m+5)V_M^2}a_{IT}\end{aligned} \tag{2.82}$$

In this example, we have

$$g(t) = -\frac{r\left[(2m+4)\theta - (m+1)\theta_f\right]}{2(2m+3)(2m+5)V_M^2}, \quad u(t) = a_{IT} \tag{2.83}$$

The optimal error dynamics with respect to the impact time error ε_t can then be selected as

$$\dot{\varepsilon}_t + \frac{K}{t_{go}}\varepsilon_t = 0 \tag{2.84}$$

Substituting Eq. (2.82) into Eq. (2.84) gives the guidance command to nullify the impact time error as

$$a_{IT} = \frac{2K(2m+3)(2m+5)V_M^2}{rt_{go}\left[(2m+4)\theta - (m+1)\theta_f\right]}\varepsilon_t \tag{2.85}$$

Then final guidance command is then given by

$$\begin{aligned} a_M = & -\frac{V_M^2}{r}\left[2(m+2)\theta + (m+1)(m+2)\theta_f\right] \\ & + \frac{2K(2m+3)(2m+5)V_M^2}{rt_{go}\left[(2m+4)\theta - (m+1)\theta_f\right]}\varepsilon_t \end{aligned} \tag{2.86}$$

Note that the proposed generalized guidance law reduce to the impact angle and time guidance law proposed in [35] when $m = 0$. From Theorem 2.1, it can be easily verified that the performance index of biased command a_{IT} is determined as

$$J = \frac{1}{2}\int_t^{t_f} \frac{r^2\left[(2m+4)\theta - (m+1)\theta_f\right]^2}{4(2m+3)^2(2m+5)^2V_M^4(t_f-\tau)^{K-1}}u^2(\tau)\,d\tau \tag{2.87}$$

Since the constant term has no effect on the pattern of the performance index, Eq. (2.87) can be reduced to

$$J = \frac{1}{2}\int_t^{t_f} \frac{r^2\left[(2m+4)\theta - (m+1)\theta_f\right]^2}{(t_f-\tau)^{K-1}}u^2(\tau)\,d\tau \tag{2.88}$$

which provides us the guideline on how to choose proper gain K to guarantee finite guidance command.

Remark 2.4 Unlike guidance law (2.65), guidance law (2.86) suffers from the singularity problem when $(2m+4)\theta = (m+1)\theta_f$. To address this problem, the biased command a_{IT} can be modified by utilizing the dead zone concept as

$$a_{IT} = \begin{cases} \dfrac{2K(2m+3)(2m+5)V_M^2}{rt_{go}\left[(2m+4)\theta - (m+1)\theta_f\right]}\varepsilon_t, & \left|(2m+4)\theta - (m+1)\theta_f\right| \geq \delta \\ \dfrac{2K(2m+3)(2m+5)V_M^2}{rt_{go}\delta \operatorname{sgn}\left[(2m+4)\theta - (m+1)\theta_f\right]}\varepsilon_t, & \left|(2m+4)\theta - (m+1)\theta_f\right| \leq \delta \end{cases} \tag{2.89}$$

where δ is a small positive constant and function sgn (x) is defined as

$$\operatorname{sgn}(x) = \begin{cases} 1, & x \geq 0 \\ -1, & x < 0 \end{cases}. \tag{2.90}$$

2.5 Simulation Results

In this section, nonlinear simulations are conducted to validate the examples shown in Sect. 2.4. For simplicity, we consider a stationary target interception scenario and the initial conditions are summarized in Table 2.3.

2.5.1 *Homing Guidance*

The simulation results of PNG with various navigation gains $N = 3, 4, 5$ are presented in Fig. 2.3, which indicates that the ZEM under PNG gradually converges to zero when the missile approaches the target. Additionally, the convergence rate increases with the increase of the navigation gain. The commanded acceleration with $N \geq 3$ converges to zero at the time of impact, as confirmed by Fig. 2.3b. This is helpful in improving the operational margins against undesired external disturbances during the endgame. We can also note that the magnitude of acceleration command during the initial phase increases with bigger value of the navigation gain. This can be attributed to the fact more control energy is consumed to increase the convergence rate of ZEM. From Fig. 2.3c, it can be observed that the PNG with $N = 3$ provides energy-optimal interception, which conforms with our analytic findings.

Table 2.3 Initial conditions for homing engagement

Parameters	Values
Missile-target initial relative range, $r(0)$	5000 m
Initial LOS angle, $\sigma(0)$	0°
Missile initial flight path angle, $\gamma_M(0)$	30°
Missile velocity, V_M	300 m/s

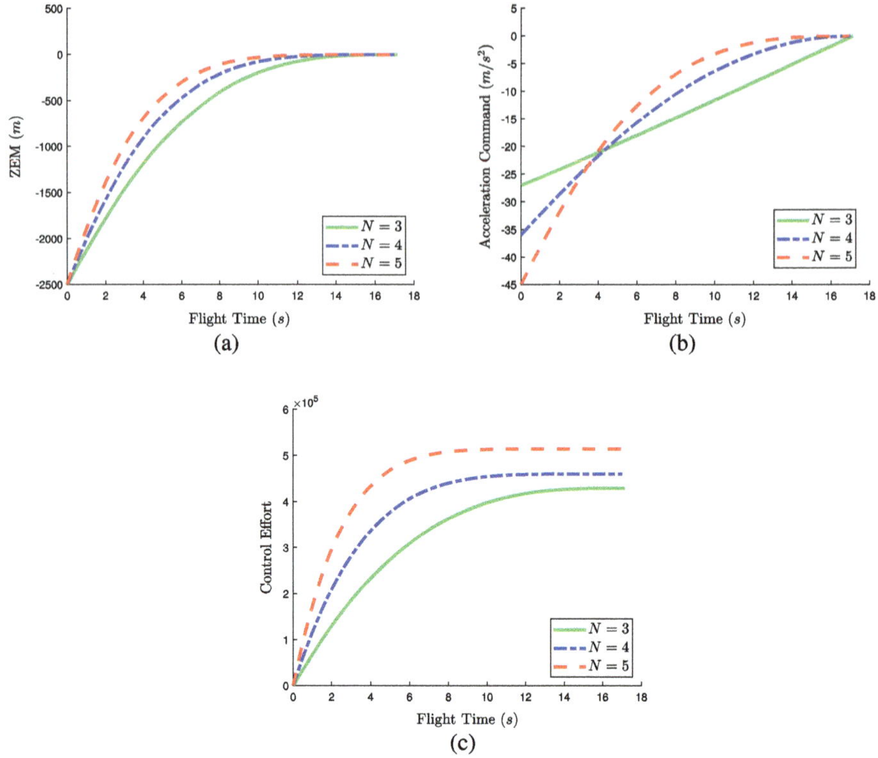

Fig. 2.3 Simulation results of PNG. **a** ZEM; **b** Acceleration command; and **c** Control effort

2.5.2 Alleviating Transition Effect

This subsection investigates the characteristics of the PNG with time-varying navigation gain (2.54) for the purpose of alleviating the transition effect. The simulation results with various design parameters $b = 0.04, 0.25, 1$ are shown in Fig. 2.4. The results indicate that the proposed guidance law ensures that the ZEM converges to zero at the time of impact and thereby guarantees accurate target interception. From weighting function (2.53), it is clear that the term bt_{go} dominate over $1/t_{go}$ during the initial time. Therefore, the magnitude of the initial guidance command reduces with the increase of the design parameter b. However, this will enforces shaper turn of the guidance command when the missile is close to the target and impose adverse effect on the control system, i.e., autopilot. Consequently, there is a tradeoff between transition effect alleviation and response delay of the autopilot system. Furthermore, increasing the design parameter b requires more energy consumption to drive the missile to converge to the collision course after a certain time period.

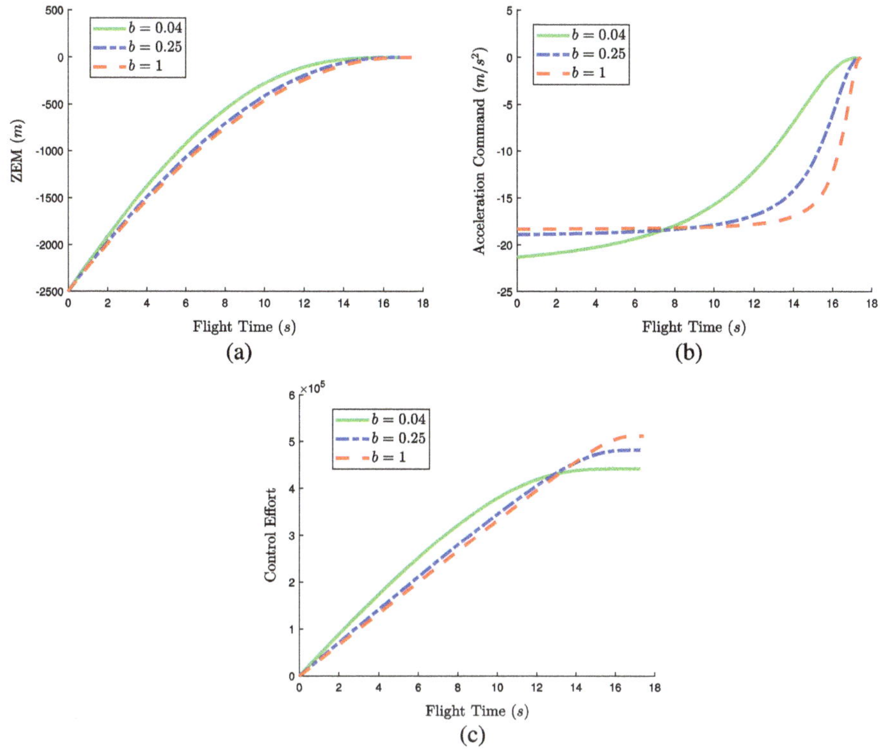

Fig. 2.4 Simulation results of guidance law to alleviate transition effect. **a** ZEM; **b** Acceleration command; and **c** Control effort

2.5.3 Shaping Aerodynamic Maneuverability

The performance of the optimal guidance law in shaping aerodynamic maneuverability is analyzed in this subsection. The simulation results obtained by PNG with time-varying navigation gain (2.57) are provided in Fig. 2.5. By reducing the value of n or increasing the value of m, the exponential term $e^{mt_{go}}$ in the denominator of weighting function (2.55) will play a dominant role over the constant term n. This will enforce the missile to consume more energy and hence to generate larger magnitude of the guidance command during the initial phase. Consequently, by using the weighting function (2.55) with properly tuned design parameters, we can shape the trajectory to account for the decrease of aerodynamic maneuverability.

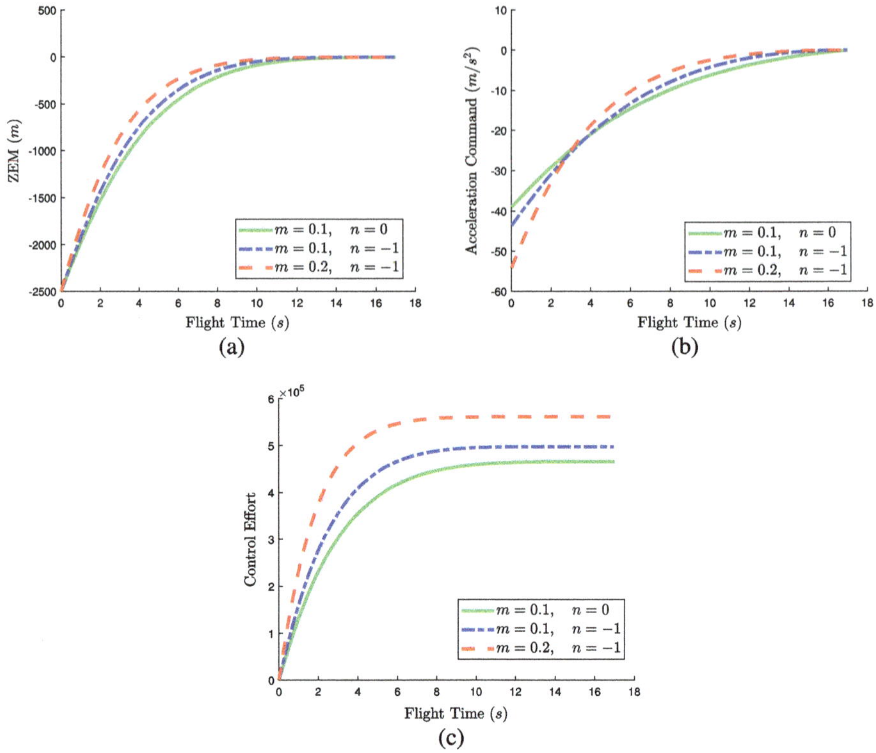

Fig. 2.5 Simulation results of guidance law to shape aerodynamic maneuverability. **a** ZEM; **b** Acceleration command; and **c** Control effort

2.5.4 Impact Time Control

The performance of generalized optimal impact time guidance law (2.65) with $t_d = 20\,\text{s}$ is numerically analyzed in this subsection. The simulation results, including interception trajectory, impact time error, acceleration command and control effort, obtained by the proposed guidance law with various guidance gains are shown in Fig. 2.6. The results indicate that the proposed guidance law successfully guide the missile to intercept the target with a desired impact time. Increasing the value of the biased gain K, the convergence rate of impact time error becomes faster at the sacrifice of more energy consumption, as shown in Fig. 2.6b and d. We can also note from Fig. 2.6c that the the guidance gain under condition $K > 2N + 1$ ensures theoretical zero terminal guidance command and this fact supports the analytical findings.

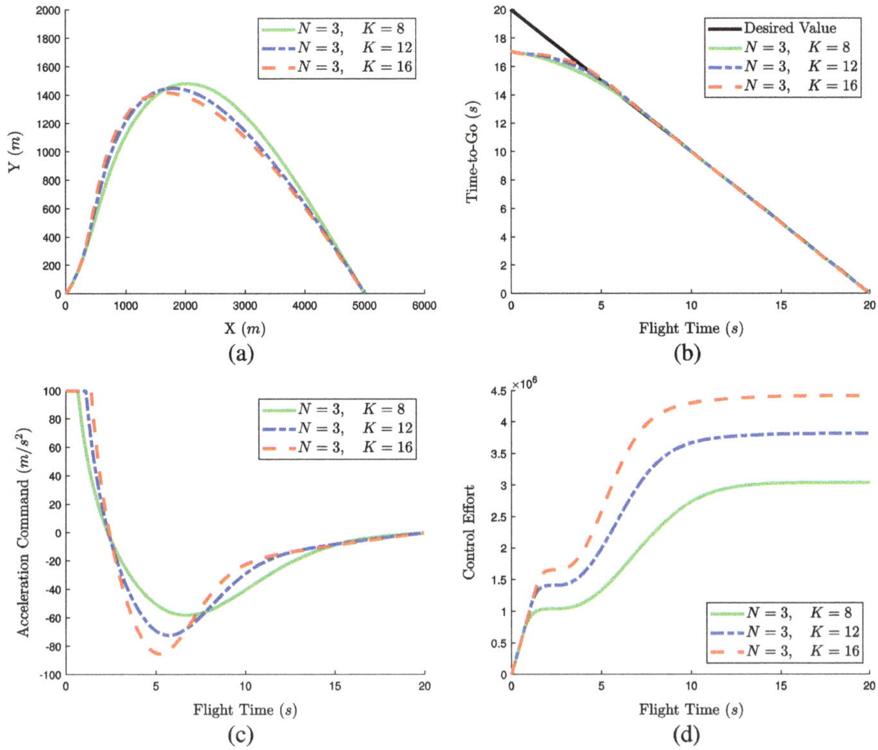

Fig. 2.6 Simulation results of impact time control guidance law. **a** Interception trajectory; **b** Time-to-go error; **c** Acceleration command; and **d** Control effort

2.5.5 Impact Angle Control

This subsection investigates the properties of generalized optimal impact angle guidance law (2.77). Without loss of generality, the desired impact angle is set as $\gamma_f = -90°$. Figure 2.7 presents the simulation results obtained from guidance law (2.77) with various guidance gains. From Fig. 2.7a and b, we can clearly note that the proposed optimal guidance law successfully guides the missile to intercept the target with a desired impact angle. Increasing the guidance gains N and K results in more curved trajectory and hence requires more control energy. The terminal guidance command converges to zero once the guidance gains satisfy condition $N > 3$, $K > 1$, as confirmed by Fig. 2.7c. The control effort consumption, demonstrated in Fig. 2.7d indicates that guidance law (2.77) with $N = 3$ and $K = 1$ provides energy-optimal interception. These results clearly conform with our analytical analysis.

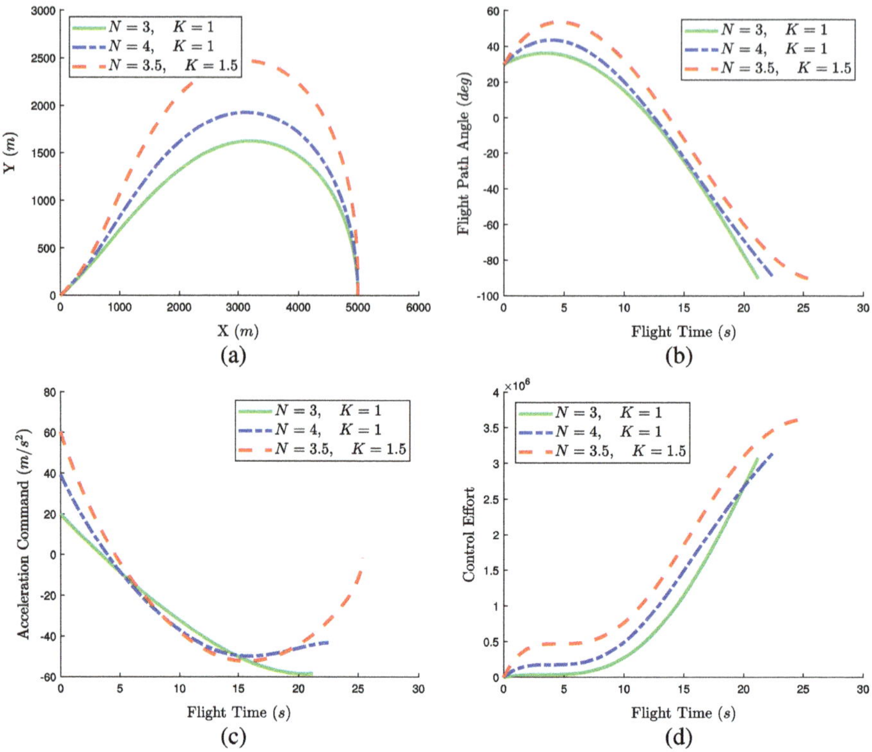

Fig. 2.7 Simulation results of impact angle control guidance law. **a** Interception trajectory; **b** Flight path angle; **c** Acceleration command; and **d** Control effort

2.5.6 *Impact Time and Angle Control*

This subsection conducts numerical simulations to investigate the performance of optimal impact time and angle guidance law (2.86). The desired impact angle and impact time are set as $\gamma_f = -90^\circ$ and $t_d = 30$ s. Figure 2.8 provides the simulation results obtained from guidance law (2.86) with various guidance gains. It follows from Fig. 2.8a–c that the missile intercepts the target with high accuracy in both impact time and impact angle control. By increasing the guidance gains, faster convergence rate of the tracking errors is recorded with a more curved interception trajectory. However, guidance law (2.86) with larger biased gain experiences longer duration of command saturation during the initial phase since more control energy is consumed to correct the collision course.

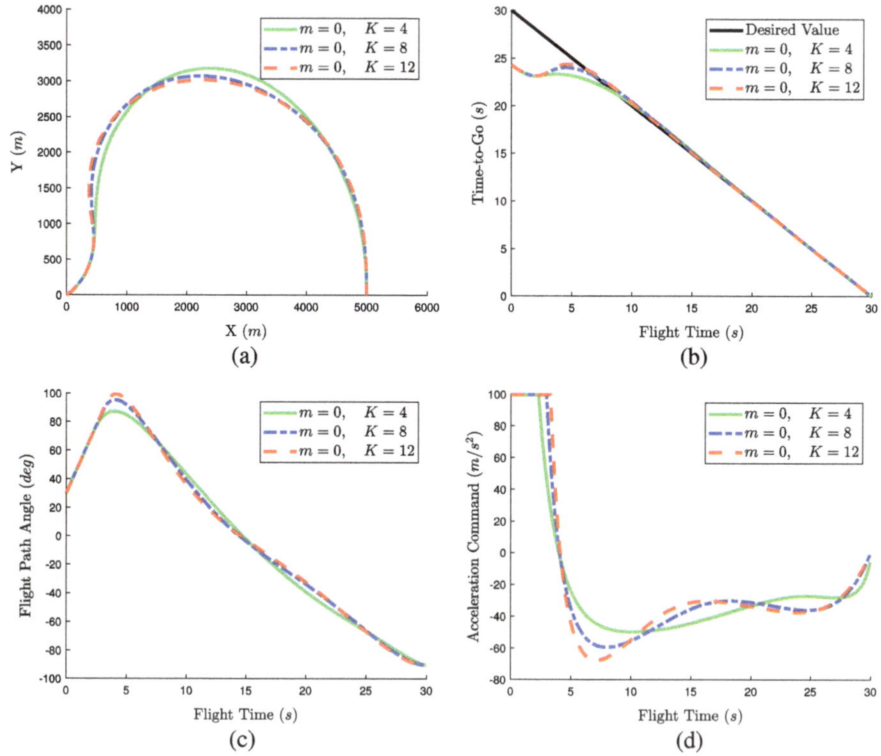

Fig. 2.8 Simulation results of impact time and angle control guidance law. **a** Interception trajectory; **b** Time-to-go error; **c** Flight path angle; and **d** Acceleration command

2.6 Summary

This chapter proposes a novel optimal error dynamics for missile guidance problems. The uniqueness of the proposed approach is that it can be applied to any operational objectives if the tracking error is suitably defined. We first reveal that the essence of missile guidance law design is a kind of finite-time tracking problem in control theory. Based on this result, a generalized error dynamics for unified missile guidance law design is proposed by solving a linear quadratic optimal control problem through Schwarz's inequality. As illustrative examples, we provide the applications of the proposed results to four well-known missile guidance problems, including homing guidance, impact angle guidance, impact time guidance and impact time and angle control.

The proposed results are believed to have an academic significance as well as a practical one. By using the proposed results, we can extend existing nonlinear guidance laws to their corresponding optimal version by exploiting advantages of the proposed optimal error dynamics. The proposed results also provide the theoretical background to evaluate the performance index of existing nonlinear missile guidance laws. As we suggest a unified approach for guidance law design, various new missile guidance laws are expected to be developed by applying the proposed approach to practical problems.

References

1. Shima T, Idan M, Golan OM (2006) Sliding-mode control for integrated missile autopilot guidance. J Guid Control Dyn 29(2):250–260
2. Idan M, Shima T, Golan OM (2007) Integrated sliding mode autopilot-guidance for dual-control missiles. J Guid Control Dyn 30(4):1081–1089
3. Dwivedi P, Bhale P, Bhattacharyya A, Padhi R (2016) Generalized estimation and predictive guidance for evasive targets. IEEE Trans Aerosp Electron Syst 52(5):2111–2122
4. He S, Lee C-H (2018) Gravity-turn-assisted optimal guidance law. J Guid Control Dyn 41(1):171–183
5. Kim M, Grider KV (1973) Terminal guidance for impact attitude angle constrained flight trajectories. IEEE Trans Aerosp Electron Syst AES–9(6):852–859
6. Kim BS, Lee JG, Han HS (1998) Biased png law for impact with angular constraint. IEEE Trans Aerosp Electr Syst 34(1):277–288
7. Lu P, Doman DB, Schierman JD (2006) Adaptive terminal guidance for hypervelocity impact in specified direction. J Guid Control Dyn 29(2):269–278
8. Erer KS, Merttopçuoglu O (2012) Indirect impact-angle-control against stationary targets using biased pure proportional navigation. J Guid Control Dyn 35(2):700–704
9. Lee C-H, Kim T-H, Tahk M-J (2013) Interception angle control guidance using proportional navigation with error feedback. J Guid Control Dyn 36(5):1556–1561
10. Kumar SR, Rao S, Ghose D (2012) Sliding-mode guidance and control for all-aspect interceptors with terminal angle constraints. J Guid Control Dyn 35(4):1230–1246
11. Kumar SR, Rao S, Ghose D (2014) Nonsingular terminal sliding mode guidance with impact angle constraints. J Guid Control Dyn 37(4):1114–1130
12. He S, Song T, Lin D (2017) Impact angle constrained integrated guidance and control for maneuvering target interception. J Guid Control Dyn 40(10):2653–2661
13. Jeon I-S, Lee J-I, Tahk M-J (2006) Impact-time-control guidance law for anti-ship missiles. IEEE Trans Control Syst Technol 14(2):260–266
14. Jeon I-S, Lee J-I, Tahk M-J (2010) Homing guidance law for cooperative attack of multiple missiles. J Guid Control Dyn 33(1):275–280
15. Kim T-H, Lee C-H, Tahk M-J, Jeon I-S (2013) Biased png law for impact-time control. Trans Jpn Soc Aeronaut Space Sci 56(4):205–214
16. Cho N, Kim Y (2016) Modified pure proportional navigation guidance law for impact time control. J Guid Control Dyn 39(4):852–872
17. Harl N, Balakrishnan SN (2012) Impact time and angle guidance with sliding mode control. IEEE Trans Control Syst Technol 20(6):1436–1449
18. Moon J, Kim K, Kim Y (2001) Design of missile guidance law via variable structure control. J Guid Control Dyn 24(4):659–664
19. Brierley SD, Longchamp R (1990) Application of sliding-mode control to air-air interception problem. IEEE Trans Aerosp Electron Syst 26(2):306–325
20. Koren A, Idan M, Golan OM et al (2008) Integrated sliding mode guidance and control for a missile with on-off actuators. J Guid Control Dyn 31(1):204
21. Yang C-D, Chen H-Y (1998) Nonlinear h_∞ robust guidance law for homing missiles. J Guid Control Dyn 21:882–890
22. Lechevin N, Rabbath CA (2004) Lyapunov-based nonlinear missile guidance. J Guid Control Dyn 27(6):1096–1102

23. Talole SE, Banavar RN (1998) Proportional navigation through predictive control. J Guid Control Dyn 21:1004–1005
24. Menon PK, Sweriduk GD, Ohlmeyer EJ (2003) Optimal fixed-interval integrated guidance-control laws for hit-to-kill missiles. In: *AIAA guidance, navigation, and control conference*, Austin, Texas. AIAA
25. Weiss G, Rusnak I (2015) All-aspect three-dimensional guidance law based on feedback linearization. J Guid Control Dyn 38(12):2421–2428
26. Zarchan P (2012) Tactical and strategic missile guidance. American Institute of Aeronautics and Astronautics
27. He S, Wang W, Lin D, Lei H (2018) Consensus-based two-stage salvo attack guidance. IEEE Trans Aerosp Electron Syst 54(3):1555–1566
28. Chen X, Wang J (2017) Nonsingular sliding-mode control for field-of-view constrained impact time guidance. J Guid Control Dyn 41(5):1214–1222
29. He S, Kim M, Song T, Lin D (2018) Three-dimensional salvo attack guidance considering communication delay. Aerosp Sci Technol 73:1–9
30. Lee C-H, Seo M-G (2018) New insights into guidance laws with terminal angle constraints. J Guid Control Dyn 41(8):1832–1837
31. Ryoo C-K, Cho H, Tahk M-J (2005) Optimal guidance laws with terminal impact angle constraint. J Guid Control Dyn 28(4):724–732
32. Ryoo C-K, Cho H, Tahk M-J (2006) Time-to-go weighted optimal guidance with impact angle constraints. IEEE Trans Control Syst Technol 14(3):483–492
33. Ohlmeyer EJ, Phillips CA (2006) Generalized vector explicit guidance. J Guid Control Dyn 29(2):261–268
34. Lee C-H, Kim T-H, Tahk M-J, Whang I-H (2013) Polynomial guidance laws considering terminal impact angle and acceleration constraints. IEEE Trans Aerosp Electron Syst 49(1):74–92
35. Tahk M-J, Shim S-W, Hong S-M, Lee C-H, Choi H-L (2018) Impact time control based on time-to-go prediction for sea-skimming anti-ship missiles. IEEE Trans Aerosp Electron Syst 54(4):2043–2052

Chapter 3
Optimal Trajectory Shaping Guidance Law with Seeker's Field-of-View Constraint

Abstract This chapter proposes a new trajectory shaping guidance law for impact time control and impact angle control with seeker's field-of-view (FOV) limit. The proposed guidance law is derived by using the concept of biased PNG (BPNG). The guidance law developed leverages a nonlinear function to ensure the boundedness of velocity lead angle to cater for seeker's FOV constraint. The finite-time convergence of the impact time error and impact angle error is theoretically analyzed. By investigating the optimality of the biased command, the physical meaning of the proposed trajectory shaping guidance law is revealed to support practical applications. Numerical simulations clearly demonstrate the effectiveness of the proposed formulation.

3.1 Introduction

Impact time control guidance enables simultaneous attack of multiple missiles and hence is a powerful strategy to counteract the close-in weapon systems of battleships. Impact angle control guidance is helpful in improving the warhead effectiveness by attacking a vulnerable spot on the target. Hence, shaping the trajectory for impact time and impact angle control is important for modern warfares and has attracted a significant level of attention in the guidance community.

By utilizing the optimal error dynamics, generalized optimal trajectory shaping guidance laws for impact time control and impact angle control are suggested in Chap. 2. From theoretical analysis and numerical simulations, it can be concluded that these trajectory shaping guidance laws require to make a detour away from the collision course to adjust the tracking error, i.e., impact time error and impact angle error. This might cause seeker's loss of target due to limited FOV constraint. For this reason, it is meaningful to design trajectory shaping guidance laws that not violate the seeker's FOV condition. Although there exist some interesting works in this field [1–4], most of them are heuristic and fail to address the optimality issue. In other words, it is difficult to analyze the physical meaning of these guidance laws and hence

S. He et al., *Optimal Guidance and Its Applications in Missiles and UAVs*, Springer Aerospace Technology, https://doi.org/10.1007/978-3-030-47348-8_3

there is a concern in practical applications. Note that analyzing the physical meaning of a guidance law is important in guaranteeing its reliability when implementing it in a real missile system.

In this chapter, a new optimal trajectory shaping guidance law against a stationary target with the consideration of seeker's FOV constraint is proposed. Similar to Chap. 2, the BPNG concept is utilized in deriving the guidance command. More specifically, the proposed guidance law consists of two main terms: the baseline PNG command and the tracking error feedback command. The error feedback term is determined by using the optimal error dynamics approach, developed in Chap. 2. A nonlinear function is leveraged to shape the error feedback term, ensuring the boundedness of velocity lead angle. Theoretical analysis reveals that the tracking error converges to zero at the time of impact and the velocity lead angle constraint is always satisfied under the proposed guidance law. Compared to optimal trajectory shaping guidance laws proposed in Chap. 2, the proposed algorithm actively constrains the seeker's FOV through a proper error dynamics and therefore extends the application of Chap. 2.

The key feature of the proposed guidance law lies in that the physical meaning can be clearly analyzed with the help of the optimal error dynamics. By formulating the corresponding function, we can easily investigate the command pattern of the proposed guidance law and the general behavior of trajectory shaping guidance laws. This result provides us theoretical guidelines on how to choose proper guidance gains to guarantee converged terminal guidance command.

3.2 Trajectory Shaping for Impact Time Control with Seeker's FOV Constraint

3.2.1 Problem Formulation

Note that the primary objective of a guidance law is to nullify the heading error such that the missile can follow a desired collision triangle to intercept the target. This condition can be mathematically formulated as

$$r\left(t_f\right) = 0 \tag{3.1}$$

where t_f denotes the final impact time.

The impact time constraint is satisfied when the final impact time t_f equals the desired impact time t_d as

$$t_f = t_d \tag{3.2}$$

In order to maintain target's lock-on, i.e., within the seeker's FOV, the seeker's look angle should not exceed its maximum permissible value. Since the angle-of-attack is usually very small during the endgame for typical anti-ship missiles, the seeker's FOV constraint can be approximated as

$$|\theta| \leq \theta_{\max} < \pi/2 \tag{3.3}$$

where $\theta_{\max}$ denotes the maximum permissible value of the velocity lead angle.

In summary, the problem considered in this section is to design an optimal guidance law to satisfy constraints (3.1)–(3.2) without violating condition (3.3). It should be pointed out that the desired impact time t_d cannot be set arbitrarily large due to seeker's FOV limit. With seeker's FOV constraint (3.3), it follows from Eq. (2.1) that

$$-V_M \leq \dot{r} \leq -V_M \cos\theta_{\max} \tag{3.4}$$

Therefore, the feasible range of achievable impact time under look angle constraint can be obtained using the formula, e.g., range over velocity, as

$$t_d \in \left(\frac{r_0}{V_M}, \frac{r_0}{V_M \cos\theta_{\max}}\right) \tag{3.5}$$

where r_0 denotes the initial value of the relative range.

3.2.2 *Impact Time Guidance Law Design*

In order to intercept the target with desired impact time t_d, the concept of BPNG is utilized in this chapter. That is, the proposed guidance command is composed of a baseline PNG part for target capture and a biased command for impact time error regulation as

$$a_M = a_{PNG} + a_{IT} \tag{3.6}$$

where a_{PNG} denotes the baseline PNG command for target interception and a_{IT} stands for the biased term to regulate the impact time error.

Define $\varepsilon_t = t_d - t_f$ as the impact time error. As derived in Chap. 2, the dynamics of ε_t under guidance command (3.6) is given by

$$\dot{\varepsilon}_t = -\frac{r\sin\theta}{(2N-1)V_M^2}a_{IT} \tag{3.7}$$

According to Theorem 2.1, the optimal desired error dynamics for a tracking error e is given as

$$\dot{e} + \frac{\rho(t)}{t_{go}}e = 0 \tag{3.8}$$

where $\rho(t) > 0$ is a nonlinear function to be designed.

Based on Eq. (3.8), consider the following desired error dynamics for system (3.7) to handle the seeker's FOV constraint.

$$\dot{\varepsilon}_t + \frac{K\phi\left(\theta/\theta_{\max}\right)}{t_{go}}\varepsilon_t = 0 \tag{3.9}$$

where $K > 0$ is the guidance gain that determines the convergence speed of the impact time error. $\phi(x)$ is a user-defined function to shape the velocity lead angle and satisfies the following condition.

Condition 3.1 The function $\phi(x)$ is defined on $[-1, 1]$ and satisfies $\phi(-1) = 0$, $\phi(1) = 0$ and $\phi(0) = 1$. Furthermore, function $\phi(x)$ monotonically increases when $x \in [-1, 0]$ and monotonically decreases when $x \in (0, 1]$.

Substituting Eq. (3.9) into Eq. (3.7) gives the biased guidance command to nullify the impact time error as

$$a_{IT} = \frac{K(2N-1)\phi\left(\theta/\theta_{\max}\right)V_m^2}{r\sin\theta t_{go}}\varepsilon_t \tag{3.10}$$

The proposed impact time control guidance law can then be obtained as

$$a_M = NV_M\dot{\sigma} + \frac{K(2N-1)\phi\left(\theta/\theta_{\max}\right)V_M^2}{r\sin\theta t_{go}}\varepsilon_t \tag{3.11}$$

It is clear that the nonlinear function $\phi(x)$ provides a way to shape the impact time error feedback command: the capacity in regulating the impact time error reduces when the magnitude of the velocity lead angle approaches its maximum permissible value $\theta_{\max}$. Furthermore, the proposed guidance law reduces to classical PNG when $|\theta| = \theta_{\max}$. Note that, when the proposed guidance law becomes PNG, the guidance law can keep satisfying seeker's FOV limit condition since the velocity lead angle starts to gradually decrease under PNG for a stationary target [5].

3.3 Analysis of Proposed Guidance Law

3.3.1 Optimality and Convergence of Impact Time Error

In this subsection, the optimality of the proposed guidance law and the convergence pattern of the impact time error will be analyzed theoretically.

Proposition 3.1 *Assuming that the error feedback gain $K\phi\left(\theta/\theta_{\max}\right)$ is instantaneous constant, then the biased guidance command a_{IT} is optimal in minimizing the performance index*

$$J = \frac{1}{2}\int_{t}^{t_f} \frac{r^2 \sin^2\theta}{\left(t_f - \tau\right)^{K\phi(\theta/\theta_{\max})-1}} a_{IT}^2 d\tau \tag{3.12}$$

Furthermore, the impact time error will converge to zero at the time of impact, i.e., $\varepsilon_t\left(t_f\right) = 0$ *under the proposed guidance law.*

Proof Denote

$$g(t) = -\frac{r \sin\theta}{(2N-1)\, V_M^2} \tag{3.13}$$

Then, according to Theorem 2.1, the biased command a_{IT}, shown in Eq. (3.10), is the solution of the following finite-time optimal regulation problem

$$\min_{a_{IT}} J = \frac{1}{2}\int_{t}^{t_f} W(\tau)\, a_{IT}^2(\tau)\, d\tau \tag{3.14}$$

subject to

$$\dot{\varepsilon}_t(t) = g(t)\, a_{IT}(t), \quad \varepsilon_t\left(t_f\right) = 0 \tag{3.15}$$

where the weighting function $W(t)$ satisfies

$$K\phi\left(\frac{\theta}{\theta_{\max}}\right) = \frac{t_{go} W^{-1}(t)\, g^2(t)}{\int_t^{t_f} W^{-1}(\tau)\, g^2(\tau)\, d\tau} \tag{3.16}$$

Under the assumption that the error feedback gain $K\phi\left(\theta/\theta_{\max}\right)$ is instantaneous constant for the purpose of analysis. Solving Eq. (3.16) gives the weighting function $W(t)$ as

$$W(t) = \frac{r^2 \sin^2\theta}{t_{go}^{K\phi(\theta/\theta_{\max})-1}} \tag{3.17}$$

Under optimal error dynamics (3.9), it is easy to verify that the closed-form solution of the impact time error is determined as

$$\varepsilon_t = \varepsilon_{t,0}\left(\frac{t_{go}}{t_f}\right)^{K\phi(\theta/\theta_{\max})} \tag{3.18}$$

where $\varepsilon_{t,0}$ denotes the initial value of the impact time error.

In the following section, we will show that $\phi\left(\theta/\theta_{\max}\right)$ is lower bounded by a small positive constant during the homing phase. Then, it follows from Eq. (3.18) that the impact time error ε_t will converge to zero at the time of impact, thus satisfying the impact time control requirement. Furthermore, the convergence rate of impact time error is determined by the guidance gain K. □

Remark 3.1 It is worth noting that the result of Proposition 3.1 is based on the approximation of the instantaneous constant for the purpose of theoretical analysis.

Therefore, the results obtained might not be exactly the same as the actual ones. However, we would say that the results obtained are enough to grasp the overall characteristics of the proposed guidance law.

3.3.2 Velocity Lead Angle Analysis

In this subsection, we will analyze the properties of velocity lead angle θ under the proposed guidance law to show that condition (3.3) is always satisfied. The results are presented in the following proposition.

Proposition 3.2 *If the initial velocity lead angle* θ_0 *satisfies* $|\theta_0| \leq \theta_{\max}$, *then the proposed guidance law ensures that* $|\theta| < \theta_{\max}$ *for all* $t > 0$.

Proof In the following, Proposition 3.2 will be proved in two steps. Step 1 will show $|\theta| \leq \theta_{\max}$ holds for all $t > 0$ if $|\theta_0| \leq \theta_{\max}$ and step 2 will subsequently demonstrate the velocity lead angle θ will never achieve the maximum permissible value $\theta_{\max}$ for all $t > 0$ under the proposed guidance law if $|\theta_0| \leq \theta_{\max}$.

Step 1. Differentiating θ with respect to time and substituting Eq. (3.11) into it yields

$$\begin{aligned}\dot{\theta} &= \dot{\gamma} - \dot{\sigma} \\ &= -\frac{(N-1)\,V_M \sin\theta}{r} + \frac{K\,(2N-1)\,\phi\,(\theta/\theta_{\max})\,V_M}{r \sin\theta t_{go}}\varepsilon_t\end{aligned} \tag{3.19}$$

For system (3.19), consider $\Omega = 0.5\theta^2$ as a Lyapunov function candidate. Taking the time derivative of V results in

$$\begin{aligned}\dot{\Omega} &= -\frac{(N-1)\,V_M\theta \sin\theta}{r} + \frac{K\,(2N-1)\,\phi\,(\theta/\theta_{\max})\,V_M\theta}{r \sin\theta t_{go}}\varepsilon_t \\ &= -\frac{(N-1)\,V_M\theta \sin\theta}{r} + \frac{a_{IT}\theta}{V_M}\end{aligned} \tag{3.20}$$

Since $\phi(-1) = 0$ and $\phi(1) = 0$, we have

$$\dot{\Omega}\,\Big|_{\theta=\pm\theta_{\max}} = -\frac{(N-1)\,V_M\theta_{\max} \sin\theta_{\max}}{r} < 0 \tag{3.21}$$

which reveals that $\Theta = \{\theta\,|\theta \leq \theta_{\max}\}$ is a positively invariant set. That is, $|\theta| \leq \theta_{\max}$ holds for all $t > 0$ if $\theta_0 \in \Theta$.

Step 2. Without loss of any generality, assume that $|\theta_0| \leq \theta_{\max} - c_1$ with c_1 being a small constant. Note that if $|\theta_0| = \theta_{\max}$, there exists a certain time $\bar{t}$ such that $|\theta\left(\bar{t}\right)| = \theta_{\max} - c_1$ since $\dot{\Omega}\,\big|_{\theta=\pm\theta_{\max}} < 0$. Next, the following two cases are considered.

Case 1: $a_{IT}\theta < 0$. Under this condition, it follows from Eq. (3.20) that $\dot{\Omega} < 0$, which means $|\theta| < \theta_{\max}$ is true.

Case 2: $a_{IT}\theta \geq 0$. If $\dot{\Omega} > 0$, there exists a small constant c_2 such that $\dot{\Omega}\big|_{|\theta|=\theta_{\max}-c_2} = 0$ since function $\dot{\Omega}$ is continuous with respect to θ and $\dot{\Omega}\big|_{\theta=\pm\theta_{\max}} < 0$. Therefore, we can readily conclude that the velocity lead angle θ is constrained by $[-\theta_{\max} + c_2, \theta_{\max} - c_2]$ for all $t > 0$. This means that the inequality $|\theta| < \theta_{\max}$ holds when $a_{IT}\theta \geq 0$.

Finally, Combining the results of Cases 1 and 2 leads to the proof of Proposition 3.2. □

Remark 3.2 Proposition 3.2 reveals that the velocity lead angle constraint (3.3) is satisfied under the proposed impact time guidance law, demonstrating that the proposed guidance law is able to handle the seeker's FOV constraint. Furthermore, Proposition 3.2 also indicates that the error feedback gain $K\phi\left(\theta/\theta_{\max}\right)$ is lower bounded by a positive constant during the homing phase.

Proposition 3.3 *The proposed guidance law enables intercepting a target with a desired impact time, i.e.,* $r\left(t_d\right) = r\left(t_f\right) = 0$.

Proof According to Proposition 3.2, one can imply that $\cos\theta \geq \cos\theta_{\max} > 0$. Then, it follows from Eq. (2.1) that r decreases monotonically. This reveals that there exists a certain time t_f such that $r\left(t_f\right) = 0$. Since Proposition 3.1 shows that $\varepsilon_t\left(t_f\right) = 0$, it can be concluded that $r\left(t_d\right) = r\left(t_f\right) = 0$. Therefore, the proposed guidance law enables intercepting a target with a desired impact time. □

The convergence of the velocity lead angle is given in the following proposition.

Proposition 3.4 *Under the proposed guidance law, the velocity lead angle* θ *will converge to zero at the time of impact, i.e.,* $\theta\left(t_f\right) = 0$.

Proof Assume that $\theta\left(t_f\right) \neq 0$, it follows from Eq. (3.20) and $\varepsilon_t\left(t_f\right) = 0$ that

$$\dot{\Omega}\big|_{t=t_f} = -(N-1)\,V_M\theta\sin\theta/r \tag{3.22}$$

As we discussed in Proposition 3.3, $r\left(t_d\right) = r\left(t_f\right) = 0$. Therefore, the condition $\dot{\Omega}\big|_{t=t_f} = -\infty$ holds. Notice that the function $\dot{\Omega}$ is continuous with respect to time. Thus, there exists a certain time $t_1, t_1 < t_f$, such that $\dot{\Omega} < 0$ holds for all $t \in \left[t_1, t_f\right]$. From Proposition 3.2, we know that the lead angle θ is uniformly bounded with $|\theta| \leq \theta_{\max}$ for all $t > 0$, which means that $\Omega\left(t_1\right)$ is bounded. With this in mind, it follows from the fact of $\dot{\Omega}\big|_{t=t_f} = -\infty$ and the continuity of $\dot{\Omega}$ that there exists a certain time t_2, $t_1 < t_2 < t_f$, such that the condition $\dot{\Omega} \leq -\Omega\left(t_1\right)/\left(t_f - t_1\right)$ holds for all $t \in \left[t_2, t_f\right]$. Therefore, we have

$$\begin{aligned}\Omega\left(t_f\right) &= \Omega\left(t_1\right) + \int_{t_1}^{t_f}\dot{\Omega}dt\\ &\leq \Omega\left(t_1\right) - \frac{\Omega\left(t_1\right)}{t_f - t_1}\left(t_f - t_1\right)\\ &\leq 0\end{aligned} \tag{3.23}$$

By the definition Ω, it is clear that $\Omega\left(t_f\right)$ is non-negative, i.e., $\Omega\left(t_f\right) \geq 0$. Then, it can be concluded that $\Omega\left(t_f\right) = 0$. Therefore, the velocity lead angle θ will converge to zero at the time of impact, i.e., $\theta\left(t_f\right) = 0$. □

3.3.3 Guidance Command Analysis

From Eq. (3.12), it can be noted that the time-to-go power order $K\phi\left(\theta/\theta_{\max}\right) - 1$ will become negative when $\phi\left(\theta/\theta_{\max}\right)$ is close to zero. This is obviously not desired to guarantee a finite guidance command when the missile approaches the target. However, Proposition 3.4 shows that the velocity lead angle θ will converge to zero at the time of impact, i.e., $\theta\left(t_f\right) = 0$, which means that $\phi\left(\theta\left(t_f\right)/\theta_{\max}\right) = 1$. This condition enables the possibility of choosing K to guarantee a finite and even zero final guidance command, thus providing enough operational margins to cope with undesired external disturbances when the missile is close to the target. From Proposition 3.1, one can conclude that the performance index of the biased command can be approximated as

$$J \approx \frac{1}{2}\int_t^{t_f} \frac{r^2\sin^2\theta}{\left(t_f - \tau\right)^{K-1}} a_{IT}^2 d\tau \tag{3.24}$$

Following similar analysis, as shown in Chap. 2, it can be readily verified that the proposed guidance law guarantees a zero final guidance command when the guidance gain satisfies condition $K - 1 > 2N$.

From Eq. (3.11), we can observe that $\theta = 0$ is a singular point, which will result in infinite guidance command. However, it is easy to verify that this singular point is trivial since the velocity lead angle $\theta \neq 0$ except for the final impact point. To see this, assume that there exists a certain time t_3, $0 < t_3 < t_f$ such that $\theta = 0$ holds for all $t \in \left[t_3, t_f\right]$, then, the missile will fly along a straight line to hit the target when $t \geq t_3$. This means that the impact time error ε_t converges to zero at $t = t_3$. However, it follows from Proposition 3.1 that the impact time error ε_t becomes zero only at the time of impact. Therefore, it can be readily concluded that $t_3 = t_f$, meaning that the point $\theta = 0$ is trivial.

3.3.4 Selection of $\phi(x)$

From previous analysis, one can note that the function $\phi(x)$ plays an important role in the proposed guidance law since it determines the convergence pattern of the impact time error and ensures the boundedness of the velocity lead angle. For this reason, this subsection will provide the details of how to choose this specific function. Any

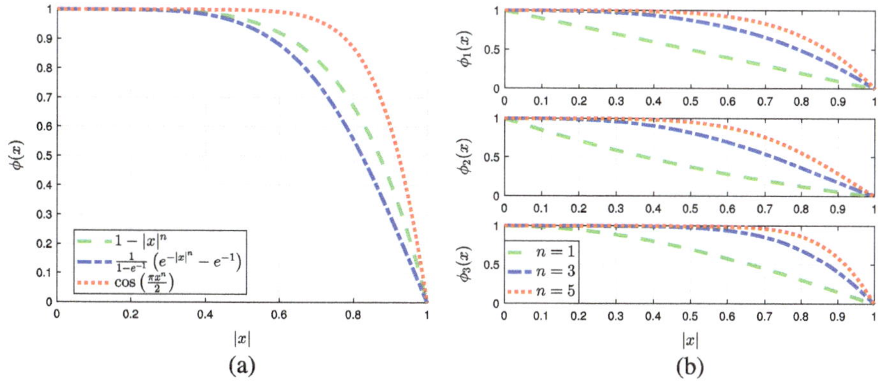

Fig. 3.1 Examples of $\phi(x)$. **a** profiles of these three different candidate functions with $n = 5$; and **b** profiles of these three different candidate functions with different n

function of $\phi(x)$ that satisfies Condition 3.1 can be utilized for implementing the proposed guidance law. Candidates of $\phi(x)$ could be

$$\phi_1(x) = 1 - |x|^n \tag{3.25}$$

$$\phi_2(x) = \frac{1}{1-e^{-1}}\left(e^{-|x|^n} - e^{-1}\right) \tag{3.26}$$

$$\phi_3(x) = \cos\left(\frac{\pi x^n}{2}\right) \tag{3.27}$$

where $n > 0$ is a design parameter to control the curvature of $\phi(x)$. Figure 3.1 shows the profiles of these three different candidate functions.

Remark 3.3 With larger n, $\phi(x)$ is closer to 1 when $|x| < 1$ and converges to zero more sharply when $|x|$ approaches 1. Therefore, the parameter n is a trade-off design between desired impact time and velocity lead angle margin: larger n provides the possibility of achieving longer desired impact time while smaller n generates more operational margins for the velocity lead angle.

Remark 3.4 Note that the form of the function $\phi(x)$ suggested in Eqs. (3.25)–(3.27) are proposed heuristically and many other meaningful candidate functions that satisfy Condition 3.1 can be utilized. It is worth pointing out the design of function $\phi(x)$ depends on the capability of the interceptor. For example, the candidate $\phi_3(x)$, shown in Eq. (3.27), is more favorable to achieve longer desired impact time as this function provides longer duration to explore the maximum look angle in the homing phase, compared with functions $\phi_1(x)$ and $\phi_2(x)$. This can be clearly confirmed by Fig. 3.1a. However, if the maneuverability of the missile is limited, candidates $\phi_1(x)$ and $\phi_2(x)$ are more favorable to save energy consumption.

3.4 Numerical Simulations

In this section, the effectiveness of the proposed impact time guidance law is demonstrated through numerical simulations, in which an anti-ship missile is considered to intercept a stationary target. The initial conditions of the considered scenario are summarized in Table 3.1. For implementing the proposed impact time guidance law, the navigation gain of the PNG command is chosen as $N = 3$ and the fixed gain of the impact time error feedback term is selected as $K = 8 > 2N + 1$ in all simulations. The nonlinear function $\phi(x)$ is set as $\phi(x) = \cos\left(\frac{\pi x^5}{2}\right)$ without any further tuning. Since the missile achievable acceleration is always bounded due to physical constraint, the acceleration command signal is upper bounded by $100\,\mathrm{m/s^2}$ in our simulations.

3.4.1 *Performance with Different Impact Times*

In this subsection, we validate the proposed guidance law with various impact times $t_d = 35\,\mathrm{s}, 37\,\mathrm{s}, 40\,\mathrm{s}, 45\,\mathrm{s}$. The maximum permissible lead angle is set as $\theta_{\max} = 50°$. Figure 3.2a compares the interception trajectory obtained from different desired impact times, showing that a longer and more curved path is required for a larger impact time. The impact time errors recorded in simulations under the considered four different conditions are less than 0.05s. This means that the proposed guidance law can successfully guide the missile to intercept the target with accurate impact time constraint. The missile acceleration command is presented in Fig. 3.2b. Clearly, more energy consumption is required during the initial phase for a larger t_d. For this reason, the duration of acceleration saturation is longer with larger desired impact time. Additionally, the guidance commands converge to zero at the time of impact in all the considered scenarios. Figure 3.2c shows the profile of velocity lead angle under the proposed guidance law. This figure reveals that the magnitude of the velocity lead angle initially increases to reduce the impact time error and finally converge to zero at the time of impact to ensure target interception.

Table 3.1 Initial conditions for homing engagement

Parameters	Values
Missile initial position	(0 m, 0 m)
Missile flight velocity	300 m/s
Missile initial flight path angle	30°
Target position	(10000 m, 0 m)

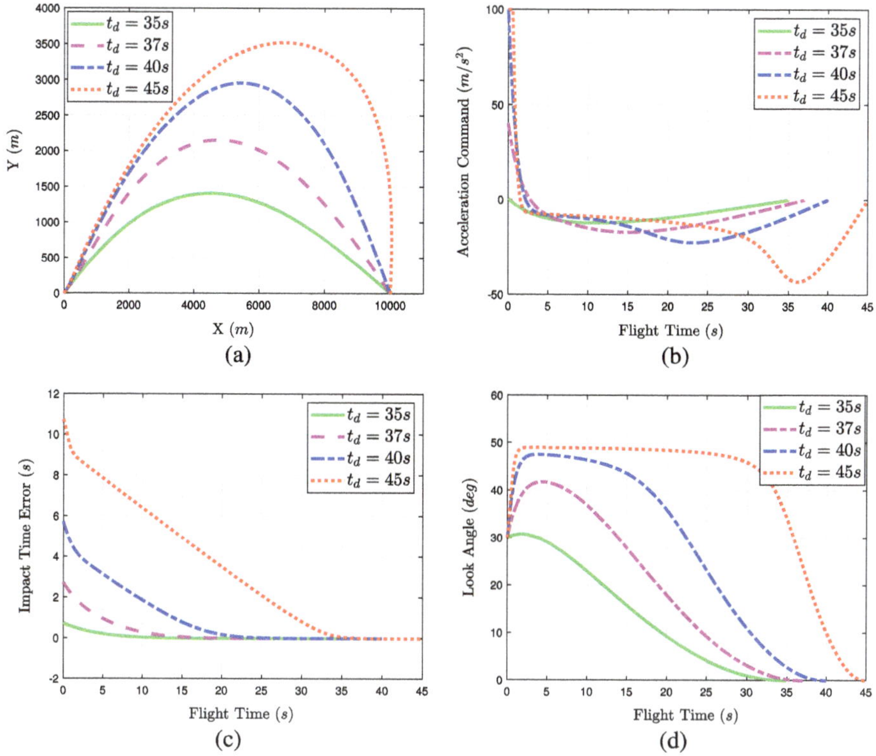

Fig. 3.2 Simulation results of the proposed guidance law with various impact times. **a** Interception trajectory; **b** Acceleration command; **c** Impact time error; and **d** Velocity lead angle

3.4.2 Performance with Different Velocity Lead Angle Constraints

In this subsection, the performance of the proposed guidance law is evaluated with various velocity lead angle constraints $\theta_{\max} = 40°, 45°, 50°, 55°$. The desired impact time is set as $t_d = 40$ s. The simulation results, including interception trajectories, missile acceleration commands, impact time error, and velocity lead angle profiles, are presented in Fig. 3.3, demonstrating that the missile successfully intercepts the target at the specific time with various lead angle constraints. This figure also reveals that the proposed guidance law tries to leverage the maximum available lead angle to reduce the impact time error during the initial period of the flight. For this reason, the interception trajectory is more curved and more control energy is required during the initial period with larger $\theta_{\max}$. With smaller $\theta_{\max}$, however, a slightly sharper turn of the acceleration command can be noted from Fig. 3.3b when the missile approaches the target. The reason of this fact is that the interceptor with smaller $\theta_{\max}$ has shorter time in regulating the lead angle to zero.

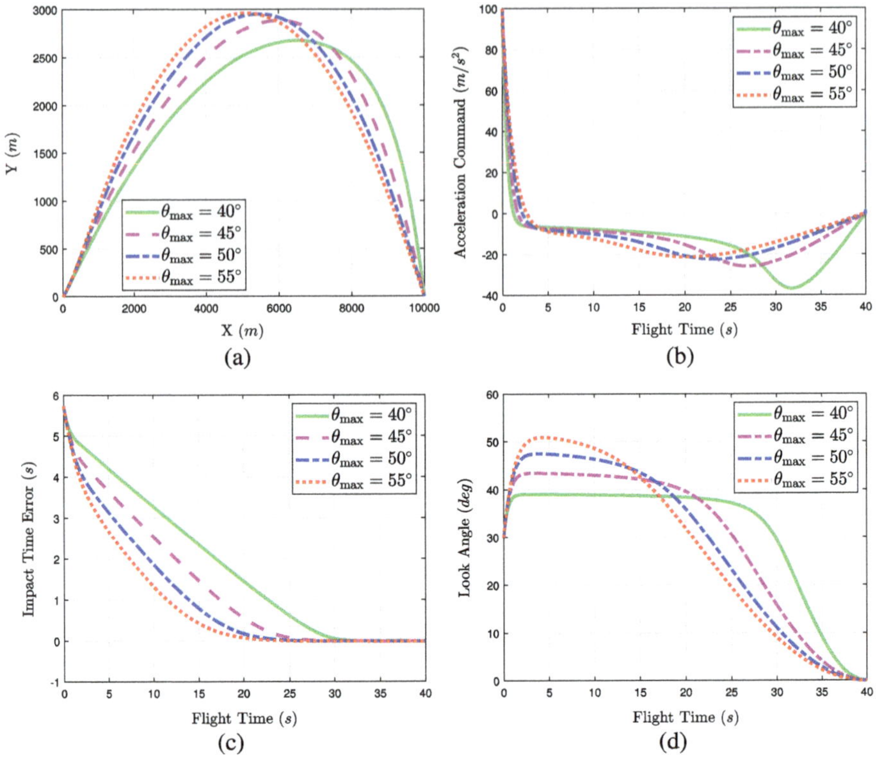

Fig. 3.3 Simulation results of the proposed guidance law with various velocity lead angle constraints. **a** Interception trajectory; **b** Acceleration command; **c** Impact time error; and **d** Velocity lead angle

3.4.3 Comparison with Other Guidance Laws

To further show the superiority of the proposed impact time guidance law, comparison simulation is conducted in this subsection. In the simulations, the desired impact time and maximum look angle are set as $t_d = 45\,\mathrm{s}$ and $\theta_{\max} = 55°$. For the purpose of comparison, the impact-time-control guidance considering the seeker's FOV limits based on the backstepping control technique [6] is implemented. The guidance command of this method is given as

$$a_M = V_M \dot{\sigma} + \frac{\dot{V}_R^d}{\sin\theta} - \frac{k_2\left(-V_M\cos\theta - V_R^d\right)}{\sin\theta} \tag{3.28}$$

where

$$V_R^d = -V_M + k_1 \mathrm{sgmf}\left(V_M\left(t_d - t\right) - r\right) \tag{3.29}$$

and the sigmoid function sgmf is selected as

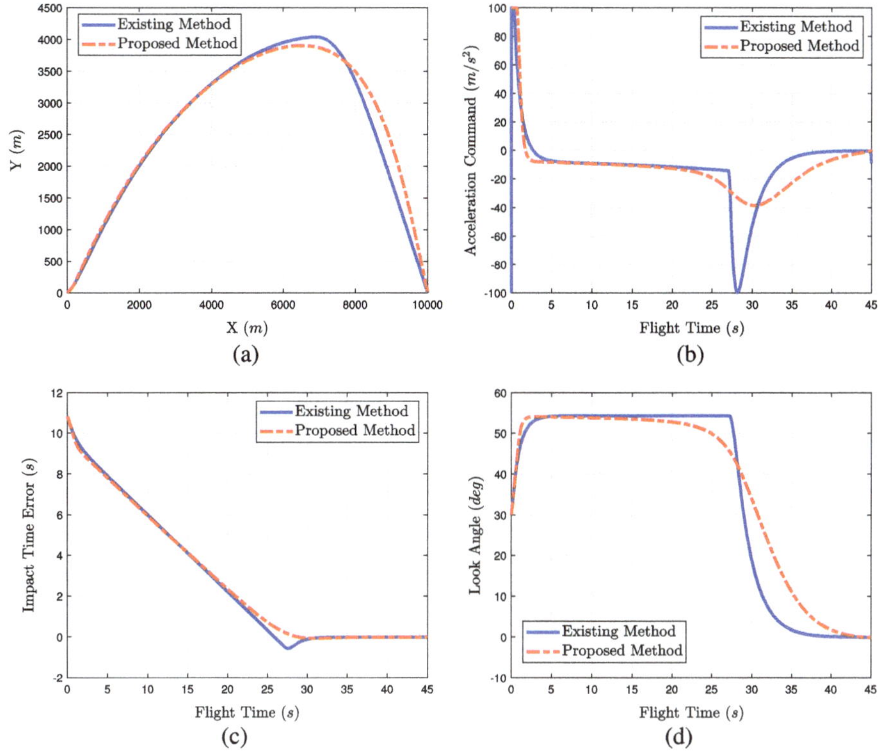

Fig. 3.4 Comparison results between the proposed method and the existing method. **a** Interception trajectory; **b** Acceleration command; **c** Impact time error; and **d** Velocity lead angle

$$\mathrm{sgmf}\,(x) = \begin{cases} -\frac{1}{2\eta^3}x^3 + \frac{3}{2\eta}x & \text{if } |x| \le \eta \\ \mathrm{sgn}\,(x) & \text{else} \end{cases} \tag{3.30}$$

where k_1, k_2, and η represents the design parameters. For fair comparison, the design parameters given in [6] are used: $k_1 = 125$, $k_2 = 1$, and $\eta = 200$. More detailed information can be found in [6].

The simulation results obtained from these two different guidance laws are presented in Fig. 3.4, which clearly reveals that all guidance laws can successfully drive the missile to intercept the target with desired impact time. However, the difference between these two guidance laws lies in the fact that the existing method shows an abrupt change of the guidance command, as can be confirmed by Fig. 3.4b. This characteristic is unfavorable for the guidance loop because it might cause the guidance loop instability [7]. The quantitative comparisons of control effort, defined as $\int_0^{t_f} a_M^2 dt$ are summarized in Table 3.2. From this table, it can be readily noted that the proposed guidance law requires less energy consumption, compared with the existing method. This is because the proposed method is developed based on the concept of optimal error dynamics.

Table 3.2 Quantitative comparisons of control effort

	Existing method	Proposed method
Control effort	30283	23554

3.5 Trajectory Shaping for Impact Angle Control with Seeker's FOV Constraint

3.5.1 *Problem Formulation*

The impact angle constraint is satisfied when the terminal flight path angle γ_{M_f} equals the desired impact angle γ_f as

$$\gamma_{M_f} = \gamma_f \tag{3.31}$$

Similarly, the seeker's look angle should not exceed its maximum permissible value to cater for the seeker's FOV constraint as

$$|\theta| \leq \theta_{\max} < \pi/2 \tag{3.32}$$

In summary, the problem considered in this section is to design an optimal guidance law to satisfy constraints (3.1), (3.31) without violating condition (3.32).

3.5.2 *Impact Angle Guidance Law Design*

Similar to Chap. 2, the BPNG concept is utilized to design an impact angle guidance law. For this reason, we formulate the guidance command in a composite form as

$$a_M = a_{PNG} + a_{IA} \tag{3.33}$$

where a_{PNG} denotes the baseline PNG command for target interception and a_{IA} stands for the biased term to regulate the impact angle error.

Define $\varepsilon_\gamma = \gamma_f - \gamma_{M_f}$ as the impact angle error. As derived in Chap. 2, the dynamics of ε_γ under guidance command (3.33) is given by

$$\dot{\varepsilon}_\gamma = \frac{a_{IA}}{(N-1)\,V_M} \tag{3.34}$$

Similar to 3.5, we choose the following desired error dynamics to handle the seeker's FOV constraint.

$$\dot{\varepsilon}_\gamma + \frac{K\phi\left(\theta/\theta_{\max}\right)}{t_{go}}\varepsilon_\gamma = 0 \tag{3.35}$$

where $K > 0$ is the guidance gain that determines the convergence speed of the impact angle error. $\phi(x)$ is a user-defined function to shape the velocity lead angle and satisfies Condition 3.1.

Substituting Eq. (3.35) into Eq. (3.34) gives the biased guidance command to nullify the impact angle error as

$$a_{IA} = -\frac{K\,(N-1)\,\phi\,(\theta/\theta_{\max})\,V_M}{t_{go}}\varepsilon_\gamma \tag{3.36}$$

The proposed impact angle control guidance law can then be obtained as

$$a_M = N V_M \dot{\sigma} - \frac{K\,(N-1)\,\phi\,(\theta/\theta_{\max})\,V_M}{t_{go}}\varepsilon_\gamma \tag{3.37}$$

which reveals that the proposed guidance law reduces to classical PNG when $|\theta| = \theta_{\max}$.

3.6 Analysis of Proposed Guidance Law

In this section, the characteristics of the proposed guidance law is analyzed to provide better insights of the working principle of impact angle control guidance laws. To begin with, the optimality of the proposed guidance law and the convergence pattern of the impact angle error is given in the following proposition.

Proposition 3.5 *Assuming that the error feedback gain $K\phi\,(\theta/\theta_{\max})$ is instantaneous constant, then the biased guidance command a_{IA} is optimal in minimizing the performance index*

$$J = \frac{1}{2}\int_t^{t_f} \frac{1}{\left(t_f-\tau\right)^{K\phi(\theta/\theta_{\max})-1}} u^2\,(\tau)\,d\tau \tag{3.38}$$

Furthermore, the impact angle error will converge to zero at the time of impact, i.e., $\varepsilon_\gamma\left(t_f\right) = 0$ under the proposed guidance law.

Proof Denote

$$g(t) = \frac{1}{(N-1)\,V_M} \tag{3.39}$$

Then, according to Theorem 2.1, the biased command a_{IA}, shown in Eq. (3.36), is the solution of the following finite-time optimal regulation problem

$$\min_{a_{IA}} J = \frac{1}{2}\int_t^{t_f} W\,(\tau)\,a_{IA}^2\,(\tau)\,d\tau \tag{3.40}$$

subject to

$$\dot{\varepsilon}_\gamma(t) = g(t)\, a_{IT}(t), \quad \varepsilon_\gamma(t_f) = 0 \tag{3.41}$$

where the weighting function $W(t)$ satisfies

$$K\phi\left(\frac{\theta}{\theta_{\max}}\right) = \frac{t_{go} W^{-1}(t)\, g^2(t)}{\int_t^{t_f} W^{-1}(\tau)\, g^2(\tau)\, d\tau} \tag{3.42}$$

Under the assumption that the error feedback gain $K\phi(\theta/\theta_{\max})$ is instantaneous constant for the purpose of analysis. Solving Eq. (3.42) gives the weighting function $W(t)$ as

$$W(t) = \frac{1}{t_{go}^{K\phi(\theta/\theta_{\max})-1}} \tag{3.43}$$

Under optimal error dynamics (3.35), it is easy to verify that the closed-form solution of the impact angle error is determined as

$$\varepsilon_\gamma = \varepsilon_{\gamma,0}\left(\frac{t_{go}}{t_f}\right)^{K\phi(\theta/\theta_{\max})} \tag{3.44}$$

where $\varepsilon_{\gamma,0}$ denotes the initial value of the impact angle error.

In Proposition 3.6, we will show that $\phi(\theta/\theta_{\max})$ is lower bounded by a small positive constant during the homing phase. Then, it follows from Eq. (3.44) that the impact angle error ε_γ will converge to zero at the time of impact, thus satisfying the impact angle control requirement. □

Proposition 3.6 *If the initial velocity lead angle θ_0 satisfies $|\theta_0| \le \theta_{\max}$, then the proposed guidance law ensures that $|\theta| < \theta_{\max}$ for all $t > 0$.*

Proof Similar to the proof of Proposition 3.2, the proof of Proposition 3.6 will be also given in two steps. Step 1 will show $|\theta| \le \theta_{\max}$ holds for all $t > 0$ if $|\theta_0| \le \theta_{\max}$ and step 2 will subsequently demonstrate the velocity lead angle θ will never achieve the maximum permissible value $\theta_{\max}$ for all $t > 0$ under the proposed guidance law if $|\theta_0| \le \theta_{\max}$.

Step 1. Differentiating θ with respect to time and substituting guidance command (3.37) into it yields

$$\begin{aligned}\dot{\theta} &= \dot{\gamma} - \dot{\sigma} \\ &= -\frac{(N-1)\, V_M \sin\theta}{r} - \frac{K(N-1)\,\phi(\theta/\theta_{\max})}{t_{go}}\varepsilon_\gamma\end{aligned} \tag{3.45}$$

For system (3.45), consider $\Omega = 0.5\theta^2$ as a Lyapunov function candidate. Taking the time derivative of V results in

$$\begin{aligned}\dot{\Omega} &= -\frac{(N-1)\,V_M\theta\sin\theta}{r} - \frac{K\,(N-1)\,\phi\,(\theta/\theta_{\max})\,\theta}{t_{go}}\varepsilon_\gamma \\ &= -\frac{(N-1)\,V_M\theta\sin\theta}{r} + \frac{a_{IA}\theta}{V_M}\end{aligned} \tag{3.46}$$

Since $\phi(-1) = 0$ and $\phi(1) = 0$, we have

$$\dot{\Omega}\,\Big|_{\theta=\pm\theta_{\max}} = -\frac{(N-1)\,V_M\theta_{\max}\sin\theta_{\max}}{r} < 0 \tag{3.47}$$

which reveals that $\Theta = \{\theta\,|\theta \le \theta_{\max}\}$ is a positively invariant set. That is, $|\theta| \le \theta_{\max}$ holds for all $t > 0$ if $\theta_0 \in \Theta$.

Step 2. Without loss of any generality, assume that $|\theta_0| \le \theta_{\max} - c_1$ with c_1 being a small constant. Note that if $|\theta_0| = \theta_{\max}$, there exists a certain time $\bar{t}$ such that $|\theta\,(\bar{t})\,| = \theta_{\max} - c_1$ since $\dot{\Omega}\,\big|_{\theta=\pm\theta_{\max}} < 0$. Next, the following two cases are considered.

Case 1: $a_{IA}\theta < 0$. Under this condition, it follows from Eq. (3.46) that $\dot{\Omega} < 0$, which means $|\theta| < \theta_{\max}$ is true.

Case 2: $a_{IA}\theta \ge 0$. If $\dot{\Omega} > 0$, there exists a small constant c_2 such that $\dot{\Omega}\,\big|_{|\theta|=\theta_{\max}-c_2} = 0$ since function $\dot{\Omega}$ is continuous with respect to θ and $\dot{\Omega}\,\big|_{\theta=\pm\theta_{\max}} < 0$. Therefore, we can readily conclude that the velocity lead angle θ is constrained by $[-\theta_{\max} + c_2, \theta_{\max} - c_2]$ for all $t > 0$. This means that the inequality $|\theta| < \theta_{\max}$ holds when $a_{IA}\theta \ge 0$.

Finally, Combining the results of Cases 1 and 2 leads to the proof of Proposition 3.6. □

Remark 3.5 Proposition 3.6 indicates that the error feedback gain $K\phi\,(\theta/\theta_{\max})$ is lower bounded by a positive constant during the homing phase and hence supports the proof of Proposition 3.5.

Remark 3.6 In light of Proposition 3.5, it can be implied that the proposed guidance law gradually converges to conventional PNG when the missile approaches the target and therefore guarantees target interception.

Notice that Proposition 3.6 reveals that the velocity lead angle θ is always bounded during the homing phase. Since the proposed guidance law converges to conventional PNG at the time of impact, we can easily verify that $\phi\left(\theta\left(t_f\right)/\theta_{\max}\right) = 1$ since PNG ensures converged velocity lead angle. This condition enables the possibility of choosing K to guarantee a finite and even zero final guidance command, thus providing enough operational margins to cope with undesired external disturbances when the missile is close to the target. From Proposition 3.5, one can conclude that the performance index of the biased command can be approximated as

$$J \approx \frac{1}{2}\int_{t}^{t_f} \frac{1}{\left(t_f - \tau\right)^{K-1}} a_{IA}^2 d\tau \tag{3.48}$$

which indicates that the biased command converges to zero at the time of impact when the guidance gain satisfies condition $K > 1$.

3.7 Numerical Simulations

In this section, nonlinear numerical simulations are conducted to validate the effectiveness of the proposed impact angle guidance law. The initial condition of the considered scenario is the same as Sect. 3.4. The design parameters of the proposed guidance law are set as $N = 4$ and $K = 2$. The magnitude of acceleration command a_M is constrained by 100 m/s^2 to cater for the physical limits of practical interceptors in simulations.

3.7.1 Performance with Different Impact Angles

In this subsection, we investigate the performance of the proposed trajectory shaping guidance law with various impact angles $\gamma_f = -60°, -90°, -120°, -150°$. The maximum permissible value of the look angle is set as $\theta_{\max} = 50°$ in the simulations. The simulation results, including interception trajectory, response curve of flight path angle, lateral acceleration command and time history of the velocity lead angle, are presented in Fig. 3.5. This figure clearly reveals that the proposed optimal trajectory shaping guidance law can successfully guide the missile to intercept the target with desired impact angles without violating the seeker's FOV constraint. From Fig. 3.5c, we can note that the magnitude of acceleration command increases significantly during the terminal homing phase when the magnitude of the desired impact angle increases. From this figure, it can be also observed that the guidance command with $N = 4$ and $K = 2$ converges to zero at the time of impact. This result clearly validate our analytical findings.

3.7.2 Performance with Different Velocity Lead Angle Constraints

In this subsection, the performance of the proposed guidance law is evaluated with various velocity lead angle constraints $\theta_{\max} = 40°, 45°, 50°, 55°$. The desired impact angle is set as $\gamma_f = -90°$. The simulation results, including interception trajectories, flight path angle, missile acceleration command, and velocity lead angle profiles, are

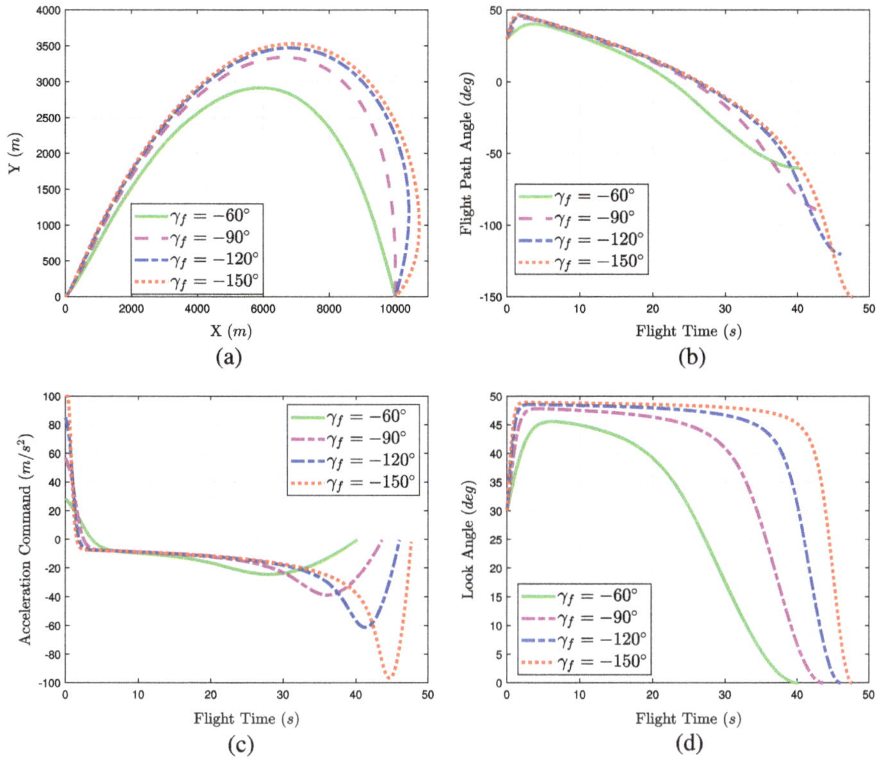

Fig. 3.5 Simulation results of the proposed guidance law with various impact angles. **a** Interception trajectory; **b** Flight path angle; **c** Acceleration command; and **d** Velocity lead angle

presented in Fig. 3.6. This figure clearly reveals that the proposed guidance law can be utilized for impact angle control with various lead angle constraints. The proposed trajectory shaping guidance law, with larger maximum permissible value of the seeker's FOV constraint, generates a more curved and longer interception trajectory. In contrast, decreasing the maximum look angle results in a sharper turn of the acceleration command when the missile is close to the target, as shown in Fig. 3.6c. This fact is obviously not desirable for the onboard autopilot system since it will exaggerate the adverse effect of the control delay.

3.7.3 Comparison with Other Guidance Laws

This subsection compares the performance of the proposed guidance law with that of [1]. When implementing the switching logic [1], the guidance command switches between the optimal optimal impact angle guidance law (2.77) and constant look angle guidance law $a_M = V_M\dot{\sigma}$ to cater for the seeker's FOV limit as

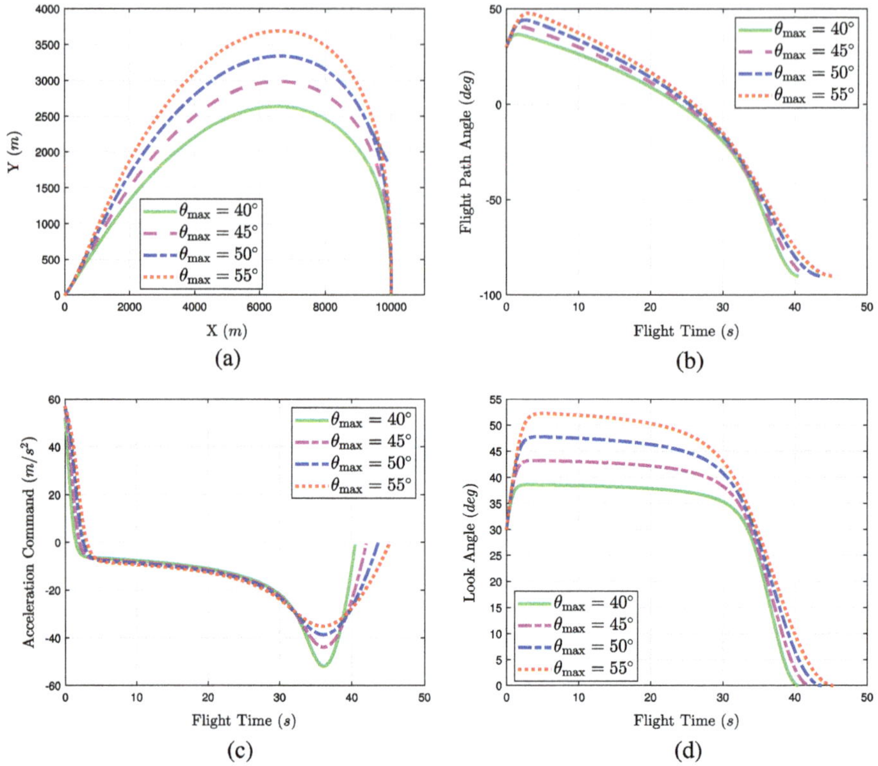

Fig. 3.6 Simulation results of the proposed guidance law with various velocity lead angle constraints. **a** Interception trajectory; **b** Flight path angle; **c** Acceleration command; and **d** Velocity lead angle

$$a_M = \begin{cases} a_{IAC}, & |\theta| < \theta_{\max} \text{ or } \dfrac{V_M \sin\theta}{r} + \dfrac{a_{IAC}}{V_M} < 0 \\ V_M \dot{\sigma}, & \text{otherwise} \end{cases} \tag{3.49}$$

where a_{IAC} denotes the optimal impact angle guidance law (2.77)

$$a_{IAC} = N V_M \dot{\sigma} - \frac{K(N-1)V_M}{t_{go}} \varepsilon_\gamma \tag{3.50}$$

For fair comparison, the guidance gains in all guidance laws are set as $N = 4$ and $K = 2$. The maximum permissible value of velocity lead and desired impact angle are set as $\theta_{\max} = 50^\circ$ and $\gamma_f = -90^\circ$.

The simulation results obtained from these two different guidance laws are presented in Fig. 3.7, which clearly reveals that all guidance laws can successfully drive the missile to intercept the target at a particular orientation. However, the switching logic [1] generates inevitable sharp command change and hence might pose adverse

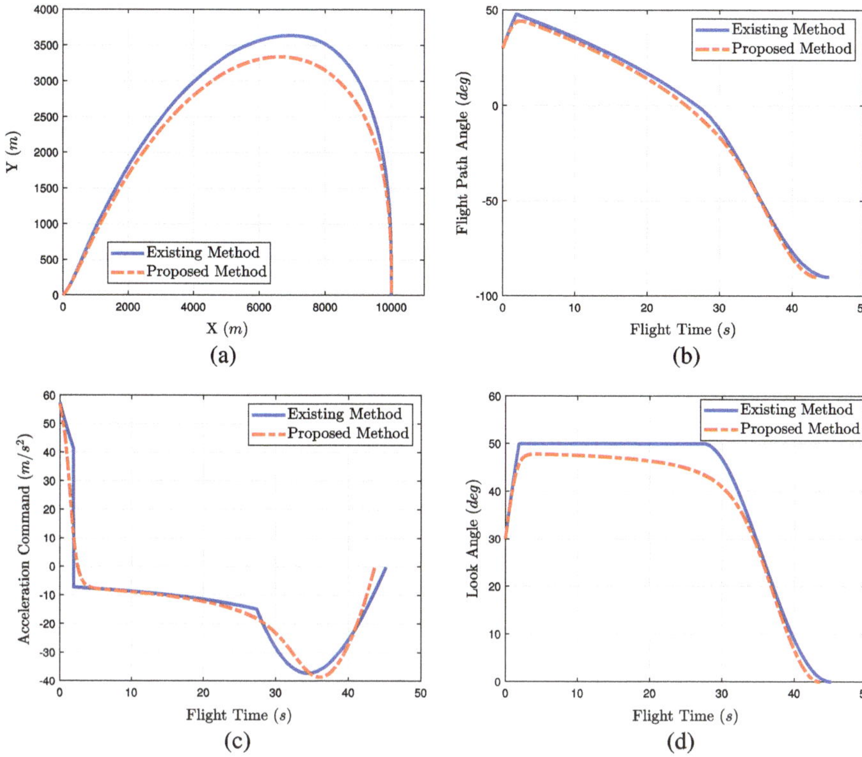

Fig. 3.7 Comparison results between the proposed method and the existing method. **a** Interception trajectory; **b** Flight path angle; **c** Impact time error; and **d** Velocity lead angle

Table 3.3 Quantitative comparisons of control effort

	Existing method	Proposed method
Control effort	21292	19401

effect on to onboard autopilot system, as can be confirmed by Fig. 3.7b. In contrast, the proposed guidance law provides smooth commanded acceleration, which is easy to be tracked by a well-designed control system. The quantitative comparisons of control effort, defined as $\int_0^{t_f} a_M^2 dt$ are summarized in Table 3.3. From this table, it can be readily noted that the proposed guidance law requires less energy consumption, compared with the existing method. This can be attributed to the fact that the proposed method is developed based on the concept of optimal error dynamics.

3.8 Summary

In this chapter, the problem of trajectory shaping guidance law design with seeker's FOV constraint against a stationary target is discussed. Both impact time and impact angle control are considered in this chapter. The proposed guidance law is composed of two parts: a baseline PNG term and a tracking error feedback command. Theoretical analysis reveals that the tracking converges to zero at the time of impact. The optimality of the biased feedback term and the behavior of the velocity lead angle are also discussed to provide better insights of the proposed guidance law.

References

1. Sang DK, Tahk M-J (2009) Guidance law switching logic considering the seeker's field-of-view limits. Proc Inst Mech Eng, Part G: J Aerosp Eng 223(8):1049–1058
2. Shim S-W, Hong S-M, Moon G-H, Tahk M-J (2018) Impact angle and time control guidance under field-of-view constraints and maneuver limits. Int J Aeronaut Space Sci 19(1):217–226
3. Saleem A, Ratnoo A (2016) Lyapunov-based guidance law for impact time control and simultaneous arrival. J Guid, Control, Dyn 39(1):164–173
4. Jeon I-S, Lee J-I (2017) Impact-time-control guidance law with constraints on seeker look angle. IEEE Trans Aerosp Electron Syst 53(5):2621–2627
5. Zarchan P (2012) Tactical and strategic missile guidance. American Institute of Aeronautics and Astronautics
6. Kim H-G, Kim HJ (2016) Impact time control guidance considering seeker's field-of-view limits. In *2016 IEEE 55th Conference on Decision and Control (CDC)*. IEEE, pp 4160–4165
7. Lee C-H, Kim T-H, Tahk M-J (2013) Interception angle control guidance using proportional navigation with error feedback. J Guid, Control, Dyn 36(5):1556–1561

Chapter 4
Linear Observability-Enhancement Optimal Guidance Law

Abstract This chapter proposes a new optimal guidance law to enhance target observability for passive guidance with bearing-only measurement. A performance index that considers terminal miss distance, control effort and target observability criterion in an integrated manner is proposed first. The proposed guidance law is then derived analytically by solving the optimization problem formulated. Under certain conditions, it is proved that the guidance law developed gradually switches from retro PNG to classical PNG as time goes. The closed-form solutions of ZEM and guidance command are also derived to provide better insights of the proposed guidance law. Nonlinear numerical simulations are conducted to support the analytical findings.

4.1 Introduction

Low-cost passive seekers are widely utilized in anti-tank/ship tactical missiles as well as projectiles when against non-maneuvering targets. It is known that the passive seeker can only measure the relative LOS angle and hence the required information for guidance law implementation needs to be extracted from angle-only measurement. The issue associated with angle-only measurement is that the target observability is highly related to the relative geometry between the target and the missile [1–6]. Lack of observability or low observability significantly degrades the estimation performance and this, in turn, poses adverse effect on the guidance performance.

Generally, the missile with a passive seeker should make a detour away from the collision course to improve target observability [3, 7–10]. This fact, obviously, contradicts with the requirement of target interception, i.e., the missile should maneuver to converge to the collision course. Hence, there is a strong motivation to develop guidance laws that have dual roles in guaranteeing certain degree of target observability and minimizing miss distance. The common idea that has been utilized in observability-enhancement guidance is to manually generate oscillating LOS motions in a heuristic manner [11–17]. The rationale behind this idea is that the target observability is proportional to the variation of LOS angle or the cumulative of the absolute value of LOS rate. One major drawback of these guidance laws

S. He et al., *Optimal Guidance and Its Applications in Missiles and UAVs*, Springer Aerospace Technology, https://doi.org/10.1007/978-3-030-47348-8_4

lies in their heuristic nature, i.e., it is difficult to analyze the physical meaning and there is a lack of systematic way to choose the design parameters for guidance law implementation.

The work in [18] suggested a linear quadratic guidance law that provides oscillating trajectory to enhance target observability. Although this algorithm has clear physical meaning provided by the cost function, it only considers the lateral relative distance in observability weighting penalty to simplify the solution. As discussed in [3, 7, 8], the relative velocity also contributes to the observability index. For this reason, it would be wise to consider these two factors in an integrated manner in the problem formulation. Another issue of the guidance law suggested in [18] is that the observability penalty weighting is constant during the terminal homing phase. Notice that observability maximization will generate weaving trajectory, which is not desirable when the interceptor is close to the target from practical point of view. Therefore, a practical way to tackle this dilemma is to maximize target observability during the initial flight phase to increase the convergence speed of the guidance filter. Additionally, the guidance law proposed in [18] is formulated in a complicated guidance command and thus it is difficult to analyze its properties theoretically.

The aim of this chapter is to propose a new optimal guidance law that can be utilized to cater for target observability requirement against non-maneuvering targets. Since the LOS rate is proportional to the ZEM, target observability can be indirectly improved by maximizing the variation of ZEM. Based on this fact, a finite-time linear quadratic optimization problem is formulated with an indefinite weighting matrix. The proposed guidance law is then derived analytically through standard optimal control theory. The resultant guidance command turns out to be the well-known PNG with a time-varying navigation gain. The proposed solution offers many elegant features:

(1) Similar to the constant-gain PNG, the proposed guidance law is simple in implementation;
(2) The proposed guidance law has clear physical meaning, providing by the closed-form solution;
(3) Compared with the optimal guidance law developed in [18], the proposed guidance requires less energy consumption and generates higher target observability.

4.2 Observability Under Proportional Navigation Guidance

In this section, we analyze the target observability under classical PNG. Since the targets' maneuverability is ignorable, compared to most types of passive missiles (e.g., anti-ship, anti-tank, anti-radiation missiles), we only consider non-maneuvering targets, i.e., $a_T = 0$, in this chapter.

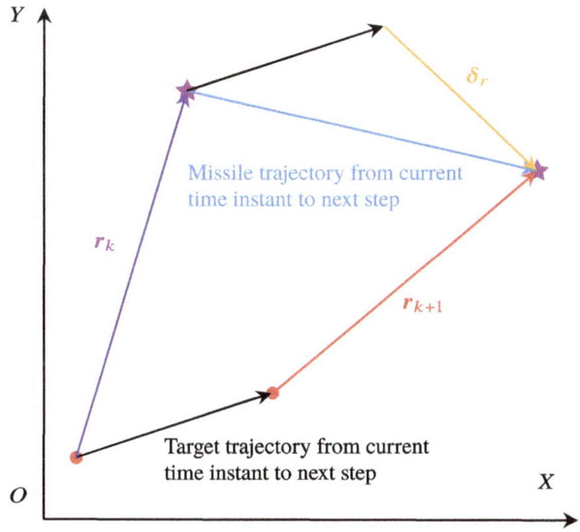

Fig. 4.1 Geometric relationship between the missile and the target between two consecutive time instants in an inertial frame. The black line represents the target movement trajectory and the blue line stands for the missile movement trajectory at current time instant. The orange line refers to the relative movement trajectory

4.2.1 Geometric Metric for Observability Analysis

This subsection suggests a metric for target observability analysis by using a geometric approach. For simplicity, we assume that the missile's heading keeps constant between two consecutive time instants. Notice that this assumption is reasonable since change in the vehicle's direction over a short interval is usually negligible. Figure 4.1 shows the geometric relationship between the missile and the target between two consecutive time instants in an inertial frame, where $\boldsymbol{r}_k$ stands for the relative position vector at time instant t_k and $\boldsymbol{\delta}_r$ represents the relative movement vector. In order to analyze target observability, we introduce a relative frame by attaching the origin at the target, as shown in Fig. 4.2, where η is the separation angle between two consecutive relative position vectors and ϵ denotes the measurement noise of the bearing angle.

Between two consecutive time instants, it is clear from Fig. 4.2 that the relative position vector $\boldsymbol{r}_{k+1}$ at next time step can be uniquely determined by triangulation if $\eta \neq 0$ and $\boldsymbol{r}_k$ is available to the missile. Note that, if $\eta = 0$, $\boldsymbol{r}_{k+1} = k_1 \boldsymbol{r}_k$, $k_1 \in \mathbb{R}^+$. Then, it is impossible to distinguish the next step relative position vector from the current step relative position vector regardless of the angle measurement. Hence, the relative maneuver should satisfy the certain conditions to guarantee target observability.

Let $\delta\rho \triangleq CE$ be the relative position estimation error at time step t_{k+1} caused by the sensor bearing measurement error ϵ. Based on the law of sines, we have

$$\triangle OCE : \frac{\delta\rho}{\sin\epsilon} = \frac{OE}{\sin[\pi - (\eta + \epsilon/2)]} = \frac{OE}{\sin(\eta + \epsilon/2)} \tag{4.1}$$

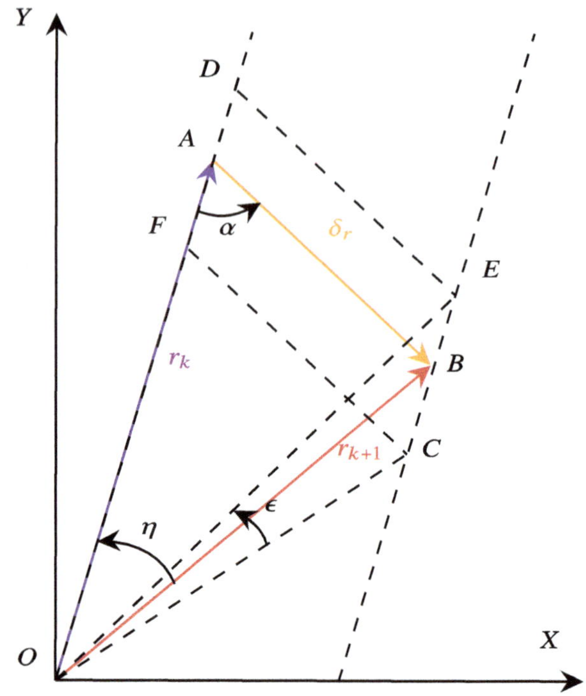

Fig. 4.2 Geometric relationship between the missile and the target between two consecutive time instants in a relative frame. The relative frame is defined by using the relative motion kinematics attaching the origin at the target position at time instant k. Line $\overline{CE}$ is parallel to line $\overline{FD}$. Both lines $\overline{DE}$ and $\overline{FC}$ are parallel to line $\overline{AB}$

$$\triangle ODE : \frac{\|\boldsymbol{\delta}_r\|}{\sin\left(\eta - \epsilon/2\right)} = \frac{OE}{\sin\alpha} \tag{4.2}$$

$$\triangle OAB : \frac{\|\boldsymbol{r}_{k+1}\|}{\sin\alpha} = \frac{\|\boldsymbol{\delta}_r\|}{\sin\eta} \tag{4.3}$$

Combining Eqs. (4.1) and (4.2), one can imply that

$$\delta\rho = \|\boldsymbol{\delta}_r\|\left(\frac{\sin\alpha\sin\epsilon}{\sin^2\eta - \sin^2\left(\epsilon/2\right)}\right) \tag{4.4}$$

where the fact that $\sin\left(a+b\right)\sin\left(a-b\right) = \sin^2 a - \sin^2 b$ is utilised in the derivation.

Assume that the sensor bearing measurement error ϵ is small, Eq. (4.4) can be reduced to

$$\delta\rho = \epsilon\,\|\boldsymbol{\delta}_r\|\left(\frac{\sin\alpha}{\sin^2\eta}\right) \tag{4.5}$$

Substituting Eq. (4.3) into Eq. (4.5) yields the error square as

$$\delta\rho^2 = \epsilon^2 \left(\frac{\|\boldsymbol{r}_{k+1}\|}{\sin \eta} \right)^2 \tag{4.6}$$

The preceding equation reveals that the relative position estimation error at next time step is characterized by the sensor measurement error ϵ, the separation angle η and the relative range $\|\boldsymbol{r}_{k+1}\|$. This provides us better insights to the influence of relative geometric configuration on system observability. Now let us assume the time separation $\Delta t = t_{k+1} - t_k$ is a small variable, we have

$$\sin \eta \approx \eta = \Delta t \dot{\sigma} (t_k) \tag{4.7}$$

where $\dot{\sigma} (t_k)$ denotes the LOS rate at time instant t_k.

The accumulated target observability from the initial time to time instant t, denoted as $\mathcal{O} (t)$, can then be readily obtained from Eqs. (4.6) and (4.7) as

$$\mathcal{O} (t) \propto \int_0^t \frac{|\dot{\sigma}|}{r} d\tau \tag{4.8}$$

Remark 4.1 Notice that the proposed geometric metric (4.8) closely resembles the expression of Fisher Information Matrix (FIM), which is commonly leveraged in observability analysis. More details of the relationship between the proposed metric and FIM can be found in Appendix A. However, it is worthy pointing out that the proposed observability metric is derived using simple geometric condition with clear physical meanings. This obviously reveals that how the missile-target relative geometry influences the estimation accuracy.

4.2.2 Target Observability Under Proportional Navigation Guidance

From Eq. (2.48), the closed-form solution of ZEM under PNG for non-maneuvering targets can be readily obtained as

$$z = \frac{z_0}{t_f^N} t_{go}^N \tag{4.9}$$

where z_0 denotes the initial ZEM.

Combining Eqs. (2.8) and (2.46), we have the relationship between ZEM and LOS rate as

$$z = V_c t_{go}^2 \dot{\sigma} \tag{4.10}$$

Substituting Eq. (4.10) into Eq. (4.9) gives the closed-form solution of the LOS rate as

$$\dot{\sigma} = \frac{z_0}{V_c t_f^N} t_{go}^{N-2} \tag{4.11}$$

which indicates that the magnitude of LOS rate under PNG with $N > 2$ exponentially decreases when the interceptor approaches the target.

From Eq. (4.8), it can be noted that target observability is proportional to the magnitude of the LOS rate and hence PNG provides low observability when implementing it in a passive guidance scenario. Motivated by this observation, the aim of this chapter is to design a dual-role optimal guidance law that guarantees zero final ZEM with observability enhancement for passive missiles.

4.3 Optimal Guidance Law for Target Observability Enhancement

In this section, we first formulate a linear quadratic optimization problem by proposing an objective function that can be utilized in the passive guidance problem. Then, the proposed guidance law is derived analytically by using the standard optimal control theory.

4.3.1 Problem Formulation

Since the relative range r is slowly varying and not adjustable for aerodynamically-controlled interceptors, a practical way to increase target observability is to maximize the cumulative of the magnitude of the LOS rate $\int |\dot{\sigma}|$, according to Eq. (4.8). Unfortunately, the issue associated with target observability enhancement is that direct optimizing the LOS rate is intractable. However, it follows from Eq. (4.10) that

$$|\dot{\sigma}| = \left| \frac{z}{V_c t_{go}^2} \right| \propto |z| \tag{4.12}$$

which means that the maximization of $\int |\dot{\sigma}|$ can be indirectly achieved by the maximization of $\int |z|$.

It is obvious that the objective of observability enhancement, i.e., maximizing the cumulative of the absolute value of ZEM, contradicts with the objective of target interception, i.e., minimizing the absolute value of ZEM. To meet these two criteria in an integrated manner, the idea here is to intentionally maximize $\int |z|$ during the initial flight period to increase target observability and then gradually force ZEM to converge to zero as time increases. To this end, we formulate the following linear quadratic optimization problem.

Problem 4.1 Find the acceleration command a_{M_σ} minimizes performance index

$$J = \frac{1}{2}cz\left(t_f\right)^2 + \frac{1}{2}\int_0^{t_f} -\left(\omega t_{go}\right)^2 z^2 + a_{M_\sigma}^2 dt \tag{4.13}$$

$$\text{subject to: } \dot{z} = -t_{go}a_{M_\sigma} \tag{4.14}$$

where $c > 0$, $\omega > 0$ are constant weighting factors.

The intuition behind choosing optimization criterion (4.13) is clear for the first and third terms: They penalize the miss distance and the command effort, respectively. The second term with negative penalty is small if target observability is high. It can be noted from Eq. (4.13) that the weighting of z^2 gradually increases from $-\left(\omega t_f\right)^2$ to 0 as time goes. This means that more penalty is placed on the objective of observability enhancement, i.e., maximizing the cumulative of the absolute value of ZEM, during the initial flight period and the observability penalty diminishes gradually when $t_{go} \to 0$. It is clear that the penalty weighting ω is of great importance to balance between target observability enhancement and control effort. A suitable condition for choosing proper ω will be presented in the next section. As shown later, the observability penalty weighting proposed also helps to remove trajectory oscillation when the interceptor is close to the target, which is beneficial to guarantee a successful interception.

Remark 4.2 Different from previous linear quadratic guidance laws, the proposed performance index (4.13) is formulated with a non-zero indefinite weighting for the purpose of constraining the transient behavior of ZEM. Notice that the optimization problem formulated is known as the indefinite linear-quadratic optimization problem [19–21] with indefinite state weighting and definite control weighting.

4.3.2 Optimal Guidance Law Design

To solve the formulated optimal control problem, consider the following Hamilton function

$$H = -\lambda t_{go}a_{M_\sigma} + \frac{1}{2}\left[-\left(\omega t_{go}\right)^2 z^2 + a_{M_\sigma}^2\right] \tag{4.15}$$

where λ denotes the costate variable.

The optimal guidance command a_{M_σ} can be uniquely determined as

$$\frac{\partial H}{\partial a_{M_\sigma}} = -\lambda t_{go} + a_{M_\sigma} = 0 \tag{4.16}$$

From Pontryagin's minimum principle, the costate equation is given by

$$-\dot{\lambda} = \frac{\partial H}{\partial z} = -\left(\omega t_{go}\right)^2 z \tag{4.17}$$

and the terminal costate satisfies

$$\lambda\left(t_f\right) = cz\left(t_f\right) \tag{4.18}$$

Define a new variable $S = \lambda/z$, evaluating the time derivative of S and substituting Eqs. (4.16) and (4.17) into it yields

$$\begin{aligned}\dot{S} &= \frac{\dot{\lambda} z - \lambda \dot{z}}{z^2} \\ &= \frac{\left(\omega t_{go}\right)^2 z^2 + \lambda^2 t_{go}^2}{z^2} \\ &= S^2 t_{go}^2 + \left(\omega t_{go}\right)^2\end{aligned} \tag{4.19}$$

Solving Eq. (4.19) with terminal constraint $S\left(t_f\right) = c$ results in

$$S = \omega \tan\left(\arctan\left(c/\omega\right) - \frac{\omega t_{go}^3}{3}\right) \tag{4.20}$$

For perfect interception with zero final ZEM, we have $c \to \infty$. Therefore, Eq. (4.20) reduces to

$$S = \omega \cot\left(\frac{\omega t_{go}^3}{3}\right) \tag{4.21}$$

Then, the optimal guidance command can be obtained as

$$a_{M_\sigma} = \lambda t_{go} = S z t_{go} = \omega t_{go} \cot\left(\frac{\omega t_{go}^3}{3}\right) z \tag{4.22}$$

Define $N = t_{go}^3 S$ as a navigation gain, the proposed guidance law can also be formulated as

$$a_{M_\sigma} = N \frac{z}{t_{go}^2} = N V_c \dot{\sigma} \tag{4.23}$$

which reveals that the proposed formulation has a similar structure as the well-known PNG with a time-varying navigation gain.

Remark 4.3 As the proposed guidance law is derived as a solution to an indefinite linear quadratic optimization problem, we need to check the optimality of the solution. From Eq. (4.16), it is clear that the necessary condition $\partial H/\partial a_{M_\sigma} = 0$ only has one unique solution $a_{M_\sigma} = \lambda t_{go}$. This means that the solution derived is globally optimal. Additionally, the second-order partial derivative of the Hamilton function with respect to a_{M_σ} is determined as $\partial^2 H/\partial a_{M_\sigma}^2 = 1 > 0$. Therefore, the proposed solution is the unique optimal solution.

Remark 4.4 Notice that guidance command (4.22) is given by a function of closing velocity, LOS rate, and time-to-go, where the time-to-go is required to calculate the navigation gain. Therefore, the proposed guidance law requires only one additional information, i.e., the time-to-go, compared to the conventional PNG. This means that the proposed guidance law is also easy for practical implementation. For the implementation, the closing velocity and the LOS rate can be obtained from a dedicated homing filter using the bearing measurement. In practice, the time-to-go information can be determined by using the conventional approximation method as $t_{go} \approx r/V_c$.

4.4 Analysis of Proposed Optimal Guidance Law

In this section, we will analyze the properties of the proposed guidance law to reveal the physical meaning of observability-enhancement optimal guidance law. First, we discuss the behavior of the navigation gain and its convergence pattern. Then, the closed-loop solution of the proposed guidance is presented. Finally, the dynamics of velocity lead angle will be discussed.

4.4.1 Behavior of Navigation Gain

In this subsection, the convergence pattern of the navigation gain N will be analyzed. We first show that the navigation gain will converge to a constant value without observability penalty or at the time of impact. Then, the transient behavior of the navigation gain will be investigated. This can provide the unique information that how the navigation gain behaves to cater for observability enhancement requirement during the terminal guidance phase. The results are provided in the following propositions.

Proposition 4.1 *The navigation gain N converges to constant value 3 without the penalty on observability criterion or at the time of impact, i.e.,* $\lim_{\omega \to 0} N = 3$, $\lim_{t_{go} \to 0} N = 3$.

Proof From the definition of the guidance gain, we have

$$N = \omega t_{go}^3 \cot\left(\frac{\omega t_{go}^3}{3}\right) \tag{4.24}$$

Without the penalty on the observability criterion, i.e., $\omega \to 0$, using L'Hospital's rule gives

$$\begin{aligned}\lim_{\omega\to 0} N &= \frac{\omega t_{go}^3 \cos\left(\frac{\omega t_{go}^3}{3}\right)}{\sin\left(\frac{\omega t_{go}^3}{3}\right)} \\ &= \frac{t_{go}^3 \cos\left(\frac{\omega t_{go}^3}{3}\right) - \omega t_{go}^3 \sin\left(\frac{\omega t_{go}^3}{3}\right)\frac{t_{go}^3}{3}}{\frac{t_{go}^3}{3}\cos\left(\frac{\omega t_{go}^3}{3}\right)} \\ &= 3\end{aligned} \tag{4.25}$$

which reveals that the proposed guidance law reduces to the optimal PNG without enforcing the observability requirement.

From Eq. (4.24), the final guidance gain is determined by L'Hospital's rule as

$$\begin{aligned}\lim_{t_{go}\to 0} N &= \frac{\omega t_{go}^3 \cos\left(\frac{\omega t_{go}^3}{3}\right)}{\sin\left(\frac{\omega t_{go}^3}{3}\right)} \\ &= \frac{3\omega t_{go}^2 \cos\left(\frac{\omega t_{go}^3}{3}\right) + \omega^2 t_{go}^3 \sin\left(\frac{\omega t_{go}^3}{3}\right) t_{go}^2}{\omega t_{go}^2 \cos\left(\frac{\omega t_{go}^3}{3}\right)} \\ &= 3\end{aligned} \tag{4.26}$$

which clearly shows that the navigation gain converges to constant value 3 at the time of impact. That is, the proposed guidance law converges to the optimal PNG when the interceptor approaches the target. Therefore, the proposed guidance law shows similar characteristics as conventional PNG. □

Proposition 4.2 *Let $k \in \mathbb{Z}$ with $\mathbb{Z}$ being the set of positive integers. The navigation gain N increases monotonically when $t_{go} \neq \left(\frac{3k\pi}{\omega}\right)^{\frac{1}{3}}$ and switches from $+\infty$ to $-\infty$ when $t = t_f - \left(\frac{3k\pi}{\omega}\right)^{\frac{1}{3}}$. Furthermore, the number of switch is given by $\left\lfloor \frac{\omega t_f^3}{3\pi} \right\rfloor$.*

Proof Evaluating the time derivative of the navigation gain yields

$$\begin{aligned}\frac{dN}{dt} &= \frac{3\omega\left[\cos\left(\frac{\omega t_{go}^3}{3}\right) - \frac{\omega t_{go}^3}{3}\sin\left(\frac{\omega t_{go}^3}{3}\right)\right]\sin\left(\frac{\omega t_{go}^3}{3}\right) - \omega^2 t_{go}^3 \cos^2\left(\frac{\omega t_{go}^3}{3}\right)}{\sin^2\left(\frac{\omega t_{go}^3}{3}\right)}\left(-t_{go}^2\right) \\ &= 3\omega t_{go}^2 \frac{\frac{\omega t_{go}^3}{3} - \sin\left(\frac{\omega t_{go}^3}{3}\right)\cos\left(\frac{\omega t_{go}^3}{3}\right)}{\sin^2\left(\frac{\omega t_{go}^3}{3}\right)}\end{aligned} \tag{4.27}$$

From simple algebra, it is easy to verify that the function $f(x) = x - \sin(x)\cos(x)$ is a monotonically increasing function. With this in mind, we have

$$f\left(\frac{\omega t_{go}^3}{3}\right) > f(0) = 0 \tag{4.28}$$

Therefore, when $t_{go} \neq \left(\frac{3k\pi}{\omega}\right)^{\frac{1}{3}}$, substituting Eq. (4.28) into Eq. (4.27) results in

$$\frac{dN}{dt} = 3\omega t_{go}^2 \frac{f\left(\frac{\omega t_{go}^3}{3}\right)}{\sin^2\left(\frac{\omega t_{go}^3}{3}\right)} > 0 \tag{4.29}$$

Denote $t_k = t_f - \left(\frac{3k\pi}{\omega}\right)^{\frac{1}{3}}$, then, it is easy to verify that

$$\lim_{t \to t_k^-} N = \omega\left(t_f - t_k\right)^3 \frac{1}{0} = +\infty \tag{4.30}$$

$$\lim_{t \to t_k^+} N = -\omega\left(t_f - t_k\right)^3 \frac{1}{0} = -\infty \tag{4.31}$$

Therefore, when $t = t_f - \left(\frac{3k\pi}{\omega}\right)^{\frac{1}{3}}$, the navigation gain switches from $+\infty$ to $-\infty$. Finally, since the solution of $\sin\left(\frac{\omega t_{go}^3}{3}\right) = 0$ is $t_{go} = \left(\frac{3k\pi}{\omega}\right)^{\frac{1}{3}}$, the number of switch, from $+\infty$ to $-\infty$, is determined by $\left\lfloor \frac{\omega t_f^3}{3\pi} \right\rfloor$. □

Remark 4.5 From Eq. (4.24), one can note that the navigation gain is only determined by the flight time t_f and weighting factor ω. Proposition 4.2 reveals that the navigation gain might exhibit abrupt change from $+\infty$ to $-\infty$ during the terminal homing phase. Consider a scenario with flight time $t_f = 10$ s and three different weighting factors $\omega_1 = 0.008$, $\omega_2 = 0.015$, $\omega_3 = 0.025$. According to Proposition 4.2, the numbers of switch of these three different navigation gains are obtained as $\left\lfloor \frac{\omega_1 t_f^3}{3\pi} \right\rfloor = 0$, $\left\lfloor \frac{\omega_2 t_f^3}{3\pi} \right\rfloor = 1$, $\left\lfloor \frac{\omega_3 t_f^3}{3\pi} \right\rfloor = 2$, respectively. This can be clearly observed from the navigation gain profile, shown in Fig. 4.3. Since using PNG with an infinite navigation gain is impractical for real implementation, we recommend to choose the weighting factor ω to satisfy $\frac{\omega t_f^3}{3} < \pi$ to avoid the abrupt change of the navigation gain. Also note that from the performance index (4.13) that larger value of ω generates higher target observability. Therefore, a desired criterion for choosing ω is $\pi/2 < \frac{\omega t_f^3}{3} < \pi$, which guarantees a bounded optimal solution. Under this condition, the navigation gain will gradually increase from negative value $\omega t_f^3 \cot\left(\frac{\omega t_f^3}{3}\right)$ to 3 as time goes. This also means that the proposed guidance law will gradually switch from retro-PNG [22], i.e., with negative navigation gain, to classical PNG, i.e., with positive navigation gain, during the terminal homing phase.

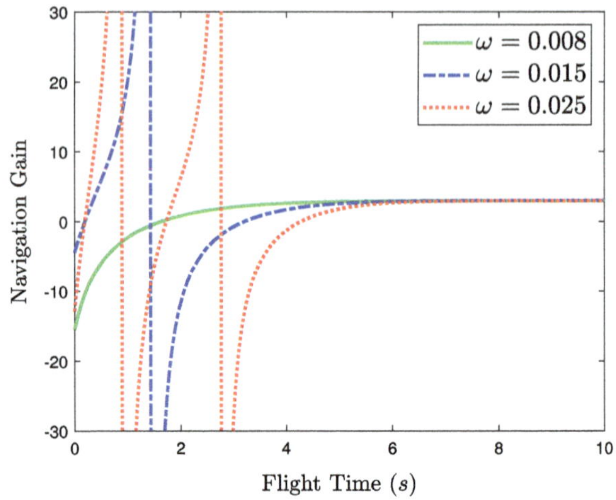

Fig. 4.3 Navigation gain with $t_f = 10\,\text{s}$ and $\omega = 0.008, 0.015, 0.025$

4.4.2 Closed-Form Solution of Proposed Guidance Law

To further provide better insights of the proposed guidance law, we seek to find its closed-form solution. More specifically, the convergence pattern of ZEM and the behavior of guidance command will be analyzed. The analytical results are presented in the following propositions.

Proposition 4.3 *Under condition $\pi/2 < \frac{\omega t_f^3}{3} < \pi$, the absolute value of ZEM $|z|$ increases monotonically to $\left|z_0/\sin\left(\frac{\omega t_f^3}{3}\right)\right|$ at time $t^* = t_f - \left(\frac{3\pi}{2\omega}\right)^{\frac{1}{3}}$ and decreases monotonically to zero until the time of impact.*

Proof Substituting Eq. (4.22) into Eq. (2.47) gives the ZEM dynamics as

$$\dot{z} = -\omega t_{go}^2 \cot\left(\frac{\omega t_{go}^3}{3}\right) z \tag{4.32}$$

Using Euler's formula $2i\cos x = e^{ix} + e^{-ix}$ and $2i\sin x = e^{ix} - e^{-ix}$ renders Eq. (4.32) to

$$\dot{z} = -\omega t_{go}^2 \frac{e^{\frac{\omega t_{go}^3}{3}i} + e^{-\frac{\omega t_{go}^3}{3}i}}{e^{\frac{\omega t_{go}^3}{3}i} - e^{-\frac{\omega t_{go}^3}{3}i}} z \tag{4.33}$$

Solving Eq. (4.33) yields

$$z = Ce^{-\frac{\omega}{3}\left(t_f^3 - t_{go}^3\right)i}\left(e^{-\frac{2\omega}{3}t_{go}^3 i} - 1\right) \tag{4.34}$$

where C is a constant determined by the initial condition.

Applying the initial condition to Eq. (4.34) gives the generic solution as

$$z = \frac{z_0}{e^{-\frac{2\omega}{3}t_f^3 i} - 1} e^{-\frac{\omega}{3}\left(t_f^3 - t_{go}^3\right)i}\left(e^{-\frac{2\omega}{3}t_{go}^3 i} - 1\right) \tag{4.35}$$

where z_0 denotes the initial ZEM.

Simplifying Eq. (4.35) through Euler's formula results in the closed-form solution of ZEM as

$$z = z_0 \frac{\sin\left(\frac{\omega t_{go}^3}{3}\right)}{\sin\left(\frac{\omega t_f^3}{3}\right)} \tag{4.36}$$

Substituting Eq. (4.36) into Eq. (4.32) gives the explicit ZEM dynamics as

$$\dot{z} = -\frac{z_0}{\sin\left(\frac{\omega t_f^3}{3}\right)} \omega t_{go}^2 \cos\left(\frac{\omega t_{go}^3}{3}\right) \tag{4.37}$$

From (4.37), one can note that if $z_0 > 0$, then $\dot{z} > 0$ when $0 \leq t < t^*$ and $\dot{z} < 0$ when $t^* < t \leq t_f$. Similarly, if $z_0 < 0$, we have $\dot{z} < 0$ when $0 \leq t < t^*$ and $\dot{z} > 0$ when $t^* < t \leq t_f$. Therefore, $|z|$ increases monotonically during $0 \leq t < t^*$ and decreases monotonically when $t^* < t \leq t_f$. The limit values of $|z|$ can be readily obtained from Eq. (4.36) as $\max|z| = |z(t^*)| = \left|z_0/\sin\left(\frac{\omega t_f^3}{3}\right)\right|$ and $\min|z| = \left|z\left(t_f\right)\right| = 0$. □

Remark 4.6 Proposition 4.3 analyzes the convergence pattern of the ZEM and demonstrates that the ZEM will converge to zero at the time of impact, guaranteeing target interception. This clearly reveals the physical meaning of observability-enhancement optimal guidance laws. Figure 4.4 shows an example of ZEM response with different weighting factor ω.

Now, let us investigate the characteristics of the guidance command.

Proposition 4.4 *Under condition* $\pi/2 < \frac{\omega t_f^3}{3} < \pi$, *the magnitude of the guidance command* $\left|a_{M_\sigma}\right|$ *achieves its maximum value at either the initial point or at time instant* $t_m = t_f - \left(\frac{1.6414}{\omega}\right)^{\frac{1}{3}}$ *as*

$$\max\left|a_{M_\sigma}\right| = \max\left\{\left|\frac{z_0}{\sin\left(\frac{\omega t_f^3}{3}\right)}\omega t_f \cos\left(\frac{\omega t_f^3}{3}\right)\right|, 1.0074\left|\frac{z_0}{\sin\left(\frac{\omega t_f^3}{3}\right)}\omega^{\frac{2}{3}}\right|\right\} \tag{4.38}$$

Furthermore, the guidance command converges to zero at the time of impact, i.e., $a_{M_\sigma}\left(t_f\right) = 0$.

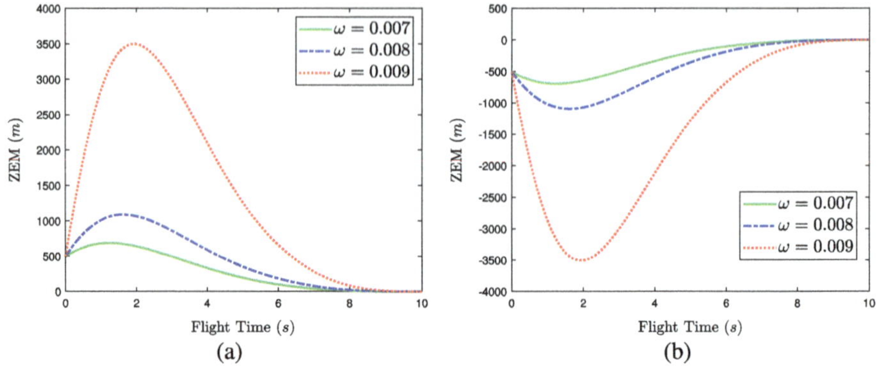

Fig. 4.4 ZEM with $t_f = 10$ s and $\omega = 0.007, 0.008, 0.009$. **a** $z_0 = 500$ m; and **b** $z_0 = -500$ m

Proof Substituting Eq. (4.36) into Eq. (4.22) gives the closed-form solution of the guidance command as

$$a_{M_\sigma} = \frac{z_0}{\sin\left(\frac{\omega t_f^3}{3}\right)} \omega t_{go} \cos\left(\frac{\omega t_{go}^3}{3}\right) \tag{4.39}$$

It is obvious from Eq. (4.39) that the guidance command converges to zero at the time of impact as $a_{M_\sigma}(t_f) = 0$. Thus, the proposed guidance law guarantees certain degree of operational margins to cope with undesired disturbances when the interceptor approaches the target. Taking the time derivative of a_{M_σ} yields

$$\dot{a}_{M_\sigma} = \frac{z_0 \omega}{\sin\left(\frac{\omega t_f^3}{3}\right)} \left[-\cos\left(\frac{\omega t_{go}^3}{3}\right) + \omega t_{go}^3 \sin\left(\frac{\omega t_{go}^3}{3}\right) \right] \tag{4.40}$$

Define function $g(x) = 3x \sin x - \cos x$ with $x \in [0, \pi]$. Solving $g\left(\frac{\omega t_{go}^3}{3}\right) = 0$ gives the solution of $\dot{a}_{M_\sigma}(t_m) = 0$ as $t = t_m$. Since $g(0) < 0$ and $g\left(\frac{\omega t_f^3}{3}\right) > 0$, one can conclude that $g(t) > 0$ when $0 \le t < t_m$ and $g(t) < 0$ when $t < t \le t_f$. Therefore, if $z_0 < 0$, $\dot{a}_{M_\sigma} < 0$ for $0 \le t < t_m$ and $\dot{a}_{M_\sigma} > 0$ for $t < t \le t_f$. This means that, for $z_0 < 0$, the guidance command monotonically decreases to $a_{M_\sigma}(t_m)$ and gradually increases to zero at the time impact. Similarly, it can be easily verified that, for $z_0 > 0$, the guidance command increases monotonically when $0 \le t < t_m$ and decreases monotonically when $t < t \le t_f$. Consequently, the guidance command takes its maximum magnitude either at the initial point or $t = t_m$ as shown in Eq. (4.38). Figure 4.5 shows an example of missile acceleration command with different weighting factor ω. □

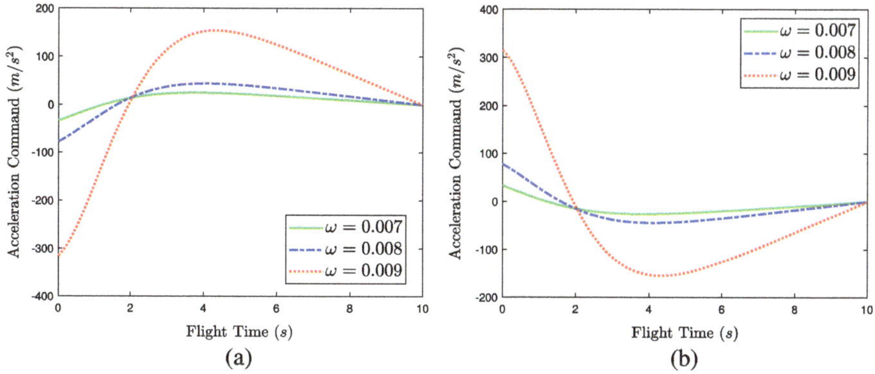

Fig. 4.5 Acceleration command with $t_f = 10\,\text{s}$ and $\omega = 0.007, 0.008, 0.009$. **a** $z_0 = 500\,\text{m}$; and **b** $z_0 = -500\,\text{m}$

Remark 4.7 Proposition 4.4 provides the information on when the acceleration command achieves its maximum magnitude during the homing phase. This information is helpful in finding the required acceleration demand with physical constraints. It is easy to verify from Eq. (4.38) that the maximum magnitude of the acceleration command is a monotonically increasing function with respect to the penalty weighting ω. Also note from Proposition 4.3 that target observability increases as the increase of ω. With these in mind, it can be concluded that the penalty weighting ω is a trade-off design parameter in real applications.

4.4.3 Behavior of Velocity Lead Angle

From previous analysis, it is clear that increasing the observability penalty weighting ω will generate more curved trajectory. Since the seeker's FOV is limited for practical interceptors, highly curved trajectory might make the look angle exceed its maximum limit and therefore lead to target loss. For this reason, this subsection will analyze the behavior of the look angle under the proposed guidance law. This is helpful in choosing the weighting parameter ω to satisfy the physical look angle constraint. For typical engagement scenarios, the angle-of-attack is usually very small. With this in mind, the look angle can be approximated by the lead angle $\theta = \gamma_M - \sigma$ in analysis. The results are given in the following proposition.

Proposition 4.5 *Under condition $\pi/2 < \frac{\omega t_f^3}{3} < \pi$, the magnitude of the look angle $|\theta|$ achieves its maximum value $\theta_{\max}$ at time instant $t_\theta = t_f - \left(\frac{3.9726}{\omega}\right)^{\frac{1}{3}}$ as*

$$\theta_{\max} = |z_0| \frac{\sin\left(\frac{\omega(t_f - t_\theta)^3}{3}\right)}{\sin\left(\frac{\omega t_f^3}{3}\right) V_m \left(t_f - t_\theta\right)} \tag{4.41}$$

Furthermore, $\theta_{\max}$ is a monotonically increasing function with respect to ω.

Proof Compared to the passive missiles, their targets, e.g., ships, tanks, etc, are slowly moving and therefore can be assumed as stationary. Under this condition, we have $\theta = -\dot{\sigma} t_{go}$ [23] and $V_c = V_m$. Then, using Eqs. (2.8) and (2.46) gives

$$\theta = -\frac{z}{V_m t_{go}} \tag{4.42}$$

Substituting Eq. (4.36) into Eq. (4.42) yields the closed-form solution of the look angle as

$$\theta = -z_0 \frac{\sin\left(\frac{\omega t_{go}^3}{3}\right)}{\sin\left(\frac{\omega t_f^3}{3}\right) V_m t_{go}} \tag{4.43}$$

Taking the time derivative of Eq. (4.43), we have

$$\dot{\theta} = z_0 \frac{\omega t_{go}^3 \cos\left(\frac{\omega t_{go}^3}{3}\right) - \sin\left(\frac{\omega t_{go}^3}{3}\right)}{\sin\left(\frac{\omega t_f^3}{3}\right) V_m t_{go}^2} \tag{4.44}$$

Define function $h(x) = 3x\cos x - \sin x$ with $x \in [0, \pi]$. Solving $h(x) = 0$ gives $x_1 = 0$ and $x_2 = 1.3241$. Since $h(0.5) > 0$ and $h(\pi) < 0$, we have $h(x) > 0$ when $0 \le x < x_2$ and $h(x) \le 0$ when $x_2 \le x \le \pi$. This means that $h\left(\frac{\omega t_{go}^3}{3}\right) \le 0$ when $0 \le t \le t_\theta$ and $h\left(\frac{\omega t_{go}^3}{3}\right) > 0$ when $t_\theta < t \le t_f$. Therefore, if $z_0 < 0$, the look angle monotonically increases from the initial time instant to t_θ and monotonically decreases thereafter. Similarly, if $z_0 \ge 0$, the look angle starts to decrease initially until t_θ and then monotonically increases when $t_\theta < t \le t_f$. Furthermore, notice from Eq. (4.42) that $\operatorname{sgn}[\theta(0)] = -\operatorname{sgn}[z_0]$. Then, it is straightforward to verify that the magnitude of the look angle $|\theta|$ achieves its maximum value at time instant $t_\theta = t_f - \left(\frac{3.9726}{\omega}\right)^{\frac{1}{3}}$ as Eq. (4.41).

Next, evaluating the partial derivative of $\theta_{\max}$ with respect to ω gives

$$\begin{aligned}\frac{\partial \theta_{\max}}{\partial \omega} &= \frac{0.9697|z_0|}{V_m} \frac{-\frac{t_f^3}{3}\left(\frac{3.9726}{\omega}\right)^{\frac{1}{3}} \cos\left(\frac{\omega t_f^3}{3}\right) + \frac{1}{3\omega}\left(\frac{3.9726}{\omega}\right)^{\frac{1}{3}} \sin\left(\frac{\omega t_f^3}{3}\right)}{\sin^2\left(\frac{\omega t_f^3}{3}\right)\left(\frac{3.9726}{\omega}\right)^{\frac{2}{3}}} \\ &= \frac{0.2041|z_0|}{V_m} \frac{-t_f^3 \omega^{\frac{1}{3}} \cos\left(\frac{\omega t_f^3}{3}\right) + \omega^{-\frac{2}{3}} \sin\left(\frac{\omega t_f^3}{3}\right)}{\sin^2\left(\frac{\omega t_f^3}{3}\right)}\end{aligned} \tag{4.45}$$

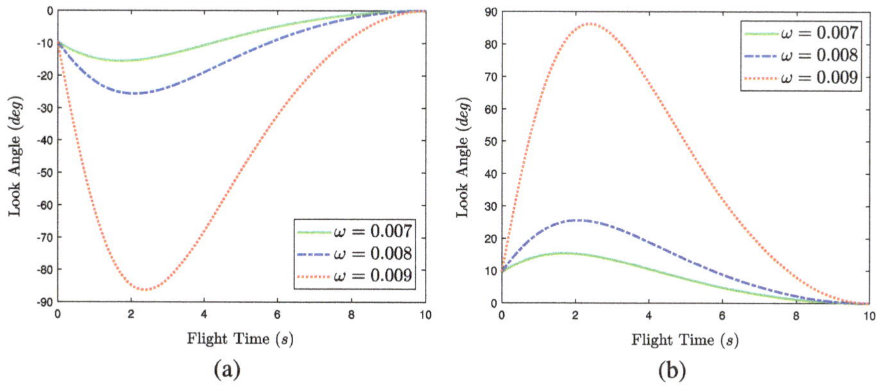

Fig. 4.6 Look angle profile with $V_m = 300\,\text{m/s}$, $t_f = 10\,\text{s}$ and $\omega = 0.007, 0.008, 0.009$. **a** $z_0 = 500\,\text{m}$; and **b** $z_0 = -500\,\text{m}$

Since $\pi/2 < \frac{\omega t_f^3}{3} < \pi$, it is clear that $\partial\theta_{\max}/\partial\omega > 0$, which implies that $\theta_{\max}$ is a monotonically increasing function with respect to ω. An example of look angle profile is presented in Fig. 4.6. □

Remark 4.8 Note that the results of Proposition 4.5 is derived based on the assumption that target is stationary. This means that the maximum magnitude of the look angle, shown in Eq. (4.41), has some approximation errors for practical scenarios if the target is moving. However, the results presented in this subsection is helpful for the initial design stage.

4.5 Simulation Results

In this section, the effectiveness of the proposed optimal guidance law is validated through nonlinear numerical simulations. We first investigate the characteristics of the proposed guidance law under different conditions. Then, the performance of the proposed guidance law is compared with that of heuristic AIM-PNG [14] and linear quadratic optimal guidance law [18]. In all simulations, we consider an air-to-surface tail-chase interception scenario. The initial conditions for the considered scenario are summarized in Table 4.1.

4.5.1 Characteristics of the Proposed Guidance Law

The simulation in this subsection investigates the characteristics of the proposed guidance law with various weighting factors $\omega = 0.0006, 0.0007, 0.0008$. Note

Table 4.1 Initial conditions for homing engagement

Parameters	Values
Missile initial position, $(x_M(0), y_M(0))$	(4143 m, 4596 m)
Missile flight velocity, V_M	300 m/s
Missile initial flight path angle, $\gamma_M(0)$	−40°
Target initial position, $(x_T(0), y_T(0))$	(8000 m, 0 m)
Target velocity, V_T	20 m/s
Target initial flight path angle, $\gamma_T(0)$	0°

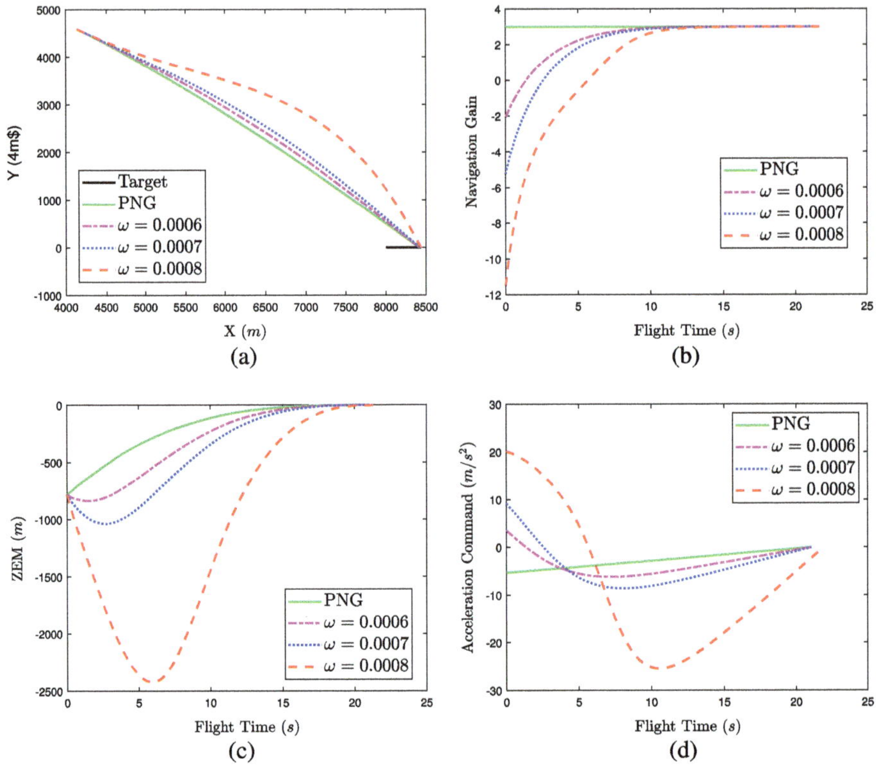

Fig. 4.7 Comparison results of the proposed guidance law with different weighting factors. **a** Interception trajectory; **b** navigation gain; **c** ZEM; and **d** acceleration command

that these three different weighting factors all satisfy the recommended condition $\pi/2 < \frac{\omega t_f^3}{3} < \pi$. For better illustration, the classical PNG is also performed in simulations. The simulation results, including interception trajectories, navigation gain profile, response curve of ZEM and guidance command, obtained from different guidance laws are provided in Fig. 4.7. The quantitative comparison results of observ-

Table 4.2 Quantitative comparison results of observability performance and control effort

	PNG	$\omega = 0.0006$	$\omega = 0.0007$	$\omega = 0.0008$
Observability index	0.0644	0.1141	0.1587	0.4512
Control effort	195.9744	345.2683	694.1029	5804.587

ability performance and control effort are summarized in Table 4.2, where the target observability index is given by $\int |\dot{\sigma}|\, dt$ and the control effort is defined as $\int_0^{t_f} a_M^2 dt$.

From Fig. 4.7a, it can be noted that successful interception is guaranteed under all guidance laws. Unsurprisingly, the proposed guidance law with higher weighting factor ω generates more curved trajectory since more penalty on the variation of ZEM is enforced in the optimization performance index. Since $\pi/2 < \frac{\omega t_f^3}{3} < \pi$, one can conclude from Propositions 4.1 and 4.2 that the navigation gain of the proposed guidance law will monotonically increase during the terminal homing phase and converge to constant value 3 at the time of impact. This fact can be clearly observed from Fig. 4.7b. From Fig. 4.7c, one can note that the ZEM under PNG monotonically converges to zero when the interceptor approaches the target. As a comparison, $|z|$ under the proposed guidance law initially increases to a certain value and then converge to zero at the time of impact to guarantee target interception. This phenomenon conforms with the analytical results shown in Proposition 4.3. Furthermore, the proposed guidance law with higher value of the weighting factor ω produces larger variation of $|z|$, thus generating higher target observability as confirmed by Table 4.2. However, increasing the weighting factor ω also increases the variation of the acceleration command during the flight period, thereby generating more control effort.

4.5.2 Comparison with Other Guidance Laws

To further show the effectiveness of the proposed optimal guidance law, the performance of the proposed guidance law is compared with that of observability-enhancement linear quadratic optimal guidance law [18]. The optimization problem formulated in [18] is given by

Problem 4.2 Find the acceleration command a_{M_σ} minimizes performance index

$$\min_{a_{m_\sigma}} J = \frac{1}{2}\bar{c}y(t_f)^2 + \frac{1}{2}\int_0^{t_f} -\bar{\omega}y^2 + a_{M_\sigma}^2 dt \tag{4.46}$$

$$\text{subject to: } \dot{y} = v, \quad \dot{v} = -a_{M_\sigma} \tag{4.47}$$

where $\bar{c} > 0$, $\bar{\omega} > 0$ are constant weighting factors.

As derived in appendix, the guidance command by solving the preceding optimization problem is given by

$$a_{M_\sigma} = [0, 1]\, \boldsymbol{N}\boldsymbol{M}^{-1}\, [y, v]^T \tag{4.48}$$

where matrices $\boldsymbol{N} \in \mathbb{R}^{2\times 2}$, $\boldsymbol{M} \in \mathbb{R}^{2\times 2}$ are given in the appendix. Different the proposed guidance law, guidance command (4.48) is quite complicated as shown in the Appendix A. For this reason, it is difficult to find the closed-form solution and analyze its physical properties theoretically.

In implementation, the required parameters of guidance law (4.48) are set as $\bar{c} = 10^{20}$, $\bar{\omega} = 0.00077$ according to the tuning rules provided in [18]. The penalty weight ω of the proposed guidance law is set as $\omega = 0.0008$ in the simulations. The simulation results, including interception trajectories, LOS rate profile, response curve of ZEM, guidance command, observability index profile and energy consumption, obtained from different guidance laws are provided in Fig. 4.8.

From Fig. 4.8a, one can note that all guidance laws successfully guide the missile to intercept the target. By introducing a negative weighting term in the optimization performance index, both guidance law (4.48) and the proposed algorithm generate an initial evasive maneuver to improve target observability. Due to this fact, the magnitude of the LOS rate under these two guidance laws initially increases to force the interceptor to make a detour from the ideal collision course. However, the LOS rate under guidance law (4.48) diverges when the missile is near the target, as shown in Fig. 4.8b. As a comparison, it is clear that the LOS rate under the proposed guidance law converges to zero at the time of impact since the proposed guidance law gradually converges to the classical PNG as we analyze before. As it is well-known that bounded LOS rate is beneficial to reduce the terminal miss distance, the proposed guidance law is expected to perform better than guidance law (4.48) in terms of interception accuracy.

It follows from Fig. 4.8c that the ZEM under the two observability-enhancement optimal guidance laws show similar patterns: the magnitude of ZEM increases initially to cater for target observability improvement and gradually converges to zero to guarantee target interception. The guidance command, depicted in Fig. 4.8d, reveals that guidance law (4.48) generates higher peak value of the acceleration command, compared with the proposed guidance law.

From Fig. 4.8e, one can clearly note that target observability under the proposed guidance law is higher than that of guidance law (4.48). This fact can be attributed to that the proposed guidance law considers the negative weighting on ZEM, i.e., the integration of lateral relative distance and lateral relative velocity, in optimization while guidance law (4.48) only utilizes the position-related term. Not surprisingly, observability-enhancement guidance laws require more energy consumption than the optimal PNG. However, the proposed guidance law still requires less control effort than guidance law (4.48), as confirmed by Fig. 4.8f. From Fig. 4.8e, f, it can be concluded that the proposed guidance law is more advantageous than the algorithm derived in [18]: the proposed guidance law requires less energy consumption while generates higher target observability.

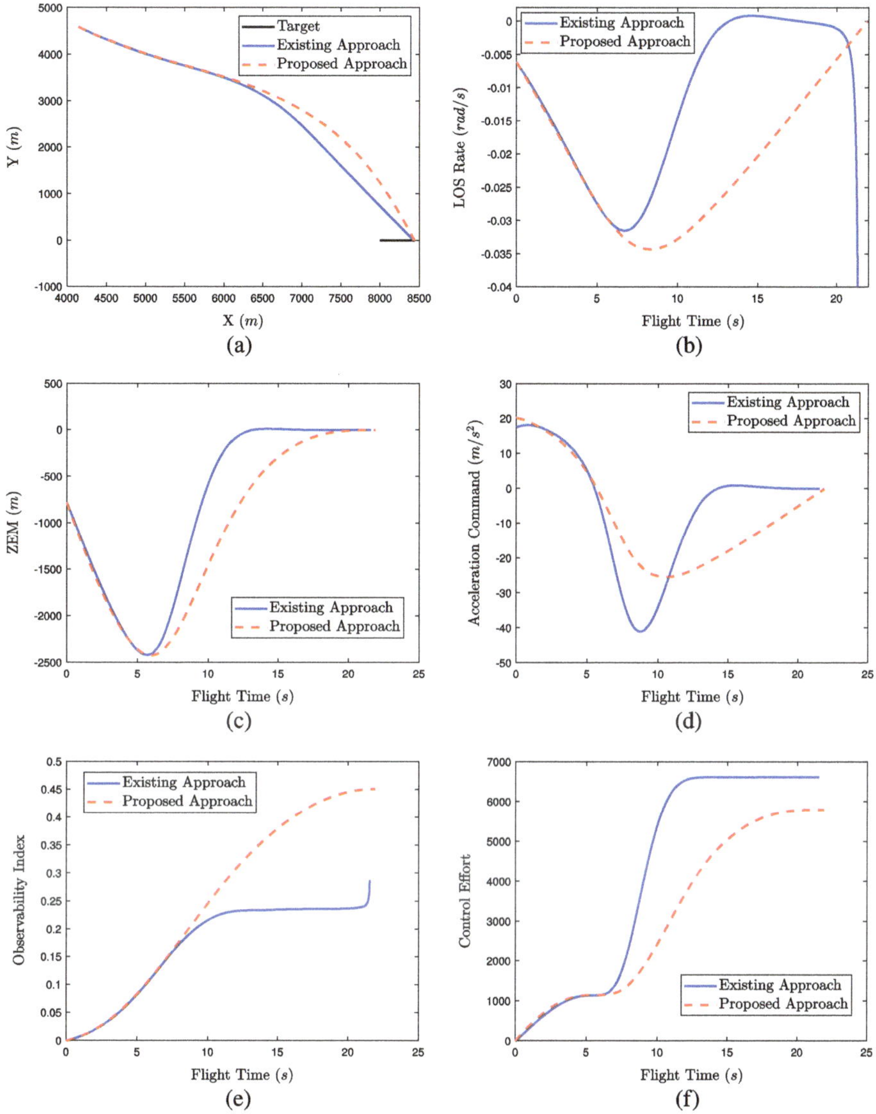

Fig. 4.8 Comparison results of different guidance laws. **a** Interception trajectory; **b** LOS rate; **c** ZEM; **d** acceleration command; **e** observability index; and **f** control effort

4.5.3 Filter-Embedded Closed-Loop Simulation

To further demonstrate the advantage of observability improvement, Unscented Kalman filter (UKF)-embedded closed-loop simulations are performed in this subsection to validate the observability enhancement property of the proposed guidance law. In the simulation, UKF in conjunction with a constant velocity model are utilized for state estimation. The system states of this model consists of four states, i.e., the down-range x, cross-range y and their corresponding rates v_x, v_y. Let $\boldsymbol{x}_k = \left[x_k, y_k, v_{x,k}, v_{y,k}\right]^T$ and $\boldsymbol{a}_{M,k} = \left[a_{M_x,k}, a_{M_y,k}\right]^T$, where a_{M_x} and a_{M_y} are the missile accelerations along the down-range and cross-range directions, respectively. The state transition of constant velocity model is determined by

$$\boldsymbol{x}_k = \boldsymbol{F}\boldsymbol{x}_{k-1} - \boldsymbol{G}\boldsymbol{a}_{M,k-1} + \boldsymbol{G}\boldsymbol{w}_{k-1} \tag{4.49}$$

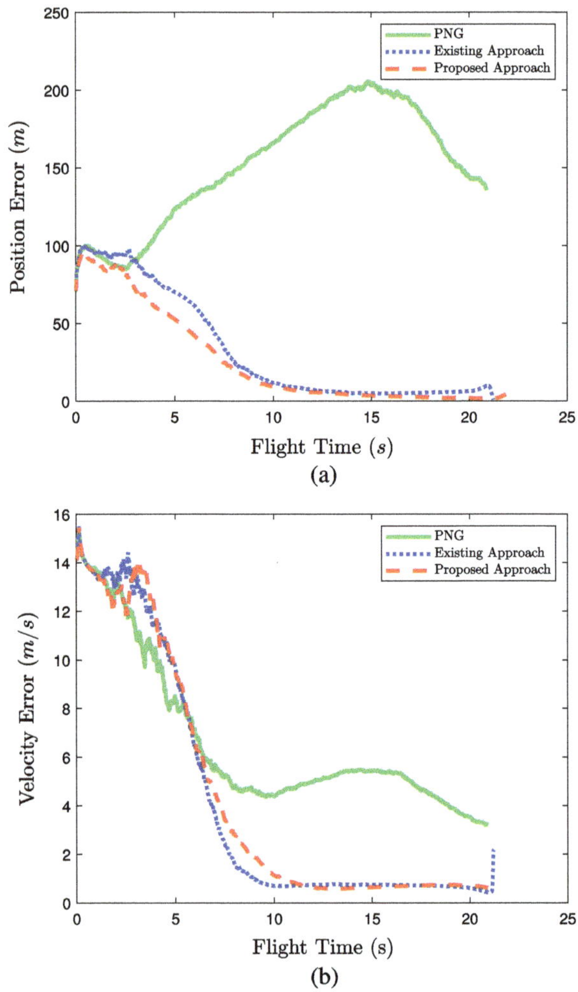

Fig. 4.9 Estimation performance comparison under different guidance laws. **a** Position RMSE; and **b** velocity RMSE

Table 4.3 Miss distance of 100 Monte-Carlo simulations

	PNG	Guidance law (4.48)	Proposed
Mean miss distance	1.864 m	0.405 m	0.331 m

with

$$F \triangleq \begin{bmatrix} \boldsymbol{I}_{2\times2} & T_s\boldsymbol{I}_{2\times2} \\ \boldsymbol{0}_{2\times2} & \boldsymbol{I}_{2\times2} \end{bmatrix}, \quad G \triangleq \begin{bmatrix} T_s^2/2\boldsymbol{I}_{2\times2} \\ T_s\boldsymbol{I}_{2\times2} \end{bmatrix} \tag{4.50}$$

where $T_s = 0.1\,\text{s}$ stands for the sampling time and $\boldsymbol{w}_k \sim \mathcal{N}\left(\cdot; 0, \boldsymbol{Q}_k\right)$ denotes the Gaussian white process noise with $\boldsymbol{Q}_k = \text{diag}\left(\sigma_v^2, \sigma_v^2\right)$ and $\sigma_v = 1\,\text{m/s}^2$. To initialize UKF, the following initial state values and initial error covariance matrix are used: $\hat{x}(0) = [4500\,\text{m}, -300\,\text{m/s}, -5000\,\text{m}, 300\,\text{m/s}]^T$, $P(0) = diag\left(\left[1000^2, 200^2, 1000^2, 200^2\right]\right)$. In simulations, we assume that the measured LOS angle is corrupted by a white noise with zero mean and standard deviation $\sigma_m = 0.2°$. The root mean square error (RMSE) is employed as a metric to evaluate the estimation performance.

The estimation performance under different guidance laws of 100 Monte-Carlo simulations is shown in Fig. 4.9. From this figure, one can clearly observe that the proposed guidance law provides more accurate position as well as velocity estimations than those of other guidance laws. The reason of this fact can be attributed to target observability improvement induced by the weighting factor ω. The average miss distance of all guidance laws is summarized in Table 4.3. From this table, one can clearly observe that the proposed outperforms other three guidance laws in terms of interception accuracy.

4.6 Summary

This chapter proposes a novel linear quadratic optimal guidance law for passive guidance to improve target observability. The resultant acceleration command turns out to be a PNG with a time-varying navigation gain. Theoretical analysis reveals that, under certain conditions, the proposed guidance law gradually switches from retro-PNG to PNG during the terminal homing phase. The magnitude of ZEM under the proposed guidance law initially increases to a certain value to enhance target observability and converges to zero at the time of impact to guarantee target interception. Nonlinear simulation results clearly validate the analytical findings of this chapter.

Appendix A. Relationship Between the Proposed Observability Metric and Fisher Information Matrix

In estimation theory, the Fisher Information Matrix (FIM), denoted as F, quantifies the amount of information that the sensor measurement carries about the unobservable variable. To analyze the relationship between the proposed geometric observability metric and FIM, we consider the following LOS angle measurement model

$$\sigma = h(\boldsymbol{p}) + \epsilon \tag{4.51}$$

where $\boldsymbol{p} = \left[p_x, p_y\right]^T$ denotes the relative position vector between the missile and the target. The notation ϵ represents a Gaussian noise with zero mean and a constant covariance Σ, i.e., $\epsilon \sim \mathcal{N}(0, \Sigma)$.

The (i, j)th element of FIM for measurement model (4.51) is given by [24]

$$F_{(i,j)} = \mathbb{E}\left[\frac{\partial}{\partial p_i} \ln f(\sigma|\boldsymbol{p}) \frac{\partial}{\partial p_j} \ln f(\sigma|\boldsymbol{p})\right] \tag{4.52}$$

where $f(\sigma|\boldsymbol{p})$ stands for the measurement likelihood and can be determined as

$$f(\sigma|\boldsymbol{p}) = \frac{1}{\sqrt{2\pi\Sigma}} \exp\left[-\frac{(\sigma - h(\boldsymbol{p}))^2}{2\Sigma}\right] \tag{4.53}$$

Under the assumption that the measurement noise is independent with the system state, the FIM can then be simplified as

$$F_{(i,j)} = \frac{1}{\Sigma} \nabla_{\boldsymbol{p}} h(\boldsymbol{p})^T \nabla_{\boldsymbol{p}} h(\boldsymbol{p}) \tag{4.54}$$

where $\nabla_{\boldsymbol{p}} h(\boldsymbol{p})$ denotes the Jacobian of the measurement model.

Based on Eq. (4.54), we can subsequently derive the increment of FIM between two consecutive time instants t_k and t_{k+1} as

$$\begin{aligned} \Delta F(t_k) &= F(t_{k+1}) - F(t_k) \\ &= \frac{1}{\Sigma} \sum_{i=k}^{k+1} \frac{1}{\|\boldsymbol{r}_k\|^2} \begin{bmatrix} \cos^2 \sigma(t_i) & -\frac{\sin[2\sigma(t_i)]}{2} \\ -\frac{\sin[2\sigma(t_i)]}{2} & \sin^2 \sigma(t_i) \end{bmatrix} \end{aligned} \tag{4.55}$$

Since FIM prescribes a lower bound of the estimation error covariance matrix, the target observability can be quantified by the determinant of FIM. The rationale behind this concept is that the maximization the determinant of FIM provides the possibility of minimizing the volume of the uncertainty ellipsoid, thus enabling the improvement of the estimation performance. Taking the determinant of $\Delta F(t_k)$ gives

$$|\Delta F(t_k)| = \frac{1}{\Sigma^2} \frac{\sin^2 \eta}{||\boldsymbol{r}_k||^2 ||\boldsymbol{r}_{k+1}||^2} \propto \frac{\sin^2 \eta}{||\boldsymbol{r}_{k+1}||^2} \tag{4.56}$$

where $\eta = \sigma(t_{k+1}) - \sigma(t_k)$. Notice that when evaluating the accumulated information from t_k to t_{k+1}, the quantity $\boldsymbol{r}_k$ is known.

Assume that the time difference between t_k and t_{k+1} is small, Eq. (4.56) reduces to

$$|\Delta F(t_k)| \propto \frac{\dot{\sigma}^2(t_k)}{||\boldsymbol{r}_k||^2} \tag{4.57}$$

The preceding equation implies that maximizing the accumulated LOS rate or minimizing the relative range helps to improve target observability. This result clearly coincides with the proposed geometric observability metric, as shown in (4.8).

Appendix B. Derivation of Guidance Command (4.48)

As this chapter utilizes a different kinematics model compared with [18], this appendix presents a brief derivation of guidance command (4.48) for the completeness of the chapter. For notational simplicity, define $\boldsymbol{x} = [y, v]^T$. Then, optimization problem (4.46) can be reformulated as

Problem 4.3 Find the acceleration command a_{M_σ} minimizes performance index

$$\min_{a_{m\sigma}} J = \frac{1}{2}\boldsymbol{x}^T(t_f)\boldsymbol{S}\boldsymbol{x}(t_f) + \frac{1}{2}\int_0^{t_f} \boldsymbol{x}^T\boldsymbol{Q}\boldsymbol{x} + a_{M_\sigma}^2 dt \tag{4.58}$$

$$\text{subject to: } \dot{\boldsymbol{x}} = \boldsymbol{A}\boldsymbol{x} + \boldsymbol{B}a_{M_\sigma} \tag{4.59}$$

where

$$\boldsymbol{S} = \begin{bmatrix} \bar{c} & 0 \\ 0 & 0 \end{bmatrix}, \quad \boldsymbol{Q} = \begin{bmatrix} -\bar{\omega} & 0 \\ 0 & 0 \end{bmatrix}, \quad \boldsymbol{A} = \begin{bmatrix} 0 & 1 \\ 0 & 0 \end{bmatrix}, \quad \boldsymbol{B} = \begin{bmatrix} 0 \\ -1 \end{bmatrix}$$

According to standard optimal control theory [25], the optimal guidance command solution is given by

$$a_{m_\sigma} = -\boldsymbol{B}^T\boldsymbol{N}\boldsymbol{M}^{-1}\boldsymbol{x} \tag{4.60}$$

where $\boldsymbol{M} = [m_{ij}] \in \mathbb{R}^{2\times 2}$ and $\boldsymbol{N} = [n_{ij}] \in \mathbb{R}^{2\times 2}$ are determined by the differential equation

$$\begin{aligned} \dot{\boldsymbol{M}} &= \boldsymbol{A}\boldsymbol{M} - \boldsymbol{B}\boldsymbol{B}^T\boldsymbol{N} \\ \dot{\boldsymbol{N}} &= -\boldsymbol{Q}\boldsymbol{M} - \boldsymbol{A}^T\boldsymbol{N} \end{aligned} \tag{4.61}$$

and terminal boundary condition

$$\boldsymbol{M}(t_f) = \boldsymbol{I}_{2\times 2}, \quad \boldsymbol{N}(t_f) = \boldsymbol{S} \tag{4.62}$$

Changing the argument from time t to time-to-go t_{go} results in the differential equation

$$\begin{aligned} \boldsymbol{M}' &= -\boldsymbol{A}\boldsymbol{M} + \boldsymbol{B}\boldsymbol{B}^T\boldsymbol{N} \\ \boldsymbol{N}' &= \boldsymbol{Q}\boldsymbol{M} + \boldsymbol{A}^T\boldsymbol{N} \end{aligned} \tag{4.63}$$

where $'$ denotes the derivative with respect to time-to-go, and initial boundary condition

$$\boldsymbol{M}(0) = \boldsymbol{I}_{2\times 2}, \quad \boldsymbol{N}(0) = \boldsymbol{S} \tag{4.64}$$

Solving matrix differential equation (4.63) with initial boundary condition (4.64) gives the elements of $\boldsymbol{M}$ and $\boldsymbol{N}$ as

$$\begin{aligned}
m_{11} &= a_1 e^{kt_{go}} + a_2 e^{-kt_{go}} + \frac{1}{2}\cos\left(kt_{go}\right) + \frac{\bar{c}}{2k^3}\sin\left(kt_{go}\right) \\
m_{12} &= -\frac{1}{4k}e^{kt_{go}} + \frac{1}{4k}e^{-kt_{go}} - \frac{1}{2k}\sin\left(kt_{go}\right) \\
m_{21} &= -ka_1 e^{kt_{go}} + ka_2 e^{-kt_{go}} + \frac{k}{2}\sin\left(kt_{go}\right) - \frac{\bar{c}}{2k^2}\cos\left(kt_{go}\right) \\
m_{22} &= \frac{1}{4}e^{kt_{go}} + \frac{1}{4}e^{-kt_{go}} + \frac{1}{2}\cos\left(kt_{go}\right) \\
n_{11} &= -k^3 a_1 e^{kt_{go}} + k^3 a_2 e^{-kt_{go}} - \frac{k^3}{2}\sin\left(kt_{go}\right) + \frac{\bar{c}}{2}\cos\left(kt_{go}\right) \\
n_{12} &= \frac{k^2}{4}e^{kt_{go}} + \frac{k^2}{4}e^{-kt_{go}} - \frac{k^2}{2}\cos\left(kt_{go}\right) \\
n_{21} &= -k^2 a_1 e^{kt_{go}} - k^2 a_2 e^{-kt_{go}} + \frac{k^2}{2}\cos\left(kt_{go}\right) + \frac{\bar{c}}{2k}\sin\left(kt_{go}\right) \\
n_{22} &= \frac{k}{4}e^{kt_{go}} - \frac{k}{4}e^{-kt_{go}} - \frac{k}{2}\sin\left(kt_{go}\right)
\end{aligned} \tag{4.65}$$

where

$$k = \bar{\omega}^{\frac{1}{4}}, \quad a_1 = \frac{1}{4} - \frac{\bar{c}}{4k^3}, \quad a_2 = \frac{1}{4} + \frac{\bar{c}}{4k^3}$$

References

1. Hepner SAR, Geering HP (1990) Observability analysis for target maneuver estimation via bearing-only and bearing-rate-only measurements. J Guid Control Dyn 13(6):977–983
2. Fonod R, Shima T (2017) Estimation enhancement by cooperatively imposing relative intercept angles. J Guid Control Dyn 40(7):1711–1725
3. Anjaly P, Ratnoo A (2018) Observability enhancement of maneuvering target with bearings-only information. J Guid Control Dyn 41(1):184–198
4. Hinson BT, Morgansen KA (2014) Observability-based optimal sensor placement for flapping airfoil wake estimation. J Guid Control Dyn 37(5):1477–1486

5. Shaferman V, Shima T (2015) Cooperative optimal guidance laws for imposing a relative intercept angle. J Guid Control Dyn 38(8):1395–1408
6. Shaferman V, Shima T (2017) Cooperative differential games guidance laws for imposing a relative intercept angle. J Guid Control Dyn 40(10):2465–2480
7. Nardone SC, Aidala VJ (1981) Observability criteria for bearings-only target motion analysis. IEEE Trans Aerosp Electron Syst AES-17(2):162–166
8. Tahk M-J, Ryu H, Song E-J (1995) Observability characteristics of angle-only measurement under proportional navigation. In: SICE'95. Proceedings of the 34th SICE annual conference. International session papers. IEEE, pp 1509–1514
9. Fogel E, Gavish M (1988) Nth-order dynamics target observability from angle measurements. IEEE Trans Aerosp Electron Syst 24(3):305–308
10. Hammel SE, Aidala VJ (1985) Observability requirements for three-dimensional tracking via angle measurements. IEEE Trans Aerosp Electron Syst AES-21(2):200–207
11. Song TL, Um TY (1996) Practical guidance for homing missiles with bearings-only measurements. IEEE Trans Aerosp Electron Syst 32(1):434–443
12. Shin H-S, Lee J-I, Cho H, Tsourdos A (2011) Trajectory modulation guidance law for anti-ship missiles. In: AIAA guidance, navigation, and control conference
13. Lee H-I, Shin H-S, Tsourdos A (2017) Weaving guidance for missile observability enhancement. IFAC-PapersOnLine 50(1):15197–15202
14. Lee H-I, Tahk M-J, Sun B-C (2001) Practical dual-control guidance using adaptive intermittent maneuver strategy. J Guid Control Dyn 24(5):1009–1015
15. Seo M-G, Tahk M-J (2015) Observability analysis and enhancement of radome aberration estimation with line-of-sight angle-only measurement. IEEE Trans Aerosp Electron Syst 51(4):3321–3331
16. Lee C-H, Kim T-H, Tahk M-J (2013) Design of impact angle control guidance laws via high-performance sliding mode control. Proc Inst Mech Eng Part G: J Aerosp Eng 227(2):235–253
17. Kim T-H, Lee C-H, Tahk M-J (2013) Time-to-go polynomial guidance with trajectory modulation for observability enhancement. IEEE Trans Aerosp Electron Syst 49(1):55–73
18. Hull DG, Speyer JL, Burris DB (1990) Linear-quadratic guidance law for dual control of homing missiles. J Guid Control Dyn 13(1):137–144
19. Rami MA, Zhou XY (2000) Linear matrix inequalities, Riccati equations, and indefinite stochastic linear quadratic controls. IEEE Trans Autom Control 45(6):1131–1143
20. Ohta H, Nikiforuk PN, Kakinuma M (1991) Use of negative weights in linear quadratic regulator synthesis. J Guid Control Dyn 14(4):791–796
21. Al-Sunni FM, Stevens BL, Lewis FL (1992) Negative state weighting in the linear quadratic regulator for aircraft control. J Guid Control Dyn 15(5):1279–1281
22. Prasanna HM, Ghose D (2012) Retro-proportional-navigation: a new guidance law for interception of high speed targets. J Guid Control Dyn 35(2):377–386
23. Lee C-H, Seo M-G (2018) New insights into guidance laws with terminal angle constraints. J Guid Control Dyn 41(8):1832–1837
24. Bishop AN, Fidan B, Anderson BDO, Doğançay K, Pathirana PN (2010) Optimality analysis of sensor-target localization geometries. Automatica 46(3):479–492
25. Bryson AE (1975) Applied optimal control: optimization, estimation and control. Taylor & Francis, Boca Raton

Chapter 5
Optimal Proportional-Integral Guidance Law

Abstract This chapter proposes a new optimal guidance law based on PI concept to reduce the sensitivity to unknown target maneuvers. By augmenting the integral ZEM as a new system state, a linear quadratic optimization problem is formulated and then the proposed guidance law is analytically derived through optimal control theory. The closed-form solution of the proposed guidance law is presented to provide better insight of its properties. Additionally, the working principle of the integral command is investigated to show why the proposed guidance law can be utilized to reduce the sensitivity to unknown target maneuvers. The analytical results reveal that the proposed optimal guidance law is exactly the same as an instantaneous direct model reference adaptive guidance law with a specified reference model. The potential significance of the obtained results is that it can provide a point of connection between PI guidance laws and adaptive guidance laws. Therefore, it allows us to have better understanding of the physical meaning of both guidance laws and provides the possibility in designing a new guidance law that takes advantages of both approaches. Finally, the performance of the guidance law developed is demonstrated by nonlinear numerical simulations with extensive comparisons.

5.1 Introduction

It follows from Chap. 2 that the optimal guidance law for maneuvering targets interception is the well-known APNG. However, the implementation of APNG requires accurate information on the target maneuver, as shown in Eq. (2.51). Unfortunately, this information is unavailable to the interceptor since target maneuver is uncertain and unpredictable in nature. To address this issue, several adaptive guidance laws have been reported by leveraging a direct or indirect adaptive law to estimate the unknown target maneuver (see [1–4] and references therein for example).

S. He et al., *Optimal Guidance and Its Applications in Missiles and UAVs*, Springer Aerospace Technology, https://doi.org/10.1007/978-3-030-47348-8_5

In the field of control theory, it is well-understood that the concept of PI control is a simple and an effective way to attenuate the effect of external disturbances. Based on this fact, guidance laws based on the PI concept were proposed in [5, 6] to reduce the sensitivity to the unknown target acceleration. However, these guidance algorithms have the following weak points:

(1) Guidance law [5] fails to address the problem of optimality in terms of a meaningful performance index;
(2) The command of guidance law [6] is not given in an explicit form and should be numerically solved at each time step;
(3) There is no rigorous explanation regarding the working principle of PI guidance laws. Notice that this information is important in ensuring confidence in the performance and reliability of the guidance law when implementing it in a real system.

Motivated by these observations, this chapter exploits the PI concept to design an optimal guidance law for maneuvering target interception. More specifically, the integral of ZEM is introduced as an augmented system state and then the proposed optimal guidance law is analytically derived by solving a linear quadratic optimization problem formulated. The key features of the proposed formulation are twofolds:

(1) The optimal PI guidance law is helpful in reducing the sensitivity against unknown target maneuvers and guarantees finite time convergence. This fact is supported by the closed-form solution of the proposed guidance law derived in this chapter.
(2) The physical meaning of the PI guidance laws is analyzed by using an inverse approach and the instantaneous linear time-invariant system concept. The results reveal that the proposed optimal PI guidance law is exactly the same as an instantaneous direct model reference adaptive guidance law with a specified reference model. This provides us better understanding of the PI guidance and establishes a theoretical link between PI guidance and adaptive guidance. Accordingly, it suggests the possibility in designing a new guidance law that exploits the benefits of both approaches.

5.2 Problem Formulation

Since target maneuver is unavailable to the missile, we consider the linearized ZEM without the target maneuver term as

$$z = y + vt_{go} \tag{5.1}$$

Differentiating Eq. (5.1) with respect to time yields

$$\dot{z} = -t_{go}a_{M\sigma} + t_{go}a_{T\sigma} \tag{5.2}$$

Again, since the target acceleration $a_{T\sigma}$ is unknown, we consider it as an external disturbance for system (5.2). By ignoring this unknown term, Eq. (5.2) reduces to

$$\dot{z} = -t_{go} a_{M\sigma} \tag{5.3}$$

In order to reduce the sensitivity against the unknown disturbance, i.e., target maneuver, we exploit the PI control concept in guidance law design. More specifically, we introduce the integration of ZEM, i.e., $\int z dt$, as an augmented state variable. For notation convenience, we define new variables x_1 and x_2 as

$$x_1 = \int z dt, \quad x_2 = z \tag{5.4}$$

By using these two variables, the differential equations that describe the augmented system can be rewritten in a compact matrix form as

$$\begin{bmatrix} \dot{x}_1 \\ \dot{x}_2 \end{bmatrix} = \begin{bmatrix} 0 & 1 \\ 0 & 0 \end{bmatrix} \begin{bmatrix} x_1 \\ x_2 \end{bmatrix} + \begin{bmatrix} 0 \\ -t_{go} \end{bmatrix} a_{M_\sigma} \tag{5.5}$$

This chapter aims to design an optimal guidance law by solving the following optimization problem.

Problem 5.1 Find guidance command a_{M_σ} that minimizes cost function

$$J = \frac{1}{2} \int_t^{t_f} R(\tau) a_{M_\sigma}^2 (\tau) d\tau \tag{5.6}$$

subject to

$$\begin{aligned} &\text{Eq. (5.5)} \\ &x_1 \left(t_f\right) = 0, \quad x_2 \left(t_f\right) = 0 \end{aligned} \tag{5.7}$$

where t_f represents the time of impact and $R(t) > 0$ is an arbitrary weighting function.

The intuition by imposing terminal constraint $x_2 \left(t_f\right) = 0$ in Problem 5.1 is clear: nullifying the ZEM provides perfect interception with zero terminal miss distance, as described in Chap. 2. The rationale by enforcing terminal constraint $x_1 \left(t_f\right) = 0$ is to minimize the cumulative ZEM so as to reduce the sensitivity to external disturbances.

5.2.1 *Preliminary*

This subsection collects a useful lemma from [7] that will be utilized in the derivation of the proposed guidance law.

Lemma 5.1 *Let $\mathcal{H}$ be a Hilbert space and $\boldsymbol{\alpha}_1, \boldsymbol{\alpha}_2, \cdots, \boldsymbol{\alpha}_n$ be a set of n linearly independent vectors in $\mathcal{H}$. If the condition $(\boldsymbol{x}, \boldsymbol{\alpha}_i) = c_i$, $i \in \{1, 2, \cdots, n\}$, with c_i being arbitrary scalars, holds among all vectors of $\mathcal{H}$, then, the one that has the minimum norm is given by*

$$\boldsymbol{x}_{\min} = \sum_{i=1}^{n} b_i \boldsymbol{\alpha}_i \tag{5.8}$$

where the coefficients b_i satisfy the condition

$$\sum_{i=1}^{n} \left(\boldsymbol{\alpha}_i, \boldsymbol{\alpha}_j\right) = b_j, \quad j \in \{1, 2, \cdots, n\} \tag{5.9}$$

5.3 Derivation of the Proposed Optimal Guidance Law

According to the linear control theory, the general solution of the system (5.7) can be expressed as

$$\begin{aligned} x_1\left(t_f\right) &= f_1 - \int_t^{t_f} h_1(\tau)\, a_M(\tau)\, d\tau \\ x_2\left(t_f\right) &= f_2 - \int_t^{t_f} h_2(\tau)\, a_M(\tau)\, d\tau \end{aligned} \tag{5.10}$$

where

$$\begin{aligned} f_1 &\triangleq x_1(t) + \left(t_f - t\right) x_2(t), & h_1 &\triangleq \left(t_f - \tau\right)^2 \\ f_2 &\triangleq x_2(t), & h_2 &\triangleq t_f - \tau \end{aligned} \tag{5.11}$$

Imposing the terminal constraints on Eq. (5.10) gives

$$\begin{aligned} f_1 &= \int_t^{t_f} h_1(\tau)\, a_{M_\sigma}(\tau)\, d\tau \\ f_2 &= \int_t^{t_f} h_2(\tau)\, a_{M_\sigma}(\tau)\, d\tau \end{aligned} \tag{5.12}$$

Introducing the weight function $R(t)$ as a slack variable in Eq. (5.12) gives

$$\begin{aligned} f_1 &= \int_t^{t_f} h_1(\tau)\, R^{-1/2}(\tau)\, R^{1/2}(\tau)\, a_{M_\sigma}(\tau)\, d\tau \\ f_2 &= \int_t^{t_f} h_2(\tau)\, R^{-1/2}(\tau)\, R^{1/2}(\tau)\, a_{M_\sigma}(\tau)\, d\tau \end{aligned} \tag{5.13}$$

Consider the Hilbert space $\mathcal{H} = \mathcal{L}^2\left[t, t_f\right]$ with the inner product $(f, g) = \int_t^{t_f} f(\tau)\, g(\tau)\, d\tau$. According to Lemma 5.1, if the guidance command a_{M_σ} is optimal in terms of minimizing performance index (5.6), then there exist two Lagrange multipliers λ_1, λ_2 such that the lateral acceleration satisfies

$$R^{1/2} a_{M_\sigma} = \lambda_1 h_1 R^{-1/2} + \lambda_2 h_2 R^{-1/2} \tag{5.14}$$

Substituting Eq. (5.14) into Eq. (5.12) results in

$$\begin{aligned} f_1 &= \lambda_1 \int_t^{t_f} h_1^2(\tau)\, R^{-1}(\tau)\, d\tau + \lambda_2 \int_t^{t_f} h_1 h_2(\tau)\, R^{-1}(\tau)\, d\tau \\ f_2 &= \lambda_1 \int_t^{t_f} h_1 h_2(\tau)\, R^{-1}(\tau)\, d\tau + \lambda_2 \int_t^{t_f} h_2^2(\tau)\, R^{-1}(\tau)\, d\tau \end{aligned} \tag{5.15}$$

which can be rewritten in a compact matrix form as

$$\begin{bmatrix} f_1 \\ f_2 \end{bmatrix} = \begin{bmatrix} g_1 & g_{12} \\ g_{12} & g_2 \end{bmatrix} \begin{bmatrix} \lambda_1 \\ \lambda_2 \end{bmatrix} \tag{5.16}$$

where

$$\begin{aligned} g_1 &= \int_t^{t_f} h_1^2(\tau) R^{-1}(\tau) d\tau \\ g_2 &= \int_t^{t_f} h_2^2(\tau) R^{-1}(\tau) d\tau \\ g_{12} &= \int_t^{t_f} h_1(\tau) h_2(\tau) R^{-1}(\tau) d\tau \end{aligned} \tag{5.17}$$

Solving Eq. (5.16) for λ_1 and λ_2 gives

$$\begin{aligned} \lambda_1 &= \frac{1}{g_1 g_2 - g_{12}^2} \left[g_2 f_1 - g_{12} f_2\right] \\ \lambda_2 &= \frac{1}{g_1 g_2 - g_{12}^2} \left[-g_{12} f_1 + g_1 f_2\right] \end{aligned} \tag{5.18}$$

The guidance command can then be obtained by substituting Eq. (5.18) into Eq. (5.14) as

$$a_{M_\sigma} = \frac{\left(h_1 g_2 - h_2 g_{12}\right) R^{-1}}{g_1 g_2 - g_{12}^2} f_1 + \frac{\left(h_2 g_1 - h_1 g_{12}\right) R^{-1}}{g_1 g_2 - g_{12}^2} f_2 \tag{5.19}$$

Using Eqs. (5.4), (5.10), and (5.11), guidance command (5.19) can be formulated in the ZEM format as

$$a_{M_\sigma} = N_1 \frac{\int z dt}{t_{go}^3} + N_2 \frac{z}{t_{go}^2} \tag{5.20}$$

where the guidance gains N_1 and N_2 are given by

$$\begin{aligned} N_1 &= \left(\frac{g_2 t_{go}^5 - g_{12} t_{go}^4}{g_1 g_2 - g_{12}^2} \right) R^{-1}(\tau) \\ N_2 &= \left(\frac{g_2 t_{go}^5 - 2 g_{12} t_{go}^4 + g_1 t_{go}^3}{g_1 g_2 - g_{12}^2} \right) R^{-1}(\tau) \end{aligned} \tag{5.21}$$

Guidance command (5.20) is the generalized optimal PI guidance law with an arbitrary weighting function. By choosing different weighting functions, one can design various optimal intercept angle guidance laws to shape the guidance command as desired. Without loss of generality, we consider a time-to-go weighting function as [8, 9]

$$R^{-1}(t) = t_{go}^N, \quad N \geq 0 \tag{5.22}$$

to shape the guidance command in this chapter. One major advantage of utilizing this weighting function is that it allows for zero final guidance command with $N > 0$ since the weighting function with $N > 0$ becomes infinite as t_{go} approaches 0. Hence, this property is beneficial in improving the operational margins to cope with external disturbances when the missile is close to the target. Substituting Eq. (5.22) into Eq. (5.21) gives the guidance gains as

$$N_1 = (N+4)(N+5), \quad N_2 = 2(N+4) \tag{5.23}$$

Then, the guidance command can be obtained by substituting Eq. (5.23) into Eq. (5.20) as

$$a_{M_\sigma} = (N+4)(N+5)\frac{\int z dt}{t_{go}^3} + 2(N+4)\frac{z}{t_{go}^2} \tag{5.24}$$

Obviously, the term $2(N+4)z/t_{go}^2$ is the well-known PNG with a navigation gain $2(N+4)$, whereas the term $(N+4)(N+5)\int zdt/t_{go}^3$ is the integral command to reduce the sensitivity to the unknown disturbances, i.e., target maneuvers. This property is totally different from previous guidance laws, where the information on the target maneuver is required for the implementation, and will be further analyzed in the following section.

Remark 5.1 For the implementation, it would be better to express the guidance command in terms of the measurable signals of on-board seekers. From Eqs. (2.8) and (5.1), we can determine the kinematics relationship between the LOS rate and the ZEM as

$$z = \dot{\sigma} V_c t_{go}^2 \tag{5.25}$$

Substituting Eq. (5.25) into Eq. (5.24) gives the explicit guidance command as

$$a_{M_\sigma} = (N+4)(N+5)\frac{\int\left(\dot{\sigma}V_c t_{go}^2\right)dt}{t_{go}^3} + 2(N+4)V_c\dot{\sigma} \tag{5.26}$$

Note that this guidance command is given by the function of closing velocity, LOS rate, and time-to-go. Therefore, the proposed guidance law requires only one additional information, i.e., the time-to-go, compared to the conventional PNG. For the implementation, the closing velocity and the LOS rate can be obtained from on-board seekers with a dedicated homing filter. In practice, the time-to-go information can be determined by using the conventional approximation method as

$$t_{go} \approx r/V_c \tag{5.27}$$

Note that although the time-to-go estimation, shown in Eq. (5.27), tends to be smaller than the true time-to-go during the initial flight period, especially when the initial heading error is large [9–12], this time-to-go estimation error gradually converges to zero as the missile approaches the target. Also, in the proposed guidance law, the time-to-go information is only used for computing the integral term, which is utilized to reduce the sensitivity against the unknown target acceleration. This implies that the time-to-go estimation error may degrade the robustness against the unknown target acceleration initially, but the degradation of final intercept accuracy due to this time-to-go estimation error is negligible in practice.

5.4 Analysis of Proposed Optimal Guidance Law

In this section, we will analyze the properties of the proposed guidance law. First, we discuss the ZEM dynamics and the closed-form solution of the guidance command to provide better understanding of the proposed guidance law. Then, the working principle of the proposed guidance law is investigated by comparing it with the adaptive guidance law.

5.4.1 Behavior of the ZEM Dynamics

Substituting Eq. (5.24) into Eq. (5.2) gives the second-order ZEM dynamics as

$$\dot{z} + N_2\frac{z}{t_{go}} + N_1\frac{\int z dt}{t_{go}^2} = t_{go}a_{T\sigma} \tag{5.28}$$

The left-hand side of Eq. (5.28) represents the characteristic equation of the ZEM dynamics under the proposed guidance law. Additionally, the right-hand side of Eq. (5.28) is the external forcing term to the ZEM dynamics. From Eq. (5.28), we can determine the natural frequency ω and damping ratio ζ of the ZEM dynamics as

$$\omega = \frac{\sqrt{N_1}}{t_{go}}, \quad \zeta = \frac{N_2}{2\sqrt{N_1}} \tag{5.29}$$

which reveals that

$$\frac{\sqrt{N_1}}{t_f} \leq \omega < \infty \tag{5.30}$$

The preceding inequality shows that the bandwidth of the guidance loop increases to infinite under the proposed guidance law when the interceptor approaches the target. Also, we can readily observe that the damping ratio is proportional to the value of N_2, which is the guidance gain of the PNG term.

5.4.2 *Closed-Form Solution of the Proposed Guidance Law*

To further analyze the characteristics of the ZEM, we seek to find its closed-form solution. It can be noted that Eq. (5.28) is a second-order Cauchy-Euler differential equation with respect to t_{go}. By considering $\int zdt = ct_{go}^n$ as a fundamental solution of Eq. (5.28), where c and n are two nonzero constants, one can readily derive the homogeneous solution as

$$\int zdt = c_1 t_{go}^{N+5} + c_2 t_{go}^{N+4} \tag{5.31}$$

where c_1, c_2 are two constants governed by the initial conditions.

Additionally, by assuming a_{T_σ} is constant and considering $\int zdt = pt_{go}^3$ as a particular solution of Eq. (5.28), where p is a nonzero constant, one can obtain the particular solution as

$$\int zdt = \frac{a_{T_\sigma}}{(N+1)(N+2)} t_{go}^3 \tag{5.32}$$

Then, combining Eqs. (5.31) and (5.32) gives the closed-form solution of the integral ZEM as

$$\int zdt = c_1 t_{go}^{N+5} + c_2 t_{go}^{N+4} + \frac{a_{T_\sigma}}{(N+1)(N+2)} t_{go}^3 \tag{5.33}$$

Applying the initial boundary conditions on the generic solution (5.33) generates

$$c_1 = -z_0 t_f^{-N-4} + \frac{a_{T_\sigma}}{N+2} t_f^{-N-2} \tag{5.34}$$

$$c_2 = z_0 t_f^{-N-3} - \frac{a_{T_\sigma}}{N+1} t_f^{-N-1} \tag{5.35}$$

where z_0 denotes the initial ZEM.

Differentiating Eq. (5.33) with respect to time gives the closed-form ZEM as

$$z = -c_1 (N+5) t_{go}^{N+4} - c_2 (N+4) t_{go}^{N+3} - \frac{3a_{T_\sigma}}{(N+1)(N+2)} t_{go}^2 \tag{5.36}$$

which clearly reveals that the ZEM can converge to zero as the missile approaches the target, $t_{go} \to 0$, even in the presence of unknown target accelerations. Therefore, unlike to previous PI guidance laws [5, 6], the proposed guidance law can rigorously guarantee the finite time convergence. Also, from Eq. (5.36), we can readily observe that the effect of unknown target acceleration on the ZEM can be mitigated as N increases.

Substituting Eq. (5.33) and Eq. (5.36) into Eq. (5.24) results in

$$\begin{aligned} a_{M_\sigma} = & - c_1 (N+4)(N+5) t_{go}^{N+2} - c_2 (N+3)(N+4) t_{go}^{N+1} \\ & + \frac{(N+4)(N-1)}{(N+1)(N+2)} a_{T_\sigma} \end{aligned} \tag{5.37}$$

which shows that the guidance command converges to a specific value at the time of impact once as

$$a_{M_\sigma}\left(t_f\right) = \frac{(N+4)(N-1)}{(N+1)(N+2)} a_{T_\sigma} \tag{5.38}$$

As shown in Eq. (5.38), the magnitude of the final guidance command is governed by N and the magnitude of target acceleration. Therefore, we can readily predict that the proposed guidance command converges to zero under $N \geq 0$ for a non-maneuvering target, i.e., $a_{T_\sigma} = 0$. Additionally, taking the derivative of a_{M_σ} with respect to time, one can imply that

$$\dot{a}_{M_\sigma} = c_1 (N+2)(N+4)(N+5) t_{go}^{N+1} + c_2 (N+1)(N+3)(N+4) t_{go}^{N} \tag{5.39}$$

Solving $\dot{a}_{M_\sigma} = 0$ for t_{go} gives

$$t_{go,m} = -\frac{c_2 (N+1)(N+3)}{c_1 (N+2)(N+5)} \tag{5.40}$$

which reveals that the guidance command takes its maximum magnitude either at the boundary points or $t_{go} = t_{go,m}$. This information is helpful in finding the required acceleration demand with physical constraints.

5.4.3 Sensitivity to Unknown Target Acceleration

This subsection reveals the working principle of the proposed guidance law, i.e., how the proposed guidance law can be leveraged to reduce the sensitivity against unknown target maneuvers. Especially, the physical meaning of the integral term in the guidance command is analyzed by introducing an alternative form of the proposed guidance law. To this end, substituting Eq. (5.24) into Eq. (5.2) gives the closed-loop ZEM dynamics as

$$\dot{z} = -t_{go}\left(N_2\frac{z}{t_{go}^2} + N_1\frac{\int z dt}{t_{go}^3}\right) + t_{go}a_{T_\sigma} \tag{5.41}$$

For notation simplicity, let us define the equivalent proportional gain k_P, the equivalent integral gain k_I, and the control input gain b, as

$$k_P = \frac{N_2}{t_{go}^2}, \quad k_I = \frac{N_1}{t_{go}^3}, \quad b = t_{go} \tag{5.42}$$

Then, by using these variables, the closed-loop ZEM dynamics shown in Eq. (5.41) can be rewritten as

$$\dot{z} = -b\left(k_P z + k_I \int z dt\right) + b a_{T_\sigma} \tag{5.43}$$

Since the gains k_P, k_I and b are time-varying due to the time-to-go, we utilize the widely-accepted instantaneous linear time-invariant system concept for analysis. In other words, we assume that k_P, k_I and b are piece-wise constants. Note that this assumption is widely-used in guidance loop analysis. Next, let us define a new variable as

$$\hat{a}_{T_\sigma} = k_I \int z dt \tag{5.44}$$

Then, the ZEM dynamics (5.43) can be rearranged using this variable as

$$\dot{z} = -b k_P z - b\hat{a}_{T_\sigma} + b a_{T_\sigma} \tag{5.45}$$

Additionally, by the instantaneous linear time-invariant system concept, the time-derivation of Eq. (5.44) can be determined as

$$\dot{\hat{a}}_{T_\sigma} = k_I z \tag{5.46}$$

In the guidance problem, since zero ZEM will lead to a perfect interception, the desired value of the ZEM can be written as $z_d = 0$. Then, by using a slack variable z_d, Eq. (5.46) can be rewritten as

$$\dot{\hat{a}}_{T_\sigma} = k_I (z - z_d) \tag{5.47}$$

From Eq. (5.47), the term $(z - z_d)$ can be considered as the ZEM tracking error between the current value and the desired value. Therefore, from Eqs. (5.45) and (5.47), one can clearly observe that the proposed optimal PI guidance law can be interpreted as an instantaneous direct model reference adaptive guidance law. In the proposed guidance law, $z_d = 0$ is considered as the reference model and k_I can be regarded as the adaptive gain. Compared to the adaptive guidance law, the term, corresponding to the adaptive gain, is given by a time-varying function of the time-to-go in the proposed guidance law. Based on this observation, we can readily predict that the integral term in the optimal PI guidance law plays the same role as the adaptive/update law in adaptive guidance law. This is why the PI guidance law can be utilized to attenuate the effect of the unknown target maneuver: the integral term, acting as an instantaneous direct adaptive law, estimates the unknown target maneuver online and the estimated signal is feedback to the guidance command to attenuate the effect of target maneuvers. In conclusion, the proposed optimal PI guidance can be regarded as a special case of adaptive guidance with a specified reference model.

Hereafter, we discuss the boundedness of the closed-loop guidance system. We define $\tilde{a}_{T\sigma} = a_{T\sigma} - \hat{a}_{T_\sigma}$ as the estimation error of the adaptive signal and consider the following Lyapunov function candidate

$$\Omega = z^2 + \frac{b}{k_I}\tilde{a}_{T\sigma}^2 \tag{5.48}$$

Assume that the target maneuver is slowly-varying with respect to time, i.e., $\dot{a}_{T\sigma} = 0$, differentiating Ω with respect to time yields

$$\begin{aligned}
\dot{\Omega} &= 2z\dot{z} + \frac{2b}{k_I}\tilde{a}_{T_\sigma}\dot{\tilde{a}}_{T_\sigma} \\
&= 2z\left(-bk_P z + b\tilde{a}_{T_\sigma}\right) - \frac{2b}{k_I}\tilde{a}_{T_\sigma}\dot{\hat{a}}_{T_\sigma} \\
&= 2z\left(-bk_P z + b\tilde{a}_{T_\sigma}\right) - 2b\tilde{a}_{T_\sigma} z \\
&= -2bk_P z^2 \\
&\leq 0
\end{aligned} \tag{5.49}$$

where the fact $bk_P > 0$ is used in the derivation. The preceding inequality proves that the ZEM and the target maneuver estimation errors are uniformly ultimately bounded. Application of Barbalat's lemma [13] reveals that the ZEM can converge to zero even in the presence of target maneuvers. From the above Lyapunov analysis, we notice that the ZEM can be upper bounded by

$$\|z\| \leq \sqrt{\Omega} \leq \sqrt{\Omega\,(0)} = \sqrt{z^2\,(0) + \frac{b}{k_I}\tilde{a}_{T\sigma}^2\,(0)} \tag{5.50}$$

This bound reveals that the effect of target maneuver on ZEM can be reduced by increasing the integral gain k_I. This coincides with the property of adaptive gain in adaptive control. Furthermore, since the integral gain k_I increases as time goes in the proposed guidance law, fast adaptation is guaranteed as the missile approaches the target. In the analysis, the results of sensitivity to target maneuvers under the proposed optimal PI guidance law are derived based on the instantaneous linear time-invariant system concept for the purpose of analysis. Therefore, the presented results may not be exactly the same as the practical ones, but the tendency of the presented results is still valid.

The findings in this subsection have potential importance. First, the results give us better understanding of how PI guidance law works in maneuvering target interception through the integral action. In practice, such information is important in ensuring confidence in the performance and reliability of the guidance law when implementing it in a real system. Second, the results can also provide a point of connection between the PI guidance and adaptive guidance. This allows us to have better understanding and utilization of both guidance laws. For example, we can design new PI guidance laws that provide better performance through appropriate choice of the reference model, like the approaches in modified adaptive control.

5.5 Simulation Results

In this section, the effectiveness of the proposed optimal PI guidance law is validated through nonlinear numerical simulations. We first investigate the fundamental characteristics of the proposed guidance law. Then, the capability of reducing the sensitivity to the unknown target maneuver is tested and compared with the benchmark PNG law and APNG. Finally, the proposed optimal PI guidance is compared with other PI guidance laws. In all simulations, we consider a fast moving target interception scenario. The initial conditions for the considered scenario are summarized in Table 5.1.

Table 5.1 Initial conditions for homing engagement

Parameters	Values
Missile initial position, $(x_M(0), y_M(0))$	(200 m, 3200 m)
Missile flight velocity, V_M	1000 m/s
Missile initial flight path angle, $\gamma_M(0)$	0°
Target initial position, $(x_T(0), y_T(0))$	(1000 m, 3100 m)
Target velocity, V_T	600 m/s
Target initial flight path angle, $\gamma_T(0)$	30°

5.5.1 Characteristics of the Proposed Guidance Law

This simulation in this subsection investigates the basic characteristics of the proposed guidance law with various guidance gains $N = 0, 1, 2, 3$. For this purpose, we assume that the target performs no evasive maneuvers in this case. Figure 5.1a compares the patterns of state variables with various guidance gains. From this figure, it can be noted that the ZEM can converge to zero at the time of impact and therefore successful interception is guaranteed. This result matches with the analytical results of the closed-form solution. The results in Fig. 5.1a also clearly reveal that the proposed guidance law with larger guidance gain N leads to faster convergence speed of the ZEM. The response of the integral ZEM $\int zdt$ with various guidance gains is presented in Fig. 5.1b, which shows that the terminal constraint $\int_0^{t_f} zdt = 0$ is satisfied by the proposed guidance law. Furthermore, Fig. 5.1b also reveals that the variation of the ZEM can be reduced by increasing the guidance gain N, leading to the improved robustness against external disturbances. Figure 5.1c provides the required guidance commands for various guidance gains. Due to physical limits, we enforce hard constraint $|a_M| \leq 400\,\mathrm{m/s^2}$ in the simulations. The corresponding control efforts, defined as $\int_0^{t_f} a_M^2(\tau)\,d\tau$, for various guidance gains are summarized in Table 5.2. One can note from Fig. 5.1b that the proposed guidance law shown with $N \geq 0$ guarantees zero guidance command at the time of impact for a non-maneuvering target as expected in the closed-form solution. However, increasing the guidance gain N requires higher acceleration command during the initial flight period, thereby generating more control effort, as shown in Table 5.2.

5.5.2 Reduced Sensitivity to Unknown Target Maneuvers

The simulation in this subsection investigates the robustness against unknown target maneuvers under the proposed optimal PI guidance law. To justify the robustness of the proposed guidance scheme, two different cases of target maneuvers are considered: (1) $a_T = 300\,\mathrm{m/s^2}$; and (2) $a_T = -300\,\mathrm{m/s^2}$. For better illustration, the

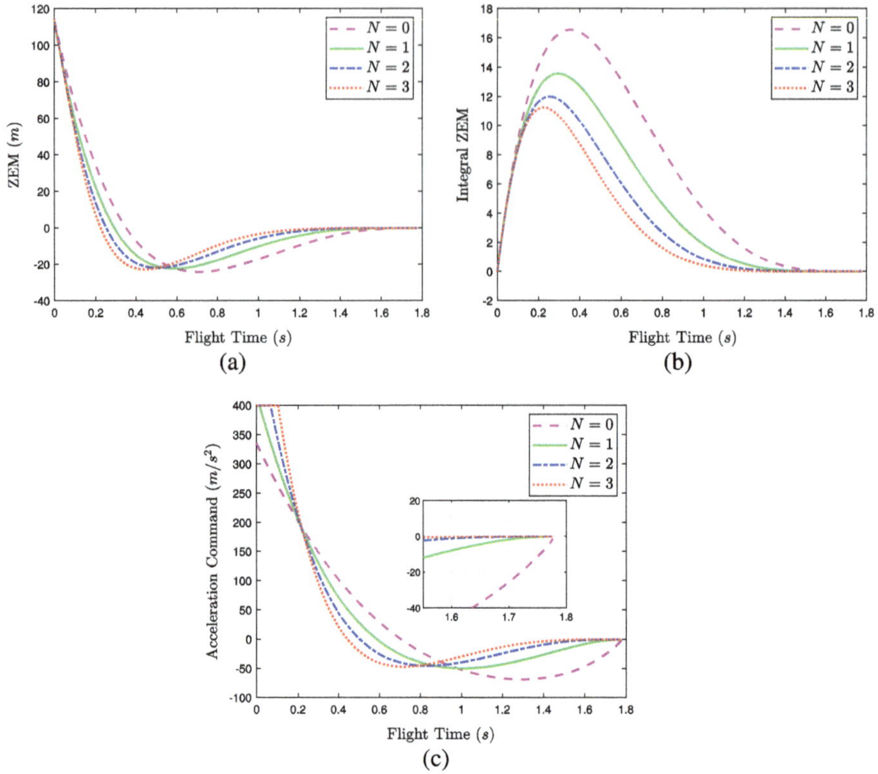

Fig. 5.1 Comparison results of different guidance laws. **a** ZEM; **b** Integral ZEM; and **c** Acceleration command

Table 5.2 Control effort for various guidance gains

	$N = 0$	$N = 1$	$N = 2$	$N = 3$
Control effort	23044	24862	27337	29576

following guidance laws are considered in the simulations for the purpose of comparison.

(1) PNG5:

$$a_{M_\sigma} = 5V_c\dot{\sigma} \tag{5.51}$$

(2) PNG10:

$$a_{M_\sigma} = 10V_c\dot{\sigma} \tag{5.52}$$

(3) APNG:

$$a_{M_\sigma} = 5V_c\dot{\sigma} + \frac{5}{2}a_{T_\sigma} \tag{5.53}$$

(4) The proposed guidance law with $N = 2$.

Note that the proposed guidance law with $N = 2$ minimizes the same performance index as PNG5 and APNG utilized in simulations. However, the implementation of the APNG law requires the information on target maneuver and hence is an ideal case.

The interception trajectories for cases 1 and 2, obtained from these four different guidance laws, are presented in Fig. 5.2a, b. The recorded miss distance and control effort are summarized in Table 5.3. Figure 5.2a, b show that both PNG10 and the proposed optimal PI guidance law successfully intercept the target in both cases with guaranteed miss distance, whereas the missile guided by PNG5 misses the target in case 2. Moreover, the recorded miss distance, provided in Table 5.3, reveals that the proposed guidance law exhibits better performance in terms of the interception accuracy. The responses of ZEM for cases 1 and 2 under different guidance laws are presented in Fig. 5.2c, d. From these two figures, it can be noted that the ZEM under the proposed optimal PI guidance law as well as the PNG10 law can converge to zero at the time of impact to guarantee successful interception and has a divergent pattern under the PNG5 law for case 2. Figure 5.2e, f provide the acceleration command of different guidance laws for both cases. Compared to the PNG10 law, the proposed guidance law requires relatively higher acceleration command at the initial flight phase but the demand command converges to a low level when the interceptor approaches the target. Furthermore, the recorded control efforts, given in Table 5.3, reveal that the proposed optimal PI guidance law requires less energy consumption than the other two laws in the two considered scenarios. From the results, shown in Table 5.3, one can note that APNG achieves the best performance in terms of miss distance and control effort. However, APNG is an ideal law that is limited to implement in real applications due to the requirement of accurate target maneuver information. Compared to two PNG laws, the performance of the proposed optimal PI guidance law is comparable to that of the APNG and it is not required target maneuver information, which demonstrates the superiority of the proposed approach.

5.5.3 Comparison with Previous PI Guidance Laws

To further show the superiority and gain insight of the proposed guidance law, comparisons with previous PI guidance laws [5, 6] are performed in a same scenario in this subsection. In [5], a heuristic PI guidance law was developed as

$$a_{M_\sigma} = N_p V_c \dot{\sigma} + N_I V_c \sigma \tag{5.54}$$

where N_p and N_I are the user-defined constant guidance gains.

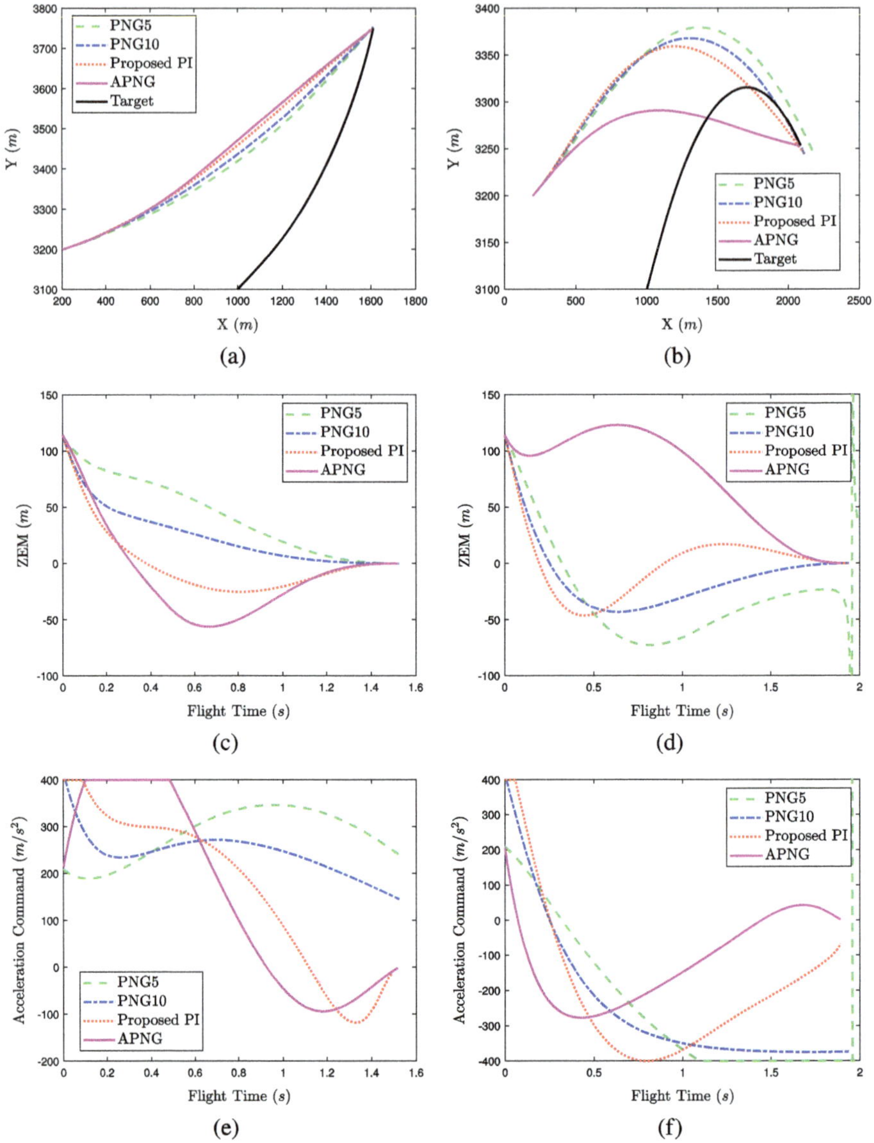

Fig. 5.2 Comparison results of different guidance laws. **a** Interception trajectory of case 2; **b** Interception trajectory of case 2; **c** ZEM of case 1; **d** ZEM of case 2; **e** Acceleration command of case 1; and **f** Acceleration command of case 2

Table 5.3 Miss distance and control effort of different guidance laws

	Scenario	PNG5	PNG10	Proposed PI	APNG
Miss distance	Case1	2.018 m	1.114 m	0.743 m	0.651 m
	Case2	30.669 m	1.554 m	0.914 m	0.663 m
Control effort	Case 1	125582	92221	84605	83511
	Case 2	200028	190457	151577	50864

Unlike [5], the authors in [6] suggested an optimal PI guidance law by utilizing the linear quadratic regulation (LQR) approach. The acceleration command of this guidance law is given by

$$a_{M_\sigma} = K_P(t)\dot{\sigma} + K_I(t)\sigma \tag{5.55}$$

where the time-varying guidance gains $K_P(t)$ and $K_I(t)$ are obtained by solving a infinite time LQR problem. Detailed derivation of this command is shown in appendix.

It follows from Eqs. (5.54) and (5.55) that PI guidance law (5.54) is a special form of the optimal PI guidance law (5.55). Therefore, we only compare the proposed PI guidance law with the optimal PI guidance law Eq. (5.55). It is known that, in LQR control, increasing the value of the diagonal elements of $\boldsymbol{Q}$ increases the penalty on the system states while increasing R helps to reduce the control effort. Therefore, we utilize different $\boldsymbol{Q}$ and R for the implementation of guidance law Eq. (5.55) to demonstrate the effectiveness of the proposed approach: (1) PI guidance 1: $\boldsymbol{Q} = 10^4\boldsymbol{I}, R = 1$; (2) PI guidance 2: $\boldsymbol{Q} = 10^6\boldsymbol{I}, R = 1$; and (3) PI guidance 3: $\boldsymbol{Q} = 10^6\boldsymbol{I}$, $R = 10$.

The simulations results, including interception trajectory, ZEM profile and guidance command with $a_T = -300\,\mathrm{m/s^2}$ are presented in Figs. 5.3. The recorded miss distance and control effort are summarized in Table 5.4. From these results, it is clear that both PI guidance law 2 and the proposed guidance law can intercept the target successfully while other two PI guidance laws miss the target due to limited maneuverability of the interceptor. Additionally, the proposed guidance law can reduce the required acceleration command when the missile approaches the target and the control effort during the entire engagement. Since kinematics (5.57) is nonlinear and time-varying, it is intractable to derive a closed-form solution for $K_P(t)$ and $K_I(t)$. Therefore, implementing optimal PI guidance law Eq. (5.55) requires solving a LQR problem at every time instant, which is time-consuming for the onboard flight computer. As a comparison, the proposed guidance law is given in an explicit form of the measured information and thus is more suitable for practical application. Another issue of the optimal PI guidance law (5.55), as stated in [6], is that there is no theoretical way to choose proper $\boldsymbol{Q}$ and R. In contrary, the guidance gain N of the proposed PI guidance law can be easily selected to achieve desired convergence rate of the ZEM by utilizing the closed-form solution, as we discussed in the previous section.

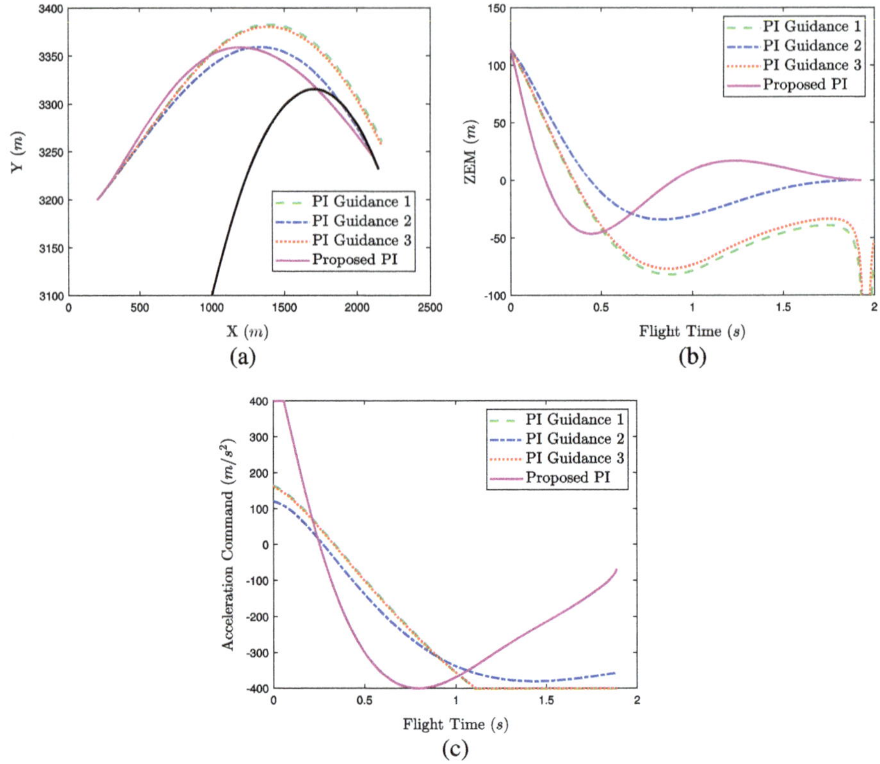

Fig. 5.3 Comparison results of different guidance laws. **a** Interception trajectory; **b** ZEM; and **c** Acceleration command

Table 5.4 Miss distance and control effort of different guidance laws

	PI Guidance 1	PI Guidance 2	PI Guidance 3	Proposed PI
Miss distance	27.127 m	0.981 m	17.003 m	0.914 m
Control effort	191581	167722	180454	151577

5.6 Summary

This chapter proposes a novel optimal guidance law using PI control concept to reduce the sensitivity to unknown target maneuvers. We first augment the integral ZEM as a new system state and formulate a finite time regulation problem for the augmented system. The analytic guidance command is derived by solving the optimization problem formulated. The closed-form solutions are also determined to investigate the properties of the proposed guidance law. By introducing an alternative form of the guidance command, the working principle of the proposed optimal PI guidance law is analyzed and the results reveal that the proposed guidance law is exactly the same as an instantaneous model reference adaptive guidance law with a specified reference model.

Appendix. Derivation of Guidance Command (5.55)

This appendix presents a brief derivation of guidance command (5.55) for the completeness of the chapter. From Eq. (2.1), we have

$$\ddot{\sigma} = -\frac{2\dot{r}}{r}\dot{\sigma} - \frac{1}{r}a_{M_\sigma} + \frac{1}{r}a_{T_\sigma} \tag{5.56}$$

By considering the unknown target maneuver as external disturbance, we ignore the unknown term a_{T_σ} in Eq. (5.56) as

$$\ddot{\sigma} = -\frac{2\dot{r}}{r}\dot{\sigma} - \frac{1}{r}a_{M_\sigma} \tag{5.57}$$

For notational simplicity, define $\boldsymbol{x} = [x_1, x_2]^T$ with $x_1 = \sigma$ and $x_2 = \dot{\sigma}$. Then, the LOS rate dynamics can be rewritten as

$$\begin{aligned} \dot{x}_1 &= x_2 \\ \dot{x}_2 &= -\frac{2\dot{r}}{r}x_2 - \frac{1}{r}a_{M_\sigma} \end{aligned} \tag{5.58}$$

Based on Eq. (5.57), we can formulate the following optimization problem.

Problem 5.2 Find guidance command a_{M_σ} that minimizes cost function

$$J = \int_0^\infty \left(\boldsymbol{x}^T \boldsymbol{Q} \boldsymbol{x} + R a_{M_\sigma}^2\right) dt \tag{5.59}$$

subject to

Eq. (5.58)

where $\boldsymbol{Q}$ and R are the user-defined positive definite weighting matrices.

The guidance command (5.55) can then be obtained by solving optimization Problem 5.2 at each step with proper weighting matrices $\boldsymbol{Q}$ and R.

References

1. Chwa D, Choi JY (2003) Adaptive nonlinear guidance law considering control loop dynamics. IEEE Trans Aerosp Electron Syst 39(4):1134–1143
2. Chwa D, Choi JY, Anavatti SG (2006) Observer-based adaptive guidance law considering target uncertainties and control loop dynamics. IEEE Trans Control Syst Technol 14(1):112–123
3. Zhou D, Biao X (2016) Adaptive dynamic surface guidance law with input saturation constraint and autopilot dynamics. J Guid Control Dyn 39(5):1155–1162
4. Cho D, Kim HJ, Tahk M-J (2016) Fast adaptive guidance against highly maneuvering targets. IEEE Trans Aerosp Electron Syst 52(2):671–680
5. Golestani M, Mohammadzaman I (2015) Pid guidance law design using short time stability approach. Aerosp Sci Technol 43:71–76
6. Evcimen Ç, Leblebicioğlu K (2013) An adaptive, optimal proportional-integral guidance for missiles. In: *AIAA guidance, navigation, and control conference*
7. Luenberger DG (1997) Optimization by vector space methods. Wiley, Hoboken
8. Zarchan P (2012) Tactical and strategic missile guidance. American Institute of Aeronautics and Astronautics
9. Ryoo C-K, Cho H, Tahk M-J (2006) Time-to-go weighted optimal guidance with impact angle constraints. IEEE Trans Control Syst Technol 14(3):483–492
10. Ratnoo A, Ghose D (2009) State-dependent riccati-equation-based guidance law for impact-angle-constrained trajectories. J Guid Control Dyn 32(1):320–325
11. Lee Y-I, Kim S-H, Lee J-I, Tahk M-J (2013) Analytic solutions of generalized impact-angle-control guidance law for first-order lag system. J Guid Control Dyn 36(1):96–112
12. Ryoo C-K, Cho H, Tahk M-J (2005) Optimal guidance laws with terminal impact angle constraint. J Guid Control Dyn 28(4):724–732
13. Khalil HK (1996) Noninear systems. Prentice-Hall, New Jersey

Chapter 6
Gravity-Turn-Assisted Optimal Guidance Law

Abstract This chapter proposes a new optimal guidance law that directly utilizes, instead of compensating, the gravity for accelerating missiles. The desired collision triangle that considers both gravity and vehicle's axial acceleration is analytically derived based on geometric conditions. The concept of instantaneous zero-effort-miss is introduced to allow for analytical guidance command derivation. The proposed optimal guidance law is derived by using the optimal error dynamics proposed in Chap. 2. The relationships of the proposed formulation with conventional PNG and guidance-to-collision (G2C) are analyzed and the results show that the proposed guidance law encompasses previously suggested approaches. The significant contribution of the proposed guidance law lies in that it ensures zero final guidance command and enables energy saving with the aid of utilizing gravity turn. Nonlinear numerical simulations clearly demonstrate the effectiveness of the proposed approach.

6.1 Introduction

The derivation of classical PNG assumes the vehicle is moving with a constant speed without the effect of gravitational acceleration. For endo-atmospheric scenarios, the vehicle's speed is only influenced by the aerodynamic forces and hence can be considered as a slowly varying variable. This means that the moving speed can be assumed as a piece-wise constant and justify the constant-speed assumption. However, exo-atmospheric interceptors, e.g., kinematic kill vehicles, require high impact energy to increase the effectiveness of direct hit and hence are usually subject to large accelerations. In this case, PNG cannot drive the missile to follow the desired straight line collision course and is far away from energy optimal [1, 2]. To address this issue, the concept of G2C is introduced to design guidance laws for accelerating missiles. The basic idea of G2C is to consider the effect of axial acceleration in deriving the collision triangle. This strategy was realized by both SMC [3] and optimal control [4] to guide the interceptor on a straight line collision course to approach the predicted interception point (PIP).

S. He et al., *Optimal Guidance and Its Applications in Missiles and UAVs*, Springer Aerospace Technology, https://doi.org/10.1007/978-3-030-47348-8_6

The issue associated with the G2C law is that it is derived under the gravity-free assumption and hence requires additional compensation term to counteract the effect of gravity in implementation. Due to the existence of the additional term, the direct compensation strategy cannot guarantee zero terminal guidance command, thereby leading to the sacrifice of operational margins to cope with undesired disturbances. Notice that direct gravity compensation might require extra energy and hence result in the decrease of impact energy. Motivated by these observations, this chapter aims to propose a new gravity-turn-assisted optimal guidance law that automatically utilizes the gravity, rather than compensating it, for accelerating missiles. The main idea to solve this problem is to regulate the ZEM, that considers both axial acceleration and gravity, to zero in a finite time, thus guiding the missile to follow a desired curved path to intercept the target. The key features of the proposed algorithm are twofolds:

(1) Unlike PNG and G2C, the resulting desired interception path defined by the collision triangle that considers both axial acceleration and gravity is a curved trajectory instead of a straight line. This fact makes analytic guidance command derivation intractable. To address this problem, a new concept, called instantaneous ZEM, is introduced and the proposed optimal guidance law is then derived by using the optimal error dynamics proposed in Chap. 2.
(2) Detailed analysis reveals that both PNG and G2C are special cases of the proposed guidance law. The advantages of the proposed approach are clear: guaranteeing zero final guidance command and saving energy without requiring extra control effort to compensate for the gravity.

6.2 Problem Formulation

This chapter considers an exo-atmospheric scenario and the interceptor is equipped with a rocket motor to provide a constant axial acceleration a_x. The missile employs an attitude controller to change the angle-of-attack α_M and consequently the axial acceleration can be leveraged to control the vehicle's heading for trajectory shaping [3]. Figure 6.1 presents the typical configuration of an exo-atmospheric interceptor. For simplicity, we assume that the attitude control system is ideal, i.e., the desired attitude is obtained with no time delays. Due to the existence of axial acceleration and gravity, the flight path angle and speed evolve according to

$$\dot{\gamma}_M = \frac{a_x \sin \alpha_M - g \cos \gamma_M}{V_M} \tag{6.1}$$

$$\dot{V}_M = a_x \cos \alpha_M - g \sin \gamma_M \tag{6.2}$$

where g stands for the gravitational acceleration, which is assumed to be constant in guidance law design since the duration of terminal guidance phase is typically very short.

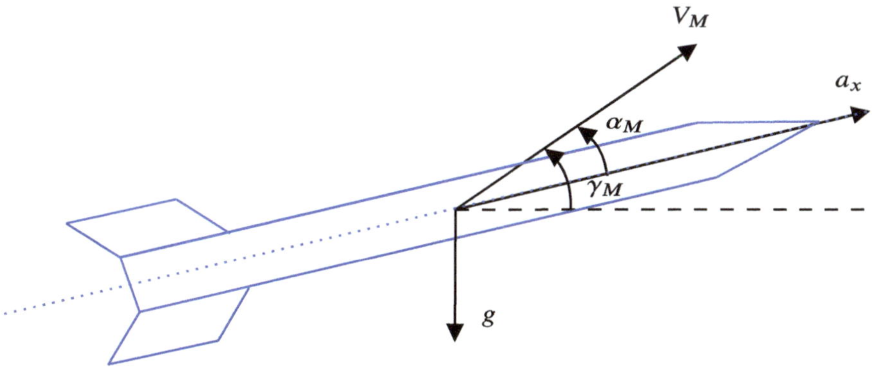

Fig. 6.1 Typical configuration of an exo-atmospheric interceptor

The control interest of exo-atmospheric interception is to design a guidance law to nullify the initial heading error such that the interceptor maintains its acceleration vector in the direction of its velocity vector thereafter. Consequently, if the gravitational effect is ignored, the missile will fly along a straight line that requires no extra control effort to the expected impact point. This characteristic is of great importance to kinematic kill vehicles since this strategy can reduce the magnitude of interceptor's angle-of-attack and hence increase the kill probability [3]. To realize the concept of direct hit, the key part is to find the closed-form solution of an ideal collision triangle for the missile that requires no extra control effort, i.e., $\alpha_M = 0$. Once we have the closed-form solution, we can easily design a guidance law that drives the missile trajectory to converge to the ideal collision triangle in finite time by utilizing the optimal error dynamics proposed in Chap. 2.

However, due to the effect of gravity, the zero-control trajectory is no longer a straight line and hence most previous guidance laws were developed under the gravity-free assumption for simplicity. In practical applications, an additional gravity-compensation term $g \cos \gamma_M$ is then leveraged to reject the effect of gravity in implementation. This straightforward approach obviously has two main drawbacks:

(1) Compensating gravity using extra term $g \cos \gamma_M$ cannot guarantee zero terminal guidance command, leading to the sacrifice of operational margins;
(2) The additional term might require additional energy consumption for gravity compensation and hence impose adverse effect on the effectiveness of direct hit due to energy loss.

In order to address these two problems, the objective of this chapter is to propose a new gravity turn-assisted optimal guidance law that automatically utilizes the gravity, rather than compensating it, for accelerating missiles.

6.3 Collision Triangle Derivation

This section derives the closed-form solution of the proposed ideal collision triangle in the presence of gravity. Once we obtain the ideal collision triangle, we can utilize the optimal control theory to design a guidance law that forces to missile to fly along the collision triangle to achieve the design goal.

Definition 6.1 The ideal motion of the interceptor is defined as the missile kinematics with zero control input.

In the derivation of the desired collision triangle, it is natural to enforce the condition of ideal motion of the interceptor, that is, $\alpha_M = 0$ for exo-atmospheric case as our goal is to make the terminal guidance command converge to zero. Under this condition, the missile's kinematics is formulated as

$$\dot{x}_M = V_M \cos \gamma_M \tag{6.3}$$

$$\dot{y}_M = V_M \sin \gamma_M \tag{6.4}$$

$$\dot{\gamma}_M = -\frac{g \cos \gamma_M}{V_M} \tag{6.5}$$

$$\dot{V}_M = a_x - g \sin \gamma_M \tag{6.6}$$

where (x_M, y_M) represents the inertial position of the interceptor.

It follows from Eqs. (6.3)–(6.6) that these four differential equations are dependent on the flight path angle γ_M and direct integration seems to be intractable. From the point of problem dimension reduction, we manually change the argument from time t to flight path angle γ_M. Then, Eqs. (6.3)–(6.6) can be reformulated as

$$\frac{dt}{d\gamma_M} = -\frac{V_M}{g \cos \gamma_M} \tag{6.7}$$

$$\frac{dx_M}{d\gamma_M} = \frac{dx_M}{dt}\frac{dt}{d\gamma_M} = -\frac{V_M^2}{g} \tag{6.8}$$

$$\frac{dy_M}{d\gamma_M} = \frac{dy_M}{dt}\frac{dt}{d\gamma_M} = -\frac{V_M^2}{g} \tan \gamma_M \tag{6.9}$$

$$\frac{dV_M}{d\gamma_M} = \frac{dV_M}{dt}\frac{dt}{d\gamma_M} = V_M \tan \gamma_M - \frac{a_x V_M}{g \cos \gamma_M} \tag{6.10}$$

Through this argument changing, we only need to solve three independent differential equations to find the analytical solution. Let $\kappa = -a_x/g$, then, Eq. (6.10) can be rewritten as

$$\frac{dV_M}{d\gamma_M} = V_M \tan\gamma_M + \frac{\kappa V_M}{\cos\gamma_M} \tag{6.11}$$

Rearranging Eq. (6.11) as

$$\frac{dV_M}{V_M} = \left(\tan\gamma_M + \frac{\kappa}{\cos\gamma_M}\right) d\gamma_M \tag{6.12}$$

Integrating both sides of Eq. (6.12) gives

$$\ln V_M\left(\gamma_M\right)\Big|_{\gamma_{M0}}^{\gamma_{Mf}} = -\ln\left|\cos\gamma_M\right|\Big|_{\gamma_{M0}}^{\gamma_{Mf}} + \kappa \ln\left|\sec\gamma_M + \tan\gamma_M\right|\Big|_{\gamma_{M0}}^{\gamma_{Mf}} \tag{6.13}$$

where γ_{Mf} and γ_{M0} denote the final and initial flight path angles, respectively.
Solving Eq. (6.13) for $V_M\left(\gamma_{Mf}\right)$ yields

$$V_M\left(\gamma_{Mf}\right) = C\left|\sec\gamma_{Mf}\right|\left|\sec\gamma_{Mf} + \tan\gamma_{Mf}\right|^{\kappa} \tag{6.14}$$

where C is the integration constant, determined by the initial conditions as

$$C = \frac{V_M(\gamma_{M0})}{\left|\sec\gamma_{M0}\right|\left|\sec\gamma_{M0} + \tan\gamma_{M0}\right|^{\kappa}} \tag{6.15}$$

Setting γ_{Mf} as γ_M and substituting Eq. (6.14) into Eqs. (6.7)–(6.9) results in

$$\begin{aligned}\frac{dt}{d\gamma_M} &= -\frac{C}{g}\sec\gamma_M\left|\sec\gamma_M\right|\left|\sec\gamma_M + \tan\gamma_M\right|^{\kappa} \\ &= -\mathrm{sgn}\left(\cos\gamma_M\right)\frac{C}{g}\sec\gamma_M\left|\sec\gamma_M + \tan\gamma_M\right|^{\kappa}\end{aligned} \tag{6.16}$$

$$\frac{dx_M}{d\gamma_M} = -\frac{C^2}{g}\sec^2\gamma_M\left|\sec\gamma_M + \tan\gamma_M\right|^{2\kappa} \tag{6.17}$$

$$\frac{dy_M}{d\gamma_M} = -\frac{C^2}{g}\sec^2\gamma_M\tan\gamma_M\left|\sec\gamma_M + \tan\gamma_M\right|^{2\kappa} \tag{6.18}$$

Following the detailed derivations shown in Appendix, we have the closed-form solution as

$$t\left(\gamma_M\right) = t\left(\gamma_{M0}\right) - \mathrm{sgn}\left(\cos\gamma_M\right)\frac{C}{g}\left[f_t\left(\gamma_M\right) - f_t\left(\gamma_{M0}\right)\right] \tag{6.19}$$

$$x_M\left(\gamma_M\right) = x_M\left(\gamma_{M0}\right) - \frac{C^2}{g}\left[f_x\left(\gamma_M\right) - f_x\left(\gamma_{M0}\right)\right] \tag{6.20}$$

$$y_M(\gamma_M) = y_M(\gamma_{M0}) - \frac{C^2}{g}\left[f_y(\gamma_M) - f_y(\gamma_{M0})\right] \tag{6.21}$$

where

$$f_t(\gamma_M) = \frac{1}{\kappa^2 - 1}(\kappa \sec\gamma_M - \tan\gamma_M)\left|\sec\gamma_M + \tan\gamma_M\right|^{\kappa} \tag{6.22}$$

$$f_x(\gamma_M) = \frac{1}{4\kappa^2 - 1}(2\kappa \sec\gamma_M - \tan\gamma_M)\left|\sec\gamma_M + \tan\gamma_M\right|^{2\kappa} \tag{6.23}$$

$$f_y(\gamma_M) = \frac{1}{4\kappa^2 - 4}\left(2\kappa \sec\gamma_M \tan\gamma_M - \tan^2\gamma_M - \sec^2\gamma_M\right)\left|\sec\gamma_M + \tan\gamma_M\right|^{2\kappa} \tag{6.24}$$

Setting $\gamma_{M0} = \gamma_M$ and $\gamma_M = \gamma_{Mf}$ in Eqs. (6.19)–(6.21) provides the trajectory of the interceptor with zero control effort as

$$t(\gamma_{Mf}) = t(\gamma_M) - \frac{C}{g}\left[f_t(\gamma_{Mf}) - f_t(\gamma_M)\right] \tag{6.25}$$

$$x_M(\gamma_{Mf}) = x_M(\gamma_M) - \frac{C^2}{g}\left[f_x(\gamma_{Mf}) - f_x(\gamma_M)\right] \tag{6.26}$$

$$y_M(\gamma_{Mf}) = y_M(\gamma_M) - \frac{C^2}{g}\left[f_y(\gamma_{Mf}) - f_y(\gamma_M)\right] \tag{6.27}$$

Since the target acceleration is usually unavailable to the missile in practice, we assume that the target leverages a gravity compensation scheme with constant moving speed in the derivation of the collision triangle. With this in mind, the terminal position of the target after t_{go} is given by

$$x_T(t_f) = x_T + V_T \cos\gamma_T t_{go} = x_M + r\cos\sigma + V_T \cos\gamma_T t_{go} \tag{6.28}$$

$$y_T(t_f) = y_T + V_T \sin\gamma_T t_{go} = y_M + r\sin\sigma + V_T \sin\gamma_T t_{go} \tag{6.29}$$

From Eq. (6.25), the time-to-go t_{go} can be formulated as

$$t_{go} = t(\gamma_{Mf}) - t(\gamma_M) = -\frac{C}{g}\left[f_t(\gamma_{Mf}) - f_t(\gamma_M)\right] \tag{6.30}$$

A perfect interception requires

$$x_M(t_f) = x_T(t_f) \tag{6.31}$$

$$y_M(t_f) = y_T(t_f) \tag{6.32}$$

Equations (6.26)–(6.32) define the ideal instantaneous collision triangle considering gravity that requires no extra control effort for the missile to intercept the target.

Remark 6.1 In previous derivations, we utilize the widely-accepted assumptions that the target adopts a gravity compensation scheme and maintains constant flying velocity. For ballistic targets with no gravity compensation, one can include the gravitational effect in target position prediction to obtain more accurate PIP. This can be easily achieved in a similar way as Eqs. (6.20)–(6.21) by setting $\kappa = 0$.

6.4 Optimal Guidance Law Design and Analysis

In this section, the concept of instantaneous ZEM considering gravity is first introduced and a new optimal guidance law is proposed to drive the instantaneous ZEM to converge to zero in finite time. We then analyze the relationships between our approach and some previous guidance laws and finally extend the proposed guidance law to intercept angle control.

6.4.1 Instantaneous Zero-Effort-Miss

Substituting Eqs. (6.26)–(6.30) into Eqs. (6.31)–(6.32) and using the relationships (6.22)–(6.24), one can obtain two coupled equalities, which are functions of the current and final flight path angles, as

$$f_1\left(\gamma_M, \gamma_{Mf}\right) = 0, \quad f_2\left(\gamma_M, \gamma_{Mf}\right) = 0 \tag{6.33}$$

Due to the complicated forms of $f_1\left(\gamma_M, \gamma_{Mf}\right)$ and $f_2\left(\gamma_M, \gamma_{Mf}\right)$, it is difficult to find the analytical solutions of γ_M and γ_{Mf}. However, the exact roots of the two coupled equations in Eq. (6.33) at each time instant during the terminal homing phase can be easily obtained through some well-known numerical algorithms, such as trust-region algorithm [5] and levenberg–marquardt algorithm [6]. Let the two roots of Eq. (6.33) at the current time be γ^* and γ_f^*, where γ^* denotes the desired current flight path angle and γ_f^* represents the desired terminal flight path angle.

Assumption 6.1 Since the flight path angle is slowly varying, we assume that γ^* is piece-wise constant, i.e., $\dot{\gamma}^* = 0$, in guidance law design.

It is well-known that forcing the missile to fly along the collision triangle requires regulating the corresponding ZEM to converge to zero. Recall that the original ZEM is defined as the final miss distance to the desired final interception course if both the pursuer and the target do not perform any maneuver from the current time instant onward [7], finding the analytic dynamics of the original ZEM is intractable for the

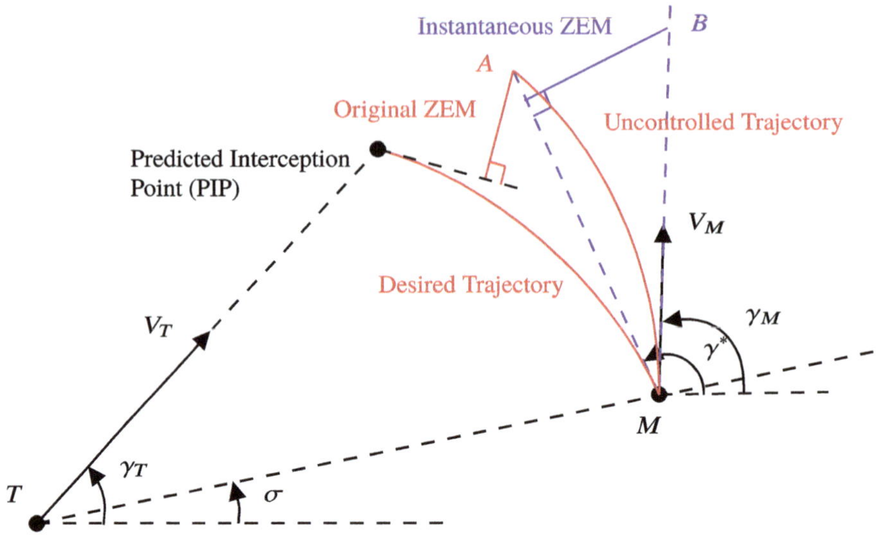

Fig. 6.2 Definition of instantaneous ZEM

curved trajectory due to gravity. To address this problem, we introduce a new concept, called instantaneous ZEM, which is defined on the basis of straight line trajectory at each time instant.

Definition 6.2 The instantaneous ZEM is defined as the final distance the missile would miss the target if the target continues along its present course and the missile follows a straight line along its current flight path angle with no further corrective maneuvers.

The geometric interpretation of the proposed instantaneous ZEM is shown in Fig. 6.2, where the curved path $M - PIP$ is the desired trajectory, $M - A$ the uncontrolled flight path from the current time onward and the straight line $M - B$ the instantaneous uncontrolled flight trajectory. According to the concept of instantaneous ZEM, the travelled length of $M - B$ is given by $V_M t_{go} + 0.5 a_x t_{go}^2$. Let $e_\gamma = \gamma_M - \gamma^*$ be the flight path angle error and z be the instantaneous ZEM. Obviously, one can note that if the instantaneous ZEM converges to zero, the flight path angle error e_γ converges to zero. This also means that the original ZEM converges to zero and therefore the interceptor will follow the desired collision triangle to hit the target. From Fig. 6.2, one can directly derive the instantaneous ZEM as

$$z = -\sin\left(e_\gamma\right)\left(V_M t_{go} + \frac{1}{2} a_x t_{go}^2\right) \tag{6.34}$$

For notation simplicity, let $a_M = a_x \sin \alpha_M$ be the virtual control input. Under small angle assumption of e_γ, the first-order time derivative of the instantaneous

ZEM can be approximated by the length of $M - B$ times the angular rotating speed $\dot{e}_\gamma$ as

$$\dot{z} = -\dot{e}_\gamma \left(V_M t_{go} + \frac{1}{2} a_x t_{go}^2 \right) = -a_M \left(t_{go} + \frac{a_x}{2V_M} t_{go}^2 \right) \tag{6.35}$$

where the kinematics equation $\dot{\gamma}_M = a_M / V_M$ and Assumption 6.1 are used in the derivation of Eq. (6.35).

Remark 6.2 Note that our collision triangle derivation automatically contains the gravity. That is, the desired flight path angle has been computed using a point mass model that already considers gravity and therefore we only need to use $\dot{\gamma}_M = a_M / V_M$ in guidance law design.

6.4.2 Optimal Guidance Law Design

In order to drive the instantaneous ZEM to converge to zero in finite time, various nonlinear control methods can be applied to the dynamics of ZEM as given in Eq. (6.35). Following the general prediction-correction concept established in Chap. 2, a guidance correction command can be determined to make a designed error dynamics to nullify the instantaneous ZEM. To this end, we choose the optimal error dynamics, proposed in Chap. 2, as

$$\dot{z} + \frac{Nz}{t_{go}} = 0 \tag{6.36}$$

where $N > 0$ is a constant guidance gain, which can be tuned to regulate the convergence rate of the instantaneous ZEM.

Substituting Eqs. (6.35) into (6.36) gives the guidance command as

$$a_M = \frac{Nz}{\left(1 + \frac{a_x}{2V_M}\right) t_{go}^2} \tag{6.37}$$

The angle-of-attack command α_M is then obtained as

$$\alpha_M = \arcsin \left(\frac{a_M}{a_x} \right) \tag{6.38}$$

According to Theorem 2.1, guidance command (6.37) is optimal in the sense of minimizing performance index

$$J = \int_t^{t_f} \frac{\left[1 + \frac{1}{2V_M} a_x \left(t_f - \tau\right)\right]^2}{\left(t_f - \tau\right)^{N-3}} a_M^2 (\tau) d\tau \tag{6.39}$$

For interceptors with fast moving velocity or short homing phase duration, the performance index can be approximated as

$$J \approx \int_t^{t_f} \frac{1}{\left(t_f - \tau\right)^{N-3}} a_M^2(\tau) d\tau \tag{6.40}$$

which indicates that the proposed guidance law with $N = 3$ provides energy-optimal interception and it is wise to choose $N \geq 3$ guarantee a zero final guidance command.

When the interceptor converges to the ideal collision course, the terminal velocity can be obtained as

$$V_{Mf} = V_M + a_x t_{go} \tag{6.41}$$

Following Eq. (6.41), the average velocity during the homing phase can be readily determined as

$$\bar{V}_M = V_M + \frac{1}{2} a_x t_{go} \tag{6.42}$$

By using the concept of average velocity, guidance command (6.37) can be also rewritten as

$$a_M = \frac{\bar{N} z}{t_{go}^2} \tag{6.43}$$

where the new guidance gain $\bar{N}$ is given by

$$\bar{N} = \frac{N V_M}{\bar{V}_M} \tag{6.44}$$

which implies that the proposed guidance law is a PNG-type guidance law with a time-varying navigation gain $\bar{N}$.

The initial and final value of the guidance gains are given by

$$\begin{aligned} \bar{N}_0 &= \lim_{t \to t_f} N(t) = \frac{N}{1 + \frac{1}{2V_{M0}} a_x t_f} \\ \bar{N}_f &= \lim_{t \to t_f} \bar{N}(t) = N \end{aligned} \tag{6.45}$$

where V_{M0} denotes the initial vehicle speed.

The influence of missile's velocity on the initial navigation gain with various axial accelerations, flight time and $N = 3$ is illustrated in Fig. 6.3. From this figure, it can be easily verified that the navigation gain gradually increases from an initial small value to N as the interceptor approaches the target.

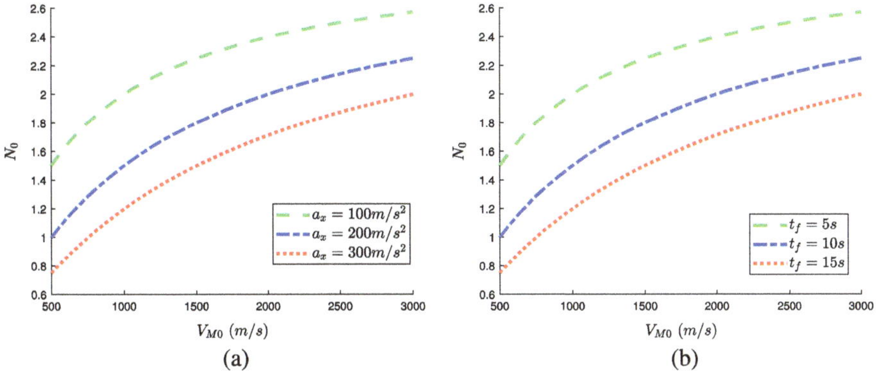

Fig. 6.3 Effect of missile's axial acceleration and flight time on the initial navigation gain. **a** Effect of missile's axial acceleration; and **b** Effect of missile's flight time

6.4.3 Relationships with Previous Guidance Laws

This subsection discusses the relationships between the proposed guidance law (6.43) and classical PNG and G2C by using the geometric approach. Figure 6.4 presents different types of ZEMs that are used in the derivation of PNG, G2C and the proposed guidance law: the length of the green straight line $A - D$ is the ZEM for PNG; the length of the blue straight line $A - C$ denotes the ZEM for G2C; and the length of the purple straight line $A - B$ stands for the proposed instantaneous ZEM. One can clearly observe from Fig. 6.4 that the desired trajectories for both PNG and G2C are straight lines from the current point to the their corresponding PIP points while the desired path of the proposed guidance law is a curved line.

6.4.3.1 Relationships with Proportional Navigation Guidance

The behind idea of PNG is to generate a lateral acceleration to nullify the ZEM so as to follow a straight line interception course. To maintain the collision triangle, it is necessary to equalize between the distances travelled by the interceptor and the target perpendicular to the LOS. In PNG formulation, the assumptions regarding no axial acceleration and gravity are required. Under these two assumptions, one can imply that

$$L_M \sin\left(\gamma_{PNG}^* - \sigma\right) - L_T \sin\left(\gamma_T - \sigma\right) = 0 \tag{6.46}$$

where γ_{PNG}^* denotes the desired current flight path angle of PNG. L_M and L_T are the vehicles' travelled distances from the current point to the PIP point.

Solving γ_{PNG}^* using Eq. (6.46) for the engagement case shown in Fig. 6.4 gives

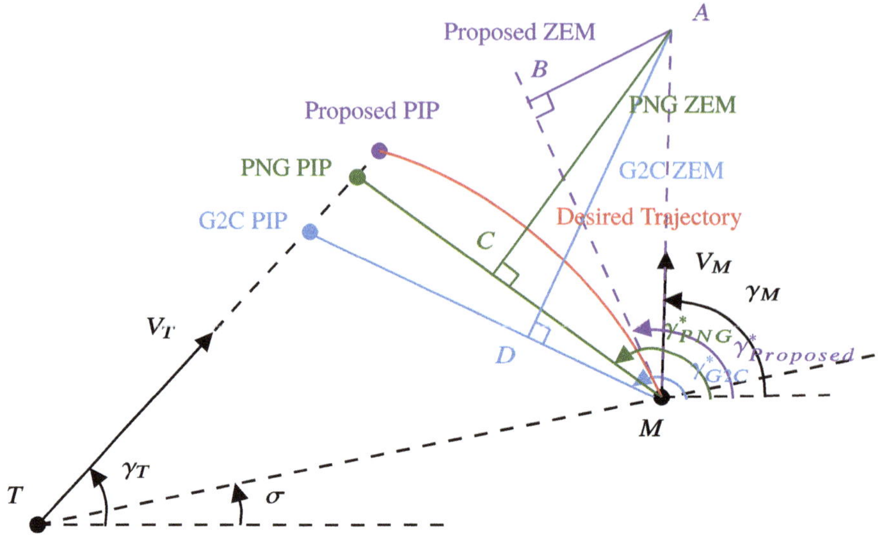

Fig. 6.4 Comparisons of different ZEMs in guidance law design

$$\begin{aligned}\gamma_{PNG}^{*} &= \pi - \arcsin\left(\frac{L_T \sin(\gamma_T - \sigma)}{L_M}\right) + \sigma \\ &= \pi - \arcsin\left(\frac{V_T \sin(\gamma_T - \sigma)}{V_M}\right) + \sigma\end{aligned} \tag{6.47}$$

Then, the ZEM dynamics of PNG can be obtained as

$$z_{PNG} = -\sin\left(e_{\gamma,PNG}\right) V_M t_{go} \tag{6.48}$$

$$\dot{z}_{PNG} = -\dot{e}_{\gamma,PNG} V_M t_{go} = -a_M t_{go} \tag{6.49}$$

where $e_{\gamma,PNG} = \gamma_M - \gamma_{PNG}^{*}$ denotes the flight path angle tracking error.

Note that the difference between the PNG ZEM and the proposed ZEM is that the desired flight path angle for the former one is derived based on the assumption of constant flight velocity under the gravity-free condition, whereas the latter one computes it under the influence of gravity for varying-speed missiles. The preceding PNG ZEM dynamics also reveals that if we remove the gravity and axial acceleration, the proposed instantaneous ZEM reduces to the PNG ZEM.

As derived in Chap. 2, the guidance command of PNG is given by

$$a_M = \frac{N z_{PNG}}{t_{go}^2} \tag{6.50}$$

Since classical PNG leverages the ZEM that is derived based on the gravity-free assumption, it requires extra term $g \cos \gamma_M$ to compensate the effect of gravity in practical implementation as

$$a_M = \frac{N z_{PNG}}{t_{go}^2} + g \cos \gamma_M \tag{6.51}$$

which indicates that the terminal guidance command of PNG cannot converge to zero due to the extra term $g \cos \gamma_M$. As a comparison, the proposed instantaneous ZEM is derived by considering the gravity, and therefore the guidance command converges to zero once the interceptor is maintained on the collision triangle.

6.4.3.2 Relationships with Guidance-to-Collision

Similar to PNG, the objective of G2C is to maintain a straight line interception course to approach the target. This means that, the distances traveled by the interceptor and the target perpendicular to the LOS is equal at any time instant during the homing phase as

$$L_M \sin\left(\gamma_{G2C}^* - \sigma\right) - L_T \sin\left(\gamma_T - \sigma\right) = 0 \tag{6.52}$$

where γ_{G2C}^* denotes the desired current flight path angle of G2C.

With constant axial acceleration, Eq. (6.52) can be reformulated as

$$\left(V_M + \frac{1}{2} a_M t_{go}\right) \sin\left(\gamma_{G2C}^* - \sigma\right) - V_T \sin\left(\gamma_T - \sigma\right) = 0 \tag{6.53}$$

Once a straight line collision course is reached and maintained after the heading error has been nulled, the time-to-go for G2C law can then be computed by [4]

$$\begin{aligned} r = {} & V_T t_{go} \cos\left(\gamma_T - \sigma\right) + V_M t_{go} \cos\left[\pi - \left(\gamma_{G2C}^* - \sigma\right)\right] \\ & + \frac{1}{2} a_M t_{go}^2 \cos\left[\pi - \left(\gamma_{G2C}^* - \sigma\right)\right] \end{aligned} \tag{6.54}$$

Solving Eq. (6.53) for γ_{G2C}^* gives

$$\gamma_{G2C}^* = \pi - \arcsin\left(\frac{V_T \sin\left(\gamma_T - \sigma\right)}{V_M + 0.5 a_M t_{go}}\right) + \sigma \tag{6.55}$$

Then, the G2C ZEM dynamics can be readily obtained as

$$z_{G2C} = -\sin\left(e_{\gamma,G2C}\right)\left(V_M t_{go} + \frac{1}{2} a_x t_{go}^2\right) \tag{6.56}$$

Table 6.1 Relationships between the proposed formulation and previous guidance laws

Guidance law	Guidance command	Terminal guidance command	Consideration
PNG	$a_M = \frac{N z_{PNG}}{t_{go}^2} + g\cos\gamma_M$	Bounded	None
G2C	$a_M = \frac{\bar{N} z_{G2C}}{t_{go}^2} + g\cos\gamma_M$	Bounded	Speed variation
Proposed	$a_M = \frac{\bar{N} z_{Proposed}}{t_{go}^2}$	Zero	Speed variation and gravity

$$\dot{z}_{G2C} = -\dot{e}_{\gamma,G2C}\left(V_M t_{go} + \frac{1}{2}a_x t_{go}^2\right) = -a_M\left(t_{go} + \frac{a_x}{2V_M}t_{go}^2\right) \tag{6.57}$$

where $e_{\gamma,G2C} = \gamma_M - \gamma_{G2C}^*$ denotes the flight path angle tracking error.

Following the same line shown in the previous subsection, one can easily derive the optimal G2C command is given by

$$a_M = \frac{\bar{N} z_{G2C}}{t_{go}^2} \tag{6.58}$$

Similarly, as the ZEM of G2C is only valid without the gravitational effect, the practical implementation of G2C requires an additional compensation term as

$$a_M = \frac{\bar{N} z_{G2C}}{t_{go}^2} + g\cos\gamma_M \tag{6.59}$$

Since the proposed guidance law utilizes gravity instead of compensating it, one can safely predict that our approach can save energy in the case of instantaneous ZEM$\leq$G2C ZEM.

Table 6.1 summarizes the relationships between the proposed guidance law and PNG as well as G2C. In conclusion, our approach is a generalized guidance law derived from the collision triangle and the gravity effect is automatically considered in the instantaneous ZEM. From the analysis, it can be noted that both PNG and G2C are special cases of the proposed guidance law. As we stated earlier, the advantage of using gravity-turn in guidance law design is to guarantee zero guidance command at the time of impact, thereby providing additional operation margins to cope with the undesired disturbances. Furthermore, by considering the gravitational effect in guidance law design, no extra energy for gravity compensation is required for guidance law implementation.

Table 6.2 Initial conditions for homing engagement

Parameters	Values
Missile-target initial relative range, $r(0)$	50 km
Initial LOS angle, $\sigma(0)$	0°
Missile initial velocity, $V_M(0)$	1500 m/s
Target velocity, V_T	3000 m/s
Interceptor's axial acceleration, a_x	20 g

6.5 Simulation Results

In this section, nonlinear simulations are performed to validate the proposed guidance law in an exo-atmospheric interception scenario. We assume the interceptor is equipped with a rocket motor that provides constant axial acceleration. The attitude of the interceptor can be adjusted to provide desired lateral acceleration to change the heading direction. The required initial conditions for a typical exo-atmospheric engagement, taken from [1, 3], are summarized in Table 6.2.

6.5.1 Characteristics of the Proposed Guidance Law

We first analyze the effect of guidance gain α on the guidance performance. In the simulations, the initial flight path angle is choose as $\gamma_M(0) = 150°$ and $\gamma_T(0) = 20°$. The simulation results, including interception trajectory, angle-of-attack command, flight path angle error, control effort, time-to-go estimation and missile velocity, with various guidance gains $N = 2, 3, 4, 5, 6$ are presented in Fig. 6.5, where the control effort is defined as $\int_0^{t_f} a_M^2(\tau)\, d\tau$. The results in this figure clearly reveal that guidance law (6.43) with larger guidance gain N provides faster convergence speed of the flight path angle error. However, large guidance gain also requires higher angle-of-attack command during the initial flight period, thereby generating more control effort. When the flight path angle tracking error is close to zero, the control effort almost remains the same and the angle-of-attack commands are also close to zero for all guidance gain cases, meaning that the interceptor is maintained on the collision triangle. Since the gravity is considered in the derivation of the collision triangle, the guidance command converges to zero at the time of impact with the increasing of the guidance gain. This characteristic is totally different from previous guidance laws that used an additional term $g \cos \gamma_M$ to compensate the gravity. From Fig. 6.5(b) and (d), we can readily note that the proposed guidance law with $N \geq 3$ guarantees zero final guidance command and $N = 3$ provides energy-optimal interception. These results support the analytic findings of previous sections.

The results in Fig. 6.5(e) our closed-form solution gives accurate estimation of the time-to-go and the estimation error converges to zero once the interceptor is

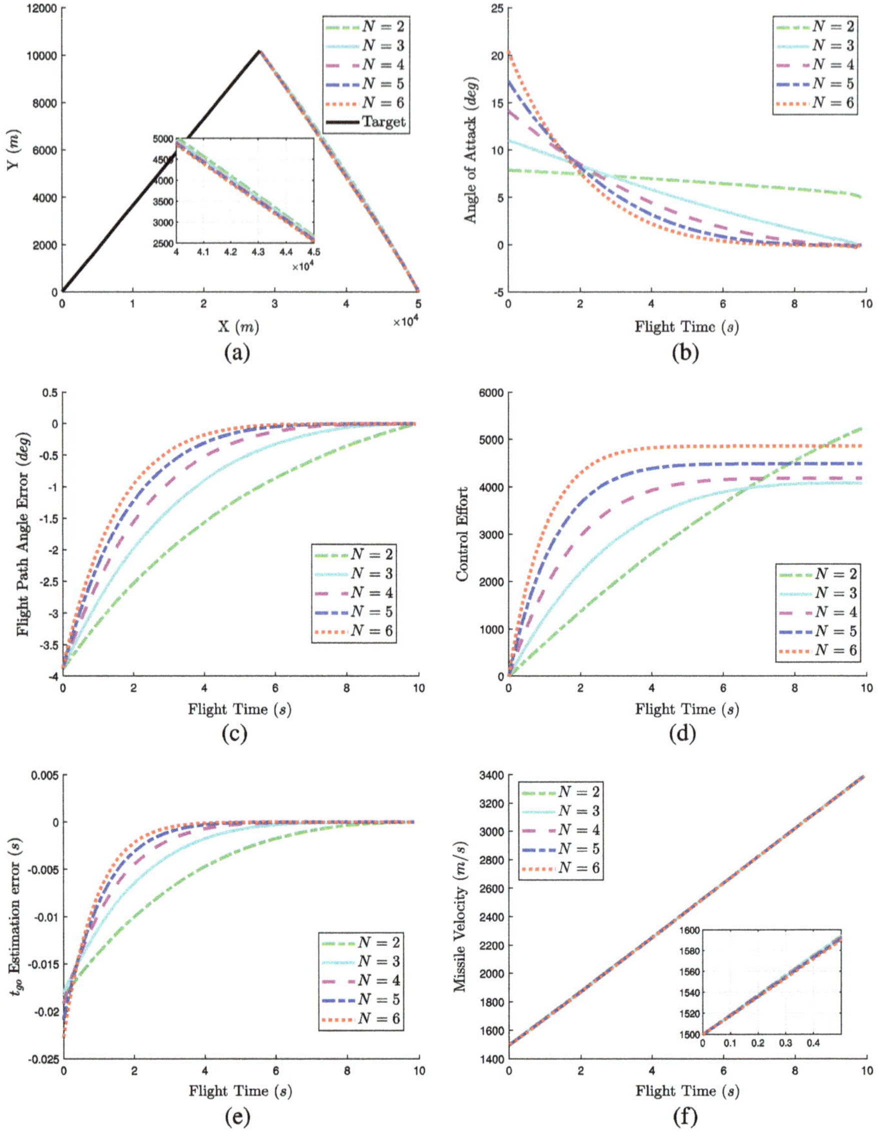

Fig. 6.5 Simulation results of the proposed guidance law with different guidance gains. **a** Interception trajectory; **b** Angle-of-attack command; **c** Flight path angle error; **d** Control effort; **e** Time-to-go estimation; and **f** Missile velocity

maintained on the desired trajectory. From the zoomed-in subfigure in Fig. 6.5(f), one can observe that the missile velocity with smaller guidance gain N increases slightly faster than that with larger guidance gain N during the initial flight phase. This can be attributed to the fact that larger guidance gain requires more control effort, i.e. larger magnitude of the angle-of-attack. Since the angle-of-attack remains very small during most of the flight period, the missile velocity in the considered scenario increases almost linearly, as shown in Fig. 6.5(f).

Now let us investigate the performance of the proposed guidance law with various interceptor's initial flight path angles. The same scenario is simulated with four different initial flight path angles: $\gamma_M(0) = 90°, 120°, 150°, 180°$ and $\gamma_T(0) = 20°$. The simulation results with guidance gain $N = 9$ are depicted in Fig. 6.6, which clearly shows that the proposed guidance law can guide the missile to successfully intercept the target in all tested scenarios. The interceptor with larger initial heading error experiences more curved trajectory to approach the target. For this reason, the duration of acceleration saturation of $\gamma_M(0) = 180°$ is longer than that of other cases. As the proposed time-to-go is calculated from the desired collision triangle, it gives an underestimation when the missile deviates from the zero control effort collision course. However, the simulation results indicate that the estimation accuracy is acceptable in most cases. From Fig. 6.6(f), it can be noted that the missile's velocity slightly deceases due to the gravitational effect when the guidance command is saturated, i.e., all control effort is leverged to nullify the instantaneous ZEM.

6.5.2 Comparison with Other Guidance Laws

To further demonstrate the superiority of the proposed approach, the performance of the new guidance law is compared to that of classical PNG and G2C in this subsection. In the simulations, we utilize an additional term $g \cos \gamma_M$ in both PNG and G2C to compensate for the gravitational effect. In order to make fair comparisons, the guidance gain for all guidance laws is set as $N = 9$.

Figure 6.7 compares the performance of these three guidance laws for intercepting a constant moving target. From this figure, it can be noted that the axial acceleration of PNG does not align with the velocity vector, thereby forcing the missile to fly along a curved path to intercept the target. Unlike PNG, G2C guides the missile to intercept the target by following a straight line after the initial heading error is nullified. As a comparison, the proposed guidance law guides the missile to fly along a slightly curved trajectory as it exploits the gravity-turn concept. It is evident from Fig. 6.7(b) that the time evolutions of angle-of-attack of all guidance laws remain bounded and the proposed guidance law guarantees zero guidance command at the time of impact. Therefore, the proposed guidance law provides more operational margins than the G2C law when the missile is close to the target. Since the proposed guidance law requires no extra term to counteract the gravity, one can clearly observe from Fig. 6.7(b) that the proposed approach enables energy saving during the terminal guidance phase. The time history of ZEM, shown in Fig. 6.7(c), reveals that all guidance laws can nullify their corresponding ZEM to zero to guarantee target interception. From Fig. 6.7(d), one can clearly observe that the terminal missile velocity under both the proposed guidance law and the G2C law than that of the PNG law, thereby enabling higher kill probability. Table 6.3 summarizes the mean miss

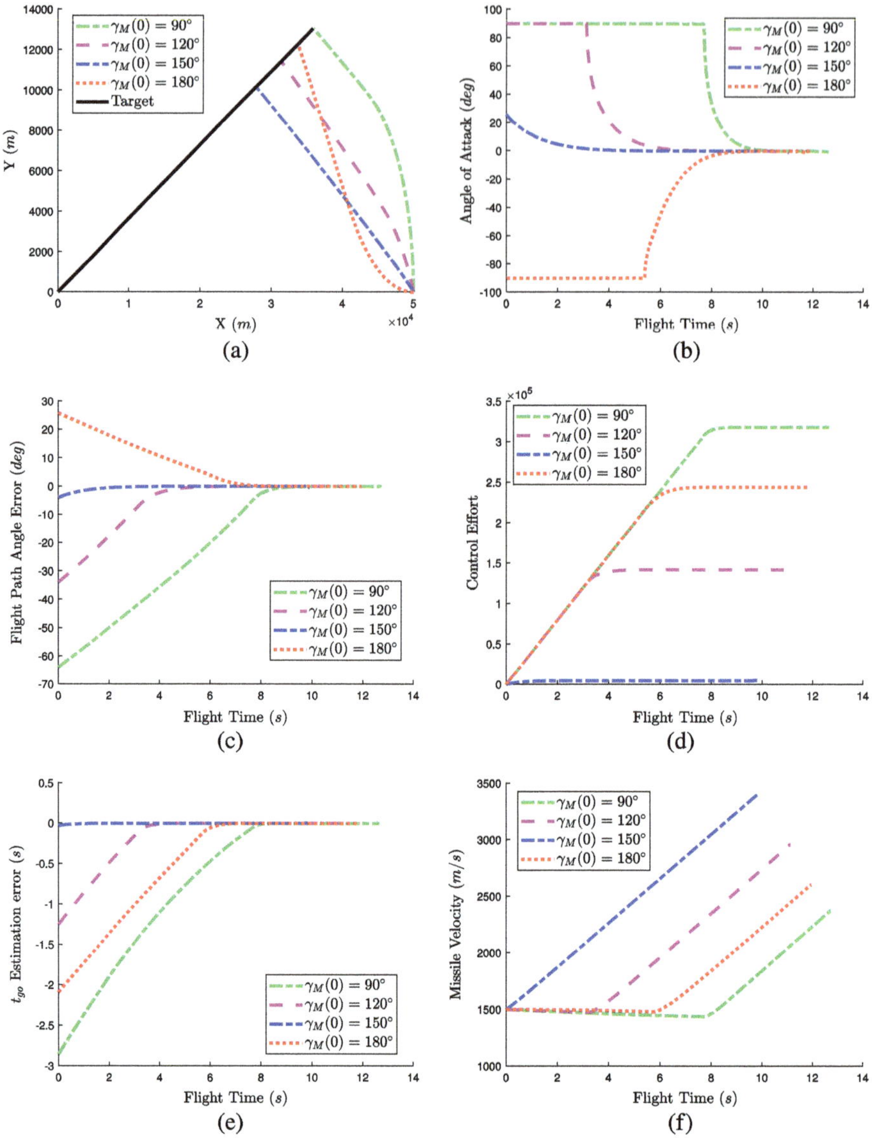

Fig. 6.6 Simulation results of the proposed guidance law with different initial flight path angles. **a** Interception trajectory; **b** Angle-of-attack command; **c** Flight path angle error; **d** Control effort; **e** Time-to-go estimation; and **f** Missile velocity

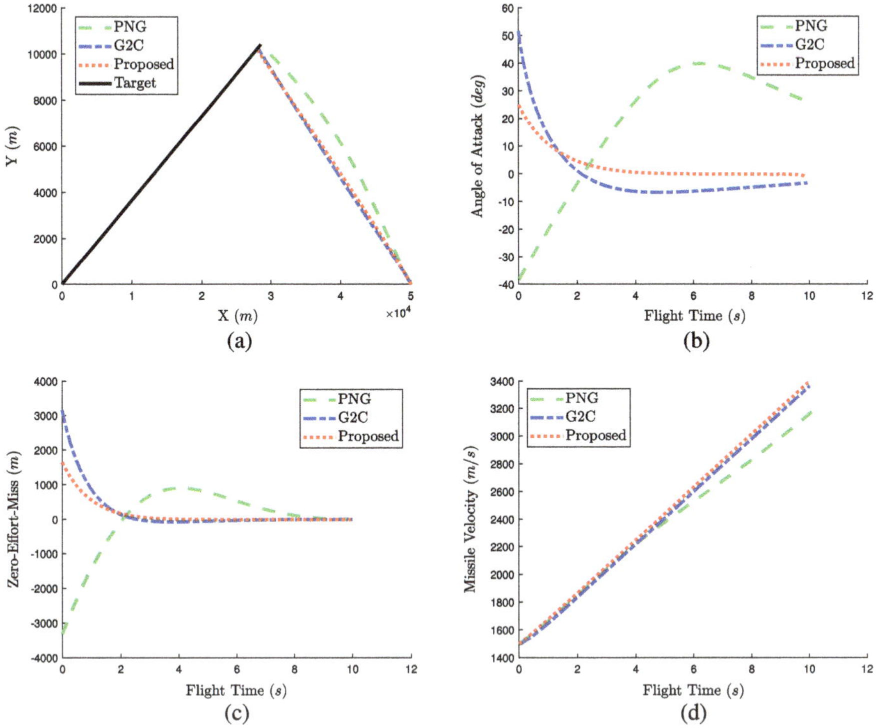

Fig. 6.7 Comparison results of exo-atmospheric interception with a non-maneuvering target. **a** Interception Trajectory; **b** Angle-of-attack command; **c** Zero-effort-miss; and **d** Missile velocity

Table 6.3 Mean miss distances of three different guidance laws

	PNG	G2C	Proposed
Mean miss distance	0.32 m	0.18 m	0.18 m

distances of these three different guidance laws in 100 Monte-Carlo simulations. It is clear that both G2C laws exhibit better homing performance than the PNG law.

6.6 Summary

The problem of optimal guidance law design for accelerating missiles in the presence of gravity is investigated in this chapter. The proposed guidance law is derived based on a new concept, called instantaneous ZEM. The key feature of the proposed guidance law lies in that it automatically leverages the gravitational acceleration. The benefits of using gravity are clear: guaranteeing zero final guidance command and saving energy. We also show that the conventional PNG and G2C are special cases of the proposed guidance law.

The proposed results are believed to have an academic significance as well as a practical one since it suggests a new way to utilize gravity in guidance law design. By using the proposed results, one can also exploit the advantages of gravity turn in midcourse guidance law design.

Appendix. Closed-Form Solution of Eqs. (6.16)–(6.18)

Noting from Eqs. (6.16)–(6.18) that the solution requires the integration of

$$h(\gamma_M)\left|\sec\gamma_M + \tan\gamma_M\right|^n \tag{6.60}$$

where $h(\gamma_M)$ is a function of secant and tangent functions. For an arbitrary function $\beta(\gamma_M)$, we have

$$\begin{aligned}
&\frac{d}{d\gamma_M}\left[\beta(\gamma_M)\left|\sec\gamma_M + \tan\gamma_M\right|^n\right] \\
&\quad = \frac{d\beta(\gamma_M)}{d\gamma_M}\left|\sec\gamma_M + \tan\gamma_M\right|^n + n\beta(\gamma_M)\sec\gamma_M\left|\sec\gamma_M + \tan\gamma_M\right|^n \\
&\quad = \left(\frac{d\beta(\gamma_M)}{d\gamma_M} + n\beta(\gamma_M)\sec\gamma_M\right)\left|\sec\gamma_M + \tan\gamma_M\right|^n
\end{aligned} \tag{6.61}$$

which reveals that the general solution of $\int h(\gamma_M)\left|\sec\gamma_M + \tan\gamma_M\right|^n d\gamma_M$ can be obtained by equalizing $h(\gamma_M)$ with $\frac{d\beta(\gamma_M)}{d\gamma_M} + n\beta(\gamma_M)\sec\gamma_M$.

Based on the properties of secant and tangent functions, the function $\beta(\gamma_M)$ in solving Eq. (6.16)–(6.18) can be formulated as a general form as

$$\beta(\gamma_M) = a_1\sec\gamma_M + a_2\tan\gamma_M + a_3\sec^2\gamma_M + a_4\tan^2\gamma_M + a_5\sec\gamma_M\tan\gamma_M \tag{6.62}$$

where a_i, $i = 1, 2, 3, 4, 5$, are unknown constant coefficients to be determined.

Differentiating Eq. (6.62) with respect to γ_M gives

$$\begin{aligned}
\frac{d\beta(\gamma_M)}{d\gamma_M} &= a_1\sec\gamma_M\tan\gamma_M + a_2\sec^2\gamma_M + 2a_3\sec^2\gamma_M\tan\gamma_M \\
&\quad + 2a_4\sec^2\gamma_M\tan\gamma_M + a_5\left(\sec\gamma_M\tan^2\gamma_M + \sec^3\gamma_M\right)
\end{aligned} \tag{6.63}$$

Substituting Eqs. (6.62) and (6.63) into Eq. (6.61) yields

$$
\begin{aligned}
&\frac{d}{d\gamma_M}\left[\beta(\gamma_M)\left|\sec\gamma_M+\tan\gamma_M\right|^n\right] \\
&\quad=\left[(na_1+a_2)\sec^2\gamma_M+(na_2+a_1)\sec\gamma_M\tan\gamma_M\right. \\
&\qquad+(na_3+a_5)\sec^3\gamma_M+(na_4+a_5)\sec\gamma_M\tan^2\gamma_M \\
&\qquad\left.+(na_5+2a_3+2a_4)\sec^2\gamma_M\tan\gamma_M\right]\left|\sec\gamma_M+\tan\gamma_M\right|^n
\end{aligned}
\tag{6.64}
$$

For Eq. (6.16), we have $h(\gamma_M)=\sec^2\gamma_M$, $n=\kappa$. Comparing the coefficients results in

$$
\begin{cases}
\kappa a_1+a_2=1\\
\kappa a_2+a_1=0\\
\kappa a_3+a_5=0\\
\kappa a_4+a_5=0\\
\kappa a_5+2a_3+2a_4=0
\end{cases}
\tag{6.65}
$$

Solving Eq. (6.65) gives the unknown coefficients as

$$
a_1=\frac{\kappa}{\kappa^2-1},\quad a_2=-\frac{1}{\kappa^2-1},\quad a_3=a_4=a_5=0 \tag{6.66}
$$

Then, the closed-form solution of Eq. (6.16) is given by

$$
t(\gamma_M)=t(\gamma_{M0})-\frac{C}{g}\left[f_t(\gamma_M)-f_t(\gamma_{M0})\right] \tag{6.67}
$$

For Eq. (6.17), we have $h(\gamma_M)=\sec^2\gamma_M$, $n=2\kappa$. Replacing κ with 2κ in Eq. (6.65) leads to the closed-form solution of Eq. (6.17) as

$$
x_M(\gamma_M)=x_M(\gamma_{M0})-\frac{C^2}{g}\left[f_x(\gamma_M)-f_x(\gamma_{M0})\right] \tag{6.68}
$$

For Eq. (6.18), we have $h(\gamma_M)=\sec^2\gamma_M\tan\gamma_M$, $n=2\kappa$. Comparing the coefficients gives the following coupled equations

$$
\begin{cases}
\kappa a_1+a_2=0\\
\kappa a_2+a_1=0\\
\kappa a_3+a_5=0\\
\kappa a_4+a_5=0\\
\kappa a_5+2a_3+2a_4=1
\end{cases}
\tag{6.69}
$$

Solving Eq. (6.69) yields

$$
a_1=a_2=0,\quad a_3=-\frac{1}{\kappa^2-4},\quad a_4=-\frac{1}{\kappa^2-4},\quad a_5=\frac{\kappa}{\kappa^2-4} \tag{6.70}
$$

which gives the closed-form solution of Eq. (6.18) as

$$y_M(\gamma_M) = y_M(\gamma_{M0}) - \frac{C^2}{g}\left[f_y(\gamma_M) - f_y(\gamma_{M0})\right] \tag{6.71}$$

References

1. Gazit R, Gutman S (1991) Development of guidance laws for a variable-speed missile. Dyn Control 1(2):177–198
2. Gazit R (1993) Guidance to collision of a variable-speed missile. In: Proceedings of the first IEEE regional conference on aerospace control systems. IEEE, Westlake Village, CA, pp 734–737
3. Shima T, Golan OM (2012) Exo-atmospheric guidance of an accelerating interceptor missile. J Franklin Inst 349(2):622–637
4. Reisner D, Shima T (2013) Optimal guidance-to-collision law for an accelerating exoatmospheric interceptor missile. J Guid Control Dyn 36(6):1695–1708
5. Conn AR, Gould NIM, Toint PL (2000) Trust region methods, vol 1. SIAM, Philadelphia, PA
6. Kanzow C, Yamashita N, Fukushima M (2004) Levenberg-marquardt methods with strong local convergence properties for solving nonlinear equations with convex constraints. J Comput Appl Math 172(2):375–397
7. Zarchan P (2012) Tactical and strategic missile guidance. American Institute of Aeronautics and Astronautics, Reston, VA

Chapter 7
Gravity-Turn-Assisted Optimal Intercept Angle Guidance Law

Abstract This chapter develops a new optimal gravity-turn-assisted intercept angle guidance law for an exo-atmospheric interceptor. The analytical guidance command is derived as a solution of a finite-time optimal regulation problem by using the Lagrange multiplier approach. The convergence of the instantaneous ZEM and intercept angle error is theoretically analyzed based on the instantaneous linear time-invariant system concept. We also reveal that existing optimal impact angle constrained guidance laws are the special cases of the proposed formulation. Numerical simulations with some comparisons clearly demonstrate the effectiveness of the proposed guidance law.

7.1 Introduction

It is known that constraining the interception angle is helpful to maintain advantageous homing engagement geometry against a target as well as to ensure a better quality of seeker measurement on a target [1–3]. For this reason, extensive efforts have been made in the area of intercept angle or impact angle guidance. Until recently, many elegant solutions of optimal intercept angle control guidance have reported in the literature. Interested readers can refer to [4–10] and references therein.

The issue with most previous optimal intercept angle guidance laws is that these algorithms were established under constant-speed and gravity-free assumption. As we explained in Chap. 6, the constant-speed assumption is might be violated for exo-atmospheric interceptors and ignoring gravity in guidance law design will reduce the terminal impact energy. Although the recent work [11] provides a solution to design intercept angle guidance law for speed-varying missiles. The resultant algorithm is limited to address the issue of optimality and fails to address the problem induced by gravity compensation.

This chapter is an attempt to solve the problem of intercept angle control by utilizing gravity for exo-atmospheric interceptions. By considering instantaneous ZEM and intercept angle as the system states, we first formulate a finite-time quadratic regulation problem with a weighted cost function. The commanded acceleration is then derived with the aid of the Lagrange multiplier approach. The relationship

S. He et al., *Optimal Guidance and Its Applications in Missiles and UAVs*, Springer Aerospace Technology, https://doi.org/10.1007/978-3-030-47348-8_7

between the proposed guidance law and previous optimal impact angle guidance laws are also analyzed to provide better understandings of the proposed approach. The key features of the proposed guidance law are summarized as follows.

(1) The gravity-turn concept is leveraged in guidance law derivation and thus the proposed algorithm requires no additional term to counteract the gravitational effect. This strategy enables energy saving and accurate interception, as confirmed by numerical simulation results.
(2) The proposed approach guarantees that the terminal angle-of-attack command converges to zero, thus providing operational margins and improves robustness against external uncertainties, especially near the interception.

7.2 Problem Formulation

In this chapter, we assume that the interceptor is equipped with a rocket motor to provide constant axial acceleration during the endgame. The attitude control provide the possibility to change the direction of the body axis for both trajectory shaping and energy management. It is known that increasing the terminal velocity is beneficial to maximize the effectiveness of direct hit in the exo-atmospheric interception scenario. For this reason, it is desirable to minimizing the magnitude of the angle-of-attack α_M since the lateral acceleration is proportional to the angle-of-attack as shown in Eq. (6.1).

From Eq. (6.2), it can be also noted that the gravity-related term $g \sin \gamma_M$ also contributes to the decrease of the moving speed. Hence, direct gravity compensation concept inevitably require more energy consumption, leading to the sacrifice of the effectiveness of direct hit and the operational margins when close to the target. In order to tackle this problem, a gravity-turn-assisted optimal guidance law that directly utilizes the gravity for exo-atmospheric interception was proposed in Chap. 6. The issue associated with utilizing gravity-turn is that the zero-control-effort collision course from the missile to the PIP is a curved line, and therefore the derivation of analytic guidance command by minimizing the ZEM is intractable. To address this issue, we introduced a new concept, called instantaneous ZEM, in Chap. 6 to allow for the derivation of analytic optimal guidance law. According to Chap. 6, the instantaneous ZEM z and its rate $\dot{z}$ can be obtained as

$$z = -e_\gamma \left(V_M t_{go} + \frac{1}{2} a_x t_{go}^2 \right) \tag{7.1}$$

$$\dot{z} = -a_M \left(t_{go} + \frac{a_x}{2V_M} t_{go}^2 \right) \tag{7.2}$$

where $e_\gamma = \gamma_M - \gamma^*$ is the heading error with γ^* being the desired current flight path angle calculated from the zero-control-effort collision triangle.

As stated before, constraining the intercept angle is desirable in terms of increasing the direct hit effectiveness as well as the kill probability for exo-atmospheric interception in practice. Consequently, the aim of this chapter is to design an optimal guidance law nullify both instantaneous ZEM and intercept angle error in a finite time. Note that the intercept angle here is defined as the missile's flight path angle at the time of impact. Let desired impact angle be denoted as γ_f and $a_M = a_x \sin \alpha_M$ be the virtual control input, the guidance problem can then be formulated as the following optimization problem.

Problem 7.1 Find optimal virtual control input a_M to minimize the performance index

$$J = \frac{1}{2} \int_t^{t_f} R(\tau) a_M^2(\tau) d\tau \tag{7.3}$$

subject to

$$\begin{aligned} \dot{z} &= -a_M \left(t_{go} + \frac{a_x}{2V_M} t_{go}^2 \right), \quad z(t_f) = 0 \\ \dot{\gamma}_M &= \frac{a_M}{V_M}, \quad \gamma_M(t_f) = \gamma_f \end{aligned} \tag{7.4}$$

where $R(t) > 0$ is an arbitrary weighting function.

Remark 7.1 Note that the collision triangle, derived in Chap. 6, automatically contains the gravity and therefore one only needs to use $\dot{\gamma}_M = a_M / V_M$ in guidance law design.

7.3 Derivation of the Optimal Intercept Angle Guidance Law

According to the linear control theory, the general solution of the guidance system (7.4) can be expressed as

$$\begin{aligned} z(t_f) &= z - \int_t^{t_f} h_1(\tau) a_M(\tau) d\tau \\ \gamma_M(t_f) &= \gamma_M - \int_t^{t_f} h_2(\tau) a_M(\tau) d\tau \end{aligned} \tag{7.5}$$

where

$$h_1 \triangleq \left(t_{go} + \frac{a_x}{2V_M} t_{go}^2 \right), \quad h_2 \triangleq -\frac{1}{V_M} \tag{7.6}$$

Imposing the terminal constraints on Eq. (7.5) gives

$$\begin{aligned} z &= \int_t^{t_f} h_1(\tau)\, a_M(\tau)\, d\tau \\ \gamma_M - \gamma_f &= \int_t^{t_f} h_2(\tau)\, a_M(\tau)\, d\tau \end{aligned} \tag{7.7}$$

Introducing the weight function $R(t)$ as a slack variable in Eq. (7.7) gives

$$\begin{aligned} z &= \int_t^{t_f} h_1(\tau)\, R^{-1/2}(\tau)\, R^{1/2}(\tau)\, a_M(\tau)\, d\tau \\ \gamma_M - \gamma_f &= \int_t^{t_f} h_2(\tau)\, R^{-1/2}(\tau)\, R^{1/2}(\tau)\, a_M(\tau)\, d\tau \end{aligned} \tag{7.8}$$

Consider the Hilbert space $\mathcal{H} = \mathcal{L}^2\left[t, t_f\right]$ with the inner product $(f, g) = \int_t^{t_f} f(\tau)\, g(\tau)\, d\tau$. According to Lemma 5.1, if the guidance command a_M is optimal in terms of minimizing performance index (7.3), then there exist two Lagrange multipliers λ_1, λ_2 such that the lateral acceleration satisfies

$$R^{1/2} a_M = \lambda_1 h_1 R^{-1/2} + \lambda_2 h_2 R^{-1/2} \tag{7.9}$$

Substituting Eq. (7.9) into Eq. (7.7) results in

$$\begin{aligned} z &= \lambda_1 \int_t^{t_f} h_1^2(\tau)\, R^{-1}(\tau)\, d\tau + \lambda_2 \int_t^{t_f} h_1 h_2(\tau)\, R^{-1}(\tau)\, d\tau \\ \gamma_M - \gamma_f &= \lambda_1 \int_t^{t_f} h_1 h_2(\tau)\, R^{-1}(\tau)\, d\tau + \lambda_2 \int_t^{t_f} h_2^2(\tau)\, R^{-1}(\tau)\, d\tau \end{aligned} \tag{7.10}$$

which can be rewritten in a compact matrix form as

$$\begin{bmatrix} z \\ \gamma_M - \gamma_f \end{bmatrix} = \begin{bmatrix} g_1 & g_{12} \\ g_{12} & g_2 \end{bmatrix} \begin{bmatrix} \lambda_1 \\ \lambda_2 \end{bmatrix} \tag{7.11}$$

where

$$\begin{aligned} g_1 &= \int_t^{t_f} h_1^2(\tau) R^{-1}(\tau) d\tau \\ g_2 &= \int_t^{t_f} h_2^2(\tau) R^{-1}(\tau) d\tau \\ g_{12} &= \int_t^{t_f} h_1(\tau) h_2(\tau) R^{-1}(\tau) d\tau \end{aligned} \tag{7.12}$$

Solving Eq. (7.11) for λ_1 and λ_2 gives

$$\begin{aligned}\lambda_1 &= \frac{1}{g_1 g_2 - g_{12}^2}\left[g_2 z - g_{12}\left(\gamma_M - \gamma_f\right)\right] \\ \lambda_2 &= \frac{1}{g_1 g_2 - g_{12}^2}\left[-g_{12} z + g_1\left(\gamma_M - \gamma_f\right)\right]\end{aligned} \quad (7.13)$$

The guidance command can then be obtained by substituting Eq. (7.13) into Eq. (7.9) as

$$a_M = \frac{\left(h_1 g_2 - h_2 g_{12}\right) R^{-1}}{g_1 g_2 - g_{12}^2} z + \frac{\left(h_2 g_1 - h_1 g_{12}\right) R^{-1}}{g_1 g_2 - g_{12}^2}\left(\gamma_M - \gamma_f\right) \quad (7.14)$$

which can also be reformulated as

$$a_M = N_1 \frac{z}{t_{go}^2} + N_2 \frac{V_M\left(\gamma_M - \gamma_f^d\right)}{t_{go}} \quad (7.15)$$

where the guidance gains are determined by

$$\begin{aligned}N_1 &= \frac{h_1 g_2 - h_2 g_{12}}{g_1 g_2 - g_{12}^2} R^{-1} t_{go}^2 \\ N_2 &= \frac{h_2 g_1 - h_1 g_{12}}{\left(g_1 g_2 - g_{12}^2\right) V_M} R^{-1} t_{go}\end{aligned} \quad (7.16)$$

Guidance command (7.16) is the generalized gravity-turn-assisted optimal intercept angle guidance law for an exo-atmospheric interceptor with an arbitrary weighting function. By choosing different weighting functions, one can design various optimal intercept angle guidance laws to shape the guidance command as desired. Without loss of generality, we consider a time-to-go weighting function as [6, 9]

$$R^{-1}(t) = t_{go}^N, \quad N \geq 0 \quad (7.17)$$

to shape the guidance command in this chapter. One major advantage of utilizing this weighting function is that it allows for zero final guidance command with $N > 0$ since the weighting function with $N > 0$ becomes infinite as t_{go} approaches 0. Hence, this property is beneficial in improving the effectiveness of direct hit in the exo-atmospheric interception scenario.

Similar to Chap. 6, we introduce the concept of average velocity to simplify the derivation process. Notice that the axial acceleration fully contributes to the increase of the moving speed once the interceptor converges to the ideal collision course. For this reason, the average speed can be approximated by the definition as

$$\bar{V}_M \triangleq \frac{1}{2}\left(V_M + V_{M_f}\right) = V_M + \frac{1}{2}a_x t_{go} \tag{7.18}$$

where the fact $V_{M_f} = V_M + a_x t_{go}$ is used. By using the average speed, the term h_1 in Eq. (7.6) can be reformulated as

$$h_1 = \frac{\bar{V}_M}{V_M} t_{go} \tag{7.19}$$

Since the V_M is a time-varying variable due to the existence of axial acceleration and gravity, finding analytical solutions of integrations in Eq. (7.12) is intractable. To address this problem, we assume the missile is flying with a constant-speed and update the moving speed at every time instant when implementing the guidance law. With this reasonable assumption, the guidance gains can be analytically derived as

$$\begin{aligned} N_1 &= (N+2)(N+3)\frac{V_M}{\bar{V}_M} \\ N_2 &= (N+1)(N+2) \end{aligned} \tag{7.20}$$

which indicates that the time-varying N_1 is scaled by the ratio of the current speed over the average speed

Substituting Eq. (7.20) into Eq. (7.15) yields the explicit guidance command as

$$a_M = (N+2)(N+3)\frac{V_M z}{\bar{V}_M t_{go}^2} + (N+1)(N+2)\frac{V_M\left(\gamma_M - \gamma_f^d\right)}{t_{go}} \tag{7.21}$$

which clearly reveals that the proposed guidance law is given by a feedback form with a time-varying gain N_1 and a fixed gain N_2.

7.4 Analysis of the Proposed Guidance Law

In this section, we first investigate the convergence of instantaneous ZEM and intercept angle error of the proposed optimal guidance law based on the concept of instantaneous linear time-invariant system. Then, the behavior of the guidance gain is discussed. Finally, the relationship between the proposed guidance law with existing optimal intercept angle guidance laws is analyzed.

7.4.1 Convergence of Instantaneous ZEM and Intercept Angle Error

According to Eq. (7.1), the heading error can be

$$e_\gamma = -\frac{z}{V_M t_{go} + \frac{1}{2} a_x t_{go}^2} = -\frac{z}{\bar{V}_M t_{go}} \tag{7.22}$$

Differentiating Eq. (7.22) with respect to time and substituting guidance command (7.21) into it gives the heading error dynamics as

$$\dot{e}_\gamma = -\frac{1}{t_{go}} \left(N_1 \frac{\bar{V}_M}{V_M} - 1 \right) e_\gamma + \frac{N_2}{t_{go}} e_I \tag{7.23}$$

where $e_I = \gamma_M - \gamma_f$ denotes the intercept angle error.

We can also obtain the dynamics of the intercept angle error dynamics as

$$\dot{e}_I = -N_1 \frac{\bar{V}_M}{V_M t_{go}} e_\gamma + N_2 \frac{e_I}{t_{go}} \tag{7.24}$$

Let $x \triangleq \left[e_\gamma, e_I \right]^T$ and $g(t) \triangleq 1/t_{go}$. Then, the error dynamics of the guidance loop can be reformulated as a compact matrix form as

$$\dot{x} = g(t) Ax \tag{7.25}$$

where

$$A \triangleq \begin{bmatrix} -[(N+2)(N+3)-1] & (N+1)(N+2) \\ -(N+2)(N+3) & (N+1)(N+2) \end{bmatrix} \tag{7.26}$$

It is clear that system (7.25) is a special linear time-varying system with $g(t)$ being a continuous positive function for all $t \geq 0$ and A being a nonsingular constant matrix. The eigenvalues of matrix A can be easily obtained as

$$\sigma_1 = -(N+2), \quad \sigma_2 = -(N+1) \tag{7.27}$$

which implies that the eigenvalues of matrix A are always negative if $N \geq 0$ and hence matrix A is Hurwitz.

Then, we can easily conclude that linear time-varying system (7.25) is globally exponentially stable by utilizing the results from [9]. Together with the fact that the origin is the equilibrium of system (7.25), one can imply that both the heading error e_γ and the intercept angle error e_I will exponentially converge to zero. This means that the proposed guidance law guarantees target interception with a specific intercept angle.

7.4.2 Behavior of Navigation Gain

From Eqs. (7.18) and (7.20), it is clear that the initial value of time-varying navigation gain N_1 is determined as

$$N_1(0) = (N+2)(N+3)\frac{1}{\left(1+\frac{a_x}{2V_{M0}}t_f\right)} \tag{7.28}$$

where t_f represents the time of impact, V_{M0} the initial velocity of the interceptor. The final value of N_1 is determined by

$$N_1\left(t_f\right) = (N+2)(N+3) \tag{7.29}$$

Through simple algebra, it is easy to verify that

$$N_1(0) < N_1(t) < N_1\left(t_f\right), \quad \forall t \in \left(0, t_f\right) \tag{7.30}$$

which reveals that the navigation gain N_1 monotonically increases and converges to constant value $(N+2)(N+3)$ at the time of impact. This provides the unique information that how the navigation gain N_1 behaves as the vehicle's speed changes during the terminal guidance phase. When the interceptor is close to the target, the average moving speed $\bar{V}_M$ gradually converges to the moving speed V_M and hence the effect of speed variation can be ignored.

By comparing Eq. (7.20) with Eq. (6.44), we have another interesting observation: the speed variation has the same effect on the instantaneous ZEM for guidance laws with and without intercept angle constraint. Therefore, it is safe to predict that classical optimal impact angle guidance laws with constant navigation gains are far away from optimal in a realistic exo-atmospheric engagement.

7.4.3 Relationship Between the Proposed Guidance Law and Previous Guidance Laws

In Chap. 6, it has been proved that the instantaneous ZEM reduces to the G2C ZEM by ignoring the gravitational effect. The dynamics of z_{G2C} is given by

$$\dot{z}_{G2C} = -a_M\left(t_{go} + \frac{a_x}{2V_M}t_{go}^2\right) \tag{7.31}$$

By using Eq. (7.31), the time-to-go weighted optimal G2C impact angle guidance (TWOG2CIAG) problem can be formulated as follows.

Problem 7.2 Find optimal virtual control input a_M to minimize the performance index

$$J = \frac{1}{2}\int_{t}^{t_f} \frac{a_M^2(\tau)}{\left(t_f - \tau\right)^N} d\tau \tag{7.32}$$

subject to

$$\begin{aligned} \dot{z}_{G2C} &= -a_M\left(t_{go} + \frac{a_x}{2V_M}t_{go}^2\right), \quad z_{G2C}\left(t_f\right) = 0 \\ \dot{\gamma}_M &= \frac{a_M}{V_M}, \quad \gamma_M\left(t_f\right) = \gamma_f \end{aligned} \tag{7.33}$$

Following the same lines as shown in Sect. 7.3, we can easily derive the analytic command of TWOG2CIAG as

$$a_M = N_1\frac{z_{G2C}}{t_{go}^2} + N_2\frac{V_M\left(\gamma_M - \gamma_f^d\right)}{t_{go}} \tag{7.34}$$

which shares the same guidance gains as the proposed guidance law.

Since z_{G2C} ignores the gravitational effect, the implementation of TWOG2CIAG requires an extra term to reject the effect of gravity as

$$a_M = N_1\frac{z_{G2C}}{t_{go}^2} + N_2\frac{V_M\left(\gamma_M - \gamma_f^d\right)}{t_{go}} + g\cos\gamma_M \tag{7.35}$$

which might require more control energy and lose some control operational margins and direct hit effectiveness.

If we ignore the influence of both gravity and axial acceleration, the instantaneous ZEM converges to the classical PNG ZEM, as demonstrated in Chap. 6. The dynamics of z_{PNG} is determined by

$$\dot{z}_{PNG} = -a_M t_{go} \tag{7.36}$$

By using Eq. (7.36), the time-to-go weighted optimal impact angle guidance (TWOIAG) problem can be formulated as follows.

Problem 7.3 Find optimal virtual control input a_M to minimize the performance index

$$J = \frac{1}{2}\int_{t}^{t_f} \frac{a_M^2(\tau)}{\left(t_f - \tau\right)^N} d\tau \tag{7.37}$$

subject to

$$\begin{aligned} \dot{z}_{PNG} &= -a_M\left(t_{go} + \frac{a_x}{2V_M}t_{go}^2\right), \quad z_{PNG}\left(t_f\right) = 0 \\ \dot{\gamma}_M &= \frac{a_M}{V_M}, \quad \gamma_M\left(t_f\right) = \gamma_f \end{aligned} \tag{7.38}$$

Following the same lines as shown in Sect. 7.3, the analytic command of TWOIAG can be easily obtained as

$$a_M = (N+2)(N+3)\frac{z_{PNG}}{t_{go}^2} + (N+1)(N+2)\frac{V_M\left(\gamma_M - \gamma_f^d\right)}{t_{go}} \tag{7.39}$$

which is an exact equivalent form of the guidance law proposed in [8, 9]. This clearly demonstrates that the guidance gains of the proposed guidance converge to those of the TWOIAG at the time of impact.

Similarly, the implementation of TWOIAG requires an extra gravity-compensation term as

$$a_M = (n+2)(n+3)\frac{z_{PNG}}{t_{go}^2} + (n+1)(n+2)\frac{V_M\left(\gamma_M - \gamma_f^d\right)}{t_{go}} + gcos\gamma_M \tag{7.40}$$

By observing the guidance commands of TWOG2CIAG and TWOIAG, it can be noted that both TWOG2CIAG and TWOIAG cannot ensure zero final guidance command due to the extra gravity-compensation term. As a comparison, the proposed guidance law utilizes the concept of instantaneous ZEM with the effect of the gravity automatically being considered in guidance law derivation. This means that no extra gravity-compensation term is required for the implementation of the proposed guidance law and zero final guidance command is theoretically guaranteed. This is beneficial to improve the operational margins to cope with unexpected disturbances and the effectiveness of direct hit during the endgame. Table 7.1 summarizes the relationships between the proposed guidance law and TWOIAG and TWOG2CIAG. We can clearly note that both TWOIAG and TWOG2CIAG are special cases of the proposed guidance law under one or two simplified assumptions.

Table 7.1 Relationships between the proposed formulation and previous guidance laws

Guidance law	Guidance command	Terminal guidance command	Consideration
TWOIAG	Eq. (7.40)	Bounded	Intercept angle
TWOG2CIAG	Eq. (7.35)	Bounded	Speed variation and intercept angle
Proposed	Eq. (7.21)	Zero	Speed variation, gravity and intercept angle

7.5 Simulation Results

In this section, nonlinear numerical simulations are performed to validate the proposed optimal guidance law in an exo-atmospheric interception scenario. The required initial conditions are summarized in Table 7.2.

7.5.1 Characteristics of the Proposed Guidance Law

This subsection analyzes the characteristics of the proposed guidance law with various design parameters $N = 0, 2, 4$. Figure 7.1a, b provide the simulation results of instantaneous ZEM and time history of flight path angle obtained from the proposed guidance law. It is clear that the guidance law with $N \geq 0$ can successfully guide the missile to intercept the target with a desired intercept angle. We can also note that the convergence rates of both instantaneous ZEM and intercept angle error become faster with the increase of the time-to-go order N. Hence, more control energy is consumed during the initial phase with a bigger value of N. This results in longer saturation time of the shear angle command at the beginning. The guidance command with different N is shown in Fig. 7.1c, which reveals that the angle-of-attack converges to zero at the time of impact when $N \geq 1$. This is consistent with what we expect. However, the guidance law with $N = 0$ cannot guarantee a finite terminal guidance command and therefore the angle-of-attack command is saturated when the missile is close to the target. Figure 7.1d compares the time history of missile velocity under the proposed guidance law with various N. The results reveal that the missile velocity keeps the initial value when all available control energy is utilized to guide the missile converge to the desired collision course. Once the collision course is achieved, the missile velocity increases almost linearly. However, the missile velocity with $N = 0$ remains constant when the missile is near the target due to command saturation.

Table 7.2 Initial conditions for homing engagement

Parameters	Values
Missile-target initial relative range, $r(0)$	50 km
Initial LOS angle, $\sigma(0)$	0°
Missile initial velocity, $V_M(0)$	2500 m/s
Missile initial flight path angle, $\gamma_M(0)$	150°
Missile axial acceleration, a_x	20 g
Target velocity, V_T	3000 m/s
Target initial flight path angle, $\gamma_T(0)$	20°
Desired Intercept Angle, γ_f	160°

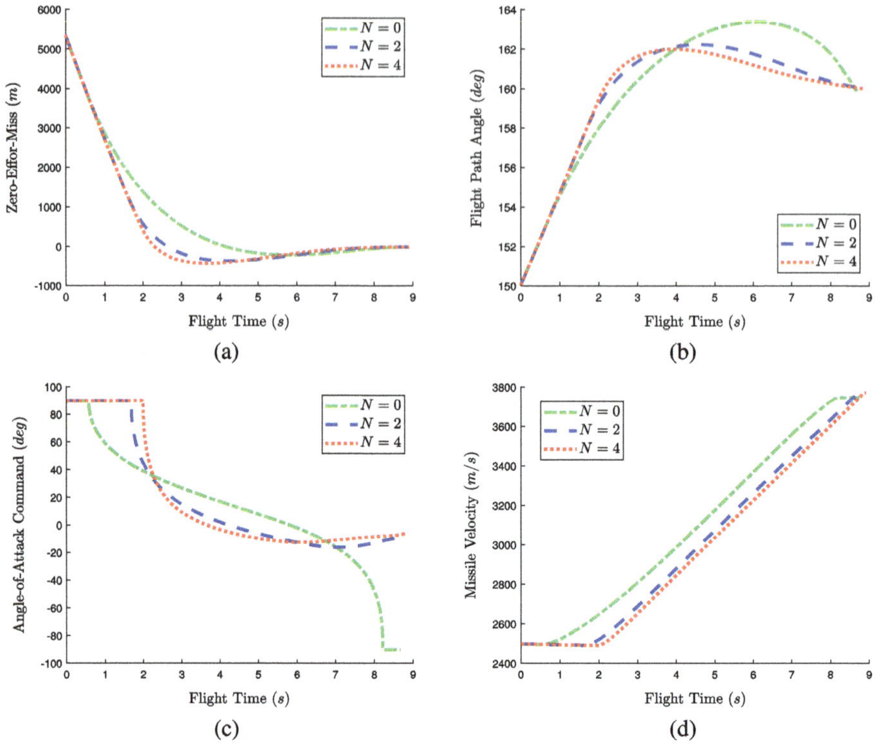

Fig. 7.1 Simulation results of the proposed guidance law with various guidance gains. **a** Interception trajectory; **b** Flight path angle; **c** Angle-of-attack command; and **d** Missile velocity profile

7.5.2 Comparison with Other Guidance Laws

The performance of the proposed guidance law is compared with that of TWOIAG and TWOG2CIAG in this subsection. To make fair comparisons, the time-to-go order of all guidance laws are chosen as $N = 3$ and an extra gravity-compensation term is utilized when implementing TWOIAG and TWOG2CIAG. The simulation results are presented in Fig. 7.2 and the quantitative comparison results are summarized in Table 7.3, where the control effort is defined as $\int_0^{t_f} a_M^2(\tau)\,d\tau$.

Figure 7.2a reveals that all tested guidance laws can successfully guide the missile to intercept the target with satisfactory performance and the difference is that TWOIAG generates more curved trajectory. However, Table 7.3 shows that the terminal miss distance under TWOIAG is much larger compared with the other two guidance laws, meaning that both TWOG2CIAG and the proposed guidance law provide more precise interception for the considered scenario.

The time history of the flight angle is presented in Fig. 7.2b and it is clear that the proposed guidance law and TWOG2CIAG have higher accuracy in driving the intercept angle to the desired value than that of TWOIAG. From Table 7.3, one

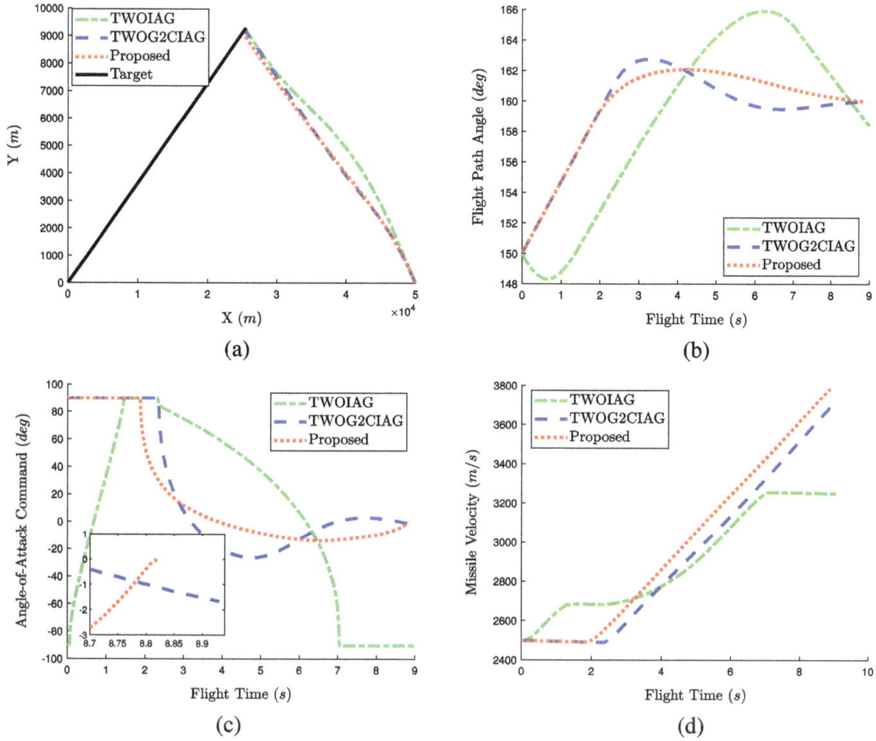

Fig. 7.2 Simulation results of the proposed guidance law with various guidance gains. **a** Interception trajectory; **b** Flight path angle; **c** Angle-of-attack command; and **d** Missile velocity profile

Table 7.3 Quantitative comparison results

	TWOIAG	TWOG2CIAG	Proposed
Miss distance	5.771 m	0.940 m	0.931 m
Intercept angle error	1.66°	0.88°	0.88°
Control effort	228575	109008	91503

can also note that the terminal impact angle error of the proposed guidance law and TWOG2CIAG is much smaller than that of TWOIAG. This further demonstrates the benefits of utilizing time-varying guidance gains in TWOG2CIAG and proposed framework rather than constant gains in TWOIAG. These results indicate that TWOIAG has limited applicability in impact angle control since it is not suitable for the scenario of accelerating missiles. The comparison of angle-of-attack commands of different guidance laws is given in Fig. 7.2c. It can be noted from this figure that the angle-of-attack command under TWOIAG saturates during the last 2 s. Since TWOG2CIAG leverages an extra gravity-compensation term in implementation, it only ensures bounded nonzero terminal guidance command. As a comparison, the proposed guidance law guarantees theoretical zero final guidance command and hence can improve the effectiveness of direct hit compared with the other guidance laws. Figure 7.2d compares the moving speed under different guidance laws. This figure shows that the proposed guidance law provides the highest terminal impact velocity compared to the other two laws, leading to the increase of final impact energy. Since the guidance command of TWOIAG saturates when the missile approaches the target, the missile velocity remains constant during that period. Compared to the proposed approach, the TWOG2CIAG requires more energy during the initial phase and only guarantee bounded terminal guidance command. In conclusion, the proposed guidance law exhibits better overall performance than the other two guidance laws in terms of control effort and guidance accuracy.

7.6 Summary

This chapter proposes a new optimal intercept angle guidance for exo-atmospheric interception. The proposed guidance law is given in a feedback form in terms of the instantaneous ZEM and intercept angle error with both time-varying and fixed guidance gains. The uniqueness of the proposed approach is that it directly uses, instead of compensating for, the gravity and also accommodate the speed-varying issue. Theoretical analysis shows that both TWOIAG and TWOG2CIAG are special cases of the proposed guidance law. Compared with TWOIAG and TWOG2CIAG, the proposed guidance law has better performance in terms of energy consumption and guidance accuracy.

It should be noted that the proposed approach can also be utilized in mid-course guidance as the angle constraint is important for guaranteeing the lock-on condition of the onboard seekers.

References

1. He S, Song T, Lin D (2017) Impact angle constrained integrated guidance and control for maneuvering target interception. J Guidance Control Dyn 40(10):2653–2661
2. Reisner D, Shima T (2013) Optimal guidance-to-collision law for an accelerating exoatmospheric interceptor missile. J Guidance Control Dyn 36(6):1695–1708
3. Shima T (2011) Intercept-angle guidance. J Guidance Control Dyn 34(2):484
4. Kim M, Grider KV (1973) Terminal guidance for impact attitude angle constrained flight trajectories. IEEE Trans Aerosp Electron Syst AES-9(6):852–859
5. Song TL, Shin SJ (1999) Time-optimal impact angle control for vertical plane engagements. IEEE Trans Aerosp Electron Syst 35(2):738–742
6. Zarchan P (2012) Tactical and strategic missile guidance. American Institute of Aeronautics and Astronautics
7. Ryoo C-K, Cho H, Tahk M-J (2005) Optimal guidance laws with terminal impact angle constraint. J Guidance Control Dyn 28(4):724–732
8. Ohlmeyer EJ, Phillips CA (2006) Generalized vector explicit guidance. J Guidance Control Dyn 29(2):261–268

9. Ryoo C-K, Cho H, Tahk M-J (2006) Time-to-go weighted optimal guidance with impact angle constraints. IEEE Trans Control Syst Technol 14(3):483–492
10. Lee C-H, Tahk M-J, Lee J-I (2013) Generalized formulation of weighted optimal guidance laws with impact angle constraint. IEEE Trans Aerosp Electron Syst 49(2):1317–1322
11. Seo M-G, Lee C-H, Tahk M-J (2017) New design methodology for impact angle control guidance for various missile and target motions. IEEE Trans Control Syst Technol

Part II
Optimal Guidance in UAV Applications

Chapter 8
Minimum-Effort Waypoint-Following Guidance Law

Abstract This chapter addresses the problem of minimum-effort waypoint-following guidance with/without arrival angle constraints of a UAV. By utilizing a linearized kinematics model, the proposed guidance laws are derived as the solutions of a linear quadratic optimal control problem with an arbitrary number of terminal boundary constraints. Theoretical analysis reveals that both optimal proportional navigation guidance and trajectory shaping guidance are special cases of the proposed guidance laws. The key feature of the proposed algorithms lies in their generic property. For this reason, the guidance laws developed can be applied to general waypoint-following missions with an arbitrary number of waypoints and an arbitrary number of arrival angle constraints. Nonlinear numerical simulations clearly demonstrate the effectiveness of the proposed formulations.

8.1 Introduction

UAVs have been successfully deployed and show great potentials in both civil and military applications. One of the fundamental enablers for UAVs to achieve high-level autonomy is to reach its destination with desired path constraints, e.g., waypoint and passing angle constraints [1–4]. Even though there exists long history in this domain, complicated numerical trajectory optimization is usually utilized to find the energy or time optimal path of a UAV [5–12]. Numerical optimization, however, requires high-computational onboard power and therefore might not be suitable to the ever-increasing small-scale UAVs. For this reason, finding analytical guidance algorithm is more beneficial for low-cost UAVs.

The most widely-accepted approach of path following guidance is the control-theoretic error-feedback-regulation concept. The basic idea of this concept is to use well-established control theories to force the trajectory tracking errors to converge to zero asymptotically or in finite time. However, the error-feedback-regulation method requires separate path planning and following modules and hence might involve iterative loops to tune these two different modules separately. Path following has also been investigated from the perspective of optimal control theory [13–16]. However, most optimal path/waypoint following guidance laws only ensure local optimality, except

S. He et al., *Optimal Guidance and Its Applications in Missiles and UAVs*, Springer Aerospace Technology, https://doi.org/10.1007/978-3-030-47348-8_8

for a few cases [17]. Although the authors in [17] revealed that the energy-optimal waypoint-following problem is equivalent to applying the optimal terminal acceleration constrained impact angle guidance to every two consecutive waypoints in conjunction with proper boundary conditions, the boundary constraints, e.g., passing angle and passing acceleration, require the numerical solution of a parameter optimization problem.

Intuitively, waypoint-following guidance can be viewed as a point-to-point guidance between two consecutive waypoints. This means that the well-developed missile guidance laws during last few decades can also be applied for UAV waypoint-following guidance. It is well-known that classical PNG with a constant navigation gain three is energy-optimal for point-to-point guidance [18–20]. If a specific arrival angle or approach angle is required, trajectory shaping guidance (TSG) becomes the energy-optimal point-to-point guidance law [21–23]. However, it is unclear whether or not the total energy consumption is optimal for the case of multiple waypoints if we simply apply these two optimal guidance laws between two consecutive points.

Motivated by the these observations, this chapter aims to propose a globally minimum-effort or energy-optimal guidance law for UAV waypoint following. By formulating a finite-time linear regulation problem, the proposed waypoint-following guidance law is analytically derived using optimal control theory based on a linear kinematics model. Theoretical analysis reveals that the proposed guidance laws reduce to classical PNG and TSG when there exists only one waypoint to be visited. The key features of the proposed guidance law are twofold:

(1) The proposed guidance law is generic: it can be applied a general waypoint-following guidance mission with an arbitrary number of waypoints and an arbitrary number of arrival angle constraint;
(2) The algorithm developed integrates path planning and following into a single step, which differs from existing error-feedback-regulation approaches. This advantage is beneficial to reduce the design complexity for initial mission analysis.

8.2 Backgrounds and Preliminaries

This chapter assumes that the UAV is equipped with a high-performance low-level flight control system that provides roll, pitch and yaw stability of the UAV as well as velocity tracking, heading and altitude hold functions. This study aims to design guidance inputs to this low-level controller for energy-optimal waypoint following in a two-dimensional geometry.

8.2.1 Nonlinear Kinematics

Consider there exist N waypoints that will be visited by the UAV. The relative geometry between the UAV and the ith waypoint is shown in Fig. 8.1, where the symbols U and W_i denote the UAV and the ith waypoint, respectively. The frame $X_I O Y_I$ is a inertial coordinate. The notation γ stands for the UAV's flight path angle. In general, the UAV speed is pre-determined according to specific mission objective and is maintained by an engine controller. For this reason, we assume that the UAV is flying with a constant velocity V. The UAV changes its direction through the lateral acceleration a. The variable a_i^σ denotes the UAV acceleration normal to the LOS direction. For simplicity, the UAV is assumed to be an ideal point mass model, i.e., the autopilot has no time delay. The variables r_i and σ_i represent the relative range and LOS angle between the UAV and the ith waypoint. Based on the principles of kinematics, the differential equations describing the engagement geometry, depicted in Fig. 8.1, are formulated as

$$
\begin{aligned}
\dot{r}_i &= -V\cos(\gamma - \sigma_i) \\
\dot{\sigma}_i &= -\frac{V\sin(\gamma - \sigma_i)}{r_i} \\
\dot{\gamma} &= \frac{a}{V}, \quad i \in [N]
\end{aligned}
\tag{8.1}
$$

where $[N] \triangleq \{1, 2, \ldots, N\}$.

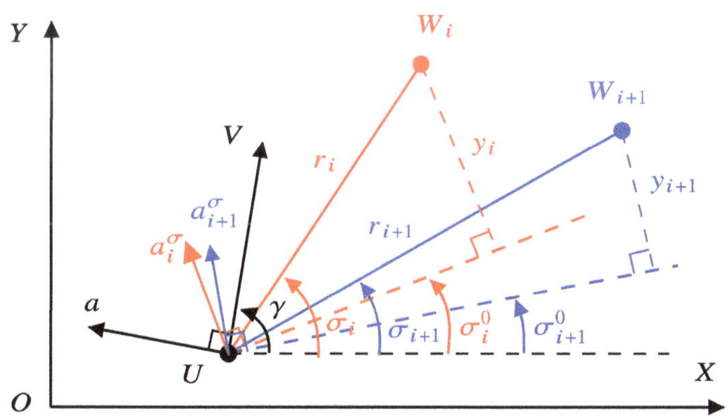

Fig. 8.1 Planar engagement geometry

8.2.2 Passing Time

Without loss of generality, assume that the N waypoints are ordered by the increase of their corresponding passing times $t_{f,i}$ as $t_{f,i} < t_{f,i+1}$. Around the ideal approaching course, i.e., when the heading error becomes error, the waypoint passing time can be approximated by

$$t_{f,i} = t + t_{go,i} \tag{8.2}$$

where $t_{go,i} \triangleq r_i / V$ denotes the remaining flight time, or the so-called time-to-go, to pass the ith waypoint.

8.2.3 Linearized Kinematics

In this chapter, optimal guidance laws will be derived based on a linearized model around the desired approaching course. For the purpose of kinematics linearization, a new frame called the reference frame $X_R O Y_R$ is introduced, as depicted in Fig. 8.1. This frame is rotated from the inertial frame by σ_0, which is the reference angle. Let $y_i = y_{W_i} - y_U$ be the relative displacement between the UAV and the ith waypoint normal to the X_R axis. In the reference frame, the engagement kinematics can be expressed as

$$\begin{aligned} \dot{y}_i &= v_i \\ \dot{v}_i &= -a_i^\sigma \cos(\sigma_i - \sigma_0) \end{aligned} \tag{8.3}$$

where v_i is the relative velocity between the UAV and the ith waypoint perpendicular to the X_R direction.

The complementary equation that describes the relationship between a_i^σ and a is given by

$$a_i^\sigma = a \cos(\gamma - \sigma_i) \tag{8.4}$$

By choosing proper σ_0, the angle $\sigma_i - \sigma_0$ can be made relatively small since the LOS angle variation is small during the flight if a guidance law works properly [24]. In practical flight, the velocity lead angle is also relatively small [25]. With these assumptions in mind, the relative kinematics between the UAV and the ith waypoint can be formulated as

$$\begin{aligned} \dot{y}_i &= v_i \\ \dot{v}_i &= -a, \end{aligned} \tag{8.5}$$

Define $\boldsymbol{x} = [y_1, v_1, y_2, v_2, \ldots, y_N, v_N]^T \in \mathbb{R}^{2N \times 1}$ as the system state vector. Then, the linearized equations of motion can be written in a compact matrix form as

$$\dot{\boldsymbol{x}} = \boldsymbol{A}\boldsymbol{x} + \boldsymbol{B}a \tag{8.6}$$

where $\boldsymbol{A} \in \mathbb{R}^{2N\times 2N}$ is a block diagonal matrix, $\boldsymbol{B} \in \mathbb{R}^{2N\times 1}$, and $\boldsymbol{C} \in \mathbb{R}^{1\times 2N}$. These three matrices are defined as

$$\boldsymbol{A} = \operatorname{diag}\left(\boldsymbol{A}_1, \boldsymbol{A}_2, \ldots \boldsymbol{A}_N\right), \quad \boldsymbol{B} = \left[\boldsymbol{B}_1^T, \boldsymbol{B}_2^T, \ldots, \boldsymbol{B}_N^T\right]^T \tag{8.7}$$

where

$$\boldsymbol{A}_i = \begin{bmatrix} 0 & 1 \\ 0 & 0 \end{bmatrix}, \quad \boldsymbol{B}_i = \begin{bmatrix} 0 \\ -1 \end{bmatrix}. \tag{8.8}$$

8.2.4 *Problem Formulation*

In practice, the energy consumption is of paramount importance for a UAV since it determines the endurance of the vehicle. For this reason, this chapter considers the following quadratic integral control effort performance index

$$J = \int_t^{t_{f,N}} a^2(\tau)\, d\tau = \begin{cases} \displaystyle\int_t^{t_{f,1}} a^2(\tau)\, d\tau + \sum_{i=1}^{N-1} \int_{t_{f,i}}^{t_{f,i+1}} a^2(\tau)\, d\tau, & t \le t_{f,1} \\ \displaystyle\int_t^{t_{f,2}} a^2(\tau)\, d\tau + \sum_{i=2}^{N-1} \int_{t_{f,i}}^{t_{f,i+1}} a^2(\tau)\, d\tau, & t \le t_{f,2} \\ \vdots & \\ \displaystyle\int_t^{t_{f,N}} a^2(\tau)\, d\tau, \quad t_{f,N-1} < t \le t_{f,N} & \end{cases} \tag{8.9}$$

It has also been shown that quantity (8.9) relates to the speed loss due to induced drag for aerodynamically-controlled vehicles, e.g., fixed-wing UAVs [26]. Therefore, minimizing the quadratic energy consumption is a worthy goal for guidance law design. The aim of this chapter is to find analytical solutions of the following two generalized optimal waypoint-following problems:

Problem 8.1 (*Optimal waypoint-following guidance problem*) Given the kinematics model (8.5), finds the guidance command a that minimizes performance index (8.9) and ensures perfect waypoint passing constraints

$$y_i\left(t_{f,i}\right) = 0, \quad i \in [N] \tag{8.10}$$

Problem 8.2 (*Optimal waypoint-following with partial flight path angle constraint guidance problem*) Given the kinematics model (8.5), finds the guidance command a that satisfies the same conditions as in Problem 1 and the additional M flight path angle constraints

$$\gamma\left(t_{f,l(j)}\right)=\gamma_{l(j)}^{d},\quad j\in[M],\quad l(j)\in[N],\quad M\leq N \tag{8.11}$$

where $l(j)$ is the index of waypoints that have specific flight path angle constraints and $\gamma_{l(j)}^{d}$ denotes the desired flight angle of the UAV when it passes the $l(j)$th waypoint.

8.3 Optimal Guidance for Waypoint-Following

8.3.1 Guidance Law Derivation

This subsection will derive an optimal guidance law to address Problem 8.1. To reduce the system order, the concept of ZEM transformation [21, 25] is used in this chapter. The ZEM between the UAV and the ith waypoint, denoted as z_i, is defined as

$$z_i=\begin{cases}\boldsymbol{C}_i\boldsymbol{\Phi}_i\left(t_{f,i},t\right)\left[y_i,v_i\right]^T, & t\leq t_{f,i}\\ z_i\left(t_{f,i}\right), & t>t_{f,i}\end{cases} \tag{8.12}$$

where $\boldsymbol{\Phi}_i\left(t_{f,i},t\right)$ denotes the transition matrix associated with matrix $\boldsymbol{A}_i$ and $\boldsymbol{C}_i$ is a constant matrix that extracts appropriate elements from the system state vector. The transition matrix $\boldsymbol{\Phi}_i\left(t_{f,i},t\right)$ is determined as

$$\boldsymbol{\Phi}_i\left(t_{f,i},t\right)=\begin{bmatrix}1 & t_{f,i}-t\\ 0 & 1\end{bmatrix} \tag{8.13}$$

To extract ZEM for each waypoint:, the matrix $\boldsymbol{C}_i$ is given by

$$\boldsymbol{C}_i=\begin{bmatrix}1 & 0 & 0 & 0\end{bmatrix} \tag{8.14}$$

Substituting Eqs. (8.13) and (8.14) into Eq. (8.12) gives the ZEM as

$$z_i=\begin{cases}y_i+t_{go,i}v_i, & t\leq t_{f,i}\\ z_i\left(t_{f,i}\right), & t>t_{f,i}\end{cases} \tag{8.15}$$

The ZEM dynamics can be obtained from Eq. (8.15) as

$$\dot{z}_i=\begin{cases}-t_{go,i}a, & t\leq t_{f,i}\\ 0, & t>t_{f,i}\end{cases} \tag{8.16}$$

With this state transformation, the system order now reduces from $2N$ to N. Problem 8.1 reduces to a problem with the same performance index given by Eq. (8.9) but reduced-order system dynamics (8.16). Notice that the ZEM z_i represents the miss distance between the UAV and the ith waypoint if, from the current time onward, the UAV will not apply any maneuver input. With this in mind, the terminal constraints of the new system states are determined as

$$z_i\left(t_{f,i}\right) = 0 \tag{8.17}$$

According to linear system theory, the solution of system (8.16) can be obtained as

$$z_i\left(t_{f,i}\right) - z_i\left(t\right) = \int_t^{t_{f,i}} -\left(t_{f,i} - \tau\right) a\left(\tau\right) d\tau, \quad t \le t_{f,i} \tag{8.18}$$

Imposing the terminal constraints (8.17) on Eq. (8.18) gives

$$z_i\left(t\right) = \int_t^{t_{f,i}} \left(t_{f,i} - \tau\right) a\left(\tau\right) d\tau, \quad t \le t_{f,i} \tag{8.19}$$

Consider the Hilbert space $\mathcal{H} = \mathcal{L}^2\left[t, t_{f,N}\right]$ with the inner product $(f, g) = \int_t^{t_{f,i}} f\left(\tau\right) g\left(\tau\right) d\tau$. According to Lemma 5.1, if the guidance command a is optimal in terms of energy minimization, then there exist N Lagrange multipliers λ_i, $i \in \{1, 2, \ldots, N\}$, such that the lateral acceleration command can be formulated as

$$a = \begin{cases} \sum_{i=1}^{N} \lambda_i \left(t_{f,i} - t\right), & t \le t_{f,1} \\ \sum_{i=2}^{N} \lambda_i \left(t_{f,i} - t\right), & t_{f,1} < t \le t_{f,2} \\ \vdots & \\ \lambda_N \left(t_{f,N} - t\right), & t_{f,N-1} < t \le t_{f,N} \end{cases} \tag{8.20}$$

The Lagrange multipliers λ_i can then be determined by introducing the expression in Eq. (8.19) and solving the resultant equations. Note that this approach can be viewed as the extension of Schwarz inequality method [27] to arbitrary number of terminal constraints. Without loss of generality, we only consider the case $t \le t_{f,1}$ in the following derivations. By substituting Eq. (8.20) into Eq. (8.19) under condition $t \le t_{f,1}$, we have

$$
\begin{aligned}
z_i(t) &= \sum_{i'=1}^{N} \lambda_{i'} \int_t^{t_{f,i}} \left(t_{f,i} - \tau\right)\left(t_{f,i'} - \tau\right) d\tau \\
&= \sum_{i'=1}^{N} \lambda_{i'} \int_t^{t_{f,\min\{i,i'\}}} \left(t_{f,i} - \tau\right)\left(t_{f,i'} - \tau\right) d\tau \\
&= \sum_{i'=1}^{N} \lambda_{i'} \int_t^{t_{f,\min\{i,i'\}}} \left(t_{f,\max\{i,i'\}} - t_{f,\min\{i,i'\}}\right)\left(t_{f,\min\{i,i'\}} - \tau\right) + \left(t_{f,\min\{i,i'\}} - \tau\right)^2 d\tau \\
&= \sum_{i'=1}^{N} \lambda_{i'} \left[\frac{t_{go,\min\{i,i'\}}^3}{3} + \frac{t_{go,\min\{i,i'\}}^2}{2}\left(t_{f,\max\{i,i'\}} - t_{f,\min\{i,i'\}}\right) \right] \\
&= \sum_{i'=1}^{N} \lambda_{i'} \left[\frac{t_{go,\min\{i,i'\}}^3}{3} + \frac{t_{go,\min\{i,i'\}}^2}{2}\left(t_{go,\max\{i,i'\}} - t_{go,\min\{i,i'\}}\right) \right] \\
&= \sum_{i'=1}^{N} \lambda_{i'} \left[\frac{t_{go,\max\{i,i'\}} t_{go,\min\{i,i'\}}^2}{2} - \frac{t_{go,\min\{i,i'\}}^3}{6} \right]
\end{aligned}
\tag{8.21}
$$

Define $\boldsymbol{\lambda} = [\lambda_1, \lambda_2, \ldots, \lambda_N]^T$ and $\boldsymbol{Z} = [z_1, z_2, \ldots, z_N]^T$ as the Lagrange multiplier vector and the ZEM vector, respectively. Then, Eq. (8.21) can be rewritten in a compact matrix form as

$$
\boldsymbol{G}\boldsymbol{\lambda} = \boldsymbol{Z} \tag{8.22}
$$

where $\boldsymbol{G} \in \mathbb{R}^{N \times N}$ is a symmetric matrix, i.e., $\boldsymbol{G}^T = \boldsymbol{G}$, and is given by

$$
\boldsymbol{G} = \begin{bmatrix}
\frac{t_{go,1}^3}{3} & \frac{t_{go,2} t_{go,1}^2}{2} - \frac{t_{go,1}^3}{6} & \cdots & \cdots & \frac{t_{go,N} t_{go,1}^2}{2} - \frac{t_{go,1}^3}{6} \\
 & \frac{t_{go,2}^3}{3} & \frac{t_{go,3} t_{go,2}^2}{2} - \frac{t_{go,2}^3}{6} & \cdots & \frac{t_{go,N} t_{go,2}^2}{2} - \frac{t_{go,2}^3}{6} \\
\ddots & & \ddots & & \\
 & \ddots & & \frac{t_{go,N-1}^3}{3} & \frac{t_{go,N} t_{go,N-1}^2}{2} - \frac{t_{go,N-1}^3}{6} \\
 & & \ddots & & \frac{t_{go,N}^3}{3}
\end{bmatrix} \tag{8.23}
$$

From Eq. (8.22), the Lagrange multiplier vector $\boldsymbol{\lambda}$ can be obtained as

$$
\boldsymbol{\lambda} = \boldsymbol{G}^{-1}\boldsymbol{Z} \tag{8.24}
$$

The guidance command for $t \leq t_{f,1}$ can then be obtained by substituting Eq. (8.24) into Eq. (8.20) as

$$
\begin{aligned}
a &= \boldsymbol{\lambda}^T \left[t_{go,1}, t_{go,2}, \ldots, t_{go,N}\right]^T \\
&= \left(\boldsymbol{G}^{-1}\boldsymbol{Z}\right)^T \left[t_{go,1}, t_{go,2}, \ldots, t_{go,N}\right]^T
\end{aligned}
\tag{8.25}
$$

Remark 8.1 Following similar procedures, the solutions for $t > t_{f,1}$ can be easily obtained. For example, when $t_{f,1} < t \leq t_{f,2}$, the ZEM vector reduces to $\boldsymbol{Z} = [Z_2, \ldots, Z_N]^T$ and the matrix $\boldsymbol{G} \in \mathbb{R}^{(N-1)\times(N-1)}$ becomes

$$\boldsymbol{G} = \begin{bmatrix} \frac{t_{go,2}^3}{3} & \frac{t_{go,3}t_{go,2}^2}{2} - \frac{t_{go,2}^3}{6} & \cdots & \cdots & \frac{t_{go,N}t_{go,2}^2}{2} - \frac{t_{go,2}^3}{6} \\ & \frac{t_{go,3}^3}{3} & \frac{t_{go,4}t_{go,3}^2}{2} - \frac{t_{go,3}^3}{6} & \cdots & \frac{t_{go,N}t_{go,3}^2}{2} - \frac{t_{go,3}^3}{6} \\ \ddots & & \ddots & & \\ & \ddots & & \frac{t_{go,N-1}^3}{3} & \frac{t_{go,N}t_{go,N-1}^2}{2} - \frac{t_{go,N-1}^3}{6} \\ & & \ddots & & \frac{t_{go,N}^3}{3} \end{bmatrix} \tag{8.26}$$

The guidance command for $t_{f,1} < t \leq t_{f,2}$ is then given by

$$a = \left(\boldsymbol{G}^{-1}\boldsymbol{Z}\right)^T \left[t_{go,2}, t_{go,3}, \ldots, t_{go,N}\right]^T \tag{8.27}$$

Remark 8.2 For the purpose of implementation, it is desired to formulate the guidance command in terms of measured signals. From Eqs. (2.8) and (2.46), the ZEM z_i can be written in an alternative form as

$$z_i = V\dot{\sigma}_i t_{go,i}^2 \tag{8.28}$$

Substituting Eq. (8.28) into Eq. (8.25) gives the guidance command in terms of measured signals $\dot{\sigma}_i$, V and r_i. This supports the practical application of the proposed guidance law. Note that although the proposed guidance law is derived based on the linearized engagement kinematics, the error generated in the linearization process can be alleviated by using Eq. (8.28) in implementation since Eq. (8.28) transforms the original linear terms into their corresponding nonlinear expressions.

Remark 8.3 It follows from Eq. (8.25) that implementing the proposed guidance law requires calculation of the inverse of matrix $\boldsymbol{G}$. Notice that the size of matrix $\boldsymbol{G}$ is proportional to the number of waypoints to be traveled. Therefore, the complexity, or computational burden, of the proposed guidance law increases for a large number of waypoints. However, compared with numerical optimization solutions, the proposed method is still much more efficient since the guidance command is explicitly given as a feedback form of the measured signals.

8.3.2 *Particular Cases*

8.3.2.1 $N = 1$

When there exists only one waypoint to be visited by the UAV, Problem 8.1 reduces to energy-optimal intercept problem. For this specific case, the guidance command (8.20) can be written as

$$a = \lambda_1 \left(t_{f,1} - t\right) \tag{8.29}$$

Under condition $N = 1$, matrix $\boldsymbol{G}$ reduces to a scalar as $\boldsymbol{G} = t_{go,1}^3/3$ and the single Lagrange multiplier can be readily obtained as $\lambda_1 = 3/t_{go,1}^3 z_1$. Substituting this expression into Eq. (8.29) gives the explicit guidance command as

$$a = \frac{3z_1}{t_{go,1}^2} \tag{8.30}$$

which coincides with the classical optimal PNG. As shown in [28], the PNG with a navigation gain three is energy optimal in the case of single waypoint. However, from the previous derivation, it is clear that simply applying PNG to every two consecutive waypoints is not energy optimal when the number of waypoints satisfies $N \geq 2$. The reason is that PNG cannot control the UAV's flight path angle and therefore the energy consumption of PNG is different from that of the proposed guidance law if the arrival angles $\gamma\left(t_{f,i}\right)$ are different.

8.3.2.2 $N = 2$

When there are two waypoints to be visited by the UAV, the guidance command (8.20) under condition $t \leq t_{f,1}$ can be written as

$$a = \lambda_1 \left(t_{f,1} - t\right) + \lambda_2 \left(t_{f,2} - t\right) \tag{8.31}$$

and matrix $\boldsymbol{G}$ becomes

$$\boldsymbol{G} = \begin{bmatrix} \frac{t_{go,1}^3}{3} & \frac{t_{go,2}t_{go,1}^2}{2} - \frac{t_{go,1}^3}{6} \\ \frac{t_{go,2}t_{go,1}^2}{2} - \frac{t_{go,1}^3}{6} & \frac{t_{go,2}^3}{3} \end{bmatrix} \tag{8.32}$$

The corresponding Lagrange multipliers can be readily obtained from Eq. (8.24) as

$$\begin{aligned} \lambda_1 &= \frac{6\left(2t_{go,2}^3 z_1 + t_{go,1}^3 z_2 - 3t_{go,1}^2 t_{go,2} z_2\right)}{t_{go,1}^3\left(t_{go,2} - t_{go,1}\right)^2\left(4t_{go,2} - t_{go,1}\right)} \\ \lambda_2 &= \frac{6\left(t_{go,1} z_1 - 3t_{go,2} z_1 + 2t_{go,1} z_2\right)}{t_{go,1}\left(t_{go,2} - t_{go,1}\right)^2\left(4t_{go,2} - t_{go,1}\right)} \end{aligned} \tag{8.33}$$

Substituting Eq. (8.33) into Eq. (8.31) gives the explicit acceleration command as

$$a = \frac{6\left(2t_{go,2}^2 z_1 - t_{go,1}t_{go,2} z_1 - t_{go,1}^2 z_2\right)}{t_{go,1}^2\left(t_{go,2} - t_{go,1}\right)\left(4t_{go,2} - t_{go,1}\right)} \tag{8.34}$$

Using Eq. (8.28), the guidance command can be reformulated in terms of measured signals as

$$a = N_1 V \dot{\sigma}_1 + N_2 V \dot{\sigma}_2 \tag{8.35}$$

where

$$\begin{aligned} N_1 &= \frac{-6\left(t_{go,1} - 2t_{go,2}\right) t_{go,2}}{\left(t_{go,1} - 4t_{go,2}\right)\left(t_{go,1} - t_{go,2}\right)} \\ N_2 &= \frac{-6t_{go,2}^2}{\left(t_{go,1} - 4t_{go,2}\right)\left(t_{go,1} - t_{go,2}\right)} \end{aligned} \tag{8.36}$$

Define $a_B = N_2 V \dot{\sigma}_2$, Eq. (8.35) can be further reduced to

$$a = N_1 V \dot{\sigma}_1 + a_B \tag{8.37}$$

which means that the proposed guidance law can be viewed as a biased PNG law with a time-varying navigation gain N_1 when $t \leq t_{f,1}$.

In the following, we will analyze the effect of the biased term a_B to provide better insights of the proposed guidance law. To this end, we assume that the navigation gain N_1 and the biased term a_B are constant in the vicinity of the first waypoint. For PNG with a constant biased term, the closed-form solution is given by [29]

$$y_1 = c_1 t_{go,1}^{N_1} + c_2 t_{go,1} + \frac{a_B}{N_1 - 2} t_{go,1}^2 \tag{8.38}$$

where c_1 and c_2 are constants determined by the initial conditions.

Since $\sigma_1 - \sigma_0 \approx y_1/r_1 \approx y_1/\left(V t_{go,1}\right)$, we have

$$\sigma_1 = \frac{c_1}{V} t_{go,1}^{N_1 - 1} + \frac{c_2}{V} + \frac{a_B}{V\left(N_1 - 2\right)} t_{go,1} + \sigma_0 \tag{8.39}$$

Differentiating Eq. (8.39) with respect to time gives the LOS rate $\dot{\sigma}_1$ as

$$\dot{\sigma}_1 = -\frac{c_1}{V}\left(N_1 - 1\right) t_{go,1}^{N_1 - 2} - \frac{a_B}{V\left(N_1 - 2\right)} \tag{8.40}$$

It follows from Eq. (8.36) that $\lim\limits_{t_{go,1} \to 0} N_1 = 3 > 2$. Hence, when the UAV approaches the first waypoint, LOS rate $\dot{\sigma}_1$ becomes

$$\lim_{t_{go,1} \to 0^+} \dot{\sigma}_1 = -\frac{a_B}{V\left(N_1 - 2\right)} \tag{8.41}$$

Therefore, the acceleration command in the vicinity of the first waypoint becomes

$$\begin{aligned}\lim_{t_{go,1}\to 0^+} a &= -\frac{N_1}{N_1-2}a_B + a_B \\ &= -\frac{2}{N_1-2}a_B \\ &= N'_2 V\dot{\sigma}_2\end{aligned} \tag{8.42}$$

where

$$N'_2 = \frac{12t_{go,2}^2}{4t_{go,2}^2 - 2t_{go,1}^2 + 4t_{go,1}t_{go,2}} \tag{8.43}$$

Since $\lim_{t_{go,1}\to 0^+} N'_2 \to 3$, the guidance command of the proposed guidance law under condition $t_{go,1} \to 0$ reduces to

$$\lim_{t_{go,1}\to 0^+} a = 3V\dot{\sigma}_2 \tag{8.44}$$

which coincides with the guidance command when $t_{f,1} < t \le t_{f,2}$. This clearly reveals that the bias term $N_2V\dot{\sigma}_2$ helps to reduce the transient effect when the UAV passes the first waypoint.

8.3.2.3 $N = 3$

When there are three waypoints to be visited by the UAV, the guidance command (8.20) under condition $t \le t_{f,1}$ can be written as

$$a = \lambda_1\left(t_{f,1}-t\right) + \lambda_2\left(t_{f,2}-t\right) + \lambda_3\left(t_{f,3}-t\right) \tag{8.45}$$

and matrix $\boldsymbol{G}$ becomes

$$\boldsymbol{G} = \begin{bmatrix} \frac{t_{go,1}^3}{3} & \frac{t_{go,2}t_{go,1}^2}{2} - \frac{t_{go,1}^3}{6} & \frac{t_{go,3}t_{go,1}^2}{2} - \frac{t_{go,1}^3}{6} \\ \frac{t_{go,2}t_{go,1}^2}{2} - \frac{t_{go,1}^3}{6} & \frac{t_{go,2}^3}{3} & \frac{t_{go,3}t_{go,2}^2}{2} - \frac{t_{go,2}^3}{6} \\ \frac{t_{go,3}t_{go,1}^2}{2} - \frac{t_{go,1}^3}{6} & \frac{t_{go,3}t_{go,2}^2}{2} - \frac{t_{go,2}^3}{6} & \frac{t_{go,3}^3}{3} \end{bmatrix} \tag{8.46}$$

Following previous derivations, the explicit guidance command is obtained as

$$\begin{aligned} a = & \frac{3t_{go,2}\left(t_{go,3}-t_{go,2}\right)\left[t_{go,1}\left(2t_{go,3}+t_{go,2}\right) - t_{go,2}\left(4t_{go,3}-t_{go,2}\right)\right]z_1}{t_{go,1}^2\left(t_{go,1}-t_{go,2}\right)\left(t_{go,3}-t_{go,2}\right)\left(2t_{go,1}t_{go,2}+t_{go,1}t_{go,3}-4t_{go,2}t_{go,3}+t_{go,2}^2\right)} \\ & + \frac{3\left(t_{go,3}-t_{go,1}\right)\left(2t_{go,3}-t_{go,2}-t_{go,1}\right)z_2 - 3\left(t_{go,1}-t_{go,2}\right)^2 z_3}{\left(t_{go,1}-t_{go,2}\right)\left(t_{go,3}-t_{go,2}\right)\left(2t_{go,1}t_{go,2}+t_{go,1}t_{go,3}-4t_{go,2}t_{go,3}+t_{go,2}^2\right)} \end{aligned} \tag{8.47}$$

Using Eq. (8.28), the guidance command can be reformulated in terms of measured signals as

$$a = N_1 V\dot{\sigma}_1 + N_2 V\dot{\sigma}_2 + N_3 V\dot{\sigma}_3 \tag{8.48}$$

where

$$\begin{aligned}
N_1 &= \frac{3t_{go,2}\left(t_{go,3}-t_{go,2}\right)\left[t_{go,1}\left(2t_{go,3}+t_{go,2}\right)-t_{go,2}\left(4t_{go,3}-t_{go,2}\right)\right]}{\left(t_{go,1}-t_{go,2}\right)\left(t_{go,3}-t_{go,2}\right)\left(2t_{go,1}t_{go,2}+t_{go,1}t_{go,3}-4t_{go,2}t_{go,3}+t_{go,2}^2\right)}\\
N_2 &= \frac{3t_{go,2}^2\left(t_{go,3}-t_{go,1}\right)\left(2t_{go,3}-t_{go,2}-t_{go,1}\right)}{\left(t_{go,1}-t_{go,2}\right)\left(t_{go,3}-t_{go,2}\right)\left(2t_{go,1}t_{go,2}+t_{go,1}t_{go,3}-4t_{go,2}t_{go,3}+t_{go,2}^2\right)}\\
N_3 &= -\frac{3t_{go,3}^2\left(t_{go,1}-t_{go,2}\right)^2}{\left(t_{go,1}-t_{go,2}\right)\left(t_{go,3}-t_{go,2}\right)\left(2t_{go,1}t_{go,2}+t_{go,1}t_{go,3}-4t_{go,2}t_{go,3}+t_{go,2}^2\right)}
\end{aligned} \tag{8.49}$$

Recall the results derived for $N = 2$, the guidance command for $t_{f,1} < t \leq t_{f,2}$ is determined as

$$a = N_2' V\dot{\sigma}_2 + N_3' V\dot{\sigma}_3 \tag{8.50}$$

where

$$\begin{aligned}
N_2' &= \frac{-6\left(t_{go,2}-2t_{go,3}\right)t_{go,3}}{\left(t_{go,2}-4t_{go,3}\right)\left(t_{go,2}-t_{go,3}\right)}\\
N_3' &= \frac{-6t_{go,3}^2}{\left(t_{go,2}-4t_{go,3}\right)\left(t_{go,2}-t_{go,3}\right)}
\end{aligned} \tag{8.51}$$

Therefore, the explicit guidance command for $N = 3$ is given by

$$a = \begin{cases} N_1 V\dot{\sigma}_1 + N_2 V\dot{\sigma}_2 + N_3 V\dot{\sigma}_3, & t \leq t_{f,1} \\ N_2' V\dot{\sigma}_2 + N_3' V\dot{\sigma}_3, & t_{f,1} < t \leq t_{f,2} \\ 3V\dot{\sigma}_3, & t_{f,2} < t \leq t_{f,3} \end{cases} \tag{8.52}$$

Following similar procedures shown in the previous subsection, the characteristics of the proposed guidance law near the waypoint can also be analyzed in a theoretical way using the concept of biased PNG.

8.4 Optimal Guidance for Waypoint-Following with Partial Flight Path Angle Constraints

8.4.1 Guidance Law Derivation

In practice, it is desirable to constrain the UAV's flight path angle as some fixed desired values at some certain waypoints for surveillance missions. For this reason, this subsection will derive an optimal guidance law to address Problem 8.2. With the

ZEM state transformation provided in the previous section, Problem 8.1 reduces to a problem with the same performance index given by Eq. (8.9) but reduced-order system dynamics (8.16). The terminal constraints are determined as

$$\begin{aligned} z_i\left(t_{f,i}\right) &= 0, \quad i \in \{1, 2, \ldots, N\} \\ \gamma\left(t_{f,l(j)}\right) &= \gamma_{l(j)}^d, \quad j \in [M], \quad l(j) \in [N], \quad M \leq N \end{aligned} \tag{8.53}$$

According to the linear system theory, the solution of ZEM and flight path angle can be obtained as

$$\begin{aligned} z_i\left(t_{f,i}\right) - z_i(t) &= \int_t^{t_{f,i}} -\left(t_{f,i} - \tau\right) a(\tau)\, d\tau, \quad t \leq t_{f,i} \\ \gamma\left(t_{f,l(j)}\right) - \gamma(t) &= \int_t^{t_{f,l(j)}} \frac{a(\tau)}{V} d\tau, \quad t \leq t_{f,l(j)} \end{aligned} \tag{8.54}$$

Imposing the terminal constraints (8.53) on Eq. (8.54) gives

$$\begin{aligned} z_i\left(t_{f,i}\right) &= \int_t^{t_{f,i}} \left(t_{f,i} - \tau\right) a(\tau)\, d\tau, \quad t \leq t_{f,i} \\ \gamma_{l(j)}^d - \gamma(t) &= \int_t^{t_{f,l(j)}} \frac{a(\tau)}{V} d\tau, \quad t \leq t_{f,l(j)} \end{aligned} \tag{8.55}$$

According to Lemma 5.1, if the guidance command a is optimal in terms of energy minimization, then there exist $N + M$ Lagrange multipliers $\lambda_i, \beta_j, i \in \{1, 2, \ldots, N\}$, $j \in \{1, 2, \ldots, M\}$, such that the lateral acceleration command can be formulated as

$$a = a_\lambda + a_\beta \tag{8.56}$$

with

$$a_\lambda = \begin{cases} \sum_{i=1}^{N} \lambda_i \left(t_{f,i} - t\right), & t \leq t_{f,1} \\ \sum_{i=2}^{N} \lambda_i \left(t_{f,i} - t\right), & t_{f,1} < t \leq t_{f,2} \\ \vdots & \\ \lambda_N \left(t_{f,N} - t\right), & t_{f,N-1} < t \leq t_{f,N} \end{cases}, \quad a_\beta = \begin{cases} \sum_{j=1}^{M} \frac{\beta_j}{V}, & t \leq t_{f,l(1)} \\ \sum_{j=2}^{M} \frac{\beta_j}{V}, & t_{f,l(1)} < t \leq t_{f,l(2)} \\ \vdots & \\ \frac{\beta_M}{V}, & t_{f,l(M-1)} < t \leq t_{f,l(M)} \end{cases} \tag{8.57}$$

where a_λ refers to the ZEM regulation term and a_β represents the flight path angle error regulation command.

Similar to previous section, we only consider the case $t \le t_{f,1}$ in the following derivations. The solutions for $t > t_{f,1}$ can be easily obtained through similar procedures. Substituting Eq. (8.56) into Eq. (8.55) under condition $t \le t_{f,1}$ results in

$$\begin{aligned} z_i(t) &= \sum_{i'=1}^{N} \lambda_{i'} \int_t^{t_{f,i}} (t_{f,i}-\tau)(t_{f,i'}-\tau)\,d\tau + \sum_{j'=1}^{M} \frac{\beta_{j'}}{V} \int_t^{t_{f,i}} (t_{f,i}-\tau)\,d\tau \\ &= \sum_{i'=1}^{N} \lambda_{i'} \int_t^{t_{f,\min\{i,i'\}}} (t_{f,i}-\tau)(t_{f,i'}-\tau)\,d\tau + \sum_{j'=1}^{M} \frac{\beta_{j'}}{V} \int_t^{t_{f,\min\{i,l(j')\}}} (t_{f,i}-\tau)\,d\tau \end{aligned} \tag{8.58}$$

$$\begin{aligned} \gamma_{l(j)}^d - \gamma &= \sum_{i'=1}^{N} \frac{\lambda_{i'}}{V} \int_t^{t_{f,l(j)}} (t_{f,i'}-\tau)\,d\tau + \sum_{j'=1}^{M} \frac{\beta_{j'}}{V^2} \int_t^{t_{f,l(j)}} d\tau \\ &= \sum_{i'=1}^{N} \frac{\lambda_{i'}}{V} \int_t^{t_{f,\min\{i',l(j)\}}} (t_{f,i'}-\tau)\,d\tau + \sum_{j'=1}^{M} \frac{\beta_{j'}}{V^2} \int_t^{t_{f,\min\{l(j),l(j')\}}} d\tau \end{aligned} \tag{8.59}$$

Evaluating the integrals in Eqs. (8.58) and (8.59) gives

$$\begin{aligned} &\int_t^{t_{f,\min\{i,i'\}}} (t_{f,i}-\tau)(t_{f,i'}-\tau)\,d\tau \\ &= \int_t^{t_{f,\min\{i,i'\}}} \left(t_{f,\max\{i,i'\}} - t_{f,\min\{i,i'\}}\right)\left(t_{f,\min\{i,i'\}}-\tau\right) + \left(t_{f,\min\{i,i'\}}-\tau\right)^2 d\tau \\ &= \frac{t_{go,\min\{i,i'\}}^3}{3} + \frac{t_{go,\min\{i,i'\}}^2}{2}\left(t_{f,\max\{i,i'\}} - t_{f,\min\{i,i'\}}\right) \\ &= \frac{t_{go,\min\{i,i'\}}^3}{3} + \frac{t_{go,\min\{i,i'\}}^2}{2}\left(t_{go,\max\{i,i'\}} - t_{go,\min\{i,i'\}}\right) \\ &= \frac{t_{go,\max\{i,i'\}} t_{go,\min\{i,i'\}}^2}{2} - \frac{t_{go,\min\{i,i'\}}^3}{6} \end{aligned} \tag{8.60}$$

$$\begin{aligned} \int_t^{t_{f,\min\{i,l(j')\}}} (t_{f,i}-\tau)\,d\tau &= \begin{cases} \int_t^{t_{f,i}} (t_{f,i}-\tau)\,d\tau, & i < l(j') \\ \int_t^{t_{f,l(j')}} \left(t_{f,l(j')} - \tau + t_{f,i} - t_{f,l(j')}\right) d\tau, & i \ge l(j') \end{cases} \\ &= \begin{cases} \frac{t_{go,i}^2}{2}, & i < l(j') \\ t_{go,i} t_{go,l(j')} - \frac{t_{go,l(j')}^2}{2}, & i \ge l(j') \end{cases} \end{aligned} \tag{8.61}$$

$$\begin{aligned} \int_t^{t_{f,\min\{i',l(j)\}}} (t_{f,i'}-\tau)\,d\tau &= \begin{cases} \int_t^{t_{f,i'}} (t_{f,i'}-\tau)\,d\tau, & i' < l(j) \\ \int_t^{t_{f,l(j)}} \left(t_{f,l(j)} - \tau + t_{f,i'} - t_{f,l(j)}\right) d\tau, & i' \ge l(j) \end{cases} \\ &= \begin{cases} \frac{t_{go,i'}^2}{2}, & i' < l(j) \\ t_{go,i'} t_{go,l(j)} - \frac{t_{go,l(j)}^2}{2}, & i' \ge l(j) \end{cases} \end{aligned} \tag{8.62}$$

$$\int_{t}^{t_{f,\min\{l(j),l(j')\}}} d\tau = t_{go,\min\{l(j),l(j')\}} \tag{8.63}$$

Define $\boldsymbol{\beta} = [\beta_1, \beta_2, \ldots, \beta_M]^T$ and $\boldsymbol{e}_\gamma = \left[\gamma_{l(1)}^d - \gamma, \gamma_{l(2)}^d - \gamma, \ldots, \gamma_{l(M)}^d - \gamma\right]^T$ as the Lagrange multiplier vector and the flight path angle error vector, respectively. Substituting Eqs. (8.60)–(8.63) into Eqs. (8.58) and (8.59) yields

$$\boldsymbol{G}\begin{bmatrix} \boldsymbol{\lambda} \\ \boldsymbol{\beta} \end{bmatrix} = \begin{bmatrix} \boldsymbol{Z} \\ \boldsymbol{e}_\gamma \end{bmatrix}, \quad \boldsymbol{G} = \begin{bmatrix} \boldsymbol{G}_1 & \boldsymbol{G}_{12} \\ \boldsymbol{G}_{21} & \boldsymbol{G}_2 \end{bmatrix} \tag{8.64}$$

where $\boldsymbol{G}_1 \in \mathbb{R}^{N \times N}$ is a symmetric matrix which is given by

$$\boldsymbol{G}_1 = \begin{bmatrix} \frac{t_{go,1}^3}{3} & \frac{t_{go,2}t_{go,1}^2}{2} - \frac{t_{go,1}^3}{6} & \cdots & \cdots & \frac{t_{go,N}t_{go,1}^2}{2} - \frac{t_{go,1}^3}{6} \\ & \frac{t_{go,2}^3}{3} & \frac{t_{go,3}t_{go,2}^2}{2} - \frac{t_{go,2}^3}{6} & \cdots & \frac{t_{go,N}t_{go,2}^2}{2} - \frac{t_{go,2}^3}{6} \\ \ddots & & \ddots & & \\ & \ddots & & \frac{t_{go,N-1}^3}{3} & \frac{t_{go,N}t_{go,N-1}^2}{2} - \frac{t_{go,N-1}^3}{6} \\ & & \ddots & & \frac{t_{go,N}^3}{3} \end{bmatrix} \tag{8.65}$$

and $\boldsymbol{G}_{12} = \boldsymbol{G}_{21}^T = \left(g_{ij}\right) \in \mathbb{R}^{N \times M}$ with elements g_{ij} being

$$g_{ij} = \begin{cases} \frac{t_{go,i}^2}{2V}, & i < l(j) \\ \frac{t_{go,i}t_{go,l(j)}}{V} - \frac{t_{go,l(j)}^2}{2V}, & i \geq l(j) \end{cases} \tag{8.66}$$

and $\boldsymbol{G}_2 \in \mathbb{R}^{M \times M}$ is a symmetric matrix which is obtained as

$$\boldsymbol{G}_2 = \frac{1}{V^2}\begin{bmatrix} t_{go,l(1)} & t_{go,l(1)} & \cdots & \cdots & t_{go,l(1)} \\ & t_{go,l(2)} & t_{go,l(2)} & \cdots & t_{go,l(2)} \\ \ddots & & \ddots & & \\ & \ddots & & t_{go,l(M-1)} & t_{go,l(M-1)} \\ & & \ddots & & t_{go,l(M)} \end{bmatrix} \tag{8.67}$$

From Eq. (8.64), the Lagrange multiplier vectors $\boldsymbol{\lambda}$ and $\boldsymbol{\beta}$ can be obtained as

$$\begin{bmatrix} \boldsymbol{\lambda} \\ \boldsymbol{\beta} \end{bmatrix} = \boldsymbol{G}^{-1}\begin{bmatrix} \boldsymbol{Z} \\ \boldsymbol{e}_\gamma \end{bmatrix} \tag{8.68}$$

The explicit guidance command for $t \leq t_{f,1}$ can then be readily obtained by substituting Eq. (8.68) into Eq. (8.56) as

$$
\begin{aligned}
a &= \boldsymbol{\lambda}^T \left[t_{go,1}, t_{go,2}, \ldots, t_{go,N} \right]^T + \boldsymbol{\beta}^T \underbrace{\left[\frac{1}{V}, \frac{1}{V}, \ldots, \frac{1}{V} \right]^T}_{M \text{ elements}} \\
&= \left(\boldsymbol{G}^{-1} \begin{bmatrix} \boldsymbol{Z} \\ \boldsymbol{e}_\gamma \end{bmatrix} \right)^T \left[t_{go,1}, t_{go,2}, \ldots, t_{go,N}, \underbrace{\frac{1}{V}, \frac{1}{V}, \ldots, \frac{1}{V}}_{M \text{ elements}} \right]^T
\end{aligned}
\tag{8.69}
$$

Remark 8.4 From Eq. (8.69), it is clear that the implementation of the proposed guidance law requires the inverse of matrix $\boldsymbol{G}$. Since matrix $\boldsymbol{G}$ is defined as a block form, the block inverse approach [30] can be applied to simplify the matrix inverse calculation as

$$
\boldsymbol{G}^{-1} = \begin{bmatrix} \left(\boldsymbol{G}_1 - \boldsymbol{G}_{12} \boldsymbol{G}_2^{-1} \boldsymbol{G}_{12}^T \right)^{-1} & -\boldsymbol{G}_1^{-1} \boldsymbol{G}_{12} \left(\boldsymbol{G}_2 - \boldsymbol{G}_{12} \boldsymbol{G}_1^{-1} \boldsymbol{G}_{12}^T \right)^{-1} \\ -\left(\boldsymbol{G}_2 - \boldsymbol{G}_{12} \boldsymbol{G}_1^{-1} \boldsymbol{G}_{12}^T \right)^{-1} \boldsymbol{G}_{12}^T \boldsymbol{G}_1^{-1} & \left(\boldsymbol{G}_2 - \boldsymbol{G}_{12} \boldsymbol{G}_1^{-1} \boldsymbol{G}_{12}^T \right)^{-1} \end{bmatrix}
\tag{8.70}
$$

Remark 8.5 Note that the proposed guidance law is generic. For this reason, it can be applied to general UAV waypoint-following guidance missions with an arbitrary number of waypoints and an arbitrary number of arrival angle constraints. However, the computational burden increases with the increase in the number of boundary constraints.

8.4.2 Particular Case: $M = N$

Let us consider the case that the UAV is required to pass multiple waypoints and every waypoint requires a specific arrival angle γ_i^d. For this specific case, the proposed guidance law is equivalent to TSG applied to every two consecutive waypoints. To validate this, let a_i and $\bar{a}_i$ be, respectively, the guidance commands of the proposed guidance law and TSG between the $(i-1)$th and the ith waypoints, with $i = 0$ being the initial point for convenience. The corresponding energy consumptions of these two different guidance laws between the $(i-1)$th and the ith waypoints are then determined as

$$
\begin{aligned}
J_i &= \int_{t_{f,i}}^{t_{f,i+1}} a_i^2(\tau)\, d\tau \\
\bar{J}_i &= \int_{t_{f,i}}^{t_{f,i+1}} \bar{a}_i^2(\tau)\, d\tau
\end{aligned}
\tag{8.71}
$$

The equivalence of these two guidance laws will be proved through mathematical induction. When there exists only one waypoint to be visited by the UAV with a specified flight path, i.e., $N = 1$, Problem 8.1 reduces to energy-optimal rendezvous problem. Then, the guidance command of the proposed guidance law, shown in Eq. (8.56), can be formulated as

$$a = \lambda_1 \left(t_{f,1} - t\right) + \frac{\beta_1}{V} \tag{8.72}$$

Note that matrix $\boldsymbol{G}$ under condition $N = 1$ becomes

$$\boldsymbol{G} = \begin{bmatrix} \frac{t_{go,1}^3}{3} & \frac{t_{go,1}^2}{2V} \\ \frac{t_{go,1}^2}{2V} & \frac{t_{go,1}}{V^2} \end{bmatrix} \tag{8.73}$$

From Eq. (8.68), we can readily solve the Lagrange multipliers as

$$\begin{bmatrix} \lambda_1 \\ \beta_1 \end{bmatrix} = \frac{12V^2}{t_{go,1}^3} \begin{bmatrix} \frac{z_1}{V^2} - \frac{t_{go,1}}{2V}\left(\gamma_1^d - \gamma\right) \\ -\frac{t_{go,1}}{2V} z_1 + \frac{t_{go,1}^2}{3}\left(\gamma_1^d - \gamma\right) \end{bmatrix} \tag{8.74}$$

Substituting Eq. (8.74) into Eq. (8.72) gives the explicit guidance command as

$$a = \frac{6z_1}{t_{go,1}^2} - \frac{2V}{t_{go,1}}\left(\gamma_1^d - \gamma\right) \tag{8.75}$$

which coincides with the classical TSG. As shown in [28], the TSG is energy optimal in the case of single waypoint with an arrival angle constraint.

By the mathematical induction concept, assume that for $N = L > 1$, we have

$$a_L = \bar{a}_L, \quad J_L = \bar{J}_L \tag{8.76}$$

which means that both guidance laws ensure the same flight path at the Lth waypoint.

Then, for $N = L + 1$, if the guidance command of TSG is different from the proposed guidance law, i.e., $a_{L+1} \neq \bar{a}_{L+1}$, we have $J_{L+1} < \bar{J}_{L+1}$ since the proposed guidance law guarantees global optimality. Notice that both guidance laws have the same boundary constraints at the $(L + 1)$th waypoint. This means that condition $a_{L+1} \neq \bar{a}_{L+1}$ violates the optimality of TSG. Therefore, we have $a_{L+1} = \bar{a}_{L+1}$ and consequently these two guidance laws provide the same level of optimality of energy consumption when $M = N$. However, for the problem with $M < N$, TSG is not applicable and applying hybrid PN/TSG to every two consecutive waypoints is obviously not energy optimal.

8.5 Numerical Simulations

In this section, nonlinear numerical simulations are performed to validate the proposed guidance law. In the considered scenario, a UAV with constant speed $V = 30\,\text{m/s}$ is required to follow 8 waypoints. The UAV initially locates at (0 m, 0 m)

Table 8.1 Inertial positions of all waypoints

Waypoint ID	Inertial position
1	(1000 m, 500 m)
2	(2000 m, 750 m)
3	(2500 m, 1000 m)
4	(4000 m, 1500 m)
5	(6000 m, 2000 m)
6	(7500 m, 1500 m)
7	(9000 m, 1000 m)
8	(11000 m, 0 m)

with an initial flight path angle of 30°. The inertial positions of all waypoints are summarized in Table 8.1.

8.5.1 *Performance of Guidance Law (8.25)*

This subsection considers the problem of minimum effort waypoint following without any arrival angle constraint to validate the effectiveness of guidance law (8.25). For the purpose of comparison, the optimal PNG is also applied to every two consecutive waypoints in the simulations. Figure 8.2a compares the UAV flight trajectories obtained from these two different guidance laws. From this figure, it is clear that both guidance laws can successfully drive the UAV to follow the desired waypoints. The maximum recorded miss distance between the UAV and the waypoint under both guidance laws is less than 0.1 m in the simulations. The flight path angle response comparisons are shown in Fig. 8.2b. As shown in this figure, the UAV guided by PNG shows sharp turns when it passes waypoints while the proposed guidance law ensures a smooth trajectory.

The acceleration commands of different guidance laws are presented in Fig. 8.2c, which shows that the guidance command under PNG exhibits discontinuity when passing waypoints. As a comparison, the proposed optimal guidance law generates a continuous command, which is more desirable for the implementation. The quantitative comparison results of energy consumption obtained from both guidance laws are summarized in Fig. 8.2d. From this figure, we can clearly observe that the proposed guidance law requires less energy consumption, compared with PNG. In the considered scenario, the proposed guidance law helps to reduce more than 40% energy consumption. Therefore, the UAV guided by the proposed approach is expected to have longer endurance than PNG.

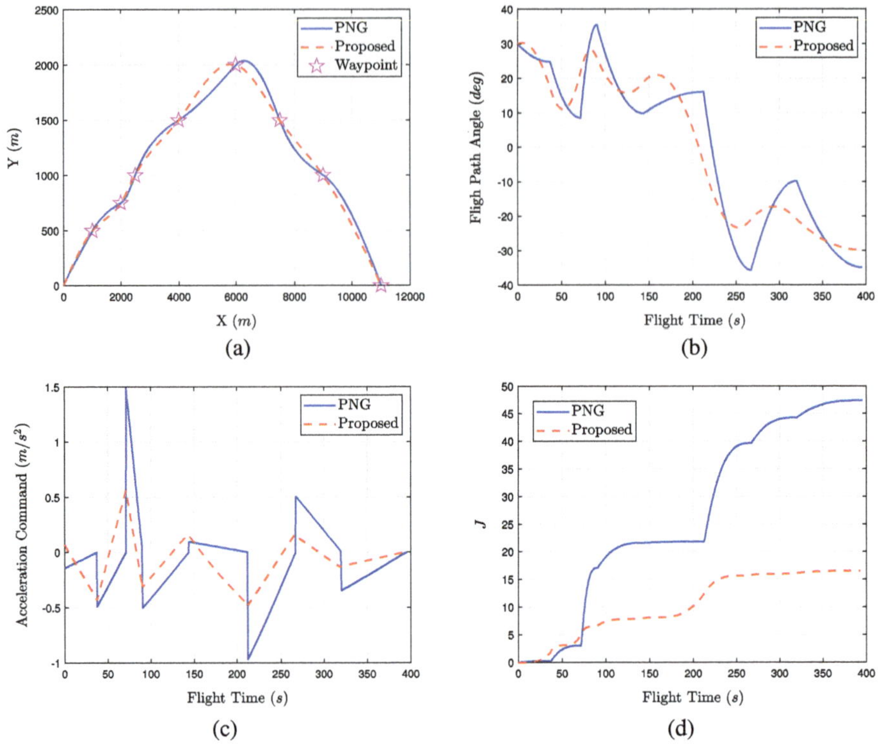

Fig. 8.2 Comparison results without angle constraint. **a** Flight trajectory; **b** Flight path angle; **c** Acceleration command; and **d** Performance index

8.5.2 Performance of Guidance Law (8.69)

This subsection considers the problem of minimum effort waypoint following with partial arrival angle constraint to validate the effectiveness of guidance law (8.69). In the scenario considered, we assume that the UAV is required to pass the 4th waypoint with desired flight path angle 0° and the 8th waypoint with desired flight path angle −90°. For comparison, the OGL is also applied to every two consecutive waypoints in the simulations. The OGL considered here is defined as

$$\text{OGL} = \begin{cases} \text{TSG}, & \text{if next waypoint requires specific arrival angle} \\ \text{PNG}, & \text{Otherwise} \end{cases}$$

The UAV flight trajectories and flight path angle profiles obtained from OGL as well as the proposed guidance law are shown in Fig. 8.3a and b, respectively. These two figures reveal that the UAV guided by both guidance laws can successfully accomplish its waypoint following mission. The recorded miss distance and flight path angle error obtained from both guidance laws are less than 0.1 m and 0.5° in our

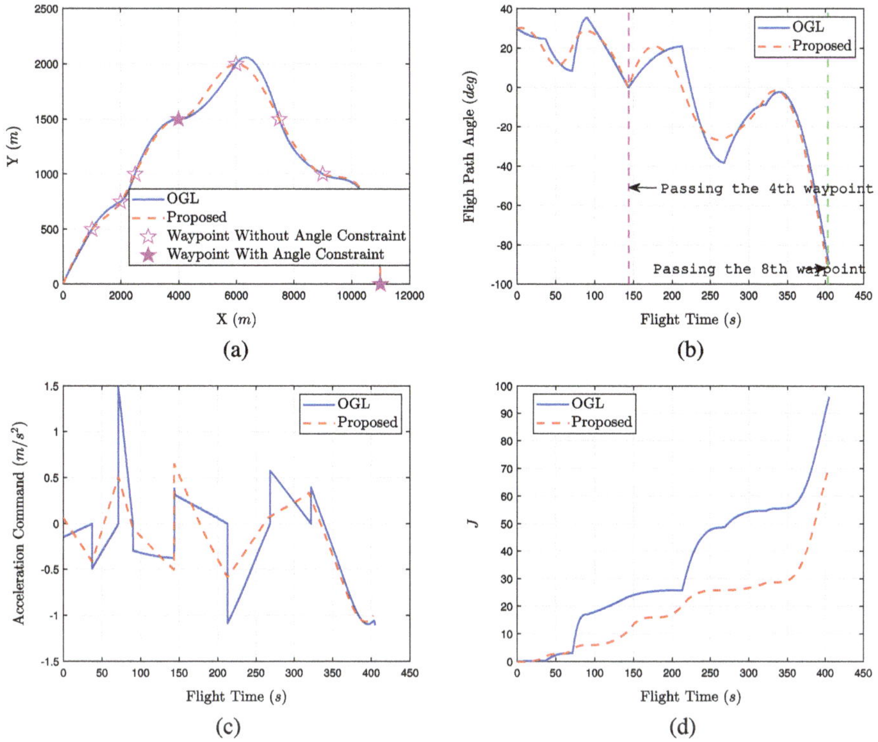

Fig. 8.3 Comparison results with angle constraint. **a** Flight trajectory; **b** Flight path angle; **c** Acceleration command; and **d** Performance index

simulations. Compare with previous scenario without any arrival angle constraint, PNG generates the same trajectory for the first three waypoints. On the other hand, the UAV flight trajectory under the proposed guidance law shows slight difference since the additional angle regulation term a_β starts to work from the beginning.

Figure 8.3c compares the UAV acceleration command generated by these two different guidance laws. Unlike the previous scenario, the proposed guidance law also shows discontinuity when the UAV passes a waypoint that requires a specific arrival angle, i.e., the 4th waypoint in the considered scenario. The reason for this fact is that the angle regulation term β_1/V disappears when $t > t_{f,4}$. After passing the 7th waypoint, there exists only one waypoint, i.e., the 8th, to be visited with a specific arrival angle constraint. For this reason, the proposed guidance law reduces to classical TSG when $t > t_{f,7}$. Therefore, the proposed guidance law and OGL are expected to share similar characteristics during that period, which is confirmed by Fig. 8.3c. The quantitative comparison results of energy consumption obtained from both guidance laws are presented in Fig. 8.3d. From this figure, it can be noted that the proposed guidance requires approximate 20% less energy consumption in the considered scenario, compared with OGL. For this reason, the proposed guidance law can be utilized to increase the endurance of the UAV.

8.6 Summary

This chapter discusses the minimum-effort waypoint-following guidance problem for a UAV. Both optimal waypoint-following guidance and optimal waypoint-following guidance with partial flight path angle constraint are rigorously derived using optimal control theory. When there exits only one waypoint to be traveled by the UAV, the proposed guidance laws reduce to the well-known PNG and TSG. The guidance law developed is generic and can be applied to an arbitrary number of waypoints with/without an arbitrary number of arrival angle constraints. Although the proposed guidance law is derived based on a linear kinematics model, nonlinear numerical simulations clearly validate the effectiveness of the proposed guidance law.

By exploiting the advantages of the proposed approach, the guidance law developed can also be applied to waypoint navigation during midcourse guidance phase of cruise missiles and sea-skimming anti-ship missiles. Another potential application of the proposed guidance law is handover between midcourse and terminal course of guided weapons, i.e., $N = 2$, $M = 0$, since the proposed guidance law can help to reduce the transient effect as proved in this chapter.

References

1. Rysdyk R (2006) Unmanned aerial vehicle path following for target observation in wind. J Guid Control Dyn 29(5):1092–1100
2. Beard RW, Ferrin J, Humpherys J (2014) Fixed wing uav path following in wind with input constraints. IEEE Trans Control Syst Technol 22(6):2103–2117
3. Kapitanyuk YA, Proskurnikov AV, Cao M (2018) A guiding vector-field algorithm for path-following control of nonholonomic mobile robots. IEEE Trans Control Syst Technol 26(4):1372–1385
4. Olavo JLG, Thums GD, Jesus TA, de Araújo Pimenta LC, Torres LAB, Palhares RM (2018) Robust guidance strategy for target circulation by controlled uav. IEEE Trans Aerosp Electron Syst 54(3):1415–1431
5. Delingette H, Hebert M, Ikeuchi K (1991) Trajectory generation with curvature constraint based on energy minimization. In: IEEE/RSJ international workshop on intelligent robots and systems' 91. Intelligence for mechanical systems, proceedings IROS'91, pp 206–211. IEEE
6. Spangelo I, Egeland O (1992) Generation of energy-optimal trajectories for an autonomous underwater vehicle. In: 1992 IEEE international conference on robotics and automation, 1992. Proceedings, pp 2107–2112. IEEE
7. Moon G, Kim Y (2003) Optimum flight path design passing through waypoints for autonomous flight control system. In: AIAA guidance, navigation, and control conference and exhibit
8. Kaplan A, Kingry N, Uhing P, Dai R (2017) Time-optimal path planning with power schedules for a solar-powered ground robot. IEEE Trans Autom Sci Eng 14(2):1235–1244

9. Zollars MD, Cobb RG, Grymin DJ (2018) Optimal path planning for suas waypoint following in urban environments. In: 2018 IEEE Aerospace Conference. IEEE
10. Ataei M, Yousefi-Koma A (2015) Three-dimensional optimal path planning for waypoint guidance of an autonomous underwater vehicle. Robot Auton Syst 67:23–32
11. Garcia G, Keshmiri S (2011) Nonlinear model predictive controller for navigation, guidance and control of a fixed-wing uav. In: AIAA guidance, navigation, and control conference
12. He S, Shin H-S, Tsourdos A (2019) Trajectory optimization for target localization with bearing-only measurement. IEEE Trans Robot 35(3):653–668
13. Frew EW, Langelaan J, Joo S (2006) Adaptive receding horizon control for vision-based navigation of small unmanned aircraft. In: American control conference, 2006, pp. 2160–2165. IEEE
14. Keviczky T, Balas GJ (2005) Flight test of a receding horizon controller for autonomous uav guidance. In: American control conference, 2005. Proceedings of the 2005, pp 3518–3523. IEEE
15. Sprinkle J, Eklund JM, Kim HJ, Sastry S (2004) Encoding aerial pursuit/evasion games with fixed wing aircraft into a nonlinear model predictive tracking controller. In: 43rd IEEE conference on decision and control, 2004. CDC, vol 3, pp 2609–2614. IEEE
16. Ick Ho Whang and Tae Won Hwang (2002) Horizontal waypoint guidance design using optimal control. IEEE Trans Aerosp Electron Syst 38(3):1116–1120
17. Ryoo C-K, Shin H-S, Tahk M-J (2010) Energy optimal waypoint guidance synthesis for antiship missiles. IEEE Trans Aerosp Electron Syst 46(1):80–95
18. Cottrell RG (1971) Optimal intercept guidance for short-range tactical missiles. AIAA J 9(7):1414–1415
19. Jeon I-S, Lee J-I (2010) Optimality of proportional navigation based on nonlinear formulation. IEEE Trans Aerosp Electron Syst 46(4):2051–2055
20. Palumbo NF, Blauwkamp RA, Lloyd JM (2010) Basic principles of homing guidance. Johns Hopkins APL Tech Digest 29(1):25–41
21. Shaferman V, Shima T (2008) Linear quadratic guidance laws for imposing a terminal intercept angle. J Guid Control Dyn 31(5):1400–1412
22. Palumbo NF, Blauwkamp RA, Lloyd JM (2010) Modern homing missile guidance theory and techniques. Johns Hopkins APL Tech Digest 29(1):42–59
23. Nesline FW (1981) Zarchan P (1981) A new look at classical vs modern homing missile guidance. J Guid Control Dyn 4(1):78–85
24. Shaoming H, Chang-Hun L (2018) Optimal proportional-integral guidance with reduced sensitivity to target maneuvers. IEEE Trans Aerosp Electron Syst
25. He S, Lee C-H (2018) Optimality of error dynamics in missile guidance problems. J Guid Control Dyn 41(7):1620–1629
26. Weiss M, Shima T, Castaneda D, Rusnak I (2017) Combined and cooperative minimum-effort guidance algorithms in an active aircraft defense scenario. J Guid Control Dyn 40(5):1241–1254
27. Lee C-H, Tahk M-J, Lee J-I (2013) Generalized formulation of weighted optimal guidance laws with impact angle constraint. IEEE Trans Aerosp Electron Syst 49(2):1317–1322
28. Zarchan P (2012) Tactical and strategic missile guidance. American Institute of Aeronautics and Astronautics
29. Lee C-H, Kim T-H, Tahk M-J (2013) Interception angle control guidance using proportional navigation with error feedback. J Guid Control Dyn 36(5):1556–1561
30. Tzon-Tzer L, Shiou S-H (2002) Inverses of 2×2 block matrices. Comput Math Appl 43(1–2): 119–129

Chapter 9
Energy-Optimal Waypoint-Following Guidance Law Considering Autopilot Dynamics

Abstract This chapter addresses the problem of energy-optimal waypoint-following guidance for a UAV with the consideration of a general autopilot dynamics model. The proposed guidance law is derived as a solution of a linear quadratic optimal control problem in conjunction with a linearized kinematics model. The algorithm developed integrates path planning and following into a single step and is able to be applied to a general waypoint-following mission. Theoretical analysis reveals that previously suggested optimal point-to-point guidance laws are special cases of the proposed approach. Nonlinear numerical simulations clearly demonstrate the effectiveness of the proposed formulations.

9.1 Introduction

In Chap. 8, we developed a generic energy-optimal waypoint-following guidance law that ensures global optimality. This guidance law, however, is formulated based on a lag-free autopilot assumption. In practice, the guidance law developed could inevitably experience performance degradation such as a significant miss distance and passing angle error associated with the waypoint, resulted from an autopilot lag in the guidance loop. Although the authors in [1] considered a first-order lag autopilot dynamics model in guidance law design for waypoint following, the resultant guidance command is not explicitly provided and requires numerical parameter optimization in practical implementation. Hence, this approach might not be suitable for low-cost UAVs with limited onboard computational capability.

As an extension of Chap. 8, an analytical energy-optimal waypoint-following guidance law in consideration of a general autopilot lag model is proposed in this chapter to support practical applications. Similar to Chap. 8, the formulation of the proposed guidance law is based on a reduced-order linearized kinematics model through terminal projection technique. The commanded acceleration is then analytically derived as a solution of a finite-time linear regulation problem using optimal control theory. Theoretical analysis reveals that classical optimal point-to-point guidance laws [2–5] are special cases of the proposed algorithm. The key features of the proposed guidance law are twofold:

S. He et al., *Optimal Guidance and Its Applications in Missiles and UAVs*, Springer Aerospace Technology, https://doi.org/10.1007/978-3-030-47348-8_9

(1) The guidance command generated by the proposed algorithm is continuous and there is no sudden change during waypoint change. This helps to reduce the transient effect when passing one waypoint.
(2) The algorithm developed considers the command response lag and thus guarantees a finite guidance command. This prevents the divergence of the guidance command near the waypoint and consequently it offers advantages of reducing the miss distances and arrival angle errors associated with the waypoints. Extensive empirical tests reveal that compensating autopilot lag will bring performance improvement for UAVs with large time lag and high speed.

9.2 Problem Formulation

In general, the UAV speed is pre-determined according to specific mission objective and is maintained by an engine controller. For this reason, we assume that the UAV is flying with a constant velocity for simplicity. In practice, the lateral acceleration a is generated by an autopilot, or flight control system, which has inevitable time delays. To account for the adverse effect of autopilot delay, it is worthy to consider the autopilot dynamics in guidance law design. For this purpose, consider the UAV autopilot dynamics as the following arbitrary-order linear equations

$$\begin{aligned} \dot{\boldsymbol{x}}_a &= \boldsymbol{A}_a \boldsymbol{x}_a + \boldsymbol{B}_a a_c \\ a &= \boldsymbol{C}_a \boldsymbol{x}_a + \boldsymbol{D}_a a_c \end{aligned} \tag{9.1}$$

where a_c represents the guidance command generated by the guidance law; $\boldsymbol{x}_a \in \mathbb{R}^{n_a \times 1}$ is the autopilot dynamics state vector; and $\boldsymbol{A}_a$, $\boldsymbol{B}_a$, $\boldsymbol{C}_a$ and $\boldsymbol{D}_a$ are some proper autopilot dynamics matrices.

Similar to Chap. 8, the proposed guidance law is derived by using a linear kinematics model. Combing Eqs. (8.5) and (9.1), the linearized model that considering the effect of autopilot dynamics can be formulated in a compact matrix form as

$$\dot{\boldsymbol{x}} = \boldsymbol{A}\boldsymbol{x} + \boldsymbol{B}a_c \tag{9.2}$$

where $\boldsymbol{x}_i = \left[y_i, v_i, \gamma, \boldsymbol{x}_a^T\right]^T \in \mathbb{R}^{(n_a+3)\times 1}$ and $\boldsymbol{x} = [\boldsymbol{x}_1, \boldsymbol{x}_2, \ldots, \boldsymbol{x}_N]^T \in \mathbb{R}^{(n_a+3)N\times 1}$ is the system state vector. $\boldsymbol{A} \in \mathbb{R}^{(n_a+3)N\times(n_a+3)N}$ is a block diagonal matrix, and $\boldsymbol{B} \in \mathbb{R}^{(n_a+3)N\times 1}$ is the control input matrix. These two matrices are defined as

$$\boldsymbol{A} = \mathrm{diag}\left(\boldsymbol{A}_1, \boldsymbol{A}_2, \ldots \boldsymbol{A}_N\right), \quad \boldsymbol{B} = \left[\boldsymbol{B}_1^T, \boldsymbol{B}_2^T, \ldots, \boldsymbol{B}_N^T\right]^T \tag{9.3}$$

where

$$A_i = \begin{bmatrix} 0 & 1 & 0 & 0_{1\times n_a} \\ 0 & 0 & 0 & -\boldsymbol{C}_a \\ 0 & 0 & 0 & \frac{\boldsymbol{C}_a}{V} \\ 0_{n_a\times 1} & 0_{n_a\times 1} & 0_{n_a\times 1} & \boldsymbol{A}_a \end{bmatrix}, \quad \boldsymbol{B}_i = \begin{bmatrix} 0 \\ -\boldsymbol{D}_a \\ \frac{\boldsymbol{D}_a}{V} \\ \boldsymbol{B}_a \end{bmatrix} \tag{9.4}$$

Similar to Chap. 8, this chapter also considers the energy-consumption performance index

$$J = \int_t^{t_{f,N}} a_c^2(\tau)\, d\tau = \begin{cases} \int_t^{t_{f,1}} a_c^2(\tau)\, d\tau + \sum_{i=1}^{N-1} \int_{t_{f,i}}^{t_{f,i+1}} a_c^2(\tau)\, d\tau, & t \leq t_{f,1} \\ \int_t^{t_{f,2}} a_c^2(\tau)\, d\tau + \sum_{i=2}^{N-1} \int_{t_{f,i}}^{t_{f,i+1}} a_c^2(\tau)\, d\tau, & t \leq t_{f,2} \\ \vdots & \\ \int_t^{t_{f,N}} a_c^2(\tau)\, d\tau, & t_{f,N-1} < t \leq t_{f,N} \end{cases} \tag{9.5}$$

In a waypoint-following mission, one path constraint is that the UAV has to visit all characteristic waypoints with zero miss distance, i.e.,

$$y_i\left(t_{f,i}\right) = 0, \quad i \in [N] \tag{9.6}$$

Additionally, the UAV might need to pass some certain waypoints with desired flight path angles for some specific purposes, such as obstacle avoidance in terrain navigation, hostile radar detection avoidance in surveillance. Hence, we also consider additional path constraint as

$$\gamma\left(t_{f,l(j)}\right) = \gamma_{l(j)}^d, \quad j \in [M], \quad l(j) \in [N], \quad M \leq N \tag{9.7}$$

where $l(j)$ stands for the index of waypoints that have specific flight path angle constraints and $\gamma_{l(j)}^d$ denotes the desired flight angle when the UAV passes the $l(j)$th waypoint.

In summary, the aim of this chapter is to find analytical guidance command solution to the following optimization problem.

Problem 9.1 Find a_c that minimizes performance index (9.5), subject to kinematics model (9.2) and path constraints (9.6), (9.7).

9.3 Guidance Law Derivation

In this section, we first present a reduced-order system using terminal projection. Then, the detailed derivation of the proposed guidance law and its implementation issue are provided.

9.3.1 Order Reduction

Before solving the optimization problem formulated, the system order is reduced first using the terminal projection technique, or the so-called zero-effort transformation [6–9]. Notice that the zero-effort transformation aims to find the terminal system state values with zero control input from the current time onward. Based on this concept, the terminal projection $\boldsymbol{Z}_i$ of system (9.2) at current time t can be readily obtained using the homogeneous solution as

$$\boldsymbol{Z}_i = \begin{cases} \boldsymbol{C}_i \boldsymbol{\Phi}_i \left(t_{f,i}, t\right) \boldsymbol{x}_i, & t \leq t_{f,i} \\ \boldsymbol{Z}_i \left(t_{f,i}\right), & t > t_{f,i} \end{cases} \tag{9.8}$$

where $\boldsymbol{\Phi}_i \left(t_{f,i}, t\right)$ denotes the transition matrix associated with matrix $\boldsymbol{A}_i$ and $\boldsymbol{C}_i$ is a constant matrix that extracts appropriate elements from the system state vector.

To satisfy terminal constraints (9.6) and (9.7), one needs to extract two variables for each waypoint: ZEM $z_{1,i}$ and zero-effort angle (ZEA) $z_{2,i}$. The ZEM and ZEA, respectively, refer to the final relative distance associated with the ith waypoint and the final flight path angle, if the UAV will not apply any control from the current time onward. For this reason, the matrix $\boldsymbol{C}_i$ is given by

$$\boldsymbol{C}_i = \begin{bmatrix} 1\ 0\ 0\ 0_{1\times n_a} \\ 0\ 0\ 1\ 0_{1\times n_a} \end{bmatrix} \tag{9.9}$$

Since

$$\dot{\boldsymbol{\Phi}}_i \left(t_{f,i}, t\right) = -\boldsymbol{\Phi}_i \left(t_{f,i}, t\right) \boldsymbol{A}_i \tag{9.10}$$

the time derivative of the terminal projection is

$$\dot{\boldsymbol{Z}}_i = \begin{cases} \boldsymbol{C}_i \boldsymbol{\Phi}_i \left(t_{f,i}, t\right) \boldsymbol{B}_i a_c, & t \leq t_{f,i} \\ \boldsymbol{0}, & t > t_{f,i} \end{cases} \tag{9.11}$$

Notice that the system order now reduces from $(n_a + 3)\, N$ to $2N$ and the terminal constraints (9.6) and (9.7) can be alternatively expressed as

$$\begin{aligned} z_{1,i}\left(t_{f,i}\right) &= 0, \quad i \in [N] \\ z_{2,l(j)}\left(t_{f,l(j)}\right) &= \gamma_{l(j)}^{d}, \quad j \in [M], \quad l(j) \in [N], \quad M \leq N \end{aligned} \tag{9.12}$$

With the zero-effort state transformations, Problem 9.1 now reduces to the following optimization problem.

Problem 9.2 Find a_c that minimizes performance index (9.5), subject to kinematics model (9.11) and path constraint (9.12).

9.3.2 General Guidance Law Solution

Define $\boldsymbol{C}_i \boldsymbol{\Phi}_i\left(t_{f,i}, t\right) \boldsymbol{B}_i = \left[-b_i(t), g_i(t)\right]^T$ for notation convenience. The dynamics of the transformed zero-effort system can be obtained using Eq. (9.11) as

$$\dot{z}_{1,i} = \begin{cases} -b_i(t)\, a_c, & t \leq t_{f,i} \\ 0, & t > t_{f,i} \end{cases} \tag{9.13}$$

$$\dot{Z}_{2,i} = \begin{cases} g_i(t)\, a_c, & t \leq t_{f,i} \\ 0, & t > t_{f,i} \end{cases} \tag{9.14}$$

The solution of differential equations (9.13) and (9.14) are given by

$$z_{1,i}\left(t_{f,i}\right) - z_{1,i}(t) = \int_t^{t_{f,i}} -b_i(\tau)\, a_c(\tau)\, d\tau, \quad t \leq t_{f,i} \tag{9.15}$$

$$z_{2,i}\left(t_{f,i}\right) - z_{2,i}(t) = \int_t^{t_{f,i}} g_i(\tau)\, a_c(\tau)\, d\tau, \quad t \leq t_{f,i} \tag{9.16}$$

Imposing terminal constraint (9.12) on Eqs. (9.15) and (9.16) gives

$$z_{1,i}(t) = \int_t^{t_{f,i}} b_i(\tau)\, a_c(\tau)\, d\tau, \quad t \leq t_{f,i} \tag{9.17}$$

$$\gamma_{l(j)}^{d} - z_{2,l(j)}(t) = \int_t^{t_{f,l(j)}} g_{l(j)}(\tau)\, a_c(\tau)\, d\tau, \quad t \leq t_{f,l(j)} \tag{9.18}$$

According to Lemma 5.1, if the guidance command a_c is optimal in terms of energy minimization, then there exist $N + M$ Lagrange multipliers λ_i, β_j, $i \in [N]$, $j \in [M]$, such that the lateral acceleration command can be formulated as

$$a_c = a_\lambda + a_\beta \tag{9.19}$$

with

$$a_\lambda = \begin{cases} \sum_{i=1}^{N} \lambda_i b_i(t), & t \le t_{f,1} \\ \sum_{i=2}^{N} \lambda_i b_i(t), & t_{f,1} < t \le t_{f,2} \\ \vdots \\ \lambda_N b_N(t), & t_{f,N-1} < t \le t_{f,N} \end{cases}, \quad a_\beta = \begin{cases} \sum_{j=1}^{M} \beta_j g_{l(j)}(t), & t \le t_{f,l(1)} \\ \sum_{j=2}^{M} \beta_j g_{l(j)}(t), & t_{f,l(1)} < t \le t_{f,l(2)} \\ \vdots \\ \beta_M g_{l(M)}(t), & t_{f,l(M-1)} < t \le t_{f,l(M)} \end{cases} \tag{9.20}$$

where a_λ refers to the ZEM regulation term and a_β represents the ZEA regulation command.

Without loss of generality, we only consider the case $t \le t_{f,1}$ in the following derivations. The solutions for $t > t_{f,1}$ can be easily obtained through similar procedures. Substituting Eq. (9.19) into Eqs. (9.17) and (9.18) under condition $t \le t_{f,1}$ results in

$$\begin{aligned} z_{1,i} &= \sum_{i'=1}^{N} \lambda_{i'} \int_t^{t_{f,i}} b_i(\tau) b_{i'}(\tau) d\tau + \sum_{j'=1}^{M} \beta_{j'} \int_t^{t_{f,i}} b_i(\tau) g_{l(j')}(\tau) d\tau \\ &= \sum_{i'=1}^{N} \lambda_{i'} \int_t^{t_{f,\min\{i,i'\}}} b_i(\tau) b_{i'}(\tau) d\tau + \sum_{j'=1}^{M} \beta_{j'} \int_t^{t_{f,\min\{i,l(j')\}}} b_i(\tau) g_{l(j')}(\tau) d\tau \end{aligned} \tag{9.21}$$

$$\begin{aligned} \gamma_{l(j)}^d - z_{2,l(j)} &= \sum_{i'=1}^{N} \lambda_{i'} \int_t^{t_{f,l(j)}} g_{l(j)}(\tau) b_{i'}(\tau) d\tau + \sum_{j'=1}^{M} \beta_{j'} \int_t^{t_{f,l(j)}} g_{l(j)}(\tau) g_{l(j')}(\tau) d\tau \\ &= \sum_{i'=1}^{N} \lambda_{i'} \int_t^{t_{f,\min\{i',l(j)\}}} g_{l(j)}(\tau) b_{i'}(\tau) d\tau \\ &\quad + \sum_{j'=1}^{M} \beta_{j'} \int_t^{t_{f,\min\{l(j),l(j')\}}} g_{l(j)}(\tau) g_{l(j')}(\tau) d\tau \end{aligned} \tag{9.22}$$

Define $\boldsymbol{\lambda} = [\lambda_1, \lambda_2, \ldots, \lambda_N]^T$, $\boldsymbol{\beta} = [\beta_1, \beta_2, \ldots, \beta_M]^T$ as two Lagrange multiplier vectors, and

$$\begin{aligned} \boldsymbol{e}_z &= \left[z_{1,1}, z_{1,2}, \ldots, z_{1,N}\right]^T \\ \boldsymbol{e}_\gamma &= \left[\gamma_{l(1)}^d - z_{2,l(1)} \gamma_{l(2)}^d - z_{2,l(2)}, \ldots, \gamma_{l(M)}^d - z_{2,l(M)}\right]^T \end{aligned} \tag{9.23}$$

as two state error vectors. Then, Eqs. (9.21) and (9.22) can be rewritten as a compact matrix form as

$$\boldsymbol{G} \begin{bmatrix} \boldsymbol{\lambda} \\ \boldsymbol{\beta} \end{bmatrix} = \begin{bmatrix} \boldsymbol{e}_z \\ \boldsymbol{e}_\gamma \end{bmatrix}, \quad \boldsymbol{G} = \begin{bmatrix} \boldsymbol{G}_1 & \boldsymbol{G}_{12} \\ \boldsymbol{G}_{21} & \boldsymbol{G}_2 \end{bmatrix} \tag{9.24}$$

where $\boldsymbol{G}_1 = \boldsymbol{G}_1^T = [g_1(i, j)] \in \mathbb{R}^{N\times N}$ with elements $g_1(i, j)$, $i \in [N]$, $j \in [N]$ being

$$g_1(i, j) = \int_t^{t_{f,\min\{i,j\}}} b_i(\tau) b_j(\tau) d\tau \tag{9.25}$$

and $\boldsymbol{G}_{12} = \boldsymbol{G}_{21}^T = [g_{12}(i, j)] \in \mathbb{R}^{N\times M}$ with elements $g_{12}(i, j)$, $i \in [N]$, $j \in [M]$ being

$$g_{12}(i, j) = \int_t^{t_{f,\min\{i,l(j)\}}} b_i(\tau) g_{l(j)}(\tau) d\tau \tag{9.26}$$

and $\boldsymbol{G}_2 = \boldsymbol{G}_2^T = [g_2(i, j)] \in \mathbb{R}^{M\times M}$ with elements $g_2(i, j)$, $i \in [M]$, $j \in [M]$ being

$$g_2(i, j) = \int_t^{t_{f,\min\{l(i),l(j)\}}} g_{l(i)}(\tau) g_{l(j)}(\tau) d\tau \tag{9.27}$$

From Eq. (9.24), the Lagrange multiplier vectors $\boldsymbol{\lambda}$ and $\boldsymbol{\beta}$ can be obtained as

$$\begin{bmatrix} \boldsymbol{\lambda} \\ \boldsymbol{\beta} \end{bmatrix} = \boldsymbol{G}^{-1} \begin{bmatrix} \boldsymbol{e}_z \\ \boldsymbol{e}_\gamma \end{bmatrix} \tag{9.28}$$

The explicit guidance command for $t \leq t_{f,1}$ can then be readily obtained by substituting Eq. (9.28) into Eq. (9.19) as

$$\begin{aligned} a_c &= \boldsymbol{\lambda}^T [b_1, b_2, \ldots, b_N]^T + \boldsymbol{\beta}^T \left[g_{l(1)}, g_{l(2)}, \ldots, g_{l(M)}\right]^T \\ &= \left(\boldsymbol{G}^{-1} \begin{bmatrix} \boldsymbol{e}_z \\ \boldsymbol{e}_\gamma \end{bmatrix}\right)^T \left[b_1, b_2, \ldots, b_N, g_{l(1)}, g_{l(2)}, \ldots, g_{l(M)}\right]^T \end{aligned} \tag{9.29}$$

Remark 9.1 It follows from Eq. (9.29) that the proposed guidance law is generic. For this reason, it can be applied to a general UAV waypoint-following guidance mission with arbitrary number of waypoints and arbitrary number of arrival angle constraints.

9.3.3 Guidance Law Implementation

From Eq. (9.29), it can be noted that the implementation of the proposed guidance law requires the information on ZEM and ZEA. This information, however, cannot be directly measured from the UAV's onboard sensors. For this reason, this subsection will transform the guidance command into an equivalent form, which is a function of the measured signals $\dot{\sigma}_i$, V, r_i, γ and $\boldsymbol{x}_a$. Under the assumption that the angle $\sigma_i - \sigma_i^0$ is small, Fig. 8.1 reveals that

$$\sigma_i - \sigma_i^0 = \frac{y_i}{r_i} \tag{9.30}$$

Notice that the state transmission matrix $\boldsymbol{\Phi}_i\left(t_{f,i}, t\right)$ can be written as

$$\boldsymbol{\Phi}_i\left(t_{f,i}, t\right) = \begin{bmatrix} 1 & t_{f,i} - t & 0 & \boldsymbol{\Phi}_{i,1}\left(t_{f,i}, t\right) \\ 0 & 1 & 0 & \boldsymbol{\Phi}_{i,2}\left(t_{f,i}, t\right) \\ 0 & 0 & 1 & \boldsymbol{\Phi}_{i,3}\left(t_{f,i}, t\right) \\ 0_{n_a \times 1} & 0_{n_a \times 1} & 0_{n_a \times 1} & \boldsymbol{\Phi}_{i,4}\left(t_{f,i}, t\right) \end{bmatrix} \tag{9.31}$$

Using Eqs. (2.8) and (9.8), the zero-effort system states can then be formulated in an alternative form as

$$z_{1,i} = \begin{cases} V\dot{\sigma}_i t_{go,i}^2 + \boldsymbol{\Phi}_{i,1}\left(t_{f,i}, t\right)\boldsymbol{x}_a, & t \le t_{f,i} \\ z_{1,i}\left(t_{f,i}\right), & t > t_{f,i} \end{cases} \tag{9.32}$$

$$z_{2,i} = \begin{cases} \gamma + \boldsymbol{\Phi}_{i,3}\left(t_{f,i}, t\right)\boldsymbol{x}_a, & t \le t_{f,i} \\ z_{2,i}\left(t_{f,i}\right), & t > t_{f,i} \end{cases} \tag{9.33}$$

Substituting Eqs. (9.32) and (9.33) into Eq. (9.29) gives the guidance command in terms of measured signals $\dot{\sigma}_i$, V, r_i, γ and $\boldsymbol{x}_a$. This supports the practical application of the proposed guidance law. Note that although the proposed guidance law is derived based on the linearized engagement kinematics, the error generated in the linearization process can be alleviated by using Eqs. (9.32), (9.33) in implementation since Eqs. (9.32), (9.33) transform the original linear terms into their corresponding nonlinear expressions [10].

9.4 Some Particular Cases

This section discusses some particular cases of the proposed guidance law. For simplicity, we consider ideal autopilot dynamics and first-order autopilot dynamics cases here.

9.4.1 Ideal Autopilot Dynamics

If the UAV's autopilot dynamics is ideal, that is,

$$\boldsymbol{A}_a = 0, \quad \boldsymbol{B}_a = 0, \quad \boldsymbol{C}_a = 0, \quad \boldsymbol{D}_a = 1 \tag{9.34}$$

Then, we have

$$z_{1,i} = \begin{cases} V\dot{\sigma}_i t_{go,i}^2, & t \le t_{f,i} \\ z_{1,i}\left(t_{f,i}\right), & t > t_{f,i} \end{cases} \tag{9.35}$$

$$z_{2,i} = \begin{cases} \gamma, & t \le t_{f,i} \\ z_{2,i}\left(t_{f,i}\right), & t > t_{f,i} \end{cases} \tag{9.36}$$

$$b_i(t) = t_{go,i}, \quad g_i(t) = \frac{1}{V} \tag{9.37}$$

The integrals in Eqs. (9.21) and (9.22) for this simple case can be easily obtained as

$$\int_t^{t_{f,\min\{i,i'\}}} b_i(\tau)\, b_{i'}(\tau)\, d\tau = \frac{t_{go,\min\{i,i'\}}^2}{6}\left(2t_{go,\min\{i,i'\}} + 3d_{i,i'}\right) \tag{9.38}$$

$$\int_t^{t_{f,\min\{i,l(j')\}}} b_i(\tau)\, g_{l(j')}(\tau)\, d\tau = \begin{cases} \frac{t_{go,i}^2}{2V}, & i \le l\left(j'\right) \\ \frac{t_{go,l(j')}^2}{2V} + \frac{d_{i,l(j')}t_{go,l(j')}}{V}, & i > l\left(j'\right) \end{cases} \tag{9.39}$$

$$\int_t^{t_{f,\min\{l(j),l(j')\}}} g_{l(j)}(\tau)\, g_{l(j')}(\tau)\, d\tau = \frac{t_{go,\min\{l(j),l(j')\}}}{V^2} \tag{9.40}$$

where $d_{i,i'} = t_{f,\max\{i,i'\}} - t_{f,\min\{i,i'\}}$.

Substituting Eqs. (9.35)–(9.40) into Eq. (9.29) gives the guidance command for $t \le t_{f,1}$ as

$$\begin{aligned} a_c &= \boldsymbol{\lambda}^T\left[c_1 t_{go,1}, c_2 t_{go,2}, \ldots, c_N t_{go,N}\right]^T + \boldsymbol{\beta}^T\underbrace{\left[\frac{1}{V}, \frac{1}{V}, \ldots, \frac{1}{V}\right]}_{M \text{ elements}}^T \\ &= \left(\boldsymbol{G}^{-1}\begin{bmatrix}\boldsymbol{e}_z \\ \boldsymbol{e}_\gamma\end{bmatrix}\right)^T \left[c_1 t_{go,1}, c_2 t_{go,2}, \ldots, c_N t_{go,N}, \underbrace{\frac{1}{V}, \frac{1}{V}, \ldots, \frac{1}{V}}_{M \text{ elements}}\right]^T \end{aligned} \tag{9.41}$$

which is the lag-free energy-optimal waypoint-following guidance law (OWFGL-0) developed in Chap. 8.

9.4.2 First-Order Autopilot Dynamics

Assume that the UAV autopilot is modeled by a first-order dynamics, that is,

$$\boldsymbol{A}_a = -\frac{1}{\tau_a}, \quad \boldsymbol{B}_a = \frac{1}{\tau_a}, \quad \boldsymbol{C}_a = 1, \quad \boldsymbol{D}_a = 0 \tag{9.42}$$

where τ_a denotes the autopilot time constant.

Then, the transition matrix is determined as

$$\boldsymbol{\Phi}_i\left(t_{f,i}, t\right) = \begin{bmatrix} 1 & t_{f,i} - t & 0 & -\tau_a^2 \phi\left(\frac{t_{f,i}-t}{\tau_a}\right) \\ 0 & 1 & 0 & \tau_a\left(e^{-\frac{t_{f,i}-t}{\tau_a}} - 1\right) \\ 0 & 0 & 1 & \frac{\tau_a}{V}\left(1 - e^{-\frac{t_{f,i}-t}{\tau_a}}\right) \\ 0 & 0 & 0 & e^{-\frac{t_{f,i}-t}{\tau_a}} \end{bmatrix} \tag{9.43}$$

which generates the following zero-effort transformations

$$z_{1,i} = \begin{cases} V\dot{\sigma}_i t_{go,i}^2 - \tau_a^2 \phi\left(\dfrac{t_{go,i}}{\tau_a}\right) a, & t \leq t_{f,i} \\ z_{1,i}\left(t_{f,i}\right), & t > t_{f,i} \end{cases} \tag{9.44}$$

$$z_{2,i} = \begin{cases} \gamma + \dfrac{\tau_a}{V}\left(1 - e^{-\frac{t_{go,i}}{\tau_a}}\right) a, & t \leq t_{f,i} \\ z_{2,i}\left(t_{f,i}\right), & t > t_{f,i} \end{cases} \tag{9.45}$$

$$b_i\left(t\right) = \tau_a \phi\left(\frac{t_{go,i}}{\tau_a}\right), \quad g_i\left(t\right) = \frac{1 - e^{-\frac{t_{go,i}}{\tau_a}}}{V} \tag{9.46}$$

where $\phi\left(x\right) \triangleq e^{-x} + x - 1$.

After some tedious but simple algebra manipulations, the integrals in Eqs. (9.21) and (9.22) are determined as

$$\begin{aligned} \int_t^{t_{f,\min\{i,i'\}}} b_i\left(\tau\right) b_{i'}\left(\tau\right) d\tau = \Bigg[& \frac{\tau_a^3}{2}\left(e^{-\frac{d_{i,i'}}{\tau_a}} - e^{-\frac{2t_{go,\min\{i,i'\}}+d_{i,i'}}{\tau_a}}\right) \\ & - \tau_a^2 t_{go,\min\{i,i'\}}\left(e^{-\frac{t_{go,\min\{i,i'\}}}{\tau_a}} + e^{-\frac{t_{go,\min\{i,i'\}}+d_{i,i'}}{\tau_a}}\right) \\ & + \tau_a^2 d_{i,i'}\left(1 - e^{-\frac{t_{go,\min\{i,i'\}}}{\tau_a}}\right) + \frac{t_{go,\min\{i,i'\}}^2}{6}\left(2t_{go,\min\{i,i'\}} + 3d_{i,i'}\right) \\ & - \tau_a t_{go,\min\{i,i'\}}^2 - \tau_a d_{i,i'} t_{go,\min\{i,i'\}} + \tau_a^2 t_{go,\min\{i,i'\}} \Bigg] \end{aligned} \tag{9.47}$$

$$\int_t^{t_{f,\min\{i,l(j')\}}} b_i(\tau)\, g_{l(j')}(\tau)\, d\tau$$
$$= \begin{cases} \frac{\tau_a}{V}\left[\tau_a - t_{go,i} + \frac{t_{go,i}^2}{2\tau_a} - \frac{\tau_a}{2} e^{-\frac{d_{i,l(j')}}{\tau_a}} - \tau_a e^{-\frac{t_{go,i}}{\tau_a}} \right. \\ \left. + t_{go,i} e^{-\frac{t_{go,i}+d_{i,l(j')}}{\tau_a}} + \frac{\tau_a}{2} e^{-\frac{2t_{go,i}+d_{i,l(j')}}{\tau_a}}\right], & i \le l(j') \\ \frac{1}{2V}\left[\tau_a^2 e^{-\frac{d_{i,l(j')}}{\tau_a}} - \tau_a^2 e^{-\frac{2t_{go,l(j')}+d_{i,l(j')}}{\tau_a}} - 2d_{i,l(j')}\tau_a \right. \\ + 2d_{i,l(j')} t_{go,l(j')} - 2\tau_a t_{go,l(j')} + t_{go,l(j')}^2 - 2\tau_a^2 e^{-\frac{t_{go,l(j')}}{\tau_a}}, & i > l(j') \\ \left. + 2\tau_a d_{i,l(j')} e^{-\frac{t_{go,l(j')}}{\tau_a}} + 2\tau_a t_{go,l(j')} e^{-\frac{t_{go,l(j')}}{\tau_a}}\right] \end{cases} \tag{9.48}$$

$$\int_t^{t_{f,\min\{l(j),l(j')\}}} g_{l(j)}(\tau)\, g_{l(j')}(\tau)\, d\tau$$
$$= \frac{1}{V^2}\left[\tau_a \phi\left(\frac{t_{go,\min\{l(j),l(j')\}}}{\tau_a}\right) + \tau_a e^{-\frac{t_{go,\min\{l(j),l(j')\}}+d_{l(j),l(j')}}{\tau_a}} \right. \tag{9.49}$$
$$\left. - \frac{\tau_a}{2} e^{-\frac{d_{l(j),l(j')}}{\tau_a}} - \frac{\tau_a}{2} e^{-\frac{2t_{go,\min\{l(j),l(j')\}}+d_{l(j),l(j')}}{\tau_a}}\right]$$

The explicit guidance command can then be readily obtained by substituting Eqs. (9.44)–(9.49) into Eq. (9.29). This guidance command is termed as first-order lag energy-optimal waypoint-following guidance law (OWFGL-1) in this chapter.

Remark 9.2 For higher-order autopilot systems, i.e., $n_a > 1$, the integrals in Eqs. (9.21) and (9.22) are algebraically complicated. However, the resultant expressions can be easily evaluated using an algebraic mathematical solver such as Mathematica®.

9.5 Relationship with Point-to-Point Optimal Guidance Laws

This section analyzes the relationship between the proposed guidance law and classical point-to-point optimal guidance laws [2–5]. For the purpose of illustration, a first-order lag autopilot system is considered here.

9.5.1 $N = 1, M = 0$

When there exists only one waypoint to be visited by the UAV without any specified flight path, the original optimization problem reduces to energy-optimal point-to-point interception problem. Then, the guidance command of the proposed guidance law, shown in Eq. (9.19), reduces to

$$a_c = \lambda_1 b_1 = \lambda_1 \tau_a \phi\left(\frac{t_{go,1}}{\tau_a}\right) \tag{9.50}$$

Note that matrix $\boldsymbol{G}$ under condition $N = 1$, $M = 0$ reduces to a scalar G as

$$G = \frac{\tau_a^3}{2}\left(1 - e^{-\frac{2t_{go,1}}{\tau_a}}\right) - 2\tau_a^2 t_{go,1} e^{-\frac{t_{go,1}}{\tau_a}} + \frac{t_{go,1}^3}{3} - \tau_a t_{go,1}^2 + \tau_a^2 t_{go,1} \tag{9.51}$$

From Eq. (9.28), we can readily solvc thc Lagrange multiplier as

$$\lambda_1 = \frac{z_{1,1}}{\frac{\tau_a^3}{2}\left(1 - e^{-\frac{2t_{go,1}}{\tau_a}}\right) - 2\tau_a^2 t_{go,1} e^{-\frac{t_{go,1}}{\tau_a}} + \frac{t_{go,1}^3}{3} - \tau_a t_{go,1}^2 + \tau_a^2 t_{go,1}} \tag{9.52}$$

Substituting Eq. (9.52) into Eq. (9.50) gives the explicit guidance command as

$$a_c = \frac{N_1 z_{1,1}}{t_{go,1}^2} \tag{9.53}$$

with

$$N_1 = \frac{\phi\left(\frac{t_{go,1}}{\tau_a}\right)}{\frac{\tau_a^2}{2t_{go,1}^2}\left(1 - e^{-\frac{2t_{go,1}}{\tau_a}}\right) - \frac{2\tau_a}{t_{go,1}} e^{-\frac{t_{go,1}}{\tau_a}} + \frac{3t_{go,1}}{\tau_a} - 1 + \frac{\tau_a}{t_{go,1}}} \tag{9.54}$$

which coincides with the energy-optimal point-to-point intercept guidance with first-order autopilot dynamics proposed in [2, 3].

9.5.2 $N = 2$, $M = 0$

To provide better understanding of the structure and nature of the proposed guidance law, we consider a special case with $N = 2$, $M = 0$, e.g., the UAV is required to visit two waypoints sequentially without any arrival angle constraints. Under this condition, the guidance command of the proposed guidance law (9.19) for $t \leq t_{f,1}$ reduces to

$$a_c = \lambda_1 b_1 + \lambda_2 b_2 = \lambda_1 \tau_a \phi\left(\frac{t_{go,1}}{\tau_a}\right) + \lambda_2 \tau_a \phi\left(\frac{t_{go,2}}{\tau_a}\right) \tag{9.55}$$

The matrix $\boldsymbol{G}$ with $N = 2$, $M = 0$ is given by

$$\boldsymbol{G} = \begin{bmatrix} g_1 & g_{12} \\ g_{21} & g_2 \end{bmatrix} \tag{9.56}$$

where

$$
\begin{aligned}
g_1 =& \frac{\tau_a^3}{2}\left(1-e^{-\frac{2t_{go,1}}{\tau_a}}\right) - 2\tau_a^2 t_{go,1} e^{-\frac{t_{go,1}}{\tau_a}} + \frac{t_{go,1}^3}{3} - \tau_a t_{go,1}^2 + \tau_a^2 t_{go,1} \\
g_{12} =& g_{21} = \frac{\tau_a^3}{2}\left(e^{-\frac{d_{1,2}}{\tau_a}} - e^{-\frac{2t_{go,1}+d_{1,2}}{\tau_a}}\right) - \tau_a^2 t_{go,1}\left(e^{-\frac{t_{go,1}}{\tau_a}} + e^{-\frac{t_{go,1}+d_{1,2}}{\tau_a}}\right) \\
&+ \tau_a^2 d_{1,2}\left(1-e^{-\frac{t_{go,1}}{\tau_a}}\right) + \frac{t_{go,1}^2}{6}\left(2t_{go,1}+3d_{1,2}\right) - \tau_a t_{go,1}^2 - \tau_a t_{go,1} d_{1,2} + \tau_a^2 t_{go,1} \\
g_2 =& \frac{\tau_a^3}{2}\left(1-e^{-\frac{2t_{go,2}}{\tau_a}}\right) - 2\tau_a^2 t_{go,2} e^{-\frac{t_{go,2}}{\tau_a}} + \frac{t_{go,2}^3}{3} - \tau_a t_{go,2}^2 + \tau_a^2 t_{go,2}
\end{aligned}
\tag{9.57}
$$

with $d_{1,2} = t_{go,2} - t_{go,1}$.

Using Eq. (9.28), the Lagrange multipliers can be formulated as

$$
\begin{aligned}
\begin{bmatrix} \lambda_1 \\ \lambda_2 \end{bmatrix} &= \begin{bmatrix} g_1 & g_{12} \\ g_{21} & g_2 \end{bmatrix}^{-1} \begin{bmatrix} z_{1,1} \\ z_{1,2} \end{bmatrix} \\
&= \frac{1}{\Lambda}\begin{bmatrix} g_2 z_{1,1} - g_{12} z_{1,2} \\ -g_{21} z_{1,1} + g_1 z_{1,2} \end{bmatrix}
\end{aligned}
\tag{9.58}
$$

where $\Lambda = g_1 g_2 - g_{12}^2$.

The guidance command for $t \le t_{f,1}$ can be readily obtained by substituting Eq. (9.58) into Eq. (9.55) as

$$
a_c = \frac{N_1 z_{1,1}}{t_{go,1}^2} + \frac{N_2 z_{1,2}}{t_{go,2}^2} \tag{9.59}
$$

where

$$
\begin{aligned}
N_1 &= \frac{\tau_a t_{go,1}^2}{\Lambda}\left[g_2 \phi\left(\frac{t_{go,1}}{\tau_a}\right) - g_{21}\phi\left(\frac{t_{go,2}}{\tau_a}\right)\right] \\
N_2 &= \frac{\tau_a t_{go,2}^2}{\Lambda}\left[g_1 \phi\left(\frac{t_{go,2}}{\tau_a}\right) - g_{12}\phi\left(\frac{t_{go,1}}{\tau_a}\right)\right]
\end{aligned}
\tag{9.60}
$$

Introducing the biased term $a_B = N_2 z_{1,2}/t_{go,2}^2$, guidance command (9.59) can then be formulated in an alternative form as

$$
a_c = \frac{N_1 z_{1,1}}{t_{go,1}^2} + a_B \tag{9.61}
$$

which can be viewed as a general ZEM shaping guidance law $N_1 z_{1,1}/t_{go,1}^2$ [10] with a biased term a_B. From Eq. (9.60), it can be easily verified that the biased term a_B gradually becomes the dominant part when the UAV approaches the first waypoint.

Recall the results derived for $N = 1$, $M = 0$, the guidance command when $t_{f,1} < t \le t_{f,2}$ for the case of $N = 2$, $M = 0$ is given by

$$a_c = \frac{N_2' z_{1,2}}{t_{go,2}^2} \tag{9.62}$$

with

$$N_2' = \frac{\phi\left(\frac{t_{go,2}}{\tau_a}\right)}{\frac{\tau_a^2}{2t_{go,2}^2}\left(1 - e^{-\frac{2t_{go,2}}{\tau_a}}\right) - \frac{2\tau_a}{t_{go,2}} e^{-\frac{t_{go,2}}{\tau_a}} + \frac{3t_{go,2}}{\tau_a} - 1 + \frac{\tau_a}{t_{go,2}}} \tag{9.63}$$

In summary, the guidance command for $N = 2$, $M = 0$ is determined as

$$a_c = \begin{cases} \frac{N_1 z_{1,1}}{t_{go,1}^2} + \frac{N_2 z_{1,2}}{t_{go,2}^2}, & t \le t_{f,1} \\ \frac{N_2' z_{1,2}}{t_{go,2}^2}, & t_{f,1} < t \le t_{f,2} \end{cases} \tag{9.64}$$

9.5.3 $N = 1$, $M = 1$

When there exists only one waypoint to be visited by the UAV with a specified flight path γ_1^d, the original optimization problem reduces to energy-optimal point-to-point rendezvous problem. Then, the guidance command of the proposed guidance law, shown in Eq. (9.19), can be formulated as

$$a_c = \lambda_1 b_1 + \beta_1 g_1 = \lambda_1 \tau_a \phi\left(\frac{t_{go,1}}{\tau_a}\right) + \beta_1 \frac{1 - e^{-\frac{t_{go,1}}{\tau_a}}}{V} \tag{9.65}$$

Under condition $N = 1$, $M = 1$, matrix $\boldsymbol{G}$ becomes

$$\boldsymbol{G} = \begin{bmatrix} g_1 & g_{12} \\ g_{21} & g_2 \end{bmatrix} \tag{9.66}$$

where

$$\begin{aligned} g_1 &= \frac{\tau_a^3}{2}\left(1 - e^{-\frac{2t_{go,1}}{\tau_a}}\right) - 2\tau_a^2 t_{go,1} e^{-\frac{t_{go,1}}{\tau_a}} + \frac{t_{go,1}^3}{3} - \tau_a t_{go,1}^2 + \tau_a^2 t_{go,1} \\ g_{12} = g_{21} &= \frac{\tau_a}{V}\left[\frac{\tau_a}{2} - t_{go,1} + \frac{t_{go,1}^2}{2\tau_a} - \tau_a e^{-\frac{t_{go,1}}{\tau_a}} + t_{go,i} e^{-\frac{t_{go,1}}{\tau_a}} + \frac{\tau_a}{2} e^{-\frac{2t_{go,1}}{\tau_a}}\right] \\ g_2 &= \frac{1}{V^2}\left[t_{go,1} + 2\tau_a e^{-\frac{t_{go,1}}{\tau_a}} - \frac{3\tau_a}{2} - \frac{\tau_a}{2} e^{-\frac{2t_{go,1}}{\tau_a}}\right] \end{aligned} \tag{9.67}$$

Using Eq. (9.28), the Lagrange multipliers can be formulated as

$$\begin{bmatrix} \lambda_1 \\ \beta_1 \end{bmatrix} = \begin{bmatrix} g_1 & g_{12} \\ g_{21} & g_2 \end{bmatrix}^{-1} \begin{bmatrix} z_{1,1} \\ \gamma_1^d - z_{2,1} \end{bmatrix} = \frac{1}{\Delta} \begin{bmatrix} g_2 z_{1,1} - g_{12}\left(\gamma_1^d - z_{2,1}\right) \\ -g_{21} z_{1,1} + g_1\left(\gamma_1^d - z_{2,1}\right) \end{bmatrix} \tag{9.68}$$

where $\Delta = g_1 g_2 - g_{12}^2$.

Substituting Eq. (9.68) into Eq. (9.54) gives the explicit guidance command as

$$a_c = \frac{K_1 z_{1,1}}{t_{go,1}^2} + \frac{K_2\left(\gamma_1^d - z_{2,1}\right)}{t_{go,1}} \tag{9.69}$$

where

$$K_1 = \frac{t_{go,1}^2}{\Delta}\left[\tau_a \phi\left(\frac{t_{go,1}}{\tau_a}\right) g_2 - \frac{1 - e^{-\frac{t_{go,1}}{\tau_a}}}{V} g_{21}\right]$$
$$K_2 = \frac{t_{go,1}}{\Delta}\left[\frac{1 - e^{-\frac{t_{go,1}}{\tau_a}}}{V} g_1 - \tau_a \phi\left(\frac{t_{go,1}}{\tau_a}\right) g_{12}\right] \tag{9.70}$$

which coincides with the energy-optimal point-to-point impact angle guidance law with first-order autopilot dynamics proposed in [4, 5].

9.6 Numerical Simulations

In this section, nonlinear numerical simulations are performed to validate the proposed guidance law. In the considered scenario, a UAV with constant speed $V = 30\,\text{m/s}$ is required to pass eight waypoints, summarized in Table 9.1. The UAV starts the mission at position (0 m, 0 m) with an initial flight path angle 30°. The UAV autopilot is modeled as a first-order lag system with time constant $\tau_a = 0.5\,\text{s}$, which is enough to identify the characteristics of a guidance law in the conservative point of view.

9.6.1 Comparison with Other Waypoint Guidance Laws

To evaluate the energy-minimization property of the guidance law developed, this subsection compares the proposed OWFGL-1 algorithm with synthetic waypoint guidance law (SGWL) [11, 12], trajectory shaping waypoint guidance law (TSWGL) [13, 14] and classical point-to-point optimal guidance laws with first-order autopilot

Table 9.1 Inertial positions of all waypoints

Waypoint ID	Inertial position
1	(1000 m, 500 m)
2	(2000 m, 750 m)
3	(3000 m, 1000 m)
4	(4000 m, 1500 m)
5	(5000 m, 1250 m)
6	(6000 m, 1750 m)
7	(7000 m, 2000 m)
8	(8000 m, 1500 m)

dynamics (P2POGL-1) [2–5]. The guidance commands of both SWGL and TSWGL are summarized in Appendix. In the simulations, P2POGL-1 is applied to every two consecutive waypoints for path following. The guidance command of P2POGL-1 is defined as

$$a_c = \begin{cases} \text{Eq.(9.69)}, & \text{if next waypoint requires specific arrival angle} \\ \text{Eq.(9.53)}, & \text{Otherwise} \end{cases}$$

We first the scenario without any passing angle constraint. The UAV flight trajectories obtained from different guidance laws are shown in Fig. 9.1a. This figure reveals that the UAV guided by all guidance laws can successfully accomplish its waypoint-following mission at some extent. The quantitative comparison results of mean miss distance over all waypoints under different guidance laws for this scenario are summarized in Table 9.2. From this table, one can note that both P2POGL-1 and the proposed guidance law outperform SWGL and TSWGL in terms of guidance accuracy. The reason is that both P2POGL-1 and the proposed algorithm consider autopilot dynamics in guidance law derivation. Compared to other guidance laws, the flight path angle under the proposed guidance law is smoother and therefore is more beneficial for practical application, as confirmed by Fig. 9.1b. Figure 9.1c compares the acceleration response generated by these four different guidance laws through the onboard autopilot. From this figure, it is clear that the acceleration response under SWGL, TSWGL and P2POGL-1 has sharp changes during waypoint switch. As a comparison, the achieved acceleration under the proposed guidance law is continuous and smooth during the entire flight period, which is more desirable for onboard flight control system. The reason of this phenomenon is clear: b_i gradually converges to zero when the UAV approaches the ith waypoint. This means that the term $\lambda_i b_i$ in the guidance command gradually diminishes when $t_{go,i} \to 0$, which helps to reduce the transient effect when passing the ith waypoint. The energy consumption, defined as $J = \int_0^{t_{f,N}} a^2(\tau)\, d\tau$, obtained from different guidance laws is presented in Fig. 9.1d. As shown in this figure, the proposed guidance law requires less energy consumption, compared with other guidance laws.

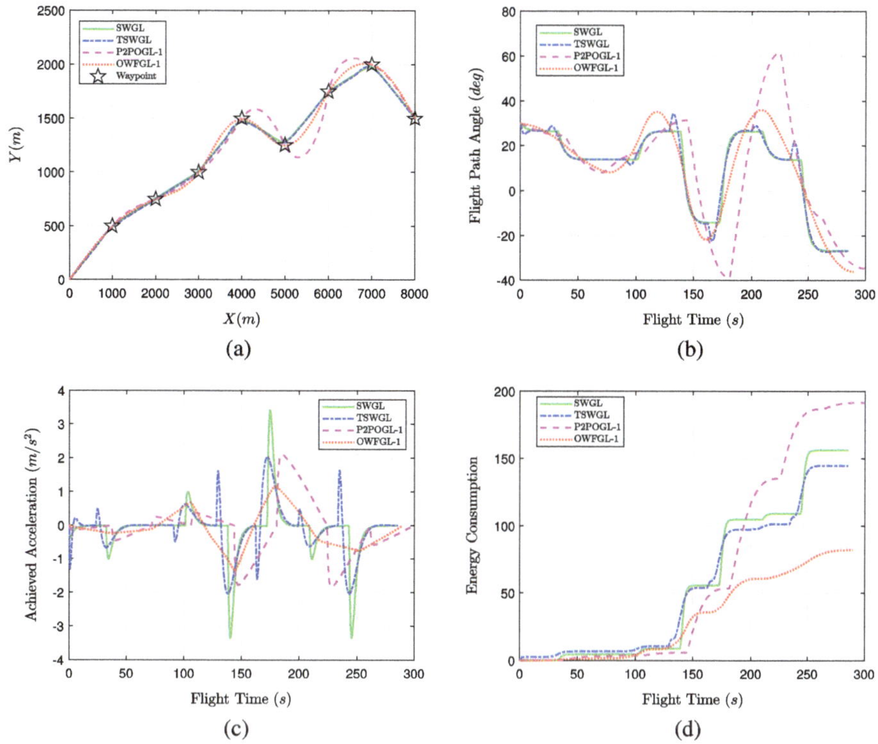

Fig. 9.1 Comparison of different guidance laws for waypoint following without flight path angle constraints. **a** Flight trajectory; **b** Flight path angle; **c** Acceleration response; and **d** Energy consumption

Table 9.2 Comparison results of mean miss distance

Metric	SWGL	TSWGL	P2POGL-1	Proposed
Mean miss distance	9.2875 m	4.2895 m	0.1806 m	0.1363 m

Now, let us analyze the characteristics of the proposed guidance law for scenarios with flight path constraint. In the simulations, we assume that the UAV is required to pass the 4th waypoint with desired flight path angle 0° and the 8th waypoint with desired flight path angle −90°. Since SWGL and TSWGL cannot control the flight path angle of the UAV, we only compare the proposed guidance law with P2POGL-1 in this scenario. The UAV flight trajectories and flight path angle profiles obtained from P2POGL-1 as well as the proposed guidance law are shown in Fig. 9.2a, b, respectively. These two figures reveal that the UAV guided by both guidance laws can successfully accomplish its waypoint-following mission. The quantitative comparison results of mean miss distance and mean passing angle over all waypoints

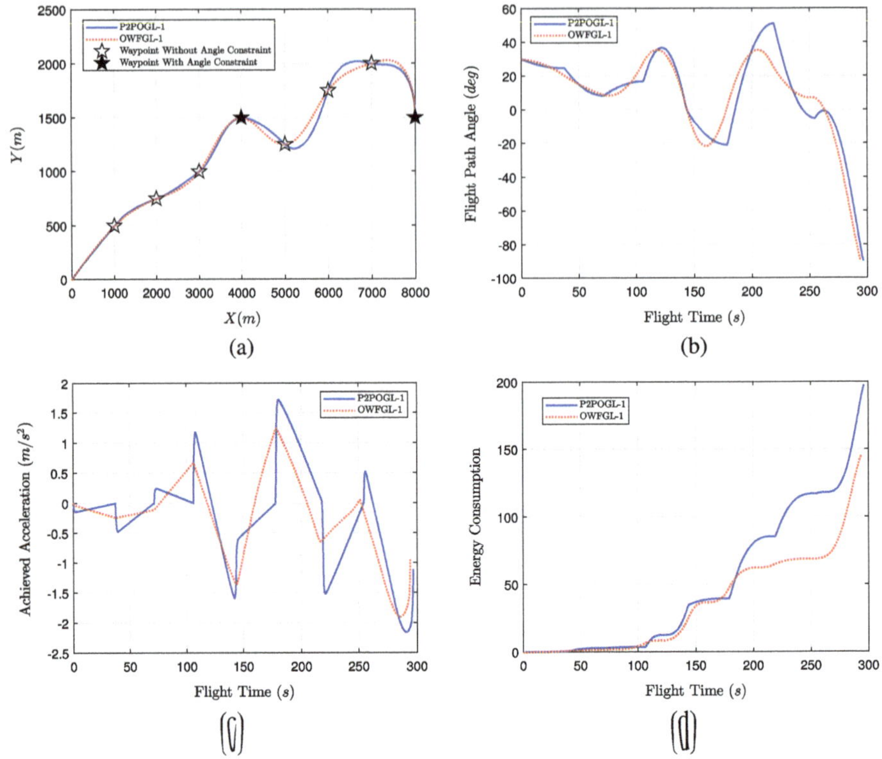

Fig. 9.2 Comparison of different guidance laws for waypoint following with partial flight path angle constraints. **a** Flight trajectory; **b** Flight path angle; **c** Acceleration response; and **d** Energy consumption

under different guidance laws for this scenario are summarized in Table 9.3. The recorded miss distance and flight path angle error obtained from both guidance laws are less than 0.2 m and 0.1° in our simulations. The achieved accelerations under both guidance laws are depicted in Fig. 9.2c. This figure demonstrates that the acceleration response under P2POGL-1 exhibits suddenly large changes when the UAV passes one waypoint due to the discontinuity of P2POGL-1 at the switching point. As a comparison, the achieved acceleration under the proposed guidance law is smoother during the entire flight period, which is more desirable for onboard flight control system. The reason can be attributed to the fact that both b_i and g_i converge to zero when the UAV approaches the ith waypoint, as deduced from in Eq. (9.46). This means that the commanded acceleration of the proposed guidance law is continuous and therefore there is no sudden command change during waypoint switch. Since both P2POGL-1 and the proposed OWFGL-1 consider autopilot dynamics in guidance law derivation, bounded guidance command can be ensured during the flight. The energy consumption obtained from different guidance laws is compared in Fig. 9.2d. As shown in this figure, the proposed guidance law requires less energy

Table 9.3 Comparison results of mean miss distance and passing angle error

Metric	P2POGL-1	Proposed
Mean miss distance	0.1901 m	0.1771 m
Mean passing angle error	0.0398°	0.0239°

consumption, compared with P2POGL-1. In the considered scenario, the proposed guidance law helps to reduce more than 25% energy consumption. Therefore, the UAV guided by the proposed approach is expected to have longer endurance than guided by P2POGL-1.

9.6.2 Effect of Autopilot Dynamics Compensation

To show the advantages of considering autopilot lag in guidance law design, this subsection compares the performance of OWFGL-1 with that of OWFGL-0 under different conditions.

We first investigate the performance of the proposed guidance law under various time lags $\tau_a = 0.5$ s, 1 s, 2 s with a fixed velocity $V = 30$ m/s. The simulation results, including UAV flight trajectories, flight angle profiles, achieved acceleration, and energy consumption, are shown in Fig. 9.3. As shown in Fig. 9.3a, b, both guidance laws successfully drive the UAV to arrive the desired waypoints at some extent. Although the recorded passing angle error in OWFGL-1 is smaller than that in OWFGL-0 in our simulations, it is fair to state that both guidance laws satisfy the arrival angle constraint in the considered scenario, as confirmed by Fig. 9.3b. Figure 9.3c indicates that the achieved acceleration of OWFGL-1 is smoother than that of OWFGL-0. Also notice from Fig. 9.3c that the guidance command of OWFGL-0 shows divergence when passing one waypoint as the autopilot lag is neglected in the guidance law derivation, especially for the 4th and the 5th waypoints in the considered scenario. Such divergence may result in a large miss distance and a significant passing angle error near the waypoint. On the other hand, the OWFGL-1 algorithm can successfully compensate the autopilot lag, and the acceleration command does not diverge as shown in Fig. 9.3c. Due to the introduced guidance command divergence, the energy consumption of OWFGL-0 is higher than that of OWFGL-1, as shown in Fig. 9.3d. The quantitative comparison results of mean miss distance and mean passing angle over all waypoints under different guidance laws for this scenario are summarized in Table 9.4. From this table, it can be concluded that the autopilot lag has adverse effects on the guidance accuracy and the performance of OWFGL-0 degrades with the increase of autopilot lag.

Now, let us investigate the performance of the proposed guidance law under various UAV speeds $V = 30$ m/s, 60 m/s, 90 m/s with a fixed autopilot lag $\tau = 0.5$ s. The comparison results, including UAV flight trajectories, flight angle profiles, achieved

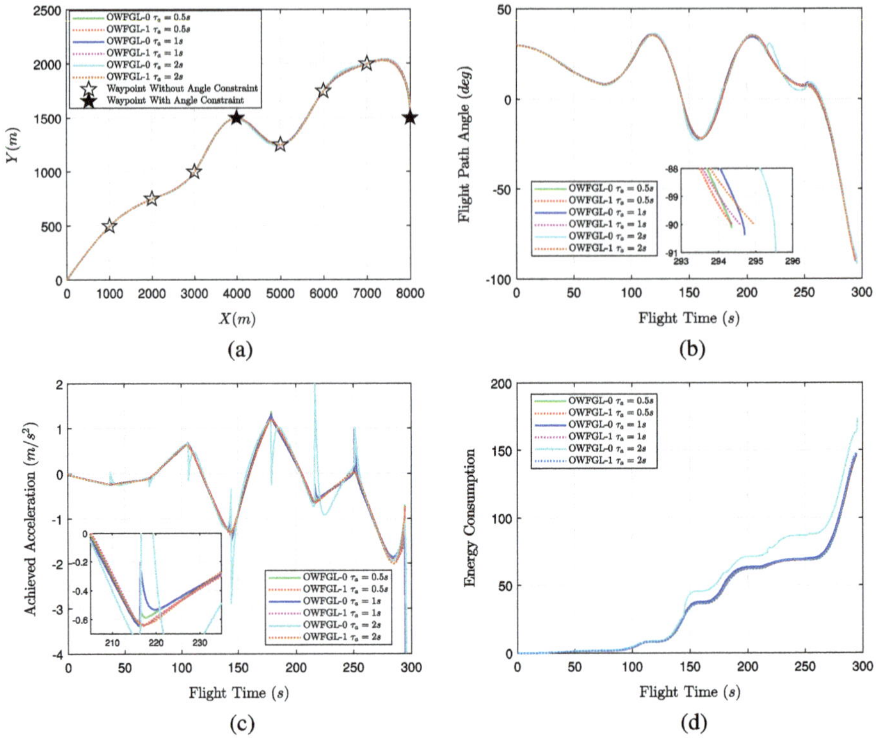

Fig. 9.3 Comparison results of the proposed guidance laws with different autopilot lags. **a** Flight trajectory; **b** Flight path angle; **c** Acceleration response; and **d** Energy consumption

Table 9.4 Comparison results of mean miss distance and passing angle error

Metric	$\tau_a = 0.5\,\text{s}$		$\tau_a = 1\,\text{s}$		$\tau_a = 2\,\text{s}$	
	OWFGL-0	OWFGL-1	OWFGL-0	OWFGL-1	OWFGL-0	OWFGL-1
Mean miss distance	0.3887 m	0.1771 m	0.9901 m	0.1819 m	2.4409 m	0.1903 m
Mean passing angle error	0.2377°	0.0239°	0.5977°	0.0951°	2.001°	0.1675°

acceleration, and energy consumption, are depicted in Fig. 9.4. From this figure, we can clearly observe that the performance of OWFGL-0 degrades with the increase of UAV speed. This can be attributed to the fact that the UAV has shorter time to initiate a response to the path change with higher flying velocity. As a comparison, the performance of OWFGL-1 remains consistent across a wide range of UAV speeds as autopilot lag is actively compensated in the guidance command. The quantitative comparison results of mean miss distance and mean passing angle over all waypoints under different guidance laws for this scenario are summarized in Table 9.5. From

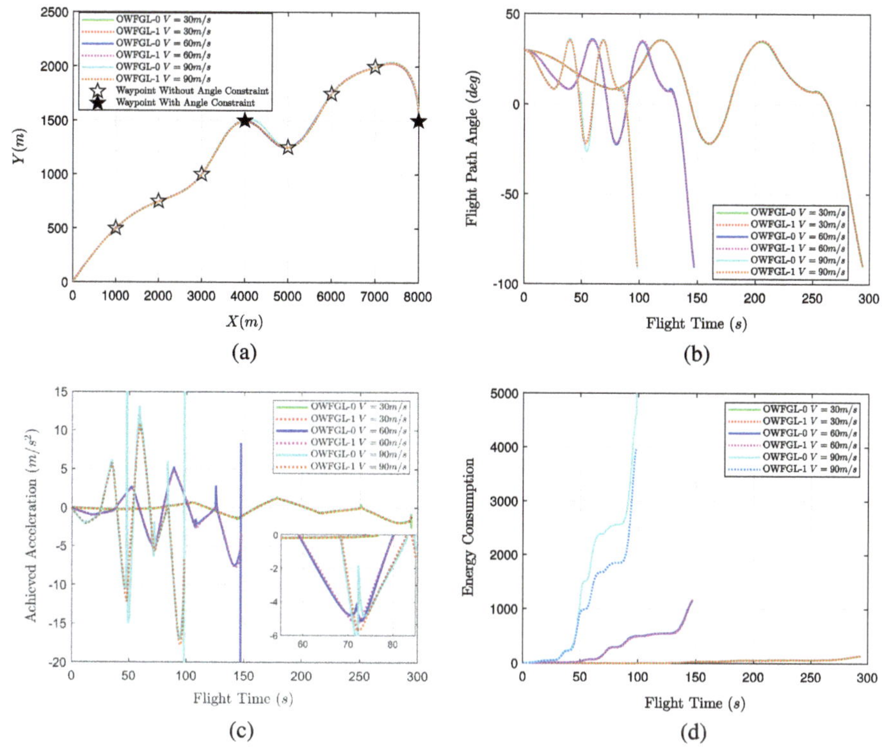

Fig. 9.4 Comparison results of the proposed guidance laws with flying speed. **a** Flight trajectory; **b** Flight path angle; **c** Acceleration response; and **d** Energy consumption

Table 9.5 Comparison results of mean miss distance and passing angle error

Metric	$V = 30\,\text{m/s}$		$V = 60\,\text{m/s}$		$V = 90\,\text{m/s}$	
	OWFGL-0	OWFGL-1	OWFGL-0	OWFGL-1	OWFGL-0	OWFGL-1
Mean miss distance	0.3887 m	0.1771 m	1.5165 m	0.2278 m	3.7727 m	0.7283 m
Mean passing angle error	0.2377°	0.0239°	1.3371°	0.1161°	2.3369°	0.2069°

this table, it can be noted that the performance of OWFGL-0 is close to OWFGL-1 for UAVs with low speed. However, the performance discrepancy between these two guidance laws drastically increases with the increase of UAV speed.

Based on the numerical simulations, it can be concluded that the proposed guidance law can effectively compensate for the autopilot lag and provide performance improvement, especially for UAVs with large time lag or high flying speed. Also, it is beneficial to compensate the autopilot delay when the range between consecutive waypoints is short due to limited response time for path change.

9.7 Summary

In this chapter, we suggest a new energy-optimal waypoint-following guidance law for a UAV with a general autopilot system. Unlike previous numerical waypoint-following guidance, the proposed approach is derived analytically through rigorous optimal control theory. We also theoretically show that the proposed guidance law encompasses previously suggested optimal point-to-point guidance laws. The proposed guidance law is generic and therefore can be applied to general waypoint-following missions. However, the computational complexity increases with the increase of number of boundary constraints. Nonlinear numerical comparisons clearly reveal that the proposed guidance law helps to increase of the endurance of the UAV and effectively compensates for the autopilot lag.

Appendix: Guidance Commands of SWGL and TSWGL

This appendix provides the guidance commands of SWGL [11, 12] and TSWGL [13, 14] for the completeness of the chapter.

Both SWGL and TSWGL requires to place a synthetic waypoint or a virtual target that travels along the flight path between designated inertial waypoints. Define (x_w, y_w) as the 2D inertial position of the synthetic waypoint. The kinematics constraint of the synthetic waypoint can then be mathematically formulated as

$$\begin{aligned} \dot{x}_w &= V_w \cos \gamma_f \\ \dot{y}_w &= V_w \sin \gamma_f \end{aligned} \tag{9.71}$$

where γ_f is the reference heading between fixed waypoints; and V_w denotes the moving speed of the synthetic waypoint, governed by

$$V_w = V \frac{R}{R*} \tag{9.72}$$

where R stands for the relative distance, or the so-called look-ahead distance, between the UAV and the synthetic waypoint; and R^* represents the desired look-ahead distance.

The desired relative distance R^* is defined by the UAV speed and time prediction horizon as

$$R^* = VT_p \tag{9.73}$$

where T_p specifies the desired time horizon that initiates a response to flight path change. This time horizon specifies the prediction of set-point and plays a similar role as the prediction horizons in model predictive control (MPC).

The SWGL applies pure pursuit missile guidance to follow the synthetic waypoint, i.e., the guidance command is given by

$$a_c = V\dot{\sigma}_w \tag{9.74}$$

where $\dot{\sigma}_w$ is the LOS rate between the UAV and the synthetic waypoint, which is determined as

$$\dot{\sigma}_w = \frac{1}{R}V_w \sin\left(\gamma_f - \sigma_w\right) - \frac{1}{R}V \sin\left(\gamma - \sigma_w\right) \tag{9.75}$$

Unlike SWGL, TSWGL employs energy-optimal trajectory shaping guidance to follow the synthetic waypoint using guidance command

$$a_c = \frac{V^2}{R}\left(6\sigma_w - 4\gamma - 2\gamma_f\right) \tag{9.76}$$

which is proved to generate the same instantaneous radius of curvature of a curved path and therefore is able to guarantee accurate path following.

References

1. Ryoo C-K, Shin H-S, Tahk M-J (2010) Energy optimal waypoint guidance synthesis for antiship missiles. IEEE Trans Aerosp Electron Syst 46(1):80–95
2. Cottrell RG (1971) Optimal intercept guidance for short-range tactical missiles. AIAA J 9(7):1414–1415
3. Zarchan P (2012) Tactical and strategic missile guidance. American Institute of Aeronautics and Astronautics
4. Ryoo C-K, Cho H, Tahk M-J (2005) Optimal guidance laws with terminal impact angle constraint. J Guidance Control Dyn 28(4):724–732
5. Ryoo C-K, Cho H, Tahk M-J (2006) Time-to-go weighted optimal guidance with impact angle constraints. IEEE Trans Control Syst Technol 14(3):483–492
6. Bryson AE (2018) Applied optimal control: optimization, estimation and control. Routledge
7. Shaferman V, Shima T (2008) Linear quadratic guidance laws for imposing a terminal intercept angle. J Guidance Control Dyn 31(5):1400–1412
8. Hayoun SY, Weiss M, Shima T (2016) A mixed l_2/l_α differential game approach to pursuit-evasion guidance. IEEE Trans Aerosp Electron Syst 52(6):2775–2788
9. Shima T, Golan OM (2007) Linear quadratic differential games guidance law for dual controlled missiles. IEEE Trans Aerosp Electron Syst 43(3):834–842
10. Lee C-H, Shin H-S, Lee J-I, Tahk M-J (2018) Zero-effort-miss shaping guidance laws. IEEE Trans Aerosp Electron Syst 54(2):693–705
11. Medagoda EDB (2010) Closed loop stability of synthetic waypoint guidance algorithm. IFAC Proc Vol 43(15):75–80
12. Medagoda EDB, Gibbens PW (2010) Synthetic-waypoint guidance algorithm for following a desired flight trajectory. J Guidance Control Dyn 33(2):601–606

13. Ratnoo A, Hayoun S, Granot A, Shima TY (2013) Path following using trajectory shaping guidance. In: AIAA Guidance, Navigation, and Control (GNC) Conference
14. Ratnoo A, Hayoun SY, Granot A, Shima T (2014) Path following using trajectory shaping guidance. J Guidance Control Dyn 38(1):106–116

Chapter 10
Optimal Integrated Waypoint Following and Obstacle Avoidance Guidance Law

Abstract This chapter addresses the problem of energy-optimal waypoint-following guidance for an UAV with the consideration of arbitrary number of obstacles. The proposed guidance law is derived as a solution of a linear quadratic optimal control problem in conjunction with convex parameter optimization. The algorithm developed integrates path following and obstacle avoidance into a single step and is able to be applied to a general waypoint-following mission. Several particular cases of the proposed guidance law are presented to provide better insights of the proposed algorithm. Nonlinear numerical simulations clearly demonstrate the effectiveness of the proposed formulations.

10.1 Introduction

To allow for online implementation with reduced computational burden, a closed-form minimum-effort waypoint following algorithm was proposed in Chap. 8 and the resultant guidance command is given as a biased PNG with time-varying navigation gains. The extension of Chap. 8 to accommodating higher-order autopilot dynamics was presented in Chap. 9 through a ZEM shaping guidance law. Notice that most small-scale UAVs are operated at low altitudes in proximity with man-made and natural structures, e.g., trees, buildings. A collision with these structures is fatal and might result in mission failure [1]. In the literature, there exist many elegant solutions for active obstacle avoidance, e.g., potential field approach [2–4], PNG [5], differential geometric apporach [6]. However, most of them are separately developed from path following or waypoint.

This chapter is an attempt to develop an energy-optimal guidance for integrated waypoint following and obstacle avoidance. As the optimization problem formulated involves inequality constraints and is difficult to solve directly, we formulate an intermediate optimization problem that relaxes the inequality constraint as equality constraint. The optimal commanded acceleration of the intermediate problem is analytically derived as a solution of a linear quadratic regulation problem. The solu-

S. He et al., *Optimal Guidance and Its Applications in Missiles and UAVs*, Springer Aerospace Technology, https://doi.org/10.1007/978-3-030-47348-8_10

tion of the original optimization problem is then obtained with the help of convex parameter optimization. The key feature of the proposed guidance law is that it integrates path following and obstacle avoidance into a single step, which differs from existing approaches. This advantage is beneficial to reduce the design complexity for initial mission analysis. Notice that the proposed guidance law is an extension of Chaps. 8 and 9 and to active obstacle avoidance.

10.2 Problem Formulation

Consider there exist N waypoints that the UAV has to visit sequentially while avoiding M static obstacles. The planner relative engagement geometry is presented in Fig. 10.1, where W_i and O_j denote the ith waypoint and the jth obstacle, respectively. The symbols in Fig. 10.1 share the same definitions as in Chap. 8 with superscript w being the waypoint and o being the obstacle. Similar to Chap. 9, we consider the following integrated kinematics and autopilot dynamics model.

$$\dot{\boldsymbol{x}}^w = \boldsymbol{A}^w \boldsymbol{x}^w + \boldsymbol{B}^w a_c \tag{10.1}$$

where $\boldsymbol{x}_i^w = \left[y_i^w, v_i^w, \boldsymbol{x}_a\right]^T \in \mathbb{R}^{(n_a+2)}$ and $\boldsymbol{x}^w = \left[\boldsymbol{x}_1^w, \boldsymbol{x}_2^w, \ldots, \boldsymbol{x}_N^w\right]^T \in \mathbb{R}^{(n_a+2)N}$. $\boldsymbol{A}^w \in \mathbb{R}^{(n_a+2)N \times (n_a+2)N}$ is a block diagonal matrix, and $\boldsymbol{B}^w \in \mathbb{R}^{(n_a+2)N}$ is the control input matrix. These two matrices are defined as

$$\boldsymbol{A}^w = \text{diag}\left(\boldsymbol{A}_1^w, \boldsymbol{A}_2^w, \ldots \boldsymbol{A}_N^w\right), \quad \boldsymbol{B}^w = \left[\left(\boldsymbol{B}_1^w\right)^T, \left(\boldsymbol{B}_2^w\right)^T, \ldots, \left(\boldsymbol{B}_N^w\right)^T\right]^T \tag{10.2}$$

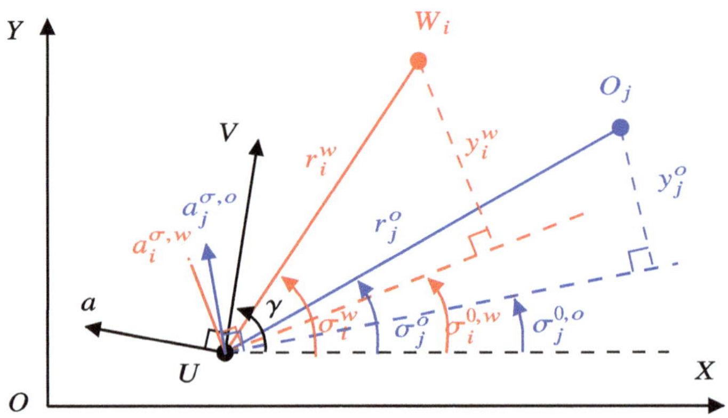

Fig. 10.1 Planar engagement geometry

with

$$A_i^w = \begin{bmatrix} 0 & 1 & 0_{1\times n_a} \\ 0 & 0 & -C_a \\ 0_{n_a\times 1} & 0_{n_a\times 1} & A_a \end{bmatrix}, \quad B_i^w = \begin{bmatrix} 0 \\ -D_a \\ B_a \end{bmatrix}, \quad i \in [N] \tag{10.3}$$

Likewise, denote $\boldsymbol{x}_j^o = \left[y_j^o, v_j^o, \boldsymbol{x}_a\right]^T \in \mathbb{R}^{(n_a+2)}$ and $\boldsymbol{x}^o = \left[\boldsymbol{x}_1^o, \boldsymbol{x}_2^o, \ldots, \boldsymbol{x}_M^o\right]^T \in \mathbb{R}^{(n_a+2)M}$. Then, the relative kinematics between the UAV and obstacles can also be linearized as

$$\dot{\boldsymbol{x}}^o = \boldsymbol{A}\boldsymbol{x}^o + \boldsymbol{B}a_c \tag{10.4}$$

where $\boldsymbol{A}^o \in \mathbb{R}^{(n_a+2)M\times(n_a+2)M}$ is a block diagonal matrix, and $\boldsymbol{B}^o \in \mathbb{R}^{(n_a+2)M}$ is the control input matrix. These two matrices are defined as

$$\boldsymbol{A}^o = \mathrm{diag}\left(\boldsymbol{A}_1^o, \boldsymbol{A}_2^o, \ldots \boldsymbol{A}_M^o\right), \quad \boldsymbol{B}^o = \left[\left(\boldsymbol{B}_1^o\right)^T, \left(\boldsymbol{B}_2^o\right)^T, \ldots, \left(\boldsymbol{B}_M^o\right)^T\right]^T \tag{10.5}$$

with

$$A_j^o = \begin{bmatrix} 0 & 1 & 0_{1\times n_a} \\ 0 & 0 & -C_a \\ 0_{n_a\times 1} & 0_{n_a\times 1} & A_a \end{bmatrix}, \quad B_j^o = \begin{bmatrix} 0 \\ -D_a \\ B_a \end{bmatrix}, \quad j \in [M] \tag{10.6}$$

To reduce the energy consumption, we consider the following cost function

$$J = \int_t^{t_{f,N}^w} a_c^2(\tau)\,d\tau = \begin{cases} \int_t^{t_{f,1}^w} a_c^2(\tau)\,d\tau + \sum_{i=1}^{N-1} \int_{t_{f,i}^w}^{t_{f,i+1}^w} a_c^2(\tau)\,d\tau, & t \le t_{f,1}^w \\ \int_t^{t_{f,2}^w} a_c^2(\tau)\,d\tau + \sum_{i=2}^{N-1} \int_{t_{f,i}^w}^{t_{f,i+1}^w} a_c^2(\tau)\,d\tau, & t \le t_{f,}^w \\ \vdots & \\ \int_t^{t_{f,N}^w} a_c^2(\tau)\,d\tau, \quad t_{f,N-1}^w < t \le t_{f,N}^w & \end{cases} \tag{10.7}$$

where $t_{f,i}^w$ stands for the time to pass the ith waypoint.

In a general waypoint following mission, one path constraint is that the UAV has to pass the ith characteristic waypoint with zero miss distance at time $t_{f,i}^w$. To cater for the requirement of obstacle avoidance, the relative distance between the UAV and the jth obstacle should be larger than a safety bound R_j^s at the predicted collision time $t_{f,j}^o$. With these in mind, we consider the following path constraints

$$y_i^w\left(t_{f,i}^w\right) = 0, \quad i \in [N] \tag{10.8}$$

$$\left|y_j^o\left(t_{f,j}^o\right)\right| \ge R_j^s, \quad j \in [M] \tag{10.9}$$

where $t_{f,i}^o$ denotes the time to reach the jth obstacle.

In summary, the aim of this chapter is to find analytical guidance command solution to the following optimization problem.

Problem 10.1 Find a_c that minimizes performance index (10.7), subject to linear kinematics (10.1), (10.4) and path constraints (10.8), (10.9).

10.3 Guidance Law Derivation

In this section, we first present a reduced-order system using terminal projection via state transmission. Then, the detailed derivation of the proposed guidance law and its implementation issue are provided.

10.3.1 Order Reduction

Similar to Chap. 9, we utilize the zero-effort transformation technique [7, 8] to reduce system order. Define z_i^w and z_j^o as ZEM distances associated with the ith waypoint and the jth obstacle, respective. Notice that the ZEM refers to the final relative distance between the UAV and the ith waypoint or the jth obstacle, if the UAV will not apply any control from the current time onward. Based on this concept, the terminal projection ZEM z_i^w and ^o_j at current time t can be readily obtained using the homogeneous solution as

$$z_i^w = \begin{cases} \boldsymbol{C}_i \boldsymbol{\Phi}_i^w \left(t_{f,i}^w, t\right) \boldsymbol{x}_i^w, & t \leq t_{f,i}^w \\ z_i^w \left(t_{f,i}^w\right), & t > t_{f,i}^w \end{cases} \tag{10.10}$$

$$z_j^o = \begin{cases} \boldsymbol{C}_j \boldsymbol{\Phi}_j^o \left(t_{f,j}^o, t\right) \boldsymbol{x}_j^o, & t \leq t_{f,j}^o \\ z_j^o \left(t_{f,j}^o\right), & t > t_{f,j}^o \end{cases} \tag{10.11}$$

where $\boldsymbol{\Phi}_i^w \left(t_f, t\right)$ and $\boldsymbol{\Phi}_j^o \left(t_f, t\right)$ denote the transition matrices associated with matrices $\boldsymbol{A}_i^w$ and $\boldsymbol{A}_i^o$, respectively. To extract the ZEM from the terminal state projection, Matrices $\boldsymbol{C}_i$ and $\boldsymbol{C}_j$ are

$$\boldsymbol{C}_i = \boldsymbol{C}_j = \left[1, 0, 0, 0_{1\times n_a}\right] \tag{10.12}$$

Since

$$\dot{\boldsymbol{\Phi}}_i^w \left(t_{f,i}^w, t\right) = -\boldsymbol{\Phi}_i^w \left(t_{f,i}^w, t\right) \boldsymbol{A}_i^w \tag{10.13}$$

$$\dot{\boldsymbol{\Phi}}_j^o \left(t_{f,j}^o, t\right) = -\boldsymbol{\Phi}_j^o \left(t_{f,j}^o, t\right) \boldsymbol{A}_j^o \tag{10.14}$$

the time evolution of the ZEM is

$$\dot{z}_i^w = \begin{cases} \boldsymbol{C}_i \boldsymbol{\Phi}_i^w \left(t_{f,i}^w, t\right) \boldsymbol{B}_i^w a_c, & t \leq t_{f,i}^w \\ 0, & t > t_{f,i}^w \end{cases} \tag{10.15}$$

$$\dot{z}_j^o = \begin{cases} \boldsymbol{C}_j \boldsymbol{\Phi}_j^o \left(t_{f,j}^o, t\right) \boldsymbol{B}_j^o a_c, & t \leq t_{f,j}^o \\ 0, & t > t_{f,j}^o \end{cases} \tag{10.16}$$

With the terminal projection, path constraints (10.8) and (10.9) can be alternatively expressed as

$$z_i \left(t_{f,i}^w\right) = 0, \quad i \in [N] \tag{10.17}$$

$$\left|z_j \left(t_{f,j}^o\right)\right| \geq R_j^s, \quad j \in [M] \tag{10.18}$$

Then, Problem 10.1 reduces to the following optimization problem.

Problem 10.2 Find a_c that minimizes performance index (10.7), subject to linear kinematics (10.1), (10.4) and path constraints (10.17), (10.18).

As it is difficult to directly derive the closed-form solution for optimization problems with inequality constraints, we formulate the following intermediate optimization problem.

Problem 10.3 Find a_c that minimizes performance index (10.7), subject to linear kinematics (10.1), (10.4) and path constraints

$$z_i \left(t_{f,i}^w\right) = 0, \quad i \in [N] \tag{10.19}$$

$$z_j \left(t_{f,j}^o\right) = z_j^d, \quad j \in [M] \tag{10.20}$$

where z_j^d is an arbitrary constant.

Notice that the solution of Problem 10.3 is a function of $\boldsymbol{Z}^d = \left[z_1^d, z_2^d, \cdots, z_M^d\right]^T$, i.e., $a_c\left(\boldsymbol{Z}^d\right)$, and $\boldsymbol{Z}^d$ is an arbitrary constant vector by definition. With this in mind, it can be concluded that the solution of Problem 10.2 can be alternatively obtained as $a_c\left(\boldsymbol{Z}^{d*}\right)$, where $\boldsymbol{Z}^{d*}$ is found by solving the parameter optimization problem

$$\boldsymbol{Z}^{d*} = \min_{\boldsymbol{Z}^d \in \boldsymbol{S}} J \tag{10.21}$$

with

$$\boldsymbol{S} = \left\{\boldsymbol{Z}^d \left| \left|z_j^d\right| \geq R_j^s, j \in [M]\right.\right\} \tag{10.22}$$

10.3.2 General Guidance Law Solution

For notation simplicity, define

$$b_i^w = \begin{cases} \boldsymbol{C}_i \boldsymbol{\Phi}_i^w \left(t_{f,i}^w, t\right) \boldsymbol{B}_i^w, & t \le t_{f,i}^w \\ 0, & t > t_{f,i}^w \end{cases} \tag{10.23}$$

$$b_j^o = \begin{cases} \boldsymbol{C}_j \boldsymbol{\Phi}_j^o \left(t_{f,j}^o, t\right) \boldsymbol{B}_j^o, & t \le t_{f,j}^o \\ 0, & t > t_{f,i}^o \end{cases} \tag{10.24}$$

Using Eqs. (10.15) and (10.16), the dynamics of the ZEM can then be rewritten as

$$\dot{z}_i^w = -b_i^w (t)\, a_c \tag{10.25}$$

$$\dot{z}_j^o = -b_j^o (t)\, a_c \tag{10.26}$$

The solution of differential equations (10.25) and (10.26) are given by

$$z_i^w \left(t_{f,i}^w\right) - z_i^w (t) = \int_t^{t_{f,i}^w} -b_i^w (\tau)\, a_c (\tau)\, d\tau \tag{10.27}$$

$$z_j^o \left(t_{f,j}^o\right) - z_j^o (t) = \int_t^{t_{f,j}^o} -b_j^o (\tau)\, a_c (\tau)\, d\tau \tag{10.28}$$

Imposing path constraints (10.19) and (10.20) on Eqs. (10.27) and (10.28) gives

$$z_i^w (t) = \int_t^{t_{f,i}^w} b_i^w (\tau)\, a_c (\tau)\, d\tau \tag{10.29}$$

$$z_j^o (t) - z_j^d = \int_t^{t_{f,j}^o} b_j^o (\tau)\, a_c (\tau)\, d\tau \tag{10.30}$$

According to Lemma 5.1, if the guidance command a_c is optimal in terms of energy minimization, then there exist $N + M$ Lagrange multipliers $\lambda_i, \beta_j, i \in [N]$, $j \in [M]$, such that the lateral acceleration command can be formulated as

$$a_c = \sum_{i=1}^{N} \lambda_i b_i^w + \sum_{j=1}^{M} \beta_j b_j^o \tag{10.31}$$

Substituting Eq. (10.31) into Eqs. (10.29) and (10.30) results in

$$z_i^w = \sum_{i'=1}^{N} \lambda_{i'} \int_t^{t_{f,i}^w} b_i^w(\tau) b_{i'}^w(\tau) d\tau + \sum_{j'=1}^{M} \beta_{j'} \int_t^{t_{f,i}^w} b_i^w(\tau) b_{j'}^o(\tau) d\tau \tag{10.32}$$

$$z_j^d - z_j^o = \sum_{i'=1}^{N} \lambda_{i'} \int_t^{t_{f,j}^o} b_j^o(\tau) b_{i'}^w(\tau) d\tau + \sum_{j'=1}^{M} \beta_{j'} \int_t^{t_{f,j}^o} b_j^o(\tau) b_{j'}^o(\tau) d\tau \tag{10.33}$$

Define $\boldsymbol{\lambda} = [\lambda_1, \lambda_2, \cdots, \lambda_N]^T$, $\boldsymbol{\beta} = [\beta_1, \beta_2, \cdots, \beta_M]^T$ as two Lagrange multiplier vectors, and $\boldsymbol{Z}^w = \left[z_1^w, z_2^w, \cdots, z_N^w\right]^T$, $\boldsymbol{Z}^o = \left[z_1^o, z_2^o, \cdots, z_M^o\right]^T$ as the ZEM vectors. Then, Eqs. (10.32) and (10.33) can be rewritten as a compact matrix form as

$$\boldsymbol{G}\begin{bmatrix} \boldsymbol{\lambda} \\ \boldsymbol{\beta} \end{bmatrix} = \begin{bmatrix} \boldsymbol{Z}^w \\ \boldsymbol{Z}^o - \boldsymbol{Z}^d \end{bmatrix}, \quad \boldsymbol{G} = \begin{bmatrix} \boldsymbol{G}_1 & \boldsymbol{G}_{12} \\ \boldsymbol{G}_{21} & \boldsymbol{G}_2 \end{bmatrix} \tag{10.34}$$

where $\boldsymbol{G}_1 = \boldsymbol{G}_1^T = [g_1(i,j)] \in \mathbb{R}^{N\times N}$ with elements $g_1(i,j)$, $i \in [N]$, $j \in [N]$ being

$$g_1(i,j) = \int_t^{t_{f,i}^w} b_i^w(\tau) b_j^w(\tau) d\tau \tag{10.35}$$

and $\boldsymbol{G}_{12} = \boldsymbol{G}_{21}^T = [g_{12}(i,j)] \in \mathbb{R}^{N\times M}$ with elements $g_{12}(i,j)$, $i \in [N]$, $j \in [M]$ being

$$g_{12}(i,j) = \int_t^{t_{f,i}^w} b_i^w(\tau) b_j^o(\tau) d\tau \tag{10.36}$$

and $\boldsymbol{G}_2 = \boldsymbol{G}_2^T = [g_2(i,j)] \in \mathbb{R}^{M\times M}$ with elements $g_2(i,j)$, $i \in [M]$, $j \in [M]$ being

$$g_2(i,j) = \int_t^{t_{f,j}^o} b_i^o(\tau) b_j^o(\tau) d\tau \tag{10.37}$$

From Eq. (10.34), the Lagrange multiplier vectors $\boldsymbol{\lambda}$ and $\boldsymbol{\beta}$ can be obtained as

$$\begin{bmatrix} \boldsymbol{\lambda} \\ \boldsymbol{\beta} \end{bmatrix} = \boldsymbol{G}^{-1}\begin{bmatrix} \boldsymbol{Z}^w \\ \boldsymbol{Z}^o - \boldsymbol{Z}^d \end{bmatrix} \tag{10.38}$$

Substituting Eq. (10.38) into Eq. (10.31) gives the analytic solution of Problem cha10prob3 as

$$\begin{aligned} a_c\left(\boldsymbol{Z}^d\right) &= \boldsymbol{\lambda}^T\left[b_1^w, b_2^w, \ldots, b_N^w\right]^T + \boldsymbol{\beta}^T\left[b_1^o, b_2^o, \ldots, b_M^o\right]^T \\ &= \left(\boldsymbol{G}^{-1}\begin{bmatrix} \boldsymbol{Z}^w \\ \boldsymbol{Z}^o - \boldsymbol{Z}^d \end{bmatrix}\right)^T \left[b_1^w, b_2^w, \ldots, b_N^w, b_1^o, b_2^o, \ldots, b_M^o\right]^T \end{aligned} \tag{10.39}$$

where the inverse of matrix $\boldsymbol{G}$ can be obtained by using the block inverse approach [9] as

$$\begin{aligned}\boldsymbol{G}^{-1} &= \begin{bmatrix} \boldsymbol{G}_{i1} & \boldsymbol{G}_{i12} \\ \boldsymbol{G}_{i12}^T & \boldsymbol{G}_{i2} \end{bmatrix} \\ &= \begin{bmatrix} \left(\boldsymbol{G}_1 - \boldsymbol{G}_{12}\boldsymbol{G}_2^{-1}\boldsymbol{G}_{12}^T\right)^{-1} & -\boldsymbol{G}_1^{-1}\boldsymbol{G}_{12}\left(\boldsymbol{G}_2 - \boldsymbol{G}_{12}\boldsymbol{G}_1^{-1}\boldsymbol{G}_{12}^T\right)^{-1} \\ -\left(\boldsymbol{G}_2 - \boldsymbol{G}_{12}\boldsymbol{G}_1^{-1}\boldsymbol{G}_{12}^T\right)^{-1}\boldsymbol{G}_{12}^T\boldsymbol{G}_1^{-1} & \left(\boldsymbol{G}_2 - \boldsymbol{G}_{12}\boldsymbol{G}_1^{-1}\boldsymbol{G}_{12}^T\right)^{-1} \end{bmatrix}\end{aligned} \tag{10.40}$$

Now, substituting solution (10.39) into performance index (10.7) gives

$$\begin{aligned} J &= \int_t^{t_{f,N}^w} \left[\sum_{i=1}^N \lambda_i b_i^w(\tau) + \sum_{j=1}^M \beta_j b_j^o(\tau)\right]^2 d\tau \\ &= \begin{bmatrix} \boldsymbol{Z}^w \\ \boldsymbol{Z}^o - \boldsymbol{Z}^d \end{bmatrix}^T \boldsymbol{G}^{-1} \begin{bmatrix} \boldsymbol{Z}^w \\ \boldsymbol{Z}^o - \boldsymbol{Z}^d \end{bmatrix} \\ &= (\boldsymbol{Z}^w)^T \boldsymbol{G}_{i1}\boldsymbol{Z}^w + (\boldsymbol{Z}^o - \boldsymbol{Z}^d)^T \boldsymbol{G}_{i12}^T \boldsymbol{Z}^w + (\boldsymbol{Z}^w)^T \boldsymbol{G}_{i12}\left(\boldsymbol{Z}^o - \boldsymbol{Z}^d\right) \\ &\quad + \left(\boldsymbol{Z}^o - \boldsymbol{Z}^d\right)^T \boldsymbol{G}_{i2}\left(\boldsymbol{Z}^o - \boldsymbol{Z}^d\right) \\ &= (\boldsymbol{Z}^w)^T \boldsymbol{G}_{i1}\boldsymbol{Z}^w + (\boldsymbol{Z}^o)^T \boldsymbol{G}_{i12}^T\boldsymbol{Z}^w + (\boldsymbol{Z}^w)^T\boldsymbol{G}_{i12}\boldsymbol{Z}^o + (\boldsymbol{Z}^o)^T\boldsymbol{G}_{i2}\boldsymbol{Z}^o \\ &\quad - \left(\boldsymbol{G}_{i2}\boldsymbol{Z}^o + \boldsymbol{G}_{i12}^T\boldsymbol{Z}^w\right)^T \boldsymbol{G}_{i2}^{-1}\left(\boldsymbol{G}_{i2}\boldsymbol{Z}^o + \boldsymbol{G}_{i12}^T\boldsymbol{Z}^w\right) \\ &\quad + \left(\boldsymbol{Z}^d - \boldsymbol{Z}^o - \boldsymbol{G}_{i2}^{-1}\boldsymbol{G}_{i12}^T\boldsymbol{Z}^w\right)^T \boldsymbol{G}_{i2}\left(\boldsymbol{Z}^d - \boldsymbol{Z}^o - \boldsymbol{G}_{i2}^{-1}\boldsymbol{G}_{i12}^T\boldsymbol{Z}^w\right) \end{aligned} \tag{10.41}$$

From Eq. (10.41), it can be noted that $\boldsymbol{Z}^{d*}$ can be obtained by

$$\boldsymbol{Z}^{d*} = \min_{\boldsymbol{Z}^d \in \boldsymbol{S}} \left(\boldsymbol{Z}^d - \boldsymbol{Z}^o - \boldsymbol{G}_{i2}^{-1}\boldsymbol{G}_{i12}^T\boldsymbol{Z}^w\right)^T \boldsymbol{G}_{i2}\left(\boldsymbol{Z}^d - \boldsymbol{Z}^o - \boldsymbol{G}_{i2}^{-1}\boldsymbol{G}_{i12}^T\boldsymbol{Z}^w\right) \tag{10.42}$$

Notice that set $\boldsymbol{S}$ can be split into two convex sets as

$$\boldsymbol{S} = \boldsymbol{S}_1 \cup \boldsymbol{S}_2 \tag{10.43}$$

with

$$\boldsymbol{S}_1 = \left\{\boldsymbol{Z}^d \left| z_j^d \geq R_j^s,\ j \in [M]\right.\right\} \tag{10.44}$$

$$\boldsymbol{S}_2 = \left\{\boldsymbol{Z}^d \left| z_j^d \leq -R_j^s,\ j \in [M]\right.\right\} \tag{10.45}$$

Then, we can alternatively solve parameter optimization problem (10.42) as

$$\boldsymbol{Z}^{d*} = \min_{\boldsymbol{Z}^d \in \left\{\boldsymbol{Z}_1^{d*}, \boldsymbol{Z}_2^{d*}\right\}} \left(\boldsymbol{Z}^d - \boldsymbol{Z}^o - \boldsymbol{G}_{i2}^{-1}\boldsymbol{G}_{i12}^T\boldsymbol{Z}^w\right)^T \boldsymbol{G}_{i2}\left(\boldsymbol{Z}^d - \boldsymbol{Z}^o - \boldsymbol{G}_{i2}^{-1}\boldsymbol{G}_{i12}^T\boldsymbol{Z}^w\right) \tag{10.46}$$

with

$$Z_1^{d*} = \min_{Z^d \in S_1} \left(Z^d - Z^o - G_{i2}^{-1} G_{i12}^T Z^w\right)^T G_{i2} \left(Z^d - Z^o - G_{i2}^{-1} G_{i12}^T Z^w\right) \quad (10.47)$$

$$Z_2^{d*} = \min_{Z^d \in S_2} \left(Z^d - Z^o - G_{i2}^{-1} G_{i12}^T Z^w\right)^T G_{i2} \left(Z^d - Z^o - G_{i2}^{-1} G_{i12}^T Z^w\right) \quad (10.48)$$

Since both S_1 and S_2 are convex sets, quadratic programming problems (10.47) and (10.48) can be efficiently solved by many existing approaches. With this in mind, the analytic guidance command for integrated waypoint following and collision avoidance with energy minimization can be readily determined as

$$a_c = \left(G^{-1} \begin{bmatrix} Z^w \\ Z^o - Z^{d*} \end{bmatrix}\right)^T \left[b_1^w, b_2^w, \ldots, b_N^w, b_1^o, b_2^o, \ldots, b_M^o\right]^T \quad (10.49)$$

Remark 10.1 Excluding the obstacle avoidance constrain (10.9), the proposed guidance law reduces to the minimum-effort waypoint-following guidance law developed in the previous two chapters.

10.3.3 Guidance Law Implementation

From Eq. (10.49), it can be noted that the implementation of the proposed guidance law requires the information on ZEM. This information, however, cannot be directly measured from the UAV's onboard sensors. For this reason, this subsection will derive an equivalent guidance command that only leverages the measured signals $\dot{\sigma}_i^w, \dot{\sigma}_i^o, V, r_i^w, r_i^o, \gamma$ and $\boldsymbol{x}_a$. Notice that the state transmission matrices $\boldsymbol{\Phi}_i^w\left(t_{f,i}^w, t\right)$ and $\boldsymbol{\Phi}_j^o\left(t_{f,j}^o, t\right)$ can be written as

$$\boldsymbol{\Phi}_i^w\left(t_{f,i}^w, t\right) = \begin{bmatrix} 1 & t_{f,i}^w - t & \boldsymbol{\Phi}_{i,1}^w\left(t_{f,i}^w, t\right) \\ 0 & 1 & \boldsymbol{\Phi}_{i,2}^w\left(t_{f,i}^w, t\right) \\ 0_{n_a \times 1} & 0_{n_a \times 1} & \boldsymbol{\Phi}_{i,3}^w\left(t_{f,i}^w, t\right) \end{bmatrix} \quad (10.50)$$

$$\boldsymbol{\Phi}_j^o\left(t_{f,j}^o, t\right) = \begin{bmatrix} 1 & t_{f,j}^o - t & \boldsymbol{\Phi}_{j,1}^o\left(t_{f,j}^o, t\right) \\ 0 & 1 & \boldsymbol{\Phi}_{j,2}^o\left(t_{f,j}^o, t\right) \\ 0_{n_a \times 1} & 0_{n_a \times 1} & \boldsymbol{\Phi}_{j,3}^o\left(t_{f,j}^o, t\right) \end{bmatrix} \quad (10.51)$$

Using Eqs. (2.8), (10.10), (10.11), (10.50) and (10.51), the ZEMs can then be formulated in alternative forms as

$$z_i^w = \begin{cases} V\dot{\sigma}_i^w \left(t_{go,i}^w\right)^2 + \boldsymbol{\Phi}_{i,1}^w \left(t_{f,i}^w, t\right) \boldsymbol{x}_a, & t \le t_{f,i}^w \\ z_i^w \left(t_{f,i}^w\right), & t > t_{f,i}^w \end{cases} \tag{10.52}$$

$$z_j^o = \begin{cases} V\dot{\sigma}_j^o \left(t_{go,j}^o\right)^2 + \boldsymbol{\Phi}_{j,1}^o \left(t_{f,j}^o, t\right) \boldsymbol{x}_a, & t \le t_{f,j}^o \\ z_j^o \left(t_{f,j}^o\right), & t > t_{f,j}^o \end{cases} \tag{10.53}$$

Substituting Eqs. (10.52) and (10.53) into Eq. (10.49) gives the guidance command in terms of measured signals $\dot{\sigma}_i^w$, $\dot{\sigma}_i^o$, V, r_i^w, r_i^o, γ and $\boldsymbol{x}_a$. This supports the practical application of the proposed guidance law.

10.4 Some Particular Cases

This section discusses some particular cases of the proposed guidance law. For simplicity, we consider ideal autopilot dynamics and first-order autopilot dynamics cases here.

10.4.1 Ideal Autopilot Dynamics

If the UAV's autopilot dynamics is ideal, that is,

$$\boldsymbol{A}_a = 0, \quad \boldsymbol{B}_a = 0, \quad \boldsymbol{C}_a = 0, \quad \boldsymbol{D}_a = 1 \tag{10.54}$$

Then, we have

$$z_i^w = \begin{cases} V\dot{\sigma}_i^w \left(t_{go,i}^w\right)^2, & t \le t_{f,i}^w \\ z_i^w \left(t_{f,i}^w\right), & t > t_{f,i}^w \end{cases} \tag{10.55}$$

$$z_j^o = \begin{cases} V\dot{\sigma}_j^o \left(t_{go,j}^o\right)^2, & t \le t_{f,j}^o \\ z_j^o \left(t_{f,j}^o\right), & t > t_{f,j}^o \end{cases} \tag{10.56}$$

$$b_i^w = t_{go,i}^w, \quad b_j^o = t_{go,j}^o \tag{10.57}$$

The integrals in Eqs. (10.32) and (10.33) for this simple case can be easily obtained as

$$\int_t^{t_{f,i}^w} b_i^w(\tau)\, b_{i'}^w(\tau)\, d\tau = \frac{\left(t_{go,\min\{i,i'\}}^w\right)^2}{6} \left(2t_{go,\min\{i,i'\}}^w + 3d_{i,i'}^w\right) \tag{10.58}$$

$$\int_t^{t_{f,i}^w} b_i^w(\tau)\, b_{j'}^o(\tau)\, d\tau = \frac{\left(t_{go,ij'}^{wo}\right)^2}{6}\left(2t_{go,ij'}^{wo} + 3d_{i,j'}^{wo}\right) \tag{10.59}$$

$$\int_t^{t_{f,j}^o} b_j^o(\tau)\, b_{j'}^o(\tau)\, d\tau = \frac{\left(t_{go,\min\{j,j'\}}^{o}\right)^2}{6}\left(2t_{go,\min\{j,j'\}}^{o} + 3d_{j,j'}^{o}\right) \tag{10.60}$$

where $d_{i,i'}^w = t_{f,\max\{i,i'\}}^w - t_{f,\min\{i,i'\}}^w, t_{go,ij'}^{wo} = \min\left\{t_{go,i}^o, t_{go,j'}^w\right\}, d_{i,j'}^{wo} = \max\left\{t_{f,i}^o, t_{f,j'}^w\right\} - \min\left\{t_{f,i}^o, t_{f,j'}^w\right\}$ and $d_{j,j'}^o = t_{f,\max\{j,j'\}}^o - t_{f,\min\{j,j'\}}^o$.

The explicit guidance command can then be readily obtained by substituting Eqs. (10.55)–(10.60) into Eq. (10.49). This guidance command is termed as lag-free energy-optimal guidance law for integrated waypoint following and obstacle avoidance in this chapter.

10.4.2 First-Order Autopilot Dynamics

Assume that the UAV autopilot is modeled by a first-order dynamics, that is,

$$\boldsymbol{A}_a = -\frac{1}{\tau_a}, \quad \boldsymbol{B}_a = \frac{1}{\tau_a}, \quad \boldsymbol{C}_a = 1, \quad \boldsymbol{D}_a = 0 \tag{10.61}$$

where τ_a denotes the autopilot time constant.

Then, the transition matrices are determined as

$$\boldsymbol{\Phi}_i^w\left(t_{f,i}^w, t\right) = \begin{bmatrix} 1 & t_{f,i}^w - t & -\tau_a^2 \phi\left(\frac{t_{f,i}^w - t}{\tau_a}\right) \\ 0 & 1 & \tau_a\left(e^{-\frac{t_{f,i^w} - t}{\tau_a}} - 1\right) \\ 0 & 0 & e^{-\frac{t_{f,i}^w - t}{\tau_a}} \end{bmatrix} \tag{10.62}$$

$$\boldsymbol{\Phi}_j^o\left(t_{f,j}^o, t\right) = \begin{bmatrix} 1 & t_{f,j}^o - t & -\tau_a^2 \phi\left(\frac{t_{f,j}^o - t}{\tau_a}\right) \\ 0 & 1 & \tau_a\left(e^{-\frac{t_{f,j}^o - t}{\tau_a}} - 1\right) \\ 0 & 0 & e^{-\frac{t_{f,j}^o - t}{\tau_a}} \end{bmatrix} \tag{10.63}$$

where $\phi(x) \triangleq e^{-x} + x - 1$.

Then, we obtain

$$z_i^w = \begin{cases} V\dot{\sigma}_i^w \left(t_{go,i}^w\right)^2 - \tau_a^2 \phi\left(\frac{t_{go,i}^w}{\tau_a}\right) a, & t \le t_{f,i}^w \\ z_i^w \left(t_{f,i}^w\right), & t > t_{f,i}^w \end{cases} \tag{10.64}$$

$$z_j^o = \begin{cases} V\dot{\sigma}_j^o \left(t_{go,j}^o\right)^2 - \tau_a^2 \phi\left(\frac{t_{go,j}^o}{\tau_a}\right) a, & t \le t_{f,j}^o \\ z_j^o \left(t_{f,j}^o\right), & t > t_{f,j}^o \end{cases} \tag{10.65}$$

$$b_i^w = \tau_a \phi\left(\frac{t_{go,i}^w}{\tau_a}\right), \quad b_j^o = \tau_a \phi\left(\frac{t_{go,j}^o}{\tau_a}\right) \tag{10.66}$$

After some tedious but simple algebra manipulations, the integrals in Eqs. (10.32) and (10.33) are determined as

$$\begin{aligned} \int_t^{t_{f,i}^w} b_i^w(\tau) b_{i'}^w(\tau)\, d\tau &= \frac{\tau_a^3}{2}\left(e^{-\frac{d_{i,i'}^w}{\tau_a}} - e^{-\frac{2t_{go,\min\{i,i'\}}^w + d_{i,i'}^w}{\tau_a}}\right) \\ &- \tau_a^2 t_{go,\min\{i,i'\}}^w \left(e^{-\frac{t_{go,\min\{i,i'\}}^w}{\tau_a}} + e^{-\frac{t_{go,\min\{i,i'\}}^w + d_{i,i'}^w}{\tau_a}}\right) \\ &+ \tau_a^2 d_{i,i'}^w \left(1 - e^{-\frac{t_{go,\min\{i,i'\}}^w}{\tau_a}}\right) - \tau_a \left(t_{go,\min\{i,i'\}}^w\right)^2 - \tau_a d_{i,i'}^w t_{go,\min\{i,i'\}}^w \\ &+ \tau_a^2 t_{go,\min\{i,i'\}}^w + \frac{\left(t_{go,\min\{i,i'\}}^w\right)^2}{6}\left(2t_{go,\min\{i,i'\}}^w + 3d_{i,i'}^w\right) \end{aligned} \tag{10.67}$$

$$\begin{aligned} \int_t^{t_{f,i}^w} b_i^w(\tau) b_{j'}^o(\tau)\, d\tau &= \frac{\tau_a^3}{2}\left(e^{-\frac{d_{i,j'}^{wo}}{\tau_a}} - e^{-\frac{2t_{go,ij'}^w + d_{i,i'}^{wo}}{\tau_a}}\right) \\ &- \tau_a^2 t_{go,ij'}^{wo}\left(e^{-\frac{t_{go,ij'}^{wo}}{\tau_a}} + e^{-\frac{t_{go,ij'}^{wo} + d_{i,j'}^{wo}}{\tau_a}}\right) \\ &+ \tau_a^2 d_{i,j'}^{wo}\left(1 - e^{-\frac{t_{go,ij'}^{wo}}{\tau_a}}\right) - \tau_a\left(t_{go,ij'}^{wo}\right)^2 - \tau_a d_{i,j'}^{wo} t_{go,ij'}^{wo} \\ &+ \tau_a^2 t_{go,ij'}^{wo} + \frac{\left(t_{go,ij'}^{wo}\right)^2}{6}\left(2t_{go,ij'}^{wo} + 3d_{i,j'}^{wo}\right) \end{aligned} \tag{10.68}$$

$$
\begin{aligned}
\int_t^{t_{f,j}^o} b_j^o(\tau)\, b_{j'}^o(\tau)\, d\tau &= \frac{\tau_a^3}{2}\left(e^{-\frac{d_{j,j'}^o}{\tau_a}} - e^{-\frac{2t_{go,\min\{j,j'\}}^o + d_{j,j'}^o}{\tau_a}}\right) \\
&- \tau_a^2 t_{go,\min\{j,j'\}}^o \left(e^{-\frac{t_{go,\min\{j,j'\}}^o}{\tau_a}} + e^{-\frac{t_{go,\min\{j,j'\}}^o + d_{j,j'}^o}{\tau_a}}\right) \\
&+ \tau_a^2 d_{j,j'}^o \left(1 - e^{-\frac{t_{go,\min\{j,j'\}}^o}{\tau_a}}\right) - \tau_a \left(t_{go,\min\{j,j'\}}^o\right)^2 - \tau_a d_{j,j'}^o t_{go,\min\{j,j'\}}^o \\
&+ \tau_a^2 t_{go,\min\{j,j'\}}^o + \frac{\left(t_{go,\min\{j,j'\}}^o\right)^2}{6}\left(2t_{go,\min\{j,j'\}}^o + 3d_{j,j'}^o\right)
\end{aligned}
\tag{10.69}
$$

The explicit guidance command can then be readily obtained by substituting Eqs. (10.64)–(10.69) into Eq. (10.49). This guidance command is termed as first-order energy-optimal guidance law for integrated waypoint following and obstacle avoidance in this chapter.

10.5 Numerical Simulations

In this section, nonlinear numerical simulations are performed to validate the proposed guidance law. In the considered scenario, a UAV with constant speed $V = 30\,\text{m/s}$ is required to sequentially visit two waypoints while avoiding two static obstacles. The inertial positions of the waypoints and obstacles are summarized in Table 10.1. The UAV starts the mission at position $(0\,m, 0\,m)$ with an initial flight path angle 30°. The UAV autopilot is modeled as a first-order lag system with time constant $\tau_a = 3\,\text{s}$, which is enough to identify the characteristics of a guidance law in the conservative point of view.

We first compare the proposed guidance algorithm with minimum-effort waypoint-following guidance law developed in Chaps. 8 and 9 to evaluate the collision avoidance property of the guidance law developed. For implementing the proposed guidance law, the safety bound is set as $R_1^s = 200\,\text{m}$, $R_2^s = 200\,\text{m}$. The UAV flight trajectories and relative distances between the UAV and obstacles obtained from these two different guidance laws are shown in Figs. 10.2a and 10.2b, respectively. As can be noted from Fig. 10.2a, both guidance laws successfully guide the UAV to sequentially visit the two selected waypoints with ignorable miss distances. How-

Table 10.1 Inertial positions of waypoints and obstacles

Waypoint ID	Inertial position	Obtacle ID	Inertial position
1	(2000 m, 750 m)	1	(1000 m, 390 m)
2	(4000 m, 3000 m)	2	(3000 m, 1500 m)

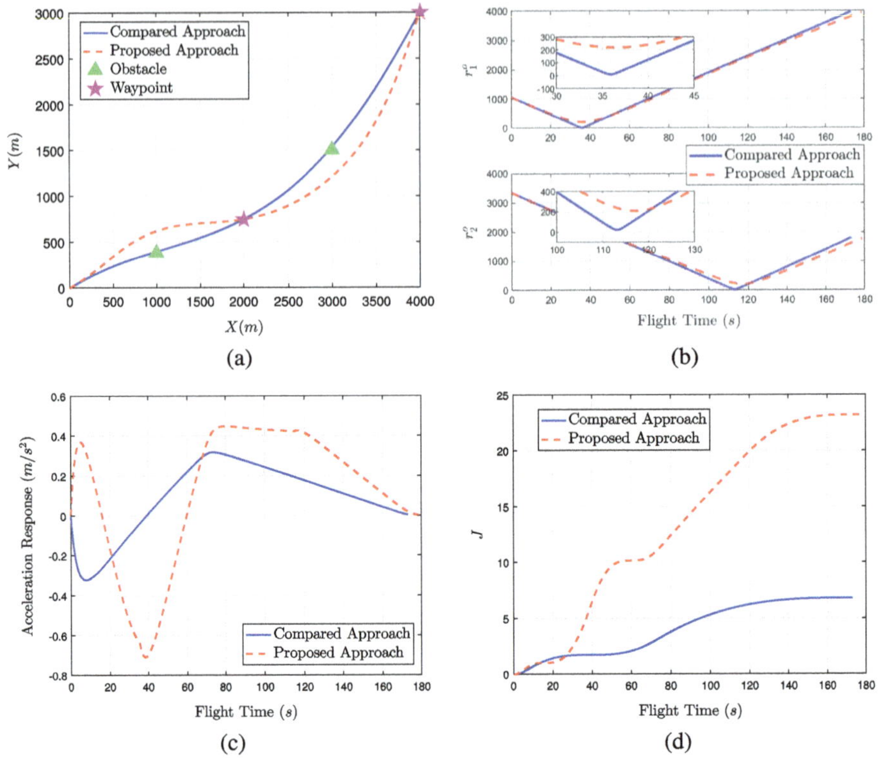

Fig. 10.2 Comparison results. **a** Flight trajectory; **b** Relative distance between the UAV and obstacles; **c** Acceleration response; and **d** Energy consumption

ever, optimal guidance law developed in Chaps. 8 and 9 cannot guarantee safe flight and the UAV might collide with potential obstacles. As a comparison, the proposed guidance law employs an active collision avoidance strategy that guides the UAV to make a detour to avoid obstacles. Figure 10.2c compares the UAV's acceleration response generated by these two different guidance laws through onboard autopilot. From this figure, it can be noted that both guidance law provides smooth acceleration response and the final guidance command converges to zero. The energy consumption, defined as $J = \int_0^{t_{f,N}^w} a^2(\tau)\, d\tau$, obtained from different guidance laws is compared in Fig. 10.2d. Not surprisingly, the proposed guidance law requires more energy consumption as additional detour maneuver is generated for collision avoidance, compared to guidance law proposed in Chaps. 8 and 9.

Now, let us investigate the effect of safety bound on the performance of the proposed guidance law. For this purpose, simulations are performed with respect to different safety bounds: (1) $R_1^s = 50\,\text{m}$, $R_2^s = 50\,\text{m}$; (2) $R_1^s = 100\,\text{m}$, $R_2^s = 150\,\text{m}$; (3) $R_1^s = 200\,\text{m}$, $R_2^s = 200\,\text{m}$; and (4) $R_1^s = 300\,\text{m}$, $R_2^s = 300\,\text{m}$. The simulation results, including the UAV's flight trajectory, relative distance between the UAV and obsta-

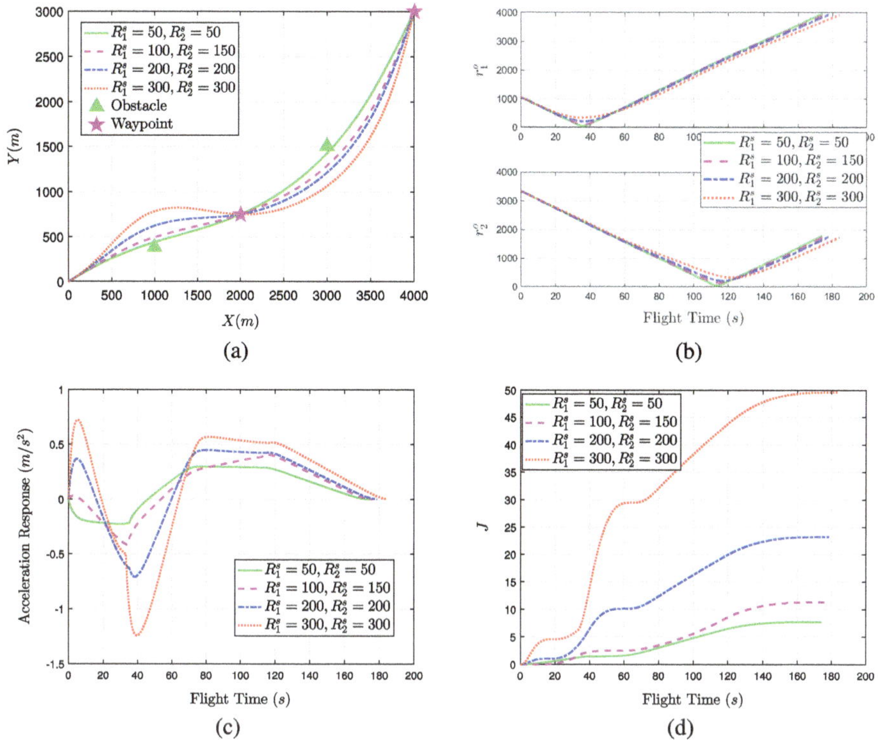

Fig. 10.3 Comparison results. **a** Flight trajectory; **b** Relative distance between the UAV and obstacles; **c** Acceleration response; and **d** Energy consumption

cles, acceleration response generated through autopilot and energy consumption, are shown in Fig. 10.3. From this figure, it can be clearly noted that the proposed guidance law with different safety bound still successfully guide the UAV to pass the selected waypoints with guaranteed safety bound. The UAV's flight trajectory becomes more curved with higher value of the radius of the safety region. This, in turn, requires higher maneuvering accelerations and more energy consumption to make a safe detour, as confirmed by Figs. 10.3c, d.

10.6 Summary

This chapter suggestes a new energy-optimal guidance law for integrated waypoint following and collision avoidance. Unlike previous numerical waypoint following guidance, the proposed approach is derived analytically through rigorous optimal control theory with the help of convex parameter optimization. The promising feature of the proposed guidance law is its global optimality and it can be applied to a general waypoint-following mission with arbitrary number of waypoints and obstacles. Nonlinear numerical comparisons clearly reveal that the proposed guidance law is able to guide the UAV to avoid potential obstacles while guaranteeing accurate waypoint following.

References

1. Mujumdar A, Padhi R (2011) Evolving philosophies on autonomous obstacle/collision avoidance of unmanned aerial vehicles. J Aerospace Comput Inf Commun 8(2):17–41
2. Koren Y, Borenstein J (1991) Potential field methods and their inherent limitations for mobile robot navigation. In: Proceedings. 1991 IEEE international conference on robotics and automation. IEEE, pp 1398–1404
3. Hamner B, Singh S, Scherer S (2006) Learning obstacle avoidance parameters from operator behavior. J Field Robot 23(11–12):1037–1058
4. Scherer S, Singh S, Chamberlain L, Elgersma M (2008) Flying fast and low among obstacles: methodology and experiments. Int J Robot Res 27(5):549–574
5. Han S-C, Bang H (2004) Proportional navigation-based optimal collision avoidance for uavs. In: 2nd international conference on autonomous robots and agents, pp 13–15
6. Mujumdar A, Padhi R (2011) Reactive collision avoidance of using nonlinear geometric and differential geometric guidance. J Guid Control Dyn 34(1):303–311
7. Arthur Earl Bryson (2018) Applied optimal control: optimization, estimation and control. Routledge
8. Shaferman V, Shima T (2008) Linear quadratic guidance laws for imposing a terminal intercept angle. J Guid Control Dyn 31(5):1400–1412
9. Tzon-Tzer L, Shiou S-H (2002) Inverses of $2\times$ 2 block matrices. Comput Math Appl 43(1–2):119–129

www.ingramcontent.com/pod-product-compliance
Ingram Content Group UK Ltd.
Pitfield, Milton Keynes, MK11 3LW, UK
UKHW022211051026
15624UKWH00001B/108
9783030473471